DATABASE SYSTEMS

DESIGN, IMPLEMENTATION, AND MANAGEMENT

PETER ROB • CARLOS CORONEL

THOMSON

COURSE TECHNOLOGY

Australia • Canada • Mexico • Singapore • Spain • United Kingdom • United States

THOMSON

COURSE TECHNOLOGY

Database Systems: Design, Implementation, and Management, Sixth Edition

by Peter Rob and Carlos Coronel

Product Manager:
Alyssa Pratt

Executive Editor:
Jennifer Locke

Developmental Editor:
Deb Kaufmann

Associate Product Manager:
Mirella Misiaszek

Editorial Assistant:
Amanda Piantedosi

Production Editor:
Melissa Panagos

Marketing Manager:
Amy Yarnevich

Cover Designer:
Rakefet Kenaan

Compositor:
GEX Publishing Services

Senior Manufacturing Coordinator:
Laura Burns

COPYRIGHT © 2004 Course Technology, a division of Thomson Learning, Inc. Thomson Learning™ is a trademark used herein under license.

Printed in the United States of America

4 5 6 7 8 9 BM 06 05

For more information, contact Course Technology, 25 Thomson Place, Boston, Massachusetts, 02210.

Or find us on the World Wide Web at: www.course.com

ALL RIGHTS RESERVED. No part of this work covered by the copyright hereon may be reproduced or used in any form or by any means—graphic, electronic, or mechanical, including photocopying, recording, taping, Web distribution, or information storage and retrieval systems—without the written permission of the publisher.

For permission to use material from this text or product, contact us by
Tel (800) 730-2214
Fax (800) 730-2215
www.thomsonrights.com

Disclaimer
Course Technology reserves the right to revise this publication and make changes from time to time in its content without notice.

ISBN 0-619-21323-x
(Student Edition)

ISBN 0-619-16033-0
(Instructor Edition)

ISBN 0-619-21372-8
(International Edition)

To Anne, who remains my best friend after 43 years of marriage. To our son, Peter William, who turned out to be the man we hoped he would be and who proved his wisdom by making Sheena our treasured daughter-in-law. To Sheena, who stole our hearts so many years ago. To our grandsons, Adam Lee and Alan Henri, who are growing up to be the fine human beings their parents are. To my mother-in-law, Nini Fontein, and to the memory of my father-in-law, Henri Fontein—their life experiences in Europe and Southeast Asia would fill a history book and they enriched my life by entrusting me with their daughter's future. To the memory of my parents, Hendrik and Hermine Rob, who rebuilt their lives after World War II's horrors, who did it again after a failed insurgency in Indonesia, and who finally found their promised land in these United States. And to the memory of Heinz, who taught me daily lessons about loyalty, uncritical acceptance, and boundless understanding. I dedicate this book to you, with love.

Peter Rob

To Victoria, my beautiful wife, who always keeps me from falling apart during the endless hours of writing: Thanks for being so caring and for being there for me during my most trying moments. To Carlos Anthony, my son, who is his father's pride and who always teaches me new things. To Gabriela Victoria, my daughter and the princess of the house. To Christian Javier, our little bundle of joy. To all my children, thanks for your laughs, your sweet voices, your beautiful smiles, and your frequent hugs. I love you; you are my Divine treasure. Finally, to my parents for their many sacrifices and encouragement.

Carlos Coronel

BRIEF CONTENTS

The following appendixes are included on the CD-ROM that accompanies this text:

CONTENTS

PART II DESIGN CONCEPTS

PART III ADVANCED DESIGN AND IMPLEMENTATION

PART V DATABASES & THE INTERNET

PART VI DATABASE ADMINISTRATION

APPENDIXES ON CD-ROM

In this sixth edition, we continue to provide a solid and practical foundation for the design, implementation, and management of database systems. We build this foundation on the notion that, while databases are very practical things, their successful creation depends on understanding the important concepts that define them. It's not easy to come up with the proper mix of theory and practice, but we are grateful that many of our previous editions' adopters and students, whose comments, suggestions, and evaluations indicate that we have largely succeeded in our quest to provide the proper balance.

THE APPROACH: AN EMPHASIS ON DESIGN

As its title suggests, *Database Systems: Design, Implementation, and Management* covers three broad aspects of database systems. However, we give special attention to database design for several important reasons:

❖ The availability of excellent database software enables even database-inexperienced people to create databases and database applications. Unfortunately, the "create without design" approach usually paves the road to any number of database disasters. In our experience, many if not most database system failures are traceable to poor design and cannot be solved with the help of even the best programmers and managers. Nor is better DBMS software likely to overcome problems created or magnified by poor design. Using an analogy, even the best bricklayers and carpenters can't create a good building from a bad blueprint.

❖ Most of the really vexing database system management problems seem to be triggered by poorly designed databases. It hardly seems worthwhile to use scarce resources to develop excellent and extensive database system management skills in order to exercise them on crises induced by poorly designed databases.

❖ Design provides an excellent means of communication. Clients are much more likely to get what they need when database system design is approached carefully and thoughtfully. In fact, clients may even discover how their organizations really function once a good database design is completed.

❖ Familiarity with database design techniques promotes one's understanding of current database technologies. For example, because data warehouses derive much of their data from operational databases, data warehouse concepts, structures, and procedures make much more sense if the operational database's structure and implementation are understood.

Because we stress the practical aspects of database design, we have covered design concepts and procedures in detail and we have made sure that the numerous end-of-chapter problems are sufficiently challenging to let students develop real and useful design skills. We also make sure that students understand the potential and actual conflicts between database design elegance, information requirements, and transaction processing speed. For example, it makes little sense to design databases that meet design elegance standards while they fail to meet end-user information requirements. Therefore, we explore the use of carefully defined trade-offs to ensure that the databases are capable of meeting end-user requirements while conforming to high design standards.

CHANGES TO THE SIXTH EDITION

In this sixth edition, we have reorganized some of the coverage to provide a better flow of the material, and we have placed even greater emphasis on proper database design. Aside from strengthening our already strong database design coverage, we have made other improvements in the sixth edition's topical coverage. Here are a few of the highlights:

- We have split the fifth edition's Chapter 1 into two chapters, **Chapter 1** "Database Systems" and **Chapter 2** "Data Models." This decision reflects user comments and suggestions and it fits our desire to improve the logical flow of the material.

- We have added a new **Chapter 7**, "Advanced SQL." SQL's advanced features such as triggers and stored procedures deserve detailed coverage to reflect our dedication to the "hands on" approach we have always promoted.

- A new **Appendix A**, "Designing Databases with Visio Professional: A Tutorial," provides a detailed tutorial on using Microsoft Visio Professional tutorial to create database designs based on the Crow's Foot notation. We believe that this tutorial adds substantial value to the student database design skill set. (And we have found that our students tend to be more interested in exploring various database design issues when they have a chance to see the effects of a proposed design change.) In any case, we believe that students will benefit from an exposure to a first-rate professional database design tool. (If you do not have access to Visio Professional, you can continue to use PowerPoint to create database designs. In fact, we have even created a PowerPoint Crow's Foot design template (available on the Student CD) to make it easier for you to develop database designs if you do not plan to use Microsoft Visio Professional.)

- A new **Appendix B**, "An Introduction to Unified Modeling Language (UML)," provides detailed coverage of UML Class diagrams to design static database components. We have used Microsoft Visio Professional's UML diagramming tool to illustrate important UML components. This appendix supports our detailed coverage of object oriented (OO) concepts in Chapter 11, "Object Oriented Databases."

- A new **Appendix C**, "Database Performance Tuning," gives practical direction for improving the performance of databases, a crucial skill in managing modern databases. In class, students typically encounter database tables that contain only a few rows, so they may not realize that a query that works just fine on a small table may be unusable when the table has thousands or millions of rows. If you don't know about performance tuning, a query executed on millions of data rows might take hours to execute; proper database tuning may cut that time to minutes or even seconds. This appendix is a perfect follow-on to Chapter 15, "Database Administration."

- Two new appendixes, **Appendix H**, "The Hierarchical Database Model," and **Appendix I**, "The Network Database Model," highlight the structural details of two older database models. Adding this coverage allows adopters to further contrast the relational and older data management approaches. These older database models contain some features that have been "rediscovered" in the current database environment, thus illustrating the notion that the more things change, the more they remain the same.

- We have moved three fifth edition chapters to sixth edition appendixes: **Appendix D**, "The University Lab: Conceptual Design," **Appendix E**, "The University Lab: Conceptual Design Verification, Logical Design, and Implementation," and **Appendix F**, "Client/Server Systems." This enables us to control the book's page count without having to sacrifice detailed coverage of several important topics.

- We have added many **new problems** and greatly improved their teaching potential. Many of these problems have multiple sections. Problem descriptions and illustrations have been strengthened considerably.

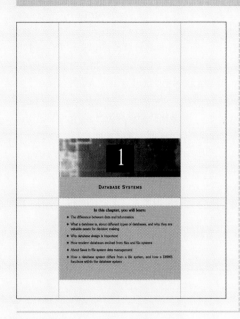

Our book's title begins with *Database Systems*. Therefore, we examine the database and design concepts covered in Chapters 1 through 5 as part of a larger whole by placing them within the systems analysis framework of Chapter 8. We believe that database designers who fail to understand that the database is a part of a larger system are likely to overlook important database design requirements. In fact, Chapter 8 provides the map for the advanced database design we develop in appendixes D and E. Within the larger systems framework, we can also explore such issues as transaction management and concurrency Control (Chapter 9), distributed database management systems (Chapter 10), data warehousing (Chapter 12), databases in electronic commerce (Chapter 14), and database administration (Chapter 15).

Database Design

The first item in the book's subtitle is *Design*, and our examination of database design is comprehensive. For example, Chapters 1 and 2 examine the development of databases and data models and illustrate the need for design. Chapter 3 lays the foundation for the relational database model, Chapter 4 provides extensive, in-depth, and practical database design coverage, and Chapter 5 is entirely devoted to critical normalization issues that affect database efficiency and effectiveness. Chapters 6 and 7 show how the database design is implemented and how the data are accessed through Structured Query Language (SQL). Chapter 8 examines the database design within the systems framework and maps the activities required to successfully design and implement the complex real-world database we develop in appendixes D and E.

Because database design is affected by real-world transactions, how data are distributed, and ever-increasing information requirements, we examine major database features that must be supported in current generation databases and models. For example, Chapter 9, "Transaction Management and Concurrency Control," focuses on the characteristics of database transactions and how they affect database integrity and consistency. Chapter 10, "Distributed Database Management Systems," focuses on data distribution, replication, and allocation. We provide detailed coverage of object oriented concepts and databases in Chapter 11, "Object Oriented Databases." In Chapter 12, "The Data Warehouse," we explore the characteristics of the databases that are used in decision support and online analytical processing. Chapter 13 covers the basic design of databases used in electronic commerce.

Implementation

The second portion of the subtitle is *Implementation*. We use Structured Query Language (SQL) in chapters 6 and 7 to show how databases are implemented and managed. Appendixes D and E demonstrate the design of a database that was fully implemented, and illustrate a wide range of implementation issues. We had to deal with conflicting design goals: design elegance, information requirements, and operational speed. Therefore, we carefully audited the initial design (Appendix D) to check its ability to meet end-user needs and to establish appropriate implementation protocols. The result of this audit yielded the final, implementable design developed in Appendix E. The special issues that are encountered in an Internet database environment are addressed in Chapter 13 ("Databases in Electronic Commerce") and in Chapter 14 ("Web Database Development").

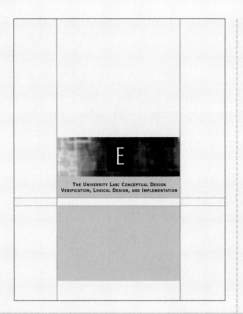

THE UNIVERSITY LAB: CONCEPTUAL DESIGN
VERIFICATION, LOGICAL DESIGN, AND IMPLEMENTATION

Management

The final portion of the subtitle is Management. We deal with database management issues in Chapter 9 ("Transaction Management and Concurrency Control"), Chapter 10 ("Distributed Database Management Systems"), and Chapter 15 ("Database Administration"). Appendix C, "Database Performance Tuning," is a valuable resource that illustrates how a DBMS manages the data retrieval operations. (In fact, you can think of Appendix C as a complement to the database administration chapter.)

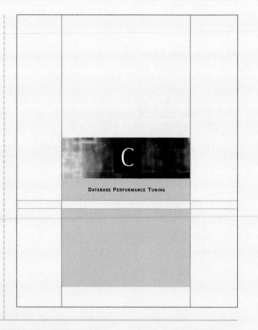

DATABASE PERFORMANCE TUNING

TEACHING DATABASE: A MATTER OF FOCUS

Given the wealth of detailed coverage, instructors may "mix and match" chapters to produce the desired coverage. Depending on where database courses fit into the curriculum, instructors may choose to emphasize database design or database management (see Figure 1.)

The hands-on nature of database design lends itself particularly well to class projects in which students use instructor-selected software to prototype a student-designed system for the end user. Several of the end-of-chapter problems are sufficiently complex to serve as projects, or the instructor may work with local businesses to give students hands-on experience. Note that some elements of the database design track are also found in the database management track. This is because it is difficult to manage database technologies that are not understood.

FIGURE 1 TOPICAL COVERAGE OPTIONS

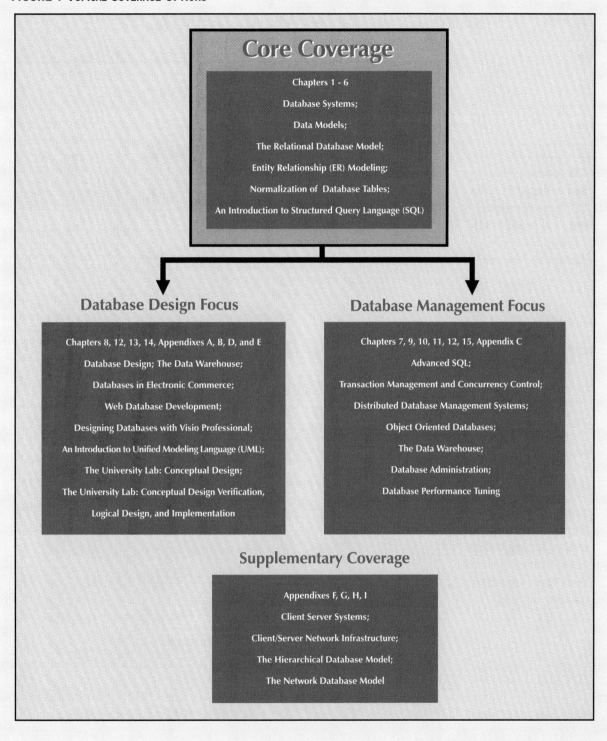

The options shown in Figure 1 serve only as a starting point. Naturally, professors will tailor their coverage based on their specific course requirements. For example, a teacher may decide to make Appendix A an outside reading assignment and use that time to cover client/server systems or object oriented databases. The latter choice would serve as a gateway to UML coverage.

CD Icons draw attention to material on the CD that accompanies the text, and provide ideas for incorporating this content into the course.

ADDITIONAL CD CONTENT

All the databases used to illustrate the material in this chapter are found on the CD bundled with this book. Check the \Databases\Student\Ch03 folder to find the databases. The database names used in the folder match the database names used in the figures. For example, the source of the tables shown in Figure 3.1 is the Ch03_TinyCollege database located in the above noted folder.

Using the STUDENT table shown in Figure 3.1, you can draw the following conclusions corresponding to the points in Table 3.1:

1. The STUDENT table shown in Figure 3.1 is perceived to be a two-dimensional structure composed of eight rows (tuples) and twelve columns. You can also describe the table as being composed of eight records and twelve attributes (fields).

2. Each row in the STUDENT table describes a single entity occurrence within the entity set. (The entity set is represented by the STUDENT table.) Note that the row (record or entity) defined by STU_NUM = 321452 defines the characteristics (attributes or fields) of a student named William C. Bowser. For example, row 4 in Figure 3.1 describes a student named Walter H. Oblonski. Similarly, row 3 describes a student named Juliette Brewer. Given the table contents, the STUDENT entity set includes eight distinct entities (rows).

NOTE

NOTE TO ORACLE USERS

If you are using Oracle DBMS, we have included SQL script files to create the tables and load the data in the database. The script files are located on the CD in the \SQL\Ch06 folder. How you connect to the Oracle database depends on how the Oracle software was installed on your server and on the access paths and methods defined and managed by the database administrator. Please follow the instructions provided by your instructor or your school's technology department.

Notes highlight important facts about the concepts introduced in the chapter.

FIGURE 14.23 **THE DELETE RECORD VALIDATION**

A variety of **four-color figures**, including ER models and implementations, tables, and illustrations, clearly illustrate difficult concepts.

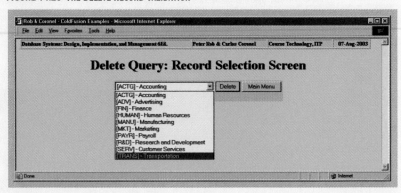

SUMMARY

Entities are the basic building blocks of a relational database. A grouping of related entities, known as an entity set, is stored in a table. Conceptually speaking, the relational table is composed of intersecting rows (tuples) and columns. Each row represents a single entity, and each column represents the characteristics (attributes) of the entities.

Keys are central to the use of relational tables. Keys define functional dependencies. That is, other attributes are dependent on the key and can, therefore, be found if the key value is known. A key can be classified according to its characteristics as a superkey, a candidate key, a primary key, a secondary key, or a foreign key.

Each table row must have a primary key. The primary key is an attribute or combination of attributes that uniquely identifies all remaining attributes found in any given row. Because a primary key must be unique, no null "values" are allowed if entity integrity is to be maintained.

Although the tables are independent, they can be linked by sharing common attributes. Thus, we can link tables by using controlled redundancy; the primary key of one table appears again as the foreign key in the table to which it is linked. To make sure that the foreign key produces the correct pointer movement, it must maintain referential integrity; that is, the foreign key must contain either values that match the primary key in the other table or contain nulls.

Relational databases can be classified according to the degree to which they support the relational algebra functions SELECT, PROJECT, JOIN, INTERSECT, UNION, DIFFERENCE, PRODUCT, and DIVIDE, and also to the degree that (entity and referential) integrity rules are enforced. A fully relational database supports all eight relational algebra functions and enforces both integrity rules. A relationally complete database lacks only the enforcement of the integrity rules. A minimally relational database supports only SELECT, PROJECT, and JOIN and does not enforce the integrity rules. The tabular database supports only the relational algebra functions supported by the minimally relational database, and it requires that the access paths to the tables be defined. The more relational the relational database is, the greater its user flexibility, and the easier it is to design its applications.

A robust **Summary** at the end of each chapter ties together the major concepts and serves as a quick review for students.

REVIEW QUESTIONS

1. What is the difference between a database and a table?
2. What does a database expert mean when (s)he observes that a database displays both entity integrity and referential integrity?
3. Why are entity integrity and referential integrity important in a database?

Review Questions challenge students to apply the skills learned in each chapter.

PROBLEMS

Use the database shown in Figure P3.1 to work Problems 1 through 7. Note that the database is composed of four tables and reflects these relationships:

- An EMPLOYEE has only one JOB_CODE, but a JOB_CODE can be held by many EMPLOYEEs.
- An EMPLOYEE can participate in many PLANs, and any PLAN can be assigned to many EMPLOYEEs.

Note also that the M:N relationship has been broken down into two 1:M relationships for which the BENEFIT table serves as the composite or bridge entity.

Problems become progressively more complex as students draw on the lessons learned from the completion of preceding problems.

SPECIAL FEATURES AND SUPPORT

We recognize that teaching database systems is hard work. If database coverage is broad, practical, and detailed, the work required includes the actual designs and the implementation of those designs through databases, tables, and relationships. To make the teacher's life easier, we have made sure that teachers are given lots of support.

1. DATABASES ARE INCLUDED

As in the previous edition, this sixth edition includes all of the database structures and table contents on the CD-ROM included with the text. This feature ensures that:

- The database designs developed in the text meet all implementation requirements

- The designs are capable of meeting the stated information requirements

- The illustrations always match the actual database contents

- Teachers do not waste their time creating the database structures and the table contents to illustrate important design and information management principles

NOTE

Although we have most frequently used Microsoft Access screenshots to present the material in the text more attractively, *Database Systems: Design, Implementation, and Management* is not an Access book. Access was used for illustration purposes because it is widely available as part of the Microsoft Office suite and because it is simple to use. If you use a different DBMS, you could easily export the Access data to that DBMS.

2. SQL SCRIPTS ARE INCLUDED

For teachers using Oracle, we have provided the SQL scripts to create and load all tables used in the SQL chapters (6 and 7). In fact, advanced SQL coverage in Chapter 7 requires the use of an Oracle DBMS. In addition, we also provide scripts for all the SQL commands illustrated in chapters 6 and 7. (Naturally, scripts for the problem solutions are only provided to the teacher.) The teacher/student can copy the SQL command from the script file to the SQL command prompt. The scripts avoid the need to manually type the SQL commands used in the chapters. In addition, the scripts eliminate delays caused by typing errors. SQL Server scripts are also included on the Student CD.

```
<HTML>
<CFQUERY NAME="SearchVendor" DATASOURCE="RobCor">
    SELECT VEN_CODE, VEN_NAME, VEN_CONTACT_NAME, VEN_ADDRESS, VEN_CITY, VEN_STATE, VEN_PH
    FROM VENDOR
    WHERE 0=0
    <CFIF #FORM.VEN_CODE# IS NOT "">
        AND VENDOR.VEN_CODE LIKE '%#FORM.VEN_CODE#%'
    </CFIF>
    <CFIF #FORM.VEN_STATE# IS NOT "ANY">
        AND VENDOR.VEN_STATE LIKE '%#FORM.VEN_STATE#%'
    </CFIF>
    ORDER BY VEN_CODE
</CFQUERY>
<HEAD>
<TITLE>Rob & Coronel - ColdFusion Examples</TITLE>
</HEAD>
<BODY BGCOLOR="LIGHTBLUE">
<H1>
<CENTER><B>Vendor Search Results</B></CENTER>
</H1>
<BR>
<CFTABLE QUERY="SearchVendor" STARTROW="1" COLSPACING="1" COLHEADERS>
<CFCOL HEADER="<B>VENDOR CODE</B>" WIDTH="14" TEXT="#VEN_CODE#">
<CFCOL HEADER="<B>VENDOR NAME</B>" WIDTH="20" TEXT="#VEN_NAME#">
<CFCOL HEADER="<B>CONTACT PERSON</B>" WIDTH="20" TEXT="#VEN_CONTACT_NAME#">
<CFCOL HEADER="<B>ADDRESS</B>" WIDTH="20" TEXT="#VEN_ADDRESS#">
<CFCOL HEADER="<B>CITY</B>" WIDTH="20" TEXT="#VEN_CITY#">
<CFCOL HEADER="<B>STATE</B>" WIDTH="2" TEXT="#VEN_STATE#">
<CFCOL HEADER="<B>PHONE</B>" WIDTH="20" TEXT="#VEN_PH#">
</CFTABLE>
<CFIF #SEARCHVENDOR.RECORDCOUNT# IS 0>
    <H2>No records were found matching your criteria </H2>
<CFELSE>
    <CFOUTPUT>Your seach returned #SearchVendor.RecordCount# record(s).</CFOUTPUT>
</CFIF>
<FORM ACTION="rc-0.cfm" METHOD="post">
    <INPUT TYPE="submit" VALUE="Main Menu ">
</FORM>
</BODY>
</HTML>
```

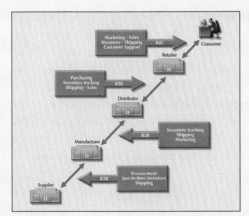

3. COLDFUSION SCRIPTS ARE INCLUDED

Chapter 14 illustrates the use of Web interfaces to access databases. ColdFusion was selected as interface software because it is a very powerful and simple Web development tool. The CD that accompanies this book includes a trial version of ColdFusion that you can install on your computer. We have included all the scripts that were used to develop the Web interfaces shown in the chapter.

4. DETAILED INSTRUCTOR'S MANUAL

Because we tackle so many problems in depth, teachers will find the Instructor's Manual especially useful. Our book emphasizes the practical aspects of the database environment, so we believe it is appropriate to show you the details of the design solution process. Also, where appropriate, we present notes about selected design procedures or alternative approaches that may be used to solve a particular problem. Finally, the book's questions and problems together with their answers and solutions are included in the Instructor's Manual. We hope that this service helps make your teaching life simpler and more rewarding.

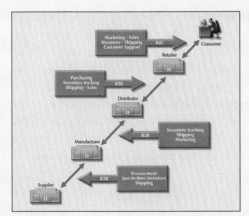

5. POWERPOINT PRESENTATIONS

To make sure that teachers do not need to create their own presentation materials, the support package includes PowerPoint presentations. These presentations follow each chapter's coverage precisely and include all appropriate illustrations.

6. VIDEO DEMONSTRATIONS

We have prepared video demonstrations in which we examine various aspects of database design, implementation, and applications development. These videos are available on the Course Technology Web site.

7. AN EXTENSIVE PROBLEM SET

As students work their way through the problem sets, the problems become progressively more complex while drawing on the lessons learned from the completion of preceding problems. The development of such comprehensive and carefully crafted problem sets makes it possible for students to gain the necessary experience to tackle real-world designs. Many of the problems are sufficiently complex to serve as cases or as class projects, especially if they are carried through the implementation and applications development stages.

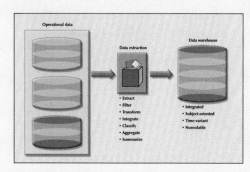

8. INCLUDES DATA WAREHOUSE AND DATA MINING COVERAGE

Chapter 12 introduces the data warehouse and how it may serve as the foundation for decision support activities. When done properly, decision support systems (DSSs) are designed to provide a computerized interface to enable business decision makers to creatively approach, analyze, and understand business problems. Given the existence of data warehouses, data mining has become a growing component of DSSs.

Because we focus on a hands-on approach to database coverage, we examine the criteria that form the basis for designing and building a data warehouse. Therefore, we are able to move beyond the basic concepts and show how the data warehouse "facts and dimensions" are used as the basis for the star schema and the multidimensional cube. Having provided practical examples, we offer students the opportunity to actually design and implement a small data warehouse.

9. DESIGNS ARE DEVELOPED TO REACH THE IMPLEMENTABLE LEVEL

As we developed database designs in our own classes, we discovered that too many students encountered problems when they tried to translate their designs into appropriate table structures and relationships. Those troubles were traceable to the fact that our relationships were sometimes displayed at the logical level, rather than at the implementation level. We have corrected that shortcoming. Optional relationships within the M:N framework were particularly troublesome; although we decomposed the M:N relationship into 1:M relationships with the help of a composite entity, some students found it difficult to migrate the optionality to the new structure. The revised entity relationship (ER) illustrations in Chapter 4, accompanied by examples, appear to have eliminated that hurdle. Teachers who choose to cover the SQL material in Chapters 6 and 7 will also be able to illustrate the impact of properly defined optional relationships by examining the SQL cascade delete and cascade update requirements.

10. COMMON GRAPHICAL MODELING DISPLAYS ARE USED

Simple Chen and Crow's Foot entity relationship diagrams (ERDs) are introduced as early as Chapter 2. Because many of the current generation CASE tools use the Crow's Foot model as the database design standard, we begin to explore the Crow's Foot ERD in greater depth in Chapter 3 and we finally make a complete switch to the Crow's Foot ERD in Chapter 4. However, some computer-based modeling tools are based on alternate ER presentation methods. Therefore, we also include coverage of the Rein85 and the IDEF1X methods in Chapter 4. By using a familiar Chen and Crow's Foot ERD as the starting point for a simple invoicing system, students find it easy to understand the alternative modeling styles.

11. SELF-DOCUMENTING NAMING CONVENTIONS

We have revised all the table structures to improve their self-documentation. Therefore, we introduce attribute naming conventions that ensure the use of descriptive attribute names, that use a prefix that ties the attribute to a table, and that make it obvious which attributes serve as foreign keys. By consistently adhering to these naming conventions, we make it easier to track the system's attributes during the design process. In addition, the use of such self-documenting naming conventions makes it much easier to develop end-user applications, especially when "drag-and-drop" techniques are used.

12. TABLE CONTENTS ARE SHOWN

When students begin their journey toward database design proficiency, some may have trouble visualizing the effect of design decisions. Seeing the table contents tends to produce those wonderful "I've got it!" moments. Examples make the more complex relationships more readily comprehensible. We have, therefore, included such examples: note particularly the expanded discussion of ternary relationships presented in Chapter 4.

13. COMPREHENSIVE TEST BANK

Because a database course requires hands-on design and implementation work, testing tends to follow that format. However, basic knowledge of concepts, implementation issues, and vocabulary may be tested with the help of a comprehensive test bank supplied to adopters. This test bank contains approximately 150 questions per chapter. The test bank includes graphics from the text and provides a mix of multiple-choice, true-false, fill-in-the-blank, and essay questions.

ACKNOWLEDGMENTS

Regardless of how many editions of this book are published, they will always rest on the solid foundation created by that first edition. We remain convinced that our work has become successful because that first edition was guided by Frank Ruggirello, a former Wadsworth Senior Editor and then Publisher. Quite aside from guiding the book's development, Frank also managed to solicit the great Peter Keen's (thankfully favorable) evaluation and subsequently convinced PK to write the foreword for the first edition. Although we sometimes found Frank to be an especially demanding taskmaster, we also found him to be a superb professional and a fine friend. We suspect Frank will still see his fingerprints all over our current work. Many thanks.

In many respects, *rewriting* a successful book is more difficult than writing the book in the first place. One of the truly difficult tasks is to decide what new approaches, topical coverage, and depth of coverage changes can or cannot fit into a book that has successfully weathered the test of the marketplace. The comments and suggestions made by the book's adopters, students, and reviewers play a major role in deciding what coverage is desirable and how that coverage is to be treated.

Some adopters became extraordinary reviewers, providing incredibly detailed and well-reasoned critiques even as they praised the book's coverage and style. Dr. David Hatherly, a superb database professional who is a Senior Lecturer in the School of Information Technology, Charles Sturt University–Mitchell, Bathhurst, Australia, made sure we knew precisely what issues led to his critiques. Better yet for us, he provided the suggestions that made it much easier to improve our topical coverage in our earlier editions. Dr. Hatherly's recommendations continue to be reflected in this sixth edition. All of his help was given freely and without prompting on our part. His efforts are much appreciated and our thanks are heartfelt.

We also owe a debt of gratitude to Professor Emil T. Cipolla, who teaches at St. Mary College. Professor Cipolla's wealth of IBM experience turned out to be a valuable resource when we tackled the embedded SQL coverage in Chapter 7.

Every technical book receives careful scrutiny by several groups of reviewers selected by the publisher. While holding them blameless for any shortcomings, we owe these reviewers many thanks for their contributions to this sixth edition:

Paul A. Seibert: North Greenville College, Neil Dunlop: Vista Community College, Samuel Sambasivam: Azusa Pacific University, Genard Catalano: Columbia College, Craig Shaw: Central Community College, Lei-da Chen: Creighton University, Linda K. Lau: Longwood University, Anita Lee-Post: University Of Kentucky, Lenore Horowitz: Schenectady County Community College, Dr. Scott L. Schneberger: Georgia State University, Tony Pollard: University of Western Sydney, Lejla Vrazalic: University of Wollongong, and David Witzany: Parkland College.

We would also like to thank the following reviewers for their efforts in helping to make the previous editions successful: Dr. Reza Barkhi, Pamplin College of Business, Virginia Polytechnic Institute and State University; Dr. Vance Cooney, Xavier University; Janusz Szczypula, Carnegie Mellon University, Dr. Ahmad Abuhejleh, University of Wisconsin, River Falls; Dr. Terence M. Baron, University of Toledo; Dr. Juan Estava, Eastern Michigan University; Dr. Kevin Gorman, University of North Carolina, Charlotte; Dr. Jeff Hedrington, University of Wisconsin, Eau Claire; Dr. Herman P. Hoplin, Syracuse University; Dr. Sophie Lee, University of Massachusetts, Boston; Dr. Michael Mannino, University of Washington; Dr. Carol Chrisman, Illinois State University; Dr. Timothy Heintz, Marquette University; Dr. Herman Hoplin, Syracuse University; Dr. Dean James, Embry-Riddle University; Dr. Constance Knapp, Pace University; Dr. Mary Ann Robbert, Bentley College; Dr. Francis J. Van Wetering, University of Nebraska; Dr. Joseph Walls, University of Southern California; Dr. Stephen C. Solosky, Nassau Community College; Dr. Robert Chiang, Syracuse University; Dr. Crist Costa, Rhode Island College; Dr. Sudesh M. Duggal, Northern Kentucky University; and Dr. Chang Koh, University of North Carolina, Greensboro.

In some respects, writing books resembles building construction: When 90% of the work seems done, 90% of the work remains to be done. Fortunately for us, we had a great team to work with. Deb Kaufmann is absolutely top of the line as a person and as a developmental editor. Her many fine professional touches have helped this sixth edition to be a grand improvement over its successful predecessor. Alyssa Pratt, our Product Manager, made sure that the team members stayed happy while they performed their many critical tasks. Alyssa also is the main contact for our all-important book reps and their professors when they have questions about the text and she has a hand in the marketing efforts. Melissa Panagos handled all of the production aspects of the book. As you can tell, she did a superb job—and she did it with panache. We were also fortunate to have an excellent copyeditor, Chris Smith. Marianne Snow and Danielle Shaw were our book's Quality Assurance testers. Such work requires extreme attention to detail and an unflagging effort to get all the testable technical material in A-OK shape. Janet Aras was this edition's Assistant Product Manager, and she handled our many supplements. Janet's work helps make life easier for both authors and adopters, so her hard work is much appreciated.

We also thank our students for their comments and suggestions. They are the reason for writing this book in the first place. Two student comments stand out in particular. "I majored in information systems for four years and I finally discovered why when I took your course." And one of our favorite comments by a former student was triggered by our question about the challenges created by a real-world information systems job: "Doc, it's just like class, only easier. You really prepared me well. Thanks!"

Last, and certainly not least, we thank our families for the solid home support. They graciously accepted that, during more than a year's worth of rewriting, there would be no free weekends, rare free nights, and even rarer free days. We owe you much and the dedication we wrote to you is but a small reflection of the important space you occupy in our hearts.

Peter Rob

Carlos M. Coronel

PART

I

DATABASE CONCEPTS

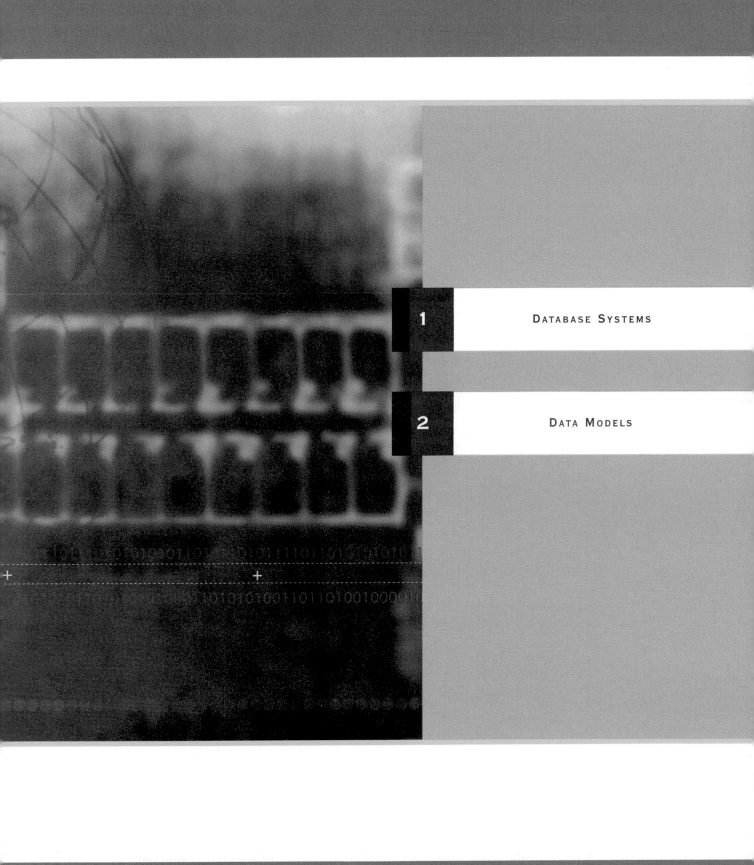

1 DATABASE SYSTEMS

2 DATA MODELS

DATABASE SYSTEMS

In this chapter, you will learn:

- ❖ The difference between data and information

- ❖ What a database is, about different types of databases, and why they are valuable assets for decision making

- ❖ Why database design is important

- ❖ How modern databases evolved from files and file systems

- ❖ About flaws in file system data management

- ❖ How a database system differs from a file system, and how a DBMS functions within the database system

Good decisions require good information, derived from raw facts known as data. Data are likely to be managed most efficiently when they are stored in a database. In this chapter you will learn what a database is, what it does, and why it yields better results than some other data management methods. You will also begin to appreciate why database design is so important.

Databases evolved from computer file systems. Although file system data management is now largely outmoded, understanding the basic characteristics of these file systems is important because they are the source of some serious data management limitations. If you understand a file system's flaws, you are more likely to understand why and how specific database properties are especially useful, and to employ these features appropriately. Finally, you will see how basic database features help eliminate most of the file system's data management shortcomings.

1.1 DATA VS. INFORMATION

To better understand what drives the design of databases, you must understand the difference between *data* and *information*. **Data** are raw facts. The word "raw" is used to indicate that the facts have not yet been *processed* to reveal their meaning. For example, suppose that ROBCOR Company tracks all sales for its two divisions through invoices. Each of the invoices contains raw facts such as these:

invoice number = 300124 invoice date = 12-JAN-2004 sales amount = $125.98

Further suppose that ROBCOR's two divisions have generated 1,380,456 and 1,453,907 invoices, respectively, between the first quarter of 1999 and the first quarter of 2004. Therefore, ROBCOR's raw data include 2,834,363 invoice numbers, 2,834,363 invoice dates, and 2,834,363 sales amounts. Given such an abundant data environment, coupled with employee data for each of the two divisions for each of these quarters, how likely is it that ROBCOR's managers can draw useful conclusions about sales productivity per employee for each of the two divisions? Examining these 2,834,363 invoices individually will merely overwhelm ROBCOR's managers. On the other hand, if they *process* these facts to yield total sales per quarter for each of the two divisions, and then divide these quarterly sales summaries by the quarterly employee count, as shown in Figure 1.1, ROBCOR's managers will have **information**, that is, data *processed* to reveal meaning. These results make it quite plain that the employees of Division 1 have greater sales productivity per employee than those of Division 2. Moreover, it's easy to see that the sales productivity gap is widening. Such information can then be used as the foundation for decision making. Data processing may be as simple as organizing the data to reveal patterns or as complex as statistical modeling to make forecasts or draw inferences.

FIGURE 1.1 **SALES PER EMPLOYEE FOR EACH OF ROBCOR'S TWO DIVISIONS**

Our era is called the "information age." This term recognizes that the production of accurate, relevant, and timely information is the key to good decision making. In turn, good decision making is the key to business survival in a global market.

Let's summarize some key points:

- Data constitute the building blocks of information.

- Information is produced by processing data.

- Information is used to reveal the meaning of data.

- Accurate, relevant, and timely information is the key to good decision making.

- Good decision making is the key to organizational survival in a global environment.

Timely and useful information requires accurate data. Such data must be generated properly, and it must be stored properly in a format that is easy to access and process. And, like any basic resource, the data environment must be managed carefully. **Data management** is a discipline that focuses on the proper generation, storage, and retrieval of data. Given the crucial role played by data, it should not surprise you that data management is a core activity for any business, government agency, service organization, or charity.

1.2 INTRODUCING THE DATABASE AND THE DBMS

Efficient data management typically requires the use of a computer database. A **database** is a shared, integrated computer structure that houses a collection of:

- End user data, that is, raw facts of interest to the end user.
- **Metadata**, or data about data, through which the data are integrated and managed.

The metadata provide a description of the data characteristics and the set of relationships that link the data found within the database. In a sense, a database resembles a very well-organized electronic filing cabinet in which powerful software, known as a *database management system*, helps manage the cabinet's contents. A **database management system (DBMS)** is a collection of programs that manages the database structure and controls access to the data stored in the database. The DBMS makes it possible to share the data in the database among multiple applications or users.

Because data are the crucial raw material from which information is derived, in our information-based society we must have a good way of managing such data. As you will discover in this book, the DBMS helps make data management much more efficient and effective. The following points are worth stressing:

- The DBMS helps create an environment in which end users have better access to more and better-managed data than they did before the DBMS became the data management standard. Such access makes it possible for end users to respond quickly to changes in their environment. The availability of data, combined with the tools that transform data into usable information, empowers end users to make quick, informed decisions that can make the difference between success and failure in the global economy.

- Wider access to well-managed data promotes an integrated view of the organization's operations and a clearer view of the "big picture." It becomes much easier to see how actions in one segment of the company affect other segments.

- The probability of data inconsistency is greatly reduced in a properly designed database that is managed through a DBMS. Better data make it possible to generate better information, on which better decisions are based.

- The DBMS makes it possible to produce quick answers to ad hoc queries. (A **query** is a question, and an **ad hoc query** is a spur-of-the-moment question.) For example, end users, when dealing with large amounts of sales data, might want quick answers to questions (ad hoc queries) such as:

 - What was the dollar volume of sales by product during the past six months?
 - What is the sales bonus figure for each of our salespeople during the past three months?
 - How many of our customers have credit balances of $3,000 or more?

The advantages of using a DBMS are not limited to the few we have just listed. In fact, you will discover many more advantages as you learn more about the technical details of databases and their proper design.

Figure 1.2 illustrates that the DBMS serves as the intermediary between the user and the database by translating user requests into the complex code required to fulfill those requests. The DBMS hides much of the database's internal complexity from the application programs that use the database. The application program might be written by a programmer using a programming language such as COBOL, Visual Basic, or C++, or it might be created through a DBMS utility program.

FIGURE 1.2 **THE DBMS MANAGES THE INTERACTION BETWEEN THE END USER AND THE DATABASE**

1.2.1 TYPES OF DATABASES

A DBMS can support many different types of databases. Databases can be classified according to the number of users, the database site location(s), and the expected type and extent of use.

The number of users determines whether the database is classified as single-user or multiuser. A **single-user database** supports only one user at a time. In other words, if user A is using the database, users B and C must wait until user A has completed his/her database work. If a single-user database runs on a personal computer, it is also called a **desktop database**. In contrast, a **multiuser database** supports multiple users at the same time. If the multiuser database supports a relatively small number of users (usually fewer than 50) or a specific department within an organization, it is called a **workgroup database**. If the database is used by the entire organization and supports many users (more than 50, usually hundreds) across many departments, the database is known as an **enterprise database**.

The database site location might also be used to classify the database. For example, a database that supports data located at a single site is called a **centralized database**. A database that supports data distributed across several different sites is called a **distributed database**. The extent to which a database can be distributed and the way in which such distribution is managed is addressed in detail in Chapter 10, "Distributed Database Management Systems."

The most popular way of classifying databases today, however, is based on how they will be used, and on the time sensitivity of the information gathered from them. For example, transactions such as product or service sales, payments, and supply purchases reflect critical day-to-day operations. Such transactions are time-critical and must be recorded accurately and immediately. A database that is primarily designed to support a company's day-to-day operations is classified as a **transactional database** or a **production database**. In contrast, a **data warehouse database** focuses primarily on the storage of data used to generate information required to make tactical or strategic decisions. Such decisions typically require extensive "data massaging" (data manipulation) to extract information from historical data to formulate pricing decisions, sales forecasts, market positioning, and so on. Because most decision support information is based on historical data, the time factor is not likely to be as critical as for transactional databases. Additionally, the data warehouse database can store complex data derived from many sources. To make it easier to retrieve such complex data, the data warehouse database structure is quite different from that of a transaction-oriented database. We cover the design, implementation, and use of data warehouse databases and their "offspring" in detail in Chapter 12, "The Data Warehouse."

> **N O T E**
>
> Most of the database design, implementation, and management issues addressed in this book are based on production (transaction) databases. Our focus on production databases is based on two considerations. First, they are the databases you most frequently encounter in common activities such as enrolling in a class, registering your car, buying a product, or making a bank deposit or withdrawal. Second, data warehouse databases derive most of their data from production databases, and if production databases are poorly designed, the data warehouse databases based on them will lose their reliability and value as well.

1.3 WHY DATABASE DESIGN IS IMPORTANT

A good database does not just happen; the structure of its contents must be designed carefully. In fact, database design is such a crucial aspect of working with databases that most of this book is dedicated to the development of good database design techniques. Even a good DBMS will perform poorly with a badly designed database.

> **N O T E**
>
> Keep in mind that the DBMS is the database software you buy commercially; you do not have the option of making design changes to it. Therefore, when we speak of **database design**, we mean the design of the database structure that will be used to store and manage data, rather than the design of the DBMS software. Once the database design is completed, the DBMS handles all the complicated activities required to translate the designer's view of the structures that are usable by the computer.

Proper database design requires the database designer to precisely identify the database's expected use. Designing a transactional database emphasizes data integrity, data consistency, and operational speed. The design of a data warehouse database recognizes the use of historical and aggregated data. Designing a database to be used in a centralized, single-user environment requires a different approach from that used in the design of a distributed, multiuser database. In this book we will emphasize the design of transactional, centralized, single-user, and multiuser databases. However, we will also examine critical issues confronting the designer of distributed and data warehouse databases in Chapter 10 and Chapter 12.

A well-designed database facilitates data management and becomes a valuable information generator. A poorly designed database will likely become a breeding ground for **redundant data**, that is, unnecessarily duplicated data. Redundant data are often the source of difficult-to-trace information errors. A database contains redundant data when the same data about the same entity are kept in different locations. For example, data redundancy exists when a customer's telephone number is stored in a customer file, a sales agent file, and an invoice file. If the customer's phone number changes, the correction might not be made in all the locations where it occurs. Therefore, the existence of redundant data can produce uncorrected data entries, and you probably won't know which value is the correct one. Reports might yield different results, depending on which version of the data was used. In short, uncontrolled data redundancies are typical of a poorly designed database.

A poorly designed database tends to generate errors that are likely to lead to bad decisions, and bad decisions can lead to the failure of an organization. Database design is simply too important to be left to luck. That's why college students study database design, why organizations of all types and sizes send personnel to database design seminars, and why database design consultants often make an excellent living. In this book, we take a practical approach to database design. If you want to learn how to design a useful database, this book will point you in the right direction. If you want to develop a thorough understanding of the outer limits of database theory, this is not the book for you.

In addition to including the material necessary to build an appropriate database vocabulary and to understand basic database concepts, we have also developed two complete applications through the logical design stage. (The **logical design** is the structural blueprint of the database.) We believe that exposure to the practical aspects of the design process will properly set the stage for your own successful designs.

1.4 THE HISTORICAL ROOTS OF THE DATABASE: FILES AND FILE SYSTEMS

Although file systems as a way of managing data are now largely obsolete, there are several good reasons for studying them in some detail:

- An understanding of the relatively simple characteristics of file systems makes the complexity of database design easier to understand.

- An awareness of the problems that plagued file systems can help you avoid such pitfalls with DBMS software.

- If you intend to convert an obsolete file system to a database system, knowledge of the file system's basic limitations will be useful.

In the recent past, a manager of almost any small organization was (and sometimes still is) able to keep track of necessary data by using a manual file system. Such a file system was traditionally composed of a collection of file folders, each properly tagged and kept in a filing cabinet. Organization of the data within the file folders was determined by the data's expected use. Ideally, the contents of each file folder were logically related. For example, a file folder in a doctor's office might contain patient data, one file folder for each patient. All the data in that file folder described only that particular patient's medical history. Similarly, a personnel manager might organize personnel data by category of employment (clerical, technical, sales, administrative, etc.). Therefore, a file folder labeled "Technical" would contain data pertaining to only those people whose duties were properly classified as technical.

As long as a data collection was relatively small and an organization's managers had few reporting requirements, the manual system served its role well as a data repository. However, as organizations grew and as reporting requirements became more complex, keeping track of data in a manual file system became more difficult. In fact, finding and using data in growing collections of file folders became such a time-consuming and cumbersome task that it became less and less likely that such data would ever generate useful information. Consider just these few questions that a retail business owner might want to answer:

- What products sold well during the past week, month, quarter, or year?

- What is the current daily, weekly, monthly, quarterly, or yearly sales dollar volume?

- How do the current period's sales compare to those of last week, last month, or last year?

- Were the various cost categories increasing, decreasing, or remaining stable during the past week, month, quarter, or year?

- Did sales show trends that could change the inventory requirements?

The list of questions tends to be long and growing!

Unfortunately, report generation from a manual file system can be slow and cumbersome. In fact, some business managers faced government-imposed reporting requirements that required weeks of intensive effort each quarter, even when a well-designed manual system was used. Consequently, the pressure built to design a computer-based system that would track data and produce required reports.

The conversion from a manual file system to a matching computer file system could be technically complex. (Because we are accustomed to today's relatively user-friendly computer interfaces, we have already forgotten how painfully hostile computers used to be!) Consequently, a new kind of professional, known as a **data processing (DP) specialist**, had to be hired or "grown" from the current staff. The DP specialist created the necessary computer file structures, often wrote the software that managed the data within those structures, and designed the application programs that produced reports based on the file data. Thus, numerous home-grown computerized file systems were born.

Initially, the computer files within the file system were similar to the manual files. A simple example of a customer data file for a small insurance company is shown in Figure 1.3. (You will discover later that the file structure shown in Figure 1.3, although typically found in early file systems, is unsatisfactory for a database. The search for solutions to structure-induced problems helped pave the way for the development of different data storage concepts and methods that are discussed later in this chapter.)

FIGURE 1.3 CONTENTS OF THE CUSTOMER FILE

C_NAME	C_PHONE	C_ADDRESS	C_ZIP	A_NAME	A_PHONE	TP	AMT	REN
Alfred A. Ramas	615-844-2573	218 Fork Rd., Babs, TN	36123	Leah F. Hahn	615-882-1244	T1	$100.00	05-Apr-2004
Leona K. Dunne	713-894-1238	Box 12A, Fox, KY	25246	Alex B. Alby	713-228-1249	T1	$250.00	16-Jun-2004
Kathy W. Smith	615-894-2285	125 Oak Ln, Babs, TN	36123	Leah F. Hahn	615-882-2144	S2	$150.00	29-Jan-2005
Paul F. Olowski	615-894-2180	217 Lee Ln., Babs, TN	36123	Leah F. Hahn	615-882-1244	S1	$300.00	14-Oct-2004
Myron Orlando	615-222-1672	Box 111, New, TN	36155	Alex B. Alby	713-228-1249	T1	$100.00	28-Dec-2004
Amy B. O'Brian	713-442-3381	387 Troll Dr., Fox, KY	25246	John T. Okon	615-123-5589	T2	$850.00	22-Sep-2004
James G. Brown	615-297-1228	21 Tye Rd., Nash, TN	37118	Leah F. Hahn	615-882-1244	S1	$120.00	25-Mar-2004
George Williams	615-290-2556	155 Maple, Nash, TN	37119	John T. Okon	615-123-5589	S1	$250.00	17-Jul-2004
Anne G. Farriss	713-382-7185	2119 Elm, Crew, KY	25432	Alex B. Alby	713-228-1249	T2	$100.00	03-Dec-2004
Olette K. Smith	615-297-3809	2782 Main, Nash, TN	37118	John T. Okon	615-123-5589	S2	$500.00	14-Mar-2004

C_NAME = Customer name
C_PHONE = Customer phone
C_ADDRESS = Customer address
C_ZIP = Customer ZIP code

A_NAME = Agent name
A_PHONE = Agent phone
TP = Insurance type
AMT = Insurance policy amount, in thousands of $
REN = Insurance renewal date

ADDITIONAL CD CONTENT

The databases used in the chapters are available on the student CD bundled with this book. Each chapter's databases are stored in a folder that bears the chapter label. For example, this chapter's databases are stored in the **Databases\Student\Ch01** folder. Throughout the book, Additional CD Content boxes will highlight related material located on the CD.

We have used a Microsoft Access database to simulate the file structures shown in Figures 1.3 and 1.4. For example, the file contents shown in Figures 1.3 and 1.4 are shown in the **Ch01_Text** database as CUSTOMER_INIT and AGENT_INIT.

The description of computer files requires a specialized vocabulary. Every discipline develops its own jargon to enable its practitioners to communicate very precisely. The basic file vocabulary summary shown in Table 1.1 will help you understand subsequent discussions more easily.

TABLE 1.1 BASIC FILE TERMINOLOGY

TERM	DEFINITION
Data	"Raw" facts, such as a telephone number, a birth date, a customer name, and a year-to-date (YTD) sales value. Data have little meaning unless they have been organized in some logical manner. The smallest piece of data that can be "recognized" by the computer is a single character, such as the letter A, the number 5, or a symbol such as /. A single character requires one byte of computer storage.
Field	A character or group of characters (alphabetic or numeric) that has a specific meaning. A field is used to define and store data.
Record	A logically connected set of one or more fields that describes a person, place, or thing. For example, the fields that constitute a record for a customer named J.D. Rudd might consist of J.D. Rudd's name, address, phone number, date of birth, credit limit, and unpaid balance.
File	A collection of related records. For example, a file might contain data about vendors of ROBCOR Company; or a file might contain the records for the students currently enrolled at Gigantic University.

Using the proper file terminology given in Table 1.1, we can identify the file components shown in Figure 1.3. Note that the CUSTOMER file shown in Figure 1.3 contains 10 records. Each record is composed of nine fields: C_NAME, C_PHONE, C_ADDRESS, C_ZIP, A_NAME, A_PHONE, TP, AMT, and REN. The 10 records are stored in a named file. Because the file in Figure 1.3 contains customer data, its filename is CUSTOMER.

Using the CUSTOMER file's contents, the DP specialist wrote programs that produced very useful reports for the sales department:

- Monthly summaries that showed the types and amounts of insurance sold by each agent. (Such reports might be used to analyze each agent's productivity.)

- Monthly checks to determine which customers must be contacted for renewal.

- Reports that analyzed the ratios of insurance types sold by each agent.

- Periodic customer contact letters designed to summarize coverage and to provide various customer relations bonuses.

As time went on, additional programs were written to produce new reports. Although it took some time to specify the report contents and to write the programs that produced the reports, the sales department manager did not miss the old manual system—using the computer saved much time and effort. The reports were impressive, and the ability to perform complex data searches yielded the information needed to make sound decisions.

Then the sales department created a file named SALES, which helped track daily sales efforts. Additional files were created as needed to produce even more useful reports. In fact, the sales department's success was so obvious that the personnel department manager demanded access to the DP specialist to automate payroll processing and other personnel functions. Consequently, the DP specialist was asked to create the AGENT file shown in Figure 1.4. The data in the AGENT file were used to write checks, keep track of taxes paid, summarize insurance coverage, and so on.

FIGURE 1.4 CONTENTS OF THE AGENT FILE

	A_NAME	A_PHONE	A_ADDRESS	ZIP	HIRED	YTD_PAY	YTD_FIT	YTD_FICA	YTD_SLS	DEP
▶	Alex B. Alby	713-228-1249	123 Toll, Nash, TN	37119	01-Nov-1998	$26,566.24	$6,641.56	$2,125.30	$132,735.75	3
	Leah F. Hahn	615-882-1244	334 Main, Fox, KY	25246	23-May-1984	$32,213.76	$8,053.44	$2,577.10	$138,967.35	0
	John T. Okon	615-123-5589	452 Elm, New, TN	36155	15-Jun-2003	$23,198.29	$5,799.57	$1,855.86	$127,093.45	2

A_NAME	= Agent name	YTD_PAY	= Year-to-date pay
A_PHONE	= Agent phone	YTD_FIT	= Year-to-date federal income tax paid
A_ADDRESS	= Agent address	YTD_FICA	= Year-to-date Social Security taxes paid
ZIP	= Agent zip code	YTD_SLS	= Year-to-date sales
HIRED	= Agent date of hire	DEP	= Number of dependents

As the number of files increased, a small file system, like the one shown in Figure 1.5, evolved. Each file in the system used its own application programs to store, retrieve, and modify data. And each file was owned by the individual or the department that commissioned its creation.

FIGURE 1.5 A SIMPLE FILE SYSTEM

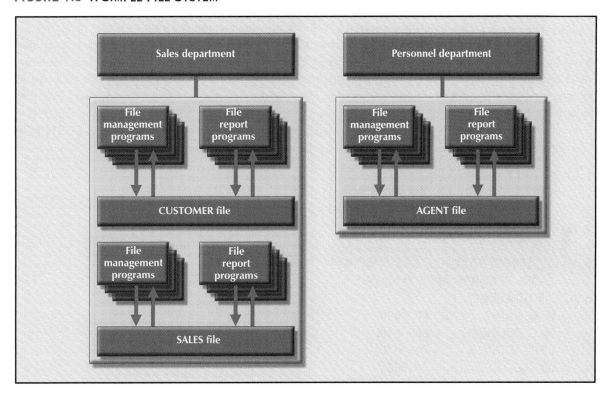

As the file system grew, the demand for the DP specialist's programming skills grew even faster. Therefore, the DP specialist was authorized to hire additional programmers. The size of the file system also required a larger, more complex computer. The new computer and the additional programming staff caused the DP specialist to spend less time programming and more time managing technical and human resources. Therefore, the DP specialist's job evolved into that of a **data processing (DP) manager**, who supervised a DP department. In spite of these organizational changes, however, the DP department's primary activity remained programming, and the DP manager inevitably spent much time as a supervising senior programmer and program troubleshooter.

1.5 PROBLEMS WITH FILE SYSTEM DATA MANAGEMENT

Although the file system method of organizing and managing data was a definite improvement over a manual system, many problems and limitations became evident in this approach, as described in this section. However, as you learn about these problems, remember that the file system served a useful purpose in data management for over two decades, a very long timespan in the computer era. Our critique serves two major purposes:

1. Understanding the shortcomings of the file system enables us to understand the development of modern databases.

2. Many of the problems we will catalog are not unique to file systems. Failure to understand such problems is likely to lead to their duplication in a database environment, even though database technology makes it relatively easy to avoid them.

The first and most glaring problem with the file system approach is that even the simplest data-retrieval task requires extensive programming in a **third-generation language (3GL)**. A 3GL requires the programmer to

specify both *what* must be done and *how* it is to be done. Examples of 3GLs include Common Business-Oriented Language (COBOL), Beginner's All-purpose Symbolic Instruction Code (BASIC), and FORmula TRANslation (FORTRAN). As you will learn in upcoming chapters, more modern databases use a **fourth-generation language (4GL)**. A 4GL allows the user to specify what must be done without specifying how it must be done.

Programming in a 3GL can be a time-consuming, high-skill activity. Because the simple file system depicted in Figure 1.5 is quite different from the way the computer physically stores the data on disk, the programmer must be familiar with the physical file structure, that is, how and where the files are stored in the computer. Therefore, every file reference in a program requires the programmer to define access paths to the data. Such access paths use complex coding to establish the precise location of the various file and system components and their data characteristics. As file systems become more complex, the access paths become difficult to manage and tend to produce system malfunctions. For example, if you were working in a COBOL environment with the CUSTOMER file illustrated in Figure 1.3, you would need to include the following lines in your program:

```
21. SELECT CUSTOMER-DATA-FILE.
22.      ASSIGN TO "C:\DATAFILES\CUSTOMER.DAT"
23.             ORGANIZATION IS RECORD SEQUENTIAL.
24. DATA DIVISION.
25. FILE SECTION.
26. FD CUSTOMER-DATA-FILE
27.      LABEL RECORDS ARE STANDARD.
28. 01 CUST-DATA-REC.
29.      05 C_NAME        PIC X(20).
30.      05 C_PHONE       PIC X(10).
31.      05 C_ADDRESS     PIC X(30).
32.      05 C_ZIP         PIC X(5).
33.      05 A_NAME        PIC X(20).
34.      05 A_PHONE       PIC X(10).
35.      05 TP            PIC X(2).
36.      05 AMT           PIC 9(3)V99.
37.      05 REN           PIC X(11).
```

Even if you don't know how to write COBOL programs, the brief excerpt shown here illustrates that the programmer must predefine the physical location of the file (line 22) and also the data access path characteristics (line 23). If the CUSTOMER data file location is changed, all the programs using it must be changed to reflect the new location.

The need to write 3GL programs to produce even the simplest reports makes ad hoc queries impossible. Harried DP specialists or DP managers who work with mature file systems often receive numerous requests for new reports. They are often forced to say that the report will be ready "next week" or even "next month." If you need the information *now*, getting it next week or next month will not serve your information needs well.

Another problem, related to the need for extensive programming, is that as the number of files in the system expands, system administration becomes difficult. Each file must have its own file management system, composed of programs that allow the user to:

- Create the file structure.
- Add data to the file.
- Delete data from the file.
- Modify the data contained in the file.
- List the file contents.

Even a simple file system of only 20 files requires $5 \times 20 = 100$ file management programs. If each of the files is accessed by 10 different reporting programs, an additional $20 \times 10 = 200$ programs must be written. Because ad hoc queries are not possible, the file reporting programs can multiply quickly. And, because each department in the organization owns its data by creating its own files, the number of files can multiply rapidly.

Making changes in an existing structure can be difficult in a file system environment. For example, changing just one field in the original CUSTOMER file (such as changing the AMT field in the preceding COBOL example to handle amounts over 999.99) requires a program that:

1. Puts the new file structure into a special portion of the computer's memory known as a buffer.

2. Opens the original file, using a different buffer.

3. Reads a record from the original file.

4. Transforms the original data to conform to the new structure's storage requirements.

5. Writes the transformed data into the new file structure.

6. Deletes the original file.

7. Modifies all the programs that use the CUSTOMER file to fit the revised file structure.

In fact, any file structure change, no matter how minor, forces modifications in all the programs that use the data in that file. Modifications are likely to produce errors (bugs), and additional time can be spent using a debugging process to find those errors. The advice "think before you do" is especially valid in the file system environment, because *all* data access programs are subject to change when even minor changes in the file structure are made.

Another fault is that security features such as effective password protection, locking out parts of files or parts of the system itself, and other measures designed to safeguard data confidentiality are difficult to program and are, therefore, often omitted in a file system environment. Even when an attempt is made to improve system and data security, the security devices tend to be limited in scope and effectiveness.

To summarize the limitations of file system data management so far:

- It requires extensive programming.

- System administration can be complex and difficult.

- It is difficult to make changes to existing structures.

- Security features are likely to be inadequate.

These limitations in turn lead to problems of structural and data dependency.

1.5.1 STRUCTURAL AND DATA DEPENDENCE

In the preceding sections you saw how a change in any file's structure, such as the addition or deletion of a field, requires the modification of all programs using that file. Such modifications are required because the file system exhibits **structural dependence**; that is, access to a file is dependent on its structure.

Even changes in file data characteristics, such as changing a field from integer to decimal, require changes in all programs that access the file. Because all data access programs are subject to change when any of the file's data characteristics change, the file system is said to exhibit **data dependence**. (Conversely, **structural independence** exists when it is possible to make changes in the database structure without affecting the application program's ability to access the data.)

The practical significance of data dependence is the difference between the **logical data format** (how the human being views the data) and the **physical data format** (how the computer "sees" the data). Any program that accesses a file system's file must not only tell the computer what to do, but also how to do it. Consequently,

each program must contain lines that specify the opening of a specific file type, its record specification, and its field definitions. Data dependence thus makes the file system extremely cumbersome from a programming and data management point of view.

1.5.2 FIELD DEFINITIONS AND NAMING CONVENTIONS

At first glance the CUSTOMER file shown in Figure 1.3 appears to have served its purpose well: requested reports usually could be generated. But suppose we want to create a customer phone directory based on the data stored in the CUSTOMER file. Storing the customer name as a single field turns out to be a liability because the directory must break up the field contents to list the last names, first names, and initials in alphabetical order. Or suppose we want to get a customer listing by area code. Including the area code in the phone number field is inefficient.

Similarly, producing a listing of customers by city is a more difficult task than is necessary. From the user's point of view, a much better (more flexible) record definition would be one that anticipates reporting requirements by breaking up fields into their component parts. Thus the CUSTOMER file's fields might be listed as shown in Table 1.2.

TABLE 1.2 SAMPLE CUSTOMER FILE FIELDS

FIELD	CONTENTS	SAMPLE ENTRY
CUS_LNAME	Customer last name	Ramas
CUS_FNAME	Customer first name	Alfred
CUS_INITIAL	Customer initial	A
CUS_AREACODE	Customer area code	615
CUS_PHONE	Customer phone	234-5678
CUS_ADDRESS	Customer street address or box number	123 Green Meadow Lane
CUS_CITY	Customer city	Murfreesboro
CUS_STATE	Customer state	TN
CUS_ZIP	Customer zip code	37130

Selecting proper field names is also important. For example, make sure that the field names are reasonably descriptive. As you examine the file structure shown in Figure 1.3, it is not obvious that the field name REN represents the customer's insurance renewal date. Using the field name CUS_RENEW_DATE would be better for two reasons. First, the prefix CUS can be used as an indicator of the field's origin, which is the CUSTOMER file. Therefore, we know that the field in question yields a CUSTOMER property. Second, the RENEW_DATE portion of the field name is more descriptive of the field's contents. With proper naming conventions, the file structure becomes self-documenting. That is, by simply looking at the field names we can determine which files the fields belong to and what information the fields are likely to contain.

Some software packages place restrictions on the length of field names, so it is wise to be as descriptive as possible within those restrictions. In addition, remember that very long field names make it difficult to fit more than a few fields on a page, thus making output spacing a problem. For example, the field name CUSTOMER_INSURANCE_RENEWAL_DATE, while being self-documenting, is less desirable than CUS_RENEW_DATE.

Another problem in Figure 1.3's CUSTOMER file is the difficulty of finding desired data efficiently. Note that the CUSTOMER file currently does not have a unique record identifier. For example, it is possible to have several

customers named John B. Smith. Consequently, the addition of a CUS_ACCOUNT field that contains a unique customer account number would be appropriate.

Although we have just criticized the field definitions and naming conventions shown in the file structure of Figure 1.3, the points we raised are not unique to file systems. Because such conventions will prove to be important later, we want you to be introduced to them early. You will revisit field definitions and naming conventions when you learn about database design in Chapter 4, "Entity Relationship (ER) Modeling," when you learn about database implementation issues in Chapter 8, "The Database Design Life Cycle," and when you see an actual database design implemented in Appendixes D and E ("The University Lab" design and implementation). Regardless of the data environment, the design—whether it involves a file system or a database—must always reflect the designer's documentation needs and the end user's reporting and processing requirements. Both types of needs are served best by adhering to proper field definitions and naming conventions.

> **NOTE**
>
> Keep in mind that no naming convention can fit all requirements for all systems. Some words or phrases are reserved for the DBMS's internal use. For example, the name ORDER generates an error in some DBMSs. Similarly, your DBMS might interpret a dash (-) as a command to subtract. Therefore, the field CUS-NAME would be interpreted as a command to subtract the NAME field from the CUS field. Because neither field exists, you would get an error message. On the other hand, CUS_NAME would work fine, because it uses an underscore.

1.5.3 DATA REDUNDANCY

The file system's structure and lack of security make it difficult to pool data. The organizational structure promotes data ownership, thus promoting the storage of the same basic data in different locations. (Database professionals use the term **islands of information** to label such scattered data locations.) Because it is unlikely that data stored in different locations will always be updated consistently, the islands of information often contain different versions of the same data. For example, in Figures 1.3 and 1.4, the agent names and phone numbers occur in both the CUSTOMER and the AGENT files. You need only one correct copy of the agent names and phone numbers. Having them occur in more than one place unnecessarily produces data redundancy. (In short, a condition of **data redundancy** exists when the data environment contains redundant—unnecessarily duplicated—data).

Uncontrolled data redundancy sets the stage for:

1. *Data inconsistency.* **Data inconsistency** exists when different and conflicting versions of the same data appear in different places. For example, suppose we change an agent's phone number or address in the AGENT file. If we forget to make corresponding changes in the CUSTOMER file, the files contain different data for the same agent. Reports yield inconsistent results, depending on which version of the data is used. Data that display data inconsistency are also referred to as data that lack **data integrity**.

 Data entry errors are much more likely to occur when complex entries (such as 10-digit phone numbers) are made in several different files and/or recur frequently in one or more files. In fact, the CUSTOMER file shown in Figure 1.3 contains just such an entry error: the third record in the CUSTOMER file has a transposed digit in the agent's phone number (615-882-2144 rather than 615-882-1244).

 It is possible to enter a nonexistent sales agent's name and phone number into the CUSTOMER file, but customers are not likely to be impressed if the insurance agency supplies the name and phone number of an agent who does not exist. And should the personnel manager allow a nonexistent agent to accrue bonuses and benefits? In fact, a data entry error such as an incorrectly spelled name or an incorrect phone number yields the same kind of data integrity problems.

2. *Data anomalies.* The dictionary defines "anomaly" as an abnormality. Ideally, a field value change should be made only in a single place. Data redundancy, however, fosters an abnormal condition by forcing field value changes in many different locations. Look at the CUSTOMER file in Figure 1.3. If agent Leah F. Hahn decides to get married and move, the agent name might change, and the address and phone will most likely change. Instead of making just a single name and/or phone/address change in a single file (AGENT), we must make the change each time that agent's name, phone number, and address occur in the CUSTOMER file, too. We could be faced with the prospect of making hundreds of corrections, one for each of the customers served by that agent! The same problem occurs when an agent decides to quit. Each customer served by that agent must be assigned a new agent. Any change in any field value must be correctly made in many places to maintain data integrity. A **data anomaly** develops when all the required changes in the redundant data are not made successfully. The data anomalies found in Figure 1.3 are commonly defined as:

- *Modification anomalies.* If agent Leah F. Hahn has a new phone number, that new number must be entered in each of the CUSTOMER file records in which Ms. Hahn's phone number is shown. In this case, only three changes must be made. In a large file system, such changes might occur in hundreds or even thousands of records. Clearly, the potential for data inconsistencies is great.

- *Insertion anomalies.* To add each new customer in the CUSTOMER file, we must also add the corresponding agent data. If we add several hundred new customers, we must also enter several hundred agent names and telephone numbers. Again, the potential for creating data inconsistencies is great.

- *Deletion anomalies.* If agent Alex B. Alby quits and is deleted from the payroll, all the customers in the CUSTOMER file refer to a nonexistent agent. To resolve this problem, we must modify all records in which Mr. Alby's name and phone number appear.

1.6 DATABASE SYSTEMS

The problems inherent in file systems make using a database system very desirable. Unlike the file system, with its many separate and unrelated files, the database consists of logically related data stored in a single logical data repository. (The "logical" label reflects the fact that, although the data repository *appears* to be a single unit to the end user, its contents may actually be physically distributed among multiple data storage facilities and/or locations.) Because the database's data repository is a single logical unit, the database represents a major change in the way end user data are stored, accessed, and managed. The database's DBMS, shown in Figure 1.6, provides numerous advantages over file system management, shown in Figure 1.5, by making it possible to eliminate most of the file system's data inconsistency, data anomalies, data dependency, and structural dependency problems. Better yet, the current generation of DBMS software stores not only the data structures, but also the relationships between those structures and the access paths to those structures, all in a central location. The current generation of DBMS software also takes care of defining, storing, and managing all the required access paths to those components.

Remember that the DBMS is just one of several crucial components of a database system. Perhaps it is even appropriate to refer to the DBMS as the database system's heart. However, just as it takes more than a heart alone to make a human being function, it takes more than a DBMS to make a database system function. In the sections that follow, you'll learn what a database system is, what its components are, and how the DBMS fits into the database system picture.

FIGURE 1.6 CONTRASTING DATABASE AND FILE SYSTEMS

1.6.1 THE DATABASE SYSTEM ENVIRONMENT

The term **database system** refers to an organization of components that define and regulate the collection, storage, management, and use of data within a database environment. From a general management point of view, the database system is composed of the five major parts shown in Figure 1.7: hardware, software, people, procedures, and data.

Let's take a closer look at the five components shown in Figure 1.7:

1. *Hardware.* Hardware refers to all the system's physical devices. The database system's main and most easily identified hardware component is the computer, which might be a microcomputer, a minicomputer, or a mainframe computer. The hardware also includes all of the computer peripherals, which are the physical devices that control computer input and output, such as keyboards, mice, modems, and printers. Hardware also includes devices that are used to connect two or more computers, thereby producing a computer network. Networks are an essential part of modern database systems, because data are likely to be accessed from a local network, from remote locations such as airplane reservation systems and automatic teller machines, or over the Internet, or all of the above.

2. *Software.* Software refers to the collection of programs used by the computers within the database system. Although the most readily identified software is the DBMS itself, to make the database system function fully it takes three types of software: operating system software, DBMS software, and application programs and utilities.

 - *Operating system software* manages all hardware components and makes it possible for all other software to run on the computers. Examples of operating system software include DOS, Microsoft Windows versions 95 / 98 / ME / NT / 2000 / XP, Linux, MacOS; UNIX, and MVS.

 - *DBMS software* manages the database within the database system. Some examples of DBMS software include Microsoft Access and SQL Server, Oracle Corporation's Oracle, and IBM's DB2.

FIGURE 1.7 THE DATABASE SYSTEM ENVIRONMENT

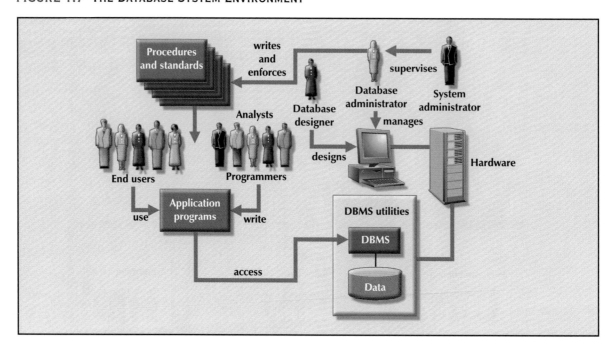

- *Application programs and utility software* are used to access and manipulate the data in the DBMS and to manage the computer environment in which data access and manipulation take place. Application programs are most commonly used to access the data found within the database to generate reports, tabulations, and other information to facilitate decision making. Utilities are the software tools used to help manage the database system's computer components. For example, all the major DBMS vendors now provide GUI interfaces to help database administrators create database structures, control database access, and monitor database operations.

3. *People.* This component includes all users of the database system. On the basis of primary job functions, we can identify five types of users in a database system: systems administrators, database administrators, database designers, systems analysts/programmers, and end users. The members of each user type perform both unique and complementary functions:

 - *Systems administrators* oversee the database system's general operations.

 - *Database administrators*, also known as **DBAs**, manage the DBMS's use and ensure that the database is functioning properly. The DBA's role is sufficiently important to warrant a detailed exploration in Chapter 15, "Database Administration."

 - *Database designers* design the database structure. They are, in effect, the database architects. If the database design is poor, even the best application programmers and the most dedicated DBAs cannot produce a useful database environment. To use an old analogy, the finest bricklayers and carpenters cannot produce a good building from a bad blueprint. Because organizations strive to optimize their data resources, the database designer's job description has expanded to cover new dimensions and growing responsibilities.

 - *Systems analysts and programmers* design and implement the application programs. They design and create the data entry screens, reports, and procedures through which end users access and manipulate the database's data.

 - *End users* are the people who use the application programs to run the organization's daily operations. For example, sales clerks, supervisors, managers, and directors are all classified as end users. High-level end users employ the information obtained from the database to make tactical and strategic business decisions.

4. *Procedures.* Procedures are the instructions and rules that govern the design and use of the database system. Procedures are a critical, although occasionally forgotten, component of the system. Procedures play an important role in a company, because they enforce the standards by which business is conducted within the organization and with customers. Procedures also are used to ensure that there is an organized way to monitor and audit both the data that enter the database and the information that is generated through the use of such data.

5. *Data.* The word "data" covers the collection of facts stored in the database. Because data are the raw material from which information is generated, the determination of which data are to be entered into the database and how such data are to be organized is a vital part of the database designer's job.

The existence of a database system adds a new dimension to an organization's management structure. Just how complex this managerial structure is depends on the organization's size, its functions, and its corporate culture. Therefore, database systems can be created and managed at quite different levels of complexity and with widely varying adherence to precise standards. For example, compare a local movie rental system with a national insurance claims system. The movie rental system might be managed by two people, the hardware used is probably a single microcomputer, the procedures are probably simple, and the data volume will tend to be low. The national insurance claims system is likely to have at least one systems administrator, several full-time DBAs, and many designers and programmers; the hardware probably includes several mainframes at multiple locations throughout the United States; the procedures are likely to be numerous, complex, and rigorous; and the data volume will tend to be very high.

In addition to the fact that different levels of database system complexity are dictated by the organizational activities and the environment within which those activities take place, managers must also take another important fact into account: database solutions must be cost-effective as well as tactically and strategically effective. Producing a million-dollar solution to a thousand-dollar problem is hardly an example of good database system selection or of good database design and management. Finally, the database technology already in use is likely to affect the selection of a database system.

1.6.2 DBMS FUNCTIONS

A DBMS performs several important functions that guarantee the integrity and consistency of the data in the database. Most of these functions are transparent to end users, and most can be achieved only through the use of a DBMS. They include data dictionary management, data storage management, data transformation and presentation, security management, multiuser access control, backup and recovery management, data integrity management, database access languages and application programming interfaces, and database communication interfaces.

1. *Data dictionary management.* The DBMS stores the definitions of the data elements and their relationships (metadata) in a **data dictionary**. In turn, all programs that access the data in the database work through the DBMS. The DBMS uses the data dictionary to look up the required data component structures and relationships, thus relieving us from having to code such complex relationships in each program. Additionally, any changes made in a database structure are automatically recorded in the data dictionary, thereby freeing us from having to modify all the programs that access the changed structure. In other words, the DBMS provides data abstraction, and it removes structural and data dependency from the system. For example, Figure 1.8 shows an example of how Microsoft Access presents the data definition for the CUSTOMER table. Note the definition of the field properties for the CUS_RENEW_DATE.

FIGURE 1.8 ILLUSTRATING METADATA WITH MICROSOFT ACCESS

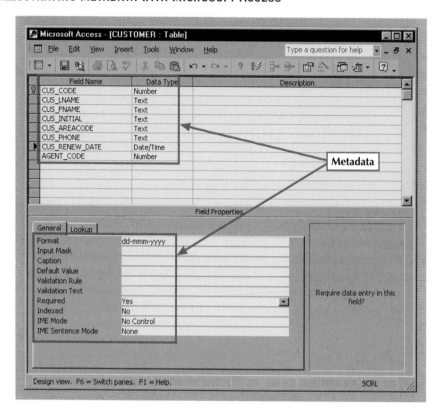

2. *Data storage management.* The DBMS creates and manages the complex structures required for data storage, thus relieving us from the difficult task of defining and programming the physical data characteristics. A modern DBMS system provides storage not only for the data, but also for related data entry forms or screen definitions, report definitions, data validation rules, procedural code, structures to handle video and picture formats, and so on. Data storage management is also important for database performance tuning. **Performance tuning** relates to the activities that make the database perform more efficiently in terms of storage and access speed. Although the user sees the database as a single data storage unit, the DBMS actually stores the database in multiple physical data files. (See Figure 1.9.) Such datafiles may even be stored on different storage media. Therefore the DBMS doesn't have to wait for one disk request to finish before the next one starts. In other words, the DBMS can fulfill database requests concurrently.

3. *Data transformation and presentation.* The DBMS transforms entered data to conform to the data structures that are required to store the data. Therefore, the DBMS relieves us of the chore of making a distinction between the data logical format and the data physical format. By maintaining **data independence**, the DBMS translates logical requests into commands that physically locate and retrieve the requested data. That is, the DBMS formats the physically retrieved data to make it conform to the user's logical expectations. In other words, a DBMS provides application programs with software independence and data abstraction. For example, imagine a enterprise database used by a multinational company. An end user in England would expect to enter data such as July 11, 2004 as "11/07/2004". In contrast, the same date would be entered in the United States as "07/11/2004". Regardless of the data presentation format, the DBMS must manage the date in the proper format for each country.

FIGURE 1.9 ILLUSTRATING DATA STORAGE MANAGEMENT WITH ORACLE

The Oracle DBA Studio Administrator GUI shows the data storage management characteristics for the ORALAB database.

4. *Security management.* The DBMS creates a security system that enforces user security and data privacy within the database. Security rules determine which users can access the database, which data items each user may access, and which data operations (read, add, delete, or modify) the user may perform. This is especially important in multiuser database systems where many users can access the database simultaneously. Chapter 15, "Database Administration," examines data security and privacy issues in greater detail. All database users are authenticated to the DBMS through the use of a user name and a password. The DBMS then uses this information to assign access priveleges to various database components such as queries and reports.

5. *Multiuser access control.* The DBMS creates the complex structures that allow multiple users to access the data. In order to provide data integrity and data consistency, the DBMS uses sophisticated algorithms to ensure that multiple users can access the database concurrently without compromising the integrity of the database. Chapter 9, "Transaction Management and Concurrency Control," covers the details of the multiuser access control.

6. *Backup and recovery management.* The DBMS provides backup and data recovery procedures to ensure data safety and integrity. Current DBMS systems provide special utilities that allow the DBA to perform routine and special backup and restore procedures. Recovery management deals with the recovery of the database after a failure, such as a bad sector in the disk or a power failure. Such capability is critical to the preservation of the database's integrity. Chapter 15 covers backup and recovery issues.

7. *Data integrity management.* The DBMS promotes and enforces integrity rules to eliminate data integrity problems, thus minimizing data redundancy and maximizing data consistency. The data relationships stored in the data dictionary are used to enforce data integrity. Ensuring data integrity is especially important in transaction-oriented database systems. Transaction and data integrity issues are addressed in Chapter 9.

8. *Database access languages and application programming interfaces.* The DBMS provides data access through a query language. A **query language** is a nonprocedural language—one that lets the user specify what must be done without having to specify how it is to be done. The DBMS's query language contains two components: a **data definition language (DDL)** and a **data manipulation language (DML)**. The DDL defines the structures in which the data are housed, and the DML allows end users to extract the data from the database. The DBMS also provides data access to programmers via procedural (3GL) languages such as COBOL, C, PASCAL, Visual Basic, and others. The DBMS also provides administrative utilities used by the DBA and the database designer to create, implement, monitor, and maintain the database.

9. *Database communication interfaces.* Current-generation DBMSs provide communications interfaces designed to allow the database to accept end-user requests within a computer network environment. For example, the DBMS might provide communications functions to access the database through the Internet, using Web browsers such as Netscape Navigator or Internet Explorer. In this environment, communications can be accomplished in several ways:

 - End users can generate answers to queries by filling in screen forms through their preferred Web browser.

 - The DBMS can automatically publish predefined reports on the Internet, using a Web format that enables any Web user to browse it.

 - The DBMS can connect to third-party systems to distribute information via e-mail or other productivity applications such as Lotus Notes.

We examine database communication interfaces in greater detail in Chapter 10, "Distributed Database Management Systems," Appendix F, "Client/Server Systems," and especially in Chapters 13 and 14, "Databases in Electronic Commerce" and "Web Database Development."

1.6.3 MANAGING THE DATABASE SYSTEM: A SHIFT IN FOCUS

As you can see, the DBMS handles many of the complex data management chores handled by the file system's DP specialist, and it performs them with far greater finesse and sophistication. Better yet, the DBMS allows those chores to be performed without the tedious and time-consuming programming required in the file management system. In addition, the DBMS provides a framework in which strict procedures and standards can be enforced. Consequently, the role of the DP specialist or the DP manager changes from an emphasis on programming to a focus on the broader aspects of managing the organization's data resources and on the administration of the complex database software itself.

Because the file system's DP manager performs broader managerial functions in a database environment, (s)he might be promoted to systems administrator. Initially, while the database environment is simple, the systems administrator performs both managerial and hands-on database administration tasks. As the database environment expands, the system administrator usually appoints one or more database administrators who evaluate database designs, maintain the database software, assign the right to use the database(s) to selected individuals, and coordinate the development of applications based on the data resources. If the data resources require the use of more than one database, the systems administrator may appoint a database administrator to administer each of the databases.

1.6.4 DATABASE DESIGN AND MODELING

The availability of a DBMS makes it possible to tackle far more sophisticated uses of the data resources, if the database is designed to make use of that available power. The kinds of data structures created within the database and the extent of the relationships among them play a powerful role in determining how effective the DBMS is. Therefore, database design becomes a crucial activity in the database environment.

You will learn in the next chapter that database design is made much simpler when you use models, which are simplified abstractions of real-world events or conditions. For example, such abstractions will enable us to explore the characteristics of entities and the relationships that can be created among such entities. If the models are not logically sound, the database designs derived from them will not deliver the database system's promise of effective information drawn from an efficient database. We will, therefore, endeavor to develop good modeling techniques in this book. *Good models yield good database designs that are the basis for good applications.* Conversely, you cannot expect to build good applications with bad database design based on poor models.

S U M M A R Y

Information is derived from data, which are usually stored in a database. To implement a database and to manage its contents you need commercial software known as a database management system (DBMS). Database design defines the database structure; the DBMS stores the facts about the structure in the database itself. The database thus contains the data you have collected and "data about data" known as metadata. Good database design is important because even a good DBMS will perform poorly with a poorly designed database.

Databases were preceded by file systems. Because file systems lack a DBMS, file management becomes difficult as the file system grows. Each file requires its own set of basic data management programs, and, because files are usually "owned" by those who commissioned them, the number of files tends to grow. Many of the files often contain redundant data, thus leading to data inconsistency, data anomalies, and a lack of data integrity. Because each file can be used by many application programs, a mature file system might have generated hundreds or even thousands of programs. Serious file system data management problems usually stem from data dependency—access to any file is dependent on data characteristic and storage formats; therefore, even a minor change in a data structure within a file requires that all the programs accessing that file must be modified, too.

Database management systems were developed to address the file system's inherent weaknesses. Rather than depositing data within independent files, a DBMS presents the database to the end user as a single data repository. This arrangement promotes data sharing, thus at least potentially eliminating the "islands of information" problem. In addition, the DBMS enforces data integrity, eliminates redundancy, and promotes data security.

The most effective database designs use models, simplified abstractions of real-world events or conditions.

K E Y T E R M S

ad hoc query	data processing (DP) manager	enterprise database
centralized database	data processing (DP) specialist	field
data	data redundancy	file
data anomaly	data warehouse database	fourth-generation language (4GL)
data definition language (DDL)	database	information
data dependence	database administrator (DBA)	islands of information
data dictionary	database design	logical data format
data inconsistency	database management system	logical design
data independence	(DBMS)	metadata
data integrity	database system	multiuser database
data management	desktop database	performance tuning
data manipulation language (DML)	distributed database	physical data format

production database	redundant data	systems administrator
query	single-user database	third-generation language (3GL)
query language	structural dependence	transactional database
record	structural independence	workgroup database

REVIEW QUESTIONS

1. Discuss each of the following terms:
 a. data
 b. field
 c. record
 d. file
2. What is data redundancy, and which characteristics of the file system can lead to it?
3. Discuss the lack of data independence in the file systems.
4. What is a DBMS, and what are its functions?
5. What is data independence, and why is it important?
6. Explain the difference between *data* and *information*.
7. Explain what data inconsistency is and why it occurs.
8. Discuss the different types of databases.
9. What are the main components of a database system environment?
10. What is metadata?
11. Explain why database design is important.
12. Discuss the differences between transaction databases and data warehouse databases.

PROBLEMS

ADDITIONAL CD CONTENT

The file structures you see in this problem set are simulated in a Microsoft Access database named **Ch01_Problems**. This database is located in the **\Databases\Student\Ch01** folder located on the student CD that accompanies this book.

Given the file structure shown in Figure P1.1, answer Problems 1 through 4.

FIGURE P1.1 THE FILE STRUCTURE FOR PROBLEMS 1–4

PROJECT_CODE	PROJECT_MANAGER	MANAGER_PHONE	MANAGER_ADDRESS	PROJECT_BID_PRICE
21-5Z	Holly B. Parker	904-338-3416	3334 Lee Rd., Gainesville, FL 37123	$16,833,460.00
25-2D	Jane D. Grant	615-898-9909	218 Clark Blvd., Nashville, TN 36362	$12,500,000.00
25-5A	George F. Dorts	615-227-1245	124 River Dr., Franklin, TN 29185	$32,512,420.00
25-9T	Holly B. Parker	904-338-3416	3334 Lee Rd., Gainesville, FL 37123	$21,563,234.00
27-4Q	George F. Dorts	615-227-1245	124 River Dr., Franklin, TN 29185	$10,314,545.00
29-2D	Holly B. Parker	904-338-3416	3334 Lee Rd., Gainesville, FL 37123	$25,559,999.00
31-7P	William K. Moor	904-445-2719	216 Morton Rd., Stetson, FL 30155	$56,850,000.00

1. How many records does the file contain, and how many fields are there per record?

2. What problem would you encounter if you wanted to produce a listing by city? How would you solve this problem by altering the file structure?

3. If you wanted to produce a listing of the file contents by last name, area code, city, state, or zip code, how would you alter the file structure?

4. What data redundancies do you detect, and how could these redundancies lead to anomalies?

5. Identify and discuss the serious data redundancy problems exhibited by the file structure shown in Figure P1.5.

FIGURE P1.5 THE FILE STRUCTURE FOR PROBLEMS 5–8

PROJ_NUM	PROJ_NAME	EMP_NUM	EMP_NAME	JOB_CODE	JOB_CHG_HOUR	PROJ_HOURS	EMP_PHONE
1	Hurricane	101	John D. Newson	EE	$85.00	13.3	653-234-3245
1	Hurricane	105	David F. Schwann	CT	$60.00	16.2	653-234-1123
1	Hurricane	110	Anne R. Ramoras	CT	$60.00	14.3	615-233-5568
2	Coast	101	John D. Newson	EE	$85.00	19.8	653-234-3254
2	Coast	108	June H. Sattlemeir	EE	$85.00	17.5	905-554-7812
3	Satellite	110	Anne R. Ramoras	CT	$62.00	11.6	615-233-5568
3	Satellite	105	David F. Schwann	CT	$26.00	23.4	653-234-1123
3	Satelite	123	Mary D. Chen	EE	$85.00	19.1	615-233-5432
3	Satellite	112	Allecia R. Smith	BE	$85.00	20.7	615-678-6879

6. Looking at the EMP_NAME and EMP_PHONE contents in Figure P1.5, what change(s) would you recommend?

7. Identify the different data sources in the file you examined in Problem 5.

8. Given your answer to Problem 7, what new files should you create to help eliminate the data redundancies found in the file shown in Figure P1.5?

9. Identify and discuss the serious data redundancy problems exhibited by the file structure shown in Figure P1.9. (The file is meant to be used as a teacher class assignment schedule. One of the many problems with data redundancy is the likely occurrence of data inconsistencies—note that two different initials have been entered for the teacher named Maria Cordoza.)

FIGURE P1.9 THE FILE STRUCTURE FOR PROBLEMS 9–10

BUILDING_CODE	ROOM_CODE	TEACHER_LNAME	TEACHER_FNAME	TEACHER_INITIAL	DAYS&TIME
KOM	204E	Williston	Horace	G	MWF 8:00-8:50
KOM	123	Cordoza	Maria	L	MWF 8:00-8:50
LDB	504	Patroski	Donald	J	TTh 1:00-2:15
KOM	34	Hawkins	Anne	W	MWF 10:00-10:50
JKP	225B	Risell	James		TTh 9:00-10:15
LDB	301	Robertson	Jeanette	P	TTh 9:00-10:15
KOM	204E	Cordoza	Maria	I	MWF 9-9:50
LDB	504	Williston	Horace	G	TTh 1:00-2:15
KOM	34	Cordoza	Maria	L	MWF 11:11:50
LDB	504	Patroski	Donald	J	MWF 2:00-2:50

10. Given the file structure shown in Figure P1.9, what problem(s) might you encounter if building KOM were deleted?

2

DATA MODELS

In this chapter, you will learn:

❖ Why data models are important

❖ About the basic data-modeling building blocks

❖ What business rules are and how they affect database design

❖ How the major data models evolved, and their advantages and disadvantages

❖ How data models can be classified by level of abstraction

P R E V I E W

In this chapter we examine data modeling. Data modeling is the first step in the database design journey, serving as a bridge between real-world objects and the database that resides in the computer.

You will discover that one of the most vexing problems of database design is that designers, programmers, and end users see data in different terms. Therefore, different views of the same data can lead to database designs that do not reflect an organization's actual operation, consequently failing to meet end-user needs and data efficiency requirements. To avoid such failures, database designers must obtain a precise description of the nature of the data and the many uses of such data within an organization. Communication among database designers, programmers, and end users should be as free of ambiguities as possible. Data modeling clarifies such communications by reducing real-world complexities to more easily understood abstractions that define entities and the relations among them.

First, you will learn some basic data modeling concepts and explore the background, characteristics, advantages, and disadvantages of the major data models: hierarchical, network, relational, entity relationship, and object oriented. Tracing the development of these database models and examining their advantages and disadvantages lays the foundation for understanding the database design and implementation issues addressed in the rest of this book, and helps you understand why this book focuses on the still-dominant relational database. You also learn how different degrees of data abstraction help to reconcile varying views of the same data.

 T W O

2.1 THE IMPORTANCE OF DATA MODELS

Traditionally, database designers relied on good judgment to help them develop a good design. Unfortunately, good judgment is often in the eye of the beholder. And, good judgment often develops after much trial and error (mostly error!). Fortunately, the development of data models, bolstered by powerful database design tools, has made it possible to substantially diminish the potential for database design error. Yet in spite of all the advances in the field of database design, we will show you that judgment has not been taken out of the design equation. At least for now, database design is both art and science.

Webster's Dictionary defines a model as a "description or analogy used to visualize something that cannot be directly observed." In other words, a **model** is an abstraction of a more complex real-world object or event. A **data model** is the relatively simple representation, usually graphical, of complex real-world data structures. The model's main function is to help us understand the complexities of the real-world environment. Within the database environment, a data model represents data structures and their characteristics, relations, constraints, and transformations.

> **N O T E**
>
> Among database practitioners, the terms *data model* and *database model* are often used interchangeably. To be precise, we will use the term *database model* to refer to a specific implementation of a *data model*.

The database designer employs data models as communications tools to facilitate the interaction among the designer, the applications programmer, and the end user. If the data model is developed well, it can foster a much improved understanding of the organization for which the database design is developed. This important aspect of data modeling was summed up neatly by a client, whose reaction was:

I created this business, I worked with this business for years, and this is the first time I've really understood how all the pieces really fit together.

A good database design is the foundation for good applications. Regardless of the skills of programmers and/or the effectiveness of application generators, *we cannot develop good applications without a good database design.* And good database design uses an appropriate data model as its foundation.

The importance of data modeling cannot be overstated. Data constitute the most basic information units employed by a system. Applications are created to manage data and to help transform data into information. But data are viewed in different ways by different people. For example, contrast the (data) view of a company manager and that of a company clerk. Although both manager and clerk work for the same company, the manager is much more likely to have an enterprise-wide view of company data than the clerk.

Even different managers view data differently. For example, a company president is likely to take a universal view of the data because (s)he must be able to tie the company's divisions to a common (database) vision. A purchasing manager in the same company is very likely to have a much more restricted view of the data, as is the company's inventory manager. In effect, each department manager works with a subset of the company's data. An inventory manager will be more concerned with inventory levels, while the purchasing manager will be more concerned with item costs and personal/business relations with the suppliers of those items.

Applications programmers have yet another view of data, being more concerned with data location, formatting, and specific reporting requirements. Basically, applications programmers translate company policies and procedures from a variety of sources into appropriate interfaces, reports, and query screens.

The different users and producers of data and information often reflect the "blind people and the elephant" analogy: the blind person who felt the elephant's trunk had quite a different view of the elephant from those who felt the elephant's leg or its tail. What we need is the ability to see the whole elephant. Similarly, a house is not a random collection of rooms; if we are going to build a house, we should first have the overall view provided by its blueprint. In the same vein, a sound data environment requires an overall database blueprint based on an appropriate data model.

If we have a good blueprint, it does not matter that an applications programmer's view of the data is different from that of the manager and/or the end user. Conversely, if we do not have a blueprint, problems are likely to ensue. For instance, an inventory management program or an order entry system may not fit into the overall set of operational requirements, thereby costing the company thousands (or even millions!) of dollars.

Keep in mind that a house blueprint is an abstraction; you cannot live in the blueprint. Similarly, the data model is an abstraction; you cannot draw the required data out of the data model. Just as you are not likely to build a good house without a blueprint, you are equally unlikely to create a good database without first selecting an appropriate data model.

2.2 DATA MODEL BASIC BUILDING BLOCKS

The basic building blocks of all data models are entities, attributes, and relationships. An **entity** is anything, such as a person, place, thing, or event, about which data are to be collected and stored. Entities may be physical objects such as customers or products. But entities may also be abstractions such as flight routes or musical concerts.

An **attribute** is a characteristic of an entity. For example, a CUSTOMER entity would be described by attributes such as customer last name, customer first name, customer phone, customer address, and customer credit limit. The attributes are the equivalent of fields in file systems.

A **relationship** describes an association among (two or more) entities. For example, there is a relationship between customers and agents that may be described as "*an agent can serve many customers and each customer may be served by one agent.* Data models use three types of relationships: one-to-many, many-to-many, and one-to-one. Database designers usually use the shorthand notations 1:M, M:N, and 1:1 for them, respectively. (Although the M:N notation is a standard label for the many-to-many relationship, the label M:M may also be used.) The following examples illustrate the distinctions among the three.

1. **One-to-many (1:M) relationship**. A painter paints many different paintings, but each one of them is painted by only one painter. Thus the painter (the "one") is related to the paintings (the "many"). Therefore, database designers label the relationship "PAINTER paints PAINTING" as 1:M. Similarly, a customer (the "one") might generate many invoices, but each invoice (the "many") is generated by only a single customer. The "CUSTOMER generates INVOICE" relationship would also be labeled 1:M.

2. **Many-to-many (M:N or M:M) relationship**. An employee might learn many job skills, and each job skill might be learned by many employees. Database designers label the relationship "EMPLOYEE learns SKILL" as M:N. Similarly, a student can take many classes, and each class can be taken by many students, thus yielding the M:N relationship label for the relationship expressed by "STUDENT takes CLASS."

3. **One-to-one (1:1) relationship**. A retail company's management structure may require that each one of its stores be managed by a single employee. In turn, each store manager—who is an employee—only manages a single store. Therefore, the relationship "EMPLOYEE manages STORE" is labeled 1:1.

Note that in the preceding discussion of relationships we have identified each relationship in both directions. That is:

1. *One* CUSTOMER can generate *many* INVOICEs.
2. Each of the *many* INVOICEs is generated by only *one* CUSTOMER.

In the next section you will learn the importance of properly identifying the nature and type of relationships.

2.3 BUSINESS RULES

How do database designers go about selecting or determining the entities, attributes, and relationships that will build a data model? They could start by thoroughly understanding the types of data in an organization, how they are used, and in what time frames. But such data and information do not, by themselves, yield the required understanding of the total business. From a database point of view, the collection of data becomes meaningful only when it reflects properly defined *business rules*. A **business rule** is a brief, precise, and unambiguous description of a policy, procedure, or principle within a specific organization's environment. In a sense, business rules are misnamed: they apply to *any* organization—a business, a government unit, a religious group, or a research laboratory—large or small—that stores and uses data to generate information.

Business rules, derived from a detailed description of an organization's operations, help to create and enforce actions within that organization's environment. Business rules must be rendered in writing and updated to reflect any change in the organization's operational environment.

Properly written business rules are used to define entities, attributes, relationships, and constraints. Any time you see relationship statements such as the ones in Section 2.2, you are seeing business rules at work. You will see the application of business rules throughout this book, especially in chapters devoted to data modeling and database design.

Keep in mind that some business rules are external to the organization. For example, all U.S. airlines operate under strict rules imposed by the federal government that necessitate the specification of numerous business rules for these airlines.

To be effective, business rules must be easy to understand and widely disseminated to ensure that every person in the organization shares a common interpretation of such rules. Business rules describe, in simple language, the main and distinguishing characteristics of the data *as viewed by the company*. Examples of business rules are:

1. A customer may make many payments on account.
2. Each payment on account is credited to only one customer.
3. A machine operator may not work more than 10 hours in any 24-hour period.
4. A business trip destination must be at least 200 miles away for an airline ticket to be purchased.
5. A training session cannot be scheduled for fewer than 10 employees or for more than 30 employees.
6. A company aircraft must have an airframe inspection every 100 flight hours.
7. A lab cannot be scheduled for demonstration purposes more than one hour per day.
8. A customer may generate many invoices.
9. Each invoice is generated by only one customer.

As you examine the preceding examples, note that these business rules establish entities, relationships, and constraints. For example, business rules 8 and 9 establish two entities, CUSTOMER and INVOICE, and a 1:M relationship between those two entities. Business rule 5 establishes a constraint, no fewer than 10 people nor more than 30 people; two entities, EMPLOYEE and TRAINING; and a relationship between EMPLOYEE and TRAINING.

The main sources of business rules are company managers, policy makers, department managers, and written documentation—such as a company's procedures, standards, or operations manuals. A faster and more direct source of business rules is direct interviews with end users. Unfortunately, because perceptions differ, end users sometimes are a less-reliable source when it comes to specifying business rules. For example, a maintenance department mechanic may believe that any mechanic may initiate a maintenance procedure, when actually only mechanics with inspection authorization may perform such a task. Such a distinction may seem trivial, but it can have major legal consequences. Although end users are crucial contributors to the development of business rules, *it pays to verify end-user perceptions*. We have discovered too often that interviews with several people who perform the same job can yield very different perceptions of what the job components are. While such a discovery may point to "management problems," that general diagnosis does not help the database designer. Given the discovery of such problems, the database designer's job is to reconcile such differences and verify the results of the reconciliation to ensure that the business rules are appropriate and accurate.

Knowing the business rules promotes the creation of an accurate data model based on how the organization actually works and what role is played by the data within that organization's operations. Consequently, the database designer must identify the organization's business rules and analyze their impact on the nature, role, and scope of data.

Business rules are essential to database design for several reasons:

- They help standardize the company's view of data.
- They constitute a communications tool between users and designers.
- They allow the designer to understand the nature, role, and scope of the data.
- They allow the designer to understand business processes.
- They allow the designer to develop appropriate relationship participation rules and constraints. (We will examine this point in detail in Chapter 4, "Entity Relationship (ER) Modeling.")

Of course you want to keep in mind that not all business rules can be modeled. For example, the business rule that specifies that "no pilot can fly more than 10 hours within any 24-hour period" cannot be modeled. However, such a business rule can be enforced by application software.

2.4 THE EVOLUTION OF DATA MODELS

The quest for better data management has led to several different models that attempt to resolve the file system's critical shortcomings. Because each data model evolved from its predecessors, we examine the major data models—hierarchical, network, relational, entity relationship, and object oriented—in roughly chronological order in this section. You will discover that many of the "new" database concepts and structures bear a remarkable resemblance to some of the "old" data model concepts and structures. (For example, you'll discover a remarkable similarity between some of the concepts of the hierarchical data model and those concepts found in the object oriented model in Chapter 11, "Object Oriented Databases.")

ADDITIONAL CD CONTENT

The hierarchical and network models are largely of historical interest, yet they do still contain some elements and features that interest current database professionals. The technical details of these two models are discussed in greater detail in Appendixes H and I, respectively, on the student CD bundled with this book. We also have devoted an entire chapter to the object oriented model (see Chapter 11, "Object Oriented Databababases.") However, given the dominant market presence of the relational model, most of the book focuses on that model.

2.4.1 THE HIERARCHICAL MODEL

North American Rockwell was the prime contractor for the Apollo project, which culminated in a moon landing in 1969. Bringing such a complex project to a successful conclusion required the management of millions of parts. Information concerning the parts was generated by a complex computer file system. However, as you learned in Chapter 1, file systems can lead to data redundancy, with all its attendant data anomalies.

When North American Rockwell began to develop its own database system, an audit of computer tapes revealed that over 60 percent of the data were redundant. The problems caused by data redundancy forced North American Rockwell to develop an alternate strategy for managing such huge data quantities. Borrowing parts of existing database concepts, they developed software known as **GUAM (Generalized Update Access Method)**, which was based on the recognition that the many smaller parts would come together as components of still larger components, and so on, until all the components came together in the final unit.

In the mid-1960s, IBM joined North American Rockwell to expand the capabilities of GUAM, replacing the computer tape medium with more up-to-date disk computer storage, which allowed the introduction of complex pointer systems.

> **NOTE**
>
> A **pointer** is a reference device that "points" to the location of specified data in the storage medium. It is analogous to the page reference in a book's index, which allows you to look up a particular topic in the alphabetically arranged index, find the topic, and then use the appropriate page reference to move to the appropriate page. Although a computer pointer system is much more complex, it works basically the same way.

The results of the joint Rockwell-IBM effort became known as the **information management system** (**IMS**). Given IBM's development and marketing clout, the IMS became the world's leading mainframe hierarchical database system in the 1970s and early 1980s.

Although the **hierarchical model** is no longer a major player in the current database market, you should understand at least a few of its characteristics, for these reasons:

- Its basic concepts form the basis for subsequent database development.
- Its limitations lead to a different way of looking at database design.
- Some of its basic concepts show up in current data models.

Basic Structure

Given its manufacturing heritage, the hierarchical model's basic logical structure is best understood when you examine a manufacturing process. For example, let's examine a somewhat simplified production process that creates a filing cabinet:

1. A filing cabinet has many components: a frame, a set of drawers, and sliding bars for those drawers.

2. A component may be composed of many smaller assemblies. For example, each drawer has a handle with a latching mechanism, a set of rollers that fits into the frame's sliding bars, and a divider blade.

3. An assembly may contain many parts. For instance, each roller is composed of a small wheel, an axle, and a brace.

4. The production process is based on data relationships that remain fixed over time. Whether a given filing cabinet model is produced today or tomorrow, the same parts are put together in the same ways to produce the same assemblies that are combined to produce the same components that are assembled in the same way to create the filing cabinet.

Tracking the parts, the assemblies, and the components we have just described is facilitated by understanding the logical process that is represented by the upside-down "tree," known as a **hierarchical structure**, shown in Figure 2.1. We have labeled the structure's components to help you understand the basic hierarchical model's vocabulary.

FIGURE 2.1 A HIERARCHICAL STRUCTURE

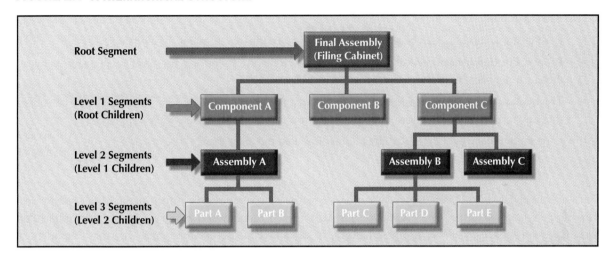

As you examine Figure 2.1, note that the user perceives the hierarchical database as a hierarchy of segments. A **segment** is the equivalent of a file system's record type. In other words, the hierarchical database is a collection of records that is logically organized to conform to the upside-down tree (hierarchical) structure shown in Figure 2.1. Within the hierarchy, the top layer (the **root**) is perceived as the parent of the segment directly beneath it. For example, in Figure 2.1, the root segment is the parent of the level 1 segments, which, in turn, are the parents of the level 2 segments, and so on. In turn, the segments below other segments are the children of the segment above them. In short:

- Each parent can have many children.
- Each child has only one parent.

In this hierarchical structure, it is easy to trace both the database's components and the 1:M relationships among them.

Keep in mind that the tree structure shown in Figure 2.1 cannot be duplicated on the computer's storage media. The computer does not "see" the logical tree structure as human beings do. Instead, the tree is defined by a path that traces the parent segments to the child segments, beginning from the left. This ordered sequencing of segments tracing the hierarchical structure is called the **hierarchical path**. For example, the hierarchical path to the segment labeled "Part D" in Figure 2.1 can be traced this way:

Final assembly → Component A → Assembly A → Part A → Part B → Component B

→ Component C → Assembly B → Part C → Part D

Note that the path traces all segments from the root, starting at the left-most segment. This "left-list" path is known as the **preorder traversal** or the **hierarchic sequence**. Given such a path, database designers must make sure that the most frequently accessed segments and their components are located closest to the left side of the tree, to ensure the most efficient data access. In Figure 2.1, for example, if Part D is the most frequently accessed and updated component, it would be wise to change the database structure to place Component C on

the left side of the level 1 segments. Then, within this relocated "branch," switch Part D to the position currently held by Part C. These changes cause the hierarchic sequence to be much shorter:

Final assembly → Component C → Assembly B → Part D

Although the hierarchical data model had its genesis in manufacturing processes, its use is not limited to that environment. It can be effective whenever you have many transactions that involve a series of 1:M relationships that remain fixed over time. For example, a bank's customer account system fits the hierarchical model well:

- Each customer account can be subject to many transactions (one-to-many relationship).

- An account transaction involves either debiting or crediting. Therefore, the basic relationship between the customer account and account transactions is fixed.

- The bank is likely to have many customer accounts. And each of the customer accounts may be subject to many transactions. Therefore, the total number of transactions is likely to be very large.

It's no surprise that many banks adopted the hierarchical data model in the 1970s.

Advantages

The hierarchical data model yielded many advantages over the file system model. *In fact, many of the hierarchical data model's features formed the foundation for current data models, and many of its database application advantages are replicated, albeit in different form, in current database environments.* The most important of these advantages are:

1. *Conceptual simplicity.* The relationship between the various layers of the model is *logically* simple. Therefore, it becomes easier to view the database conceptually, thus making its design process simpler.

2. *Database security.* Database security is provided and enforced by the DBMS. Therefore, security is enforced uniformly throughout the system, without having to rely on the efforts of individual applications programmers who may have very different ideas about the extent and type of required security.

3. *Data independence.* The DBMS creates an environment in which data independence can be maintained, thereby substantially decreasing programming effort and program maintenance. (Data independence exists when a change in a data type will be automatically cascaded throughout the database by the DBMS, thus eliminating the need to make changes in the program segments that reference the changed data type.)

4. *Database integrity.* Given the parent/child relationship, there is always a link between the parent segment and its child segment(s). Because the child segment is always automatically referenced to its parent, the hierarchical model promotes database integrity.

5. *Efficiency.* The hierarchical data model is very efficient when a database contains a large volume of data in 1:M relationships and when users require large numbers of transactions, using data having relationships that are fixed over time.

Given its inherent superiority over file systems, the hierarchical database quickly became dominant in the 1970s. Such dominance generated a large installed (mainframe) base which, in turn, created a pool of programmers who knew the systems and who developed numerous tried-and-true business applications.

Disadvantages

Although hierarchical databases and their applications still exist, the hierarchical model rapidly lost favor in the late 1970s and early 1980s, for several good reasons:

1. *Complex implementation.* Although the hierarchical model's DBMS relieves the designer and programmer of the data-dependence problems, they still must have detailed knowledge of the physical data storage characteristics. Therefore, the implementation of a database design can become very complicated.

2. *Difficult to manage.* Any changes in the database structure, such as the relocation of segments, require changes in all application programs that access the database. Therefore, database management can become a difficult chore. And, although the hierarchical structure fosters database integrity, this structure also makes it possible to perform a deletion of one segment that leads to the involuntary deletion of all the segments under it—an error that can be extremely costly!

3. *Lacks structural independence.* Structural independence exists when changes in the database structure do not affect the DBMS's ability to access the data. The hierarchical database is known as a *navigational system* because data access requires that the physical storage characteristics be known to "navigate" to the appropriate segments. Within a navigational database system, the programmer must know the access paths to relevant segments (the parent record must be accessed in order to access the child records) to retrieve data from the database. Modifications in the database structure can lead to problems with application programs that were operating correctly before the changes were made; *the benefits of data independence are thus limited by structural dependence.*

4. *Complex applications programming and use.* Because of the navigational database system's structure, applications programmers and end users must know precisely how the data are distributed physically within the database in order to access the data. Even if they understand the access paths to the data, getting to the data requires knowledge of complex pointer systems. Therefore, hierarchical databases are said to be written by programmers for programmers.

5. *Implementation limitations.* Many common relationships do not conform to the 1:M standard required by the hierarchical model. For example, consider a student enrolled at a university. Each course can contain many students, and each student can take many courses. Such a common many-to-many (M:N) relationship is difficult to implement in a hierarchical model. In addition, quite a few real-world relationships are based on a child with multiple parents. For example, the relationship shown in Figure 2.2 depicts an ORDER_LINE that has two parents: ORDER and PART.

FIGURE 2.2 **CHILD WITH MULTIPLE PARENTS**

When a customer generates an order, the order line typically contains data from both the ORDER and the PART segments:

Order: 10012 Customer: 89221 Part number: M-2W113

Such a common two-parent condition cannot be implemented easily in a hierarchical environment.

6. *Lack of standards.* Although the basic hierarchical model is incorporated in all hierarchical database software, there is no precise set of standard *concepts*, nor does the *implementation* of the model conform to a specific standard. Implementation problems were especially vexing because the hierarchical database administration component lacked a standard data definition language (DDL) to define the database components; nor did it have a data manipulation language (DML) to manipulate the database contents. Although the IMS software developed jointly by IBM and North American Rockwell became the predominant hierarchical DBMS in the late 1960s and the 1970s, many less-prominent hierarchical software programs operated outside the concept and terminology boundaries created by IMS. There-fore, moving from one hierarchical DBMS to another was difficult; that is, *portability* was limited.

In the 1970s, database professionals convened a set of meetings that culminated in the publication of a set of database standards that ultimately led to the development of alternate data models. The most prominent of these models is the network model.

2.4.2 THE NETWORK MODEL

The **network model** was created to represent complex data relationships more effectively than the hierarchical model could, to improve database performance, and to impose a database standard. The lack of database standards was troublesome to programmers and application designers because it made database designs and applications less portable. Worse, the lack of even a standard set of database *concepts* impeded the search for better data models. Disorganization seldom fosters progress.

To help establish database standards, a meeting was called in 1971 by the **Conference on Data Systems Languages (CODASYL)**. The CODASYL group was originally formed by users and computer manufacturers to create a COBOL standard. The CODASYL COBOL recommendations were subsequently accepted by the **American National Standards Institute (ANSI)** as the standard COBOL specification, thereby creating ANSI COBOL. Given its success in developing COBOL standards, the CODASYL group set out to perform a similar service in the database arena, and created the **Database Task Group (DBTG)**. The DBTG was charged to define standard specifications for an environment that would facilitate database creation and data manipulation.

The final DBTG report contained specifications for three crucial database components:

1. The network **schema**, the conceptual organization of the entire database as viewed by the database administrator. The schema includes a definition of the database name, the record type for each record, and the components that make up those records.

2. The **subschema**, which defines the portion of the database "seen" by the application programs that actually produce the desired information from the data contained within the database. The existence of subschema definitions allows all application programs to simply invoke the subschema required to access the appropriate database file(s).

3. A data management language (DML) to define the data characteristics and the data structure in order to manipulate the data.

To produce the desired standardization for each of the three components, the DBTG specified three distinct data management language components:

1. A schema data definition language (DDL), which enables the database administrator to define the schema components.

2. A subschema DDL, which allows the application programs to define the database components that will be used.

3. A DML to manipulate the database contents.

The ANSI **Standards Planning and Requirements Committee (SPARC)** augmented the database standards in 1975. All mainframe network database software was expected to conform to the DBTG report standard. Therefore, if the standards were met, designers and users would have had little trouble shifting from one commercial application to another when they were operating at the schema level. Unfortunately, the standards were often "stretched" to meet vendor software enhancements.

Basic Structure

In many respects the network model resembles the hierarchical model. For example, as in the hierarchical model, the user perceives the network database as a collection of records in 1:M relationships. However, unlike the hierarchical model, the network model allows a record to have more than one parent. Therefore, the commonly encountered relationships depicted in Figure 2.2 can be handled easily by the network model.

In network database terminology, a relationship is called a **set**. Each set is composed of at least two record types: an **owner** record that is equivalent to the hierarchical model's parent, and a **member** record that is equivalent to the hierarchical model's child. The difference between the hierarchical model and the network model is that the latter might include a condition in which a record can appear (as a member) in more than one set. In other words, a member may have several owners. *A set represents a 1:M relationship between the owner and the member.* An example of such a relationship is depicted in Figure 2.3.

FIGURE 2.3 A NETWORK DATA MODEL

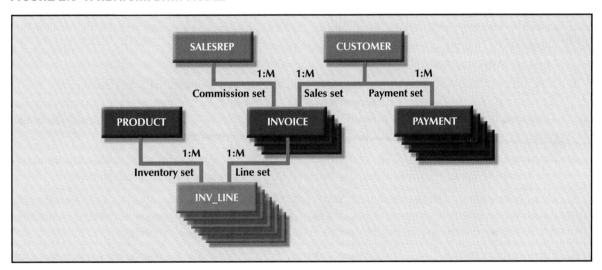

Figure 2.3 illustrates a network data model for a typical sales organization. In this model, CUSTOMER, SALESREP, INVOICE, INV_LINE, PRODUCT, and PAYMENT represent record types. As you examine Figure 2.3, note that INVOICE is "owned" by both SALESREP and CUSTOMER. Similarly, INV_LINE has two owners, PRODUCT and INVOICE. But note that the network model can still include one-owner relationships (for example, CUSTOMER makes PAYMENT) that are typical of a hierarchical data model.

The network data model in Figure 2.3 illustrates transactions that are based on a series of one-to-many relationships:

1. A SALESREP might have written many INVOICE tickets, but each INVOICE ticket was written by a single SALESREP. Therefore, there is a 1:M relationship between SALESREP and INVOICE.

2. A CUSTOMER might have made purchases on different occasions. Each time a shopping trip is made, a new INVOICE ticket is written; that is, a CUSTOMER may have generated many INVOICE tickets over

time, but each INVOICE belongs to only a single CUSTOMER. Therefore, there is a 1:M relationship between CUSTOMER and INVOICE.

3. Each product purchased is listed on one of the invoice's lines. Because a customer may buy more than one product at a time—for example, you can buy a screwdriver, a box of screws, and a can of paint during your visit to a store—each invoice can have one *or more* lines on it. But each INV_LINE occurring on a *particular* INVOICE "belongs" to that invoice. Therefore, there is a 1:M relationship between INVOICE and INV_LINE.

4. Each invoice line references a single product. But, because a store can sell many screwdrivers during any period of time, a product can appear on many different invoice lines. Therefore, there is a 1:M relationship between PRODUCT and INV_LINE.

5. A customer may make many payments over a period of time. But only one customer makes each payment. Therefore, there is a 1:M relationship between CUSTOMER and PAYMENT.

Advantages

The network data model retains many of the hierarchical data model's advantages, while it corrects or improves many of the hierarchical model's shortcomings:

1. *Conceptual simplicity*. Like the hierarchical data model, the conceptual view of the database is simple, thus promoting design simplicity.

2. *Handles more relationship types*. M:N relationships are easier to implement in the network model than in the hierarchical model. Note, for example, that the network model easily handles the relationship expressed by "a product may be sold to many customers, and a customer can buy many products" in the multiple-owner environment of Figure 2.3.

3. *Data access flexibility*. Data access flexibility is superior to that found in the hierarchical model and in file systems. An application can access an owner record and all the member records within a set. Therefore, if a member record has two or more owners, as in Figure 2.3, you can move directly from one owner to another. (In short, there is no longer a need for a preorder traversal!)

4. *Promotes database integrity*. The network data model enforces database integrity because the user must first define the owner record and then the member. (A member cannot exist without an owner.)

5. *Data independence*. The network data model achieves sufficient data independence to at least partially isolate the programs from complex physical storage details. Therefore, changes in the data characteristics do not require changes in the data access portions of the application programs.

6. *Conformance to standards*. The network database's existence is at least partially traceable to the database standards imposed in the 1970s. These standards include a DDL and a DML, thus greatly facilitating database administration and portability.

Disadvantages

Although the network data model yielded significant advantages over the hierarchical data model, it still was subject to significant disadvantages:

1. *System complexity*. Database integrity control and the efficiency with which the network model manages relationships are sometimes short-circuited by the system's complexity. Like the hierarchical model, the network model provides a navigational data access environment, in which data are accessed one record at a time. Consequently, database administrators, programmers, and end users must be very familiar with the internal structures in order to access the database. Like the hierarchical database, the network database was not designed to produce a user-friendly system.

2. *Lack of structural independence*. Given its navigational data access environment, it remains difficult to make structural changes in a network database, and some structural changes are impossible to make. If

changes are made to the database structure, all application programs must be revalidated before they can access the database. *In short, although the network model achieves data independence, it still does not produce structural independence.*

As information needs grew and as more sophisticated databases and applications were required, database design, management, and use under the network model became too cumbersome. The lack of ad hoc query capability put heavy pressure on programmers to generate the code required to produce even the simplest reports. And, although the existing databases provided limited data independence, any structural change in the database still could produce havoc in all the application programs that drew the data from the database. Many database old-timers can recall the interminable information delays caused by the hierarchical and network database environments. Because of the network data model's disadvantages, it was largely replaced by the relational data model in the 1980s.

2.4.3 THE RELATIONAL MODEL

The **relational model**, first developed by E. F. Codd (of IBM) in 1970, represented a major breakthrough for both users and designers. To use an analogy, the relational model produced an "automatic transmission" database to replace the "standard transmission" databases that preceded it. Its conceptual simplicity set the stage for a genuine database revolution.

Codd's work was considered ingenious but impractical in 1970. The relational model's conceptual simplicity was bought at the expense of computer overhead; computers at that time lacked the power to implement the relational model. Fortunately, computer power grew exponentially, as did operating system efficiency. Better yet, the cost of computers diminished rapidly even as their power grew. Today, even microcomputers, costing but a fraction of what their mainframe ancestors did, can run sophisticated relational database software such as Informix, Oracle, Ingress, DB2, and other mainframe relational software.

> **N O T E**
>
> The relational database model presented in this chapter is designed to introduce a more detailed discussion in the next chapter. In fact, the relational model is so important that it will serve as the basis for our discussions in most of the remaining chapters.

Basic Structure

The relational data model is implemented through a very sophisticated **relational database management system (RDBMS)**. The RDBMS performs the same basic functions provided by the hierarchical and network DBMS systems, plus a host of other functions that make the relational data model easier to understand and to implement.

Arguably the most important advantage of the RDBMS is its ability to let the user/designer operate in a human logical environment. The RDBMS manages all of the complex physical details. Thus, the relational database is perceived by the user to be a collection of tables in which data are stored.

Each **table** is a matrix consisting of a series of row/column intersections. Tables, also called **relations**, are related to each other by sharing a common entity characteristic. For example, the CUSTOMER table in Figure 2.4 might contain a sales agent's number that is also contained in the AGENT table.

> **A D D I T I O N A L C D C O N T E N T**
>
> This chapter's databases are stored in the Databases\Student\Ch02 folder on the student CD. For example, the CUSTOMER and AGENT table contents shown in Figure 2.4 are found in the database named Ch02_InsureCo.

FIGURE 2.4 LINKING RELATIONAL TABLES

Database name: Ch02_InsureCo Table name: AGENT (first six attributes)

AGENT_CODE	AGENT_LNAME	AGENT_FNAME	AGENT_INITIAL	AGENT_AREACODE	AGENT_PHONE
501	Alby	Alex	B	713	228-1249
502	Hahn	Leah	F	615	882-1244
503	Okon	John	T	615	123-5589

Link through AGENT_CODE

Table name: CUSTOMER

CUS_CODE	CUS_LNAME	CUS_FNAME	CUS_INITIAL	CUS_AREACODE	CUS_PHONE	CUS_RENEW_DATE	AGENT_CODE
10010	Ramas	Alfred	A	615	844-2573	05-Apr-2004	502
10011	Dunne	Leona	K	713	894-1238	16-Jun-2004	501
10012	Smith	Kathy	W	615	894-2285	29-Jan-2005	502
10013	Olowski	Paul	F	615	894-2180	14-Oct-2004	502
10014	Orlando	Myron		615	222-1672	28-Dec-2004	501
10015	O'Brian	Amy	B	713	442-3381	22-Sep-2004	503
10016	Brown	James	G	615	297-1228	25-Mar-2004	502
10017	Williams	George		615	290-2556	17-Jul-2004	503
10018	Farriss	Anne	G	713	382-7185	03-Dec-2004	501
10019	Smith	Olette	K	615	297-3809	14-Mar-2004	503

The common link between the CUSTOMER and AGENT tables thus enables us to match the customer to his/her sales agent, even though the customer data are stored in one table and the sales representative data are stored in another table. For example, we can easily determine that customer Dunne's agent is Alex Alby, because for customer Dunne the CUSTOMER table's AGENT_CODE is 501, which matches the AGENT table's AGENT_CODE for agent Alex Alby. Although the tables are completely independent of one another, we can easily connect the data between tables. The relational model thus provides a minimum level of controlled redundancy to eliminate most of the redundancies commonly found in file systems.

By linking the AGENT and the CUSTOMER through the AGENT_CODE, we can determine that agent Leah Hahn (AGENT_CODE = 502) works with customers Ramas (10010), Smith (10012), Olowski (10013), and Brown (10016). We can also determine that customer Dunne's agent is Alex Alby, and so on.

The relationship type (1:1, 1:M, or M:N) is often shown in a relational schema, an example of which is depicted in Figure 2.5. A **relational schema** is a visual representation of the relational database's entities, the attributes within those entities, and the relationships between those entities.

As you examine Figure 2.5, note that the relational schema shows the connecting fields (in this case, AGENT_CODE) and the relationship type, 1:M. Microsoft Access, the database software application used to generate Figure 2.5, employs the ∞ symbol to indicate the "many" side. In this example, the CUSTOMER represents the "many" side, because an AGENT can have many CUSTOMERs. The AGENT represents the "1" side, because each CUSTOMER has only one AGENT.

A relational table stores a collection of related entities. In this respect, the relational database table resembles a file. But there is one crucial difference between a table and a file: a table yields complete data and structural independence because it is a purely logical structure. How the data are physically stored in the database is of no concern to the user or the designer; the perception is what counts here. And this property of the relational data model, explored in depth in the next chapter, became the source of a real database revolution.

FIGURE 2.5 A RELATIONAL SCHEMA

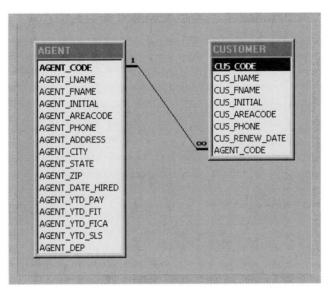

Advantages

Like the hierarchical and network databases, the relational database is a single data repository in which data independence is maintained. However, the relational data model adds significant advantages:

1. *Structural independence.* Because the relational data model does not use a navigational data access system, data access paths are irrelevant to relational database designers, programmers, and end users. Changes in the relational database structure do not affect the DBMS's data access in any way. Therefore, the relational data model achieves the structural independence not found in the hierarchical and network data models. (Recall that structural independence exists when it is possible to make changes in the database structure without affecting the application programs' ability to access the data.) In contrast to the relational database, any change in the hierarchical database's tree structure or in the network database's sets will affect the data access paths, thus requiring changes in all data access programs.

2. *Improved conceptual simplicity.* Although the hierarchical and network data models were conceptually much simpler than the file management systems they replaced, the relational data model is even simpler at the conceptual level. Because the relational data model gives us the luxury of ignoring physical data storage characteristics, we can concentrate on the *logical* view of the database. That is, we can focus on the human perception of data storage rather than on the often difficult-to-comprehend manner in which the computer "sees" the same data.

3. *Easier database design, implementation, management, and use.* Because the relational model achieves both data independence and structural independence, it becomes much easier to design the database and to manage its contents.

4. *Ad hoc query capability.* One of the reasons for the relational data model's rise to dominance in the database market is its very powerful and flexible query capability. For most relational database software, the query language is **Structured Query Language (SQL)**. In fact, we have heard the sentiment, "If it doesn't use SQL, it's not relational," expressed in professional meetings as well as in popular computer magazines such as *InfoWeek*. Given its importance, we will devote two chapters to SQL: Chapter 6, "An Introduction to Structured Query Language (SQL)," and Chapter 7, "Advanced SQL."

SQL makes pure ad hoc queries a reality; it is a fourth-generation language (4GL) that allows the user to specify what must be done without specifying how it must be done. The RDBMS uses SQL to translate the user query into the technical code required to retrieve the requested data. Consequently, the relational database SQL language makes it possible to retrieve data with far less effort than any other database or file environment.

Keep in mind that any SQL-based relational database application involves three parts: a user interface, a set of tables within the database, and the SQL "engine." The interface might include menus, query operations, and report generators. Basically, the interface allows the end user to interact with the data. Each interface is a product of the software vendor's idea of meaningful interaction with the data. You can also design your own customized interface with the help of application generators that are now standard fare in the database software arena. The database tables simply store the data.

Largely hidden from the end user, the SQL engine does the tough database jobs. Within the RDBMS, SQL is used to create table structures, maintain the data dictionary and the system catalog, channel database table access and maintenance, and translate user requests into a format that the computer can handle. The database system's maintenance is the crucial task that the RDBMS handles largely without the end user's knowledge.

5. *A powerful database management system.* A good RDBMS is a much more complex piece of software than the DBMS found in the hierarchical and network databases. Its complexity stems from the fact that it performs far more (and more complex) tasks for both the system's designers and the users. Consequently, a good RDBMS makes it possible to hide a system's (physical) complexity from both the database designer and the end user.

Because the RDBMS performs the unseen hardware tasks, we need not focus on the physical aspects of the database. Instead, we will concentrate our efforts on the logical portion of the relational database and its design in the chapters that follow.

Disadvantages

The relational database's substantial advantages over the hierarchical and network databases are gained at the cost of some disadvantages:

1. *Substantial hardware and system software overhead.* The same RDBMS that hides most of the system's complexity also necessitates its substantial hardware and causes its operating system overhead. It simply takes a more powerful computer to perform all the RDBMS-assigned tasks. Consequently, early relational database systems tended to be slower than the other contemporary database systems. However, with the rapid growth of hardware capability and the constant stream of operating system improvements, that "slow" label is fading.

2. *Can facilitate poor design and implementation.* In a sense, the relational environment's easy-to-use asset is also its liability. Relational software, especially at the microcomputer level, is so simple to use that relatively untrained people find it easy to generate handy reports and queries without giving much thought to the need to design a proper database. As the database grows, lack of proper design slows down the system and produces the data anomalies found in file systems.

3. *May promote "islands of information" problems.* Because the relational data model is so easy to use, too many people find it easy to create their own database subsets and applications. Although end-user autonomy is desirable when the end users query a common database, such autonomy may also lead to the development of database subsets that are "owned" by divisions or individuals. The proliferation of such database subsets may produce the same "islands of information" problem that was characteristic of the file management system, with the same tendency to promote inconsistent data, thus causing problems in information generation and validation.

Because the relational database's disadvantages are relatively minor when compared to its substantial advantages, it has become the dominant model for production databases. However, the ever-increasing complexity of the data environment kept database professionals searching for alternate data models, especially for models with a greater *visual* component.

2.4.4 THE ENTITY RELATIONSHIP MODEL

The relational model's conceptual simplicity made it possible to expand the database's scope. Therefore, the introduction of relational database technology triggered a demand for more (and more complex) transactions and information. In turn, the rapidly increasing transaction and information requirements created the need for more complex database implementation structures, thus creating the need for more effective database design tools. (Building a skyscraper requires more detailed design activities than building a doghouse.)

Complex design activities require conceptual simplicity to yield successful results. Although the relational model was a vast improvement over the hierarchical and network models, it still lacked the features that might make it an effective database *design* tool. Therefore, the search was on for a data model that was conceptually simpler. Because it is easier to examine structures graphically than to describe them in text form, database designers find it desirable to use a graphical tool in which entities and their relationships can be pictured.

Although many different types of data-modeling tools have been developed during the past decade, the **entity relationship (ER) model** or **ERM** is a widely accepted and adapted graphical tool for data modeling. Peter Chen first introduced the ER data model in 1976 in his landmark paper "The Entity-Relationship Model: Toward a Unified View of Data" (*ACM Transactions on Database Systems* 1:1, March 1976). The ERM yielded a graphical representation of entities and their relationships in a database structure. It is precisely this graphic data representation that popularized the use of the ER diagrams as a tool for conceptual-level data modeling. Better yet, the ER model *complemented* the relational data model concepts, thus providing the foundation for a tightly structured database design environment to ensure the proper design of relational databases.

Basic Structure

ER models are normally represented in an **entity relationship diagram (ERD)**, which uses graphic representations to model the database components.

> **NOTE**
>
> Because this chapter's objective is to simply introduce data modeling concepts, we present a simplified ERD discussion in this section. You will have the opportunity to learn additional ERD details in Chapter 3, "The Relational Database Model," and you will learn how to actually design databases with the help of ERDs in Chapter 4, "Entity Relationhip (ER) Modeling."

The ER model is based on the following components:

- Earlier in this chapter we defined an entity as anything about which data are to be collected and stored. An entity is represented in the ERD by a rectangle, also known as an entity box. The name of the entity, a noun, is written in the center of the rectangle. The entity name is generally written in capital letters and is written in the singular: PAINTER rather than PAINTERS, or EMPLOYEE rather than EMPLOYEES. Normally, when applying the ERD to the relational model, an entity is mapped to a relational table. Each row in the relational table is known as an **entity instance** or **entity occurrence** in the ER model.

NOTE

A collection of like entities is known as an **entity set**. For example, you can think of the AGENT file in Figure 2.4 as a collection of three agents (*entities*) in the AGENT *entity set*. Technically speaking, the ERD depicts entity *sets*. Unfortunately, ERD designers use "entity" as a substitute for "entity set," and we will conform to that established practice when we discuss any ERD and its components.

Each entity is described by a set of *attributes* that describe particular characteristics of the entity. For example, the entity EMPLOYEE will have attributes such as a Social Security number, a last name, and a first name. (We will show you in Chapter 4 how attributes are included in the ERD.)

- *Relationships* describe associations among data. Most relationships describe associations between two entities. When we introduced the basic data model components, we illustrated three types of relationships possible among data: one-to-many (1:M), many-to-many (M:N), and one-to-one (1:1). ERD modelers use the term **connectivity** to label the types of relationships. (The connectivities are written next to each entity box.) Relationships are represented by a diamond connected to the related entities through a relationship line. The name of the relationship, an active or passive verb, is written inside the diamond. For example, each of the company's DEPARTMENTs *has* many EMPLOYEEs. And a PAINTER *paints* many PAINTINGs.

Figure 2.6 shows some basic ERDs that illustrate these relationships and connectivities.

FIGURE 2.6 RELATIONSHIPS: THE BASIC CHEN ERD

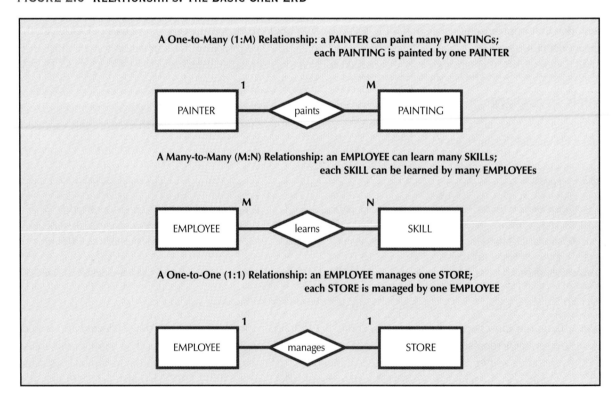

The ERD shown in Figure 2.6 is based on the so-called **Chen model**, described in Peter Chen's landmark paper. Although the entities and relationships are shown in a horizontal format in Figure 2.6, they also may be oriented vertically. The entity location and the order in which the entities are presented are immaterial—just remember to always read a 1:M relationship from the "1" side to the "M" side.

A more current version of the ERD is the **Crow's Foot model**, shown in Figure 2.7. The label "Crow's Foot" is derived from the three-pronged symbol used to represent the "many" side of the relationship. Note also that the Crow's Foot model eliminates the diamond to indicate the relationship. Instead, the Crow's Foot model simply places the relationship name in the relationship line.

FIGURE 2.7 RELATIONSHIPS: THE BASIC CROW'S FOOT ERD

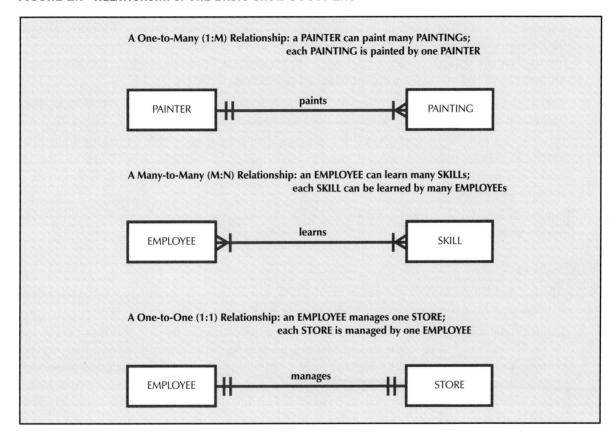

As you examine the basic Crow's Foot ERD in Figure 2.7, note that the connectivities are represented by symbols. For example, the "1" is represented by a short line segment and the "M" is represented by the three-pronged "crow's foot." Like the Chen ERD, the entities and relationships may be presented horizontally or vertically. And, again like the Chen ERD, the location and the order in which the entities are presented in a Crow's Foot ERD are immaterial—just remember to always read a 1:M relationship from the "1" side to the "M" side.

We have identified the ERDs shown in Figure 2.6 and 2.7 as *basic* ERDs. You will learn in Chapter 4 that there is much more to the Crow's Foot ERD than we have shown you in this brief introduction. For example, whether the relationship lines are solid—as we drew them in Figure 2.7—or dashed will have important consequences. And you will discover that the dual short line segments in Figure 2.7 also convey much about the nature of the relationships. However, the basic illustrations we have used are sufficient to get you started in examining the data environment graphically. To paraphrase an old Chinese proverb: even a journey of a thousand miles begins with a single step.

Although we will use the basic Chen ERD to illustrate many of our discussions, the Crow's Foot ERD will be used as our design standard in the following chapters. Most database modeling tools let you select the Crow's Foot option. In fact, we will use Microsoft's Visio Professional software to generate the Crow's Foot designs you will see in Chapter 4 and all subsequent chapters. However, you should know how to use either model.

Advantages

Human beings tend to operate most efficiently in a graphic environment. (Reading a description of a particular house takes more time and yields a lower probability of correct identification than looking at its picture.) Therefore, it is not surprising that the entity relationship model's advantages are difficult to separate from those of the entity relationship diagram (ERD) that depicts it. In fact, database designers often treat the ER model and the ERD as two sides of the same coin. In any case, we can identify at least the following advantages of the ER model:

1. *Exceptional conceptual simplicity.* All data models yield a much better logical view of the data than the file management systems. However, the ER model yields a particularly easily viewed and understood conceptual view of a database's main entities and their relationships. Therefore, complex database designs are much more easily created and managed than they otherwise would be.

2. *Visual representation.* The ER model gives database designers, programmers, and end users an easily understood *visual* representation of the data and the data relationships.

3. *Effective communication tool.* The ER model allows the database designer to capture different views of data as seen by database designers, programmers, managers, and end users. Therefore, the ER model is a particularly effective communications tool that integrates the different data views into a common framework.

4. *Integrated with the relational data model.* The ER model is well integrated with the relational data model. Such integration helps make relational database design a very structured process.

Disadvantages

Despite the fact that the ER model's representation, the ERD, has become the favorite tool in the production database design arsenal, database designers still have some reservations about its use:

1. *Limited constraint representation.* The model easily displays constraints that are directly tied to connectivities. For example, the constraint "a professor may teach as few as no classes (a research professor) but no more than four classes" can be depicted easily. Unfortunately, there are many important data constraints that cannot be modeled. For example, constraints such as "the student grade point average ranges between 0.0 and 4.0" and "a pilot may not be scheduled for more than 10 consecutive hours of duty time" cannot be represented. Instead, these constraints must be handled at the application level.

2. *Limited relationship representation.* Relationships are represented as occurring *between entities*. Therefore, relationships between attributes *within* entities cannot be represented. For example, there is no way to display the relationship between a student's classification and his or her completed credit hours. In addition, when entities have multiple relationships with other entities, the meaning of those relationships may become obscured.

3. *No data manipulation language.* Proponents of the relational model usually point to the lack of data manipulation commands in the ER model. Lacking such data manipulation commands, the ER model is not "complete."

4. *Loss of information content.* The basic Chen model tends to become "crowded" when attributes are represented. Therefore, database designers usually avoided attribute mapping, thus decreasing the model's informational content. Modifications to the Chen model and the emergence of the Crow's Foot model make it possible to store the attribute names and even basic attribute characteristics within the entity rectangle, so this criticism is much less valid than it used to be.

The ER model's disadvantages occasionally make it difficult to model the complex data types and relationships that are becoming increasingly common in our database environment. Yet, in spite of its shortcomings, the ER model's modeling tool, the ERD, has survived and prospered. In fact, vendors have made many extensions to the basic ERD presentation and it is still the dominant production database design tool. Nevertheless, the search for better data-modeling tools continues as the data environment continues to evolve.

2.4.5 THE OBJECT ORIENTED MODEL

Increasingly complex real-world problems demonstrated a need for a different data model that more closely represented the real world. The first of these models was the **semantic data model (SDM)**, developed by M. Hammer and D. McLeod and published in 1981 in their "Database Description with SDM: A Semantic Database Model" (*ACM Transactions on Database Systems* 6:3).

The SDM modeled both data *and their relationships* in a single structure known as an **object**. Because its basic modeling structure is an object, the SDM is said to be an **object oriented data model (OODM)**. In turn, the OODM becomes the basis for the **object oriented database management system (OODBMS)**.

> **N O T E**
>
> In this chapter we introduce only basic OO (object oriented) concepts to help you understand the impact of those concepts on data modeling and database management systems. You'll have a chance to examine object-orientation concepts and principles in detail in Chapter 11, "Object Oriented Databases."

An OODM reflects a very different way to define and use entities. Like the relational model's entity, an object is described by its factual content. But, quite *unlike* an entity, an object includes information about relationships between the facts within the object, as well as information about its relationships with other objects. Therefore, the facts within the object are given greater *meaning*.

> **N O T E**
>
> The word *semantics* indicates *meaning*. Therefore, in object oriented database terms, the object is said to have semantic content. That's why Hammer and McLeod named their creation a *semantic* database model.

Subsequent OODM development allowed an object to also contain all operations that can be performed on it, such as changing its data values, finding a specific data value, and printing data values. Because objects include data, various types of relationships, and operational procedures, the object becomes self-contained, thus making the object—at least potentially—a basic building block for autonomous structures.

Although object oriented (OO) concepts have been used since the 1970s, at least four developments boosted the object oriented data model's popularity in the 1990s:

1. The growing cost of generating and maintaining increasingly complex computer programs put a premium on code reusability. The most efficient way to foster code reusability is to make sure that programs are written in self-contained modules that perform specific functions. Such modules can then be "stacked" in different ways to create a wide range of applications. The object oriented data model's characteristics turned out to be particularly promising in the modular concept arena.

2. Modern databases are likely to include graphics, video, and sound in addition to text and numbers. Increasingly complex data types and system requirements became difficult to manage with a traditional RDBMS. Because complex data types are more easily modeled with the object oriented data model, this model appeared to meet such complex data requirements.

3. As the data environment became more complex, it became possible to support increasingly sophisticated transaction and information requirements. For example, a trucking company might now find it possible to generate ad hoc detail maps on which the exact location of its trucks, each truck's destination, its contents, and its estimated time of arrival would be shown. An engineering firm might want to generate exploded views of a product to troubleshoot that product's problems while the customer is on the phone.

4. Ever-increasing computing power made it possible to support the large computing overhead required to support complex object oriented data, processes, and procedures.

> **NOTE**
>
> The object oriented data model has its roots in work done in the 1970s at Xerox's Palo Alto Research Center. Xerox's research yielded the many computer breakthroughs that we take for granted today: graphical user interfaces, the computer mouse, e-mail, and object oriented programming languages.

Basic Structure

The object oriented data model is based on the following components:

- An object is an abstraction of a real-world entity. In general terms, an object may be considered equivalent to an ER model's entity. More precisely, an object represents only one individual occurrence of an entity. (The object's *semantic* content is defined through several of the items in this list.)

- Attributes describe the properties of an object. For example, a PERSON object includes the attributes Name, Social Security Number, and Date of Birth.

- Objects that share similar characteristics are grouped in classes. A **class** is a collection of similar objects with shared structure (attributes) and behavior (methods). In a general sense, a class resembles the ER model's entity *set*. However, a class is different from an entity set in that it contains a set of procedures known as *methods*. A class's **method** represents a real-world action such as *finding* a selected PERSON's name, *changing* a PERSON's name, or *printing* a PERSON's address. In other words, methods are the equivalent of *procedures* in traditional programming languages. In object oriented terms, methods define an object's *behavior*. (Some variants of the object oriented data model—such as the semantic object model—do not include methods in their representation.)

- Classes are organized in a **class hierarchy**. The class hierarchy resembles an upside-down tree in which each class has only one parent. For example, the CUSTOMER class and the EMPLOYEE class share a parent PERSON class. (Note the similarity to the hierarchical data model in this respect!)

- **Inheritance** is the ability of an object within the class hierarchy to inherit the attributes and methods of the classes above it. For example, we can create two classes, CUSTOMER and EMPLOYEE, as subclasses from the class PERSON. In this case CUSTOMER and EMPLOYEE will inherit all attributes and methods from PERSON.

To illustrate the difference between the OO model and the ER model, let's examine their graphic representations in the simple invoicing problem shown in Figure 2.8.

As you examine Figure 2.8, note that:

- The OO data model represents an object as a box; all of the object attributes and relationships to other objects are included within the object box. The object representation of the INVOICE includes all related objects within the *same* object box. Note that the connectivities (1 and M) indicate the relationship of the related objects to the INVOICE. For example, the "1" next to the CUSTOMER object indicates that each INVOICE is related to only one CUSTOMER. The "M" written next to the LINE object indicates that each INVOICE contains many LINEs.

- The ER model uses three separate entities and two relationships to represent an invoice transaction. Because customers can buy more than one item at a time, each invoice references one or more lines, one item per line! And, because invoices are generated by customers, the data-modeling requirements include a customer entity and a relationship between the customer and the invoice.

FIGURE 2.8 A COMPARISON OF THE OO MODEL AND THE ER MODEL

Advantages

The OO data model has several important advantages over the ER model:

1. *Adds semantic content.* The addition of semantic content to the data model gives the data greater meaning. For example, the OO data model in Figure 2.8 shows the *relationships* between the customer and the invoice and between the invoice and the invoice lines within the invoice object. That information is *in addition* to the attributes found in the invoice object.

2. *Visual presentation includes semantic content.* Like the ERD, the OODM models relationships visually. However, the OODM includes the semantic content within the object's visual representation, thus making it easier to visualize much more complex relationships within and between objects.

3. *Database integrity.* Like the hierarchical data model, the OODM uses inheritance to protect database integrity. However, the OODM's objects include more relationship types and more complex relationships.

4. *Both structural and data independence.* The OODM's object autonomy ensures both structural and data independence.

Disadvantages

In spite of the OO model's impressive strengths, it has failed to win the battle against the relational model. There are many reasons for this failure, but the following disadvantages of the OODM have contributed significantly:

1. *Slow pace of OODM standards development.* Object oriented standards have evolved very slowly compared to other data model standards. Therefore, in the rush to bring object oriented products to market, vendors supplied OODBM products with proprietary features, thus creating imcompatibilities among vendors. For example, there is no standard data access method comparable to SQL in the relational model. This shortcoming creates problems when data are accessed from multiple sources.

2. *Complex navigational data access.* The OO data access method resembles the navigational style of the hierarchical and network models.

3. *Steep learning curve.* Given the lack of widely accepted standards and the difficulties imposed by the navigational data access style, the object oriented model's learning curve tends to be steeper than the relational model's. Although we use objects with ease—we drag and drop Windows objects with abandon

and with no thought about the actions precipitated by those movements—the implementation of OO databases is quite a different story. Objects are complex things, and because they are able to contain so much semantic content, it is difficult to design and implement them properly. The programmer-oriented roots of object oriented systems make end users perceive them as difficult to operate, too.

4. *High system overhead slows transactions.* The system complexity tends to be far greater than that of the relational model. Therefore, the OODM's implementation requires substantial hardware and operating system overhead. The complexity of the environment and the high systems requirements tend to slow down transactions. And, because transactions are the lifeblood of the production database, *slow* transactions are simply unacceptable.

5. *Lack of market penetration.* Relational database vendors increasingly adopted OO features within their products. Given the fact that relational databases already occupied the major database market share, the inclusion of OO features in relational databases made it difficult to make a case for the adoption of OO databases.

2.4.6 OTHER MODELS

Another semantic data model was developed in response to the increasing complexity of applications—the **extended relational data model (ERDM)**. The ERDM, championed by many relational database researchers, constitutes the relational model's response to the OODM challenge. This model includes many of the OO model's best features within an inherently simpler relational database structural environment. That's why a DBMS based on the ERDM is often described as an **object/relational database management system (O/RDBMS)**.

With the relational database's huge installed base and the emergence of the ERDM, the OODM faces an uphill battle. However, the OODM's basic concepts have become the basis of a wide variety of database systems analysis and design procedures. In addition, the basic OO approach has been adopted by many application generators and other development tools.

The OODM-ERDM battle for dominance in the database marketplace seems remarkably similar to the one waged by the hierarchical and network models against the relational model more than two decades ago. The OODM and ERDM are similar in the sense that each attempts to address the demand for more semantic information to be incorporated into the model. However, the OODM and the ERDM differ substantially both in underlying philosophy and in the nature of the problem to be addressed. Although the ERDM includes a strong semantic component, it is primarily based on the relational data model's concepts. In contrast, the OODM is wholly based on the OO and semantic data model concepts. The ERDM is primarily geared to business applications, while the OODM tends to focus on very specialized engineering and scientific applications. In the database arena, the most likely scenario appears to be an ever-increasing merging of OO and relational data model concepts and procedures.

Although the ERDM label has frequently been used in the database literature to describe the—quite successful— relational data model's response to the OODM challenge, C. J. Date objects to the ERDM label for the following reasons (set forth in "Back to the Relational Future" *http://www.dbsummit.com/or/date.html*):

- The useful contribution of the object model is its ability to let users define their own—and often very complex—data types. However, mathematical structures known as "domains" in the relational model also provide this ability. Therefore, a relational DBMS that properly supports such domains greatly diminishes the reason for using the object model. Given proper support for domains, relational data models are quite capable of handling the complex data encountered in time series, engineering design, office automation, financial modeling, and so on. Because the relational model can support complex data types, the notion of an "extended relational data model" or ERDM is "extremely inappropriate and inaccurate" and "it should be firmly resisted." (The capability that is supposedly being extended is already there!)

- Even the label object/relational data model (O/RDM) is not quite accurate, because the relational data model's domain is not an object model structure. However, there are already quite a few O/R products—also known as **universal database servers**—on the market. Therefore, Date concedes that we are probably stuck with the O/R label. In fact, Date believes that "an O/R system is in everyone's future." More precisely, Date argues that a true O/R system would be "nothing more nor less than a true relational system—which is to say, a system that supports the relational model, with all that such support entails."

C. J. Date concludes his discussion by observing that "We need do nothing to the relational model to achieve object functionality. (Nothing, that is, except implement it, something that doesn't yet seem to have been tried in the commercial world.)"

Because C. J. Date is generally considered to be one of the world's leading database thinkers and innovators, his observations cannot be easily dismissed. In any case, regardless of the label that is used to tag the relational data model's growing capabilities, it seems clear that the relational data model is likely to maintain its database market dominance for some time. We believe, therefore, that our continued emphasis on the relational data model is appropriate.

2.4.7 DATA MODELS: A SUMMARY

The evolution of database management systems has always been driven by the search for new ways of modeling increasingly complex real-world data. A summary of the most commonly recognized data models is shown in Figure 2.9.

FIGURE 2.9 THE DEVELOPMENT OF DATA MODELS

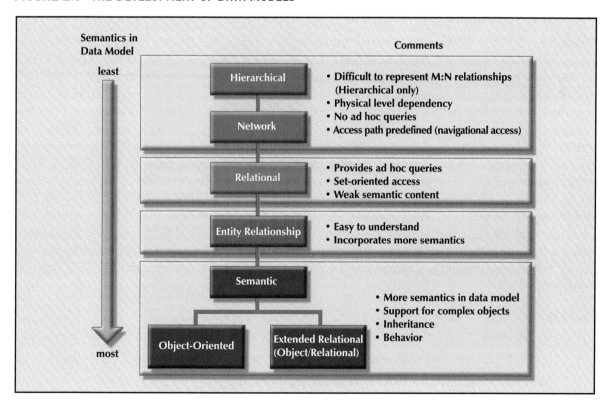

Each new data model capitalized on the shortcomings of previous models. The network model replaced the hierarchical model because the former made it much easier to represent complex (many-to-many) relationships. In turn, the relational model offered several advantages over the hierarchical and network models through its simpler data representation, superior data independence, and relatively easy-to-use query language. Although the debate about the relative merits of the older hierarchical and network models, and the newer relational model lasted several years, the relational model emerged as the dominant data model for business applications. The OO and ERDM have a high level of semantic content, provide support for complex and large objects, and include inheritance and behavior in their objects. (C. J. Date would argue that a properly implemented relational model would also have such characteristics and capabilities.) Although the OO and ERDM have gained a substantial foothold in the battle for market dominance, their attempts to dislodge the relational model from its position of market dominance have not been succesful.

In the evolution of data models, we can determine some common characteristics that data models must have in order to be widely accepted:

1. A data model must show some degree of conceptual simplicity without compromising the semantic completeness of the database. *It does not make sense to have a data model that is more difficult to conceptualize than the real world.*

2. A data model must represent the real world as closely as possible. This goal is more easily realized by adding more semantics to the model's data representation. (Semantics concern the dynamic data behavior, while the data representation constitutes the static aspect of the real-world scenario.)

3. The representation of the real-world transformations (behavior) must be in compliance with the consistency and integrity characteristics of any data model.

Database Models and the Internet

If there is a constant in the database arena, it is that database strategies are always in a state of flux. The use of the Internet as a prime business tool has drastically changed the role and scope of the database market. In fact, the Internet's impact on the database market has generated new database product strategies in which the OODM and ERDM-O/RDM have taken a backseat to Internet database access development. Therefore, instead of witnessing an OODM vs. ERDM-O/RDM data modeling duel, we now see vendors focusing their development efforts on creating database products that interface efficiently and easily with the Internet. The focus on effective Internet interfacing makes the underlying data model less important to the end user. This development, too, favors the already dominant relational database model. Change is always a challenge. Why try to deal with interface issues and database issues simultaneously? It is unnecessary because the relational database now incorporates so many of the features that would seem to give the OODM and ERDM-O/RDM a technical edge. (The market seems to support C. J. Date's argument well: a true O/R system would be "nothing more nor less than a true relational system ….") In any case, the Internet trumps all other aspects of data storage and access—thus forcing a focus on high levels of systems integration and development through new Internet-age technologies that we will examine in detail in Chapters 13, "Databases in Electronic Commerce," and 14, "Web Database Development."

The most successful "Internet age" databases seem to be those characterized by:

- Flexible, efficient, and secure Internet access that is easily used, easily developed, and easily supported.

- Support for complex data types and relationships. (Complex data types include video and sound.)

- Seamless interfacing with multiple data sources and structures.

- Relative conceptual simplicity, to make database design and implementation less cumbersome in a very complex data environment.

- An abundance of available database design, implementation, and application development tools.
- A powerful DBMS graphical user interface (GUI) to help make the DBA's job easier.

Clearly, if the database fits well into the Internet picture, its precise modeling heritage does not matter much. That's one good reason the relational model has prospered by borrowing components from other data models. For example, Oracle Corporation's Oracle9i database contains object oriented components *within a relational database structure*, as does IBM's current DB2 version. (Perhaps the most crucial barrier to OODM dominance is that the benefits of using the Internet—and creating business value—are more easily measured than those provided by pure object oriented features. Therefore, OO concepts are mostly applied to support Internet usability, rather than to develop databases that rely on OO concepts.)

Yet, in spite of the fact that most current database *application* development is now geared toward the Internet, keep in mind that all Internet business transactions ultimately interface with a database of some kind. If that database is badly designed, the best Internet interface won't fix the database structural problems, and the Internet applications will fail. That's why we focus on database design and on the issues that affect such design. The Internet coverage in Chapters 13 and 14 simply illustrates that a new data access dimension has been added to the underlying database.

Table 2.1 summarizes the advantages and disadvantages of the various database models.

2.5 DEGREES OF DATA ABSTRACTION

Another way of classifying data models is by levels or degrees of data abstraction. To illustrate what we mean by data abstraction, consider the example of automotive design. A car designer begins by drawing the *concept* of the car that is to be produced. Next, engineers design the details that help transfer the basic concept into a structure that can be produced. Finally, the engineering drawings are translated into production specifications to be used on the factory floor. As you can see, the process of producing the car begins at a high level of abstraction and proceeds to an ever-increasing level of detail. The factory floor process cannot proceed unless the engineering details are properly specified, and the engineering details could not exist without the basic conceptual framework created by the designer. Designing a usable database follows the same basic process. That is, a database designer starts with an abstract view of the overall data environment and adds details as the design comes closer to implementation. Using levels of abstraction can also be very helpful in integrating multiple (and sometimes conflicting) views of the data as seen at different levels of an organization.

In the early 1970s, the American National Standards Institute/Standards Planning and Requirements Committee (ANSI/SPARC) classified data models according to their degree of abstraction: conceptual, external, and internal (see Figure 2.10). As you examine Figure 2.10, note that we have expanded the ANSI/SPARC classification by adding a *physical* model to explicitly address DBMS-specific implementation issues of the internal model. In addition, to better represent the *designer's* view of the data, we have explicitly recognized the connection between the external model and the internal model.

Successful database design is first and foremost based on end-user requirements. However, while the designer's activities are based on extensive interaction with the end users, it is the designer who must make the technical decisions that govern the design's structure. Because we focus on the technical aspects of database design, we tend to emphasize the designer's view of the data. To use an analogy, if you want to design a successful car, you'll have to spend a lot of time with the people who are expected to buy and drive that car. But you do not ask the drivers to determine the firing order of the pistons, or the best casting method for the engine block.

TABLE 2.1 ADVANTAGES AND DISADVANTAGES OF VARIOUS DATABASE MODELS

DATA MODEL	DATA INDEPENDENCE	STRUCTURAL INDEPENDENCE	ADVANTAGES	DISADVANTAGES
Hierarchical	Yes	No	1. Promotes data sharing 2. Parent/Child relationship promotes conceptual simplicity 3. Parent/Child relationship promotes data integrity 4. Efficient with 1:M relationships	1. Navigational systems yield complex design, implementation, application development, use, and management 2. Implementation limitations (No multiparent or M:N relationships) 3. No data definition or data manipulation language in the DBMS 4. Lack of standards
Network	Yes	No	1. Conceptual simplicity at least equal to that of the hierarchical model 2. Handles more relationship types such as M:N and multiparent 3. Owner/Member relationship promotes data integrity 4. Conformance to standards 5. Includes data definition language (DDL) and data manipulation language (DML) in DBMS	1. System complexity limits efficiency—still a navigational system 2. Navigational systems yield complex design, implementation, application development, use, and management
Relational	Yes	Yes	1. Tabular view substantially improves conceptual simplicity, thereby promoting easier database design, implementation, management, and use 2. Ad hoc query capability based on SQL 3. Powerful RDBMS improves implementation and management simplicity	1. The RDBMS that makes the system easy to use and manage requires substantial hardware and system software overhead 2. Conceptual simplicity gives relatively untrained people the tools to use a good system poorly 3. May promote "islands of information" problems as individuals and departments find it easy to develop their own applications
Entity Relationship	Yes	Yes	1. Visual modeling yields exceptional conceptual simplicity 2. Visual representation makes it an effective communication tool 3. Integrated with dominant relational model	1. Limited constraint representation 2. Limited relationship representation 3. No data manipulation language 4. Loss of information content when attributes are removed from entities to avoid crowded displays (This limitation has been addressed in subsequent graphic versions)
Object Oriented	Yes	Yes	1. Adds semantic content 2. Visual representation includes semantic content 3. Inheritance promotes data integrity	1. Slow development of standards caused vendors to supply their own enhancements, thus eliminating a widely accepted standard 2. Complex navigational system 3. Steep learning curve 4. High system overhead slows transactions

Note: All databases assume the use of a common data pool within the database. Therefore, all database models promote data sharing, thus at least potentially eliminating the "islands of information" problem.

FIGURE 2.10 DATA ABSTRACTION LEVELS

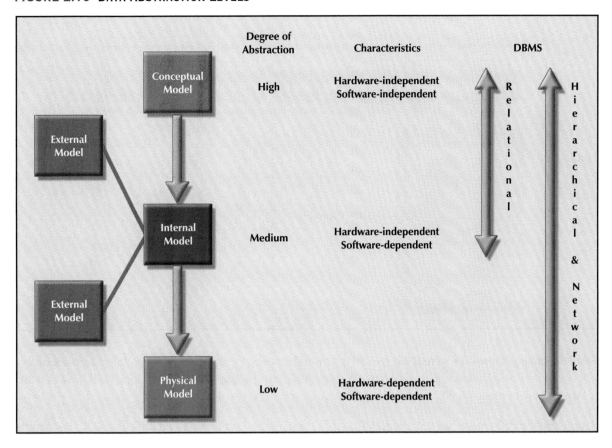

2.5.1 THE CONCEPTUAL MODEL

The **conceptual model** represents a global view of the database. It is an enterprise-wide representation of data as viewed by high-level managers. Therefore, it is located at the abstraction apex in Figure 2.10. The conceptual model is the basis for the identification and description of the main data objects, avoiding details. The most widely used conceptual model is the entity relationship (ER) model. You should recall that the ER model is illustrated with the help of the entity relationship diagram (ERD), which is, in effect, the basic database blueprint. We use the ERD to graphically *represent* the conceptual model.

To illustrate the conceptual model's use, let us examine the data environment of Tiny College. Tiny College's main objects are its students, professors, courses, classes, and classrooms. These objects are the main entities for which data are collected and stored.

> **NOTE**
>
> Earlier in this chapter you learned that an *entity set* is a collection of related entities. For example, the set of students at Tiny College can be properly referred to as the *entity set* named STUDENT. We also noted that the data-modeling vocabulary uses *entity* to label what is technically an *entity set*. That is, the data modeler refers to the set of students at Tiny College as the *entity* named STUDENT. If you do not keep this vocabulary quirk in mind, you may become confused when we describe the relationships among entities.

The Tiny College entities named STUDENT, PROFESSOR, COURSE, CLASS, and ROOM, as well as some sample STUDENT attributes, are displayed in Figure 2.11.

FIGURE 2.11 TINY COLLEGE ENTITIES

Given the entities shown in Figure 2.11, we can define and describe the relationships (also called *associations* or *interactions*) among those entities. Recall that a relationship can be classified as one-to-one (1:1), one-to-many (1:M), or many-to-many (M:N).

Having identified the entities, we use a conceptual model, graphically represented by an ERD (Figure 2.12), to relate one entity to another. As you examine the ERD in Figure 2.12, note that the relationships are described by the verbs *generates, teaches, contains,* and *is used for.* That is, a COURSE generates a CLASS, a PROFESSOR teaches a CLASS, a CLASS contains STUDENTs, and a ROOM is used for a CLASS.

> **N O T E**
>
> Our use of CLASS and COURSE entities reflects the real-world use of these labels. Students use a *class* schedule to enroll each semester or quarter. Although each of the classes listed in the class schedule represents a section of a course, the schedule is *not* named a section schedule.

Note also that the entity relationships in Figure 2.12 can be classified 1:M or M:N. For example, a PROFESSOR may teach many CLASSes, but each CLASS is taught by only one PROFESSOR; that is, there is a 1:M relationship between PROFESSOR and CLASS. Similarly, a CLASS may contain many STUDENTs, and each student may take many CLASSes, thus creating an M:N relationship between STUDENT and CLASS. Each COURSE may generate many CLASSes, but each CLASS references a single COURSE. For example, there may be several classes (sections) of a database course having a course code of CIS-420. One of those classes may be offered on MWF from 8:00 a.m. to 8:50 a.m., another may be offered on MWF from 1:00 p.m. to 1:50 p.m., while a third may be offered on Thursdays from 6:00 p.m. to 8:40 p.m., yet all three classes have the course code CIS-420. Finally, a CLASS requires one ROOM, but a ROOM may be scheduled for many CLASSes, because each classroom may be used for several classes: one at 9:00 a.m., one at 11:00 a.m., and so on. In other words, there is a 1:M relationship between ROOM and CLASS.

FIGURE 2.12 A CONCEPTUAL MODEL FOR TINY COLLEGE

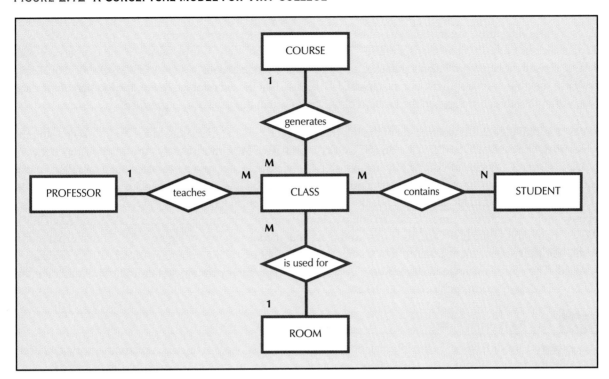

The conceptual model yields some very important advantages. First, it provides a relatively easily understood bird's-eye (macro level) view of the data environment. For example, you can get a summary of Tiny College's data environment by examining the conceptual model presented in Figure 2.12.

The DBMS only needs a name for the relationship—it does not care how a relationship is described or how it is read. However, human beings tend to prefer a relationship description that favors common language conventions. For this reason, 1:M relationships are read from the "1" side to the "M" side, regardless of how and in what position the entity boxes are presented. Therefore, you read the relationship between CLASS and ROOM in Figure 2.12 as "a ROOM *is used for* many CLASSes," and the relationship between PROFESSOR and CLASS as "a PROFESSOR *teaches* many CLASSes."

Second, the conceptual model is independent of both software and hardware. **Software independence** means that the model does not depend on the DBMS software used to implement the model. **Hardware independence** means that the model does not depend on the hardware used in the implementation of the model. Therefore, changes in either the hardware or the DBMS software will have no effect on the database design at the conceptual level.

2.5.2 THE INTERNAL MODEL

Once a specific DBMS has been selected, the internal model adapts the conceptual model to the DBMS. The **internal model** is the representation of the database as "seen" by the DBMS. In other words, the internal model requires the designer to match the conceptual model's characteristics and constraints to those of the selected implementation model. Because the internal model depends on the existence of specific database software, it is said to be software-dependent. Therefore, a change in the DBMS software requires that the internal model be changed to fit the characteristics and requirements of the DBMS.

The development of a detailed internal model is especially important to database designers who work with hierarchical or network models, because these models require very precise specification of data storage location and data access paths. In contrast, the relational model requires less detail in its internal model because most RDBMSs handle data access path definition *transparently*—that is, the designer need not be aware of the data access path details. Nevertheless, even relational database software usually requires data storage location specification, especially in a mainframe environment. For example, DB2 requires that the data storage group, the location of the database within the storage group, and the location of the tables within the database be specified.

In the case of Figure 2.12, we implement the internal model by creating the database for Tiny College using the tables PROFESSOR, COURSE, CLASS, STUDENT, and ROOM. Unfortunately, the relationship between STUDENT and CLASS is M:N. The M:N relationship is undesirable, because it produces the data redundancies you learned about in Chapter 1. (You will learn exactly how and why such redundancies occur in Chapter 4, "Entity Relationship (ER) Modeling.") In addition, the enrollment of students in classes is an activity that requires its own entity. Therefore, we must create an entity to implement the relationship between CLASS and STUDENT. We will name this entity ENROLL, as shown in the internal model segment in Figure 2.13. The internal model is still hardware-independent because it is unaffected by the choice of the computer on which the software is installed. Therefore, a change in storage devices or even a change in operating system will not affect the internal model's design requirements.

2.5.3 THE EXTERNAL MODEL

The **external model**, based on the internal model, is the end users' view of the data environment. By end users we mean the people who use the application programs as well as those who designed and implemented them. End users usually operate in an environment in which an application has a specific business unit focus. Businesses are generally divided into several business units such as sales, finance, and marketing. Each business unit is subject to specific constraints and requirements, and each one uses a data subset. Therefore, applications programmers working within those business units view their data subsets as separate from, or external to, the internal model from which those subsets were derived.

The applications programmer's working environment thus requires that the modeler subdivide the set of requirements and constraints into functional modules that can be examined within the framework of their external models. For example, the internal model depicted in Figure 2.13 is broken up into two functional modules that yield the two external models. Each of the external models summarizes the theater of operations for each applications programmer. Each external model includes the appropriate entities, relationships, processes, and constraints imposed by the business unit. Note that each of the external models in Figure 2.14 represents a subset of the internal model depicted in Figure 2.13. Also note that, *although the application views are isolated from each other, each view shares a common entity with the other view*. For example, programmers Jim and Anne share the entity CLASS. A good design should consider such relationships between views and provide the programmers with a set of restrictions that govern common entities. The external models for Tiny College may be represented as shown in Figure 2.14.

FIGURE 2.13 A DIVISION OF AN INTERNAL MODEL INTO EXTERNAL MODELS

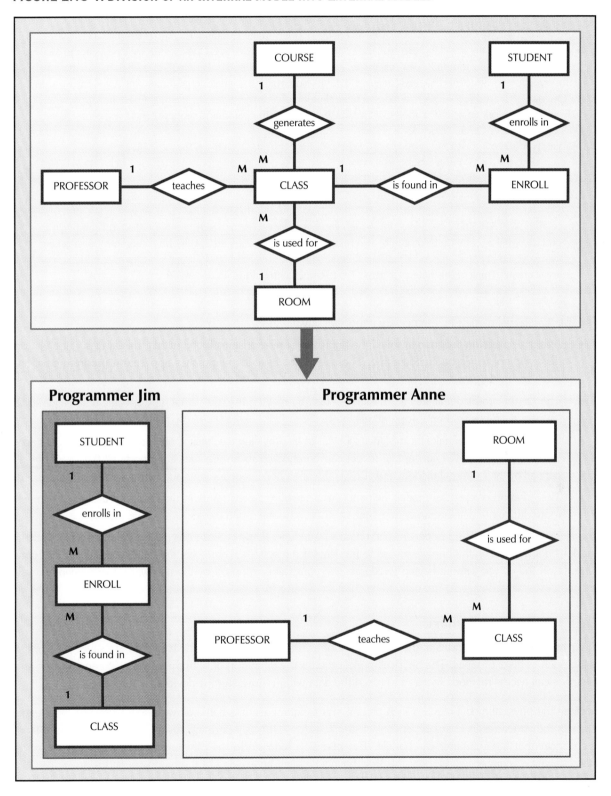

N O T E

In Section 2.4.3 we mentioned that SQL is the relational query language standard. As you examine Figure 2.14, keep in mind that the SQL language uses the command CREATE VIEW to create the end user views in the relational database environment. For example, once you have learned the SQL material in Chapter 6, "Introduction to Structured Query Language (SQL)," you will know that programmer Anne would use the ROOM, CLASS, and PROFESSOR entities in her definition of a view. Therefore, she might base her applications development work on a view named CLASS_VIEW, created by the command sequence:

 CREATE VIEW CLASS_VIEW AS
 SELECT (CLASS_ID, CLASS_NAME, PROF_NAME, CLASS_TIME, ROOM_ID)
 FROM CLASS, PROFESSOR, ROOM
 WHERE CLASS.PROF_ID = PROFESSOR.PROF_ID AND
 CLASS.ROOM_ ID = ROOM.ROOM_ID;

(We have, of course, assumed that the ROOM, CLASS, and PROFESSOR tables exist, as well as the attributes used in the CREATE VIEW command.) Although you have not yet learned SQL, we hope that you can begin to appreciate what a useful tool SQL can be in a real-world environment. Perhaps this brief example will also help convince you that knowledge of SQL is desirable if you want to learn how to implement the database designs you will create. Think of this note as a "teaser" for "coming attractions" in Chapter 6.

The use of external models has some very important advantages:

- If each application program included the entire set of relationships found within the database, application development would be cumbersome. The use of database subsets makes application program development much simpler.

- The use of subsets facilitates the designer's task by making it easier to identify specific data required to support each business unit's operations.

- External models also make the designer's job easier by providing feedback about the conceptual model's adequacy. Specifically, the conceptual model can be checked to ensure that it supports all the processes as defined by their external models, as well as all the operational requirements and constraints (see Figure 2.14).

- The creation of external models helps to ensure *security* constraints in the database design. It is much more difficult to damage an entire database if each business unit works with only a subset.

The external model is DBMS-dependent and hardware-independent, because it is designed for a specific DBMS but does not depend on the hardware used to implement the system. Therefore, the database designer can concentrate on the design without having to address the constraints imposed by the hardware. Consequently, well-designed external schemas do not need to be changed when new physical devices are installed.

2.5.4 THE PHYSICAL MODEL

The **physical model** operates at the lowest level of abstraction, describing the way data are saved on storage media such as disks or tapes. The physical model requires the definition of both the physical storage devices and the (physical) access methods required to reach the data within those storage devices.

Because the physical model requires such precise task assignments, it is both software- and hardware-dependent. The storage structures used are dependent on the software (DBMS, operating system) and on the type of storage devices that the computer can handle. The precision required in the physical model's definition demands that database designers who work at this level have a detailed knowledge of the hardware and software used to implement the database design.

FIGURE 2.14 THE EXTERNAL MODELS FOR TINY COLLEGE

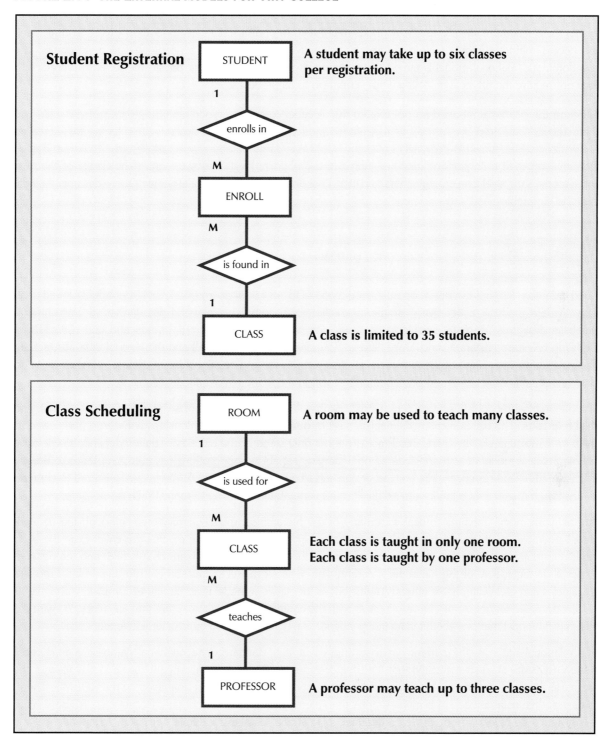

Hierarchical models—and, to a lesser extent, the network model—force the database designer to take the details of the physical model's data storage requirements into account. In both the hierarchical and network environments, the selection of specific access methods might help to optimize database operations, while poorly specified access methods might help to destroy the effectiveness of a good conceptual design. However, the now-dominant relational model is aimed largely at the logical rather than the physical level, and, therefore, it does not require the physical modeling details common to its predecessors.

Although the relational model does not require the designer to be concerned about the data's physical storage characteristics, the *implementation* of a relational model may require physical-level fine-tuning for increased performance. Fine-tuning is especially important when very large databases are installed in a mainframe environment. Yet even such performance fine-tuning at the physical level does not require knowledge of the relational model's physical data storage characteristics. In short, the relational model is much more designer friendly than the hierarchical and network models.

S U M M A R Y

Good database design is important because even a good DBMS will perform poorly with a poorly designed database. A data model is a (relatively) simple abstraction of a complex real-world data-gathering environment. Database designers use data models to communicate with applications programmers and end users. Data-modeling requirements are a function of different data views (global vs. local) and the level of data abstraction. The basic data modeling components are: entities, attributes, and relationships. Business rules are used to identify and define the basic modeling components within a specific real-world environment.

The hierarchical model is based on a tree structure that is composed of a root segment, parent segments, and child segments. The hierarchical model depicts a set of one-to-many (l:M) relationships between a parent and its children. Unfortunately, the hierarchical model returns only some of the benefits of data independence because the hierarchical model does not include ad hoc querying capability; applications programming tends to require extensive effort and resources. Hierarchical design can also be difficult; thus DBMS efficiency is hard to capture. In addition, a multiparent relationship is difficult to implement, as is a many-to-many (M:N) relationship.

The network model attempts to deal with many of the hierarchical model's limitations. Although the network model resembles the hierarchical model, its structure easily accommodates the multiparent child. Even the basic model components are similar: the network member is the equivalent of the hierarchical child, and the network owner is the equivalent of the hierarchical parent. The network model shares some of the hierarchical model's efficiencies and greatly improves others. As efficient as the network model is, its structural complexity often limits its effectiveness: designers must be very familiar with the database structure to capture that efficiency. As with the hierarchical model that preceded it, data independence is not accompanied by structural independence.

The relational model is the current database implementation standard. The relational database management system (RDBMS) is so sophisticated that the user/designer need only be concerned with the logical view of the database; the details of physical storage, access paths, and data structures are managed by the RDBMS. Therefore, relational database design is much simpler than hierarchical or network design. Because the RDBMS hides the system's complexity from the user/designer, a relational database exhibits both data independence and structural independence. Therefore, data management becomes much easier than with the earlier models.

Increasingly complex databases created a need for a more easily interpreted graphic conceptual environment. Tools developed to address this need include the entity relationship diagram (ERD), which is based on the entity relationship (ER) model. The ER model is particularly valuable because it is based on a visual presentation of data and their relationships. In addition, the ER model allows the database designer to capture different views of the data as seen by database designers, programmers, and end users and to integrate them into a common framework.

The ER model spawned a new generation of database design tools; the most widely used of these tools is the entity relationship diagram (ERD). Because the ER model integrates well with the still-dominant relational model, the ERM and the ERDs based on it are powerful instruments that help make the design of relational databases a very structured process.

Current real-world data and information problems tend to be complex. Information requirements are more extensive and include the use of data types such as pictures, sound, and video. The basic modeling structure of the object oriented model is the object. An object resembles an entity in that it includes the facts that define it. But, unlike an entity, the object also includes information about relationships between the facts as well as relationships with other objects, thus giving its data more meaning. Therefore, the object has greater semantic content than the entities used in earlier data models.

In spite of the potential benefits of the OO data model, its slow development of widely accepted standards, difficult navigational data access environment, and relatively slow transaction speeds caused by high system overhead have prevented its wide acceptance as a database standard. Instead, the relational model has adopted many OO extensions to become the extended relational data model (ERDM). At this point, the OODM is largely used in specialized engineering and scientific applications, while the ERDM is primarily geared to business applications. Although the most likely scenario is an increasing merger of OODM and ERDM technologies, both are overshadowed by the need to develop Internet access strategies for databases.

Data modeling requirements are a function of different data views (global vs. local) and the level of data abstraction. Four levels of data abstraction are illustrated in Table 2.2.

TABLE 2.2 **LEVELS OF DATA ABSTRACTION**

MODEL	DEGREE OF ABSTRACTION		DATA MODEL	FOCUS	INDEPENDENT OF
Conceptual	High		Entity	Global	Hardware and software
External		↕	ER components	Subset	Hardware
Internal			Relational and others	Global	Hardware
Physical	Low		Physical storage methods	N/A	Neither hardware nor software

K E Y T E R M S

American National Standards
 Institute (ANSI)
attribute
business rule
Chen model
class
class hierarchy
conceptual model
Conference On Data Systems
 Languages (CODASYL)
connectivity
Crow's Foot model
data model
Database Task Group (DBTG)

entity
entity instance
entity occurrence
entity relationship diagram (ERD)
entity relationship (ER) model
entity set
ERM
extended relational data model
 (ERDM)
external model
Generalized Update Access
 Method (GUAM)
hardware independence
hierarchic sequence

hierarchical model
hierarchical path
hierarchical structure
information management
 system (IMS)
inheritance
internal model
many-to-many (M:N or M:M)
 relationship
member (network database)
method
model
network model
object

object oriented data model
 (OODM)
object oriented database manage-
 ment system (OODBMS)
object/relational database manage-
 ment system (O/RDBMS)
one-to-many (1:M) relationship
one-to-one (1:1) relationship
owner (network database)
physical model
pointer

preorder traversal
relation
relational database management
 system (RDBMS)
relational model
relational schema
relationship
root
schema
segment (parent and child)

semantic data model (SDM)
set (owner/member)
software independence
Standards Planning and
 Requirements Committee
 (SPARC)
Structured Query Language (SQL)
subschema
table
universal database server

R E V I E W Q U E S T I O N S

1. Discuss the importance of data modeling.

2. What is a business rule and what is its purpose in data modeling?

3. How does the hierarchical database structure address the problem of data redundancy?

4. What do each of the following acronyms represent, and how is each related to the birth of the network data model?
 a. CODASYL
 b. SPARC
 c. ANSI
 d. DBTG

5. What three languages were adopted by the DBTG to standardize the basic network data model, and why was such standardization important to users and designers?

6. Describe the basic features of the relational data model, and discuss their importance to the end user and the designer.

7. Explain how the entity relationship (ER) model helped produce a more structured relational database design environment.

8. Use the scenario described by "A customer can make many payments, but each payment is made by only one customer" as the basis for an entity relationship diagram (ERD) presentation.

9. Why is an object said to have greater semantic content than an entity?

10. What is the difference between an object and a class in the object oriented data model (OODM)?

11. How would you model Question 8 with an OODM? (Use Figure 2.8 as your guide.)

12. What is an ERDM, and what role does it play in the modern (production) database environment?

13. Compare file system data management with the five data models discussed in this chapter, in terms of data and structural independence.

14. What is a relationship, and what three types of relationships exist?

15. Give an example of each of the three types of relationships.

16. What is a table and what role does it play in the relational model?

17. What is a relational schema? Give an example.

18. What is structural independence and what are its advantages?

19. What are the advantages of using a 4GL as opposed to a 3GL?

20. What is connectivity?
 a. Use a Chen ERD to illustrate connectivity.
 b. Use a Crow's Foot ERD to illustrate connectivity.

P R O B L E M S

Use the contents of Figure 2.4 to work Problems 1–6.

1. Write the relationship statement that governs the relationship between AGENT and CUSTOMER.

2. Given the relationship you wrote in Problem 1, create the basic Chen ERD. Make sure that you identify the connectivities.

3. Given the relationship you wrote in Problem 1, create the basic Crow's Foot ERD.

4. If the relationship between AGENT and CUSTOMER were implemented in a hierarchical model, what would the hierarchical structure look like? Label the structure fully, identifying the root segment and the level 1 segment.

5. If the relationship between AGENT and CUSTOMER were implemented in a network model, what would the network model look like? (Identify the set.)

6. Using the ERD you drew in Problem 3 as your guide, create the equivalent OO model.

Using Figure P2.7 as your guide, work Problems 7–9. The DealCo relational schema shows the initial entities and attributes for the DealCo stores, located in two regions of the country.

FIGURE P2.7 THE DEALCO RELATIONAL SCHEMA

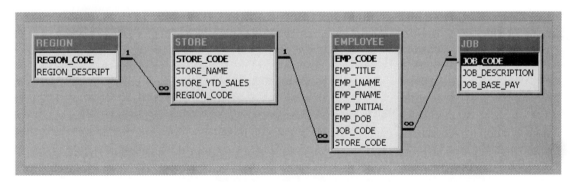

7. Identify each relationship type and write all the relationships.

8. Draw the basic Chen ERD that corresponds to the relational schema.

9. Convert the Chen ERD you created in Problem 8 to the basic Crow's Foot ERD.

Using Figure P2.10 as your guide, work Problems 10–14. The Tiny College relational schema shows the initial entities and attributes for the Tiny College university.

FIGURE P2.10 THE TINY COLLEGE RELATIONAL SCHEMA

10. Identify each relationship type and write all the relationships.

11. Draw the basic Chen ERD that corresponds to the relational schema.

12. Convert the Chen ERD you created in Problem 11 to the basic Crow's Foot ERD.

13. Create the network model that reflects the entities and relationships you identified in the relational schema.

14. Create the object oriented model that reflects the entities and relationships you identified in the relational schema.

15. Using the hierarchical representation shown in Figure P2.15, answer a, b, and c.

FIGURE P2.15 THE HIERARCHICAL STRUCTURE FOR THE ARTIST DATABASE

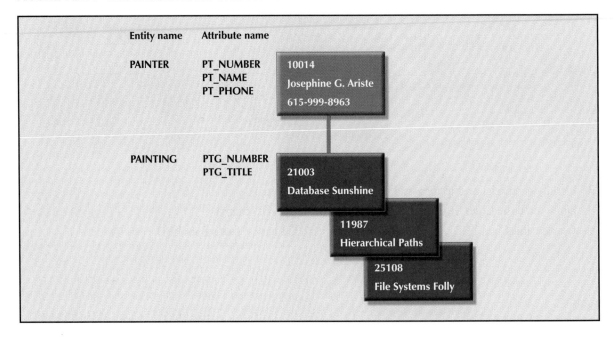

a. Identify the segment types.

b. Identify the components that are equivalent to the file system's fields.

c. Describe the hierarchical path for the occurrence of the third PAINTING segment.

16. The hierarchical diagram shown in Figure P2.16 depicts a single record occurrence of a patient named Judy D. Johanssen. Typically, a patient staying in a hospital receives medications that have been ordered by a particular doctor. Because the patient often receives several medications per day, there is a 1:M relationship between PATIENT and ORDER. Similarly, each order can include several medications, creating a 1:M relationship between ORDER and MEDICATION.

Given the structure shown in Figure P2.16:

a. Identify the segment types.

b. Identify the components that are equivalent to the file system's fields.

c. Describe the hierarchical path for the occurrence of the second MEDICATION segment.

FIGURE P2.16 THE HIERARCHICAL STRUCTURE FOR THE MEDCLINIC DATABASE

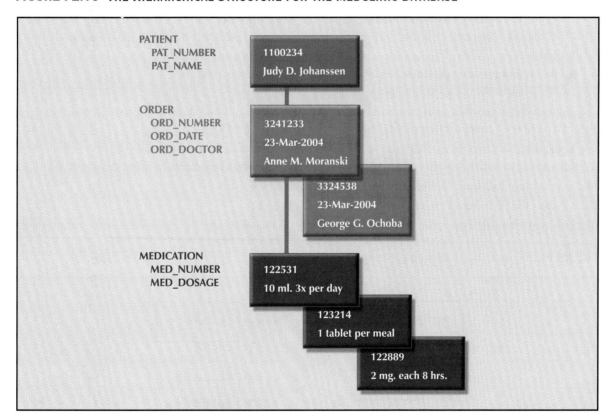

17. Expand the model in Problem 16 to include a DOCTOR segment, then draw its hierarchical structure. (Identify all segments.) (*Hint*: A patient can have several doctors assigned to his or her case, but the patient named Judy D. Johanssen occurs only once in each of those doctors' records.)

18. Suppose you want to write a report that shows:
 a. All patients treated by each doctor.
 b. All doctors who treated each patient.

 Evaluate the hierarchical structure you drew in Problem 17 in terms of its search efficiency in producing the report.

19. The PYRAID company wants to track each PART used in each specific piece of EQUIPMENT; each PART is bought from a specific SUPPLIER. Using this description, draw the network structure and identify the sets for the PYRAID company database. (*Hint*: A piece of equipment is composed of many parts, but each part is used in only one specific piece of equipment. A supplier can supply many parts, but each part has been supplied by only one supplier.)

20. United Broke Artists (UBA) is a broker for not-so-famous painters. UBA maintains a small network database to track painters, paintings, and galleries. Using PAINTER, PAINTING, and GALLERY, write the network structure and identify appropriate sets within the UBA database.

 (*Hint 1*: A painting is painted by a particular artist, and that painting is exhibited in a particular gallery.)
 (*Hint 2*: A gallery can exhibit many paintings, but each painting can be exhibited in only one gallery. Similarly, a painting is painted by a single painter, but each painter can paint many paintings.)

21. If you decide to convert the network database in Problem 20 to a relational database:
 a. What tables would you create, and what would the table components be?
 b. How might the (independent) tables be related to one another?

22. Convert the network database model in Figure 2.3 into a design for a relational database model, using an ERD. Show all entities, relationships, and connectivities.

23. Using the ERD from Problem 22, create the relational schema. (Create an appropriate collection of attributes for each of the entities. Make sure you use the appropriate naming conventions to name the attributes!)

24. Convert the ERD from Problem 22 into a basic OODM.

25. Convert the ERD from Problem 22 into a Crow's Foot model.

26. Describe the relationships depicted in the ERD shown in Figure P2.26.

FIGURE P2.26 THE CHEN ERD FOR PROBLEM 26

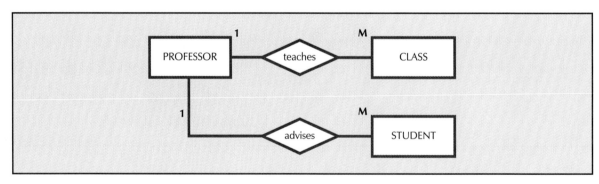

27. Convert the ERD from Problem 26 into a Crow's Foot model.

28. Describe the relationships shown in the ERD in Figure P2.28.

FIGURE P2.28 THE CROW'S FOOT ERD FOR PROBLEM 28

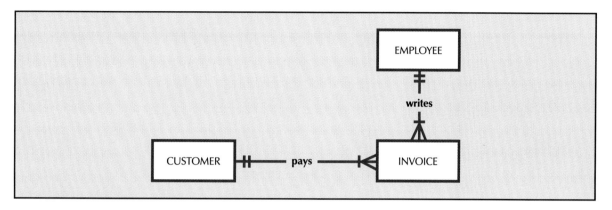

29. Convert the ERD from Problem 28 into a Chen model.

30. Create a Chen ERD and a Crow's Foot ERD for each of the following descriptions. (*Note*: The word "many" merely means "more than one" in the database modeling environment.)

 a. Each of the MegaCo Corporation's divisions is composed of many departments. Each of those departments has many employees assigned to it, but each employee works for only one department. Each department is managed by one employee, and each of those managers can manage only one department at a time.

 b. During some period of time, a customer can rent many videotapes from the BigVid store. Each of the BigVid's videotapes can be rented to many customers during that period of time.

 c. An airliner can be assigned to fly many flights, but each flight is flown by only one airliner.

 d. The KwikTite Corporation operates many factories. Each factory is located in a region. Each region can be "home" to many of KwikTite's factories. Each factory employs many employees, but each of those employees is employed by only one factory.

 e. An employee may have earned many degrees, and each degree may have been earned by many employees.

PART
II

DESIGN CONCEPTS

3

THE RELATIONAL DATABASE MODEL

In this chapter, you will learn:

- ❖ That the relational database model takes a logical view of data

- ❖ That the relational model's basic components are entities, attributes, and relationships among entities

- ❖ How entities and their attributes are organized into tables

- ❖ About relational database operators, the data dictionary, and the system catalog

- ❖ How data redundancy is handled in the relational database model

- ❖ Why indexing is important

In Chapter 2, "Data Models," you learned that the relational data model's structural and data independence allow us to examine the model's logical structure without considering the physical aspects of data storage and retrieval. You also learned that the entity relationship (ER) model may be used to depict entities and their relationships graphically through an entity relationship diagram (ERD). In this chapter, you will learn some important details about the relational model's logical structure. You will also learn more about how the ERD may be used to design a relational database.

You will learn how the relational database's basic data components—entities and their attributes—fit into a logical construct known as a table. You will discover that one important reason for the relational database model's simplicity is that its tables can be treated as logical rather than physical units. You will also learn how the independent tables within the database may be related to one another.

After learning about tables, their components, and their relationships, you will be introduced to the basic concepts that shape the design of tables. Because the table is such an integral part of relational database design, you also will learn the characteristics of well-designed and poorly designed tables.

Finally, we introduce you to some basic concepts that will become your gateway to the next few chapters. For example, you will examine different kinds of relationships and how such relationships might be handled in the relational database environment.

THREE

3.1 A LOGICAL VIEW OF DATA

In Chapter 1, "Database Systems," you learned that a database stores and manages both data and metadata. You also learned that the DBMS manages and controls access to the data and the database structure. Such an arrangement—placing the DBMS between the application and the database—eliminates most of the file system's inherent limitations. The result of such flexibility, however, is a far more complex physical structure. In fact, the database structures required by both the hierarchical and network database models often become complicated enough to diminish efficient database design. The relational data model changed all that by allowing the designer to focus on the logical representation of the data and its relationships, rather than on the physical storage details. To use an automotive analogy, the relational database uses an automatic transmission to relieve us of the need to manipulate clutch pedals and gearshifts. In short, the relational model enables us to view data *logically* rather than *physically*.

The practical significance of taking the logical view is that it reminds us of the simpler file concept of data storage. Although the use of a table, quite unlike that of a file, has the advantages of structural and data independence, a table does resemble a file from a conceptual point of view. Because we can think of related records as being stored in independent tables, the relational database model is much easier to understand than its hierarchical and network database predecessors. Greater logical simplicity tends to yield simpler and more effective database design methodologies.

Because the table plays such a prominent role in the relational model, it deserves a closer look. Therefore, we will begin our discussion by exploring the details of table structure and contents.

3.1.1 TABLES AND THEIR CHARACTERISTICS

The logical view of the relational database is facilitated by the creation of data relationships based on a (logical) construct known as a table. A table is perceived as a two-dimensional structure composed of rows and columns. As far as the table's user is concerned, *a table contains a group of related entities*—that is, an entity set; for that reason, the terms *entity set* and *table* are often used interchangeably. A table is also called a *relation* because the relational model's creator, E. F. Codd, used the term relation as a synonym for table. You can think of a table as a persistent relation, that is, a relation whose contents can be permanently saved for future use.

> **NOTE**
>
> The word *relation*, also known as a *dataset* in Microsoft Access, is based on the mathematical set theory from which Codd derived his model. Because the relational model is based on links that allow us to use entity relationships, many database users incorrectly assume that the term "relation" refers to such relationships. Many then incorrectly conclude that only the relational model can permit the use of relationships.

You will discover that the table view of data makes it easier to spot and define entity relationships, thereby greatly simplifying the task of database design. The characteristics of a relational table are summarized in Table 3.1.

TABLE 3.1 CHARACTERISTICS OF A RELATIONAL TABLE

1 A table is perceived as a two-dimensional structure composed of rows and columns.
2 Each table row (**tuple**) represents a single entity occurrence within the entity set.
3 Each table column represents an attribute, and each column has a distinct name.
4 Each row/column intersection represents a single data value.

TABLE 3.1 CHARACTERISTICS OF A RELATIONAL TABLE (CONTINUED)

5	All values in a column must conform to the same data format. For example, if the attribute is assigned an integer data format, all values in the column representing that attribute must be integers.
6	Each column has a specific range of values known as the **attribute domain**.
7	The order of the rows and columns is immaterial to the DBMS.
8	Each table must have an attribute or a combination of attributes that uniquely identifies each row.

N O T E

Relational database terminology is very precise. Unfortunately, file system terminology sometimes creeps into the database environment. Thus, rows are sometimes referred to as *records*, and columns are sometimes labeled as *fields*. Occasionally, tables are labeled *files*. Technically speaking, this substitution of terms is not always appropriate; the database table is a logical rather than a physical concept, and the terms file, record, and field describe physical concepts. Nevertheless, as long as you recognize that the table is actually a logical rather than a physical construct, you may (at the conceptual level) think of table rows as records and of table columns as fields. In fact, many database software vendors still use this familiar file system terminology.

We'll use the tables shown in Figure 3.1 to illustrate the characteristics listed in Table 3.1.

FIGURE 3.1 STUDENT TABLE ATTRIBUTE VALUES

Database name: Ch03_TinyCollege
Table name: STUDENT

STU_NUM	STU_LNAME	STU_FNAME	STU_INIT	STU_DOB	STU_HRS	STU_CLASS
321452	Bowser	William	C	12-Feb-1972	42	So
324257	Smithson	Anne	K	15-Nov-1977	81	Jr
324258	Brewer	Juliette		23-Aug-1966	36	So
324269	Oblonski	Walter	H	16-Sep-1973	66	Jr
324273	Smith	John	D	30-Dec-1955	102	Sr
324274	Katinga	Raphael	P	21-Oct-1976	114	Sr
324291	Robertson	Gerald	T	08-Apr-1970	120	Sr
324299	Smith	John	B	30-Nov-1983	15	Fr

STUDENT table, continued →

STU_GPA	STU_TRANSFER	DEPT_CODE	STU_PHONE	PROF_NUM
2.84	No	BIOL	2134	205
3.27	Yes	CIS	2256	222
2.26	Yes	ACCT	2256	228
3.09	No	CIS	2114	222
2.11	Yes	ENGL	2231	199
3.15	No	ACCT	2267	228
3.87	No	EDU	2267	311
2.92	No	ACCT	2315	230

STU_HRS	= Credit hours earned	**STU_GPA**	= Grade point average	
STU_CLASS	= Student classification	**STU_PHONE**	= 4-digit campus phone extension	
STU_DOB	= Student date of birth	**PROF_NUM**	= Number of the professor who is the student's advisor	

A D D I T I O N A L C D C O N T E N T

All the databases used to illustrate the material in this chapter are found on the CD bundled with this book. Check the **\Databases\Student\Ch03** folder to find the databases. The database names used in the folder match the database names used in the figures. For example, the source of the tables shown in Figure 3.1 is the **Ch03_TinyCollege** database located in the above noted folder.

Using the STUDENT table shown in Figure 3.1, you can draw the following conclusions corresponding to the points in Table 3.1:

1. The STUDENT table shown in Figure 3.1 is perceived to be a two-dimensional structure composed of eight rows (tuples) and twelve columns. You can also describe the table as being composed of eight records and twelve attributes (fields).

2. Each row in the STUDENT table describes a single entity occurrence within the entity set. (The entity set is represented by the STUDENT table.) Note that the row (record or entity) defined by STU_NUM = 321452 defines the characteristics (attributes or fields) of a student named William C. Bowser. For example, row 4 in Figure 3.1 describes a student named Walter H. Oblonski. Similarly, row 3 describes a student named Juliette Brewer. Given the table contents, the STUDENT entity set includes eight distinct entities (rows).

3. Each column represents an attribute, and each column has a distinct name.

N O T E

LIMITS ON TABLE AND COLUMN NAMES

Most current-generation database software can handle long, descriptive names. However, because it is sometimes necessary to "mix and match" software in a complex database environment, some of the older DBMSs might impose the following naming constraints:

 1. Table names are restricted to eight characters.

 2. Column (attribute) names are limited to 10 characters.

 3. Column names cannot begin with a digit.

In addition, some database software places restrictions on the use of spaces or symbols such as -, *, ., or /. For example, while Microsoft Access finds a name such as PRICE-DISCOUNT acceptable, other software might yield an error message and would find PRICE_DISCOUNT [note the use of the underscore (_) rather than the dash (-)] a better choice. In fact, some DBMS software would interpret PRICE-DISCOUNT as "subtract the value in the DISCOUNT attribute from the value in the PRICE attribute." And an attribute name such as FIRST.NAME might be interpreted as "the NAME attribute located in a table named FIRST." Therefore, we suggest that you consult your DBMS manual to determine the naming limitations imposed by your database software.

4. Each row/column intersection contains only a single data value.

5. All the values in a column match the entity's attribute characteristics. For example, the grade point average (STU_GPA) column contains only STU_GPA entries for each of the table rows. Data must be classified according to their format and function. Although various DBMSs can support different data types, most support at least the following:

 a. *Numeric.* Numeric data are data on which you can perform meaningful arithmetic procedures. For example, STU_HRS and STU_GPA in Figure 3.1 are numeric attributes. On the other hand, STU_PHONE is not a numeric attribute, because adding or subtracting phone numbers does not yield an arithmetically meaningful result.

b. *Character.* Character data, also known as *text* data or *string* data, can contain any character or symbol not intended for mathematical manipulation. In Figure 3.1 for example, STU_LNAME, STU_FNAME, STU_INIT, STU_CLASS, and STU_PHONE are character, text, or string attributes.

c. *Date.* Date attributes contain calendar dates stored in a special format known as a Julian date format. Although the physical storage of the Julian date is immaterial to the user and designer, the Julian date format allows us to perform a special kind of arithmetic known as Julian date arithmetic. Using Julian date arithmetic, we can determine the number of days that have elapsed between 12-May-1997 and 20-Mar-2004 by simply subtracting 12-May-1997 from 20-Mar-2004. In Figure 3.1 STU_DOB can properly be classified as a date attribute. Most relational database software packages support (Julian) date formats.

While the database's internal date format is likely to be Julian, many different *presentation* formats are available. For example, if you use Microsoft Access as we have done in Figure 3.1, you could have shown Mr. Bowser's date of birth (STU_DOB) in a long date format (Saturday, February 12, 1972), the more commonly used medium date format (12-Feb-72), or a short date format (2/12/72). Most relational DBMSs, including Microsoft Access, even allow you to define your own date presentation format. For instance, Access and Oracle users might specify the "dd-mmm-yyyy" date format to show the first STU_DOB value in Figure 3.1 as 12-Feb-1972. (As you can tell by examining the STU_DOB values in Figure 3.1, we selected the "dd-mmm-yyyy" format to present the output.)

N O T E

The date attribute, as described here, is an example of an abstract data type. For now, an **abstract data type** can be defined as a set of similar objects, in this case dates, that have:

- An encapsulated data representation. The storage details are transparent (invisible) to the user. The user does not need to know whether the data are stored internally as a 10-character string (dd/mmm/yyyy) or as an integer giving the number of days since the year 0.

- A set of methods—a set of allowable operations. For dates, the allowable operations might be to:
 - Compare two dates.
 - Find the difference in the number of days between two dates.
 - Generate a date by adding or subtracting a number of days from a given date.
 - Convert a date from its internal representation to a presentation format such as 20-Feb-2004 or 02/20/2004.

d. *Logical.* Logical data can have only a true or false (yes or no) condition. For example, is a student a junior college transfer? In Figure 3.1, the STU_TRANSFER attribute uses a logical data format. Most, but not all, relational database software packages support the logical data format. (Microsoft Access uses the label "Yes/No data type" to indicate a logical data type.)

6. The column's range of permissible values is known as its **domain**. Because the STU_GPA values are limited to the range 0–4, inclusive, the domain is [0,4].

7. The order of rows and columns is immaterial to the user.

8. Each table must have a primary key. In general terms, the **primary key (PK)** is an attribute (or a combination of attributes) that uniquely identifies any given entity (row). In this case, STU_NUM (the student number) is the primary key. Using the data presented Figure 3.1, observe that a student's last name (STU_LNAME) would not be a good primary key because it is possible to find several students whose last name is Smith. Even the combination of the last name and first name (STU_FNAME) would not be an appropriate primary key because, as Figure 3.1 shows, it is quite possible to find more than one student named John Smith.

3.2 KEYS

A **key** consists of one or more attributes that determine other attributes. (For example, an invoice number identifies all the invoice attributes, such as the invoice date and the customer name.) One type of key, the primary key, has already been introduced. Given the structure of the STUDENT table shown in Figure 3.1, defining and describing the primary key seems simple enough. However, because the primary key plays such an important role in the relational environment, we will examine the primary key's properties more carefully. There are several other kinds of keys that warrant attention. In this section you also will become acquainted with superkeys, candidate keys, and secondary keys.

The key's role is based on a concept known as **determination**. In the context of a database table, the statement "A determines B" indicates that if you know the value of attribute A you can look up (determine) the value of attribute B. For example, knowing the STU_NUM in the STUDENT table (see Figure 3.1) means that you are able to look up (determine) that student's last name, grade point average, phone number, and so on. The shorthand notation for "A determines B" is A → B. If A determines B, C, and D, you write A → B, C, D.

Therefore, using the attributes of the STUDENT table in Figure 3.1, we can represent the statement "STU_NUM determines STU_LNAME" by writing:

STU_NUM → STU_LNAME

In fact, the STU_NUM value in the STUDENT table determines all of the student's attribute values. For example, we can write:

STU_NUM → STU_LNAME, STU_FNAME, STU_INIT

and

STU_NUM → STU_LNAME, STU_FNAME, STU_INIT, STU_DOB, STU_TRANSFER

In contrast, STU_NUM is not determined by STU_LNAME, because it is quite possible for several students to have the last name Smith.

The principle of determination is very important, because it is used in the definition of a central relational database concept known as functional dependence. The term **functional dependence** can be defined most easily this way: *The attribute B is functionally dependent on A if A determines B.* More precisely:

> **N O T E**
>
> The attribute B is functionally dependent on the attribute A if each value in column A determines *one and only one* value in column B.

Using the contents of the STUDENT table in Figure 3.1, it is appropriate to say that STU_PHONE is functionally dependent on STU_NUM. For example, the STU_NUM value 321452 determines the STU_PHONE value 2134. On the other hand, STU_NUM is not functionally dependent on STU_PHONE, because the STU_PHONE value 2267 is associated with two STU_NUM values: 324274 and 324291. (Apparently, some students share a phone.) Similarly, the STU_NUM value 324273 determines the STU_LNAME value Smith. But the STU_NUM value is not functionally dependent on STU_LNAME, because more than one student may have the last name "Smith."

The functional dependence definition can be generalized to cover the case in which the determining attribute values occur more than once in a table. Functional dependence can then be defined this way:[1]

> **NOTE**
>
> Attribute A determines attribute B (that is, B is functionally dependent on A) if all the rows in the table that agree in value for attribute A must also agree in value for attribute B.

Be careful when defining the dependency's direction. For example, Gigantic State University determines its student classification as shown in Table 3.2.

TABLE 3.2 STUDENT CLASSIFICATION

HOURS COMPLETED	CLASSIFICATION
Less then 30	Fr
30–59	So
60–89	Jr
90 or more	Sr

Given the contents of Table 3.2, the student classification is dependent on the hours completed. Therefore, we can write:

STU_HRS → STU_CLASS

But the specific number of hours is not dependent on the classification. It is quite possible to find a junior with 62 completed hours or one with 84 completed hours. In other words, the classification (STU_CLASS) will not determine one and only one value for completed hours (STU_HRS).

Keep in mind that it might take more than a single attribute to define functional dependence; that is, a key may be composed of more than one attribute. Such a multiattribute key is known as a **composite key**.

Any attribute that is part of a key is known as a **key attribute**. For instance, in the STUDENT table the student's last name would not be sufficient to serve as a key. On the other hand, the combination of last name, first name, initial, and home phone is very likely to produce unique matches for the remaining attributes. For example, we can write:

STU_LNAME, STU_FNAME, STU_INIT, STU_PHONE → STU_HRS, STU_CLASS

or

STU_LNAME, STU_FNAME, STU_INIT, STU_PHONE → STU_HRS, STU_CLASS, STU_GPA

or

STU_LNAME, STU_FNAME, STU_INIT, STU_PHONE → STU_HRS, STU_CLASS, STU_GPA, STU_DOB

[1] *ISO-ANSI Working Draft Database Language/SQL Foundation (SQL3)*, Part 2, August 29, 1994. This source was provided through the courtesy of Dr. David Hatherly.

Given the possible existence of a composite key, we can further refine the notion of functional dependence by specifying **full functional dependence**:

> **N O T E**
>
> If the attribute (B) is functionally dependent on a composite key (A) but not on any subset of that composite key, the attribute (B) is fully functionally dependent on (A).

Within the broad key classification, several specialized keys can be defined. For example, a **superkey** is any key that uniquely identifies each entity. In short, the superkey functionally determines all of the entity's attributes. Given the structure of our STUDENT table, the superkey may be any of the following:

STU_NUM

STU_NUM, STU_LNAME

STU_NUM, STU_LNAME, STU_INIT

In fact, STU_NUM, with or without additional attributes, can be a superkey, even when the additional attributes are redundant.

A **candidate key** can be described as a superkey without redundancies. Using this distinction, we note that the composite key

STU_NUM, STU_LNAME

is a superkey, but it is not a candidate key because STU_NUM by itself is a candidate key! The combination

STU_LNAME, STU_FNAME, STU_INIT, STU_PHONE

might also be a candidate key, as long as you discount the possibility that two students may share the same last name, first name, initial, and phone number.

If we had included the student's Social Security number as one of the attributes in the STUDENT table in Figure 3.1—perhaps naming it STU_SOCSECNUM—both it and STU_NUM would have been candidate keys, because either one would uniquely identify each student. In that case, the selection of STU_NUM as the primary key would be driven by the designer's choice or by end user requirements. Note, incidentally, that a primary key is a superkey as well as a candidate key.

Within a table, each primary key value must be unique to ensure that each row is uniquely identified by the primary key. When this is the case, the table is said to exhibit **entity integrity**. To maintain entity integrity, a **null** "value" (that is, no data entry at all) is not permitted in the primary key.

> **N O T E**
>
> A null does NOT mean a zero or a space. Pressing the keyboard's Spacebar creates a blank (or a space). A null is created when you press the keyboard's Enter key without making a prior entry of any kind.

Nulls can *never* be part of a primary key and they should be avoided—to the greatest extent possible—in other attributes, too. There are rare cases in which nulls cannot be reasonably avoided when you are working with non-key attributes. For example, one of an EMPLOYEE table's attributes is likely to be the EMP_INITIAL. However, some employees—like this author—do not have a middle initial. Therefore, some of the EMP_INITIAL "values"

may be null. You will also discover later in this section that there may be situations in which a null exists because of the nature of the relationship between two entities. In any case, even if nulls cannot always be avoided, they must be used sparingly. In fact, the existence of nulls in a table is often an indication of poor database design skills.

Nulls, if used improperly, can create problems because they can have many different meanings. For example, a null can represent:

- An unknown attribute value.

- A known, but missing, attribute value.

- A "not applicable" condition.

Depending on the sophistication of the application development software, nulls can create problems when we use functions such as COUNT, AVERAGE, and SUM. In addition, nulls can create logical problems when relational tables are linked.

Controlled redundancy makes the relational database work. Tables within the database share common attributes that enable us to link the tables together. Note, for example, that the PRODUCT and VENDOR tables in Figure 3.2 share a common attribute named VEND_CODE. And note that the PRODUCT table's VEND_CODE value 232 occurs more than once, as does the VEND_CODE value 235. Because the PRODUCT table is related to the VENDOR table through these VEND_CODE values, the multiple occurrence of these values is *required* to make the 1:M relationship between VENDOR and PRODUCT work. Each VENDOR table VEND_CODE value is unique—the VENDOR is the "1" side in the VENDOR-PRODUCT relationship. But any given VENDOR table's VEND_CODE value may occur more than once in the PRODUCT table, thus providing evidence that PRODUCT is the "M" side of the VENDOR-PRODUCT relationship.) In database terms, the multiple occurrences of the VEND_CODE values in the PRODUCT table are not redundant, because they are *required* to make the relationship work. You should recall from Chapter 2, "Data Models," that data redundancy exists only when there is *unnecessary* duplication of attribute values.

FIGURE 3.2 **AN EXAMPLE OF A SIMPLE RELATIONAL DATABASE**

Table name: PRODUCT Database name: Ch03_SaleCo
Primary key: PROD_CODE
Foreign key: VEND_CODE

PROD_CODE	PROD_DESCRIPT	PROD_PRICE	PROD_ON_HAND	VEND_CODE
001278-AB	Claw hammer	$12.95	23	232
123-21UUY	Houselite chain saw, 16-in. bar	$189.99	4	235
QER-34256	Sledge hammer, 16-lb. head	$18.63	6	231
SRE-657UG	Rat-tail file	$2.99	15	232
ZZX/3245Q	Steel tape, 12-ft. length	$6.79	8	235

link

Table name: VENDOR
Primary key: VEND_CODE
Foreign key: none

VEND_CODE	VEND_CONTACT	VEND_AREACODE	VEND_PHONE
230	Shelly K. Smithson	608	555-1234
231	James Johnson	615	123-4536
232	Annelise Crystall	608	224-2134
233	Candice Wallace	904	342-6567
234	Arthur Jones	615	123-3324
235	Henry Ortozo	615	899-3425

As you examine Figure 3.2, note that the VEND_CODE value in one table can be used to point to the corresponding value in the other table. For example, the VEND_CODE value 235 in the PRODUCT table points to

vendor Henry Ortozo in the VENDOR table. Consequently, we discover that the product "Houselite chain saw, 16-in. bar" is delivered by Henry Ortozo, and that he can be contacted by calling 615-899-3425. The same connection can be made for the product "Steel tape, 12-ft. length" in the PRODUCT table.

Remember our naming convention—we used the prefix PROD in Figure 3.2 to indicate that the attributes "belong" to the PRODUCT table. Therefore, the prefix VEND in the PRODUCT table's VEND_CODE indicates that VEND_CODE points to some other table in the database. In this case, the VEND prefix is used to point to the VENDOR table in the database.

The link between the PRODUCT and VENDOR tables in Figure 3.2 can also be represented by the relational schema shown in Figure 3.3. In this case, the link is indicated by the line that connects the VENDOR and PRODUCT tables.

FIGURE 3.3 THE RELATIONAL SCHEMA FOR THE CH03_SaleCo DATABASE

Note that the link in Figure 3.3 is the equivalent of the relationship line in an ERD. This link is created when two tables share an attribute with common values. More specifically, the primary key of one table (VENDOR) appears again as the *foreign key* in a related table (PRODUCT.) A **foreign key (FK)** is an attribute whose values match the primary key values in the related table. For example, in Figure 3.2 the VEND_CODE is the primary key in the VENDOR table, and it occurs as a foreign key in the PRODUCT table. Because the VENDOR table is not linked to a third table, the VENDOR table shown in Figure 3.2 does not contain a foreign key.

If the foreign key contains either matching values or nulls, the table(s) that make use of such a foreign key is (are) said to exhibit *referential integrity*. In other words, **referential integrity** means that if the foreign key contains a value, that value refers to an existing valid tuple (row) in another relation. Note that referential integrity is maintained between the PRODUCT and VENDOR tables shown in Figure 3.3.

Finally, a **secondary key** is defined as a key that is used strictly for data retrieval purposes. Suppose customer data are stored in a CUSTOMER table in which the customer number is the primary key. Do you suppose that most customers will remember their number? Data retrieval for a customer can be facilitated when the customer's last name and phone number are used. In this case, the primary key is the customer number; the secondary key is the combination of the customer's last name and phone number. Keep in mind that a secondary key does not necessarily yield a unique outcome. For example, a customer's last name and home telephone number could yield several matches if there are several Smith family members living at a residence with only one phone line. Similarly, the combination last name and zip code could yield dozens of matches, which can then be "combed" for a specific match.

A secondary key's effectiveness in narrowing down a search depends on how restrictive it is. For instance, although the secondary key CUS_CITY is legitimate from a database point of view, the attribute values "New York" or "Sydney" are not likely to produce a usable return, unless you want to examine millions of possible matches. (Of course, CUS_CITY is a better secondary key than CUS_COUNTRY.)

Table 3.3 summarizes the different relational database table keys.

TABLE 3.3 **RELATIONAL DATABASE KEYS**

KEY TYPE	DEFINITION
Superkey	An attribute (or combination of attributes) that uniquely identifies each entity in a table.
Candidate key	A minimal superkey. A superkey that does not contain a subset of attributes that is itself a superkey.
Primary key	A candidate key selected to uniquely identify all other attribute values in any given row. Cannot contain null entries.
Secondary key	An attribute (or combination of attributes) used strictly for data retrieval purposes.
Foreign key	An attribute (or combination of attributes) in one table whose values must either match the primary key in another table or be null.

3.3 INTEGRITY RULES REVISITED

You will discover that the relational database integrity rules are very important to good database design. Many (but by no means all) relational database management systems (RDBMSs) enforce integrity rules automatically. However, it is much safer to make sure that your applications design conforms to the entity and referential integrity rules mentioned in this chapter. These rules are summarized in Table 3.4.

TABLE 3.4 **INTEGRITY RULES**

ENTITY INTEGRITY	DESCRIPTION
Requirement	All primary key entries are unique, and no part of a primary key may be null.
Purpose	Guarantees that each entity will have a unique identity and ensures that foreign key values can properly reference primary key values.
Example	No invoice can have a duplicate number, nor can it be null. In short, all invoices are uniquely identified by their invoice number.
REFERENTIAL INTEGRITY	**DESCRIPTION**
Requirement	A foreign key may have either a null entry—as long as it is not a part of its table's primary key—or an entry that matches the primary key value in a table to which it is related. (Every non-null foreign key value *must* reference an *existing* primary key value.)
Purpose	Makes it possible for an attribute NOT to have a corresponding value, but it will be impossible to have an invalid entry. The enforcement of the referential integrity rule makes it impossible to delete a row in one table whose primary key has mandatory matching foreign key values in another table.
Example	A customer might not (yet) have an assigned sales representative (number), but it will be impossible to have an invalid sales representative (number).

The integrity rules summarized in Table 3.4 are illustrated in Figure 3.4.

FIGURE 3.4 AN ILLUSTRATION OF INTEGRITY RULES

Table name: CUSTOMER Database name: Ch03_InsureCo
Primary key: CUS_CODE
Foreign key: AGENT_CODE

CUS_CODE	CUS_LNAME	CUS_FNAME	CUS_INITIAL	CUS_AREACODE	CUS_PHONE	CUS_RENEW_DATE	AGENT_CODE
10010	Ramas	Alfred	A	615	844-2573	12-Mar-02	502
10011	Dunne	Leona	K	713	894-1238	23-May-02	501
10012	Smith	Kathy	W	615	894-2285	05-Jan-03	502
10013	Olowski	Paul	F	615	894-2180	20-Sep-02	
10014	Orlando	Myron		615	222-1672	04-Dec-02	501
10015	O'Brian	Amy	B	713	442-3381	29-Aug-02	503
10016	Brown	James	G	615	297-1228	01-Mar-03	502
10017	Williams	George		615	290-2556	23-Jun-02	503
10018	Farriss	Anne	G	713	382-7185	09-Nov-02	501
10019	Smith	Olette	K	615	297-3809	18-Feb-03	503

Table name: AGENT
Primary key: AGENT_CODE
Foreign key: none

AGENT_CODE	AGENT_AREACODE	AGENT_PHONE	AGENT_LNAME	AGENT_YTD_SLS
501	713	228-1249	Alby	$1,735,453.75
502	615	882-1244	Hahn	$4,967,003.28
503	615	123-5589	Okon	$3,093,980.41

Note the following features of Figure 3.4:

1. *Entity integrity.* The CUSTOMER table's primary key is CUS_CODE. The CUSTOMER primary key column has no null entries, and all entries are unique. Similarly, the AGENT table's primary key is AGENT_CODE, and this primary key column also is free of null entries.

2. *Referential integrity.* The CUSTOMER table contains a foreign key AGENT_CODE, which links entries in the CUSTOMER table to the AGENT table. The CUS_CODE row that is identified by the (primary key) number 10013 contains a null entry in its AGENT_CODE foreign key because Mr. Paul F. Olowski does not (yet) have a sales representative assigned to him. The remaining AGENT_CODE entries in the CUSTOMER table all match the AGENT_CODE entries in the AGENT table.

To avoid nulls, some designers use special codes, known as **flags**, to indicate the absence of some value. Using Figure 3.4 as an example, the code -99 could be used as the AGENT_CODE entry of the fourth row of the CUSTOMER table to indicate that customer Paul Olowski does not yet have an agent assigned to him. If such a flag is used, the AGENT table must contain a dummy row with an AGENT_CODE value of -99. Thus the AGENT table's first record might contain the values shown in Table 3.5.

TABLE 3.5 A DUMMY VARIABLE VALUE USED AS A FLAG

AGENT_CODE	AGENT_AREA CODE	AGENT_PHONE	AGENT_LNAME	AGENT_YTD_SALES
-99	000	000-0000	None	$0.00

In Chapter 4, "Entity Relationship (ER) Modeling," we will discuss several ways in which nulls may be handled.

3.4 RELATIONAL DATABASE OPERATORS

The degree of relational completeness can be defined by the extent to which relational algebra is supported. **Relational algebra** defines the theoretical way of manipulating table contents using the eight relational operators: SELECT, PROJECT, JOIN, INTERSECT, UNION, DIFFERENCE, PRODUCT, and DIVIDE. (You will learn

in Chapter 6, "An Introduction to Structured Query Language (SQL)," how SQL commands can be used to accomplish relational algebra tasks.)

N O T E

To be considered minimally relational, the DBMS must support the key relational operators SELECT, PROJECT, and JOIN. Very few DBMSs are capable of supporting all eight relational operators.

The use of relational algebra operators on existing tables (relations) produces new relations. There is no need to examine the mathematical definitions, properties, and characteristics of these relational algebra operators. However, their *use* can easily be illustrated as follows:

1. UNION combines all rows from two tables, excluding duplicate rows. The tables must have the same attribute characteristics (the columns and domains must be identical) to be used in the UNION. When two or more tables share the same columns and domains, they are said to be **union-compatible**. The effect of a UNION is shown in Figure 3.5.

FIGURE 3.5 **UNION**

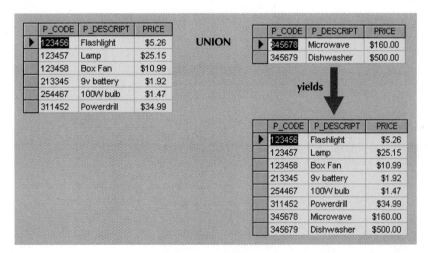

2. INTERSECT yields only the rows that appear in both tables. As was true in the case of UNION, the tables must be union-compatible to yield valid results. For example, you cannot use INTERSECT if one of the attributes is numeric and one is character-based. The effect of an INTERSECT is shown in Figure 3.6.

FIGURE 3.6 **INTERSECT**

3. DIFFERENCE yields all rows in one table that are not found in the other table; that is, it subtracts one table from the other. As was true in the case of UNION, the tables must be union-compatible to yield valid results. The effect of a DIFFERENCE is shown in Figure 3.7.

FIGURE 3.7 DIFFERENCE

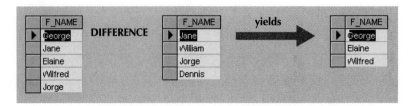

4. PRODUCT yields all possible pairs of rows from two tables—also known as the Cartesian product. Therefore, if one table has 6 rows and the other table has 3 rows, the PRODUCT yields a list composed of 6 × 3 = 18 rows. The effect of a PRODUCT is shown in Figure 3.8.

FIGURE 3.8 PRODUCT

P_CODE	P_DESCRIPT	PRICE
123456	Flashlight	$5.26
123457	Lamp	$25.15
123458	Box Fan	$10.99
213345	9v battery	$1.92
254467	100W bulb	$1.47
311452	Powerdrill	$34.99

PRODUCT

STORE	AISLE	SHELF
23	W	5
24	K	9
25	Z	6

yields

P_CODE	P_DESCRIPT	PRICE	STORE	AISLE	SHELF
123456	Flashlight	$5.26	23	W	5
123456	Flashlight	$5.26	24	K	9
123456	Flashlight	$5.26	25	Z	6
123457	Lamp	$25.15	23	W	5
123457	Lamp	$25.15	24	K	9
123457	Lamp	$25.15	25	Z	6
123458	Box Fan	$10.99	23	W	5
123458	Box Fan	$10.99	24	K	9
123458	Box Fan	$10.99	25	Z	6
213345	9v battery	$1.92	23	W	5
213345	9v battery	$1.92	24	K	9
213345	9v battery	$1.92	25	Z	6
311452	Powerdrill	$34.99	23	W	5
311452	Powerdrill	$34.99	24	K	9
311452	Powerdrill	$34.99	25	Z	6
254467	100W bulb	$1.47	23	W	5
254467	100W bulb	$1.47	24	K	9
254467	100W bulb	$1.47	25	Z	6

5. SELECT yields values for all rows found in a table. SELECT can be used to list either all of the row values, or it can yield only those row values that match a specified criterion. In other words, SELECT yields a horizontal subset of a table. The effect of a SELECT is shown in Figure 3.9.

FIGURE 3.9 **SELECT**

6. PROJECT yields all values for selected attributes. In other words, PROJECT yields a vertical subset of a table. The effect of a PROJECT is shown in Figure 3.10.

FIGURE 3.10 **PROJECT**

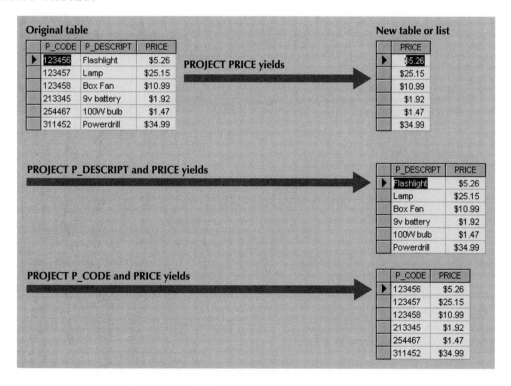

7. JOIN allows us to combine information from two or more tables. JOIN is the real power behind the relational database, allowing the use of independent tables linked by common attributes. We will use the CUSTOMER and AGENT tables shown in Figure 3.11 to illustrate several types of joins.

FIGURE 3.11 TWO TABLES THAT WILL BE USED IN JOIN ILLUSTRATIONS

Table name: CUSTOMER

CUS_CODE	CUS_LNAME	CUS_ZIP	AGENT_CODE
1132445	Walker	32145	231
1217782	Adares	32145	125
1312243	Rakowski	34129	167
1321242	Rodriguez	37134	125
1542311	Smithson	37134	421
1657399	Vanloo	32145	231

Table name: AGENT

AGENT_CODE	AGENT_PHONE
125	6152439887
167	6153426778
231	6152431124
333	9041234445

A **natural JOIN** links tables by selecting only the rows with common values in their common attribute(s). A natural join is the result of a three-stage process:

a. First, a PRODUCT of the tables is created, yielding the results shown in Figure 3.12.

FIGURE 3.12 NATURAL JOIN, STEP 1: PRODUCT

CUS_CODE	CUS_LNAME	CUS_ZIP	CUSTOMER.AGENT_CODE	AGENT.AGENT_CODE	AGENT_PHONE
1132445	Walker	32145	231	125	6152439887
1132445	Walker	32145	231	167	6153426778
1132445	Walker	32145	231	231	6152431124
1132445	Walker	32145	231	333	9041234445
1217782	Adares	32145	125	125	6152439887
1217782	Adares	32145	125	167	6153426778
1217782	Adares	32145	125	231	6152431124
1217782	Adares	32145	125	333	9041234445
1312243	Rakowski	34129	167	125	6152439887
1312243	Rakowski	34129	167	167	6153426778
1312243	Rakowski	34129	167	231	6152431124
1312243	Rakowski	34129	167	333	9041234445
1321242	Rodriguez	37134	125	125	6152439887
1321242	Rodriguez	37134	125	167	6153426778
1321242	Rodriguez	37134	125	231	6152431124
1321242	Rodriguez	37134	125	333	9041234445
1542311	Smithson	37134	421	125	6152439887
1542311	Smithson	37134	421	167	6153426778
1542311	Smithson	37134	421	231	6152431124
1542311	Smithson	37134	421	333	9041234445
1657399	Vanloo	32145	231	125	6152439887
1657399	Vanloo	32145	231	167	6153426778
1657399	Vanloo	32145	231	231	6152431124
1657399	Vanloo	32145	231	333	9041234445

b. A SELECT is performed on the output of Step (a) to yield only the rows for which the AGENT_CODE values are equal. The common column(s) are referred to as the **join column(s)**. Step (b) yields the results shown in Figure 3.13.

FIGURE 3.13 NATURAL JOIN, STEP 2: SELECT

CUS_CODE	CUS_LNAME	CUS_ZIP	CUSTOMER.AGENT_CODE	AGENT.AGENT_CODE	AGENT_PHONE
1217782	Adares	32145	125	125	6152439887
1321242	Rodriguez	37134	125	125	6152439887
1312243	Rakowski	34129	167	167	6153426778
1132445	Walker	32145	231	231	6152431124
1657399	Vanloo	32145	231	231	6152431124

c. A PROJECT is performed on the results of Step (b) to yield a single copy of each attribute, thereby eliminating duplicate columns. Step (c) yields the output shown in Figure 3.14.

FIGURE 3.14 NATURAL JOIN, STEP 3: PROJECT

CUS_CODE	CUS_LNAME	CUS_ZIP	AGENT_CODE	AGENT_PHONE
1217782	Adares	32145	125	6152439887
1321242	Rodriguez	37134	125	6152439887
1312243	Rakowski	34129	167	6153426778
1132445	Walker	32145	231	6152431124
1657399	Vanloo	32145	231	6152431124

The final outcome of a natural JOIN yields a table that does not include unmatched pairs and provides only the copies of the matches.

Note a few crucial features of the natural JOIN operation:

- If no match is made between the table rows, the new table does not include the unmatched row. In this case, neither AGENT_CODE 421 nor the customer whose last name is Smithson is included. Smithson's AGENT_CODE 421 does not match any entry in the AGENT table.

- The column on which we made the JOIN—that is, AGENT_CODE—occurs only once in the new table.

- If the same AGENT_CODE were to occur several times in the AGENT table, a customer would be listed for each match. For example, if the AGENT_CODE 167 were to occur three times in the AGENT table, the customer named Rakowski, who is associated with AGENT_CODE 167, would occur three times in the resulting table. (A good AGENT table cannot, of course, yield such a result because it would contain unique primary key values.)

Another form of JOIN, known as **equijoin**, links tables on the basis of an equality condition that compares specified columns of each table. The outcome of the equijoin does not eliminate duplicate columns, and the condition or criterion to join the tables must be explicitly defined. The equijoin takes its name from the equality comparison operator (=) used in the condition. If any other comparison operator is used, the join is called a **theta join**.

In an outer join the matched pairs would be retained and any unmatched values in the other table would be left null. More specifically, if we produce an outer join for tables CUSTOMER and AGENT, two scenarios are possible:

A **left outer join** yields all the rows in the CUSTOMER table, including those that do not have a matching value in the AGENT table. An example of such a join is shown in Figure 3.15.

FIGURE 3.15 LEFT OUTER JOIN

CUS_CODE	CUS_LNAME	CUS_ZIP	AGENT_CODE	AGENT_PHONE
1217782	Adares	32145	125	6152439887
1321242	Rodriguez	37134	125	6152439887
1312243	Rakowski	34129	167	6153426778
1132445	Walker	32145	231	6152431124
1657399	Vanloo	32145	231	6152431124
1542311	Smithson	37134	421	

A **right outer join** yields all the rows in the AGENT table, including those that do not have matching values in the CUSTOMER table. An example of such a join is shown in Figure 3.16.

FIGURE 3.16 RIGHT OUTER JOIN

CUS_CODE	CUS_LNAME	CUS_ZIP	AGENT_CODE	AGENT_PHONE
1217782	Adares	32145	125	6152439887
1321242	Rodriguez	37134	125	6152439887
1312243	Rakowski	34129	167	6153426778
1132445	Walker	32145	231	6152431124
1657399	Vanloo	32145	231	6152431124
			333	9041234445

Outer joins are especially useful when you are trying to determine what value(s) in related tables cause referential integrity problems—which are created when foreign key values do not match the primary key values in the related table(s). In fact, if you are asked to convert large spreadsheets or other non-database data into relational database tables, you will discover that the outer joins will save you vast amounts of time and uncounted headaches when you encounter referential integrity errors after the conversions. We certainly have learned to be thankful for the existence of outer joins in our database work!

You may wonder why the outer joins are labeled *left* and *right*. The labels refer to the order in which the tables are listed in the SQL command. We will explore such joins in Chapter 7, "Advanced SQL."

8. DIVIDE requires the use of one single-column table and one two-column table. Using the example shown in Figure 3.17, note that:

 a. Table 1 is "divided" by Table 3. Tables 1 and 2 both contain the column CODE but do not share LOC.

 b. To be included in the resulting Table 3, a value in the unshared column (LOC) must be associated (in the dividing Table 2) with *every* value in Table 1.

 c. The only value associated with both A and B is 5.

FIGURE 3.17 DIVIDE

3.5 THE DATA DICTIONARY AND THE SYSTEM CATALOG

The **data dictionary** is used to provide a detailed accounting of all tables found within the user/designer-created database. Thus the data dictionary contains (at least) all the attribute names and characteristics for each table in the system. In short, the data dictionary contains metadata—data about data. Using the small database presented in Figure 3.4, we might picture its data dictionary as shown in Table 3.6.

TABLE 3.6 A SAMPLE DATA DICTIONARY

TABLE NAME	ATTRIBUTE NAME	CONTENTS	TYPE	FORMAT	RANGE	REQUIRED	PK Or FK	FK REFERENCED TABLE
CUSTOMER	CUS_CODE	Customer acct. code	CHAR(5)	99999	10000-99999	Y	PK	
	CU_LNAME	Customer last name	VCCHAR(20)	Xxxxxxxx		Y		
	CUS_FNAME	Customer first name	VCHAR(20	Xxxxxxxx	100-999	Y		
	CUS_INITIAL	Customer initial	CHAR(1)	X				
	CUS_RENEW_DATE	Customer insurance renewal date	DATE	dd-mmm-yyyy				
	AGENT_CODE	Agent code	CHAR(3)	999			FK	AGENT
AGENT	AGENT_CODE	Agent code	CHAR(3)	999		Y	PK	
	AGENT_AREACODE	Agent area code	CHAR(3)	999	0.00-9,999,999.99	Y		
	AGENT_PHONE	Agent telephone number	CHAR(8)	999-9999		Y		
	AGENT_LNAME	Agent last name	VCHAR(20)	Xxxxxxxx		Y		
	AGENT_YTD_SALES	Agent year-to-date sales	NUMBER(9,2)	9,999,999.99		Y		

FK = Foreign key

PK = Primary key

CHAR = Fixed character length data (1–255 characters)

VARCHAR = Variable character length data (1–2,000 characters)

NUMBER = Numeric data. NUMBER(9,2) is used to specify numbers with two decimal places and up to nine digits, including the decimal places. Some RDBMSs permit the use of a MONEY or CURRENCY data type.

> **N O T E**
>
> The data dictionary in Table 3.6 is an example of the *human* view of the entities, attributes, and relationships. The purpose of such a data dictionary is to ensure that all members of database design and implementation teams use the same table and attribute names and characteristics. The DBMS's internally stored data dictionary contains additional information about relationship types, entity and referential integrity checks and enforcement, and index types and components. This additional information is generated during the database implementation stage.

The data dictionary is sometimes described as "the database designer's database" because it records the design decisions about tables and their structures.

> **N O T E**
>
> Telephone area codes are always composed of digits 0–9. Because area codes are not used arithmetically, they are most efficiently stored as character data. Also, the area codes are always composed of three digits. Therefore, the area code data type is defined as CHAR(3). On the other hand, names do not conform to some standard length. Therefore, the customer first names are defined as VARCHAR(20), thus indicating that up to 20 characters may be used to store such names. Character data are shown as left-justified.

Like the data dictionary, the **system catalog** contains metadata. The system catalog can be described as a very detailed system data dictionary that describes all objects within the database, including data about table names, the table's creator and creation date, the number of columns in each table, the data type corresponding to each column, index filenames, index creators, authorized users, access privileges, and so on. Because the system catalog contains all the required data dictionary information, the terms "system catalog" and "data dictionary" are often used interchangeably. In fact, current relational database software generally provides only a system catalog, from which the designer's data dictionary information may be derived. The system catalog is actually a system-created database whose tables store the user/designer-created database characteristics and contents. Therefore, the system catalog tables can be queried just like any user/designer-created table.

In effect, the system catalog automatically produces database documentation. As new tables are added to the database, such documentation also allows the RDBMS to check for and eliminate homonyms and synonyms. In general terms, **homonyms** are similar-sounding words with different meanings, such as *boar* and *bore*, or identically spelled words with different meanings, such as *fair* (meaning "just") and *fair* (meaning "festival"). In a database context, the word "homonym" indicates the use of the same attribute name to label different attributes. For example, we might use C_NAME to label a customer name attribute in a CUSTOMER table and also use C_NAME to label a consultant name attribute in a CONSULTANT table. To lesson confusion, you should avoid database homonyms; the data dictionary is very useful in this regard.

In a database context, a **synonym** is the opposite of a homonym and indicates the use of different names to describe the same attribute. For example, *car* and *auto* refer to the same object. Synonyms must be avoided. You will discover why using synonyms is a bad idea when you get a chance to correct the use of synonyms in Problem 33 at the end of the chapter.

3.6 RELATIONSHIPS WITHIN THE RELATIONAL DATABASE

You already know that relationships are classified as "one-to-one" (1:1), "one-to-many" (1:M), and "many-to-many" (M:N or M:M). In this section we will explore these relationships further to help you apply them properly when you start developing database designs. We fill focus our attention on the following points:

- The 1:M relationship is the relational modeling ideal. Therefore, this relationship type should be the norm in any relational database design.

- M:N relationships must be avoided because they lead to data redundancies. (We will show you later in this section how any M:N relationship can be changed into two 1:M relationships.)

- The 1:1 relationship should be rare in any relational database design.

3.6.1 THE 1:1 RELATIONSHIP

As the 1:1 label implies, in this relationship one entity can be related to only one other entity, and vice versa. For example, one department chair—a professor—can chair only one department, and one department can have only one department chair. The entities PROFESSOR and DEPARTMENT thus exhibit a 1:1 relationship. (You might properly argue that not all professors chair a department and professors cannot be *required* to chair a department. That is, the relationship between these two entities is optional. However, at this stage of the discussion, we'll focus our attention on the basic 1:1 relationship. We will address optional relationships in Chapter 4.) The basic 1:1 relationship is modeled in Figure 3.18 and its implementation is shown in Figure 3.19.

FIGURE 3.18 THE 1:1 RELATIONSHIP BETWEEN PROFESSOR AND DEPARTMENT

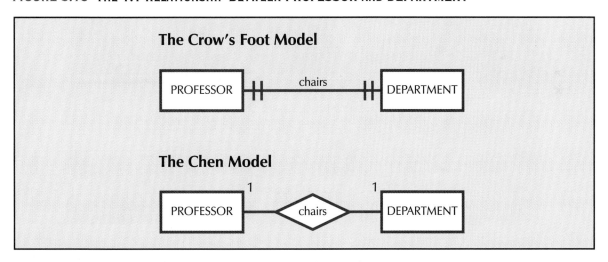

As you examine the tables in Figure 3.19, note that each professor is a Tiny College employee. Therefore, the professor identification is through the EMP_NUM. (However, note that not all employees are professors—there's another optional relationship.)

ADDITIONAL CD CONTENT

If you open the **Ch03_TinyCollege** database on the student CD, you'll see that the STUDENT and CLASS entities still use PROF_NUM as their foreign key. Because PROF_NUM and EMP_NUM are labels for the same attribute, this is an example of the use of synonyms—different names for the same attribute. These synonyms will be eliminated in Chapter 4 as we continue to improve the Tiny College database.

FIGURE 3.19 THE IMPLEMENTED 1:1 RELATIONSHIP BETWEEN PROFESSOR AND DEPARTMENT

Table name: PROFESSOR
Primary key: EMP_NUM
Foreign key: DEPT_CODE

Database name: Ch03_TinyCollege

EMP_NUM	DEPT_CODE	PROF_OFFICE	PROF_EXTENSION	PROF_HIGH_DEGREE
103	HIST	DRE 156	6783	Ph.D.
104	ENG	DRE 102	5561	MA
105	ACCT	KLR 229D	8665	Ph.D.
106	MKT/MGT	KLR 126	3899	Ph.D.
110	BIOL	AAK 160	3412	Ph.D.
114	ACCT	KLR 211	4436	Ph.D.
155	MATH	AAK 201	4440	Ph.D.
160	ENG	DRE 102	2248	Ph.D.
162	CIS	KLR 203E	2359	Ph.D.
191	MKT/MGT	KLR 409B	4016	DBA
195	PSYCH	AAK 297	3550	Ph.D.
209	CIS	KLR 333	3421	Ph.D.
228	CIS	KLR 300	3000	Ph.D.
297	MATH	AAK 194	1145	Ph.D.
299	ECON/FIN	KLR 284	2851	Ph.D.
301	ACCT	KLR 244	4683	Ph.D.
335	ENG	DRE 208	2000	Ph.D.
342	SOC	BBG 208	5514	Ph.D.
387	BIOL	AAK 230	8665	Ph.D.
401	HIST	DRE 156	6783	MA
425	ECON/FIN	KLR 284	2851	MBA
435	ART	BBG 185	2278	Ph.D.

Table name: DEPARTMENT
Primary key: DEPT_CODE
Foreign key: EMP_NUM

DEPT_CODE	DEPT_NAME	SCHOOL_CODE	EMP_NUM	DEPT_ADDRESS	DEPT_EXTENSION
ACCT	Accounting	BUS	114	KLR 211, Box 52	3119
ART	Fine Arts	A&SCI	435	BBG 185, Box 128	2278
BIOL	Biology	A&SCI	387	AAK 230, Box 415	4117
CIS	Computer Info. Systems	BUS	209	KLR 333, Box 56	3245
ECON/FIN	Economics/Finance	BUS	299	KLR 284, Box 63	3126
ENG	English	A&SCI	160	DRE 102, Box 223	1004
HIST	History	A&SCI	103	DRE 156, Box 284	1867
MATH	Mathematics	A&SCI	297	AAK 194, Box 422	4234
MKT/MGT	Marketing/Management	BUS	106	KLR 126, Box 55	3342
PSYCH	Psychology	A&SCI	195	AAK 297, Box 438	4110
SOC	Sociology	A&SCI	342	BBG 208, Box 132	2008

Although the preceding example indicates that 1:1 relationships can be used in a relational database environment, *the existence of a 1:1 relationship often means that the entity components were not defined properly*. Additionally, it could indicate that the two entities actually belong in the same table! However, there are times when 1:1 relationships are appropriate. The preceding "PROFESSOR chairs DEPARTMENT" example illustrates a defensible 1:1 relationship. (But even this kind of 1:1 relationship can be avoided by creating a JOB table that contains all job codes, their descriptions, and other job-related details, and then using a JOB_CODE foreign key in the EMPLOYEE table to keep track of which employee chairs which department.)

Rare as 1:1 relationships ought to be, there are conditions that absolutely *require* their use. For example, suppose you manage the database for a company that employs pilots, accountants, mechanics, clerks, salespeople, service personnel, and so on. Pilots have many attributes that the other employees don't have, such as licenses, medical certificates, flight experience records, dates of flight proficiency checks, and proof of required periodic medical checks. If you put all the pilot-specific attributes in the EMPLOYEE table, you will have several nulls in that employee table for all employees who are not pilots. To avoid the proliferation of nulls, it is better to split

the pilot attributes into a separate table (PILOT) that is linked to the EMPLOYEE table in a 1:1 relationship. Because pilots have many attributes that are shared by all employees—such as name, date of birth, and date of first employment—those attributes would be stored in the EMPLOYEE table.

ADDITIONAL CD CONTENT

If you take a look at the Ch03_AviaCo database on the student CD, you will see the implementation of the PILOT to EMPLOYEE relationship. We will examine this type of relationship in detail in Chapter 4.

3.6.2 THE 1:M RELATIONSHIP

The 1:M relationship is the relational database norm. To see how such a relationship is modeled and implemented, let's take another look at the PAINTER paints PAINTING example we used in Chapter 2. Compare the data model in Figure 3.20 with its implementation in Figure 3.21.

FIGURE 3.20 THE 1:M RELATIONSHIP BETWEEN PAINTER AND PAINTING

FIGURE 3.21 THE IMPLEMENTED 1:M RELATIONSHIP BETWEEN PAINTER AND PAINTING

Table name: PAINTER
Primary key: PAINTER_NUM
Foreign key: none
Database name: Ch03_Museum

	PAINTER_NUM	PAINTER_LNAME	PAINTER_FNAME	PAINTER_INITIAL
	123	Ross	Georgette	P
	126	Itero	Julio	G

Table name: PAINTING
Primary key: PAINTING_NUM
Foreign key: PAINTER_NUM

	PAINTING_NUM	PAINTING_TITLE	PAINTER_NUM
	1338	Dawn Thunder	123
	1339	Vanilla Roses To Nowhere	123
	1340	Tired Flounders	126
	1341	Hasty Exit	123
	1342	Plastic Paradise	126

As you examine the PAINTER and PAINTING table contents in Figure 3.21, note the following features:

- Each painting is painted by one and only one painter, but each painter could have painted many paintings. Note that painter 123 (Georgette P. Ross) has three paintings stored in the PAINTING table.

- There is only one row in the PAINTER table for any given row in the PAINTING table, but there may be many rows in the PAINTING table for any given row in the PAINTER table.

Now take another look at Figure 3.21 to see how the data model portrays the 1:M relationship that is implemented in the Ch03_Museum database.

The 1:M relationship is found in any database environment. Students in a typical college or university will discover that each COURSE can generate many CLASSes, but each CLASS refers to only one COURSE. For example, an Accounting II course might yield two classes: one offered on Monday, Wednesday, and Friday (MWF) from 10:00 a.m. to 10:50 a.m., and one offered on Thursday (Th) from 6:00 p.m. to 8:40 p.m. Therefore, the 1:M relationship between COURSE and CLASS might be described this way:

- Each COURSE can have many CLASSes, but each CLASS references only one COURSE.

- There will be only one row in the COURSE table for any given row in the CLASS table, but there can be many rows in the CLASS table for any given row in the COURSE table.

- Figure 3.22 maps the ER model for the 1:M relationship between COURSE and CLASS.

FIGURE 3.22 THE 1:M RELATIONSHIP BETWEEN COURSE AND CLASS

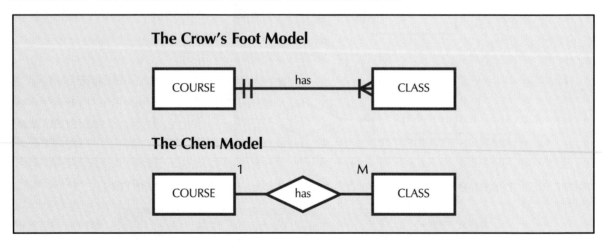

The 1:M relationship between COURSE and CLASS is further illustrated in Figure 3.23.

Using Figure 3.23, let's review some important terminology. Note that CLASS_CODE in the CLASS table uniquely identifies each row. Therefore, we have chosen CLASS_CODE to be the primary key. However, the combination CRS_CODE and CLASS_SECTION will also uniquely identify each row in the class table. In other words, the *composite key* that is composed of CRS_CODE and CLASS_SECTION forms a *candidate key*.

The one-to-many (1:M) relationship is easily implemented in the relational model. *Just make sure that the primary key of the one is included as the foreign key in the table of the many.* Note in Figure 3.21, for example, that the PAINTER table's primary key, PAINTER_NUM, is included in the PAINTING table as a foreign key. Similarly, in Figure 3.23, the COURSE table's primary key, CRS_CODE, is included in the CLASS table as a foreign key.

FIGURE 3.23 THE IMPLEMENTED 1:M RELATIONSHIP BETWEEN COURSE AND CLASS

Table name: COURSE
Primary key: CRS_CODE
Foreign key: none Database name: Ch03_TinyCollege

	CRS_CODE	DEPT_CODE	CRS_DESCRIPTION	CRS_CREDIT
+	ACCT-211	ACCT	Accounting I	3
+	ACCT-212	ACCT	Accounting II	3
+	CIS-220	CIS	Intro. to Microcomputing	3
+	CIS-420	CIS	Database Design and Implementation	4
+	QM-261	CIS	Intro. to Statistics	3
+	QM-362	CIS	Statistical Applications	4

Table name: CLASS
Primary key: CLASS_CODE
Foreign key: CRS_CODE

	CLASS_CODE	CRS_CODE	CLASS_SECTION	CLASS_TIME	CLASS_ROOM	PROF_NUM
+	10012	ACCT-211	1	MWF 8:00-8:50 a.m.	BUS311	105
+	10013	ACCT-211	2	MWF 9:00-9:50 a.m.	BUS200	105
+	10014	ACCT-211	3	TTh 2:30-3:45 p.m.	BUS252	342
+	10015	ACCT-212	1	MWF 10:00-10:50 a.m.	BUS311	301
+	10016	ACCT-212	2	Th 6:00-8:40 p.m.	BUS252	301
+	10017	CIS-220	1	MWF 9:00-9:50 a.m.	KLR209	228
+	10018	CIS-220	2	MWF 9:00-9:50 a.m.	KLR211	114
+	10019	CIS-220	3	MWF 10:00-10:50 a.m.	KLR209	228
+	10020	CIS-420	1	W 6:00-8:40 p.m.	KLR209	162
+	10021	QM-261	1	MWF 8:00-8:50 a.m.	KLR200	114
+	10022	QM-261	2	TTh 1:00-2:15 p.m.	KLR200	114
+	10023	QM-362	1	MWF 11:00-11:50 a.m.	KLR200	162
+	10024	QM-362	2	TTh 2:30-3:45 p.m.	KLR200	162

3.6.3 THE M:N RELATIONSHIP

A many-to-many (M:N) relationship is a more troublesome proposition in the relational environment. However, you will discover that the M:N relationship can be implemented by breaking it up to produce a set of 1:M relationships.

To explore the many-to-many (M:N) relationship, let's examine a rather typical college environment in which each STUDENT can take many CLASSes and each CLASS can contain many STUDENTs. The ERD model in Figure 3.24 shows this M:N relationship.

FIGURE 3.24 THE ERD'S M:N RELATIONSHIP BETWEEN STUDENT AND CLASS

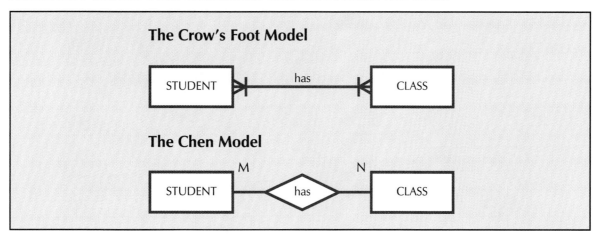

Note the features of the ER model in Figure 3.24:

- Each CLASS can have many STUDENTs, and each STUDENT can take many CLASSes.
- There can be many rows in the CLASS table for any given row in the STUDENT table, and there can also be many rows in the STUDENT table for any given row in the CLASS table.

To examine the M:N relationship more closely, let us imagine a small college with two students, each of whom takes three classes. Table 3.7 shows the enrollment data for the two students.

TABLE 3.7 SAMPLE STUDENT ENROLLMENT DATA

STUDENT'S LAST NAME	SELECTED CLASSES
Bowser	Accounting 1, ACCT-211, code 10014 Intro. to Microcomputing, CIS-220, code 10018 Intro. To Statistics, QM-261, code 10021
Smithson	Accounting 1, ACCT-211, code 10014 Intro. to Microcomputing, CIS-220, code 10018 Intro. To Statistics, QM-261, code 10021

Although the M:N relationship is logically reflected in Figure 3.24, it should *not* be implemented as shown in Figure 3.25, for two good reasons:

- The tables create many redundancies. For example, note that the STU_NUM values occur many times in the STUDENT table. In a real-world situation, additional student attributes such as address, classification, major, and home phone would also be contained in the STUDENT table, and each of these attribute values would be repeated in each of the records shown here. Similarly, the CLASS table contains many duplications: each student taking the class generates a CLASS record. The problem would be even worse if the CLASS table included such attributes as credit hours and course description. Such redundancies lead to the anomalies discussed in Chapter 1.
- Given the structure and contents of these two tables, the relational operations become very complex and are likely to lead to system efficiency errors and output errors.

FIGURE 3.25 THE M:N RELATIONSHIP BETWEEN STUDENT AND CLASS

Table name: STUDENT
Primary key: STU_NUM Database name: Ch03_CollegeTry
Foreign key: none

STU_NUM	STU_LNAME	CLASS_CODE
321452	Bowser	10014
321452	Bowser	10018
321452	Bowser	10021
324257	Smithson	10014
324257	Smithson	10018
324257	Smithson	10021

Table name: CLASS
Primary key: CLASS_CODE
Foreign key: STU_NUM

CLASS_CODE	STU_NUM	CRS_CODE	CLASS_SECTION	CLASS_TIME	CLASS_ROOM	PROF_NUM
10014	321452	ACCT-211	3	TTh 2:30-3:45 p.m.	BUS252	342
10014	324257	ACCT-211	3	TTh 2:30-3:45 p.m.	BUS252	342
10018	321452	CIS-220	2	MWF 9:00-9:50 a.m.	KLR211	114
10018	324257	CIS-220	2	MWF 9:00-9:50 a.m.	KLR211	114
10021	321452	QM-261	1	MWF 8:00-8:50 a.m.	KLR200	114
10021	324257	QM-261	1	MWF 8:00-8:50 a.m.	KLR200	114

Fortunately, we can easily avoid the problems inherent in the many-to-many (M:N) relationship by creating a **composite entity** or **bridge entity**. Because such a table is used to link the tables that originally were related in a M:N relationship, the composite entity structure includes—as foreign keys—*at least* the primary keys of the tables that are to be linked. The database designer has two main options to define a composite table's primary key: either use the combination of those foreign keys or create a new primary key.

Remember that each entity in the ER model is represented by a table. Therefore, we can create the (composite) ENROLL table shown in Figure 3.26 to link the tables CLASS and STUDENT. In this example, the ENROLL table's primary key is the combination of its foreign keys CLASS_CODE and STU_NUM. But the designer could have decided to create a single-attribute new primary key such as ENROLL_LINE, using a different line value to identify each ENROLL table row uniquely. (Microsoft Access users might use the *Autonumber* data type to generate such line values automatically.)

FIGURE 3.26 CONVERTING THE M:N RELATIONSHIP INTO TWO 1:M RELATIONSHIPS

Table name: STUDENT
Primary key: STU_NUM
Foreign key: none

Database name: Ch03_CollegeTry2

STU_NUM	STU_LNAME
321452	Bowser
324257	Smithson

Table name: ENROLL
Primary key: CLASS_CODE + STU_NUM
Foreign key: CLASS_CODE, STU_NUM

CLASS_CODE	STU_NUM	ENROLL_GRADE
10014	321452	C
10014	324257	B
10018	321452	A
10018	324257	B
10021	321452	C
10021	324257	C

Table name: CLASS
Primary key: CLASS_CODE
Foreign key: CRS_CODE

CLASS_CODE	CRS_CODE	CLASS_SECTION	CLASS_TIME	CLASS_ROOM	PROF_NUM
10014	ACCT-211	3	TTh 2:30-3:45 p.m.	BUS252	342
10018	CIS-220	2	MWF 9:00-9:50 a.m.	KLR211	114
10021	QM-261	1	MWF 8:00-8:50 a.m.	KLR200	114

Because the ENROLL table in Figure 3.26 links two tables, STUDENT and CLASS, it is also called a **linking table**. In other words, a linking table is the implementation of a composite entity.

NOTE

In addition to the linking attributes, the (composite) ENROLL table can also contain such relevant attributes as the GRADE earned in the course. In fact, a composite table can contain any number of attributes that the designer wants to track. Keep in mind that the composite entity, *although it is implemented as an actual table*, is *conceptually* a logical entity that was created as a means to an end: to eliminate the potential for multiple redundancies in the original M:N relationship.

The linking table (ENROLL) shown in Figure 3.25 yields the required M:N to 1:M conversion. Observe that the composite entity represented by the ENROLL table must contain at least the primary keys of the CLASS and STUDENT tables (CLASS_CODE and STU_NUM respectively) for which it serves as a connector. And especially note that the STUDENT and CLASS tables now contain only one row per entity. The linking ENROLL table contains multiple occurrences of the foreign key values, but such controlled redundancies are incapable of producing anomalies as long as referential integrity is enforced. Additional attributes may be assigned as needed. In this case, ENROLL_GRADE is selected to satisfy a reporting requirement. Also note that the ENROLL table's primary key consists of the two attributes CLASS_CODE and STU_NUM because it takes both the class code and the student number to define a particular student's grade. Naturally, the conversion is reflected in the entity relationship model, too. The revised relationship is shown in Figure 3.27.

FIGURE 3.27 CHANGING THE M:N RELATIONSHIP TO TWO 1:M RELATIONSHIPS

As you examine Figure 3.27, note that the (composite) entity named ENROLL represents the linking table between STUDENT and CLASS. The Chen ERD uses a diamond within a rectangle to indicate the existence of a composite entity.

We illustrated the 1:M relationship between COURSE and CLASS in Figure 3.23. With the help of this relationship, we can increase the amount of available information even as we control the database's redundancies. Thus we can expand Figure 3.27 to include the 1:M relationship between COURSE and CLASS shown in Figure 3.28. Note that the model is able to handle multiple sections of a CLASS, while controlling redundancies by making sure that all the COURSE data common to each CLASS are kept in the COURSE table.

FIGURE 3.28 THE EXPANDED ENTITY RELATIONSHIP MODEL

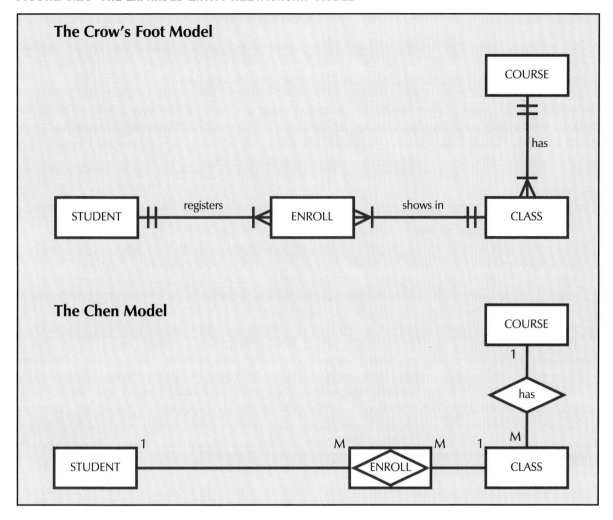

The relational schema that corresponds to the entity relationship diagram in Figure 3.28 is shown in Figure 3.29.

FIGURE 3.29 THE RELATIONAL SCHEMA FOR THE CH03_TINYCOLLEGE DATABASE

We will examine the ERD in greater detail in Chapter 4 to show you how it is used to design more-complex databases. The ERD will also be used as the basis for the development and implementation of a realistic database design in Appendixes D and E for a University Computer Lab.

3.7 DATA REDUNDANCY REVISITED

In Chapter 1 you learned that data redundancy leads to data anomalies. Such anomalies can destroy the effectiveness of the database. You also learned that the relational database makes it possible to control data redundancies by using common attributes that are shared by tables, called foreign keys.

The proper use of foreign keys is crucial to exercising data redundancy control. However, it is worth emphasizing that, in the strictest sense, the use of foreign keys does not eliminate data redundancies, because the foreign key values can be repeated many times. Nevertheless, the proper use of foreign keys *minimizes* data redundancies, thus minimizing the chance that destructive data anomalies will develop.

> **N O T E**
>
> The real test of redundancy is *not* how many copies of a given attribute are stored, *but whether or not the elimination of an attribute will eliminate information*. Therefore, if you delete an attribute and the original information can still be generated through relational algebra, the inclusion of that attribute would be redundant. Given this view of redundancy, proper foreign keys are clearly not redundant in spite of their multiple occurrences in a table. However, even when you use this less restrictive view of redundancy, keep in mind that *controlled* redundancies are often designed into the system to ensure transaction speed and/or information requirements. Exclusive reliance on relational algebra to produce required information may lead to elegant designs that fail the test of practicality.

You will learn in Chapter 4 that database designers must reconcile three often contradictory requirements: design elegance, processing speed, and information requirements. And you will learn in Chapter 12, "The Data Warehouse," that proper data warehousing design even requires carefully defined and controlled data redundancies to function properly. In short, regardless of how you describe data redundancies, their potential for damage is limited by their proper implementation and careful control.

As important as data redundancy control is, there are times when we must actually increase the level of data redundancy to make the database serve crucial information purposes. For example, let's examine a small invoicing system. The system includes the CUSTOMER, who may buy one or more PRODUCTs, thus generating an INVOICE. Because a customer may buy more than one product at a time, an invoice may contain several invoice LINEs, each providing details about the purchased product. The PRODUCT table should contain the product price to provide a consistent pricing input for each product that appears on the invoice. The tables that are part of such a system are shown in Figure 3.30. The system's relational schema is shown in Figure 3.31.

As you examine the tables in the invoicing system in Figure 3.30 and the relationships depicted in the relational schema in Figure 3.31, note that you can keep track of typical sales information. For example, by tracing the relationships among the four tables, you discover that customer 10014 (Myron Orlando) bought two items on March 8, 2004 that were written to invoice number 1001: one Houselite chain saw with a 16-inch bar and two rat-tail files. (*Note*: Check the first two rows in the LINE table, then match the two PROD_CODE values in LINE with the PROD_CODE values in PRODUCT. Trace the LINE table's INV_NUMBER to the INVOICE table and the INVOICE table's line number 1001 to the CUSTOMER table via CUS_CODE.) Application software will be used to write the correct bill by multiplying each invoice line item's LINE_UNITS by its LINE_PRICE, adding the results, applying appropriate taxes, and so on. Later, other application software might use the same technique to write sales reports that track and compare sales by week, month, or year.

FIGURE 3.30 **A SMALL INVOICING SYSTEM**

Table name: **CUSTOMER**
Primary key: **CUS_CODE** Database name: **Ch03_SaleCo**
Foreign key: none

		CUS_CODE	CUS_LNAME	CUS_FNAME	CUS_INITIAL	CUS_AREACODE	CUS_PHONE
▶	+	10010	Ramas	Alfred	A	615	844-2573
	+	10011	Dunne	Leona	K	713	894-1238
	+	10012	Smith	Kathy	W	615	894-2285
	+	10013	Olowski	Paul	F	615	894-2180
	+	10014	Orlando	Myron		615	222-1672
	+	10015	O'Brian	Amy	B	713	442-3381
	+	10016	Brown	James	G	615	297-1228
	+	10017	Williams	George		615	290-2556
	+	10018	Farriss	Anne	G	713	382-7185
	+	10019	Smith	Olette	K	615	297-3809

Table name: **INVOICE**
Primary key: **INV_NUMBER**
Foreign key: **CUS_CODE**

		INV_NUMBER	CUS_CODE	INV_DATE
▶	+	1001	10014	08-Mar-04
	+	1002	10011	08-Mar-04
	+	1003	10012	08-Mar-04
	+	1004	10011	09-Mar-04

Table name: **LINE**
Primary key: **INV_NUMBER + LINE_NUMBER**
Foreign keys: **INV_NUMBER, PROD_CODE**

	INV_NUMBER	LINE_NUMBER	PROD_CODE	LINE_UNITS	LINE_PRICE
▶	1001	1	123-21UUY	1	$189.99
	1001	2	SRE-657UG	3	$2.99
	1002	1	QER-34256	2	$18.63
	1003	1	ZZX/3245Q	1	$6.79
	1003	2	SRE-657UG	1	$2.99
	1003	3	001278-AB	1	$12.95
	1004	1	001278-AB	1	$12.95
	1004	2	SRE-657UG	2	$2.99

Table name: **PRODUCT**
Primary key: **PROD_CODE**
Foreign key: none

		PROD_CODE	PROD_DESCRIPT	PROD_PRICE	PROD_ON_HAND	VEND_CODE
▶	+	001278-AB	Claw hammer	$12.95	23	232
	+	123-21UUY	Houselite chain saw, 16-in. bar	$189.99	4	235
	+	QER-34256	Sledge hammer, 16-lb. head	$18.63	6	231
	+	SRE-657UG	Rat-tail file	$2.99	15	232
	+	ZZX/3245Q	Steel tape, 12-ft. length	$6.79	8	235

As you examine the sales transactions in Figure 3.30, you might reasonably suppose that the product price billed to the customer is derived from the PRODUCT table, because that's where the product data are stored. *But why does that same product price occur again in the LINE table? Isn't that a data redundancy?* Indeed it is. But this time, *the redundancy is crucial to the system's success.* **Copying the product price from the PRODUCT table to the LINE table means that it is possible to maintain the historical accuracy of the transactions.** Suppose, for instance, that we fail to write the LINE_PRICE in the LINE table and that we use the PROD_PRICE (product price) from the PRODUCT table to calculate the sales revenue. Now suppose that the PRODUCT table's PROD_PRICE changes. This price change will be properly reflected

in all current sales revenue calculations. Unfortunately, the calculations of past sales revenues will now also reflect the new product price that was not in effect when the transaction took place! Therefore, the revenue calculations for all past transactions will be incorrect, thus eliminating the possibility of making proper sales comparisons over time. On the other hand, if the price data are copied from the PRODUCT table and stored with the transaction in the LINE table, that price will always accurately reflect the transaction that took place *at that time*. You will discover that such planned redundancies are common in good database design.

FIGURE 3.31 THE RELATIONAL SCHEMA FOR THE INVOICING SYSTEM

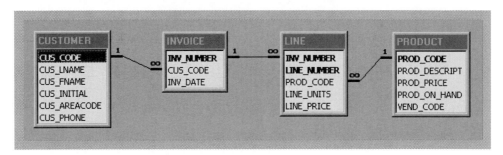

Finally, you might wonder why we used the LINE_NUMBER attribute in the LINE table in Figure 3.30. Wouldn't the combination of INV_NUMBER and PROD_CODE be a sufficient composite primary key—and, therefore, isn't the LINE_NUMBER redundant? Yes, the LINE_NUMBER is redundant, but this redundancy is quite commonly created by invoicing software that generates such line numbers automatically. In this case, the redundancy is not necessary. But given its automatic generation, the redundancy is not a source of anomalies. The inclusion of LINE_NUMBER also adds another benefit: the order of the retrieved invoicing data will always match the order in which the data were entered. If product codes are used as part of the primary key, indexing will arrange these product codes as soon as the invoice is completed and the data are stored. You can imagine the potential confusion when a customer calls and says, "The second item on my invoice has an incorrect price," and you are looking at an invoice whose invoice lines have a different order from those on the customer's copy!

3.8 INDEXES

Suppose you want to locate a particular book in your library. Would it make sense to search for the book by looking through *every* book in the library until you find that book? Of course not; you use the library's catalog, which is indexed by title, topic, and author. The index (in either a manual or a computer system) "points" you to the book's location, thereby making retrieval of the book a quick and simple matter. An **index** is an orderly arrangement used to logically access rows in a table.

Or suppose you want to find a topic, such as "ER model," in this book. Would it make sense to read through every page in the book until you stumble across the topic? Of course not; it is much simpler to go to the book's index, look up the phrase ER model, and read the page references that point you to the appropriate page(s). In each case, an index is used to locate a needed item quickly.

Indexes in the relational database environment work like the indexes described in the preceding paragraphs. From a conceptual point of view, an index is composed of an index key and a set of pointers. The **index key** is, in effect, the index's reference point. More formally, an index is an ordered arrangement of keys and pointers. Each key points to the location of the data identified by the key.

For example, suppose that we want to look up all the paintings created by a given painter in the Ch03_MUSEUM database in Figure 3.21. Without an index, we must read each row in the PAINTING table and see if the PAINTER_NUM matches the requested painter. However, if we index the PAINTER table, using the index key

PAINTER_NUM, we merely need to look up the appropriate PAINTER_NUM in the index and find the matching pointers. Conceptually speaking, the index would resemble the presentation depicted in Figure 3.32.

FIGURE 3.32 COMPONENTS OF AN INDEX

As you examine Figure 3.31 and compare it to the Ch03_Museum database tables shown in Figure 3.21, note that the first PAINTER_NUM index key value (123) is found in records 1, 2, and 4 of the PAINTING table in Figure 3.21. And the second PAINTER_NUM index key value (126) is found in records 3 and 5 of the PAINTING table in Figure 3.21.

DBMSs use indexes for many different purposes. You just learned that an index can be used to retrieve data more efficiently. But indexes can also be used by a DBMS to retrieve data ordered by a specific attribute or attributes. For example, creating an index on the customer's last name will allow us to retrieve the customer data alphabetically ordered by the customer's last name. Also, an index key can be composed of one or more attributes. For example, in Figure 3.30, you can create an index on VEND_CODE and PROD_CODE to retrieve all products rows ordered by vendor, and within vendor, ordered by product.

Indexes play an important role in DBMSs for the implementation of primary keys. When you define a table's primary key, the DBMS will automatically create a unique index on the primary key column(s) you declared. For example, in Figure 3.30, when you declare CUS_CODE to be the primary key of the CUSTOMER table, the DBMS will automatically create a unique index on that attribute. A **unique index**, as its name implies, is an index in which the index key can only have one pointer value (row) associated with it. (The index in Figure 3.32 is not a unique index, because the PAINTER_NUM has multiple pointer values associated with it. For example, painter number 123 points to three rows—1, 2, and 4—in the PAINTING table.)

A table can have many indexes, but each index is associated with only one table. The index key can have multiple attributes (composite index).

NOTE

Creating an index is easy. You will learn in Chapter 6 that a simple SQL command will produce any required index.

S U M M A R Y

Entities are the basic building blocks of a relational database. A grouping of related entities, known as an entity set, is stored in a table. Conceptually speaking, the relational table is composed of intersecting rows (tuples) and columns. Each row represents a single entity, and each column represents the characteristics (attributes) of the entities.

Keys are central to the use of relational tables. Keys define functional dependencies. That is, other attributes are dependent on the key and can, therefore, be found if the key value is known. A key can be classified according to its characteristics as a superkey, a candidate key, a primary key, a secondary key, or a foreign key.

Each table row must have a primary key. The primary key is an attribute or combination of attributes that uniquely identifies all remaining attributes found in any given row. Because a primary key must be unique, no null "values" are allowed if entity integrity is to be maintained.

Although the tables are independent, they can be linked by sharing common attributes. Thus, we can link tables by using controlled redundancy; the primary key of one table appears again as the foreign key in the table to which it is linked. To make sure that the foreign key produces the correct pointer movement, it must maintain referential integrity; that is, the foreign key must contain either values that match the primary key in the other table or contain nulls.

Relational databases can be classified according to the degree to which they support the relational algebra functions SELECT, PROJECT, JOIN, INTERSECT, UNION, DIFFERENCE, PRODUCT, and DIVIDE, and also to the degree that (entity and referential) integrity rules are enforced. A fully relational database supports all eight relational algebra functions and enforces both integrity rules. A relationally complete database lacks only the enforcement of the integrity rules. A minimally relational database supports only SELECT, PROJECT, and JOIN and does not enforce the integrity rules. The tabular database supports only the relational algebra functions supported by the minimally relational database, and it requires that the access paths to the tables be defined. The more relational the relational database is, the greater its user flexibility, and the easier it is to design its applications.

If you at least use a relationally complete database, its RDBMS performs much work behind the scenes. For example, if you create a database, the RDBMS automatically produces a structure to house a data dictionary and the system catalog for your database. Each time you create a new table within the database, the RDBMS updates the data dictionary and system catalog, thereby providing the database documentation. As you define the attributes for the database tables, a fully relational RDBMS checks for (and prevents) the improper use of homonyms and synonyms, and it enforces the integrity rules.

Once you know the relational database basics, you can concentrate on design. Good design begins by identifying appropriate entities and attributes and the relationships among the entities. Such relationships (1:1, 1:M, and M:N) can be represented by models known as entity relationship (ER) models. The use of ER models allows us to create and evaluate simple logical design. The 1:M relationships are most easily incorporated in a good design; just make sure that the primary key of the one is included in the table of the many. The best designs use 1:M relationships exclusively. Fortunately, 1:1 and M:N relationships can be converted to 1:M relationships.

Data retrieval speed can be increased dramatically by using indexes. Indexes may be created very easily with the help of SQL commands.

KEY TERMS

abstract data type	full functional dependence	primary key (PK)
attribute domain	functional dependence	referential integrity
bridge entity	homonym	relational algebra
candidate key	index	right outer join
composite entity	index key	secondary key
composite key	join column(s)	superkey
data dictionary	key	synonym
determination	key attribute	system catalog
domain	left outer join	theta join
entity integrity	linking table	tuple(s)
equijoin	natural join	union-compatible
flags	null	unique index
foreign key (FK)		

REVIEW QUESTIONS

1. What is the difference between a database and a table?

2. What does a database expert mean when (s)he observes that a database displays both entity integrity and referential integrity?

3. Why are entity integrity and referential integrity important in a database?

4. A database user manual notes that, "The file contains two hundred records, each one of which contains nine fields." Use appropriate relational database terminology to "translate" the preceding statement.

5. Use the small database shown in Figure Q3.5 to illustrate the difference between a natural JOIN, an equiJOIN, and an outer JOIN.

FIGURE Q3.5 THE CH03_COLLEGEQUE DATABASE TABLES

Table name: STUDENT Database name: Ch03_CollegeQue

STU_CODE	PROF_CODE
100278	
128569	2
512272	4
531235	2
531268	
553427	1

Table name: PROFESSOR

PROF_CODE	DEPT_CODE
1	2
2	6
3	6
4	4

A D D I T I O N A L C D C O N T E N T

All the databases used in the questions and problems are found on the CD bundled with this book. Check the **\Databases\Student\Ch03** folder to find the databases. The database names used in the folder match the database names used in the figures. For example, the source of the tables shown in Figure Q3.5 is the **Ch03_CollegeQue** database located in the above named folder.

6. Draw the basic entity relationship diagram for the database shown in Figure Q3.5.
7. Draw the relational schema for the database shown in Figure Q3.5.
8. Suppose that you have the entity relationship model shown in Figure Q3.8.

FIGURE Q3.8 THE CROW'S FOOT ERD FOR QUESTION 8

During some time interval, a DRIVER can drive many different TRUCKs and any TRUCK can be driven by many DRIVERS

How would you convert this model into an entity relationship model that displays only 1:M relationships? (Make sure that you draw the revised entity relationship model.)

9. What are homonyms and synonyms, and why should they be avoided in database design?
10. How would you implement a l:M relationship in a database composed of two tables? Give an example.
11. Identify and describe the components of the table shown in Figure Q3.11, using correct terminology. Use your knowledge of naming conventions to identify the table's probable foreign key(s).

FIGURE Q3.11 THE CH03_NOCOMP DATABASE EMPLOYEE TABLE

Table name: EMPLOYEE Database name: Ch03_NoComp

EMP_NUM	EMP_LNAME	EMP_INITIAL	EMP_FNAME	DEPT_CODE	JOB_CODE
11234	Friedman	K	Robert	MKTG	12
11238	Olanski	D	Delbert	MKTG	12
11241	Fontein		Juliette	INFS	5
11242	Cruazona	J	Maria	ENG	9
11245	Smithson	B	Bernard	INFS	6
11248	Washington	G	Oleta	ENGR	8
11256	McBride		Randall	ENGR	8
11257	Kachinn	D	Melanie	MKTG	14
11258	Smith	W	William	MKTG	14
11260	Ratula	A	Katrina	INFS	5

12. Suppose that you are using the database composed of the two tables shown in Figure Q3.12.

FIGURE Q3.12 THE CH03_THEATER DATABASE TABLES

Table name: DIRECTOR **Database name: Ch03_Theater**

		DIR_NUM	DIR_LNAME	DIR_DOB
▶	+	100	Broadway	12-Jan-65
	+	101	Hollywoody	18-Nov-53
	+	102	Goofy	21-Jun-62

Table name: PLAY

	PLAY_CODE	PLAY_NAME	DIR_NUM
▶	1001	Cat On a Cold, Bare Roof	102
	1002	Hold the Mayo, Pass the Bread	101
	1003	I Never Promised You Coffee	102
	1004	Silly Putty Goes To Washington	100
	1005	See No Sound, Hear No Sight	101
	1006	Starstruck in Biloxi	102
	1007	Stranger In Parrot Ice	101

a. Identify the primary keys.

b. Identify the foreign keys.

c. Draw the entity relationship model.

d. Draw the relational schema to show the relationship between DIRECTOR and PLAY.

e. Suppose you wanted quick lookup capability to get a listing of all the plays directed by a given director. Which table would be the basis for the INDEX table, and what would be the index key?

f. What would be the conceptual view of the INDEX table that is described in Part d? Depict the contents of the (conceptual) INDEX table.

PROBLEMS

Use the database shown in Figure P3.1 to work Problems 1 through 7. Note that the database is composed of four tables and reflects these relationships:

• An EMPLOYEE has only one JOB_CODE, but a JOB_CODE can be held by many EMPLOYEEs.

• An EMPLOYEE can participate in many PLANs, and any PLAN can be assigned to many EMPLOYEEs.

Note also that the M:N relationship has been broken down into two 1:M relationships for which the BENEFIT table serves as the composite or bridge entity.

FIGURE P3.1 THE CH03_BENECO DATABASE TABLES

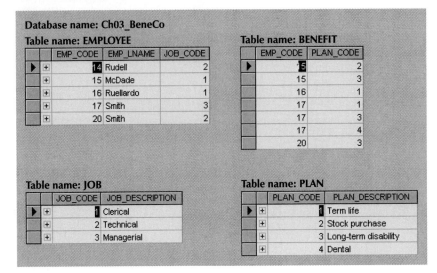

1. For each table in the database, identify the primary key and the foreign key(s). If a table does not have a foreign key, write None in the assigned space.

TABLE	PRIMARY KEY	FOREIGN KEY(S)
EMPLOYEE		
BENEFIT		
JOB		
PLAN		

2. Draw the entity relationship diagram for the relationship between EMPLOYEE and JOB.
3. Draw the relational schema for the relationship between EMPLOYEE and JOB.
4. Do the tables exhibit entity integrity? Answer Yes or No, and then explain your answer.

EXPLANATION	ENTITY INTEGRITY	EXPLANATION
EMPLOYEE		
BENEFIT		
JOB		
PLAN		

5. Do the tables exhibit referential integrity? Answer Yes or No, and then explain your answer. Write NA (Not Applicable) if the table does not have a foreign key.

TABLE	REFERENTIAL INTEGRITY	EXPLANATION
EMPLOYEE		
BENEFIT		
JOB		
PLAN		

6. Draw the entity relationship diagram to show the relationships among EMPLOYEE, BENEFIT, JOB, and PLAN.

7. Draw the relational schema to show the relationships among EMPLOYEE, BENEFIT, JOB, and PLAN.

Use the database shown in Figure P3.8 to answer Problems 8 through 15.

FIGURE P3.8 THE CH03_StoreCo DATABASE TABLES

Table name: EMPLOYEE Database name: Ch03_StoreCo

EMP_CODE	EMP_TITLE	EMP_LNAME	EMP_FNAME	EMP_INITIAL	EMP_DOB	STORE_CODE
1	Mr.	Williamson	John	W	21-May-64	3
2	Ms.	Ratula	Nancy		09-Feb-69	2
3	Ms.	Greenboro	Lottie	R	02-Oct-61	4
4	Mrs.	Rumpersfro	Jennie	S	01-Jun-71	5
5	Mr.	Smith	Robert	L	23-Nov-59	3
6	Mr.	Renselaer	Cary	A	25-Dec-65	1
7	Mr.	Ogallo	Roberto	S	31-Jul-62	3
8	Ms.	Johnsson	Elizabeth	I	10-Sep-68	1
9	Mr.	Eindsmar	Jack	W	19-Apr-55	2
10	Mrs.	Jones	Rose	R	06-Mar-66	4
11	Mr.	Broderick	Tom		21-Oct-72	3
12	Mr.	Washington	Alan	Y	08-Sep-74	2
13	Mr.	Smith	Peter	N	25-Aug-64	3
14	Ms.	Smith	Sherry	H	25-May-66	4
15	Mr.	Olenko	Howard	U	24-May-64	5
16	Mr.	Archialo	Barry	V	03-Sep-60	5
17	Ms.	Grimaldo	Jeanine	K	12-Nov-70	4
18	Mr.	Rosenberg	Andrew	D	24-Jan-71	4
19	Mr.	Rosten	Peter	F	03-Oct-68	4
20	Mr.	Mckee	Robert	S	06-Mar-70	1
21	Ms.	Baumann	Jennifer	A	11-Dec-74	3

Table name: STORE

STORE_CODE	STORE_NAME	STORE_YTD_SALES	REGION_CODE	EMP_CODE
1	Access Junction	$1,003,455.76	2	8
2	Database Corner	$1,421,987.39	2	12
3	Tuple Charge	$986,783.22	1	7
4	Attribute Alley	$944,568.56	2	3
5	Primary Key Point	$2,930,098.45	1	15

Table name: REGION

REGION_CODE	REGION_DESCRIPT
1	East
2	West

8. For each table, identify the primary key and the foreign key(s). If a table does not have a foreign key, write None in the assigned space.

TABLE	PRIMARY KEY	FOREIGN KEY(S)
EMPLOYEE		
STORE		
REGION		

9. Do the tables exhibit entity integrity? Answer Yes or No, and then explain your answer.

TABLE	ENTITY INTEGRITY	EXPLANATION
EMPLOYEE		
STORE		
REGION		

10. Do the tables exhibit referential integrity? Answer Yes or No, and then explain your answer. Write NA (Not Applicable) if the table does not have a foreign key.

TABLE	REFERENTIAL INTEGRITY	EXPLANATION
EMPLOYEE		
STORE		
REGION		

11. Describe the type(s) of relationship(s) between STORE and REGION.
12. Draw the entity relationship diagram for the relationship between STORE and REGION.
13. Draw the relational schema for the relationship between STORE and REGION.
14. Describe the type(s) of relationship(s) between EMPLOYEE and STORE. (*Hint*: Each store employs many employees, one of whom manages the store.)
15. Draw the entity relationship diagram to show the relationships among EMPLOYEE, STORE, and REGION.
16. Draw the relational schema to show the relationships among EMPLOYEE, STORE, and REGION.

Use the database shown in Figure P3.17 to answer Problems 17 through 23.

FIGURE P3.17 THE CH03_CHEAPCO DATABASE TABLES

Table name: PRODUCT Database name: Ch03_CheapCo

PROD_CODE	PROD_DESCRIPTION	PROD_STOCK_DATE	PROD_ON_HAND	PROD_PRICE	VEND_CODE
12-WW/P2	7.25-in. power saw blade	07-Apr-04	12	$1.19	123
1QQ23-55	2.5-in. wood screw, 100	19-Mar-04	123	$4.49	123
231-78-W	PVC pipe, 3.5-in., 8 ft.	07-Dec-03	45	$8.87	121
33564/U	Rat-tail file, 0.125-in., fine	08-Mar-04	18	$1.19	123
AR/3/TYR	Cordless drill, 0.25-in.	29-Nov-03	8	$45.99	121
DT-34-WW	Phillips screwdriver pack	20-Dec-03	11	$23.29	123
EE3-67/W	Sledge hammer, 12 lb.	25-Feb-04	9	$17.99	121
ER-56/DF	Houselite chain saw, 16-in.	28-Dec-03	7	$235.49	125
FRE-TRY9	Jigsaw, 12-in blade	12-Aug-03	67	$1.45	125
SE-67-89	Jigsaw, 8-in. blade	11-Oct-03	34	$1.35	125
ZW-QR/AV	Hardware cloth, 0.25-in.	23-Apr-04	14	$12.99	123
ZX-WR/FR	Claw hammer	01-Mar-04	15	$8.95	121

Table name: VENDOR

VEND_CODE	VEND_NAME	VEND_CONTACT	VEND_AREACODE	VEND_PHONE
120	BargainSnapper, Inc.	Melanie T. Travis	615	899-1234
121	Cut'nGlow Co.	Henry J. Olero	615	342-9896
122	Rip & Rattle Supply Co.	Anne R. Morrins	901	225-1127
123	Tools 'R Us	Juliette G. McHenry	615	546-7894
124	Trowel & Dowel, Inc.	George F. Frederick	901	453-4567
125	Bow & Wow Tools	Bill S. Sedwick	904	324-9988

17. For each table, identify the primary key and the foreign key(s). If a table does not have a foreign key, write None in the assigned space.

TABLE	PRIMARY KEY	FOREIGN KEY(S)
PRODUCT		
VENDOR		

18. Do the tables exhibit entity integrity? Answer Yes or No, and then explain your answer.

TABLE	ENTITY INTEGRITY	EXPLANATION
PRODUCT		
VENDOR		

19. Do the tables exhibit referential integrity? Answer Yes or No, and then explain your answer. Write NA (Not Applicable) if the table does not have a foreign key.

TABLE	REFERENTIAL INTEGRITY	EXPLANATION
PRODUCT		
VENDOR		

20. Draw the entity relationship diagram for this database.

21. Draw the relational schema for this database.

22. Create the data dictionary for this database.

Use the database shown in Figure P3.23 to answer Problems 23 through 28.

FIGURE P3.23 THE CH03_TRANSCO DATABASE TABLES

Table name: TRUCK Database name: Ch03_TransCo

TRUCK_NUM	BASE_CODE	TYPE_CODE	TRUCK_MILES	TRUCK_BUY_DATE	TRUCK_SERIAL_NUM
1001	501	1	32,123.5	23-Sep-03	AA-322-12212-W11
1002	502	1	76,984.3	05-Feb-02	AC-342-22134-Q23
1003	501	2	12,346.6	11-Nov-03	AC-445-78656-Z99
1004		1	2,894.3	06-Jan-04	WQ-112-23144-T34
1005	503	2	45,673.1	01-Mar-03	FR-998-32245-W12
1006	501	2	193,245.7	15-Jul-00	AD-456-00845-R45
1007	502	3	32,012.3	17-Oct-01	AA-341-96573-Z84
1008	502	3	44,213.6	07-Aug-02	DR-559-22189-D33
1009	503	2	10,932.9	12-Feb-04	DE-887-98456-E94

Table name: BASE

BASE_CODE	BASE_CITY	BASE_STATE	BASE_AREA_CODE	BASE_PHONE	BASE_MANAGER
501	Murfreesboro	TN	615	123-4567	Andrea D. Gallager
502	Lexington	KY	568	234-5678	George H. Delarosa
503	Cape Girardeau	MO	456	345-6789	Maria J. Talindo
504	Dalton	GA	901	456-7890	Peter F. McAvee

Table name: TYPE

TYPE_CODE	TYPE_DESCRIPTION
1	Single box, double-axle
2	Single box, single-axle
3	Tandem trailer, single-axle

23. For each table, identify the primary key and the foreign key(s). If a table does not have a foreign key, write None in the assigned space.

TABLE	PRIMARY KEY	FOREIGN KEY(S)
TRUCK		
BASE		
TYPE		

24. Do the tables exhibit entity integrity? Answer Yes or No, and then explain your answer.

TABLE	ENTITY INTEGRITY	EXPLANATION
TRUCK		
BASE		
TYPE		

25. Do the tables exhibit referential integrity? Answer Yes or No, and then explain your answer. Write NA (Not Applicable) if the table does not have a foreign key.

TABLE	REFERENTIAL INTEGRITY	EXPLANATION
TRUCK		
BASE		
TYPE		

26. Identify the TRUCK table's candidate key(s).
27. For each table, identify a superkey and a secondary key.

TABLE	SUPERKEY	SECONDARY KEY
TRUCK		
BASE		
TYPE		

28. Draw the entity relationship diagram for this database.
29. Draw the relational schema for this database.

Use the database shown in Figure P3.30 to answer Problems 30 through 33.

FIGURE P3.30A THE CH03_AVIACO DATABASE TABLES

Database name: Ch03_AviaCo
Table name: CHARTER (Part 1: first six attributes)

CHAR_TRIP	CHAR_DATE	CHAR_PILOT	CHAR_COPILOT	AC_NUMBER	CHAR_DESTINATION
10001	05-Feb-04	104		2289L	ATL
10002	05-Feb-04	101		2778V	BNA
10003	05-Feb-04	105	109	4278Y	GNV
10004	06-Feb-04	106		1484P	STL
10005	06-Feb-04	101		2289L	ATL
10006	06-Feb-04	109		4278Y	STL
10007	06-Feb-04	104	105	2778V	GNV
10008	07-Feb-04	106		1484P	TYS
10009	07-Feb-04	105		2289L	GNV
10010	07-Feb-04	109		4278Y	ATL
10011	07-Feb-04	101	104	1484P	BNA
10012	08-Feb-04	101		2778V	MOB
10013	08-Feb-04	105		4278Y	TYS
10014	09-Feb-04	106		4278Y	ATL
10015	09-Feb-04	104	101	2289L	GNV
10016	09-Feb-04	109	105	2778V	MQY
10017	10-Feb-04	101		1484P	STL
10018	10-Feb-04	105	104	4278Y	TYS

The destinations are indicated by standard three-letter airport codes. For example,
STL = St. Louis, MO ATL = Atlanta, GA BNA = Nashville, TN, and so on.

Table name: CHARTER (Part 2: next four attributes)

CHAR_TRIP	CHAR_DISTANCE	CHAR_HOURS_FLOWN	CHAR_HOURS_WAIT	CHAR_FUEL_GALLONS
10001	936.0	5.1	2.2	354.1
10002	320.0	1.6	0	72.6
10003	1,574.0	7.8	0	339.8
10004	472.0	2.9	4.9	97.2
10005	1,023.0	5.7	3.5	397.7
10006	472.0	2.6	5.2	117.1
10007	1,574.0	7.9	0	348.4
10008	644.0	4.1	0	140.6
10009	1,574.0	6.6	23.4	459.9
10010	998.0	6.2	3.2	279.7
10011	352.0	1.9	5.3	66.4
10012	884.0	4.8	4.2	215.1
10013	644.0	3.9	4.5	174.3
10014	936.0	6.1	2.1	302.6
10015	1,645.0	6.7	0	459.5
10016	312.0	1.5	0	67.2
10017	508.0	3.1	0	105.5
10018	644.0	3.8	4.5	167.4

CHAR_HOURS = number of flight hours
CHAR_FUEL_GALLONS = gallons of fuel used during the trip
CHAR_DISTANCE = roundtrip distance in miles

FIGURE P3.30B THE CH03_AVIACO DATABASE TABLES (CONTINUED)

Table name: CHARTER (Part 3: remaining attributes, primary key values repeated to see the connection to the first and second parts more easily.)

CHAR_TRIP	CHAR_OIL_QTS	CUS_CODE
10001	1	10011
10002	0	10016
10003	2	10014
10004	1	10019
10005	2	10011
10006	0	10017
10007	2	10012
10008	1	10014
10009	0	10017
10010	0	10016
10011	1	10012
10012	0	10010
10013	1	10011
10014	0	10017
10015	2	10016
10016	0	10011
10017	0	10014
10018	0	10017

Table name: AIRCRAFT

AC_NUMBER	MOD_CODE	AC_TTAF	AC_TTEL	AC_TTER
1484P	PA23-250	1,833.1	1,833.1	101.8
2289L	C-90A	4,243.8	768.9	1,123.4
2778V	PA31-350	7,992.9	1,513.1	789.5
4278Y	PA31-350	2,147.3	622.1	243.2

AC_TTAF = Aircraft total time, airframe (hours)
AC_TTEL = Total time, left engine (hours)
AC_TTER = Total time, right engine (hours)

In a fully developed database system, such attribute values would be updated by application software when the CHARTER table entries are posted.

Table name: MODEL

MOD_CODE	MOD_MANUFACTURER	MOD_NAME	MOD_SEATS	MOD_CHG_MILE
C-90A	Beechcraft	KingAir	8	$2.67
PA23-250	Piper	Aztec	6	$1.93
PA31-350	Piper	Navajo Chieftain	10	$2.35

Customers are charged per round-trip mile, using the MOD_CHG_MILE rate. The MOD_SEAT gives the total number of seats in the airplane, including the pilot and copilot seats. Therefore, a PA31-350 trip that is flown by a pilot and a copilot has six passenger seats available.

FIGURE P3.30C THE CH03_AViaCo DATABASE TABLES (CONTINUED)

Table name: PILOT

		EMP_NUM	PIL_LICENSE	PIL_RATINGS	PIL_MED_TYPE	PIL_MED_DATE	PIL_PT135_DATE
▶	+	101	ATP	ATP/SEL/MEL/Instr/CFII	1	20-Jan-04	11-Jan-04
	+	104	ATP	ATP/SEL/MEL/Instr	1	18-Dec-03	17-Jan-04
	+	105	COM	COMM/SEL/MEL/Instr/CFI	2	05-Jan-04	02-Jan-04
	+	106	COM	COMM/SEL/MEL/Instr	2	10-Dec-03	02-Feb-04
	+	109	COM	ATP/SEL/MEL/SES/Instr/CFII	1	22-Jan-04	15-Jan-04

The pilot licenses shown in the PILOT table include the ATP = Airline Transport Pilot and COM = Commercial Pilot. Businesses that operate "on demand" air services are governed by Part 135 of the Federal Air Regulations (FARs) that are enforced by the Federal Aviation Administration (FAA). Such businesses are known as "Part 135 operators." Part 125 operations require that pilots successfully complete flight proficiency checks each six months. The "Part 135" flight proficiency check date is recorded in PIL_PT135_DATE. To fly commercially, pilots must have at least a commercial license and a 2nd class medical certificate (PIL_MED_TYPE = 2.)

The PIL_RATINGS include1

SEL = Single Engine, Land MEL = Multi-engine Land
SES = Single Engine (Sea) Instr. = Instrument
CFI = Certified Flight Instructor CFII = Certified Flight Instructor, Instrument

Table name: EMPLOYEE

		EMP_NUM	EMP_TITLE	EMP_LNAME	EMP_FNAME	EMP_INITIAL	EMP_DOB	EMP_HIRE_DATE
▶	+	100	Mr.	Kolmycz	George	D	15-Jun-42	15-Mar-88
	+	101	Ms.	Lewis	Rhonda	G	19-Mar-65	25-Apr-86
	+	102	Mr.	Vandam	Rhett		14-Nov-58	18-May-93
	+	103	Ms.	Jones	Anne	M	11-May-74	26-Jul-99
	+	104	Mr.	Lange	John	P	12-Jul-71	20-Aug-90
	+	105	Mr.	Williams	Robert	D	14-Mar-75	19-Jun-03
	+	106	Mrs.	Duzak	Jeanine	K	12-Feb-68	13-Mar-89
	+	107	Mr.	Diante	Jorge	D	01-May-75	02-Jul-97
	+	108	Mr.	Wiesenbach	Paul	R	14-Feb-66	03-Jun-93
	+	109	Ms.	Travis	Elizabeth	K	18-Jun-61	14-Feb-04
	+	110	Mrs.	Genkazi	Leighla	W	19-May-70	29-Jun-90

Table name: CUSTOMER

		CUS_CODE	CUS_LNAME	CUS_FNAME	CUS_INITIAL	CUS_AREACODE	CUS_PHONE	CUS_BALANCE
▶	+	10010	Ramas	Alfred	A	615	844-2573	$0.00
	+	10011	Dunne	Leona	K	713	894-1238	$0.00
	+	10012	Smith	Kathy	W	615	894-2285	$896.54
	+	10013	Olowski	Paul	F	615	894-2180	$1,285.19
	+	10014	Orlando	Myron		615	222-1672	$673.21
	+	10015	O'Brian	Amy	B	713	442-3381	$1,014.56
	+	10016	Brown	James	G	615	297-1228	$0.00
	+	10017	Williams	George		615	290-2556	$0.00
	+	10018	Farriss	Anne	G	713	382-7185	$0.00
	+	10019	Smith	Olette	K	615	297-3809	$453.98

ROBCOR is an aircraft charter company that supplies on-demand charter flight services, using a fleet of four aircraft. Aircraft are always identified by a (unique) registration number. Therefore, the aircraft registration number is an appropriate primary key for the AIRCRAFT table.

The nulls in the CHARTER table's CHAR_COPILOT column indicate that a copilot is not necessarily required for some charter trips or for some aircraft. (Federal Aviation Administration rules require a copilot on jet aircraft and on aircraft having a gross take-off weight over 12,500 pounds. None of the aircraft in the AIRCRAFT table is governed by this requirement; however, some customers may require the presence of a copilot for insurance reasons.) All charter trips are recorded in the CHARTER table.

N O T E

Earlier in the chapter we stated that it is best to avoid homonyms and synonyms wherever possible. In this problem, both the pilot and the copilot are pilots in the PILOT table, but we cannot use EMP_NUM for both in the CHARTER table. Therefore, we used the synonyms CHAR_PILOT and CHAR_COPILOT in the CHARTER table.

Although this "solution" works in this case, it is very restrictive and it generates nulls when a copilot is not required. Worse, such nulls proliferate as crew requirements change. For example, if the AviaCo charter company grows and starts using larger aircraft, crew requirements may increase to include flight engineers and load masters. The CHARTER table would then have to be modified to include the additional crew assignments by adding such attributes as CHAR_FLT_ENGINEER and CHAR_LOADMASTER to the CHARTER table. Given this change, each time a smaller aircraft flies a charter trip without the number of crew members required in larger aircraft, the "missing" crew members yield additional nulls in the CHARTER table.

You will have a chance to correct these design shortcomings in Problem 33. This problem illustrates two important points:

1. Don't use synonyms. If your design requires the use of synonyms, revise the design!

2. To the greatest possible extent, design the database to accommodate growth without requiring structural changes in the database tables. Plan ahead—and try to anticipate the effects of change on the database.

30. For each table, where possible, identify:
 a. The primary key
 b. A superkey
 c. A candidate key
 d. The foreign key(s)
 e. A secondary key

31. Create the entity relationship diagram. (*Hint*: Look at the table contents. You will discover that an AIRCRAFT can fly many CHARTER trips, but each CHARTER trip is flown by one AIRCRAFT. Similarly, you will discover that a MODEL references many AIRCRAFT, but each AIRCRAFT references a single MODEL, and so on.)

32. Create the relational schema.

33. Modify the ERD you created in Problem 31 to eliminate the problems created by the use of synonyms. (*Hint*: Modify the CHARTER table structure by eliminating the CHAR_PILOT and CHAR_COPILOT attributes, and then create a composite table named CREW to link the CHARTER and EMPLOYEE tables. Some crew members, such as flight attendants, may not be pilots. That's why the EMPLOYEE table enters into this relationship.)

34. Draw the relational schema for the design you revised in Problem 33. (After you have had a chance to revise the design, your teacher will be able to show you the results of the design change, using a copy of the revised database named **Ch03_AviaCo_2**.)

4

ENTITY RELATIONSHIP (ER) MODELING

In this chapter, you will learn:

❖ How relationships between entities are defined and refined, and how such relationships are incorporated into the database design process

❖ How ERD components affect database design and implementation

❖ How to interpret the modeling symbols for the four most popular ER modeling tools

❖ That real-world database design often requires that you reconcile conflicting goals

In this chapter, we expand our coverage of the data modeling aspect of database design. Data modeling is the first step in the database design journey, serving as a bridge between real-world objects and the database model that is implemented in the computer. Therefore, we cannot overstate the importance of the data modeling details, expressed graphically through entity relationship diagrams (ERDs).

Most of the basic concepts and definitions used in the ER model were introduced in Chapter 3. For example, the basic components of entities and relationships and their representation should now be familiar fare. This chapter goes much deeper and broader into analyzing the graphic depiction of relationships among the entities and shows you how these depictions help you to summarize the wealth of data required to implement a successful design.

We will—at least initially—continue to show you the Crow's Foot model in tandem with the Chen model so you become well acquainted with the two major modeling tools. However, we will shift to the Crow's Foot model once the modeling basics have been covered. Because this book focuses on the practical aspects of the database environment, and because the Crow's Foot model has excellent implementation features and is available in many database design tools, the Crow's Foot model will be the standard for most of the designs you will see in this and subsequent chapters.

We will also briefly show you the basics of two additional models that can be used to create ERDs: Rein85 and IDEF1X. We will then compare the four models as they are applied to a single design, and show you how a more complex design is created using the Crow's Foot model. Finally, we will illustrate how conflicting goals can be a challenge in database design, and may require you to make design compromises.

 F O U R

4.1 THE ENTITY RELATIONSHIP (ER) MODEL

You should remember from Chapters 2 and 3 ("Data Models" and "The Relationship Database Model," respectively) that the ER model forms the basis of an ER diagram, or ERD. The ERD represents the conceptual database as viewed by the end user. ERDs depict the ER model's three main components: entities, attributes, and relationships. Because an entity represents a real-world object, the words "entity" and "object" are often used interchangeably. Thus, the entities (objects) of the Tiny College database design we develop in this chapter will include students, classes, teachers, and classrooms. The order in which the ERD components are covered in the chapter is dictated by the way the modeling tools are used to develop ERDs that can form the basis for successful database design and implementation.

4.1.1 ENTITIES

In Chapter 2 you learned that an entity at the ER modeling level actually refers to the *entity set* and not to a single entity occurrence. In other words, the word *entity* in the ER model corresponds to a table and not to a row in the relational environment. The ER model refers to a specific table row as an *entity instance* or *entity occurrence*. In both the Chen and Crow's Foot models, an entity is represented by a rectangle containing the entity's name. The entity name, a noun, is usually written in capital letters.

The Crow's Foot ERDs you see in this chapter adhere to the generally accepted Crow's Foot modeling standards. These standards are reflected in any commercial database modeling software that uses Crow's Foot modeling capability. However, although the software details do not vary significantly from one software vendor to another, most of the software that generates Crow's Foot ERDs lets you select various presentation format options. For example, the entity name may be boldface, the entity name box may be shown in color, and so on. In fact, the Microsoft Visio Professional software (which we have used for our Crow's Foot ERDs) lets you select any color combination, font, and embedded graphics combination that suits your needs.

ADDITIONAL CD CONTENT

We used Microsoft Visio Professional to generate the Crow's Foot ERDs in this and subsequent chapters. We have placed a Visio Professional tutorial in Appendix A (located on the CD that is bundled with this book) to make sure that you will be able to generate your own Crow's Foot ERDs in a truly professional way.

4.1.2 ATTRIBUTES

You should also recall from Chapter 2 that attributes are characteristics of entities. For example, the STUDENT entity includes the attributes STU_LNAME, STU_FNAME, and STU_INITIAL. In the original Chen model, attributes are represented by ovals and are connected to the entity rectangle with a line. Each oval contains the name of the attribute it represents. In the Crow's Foot model, the attributes are simply written in the attribute box below the entity rectangle. (See Figure 4.1.) Because the Chen representation is rather space-consuming, some Chen model software vendors have adopted the Crow's Foot style attribute display.

As you examine Figure 4.1, note that the boldface attributes in the Crow's Foot model indicate that a data entry will be required when the entity is implemented. In this example, STU_LNAME and STU_FNAME require data entries, because we assume that all students have a last name and a first name. But students might not have a middle name and perhaps they do not (yet) have a phone number and an e-mail address. Therefore, these attributes are not presented in boldface in the entity box.

FIGURE 4.1 THE ATTRIBUTES OF THE STUDENT ENTITY

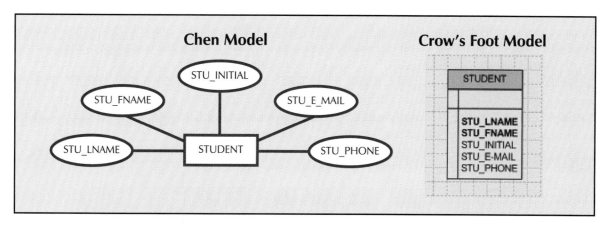

ADDITIONAL CD CONTENT

See how to create a Crow's Foot entity and how to define the attribute characteristics using Microsoft Visio Professional in Appendix A on the student CD.

Domains

Attributes have a domain. As you learned in Chapter 3, a *domain* is the attribute's set of possible values. For example, the domain for the (numeric) attribute grade point average (GPA) is written (0,4) because the lowest possible GPA value is 0, and the highest possible value is 4. The domain for the (character) attribute SEX consists of only two possibilities, M or F (or some other equivalent code). The domain for a company's date of hire attribute consists of all dates that fit into a range (for example, company startup date to current date).

Attributes may share a domain. For instance, a student address and a professor address share the same domain of all possible addresses. In fact, the data dictionary may let a newly declared attribute inherit the characteristics of an existing attribute, if the same attribute name is used. For example, the PROFESSOR and STUDENT entities may each have an attribute named ADDRESS.

Primary Keys

Primary keys (key attributes) are underlined in the ER diagram. Key attributes are also underlined in a frequently used table structure shorthand using the format:

TABLE NAME (**KEY_ATTRIBUTE 1**, ATTRIBUTE 2, ATTRIBUTE 3 . . . ATTRIBUTE K)

For example, a CAR entity may be represented by:

CAR (**CAR_VIN**, MOD_CODE, CAR_YEAR, CAR_COLOR)

(VIN is the standard acronym for vehicle identification number.)

Given the naming conventions we discussed in Chapters 2 and 3, it is easy to see that the attribute named MOD_CODE is likely to be a foreign key that is used to connect the car table to some other table. In this case, the other table, perhaps named MODEL, probably contains model-specific attributes such as MOD_TYPE to store descriptive entries such as "2-door sports coupe" or "4-door sedan" or "station wagon."

Ideally, a primary key is composed of only a single attribute. For example, the table in Figure 4.2 uses a single-attribute primary key named CLASS_CODE. However, it is possible to use a composite key, that is, a primary key composed of more than one attribute. For instance, the Tiny College database administrator may decide to identify each CLASS entity instance (occurrence) by using a composite primary key composed of the combination of CRS_CODE and CLASS_SECTION, instead of using CLASS_CODE. Either approach uniquely identifies each entity instance. Given the current structure of the CLASS table shown in Figure 4.2, CLASS_CODE is the primary key, and the combination of CRS_CODE and CLASS_SECTION is a proper candidate key. If the CLASS_CODE attribute is deleted from the CLASS entity, the candidate key (CRS_CODE and CLASS_SECTION) becomes an acceptable composite primary key.

FIGURE 4.2 THE CLASS TABLE (ENTITY) COMPONENTS AND CONTENTS

Table name: CLASS Database name: Ch04_TinyCollege

CLASS_CODE	CRS_CODE	CLASS_SECTION	CLASS_TIME	CLASS_ROOM	PROF_NUM
10012	ACCT-211	1	MWF 8:00-8:50 a.m.	BUS311	105
10013	ACCT-211	2	MWF 9:00-9:50 a.m.	BUS200	105
10014	ACCT-211	3	TTh 2:30-3:45 p.m.	BUS252	342
10015	ACCT-212	1	MWF 10:00-10:50 a.m.	BUS311	301
10016	ACCT-212	2	Th 6:00-8:40 p.m.	BUS252	301
10017	CIS-220	1	MWF 9:00-9:50 a.m.	KLR209	228
10018	CIS-220	2	MWF 9:00-9:50 a.m.	KLR211	114
10019	CIS-220	3	MWF 10:00-10:50 a.m.	KLR209	228
10020	CIS-420	1	W 6:00-8:40 p.m.	KLR209	162
10021	QM-261	1	MWF 8:00-8:50 a.m.	KLR200	114
10022	QM-261	2	TTh 1:00-2:15 p.m.	KLR200	114
10023	QM-362	1	MWF 11:00-11:50 a.m.	KLR200	162
10024	QM-362	2	TTh 2:30-3:45 p.m.	KLR200	162

NOTE

Remember from Chapter 3 that we made a commonly accepted distinction between COURSE and CLASS. A CLASS constitutes a specific COURSE offering and is defined by the course description and its section. Thus a professor who teaches Database I, Section 2; Database I, Section 5; Database I, Section 8; and Spreadsheet II, Section 6, teaches two courses (Database I and Spreadsheet II), but teaches four classes! Typically, the COURSE offerings are printed in a course catalog, while the CLASS offerings are printed in each enrollment period's (semester, trimester, or quarter) class schedule.

If the CLASS_CODE in Figure 4.2 is used as the primary key, the CLASS entity may be represented in short-hand form by:

CLASS (**CLASS_CODE**, CRS_CODE, CLASS_SECTION, CLASS_TIME, CLASS_ROOM, PROF_NUM)

On the other hand, if CLASS_CODE is deleted and the composite primary key is the combination of CRS_CODE and CLASS_SECTION, the CLASS entity may be represented by:

CLASS (**CRS_CODE**, **CLASS_SECTION**, CLASS_TIME, CLASS_ROOM, PROF_NUM)

Note that *both* key attributes are underlined in this entity notation.

Composite and Simple Attributes

Attributes are classified as simple or composite. A **composite attribute**, not to be confused with a composite key, is an attribute that can be further subdivided to yield additional attributes. For example, the attribute ADDRESS can be subdivided into street, city, state, and zip code. Similarly, the attribute PHONE_NUMBER can be subdivided into area code and exchange number. A **simple attribute** is an attribute that cannot be subdivided. For example, age, sex, and marital status would be classified as simple attributes. To facilitate detailed queries, it is usually appropriate to change composite attributes into a series of simple attributes.

Single-Valued Attributes

A **single-valued attribute** is an attribute that can have only a single value. For example, a person can have only one Social Security number, and a manufactured part can have only one serial number. *Keep in mind that a single-valued attribute is not necessarily a simple attribute.* For instance, a part's serial number, such as SE-08-02-189935, is single-valued, but it is a composite attribute because it can be subdivided into the region in which the part was produced, SE, the plant within that region, 08, the shift within the plant, 02, and the part number, 189935.

Multivalued Attributes

Multivalued attributes are attributes that can have many values. For instance, a person may have several college degrees, or a household may have several different phones, each with its own number. Similarly, a car's color may be subdivided into colors of roof, body, and trim. In the Chen ER model, the multivalued attributes are shown by a double line connecting the attribute to the entity. The Crow's Foot model does not identify multivalued attributes. The ER diagram in Figure 4.3 contains all the components we have introduced thus far. As you examine Figure 4.3, note that CAR_VIN is the primary key, and CAR_COLOR is a multivalued attribute of the CAR entity. (We will assume that each car is identified by a unique vehicle identification number, or CAR_VIN.)

FIGURE 4.3 A MULTIVALUED ATTRIBUTE IN AN ENTITY

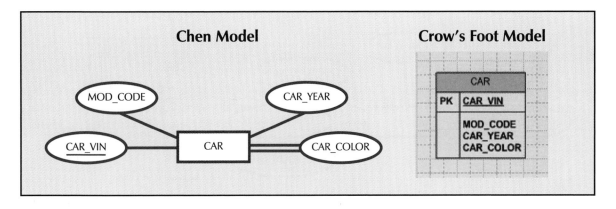

> **N O T E**
>
> If you examine the Crow's Foot model in Figure 4.3, you will see that we have typed the CAR entity's foreign key (FK) as MOD_CODE. This was done to illustrate the attribute similarities between the Chen model's attribute presentation and that of the Crow's Foot model in a free-standing entity—that is, an entity that has not (yet) been related to another entity. Actually, proper use of the Crow's Foot model generated by the Microsoft Visio Professional software—or any other database modeling software likely to be used by professional database designers—will produce the FK automatically when the relationship is defined. In addition, the software will label that FK appropriately and write that FK's implementation details in a data dictionary. Therefore, when you use Microsoft Visio Professional, *never type the FK attribute yourself*; let the software handle this task when the relationship between the entities is defined. (We show you how that's done in the Visio Professional tutorial located on your CD in Appendix A. But before you experiment with the Visio Professional software, study the text through section 4.1.7 to learn more about various types of relationships.)
>
> Keep this note in mind whenever you see an unlabeled FK in an entity. (You will see an example of a properly defined and labeled FK—generated by Microsoft Visio Professional—in Figure 4.5.)

Resolving Multivalued Attribute Problems

Although the conceptual model can handle M:N relationships and multivalued attributes, *you should not implement them in the relational DBMS*. So, if multivalued attributes exist, the designer must decide on one of two possible courses of action:

1. Within the original entity, create several new attributes, one for each of the original multivalued attribute's components. For example, we can split the CAR entity's attribute CAR_COLOR to create the new attributes CAR_TOPCOLOR, CAR_BODYCOLOR, and CAR_TRIMCOLOR, shown in Figure 4.4, and assign them to the CAR entity.

FIGURE 4.4 SPLITTING THE MULTIVALUED ATTRIBUTE INTO NEW ATTRIBUTES

Although this solution seems to work, its adoption can lead to major structural problems in the table. For example, if additional color components—such as a logo color—are added for some cars, the table structure must be modified to accommodate the new color section. In that case, cars that do not have such color sections either generate nulls for the nonexisting components or their color entries for those sections are "N/A" to indicate "not applicable." (Imagine how the solution in Figure 4.4—splitting a multivalued attribute into new attributes—would cause problems when it is applied to an employee entity containing employee degrees and certifications. If some employees have 10 degrees and certifications while most have fewer or none, the number of degree/certification attributes would number 10 and most

of those attribute values would be null for most of the employees.) In short, although we have seen solution 1 applied, it is not an acceptable solution.

2. Create a new entity composed of the original multivalued attribute's components (see Figure 4.5). The new (independent) COLOR entity is then related to the original CAR entity in a 1:M relationship. Note that such a change allows the designer to define color for different sections of the car. See Table 4.1.

TABLE 4.1 COMPONENTS OF THE MULTIVALUED ATTRIBUTE

SECTION	COLOR
Top	White
Body	Blue
Trim	Gold
Interior	Blue

Using the approach illustrated in Table 4.1, we even get a fringe benefit: we are now able to assign as many colors as necessary, without having to change the table structure. Note that the ER model in Figure 4.5 reflects the components listed in Table 4.1. This is the preferred way to deal with multivalued attributes.

FIGURE 4.5 A NEW ENTITY SET COMPOSED OF A MULTIVALUED ATTRIBUTE'S COMPONENTS

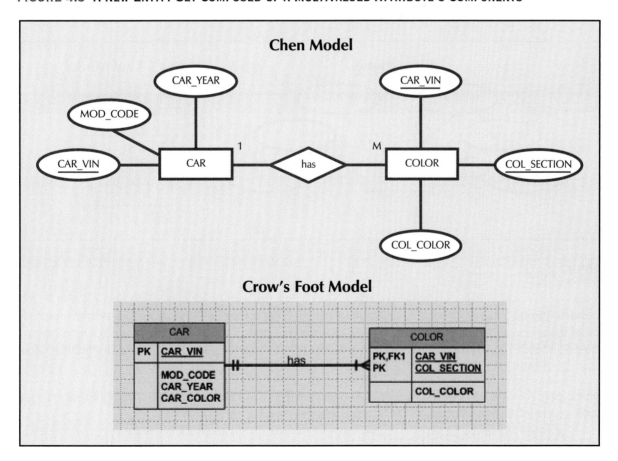

As you examine the Visio-generated Crow's Foot ERD in Figure 4.5, note that the software labeled the primary key (PK) attribute and underlined it when the PK was declared. The foreign key (FK) was labeled by the software when the relationship between CAR and COLOR was defined. (You will learn about different types of relationships and their effects on the PK/FK definitions in sections 4.1.3 through 4.1.7.) Note also that the COLOR entity's CAR_VIN is part of its PK and it is also the FK to CAR.

Derived Attributes

Finally, an attribute may be classified as a derived attribute. A **derived attribute** is an attribute whose value may be calculated (derived) from other attributes. The derived attribute need not be physically stored within the database; instead, it can be derived by using an algorithm. For example, an employee's age, EMP_AGE, may be found by computing the integer value of the difference between the current date and the EMP_DOB. If you use MS Access, you would use INT((DATE() – EMP_DOB)/365).

If you use Oracle, you would use SYSDATE, instead of DATE(). (We are assuming, of course, that the EMP_DOB was stored in the Julian date format.) Similarly, the total cost of an order can be derived by multiplying the quantity ordered by the unit price. Or the estimated average speed can be derived by dividing trip distance by the time spent en route. A derived attribute is indicated in the Chen model by a dashed line connecting the attribute and the entity (see Figure 4.6). The Crow's Foot model does not have a special display format for the derived attribute. The decision to store derived attributes in database tables depends on the processing requirements and the constraints placed on a particular application. The designer should be able to balance the design in accordance with such constraints.

FIGURE 4.6 DEPICTION OF A DERIVED ATTRIBUTE

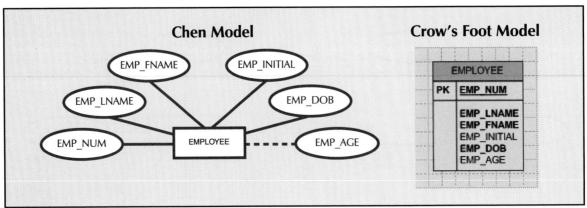

4.1.3 RELATIONSHIPS

A *relationship* is an association between entities. The entities that participate in a relationship are also known as **participants**. You should recall from Chapter 2 that each relationship is identified by a name that is descriptive of the relationship. The relationship name is an active or a passive verb. For example, a STUDENT *takes* a CLASS, a PROFESSOR *teaches* a CLASS, a DEPARTMENT *employs* a PROFESSOR, a DIVISION *is managed by* an EMPLOYEE, and an AIRCRAFT *is flown by* a CREW.

Relationships between entities always operate in both directions. That is, to define the relationship between the entities named CUSTOMER and INVOICE, you would specify that:

1. A CUSTOMER may generate many INVOICEs.
2. Each INVOICE is generated by one CUSTOMER.

Because you know both directions of the relationship between CUSTOMER and INVOICE, it is easy to see that this relationship can be classified as 1:M.

The relationship classification is difficult to establish if you only know one side of the relationship. For example, if you specify that:

1. A DIVISION is managed by one EMPLOYEE

 you don't know if the relationship is 1:1 or 1:M. Therefore, you should ask the question: "Can an employee manage more than one division?" If the answer to that question is "yes," the relationship is 1:M, and the second part of the relationship is then written as:

2. An EMPLOYEE may manage many DIVISIONs

 If an employee cannot manage more than one division, the relationship is 1:1, and the second part of the relationship is then written as:

2 An EMPLOYEE may manage only one DIVISION

4.1.4 CONNECTIVITY AND CARDINALITY

You learned in Chapter 2 that entity relationships may be classified as one-to-one, one-to-many, and many-to-many, and you also learned how such relationships were depicted in the Chen and Crow's Foot models. The term *connectivity* is used to describe the relationship classification.

Cardinality expresses the specific number of entity occurrences associated with one occurrence of the related entity. In the Chen model, cardinality is indicated by placing the appropriate numbers beside the entities, using the format (x,y). The first cardinality value represents the minimum value, while the second cardinality value represents the maximum value. In the Crow's Foot model, the cardinality is indicated through the use of symbols. Figure 4.7 illustrates how these models show connectivity and cardinality.

Knowing the minimum and maximum number of entity occurrences is very useful at the application software level. For example, Tiny College may want to ensure that a class cannot be taught unless it has at least 10 students enrolled. Similarly, if the classroom can only hold 30 students, the application software should use that cardinality to limit the enrollment in that class. However, do keep in mind that the DBMS cannot handle the implementation of the cardinalities at the table level—that capability is provided by the application software or by triggers. You will learn how to create and execute triggers in Chapter 7, "Advanced SQL."

As you examine the Chen model in Figure 4.7, keep in mind that the Chen cardinalities represent the number of occurrences in the *related* entity. For example, the cardinality (1,4) written next to the PROFESSOR entity in the "PROFESSOR teaches CLASS" relationship indicates that the PROFESSOR table's foreign key value occurs at least once and no more than four times in the CLASS table. If the cardinality had been written as (1,N), there would be no upper limit to the number of classes a professor might teach. Similarly, the cardinality (1,1) written next to the CLASS entity indicates that each class is taught by one and only one professor. (Each entity occurrence—that is, a row—in the PROFESSOR table can occur once and only once for each professor.)

The Crow's Foot model does not display a numeric range of specific cardinality values in the ERD. The Visio Professional software with which we generate the ERDs conforms to that standard. For example, note that the Chen cardinality (1,4) shows the "4" as a "many" symbol in the Crow's Foot portion of Figure 3.7. Also, unlike the Chen model, which displays the number of occurrences in the *related* table, the Crow's Foot model's cardinalities refer to the number of entity occurrences in the *originating* table. (Compare the two presentations in Figure 4.7.) Although the Crow's Foot ERD does not show the actual numbers in the cardinality segments, such numbers can be specified when you define the relationship. These cardinality numbers become part of the data dictionary when you implement the Crow's Foot ERD.

FIGURE 4.7 CONNECTIVITY AND CARDINALITY IN AN ERD

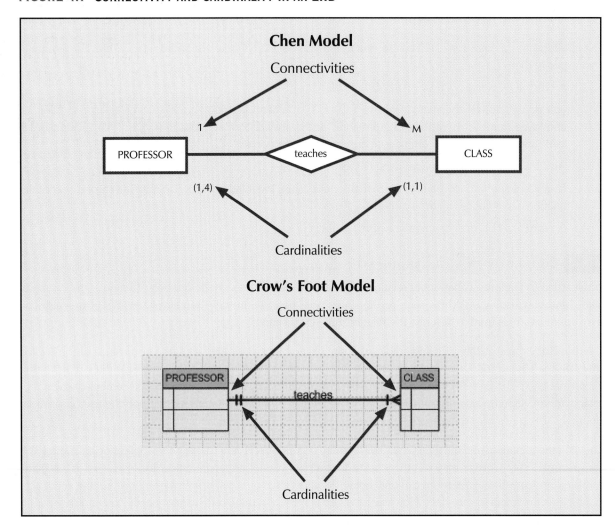

Connectivities and cardinalities are established by very concise statements known as *business rules*. (We introduced the business rules to you in Chapter 2.) Such rules, derived from a precise and detailed description of an organization's data environment, also establish the ER model's entities, attributes, relationships, connectivities, and constraints. Because business rules define the ER model's components, making sure that all appropriate business rules are identified becomes a very important part of a database designer's job.

ADDITIONAL CD CONTENT

Because the careful definition of complete and accurate business rules is crucial to good design, we examine their derivation in detail in Appendix D, "The University Lab: Conceptual Design." The modeling skills you are learning in this chapter are applied when we develop a real database design in Appendix D and edit it to its implementable stage in Appendix E, "The University Lab: Conceptual Design Verification, Logical Design, and Implementation."

4.1.5 RELATIONSHIP STRENGTH

It is worth remembering that relationships are established between entities. To understand relationships, you must be aware of the characteristics of the entities that help define those relationships. Therefore, before we can explore the details of relationship classification, we should briefly examine how existence dependence fits into the relationship framework.

Existence Dependence

If an entity's existence depends on the existence of one or more other entities, it is said to be **existence-dependent**. For example, if an XYZ Corporation employee wants to claim one or more dependents for tax withholding purposes, the relationship "EMPLOYEE claims DEPENDENT" would be appropriate. In this case, the DEPENDENT entity is clearly existence-dependent on the EMPLOYEE entity, because it is impossible for the dependent to exist apart from the EMPLOYEE in the XYZ Company's database.

If an entity can exist apart from one or more related entities, it is said to be **existence-independent** (sometimes designers refer to such an entity as a *strong* entity). For example, suppose that the XYZ Corporation uses parts to produce its products. Further suppose that some of these parts are produced "in-house" and other parts are bought from vendors. Given this scenario, it is quite possible for a PART to exist independently from a VENDOR in the relationship "PART is supplied by VENDOR." (After all, at least some of the parts are not supplied by a vendor.) Therefore, PART is existence-independent from VENDOR.

Weak (Non-Identifying) Relationships

If one entity is not existence-independent on another entity, the relationship between them is described as a **weak relationship**, also known as a **non-identifying relationship**. From a database design perspective, a weak relationship exists if the PK of the related entity does not contain a PK component of the parent entity. For example, suppose that we define the definition of the COURSE and CLASS entities as

COURSE(**CRS_CODE**, DEPT_CODE, CRS_DESCRIPTION, CRS_CREDIT)

CLASS(**CLASS_CODE**, CRS_CODE, CLASS_SECTION, CLASS_TIME, . . . etc.)

In this case, a weak relationship exists between COURSE and CLASS, because the CLASS_CODE is the CLASS entity's PK, while the CRS_CODE in CLASS is only an FK. In this case, the CLASS PK did not inherit the PK component from the COURSE entity.

Figure 4.8 shows that the Crow's Foot model depicts a weak relationship by placing a dashed relationship line between the related entities. The Chen model does not make a distinction between weak and strong relationships.

FIGURE 4.8 A WEAK (NON-IDENTIFYING) RELATIONSHIP BETWEEN COURSE AND CLASS

N O T E

If you are used to looking at relational schemas such as the ones produced by Microsoft Access, you expect to see the relationship line *in the relational schema* drawn from the PK to the FK. However, the relational schema convention is not necessarily reflected in the ERD. In an ERD, the focus is on the entities and the relationships between them, rather than how those relationships are anchored graphically. In fact, if design tools such as Visio Professional are used properly, the FKs are established *after* the relationship between the entities has been defined, so it is impossible to anchor the relationship line on the FK until *after* that FK has been created by the software. (This feature ensures that the FK attribute characteristics always properly match the attribute characteristics of the PK to which the FK points.) You can, of course, decide to move the relationship line anchor points after the update has been completed—*but that decision clearly reflects choice, rather than necessity.* You will discover that the placement of the relationship lines in a complex ERD that includes both horizontally and vertically placed entities is largely dictated by the designer's decision to improve the "readability" of the design. (Remember that the ERD is used for communication between the designer(s) and end users.) For this reason, you will see that we tend to place the relationship line anchor points on the FK/PK "connection" only when it does not interfere with the ERD's readability.

An example of the tables that exist in a weak relationship is shown in Figure 4.9. (We have added the ROOM_CODE and the PROF_NUM in the CLASS table to let you know that the class must be taught by a professor and must be taught in a room.) The ERD in Figure 4.8—on which the implementation in Figure 4.9 is based—does not include the CLASS table FKs. These FKs are not created until the ROOM and PROFESSOR entities exist and the relationships between these two entities and the CLASS entity exist. To show the weak relationship between COURSE and CLASS, only those two entities are shown in the ERD.

FIGURE 4.9 A WEAK RELATIONSHIP BETWEEN COURSE AND CLASS

Table name: COURSE **Database name: Ch04_TinyCollege**

CRS_CODE	DEPT_CODE	CRS_DESCRIPTION	CRS_CREDIT
ACCT-211	ACCT	Accounting I	3
ACCT-212	ACCT	Accounting II	3
CIS-220	CIS	Intro. to Microcomputing	3
CIS-420	CIS	Database Design and Implementation	4
MATH-243	MATH	Mathematics for Managers	3
QM-261	CIS	Intro. to Statistics	3
QM-362	CIS	Statistical Applications	4

Table name: CLASS

CLASS_CODE	CRS_CODE	CLASS_SECTION	CLASS_TIME	ROOM_CODE	PROF_NUM
10012	ACCT-211	1	MWF 8:00-8:50 a.m.	BUS311	105
10013	ACCT-211	2	MWF 9:00-9:50 a.m.	BUS200	105
10014	ACCT-211	3	TTh 2:30-3:45 p.m.	BUS252	342
10015	ACCT-212	1	MWF 10:00-10:50 a.m.	BUS311	301
10016	ACCT-212	2	Th 6:00-8:40 p.m.	BUS252	301
10017	CIS-220	1	MWF 9:00-9:50 a.m.	KLR209	228
10018	CIS-220	2	MWF 9:00-9:50 a.m.	KLR211	114
10019	CIS-220	3	MWF 10:00-10:50 a.m.	KLR209	228
10020	CIS-420	1	W 6:00-8:40 p.m.	KLR209	162
10021	QM-261	1	MWF 8:00-8:50 a.m.	KLR200	114
10022	QM-261	2	TTh 1:00-2:15 p.m.	KLR200	114
10023	QM-362	1	MWF 11:00-11:50 a.m.	KLR200	162
10024	QM-362	2	TTh 2:30-3:45 p.m.	KLR200	162
10025	MATH-243	1	Th 6:00-8:40 p.m.	DRE155	325

ADDITIONAL CD CONTENT

All the databases used to illustrate the material in this chapter are found on the CD bundled with this book. Check the Databases\Student\Ch04 folder to find the databases. The database names used in the folder match the database names used in the figures.

Strong (Identifying) Relationships

A **strong relationship**, also known as an **identifying relationship**, exists when the related entities are existence-dependent. From a database design perspective, a strong relationship between two entities exists whenever the PK of the related entity contains a PK component of the parent entity. For example, the definitions of the COURSE and CLASS entities

COURSE(**CRS_CODE**, DEPT_CODE, CRS_DESCRIPTION, CRS_CREDIT)

CLASS(**CRS_CODE**, **CLASS_SECTION**, CLASS_TIME, . . . etc.)

indicate that a strong relationship exists between COURSE and CLASS, because the CLASS entity's composite PK is composed of CRS_CODE + CLASS_SECTION. (Note that the CRS_CODE in CLASS is *also* the FK to the COURSE entity.)

Figure 4.10 shows that the Crow's Foot model depicts the strong (identifying) relationship with a solid line between the entities. Whether the relationship between COURSE and CLASS is strong or weak depends on how the CLASS entity's primary key is defined.

Keep in mind that *the order in which the tables are created and loaded is very important*. For example, in the "COURSE generates CLASS" relationship, the COURSE table must be created before the CLASS table. After all, it would not be acceptable to have the CLASS table's foreign key reference a COURSE table that did not yet exist. In some DBMSs, this sequencing problem does not crop up until the data are loaded into the tables. In fact, *you must always load the "1" side first in a 1:M relationship, to avoid the possibility of referential integrity errors*, regardless of whether the relationships are weak or strong.

As you examine Figure 4.10, you may wonder what the "O" symbol next to the CLASS entity signifies. You will discover the meaning of that cardinality in the next section, 4.1.6, "Relationship Participation."

Remember that the nature of the relationship is often determined by the database designer, who must use professional judgment to determine which relationship type and strength best suit the database transaction, efficiency, and information requirements. We will emphasize this point often and in detail!

4.1.6 RELATIONSHIP PARTICIPATION

Participation in an entity relationship is either optional or mandatory. Participation is optional if one entity occurrence does not *require* a corresponding entity occurrence in a particular relationship. For example, in the "COURSE generates CLASS" relationship, we noted that at least some courses may not generate a class. In other words, an entity occurrence (row) in the COURSE table does not necessarily require the existence of a corresponding entity occurrence in the CLASS table. (Remember that each entity is implemented as a table.) Therefore, the CLASS entity is considered to be *optional* to the COURSE entity. In the Chen and Crow's Foot models, an optional relationship between entities is shown by drawing a small circle (O) on the side of the optional entity, as shown in Figure 4.10. The existence of an *optionality* indicates that the minimum cardinality is 0 for the optional entity. (The term **optionality** is used to label any condition in which one or more optional relationships exist.)

FIGURE 4.10 **A Strong (Identifying) Relationship Between COURSE and CLASS**

Table name: COURSE **Database name: Ch04_TinyCollege_Alt**

	CRS_CODE	DEPT_CODE	CRS_DESCRIPTION	CRS_CREDIT
▶ ⊞	ACCT-211	ACCT	Accounting I	3
⊞	ACCT-212	ACCT	Accounting II	3
⊞	CIS-220	CIS	Intro. to Microcomputing	3
⊞	CIS-420	CIS	Database Design and Implementation	4
⊞	MATH-243	MATH	Mathematics for Managers	3
⊞	QM-261	CIS	Intro. to Statistics	3
⊞	QM-362	CIS	Statistical Applications	4

Table name: CLASS

	CRS_CODE	CLASS_SECTION	CLASS_TIME	ROOM_CODE	PROF_NUM
▶	ACCT-211	1	MWF 8:00-8:50 a.m.	BUS311	105
	ACCT-211	2	MWF 9:00-9:50 a.m.	BUS200	105
	ACCT-211	3	TTh 2:30-3:45 p.m.	BUS252	342
	ACCT-212	1	MWF 10:00-10:50 a.m.	BUS311	301
	ACCT-212	2	Th 6:00-8:40 p.m.	BUS252	301
	CIS-220	1	MWF 9:00-9:50 a.m.	KLR209	228
	CIS-220	2	MWF 9:00-9:50 a.m.	KLR211	114
	CIS-220	3	MWF 10:00-10:50 a.m.	KLR209	228
	CIS-420	1	W 6:00-8:40 p.m.	KLR209	162
	MATH-243	1	Th 6:00-8:40 p.m.	DRE155	325
	QM-261	1	MWF 8:00-8:50 a.m.	KLR200	114
	QM-261	2	TTh 1:00-2:15 p.m.	KLR200	114
	QM-362	1	MWF 11:00-11:50 a.m.	KLR200	162
	QM-362	2	TTh 2:30-3:45 p.m.	KLR200	162

NOTE

Remember that the burden of establishing the relationship is always placed on the entity that contains the foreign key. In most cases, this will be the entity on the "many" side of the relationship.

Relationship participation is **mandatory** if one entity occurrence *requires* a corresponding entity occurrence in a particular relationship. If no optionality symbol is depicted with the entity, that entity exists in a mandatory relationship with the related entity. The existence of a mandatory relationship indicates that the minimum cardinality is 1 for the mandatory entity.

NOTE

You may be tempted to conclude that relationships are weak when they are created between optional entities, and strong when they are created between mandatory entities. However, that conclusion is not warranted. Always keep in mind that relationship participation and relationship strength do not describe the same thing. It is quite possible for you to encounter a strong relationship when one entity is optional to another. For example, the relationship between EMPLOYEE and DEPENDENT is clearly a strong one, but DEPENDENT is just as clearly optional to EMPLOYEE. After all, you cannot *require* employees to have dependents. And it is just as possible for a weak relationship to be established when one entity is mandatory to another. *The relationship strength depends on how the PK of the related entity is formulated, while the relationship participation depends on how the business rule is written.* For example, the business rules

Each part must be supplied by a vendor

and

A part may or may not be supplied by a vendor

create different optionalities for the same entities! Failure to understand this distinction may lead to poor design decisions that cause major problems when table rows are inserted or deleted.

Because relationship participation turns out to be a very important component of the database design process, let's examine a few more scenarios. Suppose that Tiny College employs some professors who conduct research without teaching classes. If we examine the "PROFESSOR teaches CLASS" relationship, it is quite possible for a PROFESSOR not to teach a CLASS. Therefore, CLASS is *optional* to PROFESSOR. On the other hand, a CLASS must be taught by a PROFESSOR. Therefore, PROFESSOR is *mandatory* to CLASS. Note that the Chen model in Figure 4.11 shows the cardinality next to PROFESSOR to be (0,3), thus indicating that a professor may teach no classes at all or as many as three classes. The Crow's Foot model does not specify the exact cardinality limits. Instead, the Crow's Foot model indicates a minimum cardinality of 0 and an unspecified maximum cardinality (M). The Chen model also indicates the number of entity occurrences in the related table. Therefore, the Chen model shows that the CLASS table will have as few as 0 references to PROFESSOR or as many as three. And each CLASS table row will reference one and only one PROFESSOR row—assuming that each class is taught by one and only one professor.

FIGURE 4.11 AN OPTIONAL CLASS ENTITY IN THE RELATIONSHIP PROFESSOR TEACHES CLASS

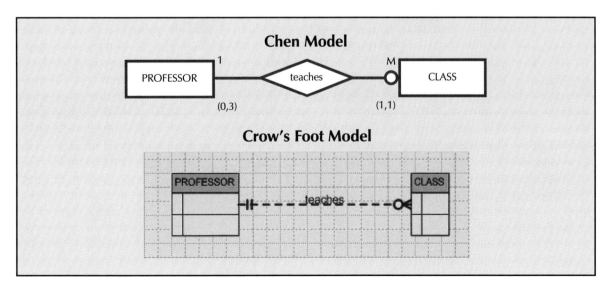

Failure to understand the distinction between *mandatory* and *optional* participation in relationships may yield designs in which awkward (and unnecessary) temporary rows (entity instances) must be created just to accommodate the creation of required entities. Therefore, it is important to clearly understand the concepts of mandatory and optional participation.

It is also important to understand that the semantics of a problem may determine the type of participation in a relationship. For example, suppose that Tiny College offers several courses; each course has several classes. Note again the distinction between *class* and *course* in this discussion: a CLASS constitutes a specific offering (or section) of a COURSE. (Typically, courses are listed in the university's course catalog, while classes are listed in the class schedules that students use to register for their classes.)

Analyzing the CLASS entity's contribution to the "COURSE generates CLASS" relationship, it is easy to see that a CLASS cannot exist without a COURSE. Therefore, we conclude that the COURSE entity is *mandatory* in the relationship. But we may write two scenarios for the CLASS entity, shown in Figures 4.12 and 4.13. The different scenarios are a function of the semantics of the problem; that is, they depend on how the relationship is defined.

1. *CLASS is optional.* It is possible for the department to create the entity COURSE first, and then create the CLASS entity after making the teaching assignments. In the real world, such a scenario is likely; there may be courses for which sections (classes) have not yet been defined. In fact, some courses are taught only once a year and do not generate classes each semester.

FIGURE 4.12 **CLASS IS OPTIONAL TO COURSE**

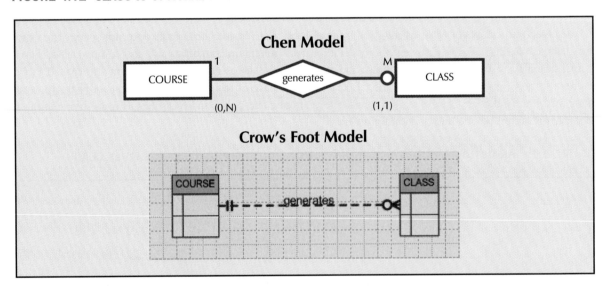

2. *CLASS is mandatory.* This condition is created by the constraint that is imposed by the semantics of the statement "Each COURSE generates one or more CLASSes." In ER terms, each COURSE in the "generates" relationship must have at least one CLASS. Therefore, a CLASS must be created as the COURSE is created, in order to comply with the semantics of the problem.

FIGURE 4.13 **COURSE** AND **CLASS** IN A **MANDATORY RELATIONSHIP**

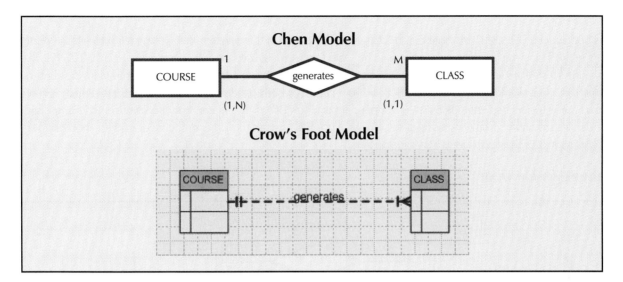

Keep in mind the practical aspects of the scenario presented in Figure 4.13. Given the semantics of this relationship, the system should not accept a course that is not associated with at least one section. Is such a rigid environment desirable from an operational point of view? For example, when we create a new COURSE, the database first updates the COURSE table, thereby inserting a COURSE entity that does not yet have a CLASS associated with it. Naturally, the apparent problem seems to be solved when CLASS entities are inserted into the corresponding CLASS table. However, given the semantics of the mandatory relationship, the system will be in temporary violation of the business rule constraint. For practical purposes, it would be desirable to classify the CLASS as optional, in order to produce a more flexible design.

Finally, as you examine the scenarios presented in Figures 4.12 and 4.13, keep the role of the DBMS in mind. To maintain data integrity, the DBMS must ensure that the "many" side (CLASS) is associated with a COURSE through the foreign key rules. The burden of maintaining data integrity is assigned to the many side rather than to the one side. The DBMS handles the data integrity constraints imposed by the business rules. (A few DBMS implementations, such as Microsoft's SQL Server and Oracle, allow the definition of business rules at the DBMS level.)

4.1.7 RELATIONSHIP STRENGTH AND WEAK ENTITIES

In Section 4.1.5 we introduced the notion of existence dependence and classified the strength of relationships into weak (non-identifying) and strong (identifying). You learned that strong relationships have an effect on how related entities are depicted. In database design terms, the existence of a strong relationship between a parent entity and its related entity or entities is associated with weak entities. Because weak entities may play an important role in the implementation of database design, let's examine them more closely.

A **weak entity** is one that meets two conditions:

1. It is existence-dependent; that is, it cannot exist without the entity with which it has a relationship.
2. It has a primary key that is partially or totally derived from the parent entity in the relationship.

For example, a company insurance policy may insure an employee and his/her dependents. For the purpose of describing an insurance policy, an EMPLOYEE may or may not have a DEPENDENT, but the DEPENDENT must be associated with an EMPLOYEE. Moreover, the DEPENDENT cannot exist without the EMPLOYEE; that is, you cannot get insurance coverage at the XYZ Corporation as a *dependent* unless you happen to be a

dependent of an employee working for the XYZ Corporation! DEPENDENT is the weak entity in the relationship "EMPLOYEE has DEPENDENT."

As you examine Figure 4.14, note that the Chen model identifies the weak entity by using a double-walled entity rectangle. The Crow's Foot model generated by Visio Professional uses the relationship line and the PK/FK designation to indicate whether or not the related entity is weak. A strong (identifying) relationship indicates that the related entity is weak, because such a relationship means that both conditions for the weak entity definition have been met—the related entity is existence-dependent and the PK of the related entity contains a PK component of the parent entity. (Some versions of the Crow's Foot ERD depict the weak entity by drawing a short line segment in each of the four weak entity box corners.)

FIGURE 4.14 A WEAK ENTITY IN AN ERD

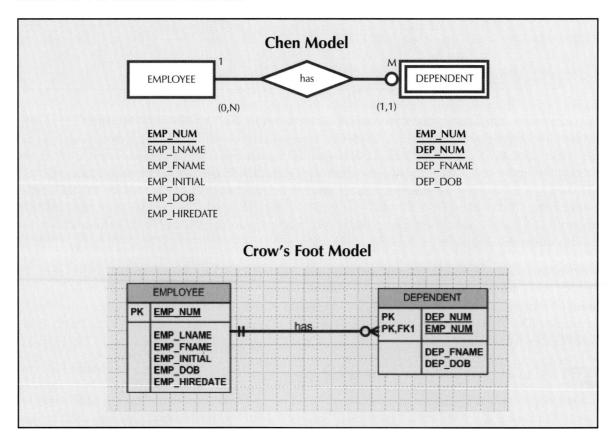

Remember that the weak entity inherits all or part of its primary key from its strong counterpart. For example, at least part of the DEPENDENT entity's key shown in Figure 4.14 was inherited from the EMPLOYEE entity:

EMPLOYEE (**EMP_NUM**, EMP_LNAME, EMP_FNAME, EMP_INITIAL, EMP_DOB, EMP_HIREDATE)

DEPENDENT (**EMP_NUM**, **DEP_NUM**, DEP_FNAME, DEP_DOB)

Figure 4.15 illustrates the implementation of the relationship between the weak entity (DEPENDENT) and its parent or strong counterpart (EMPLOYEE). Note that DEPENDENT's primary key is composed of two attributes: EMP_NUM and DEP_NUM, and that EMP_NUM was inherited from EMPLOYEE. With the help of this relationship, you can determine that:

Jeanine J. Callifante claims two dependents, Annelise and Jorge.

William K. Smithson, Lydia Chen, Melanie Johnson, and Peter O'Donnell claim no dependents.

Herman H. Washington claims one dependent, Suzanne.

Jorge Ortega claims one dependent, Carlos.

Barbara Brzenski claims three dependents, Michael, George, and Katherine.

FIGURE 4.15 A WEAK ENTITY IN A STRONG RELATIONSHIP

Table name: EMPLOYEE Database name: Ch04_ShortCo

EMP_NUM	EMP_LNAME	EMP_FNAME	EMP_INITIAL	EMP_DOB	EMP_HIREDATE
1001	Callifante	Jeanine	J	12-Mar-64	25-May-97
1002	Smithson	William	K	23-Nov-70	28-May-97
1003	Washington	Herman	H	15-Aug-68	28-May-97
1004	Chen	Lydia	B	23-Mar-74	15-Oct-98
1005	Johnson	Melanie		28-Sep-66	20-Dec-98
1006	Ortega	Jorge	G	12-Jul-79	05-Jan-02
1007	O'Donnell	Peter	D	10-Jun-71	23-Jun-02
1008	Brzenski	Barbara	A	12-Feb-70	01-Nov-03

Table name: DEPENDENT

EMP_NUM	DEP_NUM	DEP_FNAME	DEP_DOB
1001	1	Annelise	05-Dec-97
1001	2	Jorge	30-Sep-02
1003	1	Suzanne	25-Jan-04
1006	1	Carlos	25-May-01
1008	1	Michael	19-Feb-95
1008	2	George	27-Jun-98
1008	3	Katherine	18-Aug-03

Keep in mind that *it is the database designer who usually determines whether or not an entity can be described as weak based on the business rules.* For example, an examination of the relationship between COURSE and CLASS in Figure 4.9 may cause you to conclude that CLASS is a weak entity to COURSE. After all, if you examine the CLASS rows in Figure 4.9, it seems clear that a CLASS cannot exist without a COURSE, so there is existence dependency. For example, a student cannot enroll in the Accounting I class ACCT-211, Section 3 (CLASS_CODE 10014) unless there is an ACCT_211 course. However, note that the CLASS table's primary key is CLASS_CODE, which is *not* derived from the COURSE parent entity. That is, we may represent CLASS by:

CLASS (**CLASS_CODE**, CRS_CODE, CLASS_SECTION, CLASS_TIME, ROOM_CODE, PROF_NUM)

The second weak entity requirement has not been met and, therefore, the CLASS entity in Figure 4.9 may not be classified as weak by definition. On the other hand, if the CLASS entity's primary key had been defined as a composite key, composed of the combination CRS_CODE and CLASS_SECTION, we could represent CLASS by:

CLASS (**CRS_CODE**, **CLASS_SECTION**, CLASS_TIME, ROOM_CODE, PROF_NUM)

In this case, illustrated in Figure 4.10, the CLASS primary key is partially derived from COURSE, because CRS_CODE is the COURSE table's primary key. Given this decision, CLASS is a weak entity by definition. (In Visio Professional Crow's Foot terms, the relationship between COURSE and CLASS is classified as strong, or identifying.) In any case, CLASS is always existence-dependent on COURSE, *whether or not it is defined to be weak.*

4.1.8 RELATIONSHIP DEGREE

A relationship's **degree** indicates the number of associated entities or participants. A **unary relationship** exists when an association is maintained within a single entity. A **binary relationship** exists when two entities are associated. A **ternary relationship** exists when three entities are associated. Although higher degrees exist, they are rare and are not specifically named. (For example, an association of four entities is simply described as a *four-entity relationship*.) We have shown a few types of relationships in Figure 4.16. (The Crow's Foot model will not let you implement M:N relationships, so we have shown the unary relationship by using a second copy of the COURSE entity. This display also properly recognizes that the relationship is defined in both directions: many COURSEs may be prerequisites to other COURSEs and a COURSE may require many COURSE prerequisites.)

As you compare the Chen and Crow's Foot ERDs in Figure 4.16, note that the implementation focus of the Crow's Foot model requires that you translate the CFR (contributor-fund-recipient) *relationship* into the CFR *entity* through which the relationship is implemented. The conceptual orientation of the Chen model makes it easier to see that the CFR is, in fact, a ternary *relationship*.

In the case of the unary relationship shown in Figure 4.16, a course within the COURSE entity is a prerequisite for another course within that entity. In this case, the existence of a course prerequisite means that a COURSE requires a COURSE—that is, COURSE has a relationship with itself. Such a relationship is known as a **recursive relationship**.

> **N O T E**
>
> Binary relationships are the most common. In fact, to simplify the conceptual design, most higher-order (ternary and higher) relationships are decomposed into appropriate equivalent binary relationships whenever possible.

FIGURE 4.16 THREE TYPES OF RELATIONSHIPS

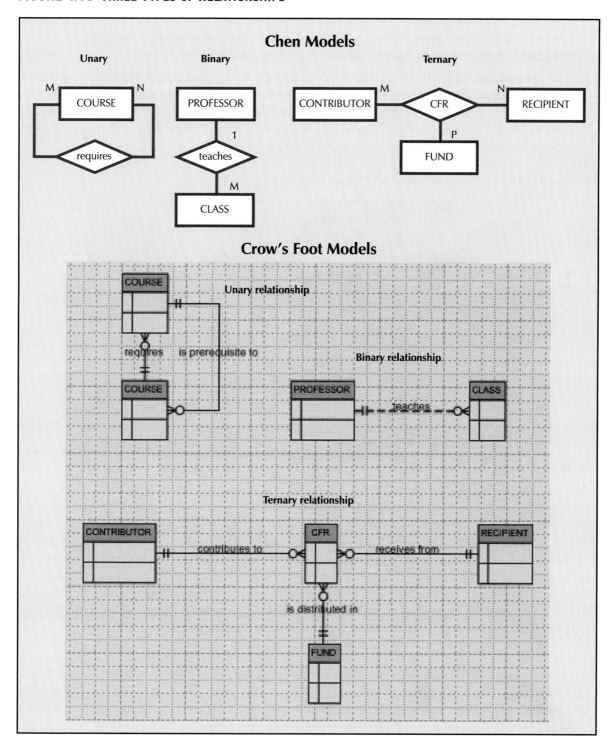

Although most relationships are binary, the use of ternary and higher-order relationships does allow the designer some latitude regarding the semantics of a problem. For example, note the relationships (and their consequences!) in Figure 4.17:

1. People or institutions in the CONTRIBUTOR group donate money to a special research FUND. (The research funds may be placed into several categories. For example, if the fund is designed to support medical research, the categories may include cancer, AIDS, heart disease, and so on.)

2. The researchers who are found in the RECIPIENT entity are funded through the FUND contents.

3. The relationships are all classified as M:N. For example, contributors can make donations to many funds. A fund can have many contributors. A fund may support many researchers, who become the fund's recipients, and a researcher may draw support from many funds.

4. The three entities CONTRIBUTOR, RECIPIENT, and FUND are related through the ternary relationship we have labeled CFR. Because a single verb cannot express a ternary relationship, we elected to use the acronym CFR to indicate its role in defining the relationships among the <u>C</u>ONTRIBUTOR, the <u>F</u>UND, and the <u>R</u>ECIPIENT. Thus the CFR entity enables the database end user to match the fund's contributors with the recipients of the fund, as shown in Figure 4.17.

FIGURE 4.17 THE IMPLEMENTATION OF A TERNARY RELATIONSHIP

Database name: Ch04_MedCo

Table name: CONTRIBUTOR

CONTRIB_ID	CONTRIB_LNAME
C1	Brown
C2	Iglesas
C3	Smith

Table name: FUND

FUND_ID	FUND_NAME	CONTRIB_ID	FUND_AMOUNT
F1	Heart	C1	$50,000.00
F1	Heart	C2	$10,000.00
F2	Cancer	C1	$10,000.00
F2	Cancer	C2	$5,000.00
F2	Cancer	C3	$10,000.00

Table name: RECIPIENT

REC_ID	REC_TYPE
R1	Rogers
R2	Chen
R3	Oshanski

Table name: CFR

FUND_ID	CON_ID	REC_ID	CFR_AMOUNT
F1	C1	R2	$30,000.00
F1	C1	R3	$20,000.00
F1	C2	R2	$10,000.00
F2	C1	R1	$10,000.00
F2	C2	R1	$5,000.00

As you examine the table contents in Figure 4.17, note that it is possible to track all transactions. For instance, researcher R1, specializing in cancer research, has received a total of $15,000, divided as follows: $10,000 from contributor C1 (Brown) and $5,000 from contributor C2 (Iglesas). Researcher R2, specializing in heart research, received $40,000: $30,000 of which came from contributor C1 and $10,000 of which came from contributor C2. In addition, the CFR (ternary) relationship allows us to relate the available funds to the contributors. For example, the table contents indicate that contributor C1 donated $50,000 to heart research, $30,000 of which was received by researcher R2, while the remaining $20,000 was received by researcher R4. Note that the Crow's Foot model in Figure 4.16 already shows that the ternary relationship must be implemented as an entity, thus making it easier to trace the transaction issues we have just discussed.

The ternary relationship described here has an effect on the type and extent of the data available to a database query. For example, if 500 people make donations to the heart portion of the fund and we do not use a ternary relationship, there is no way to identify the specific source of those funds when the recipient receives a heart grant. Therefore, if it is necessary to keep track of specific contributors to the fund and the recipients of those funds, the ternary relationship is necessary. As the preceding example demonstrates, you should keep in mind that a ternary relationship is not always equivalent to several 1:M relationships. Consequently, the designer must consider the semantics of the problem that is being addressed; desirable simplifications may not match end user needs!

Recursive Relationships

As we mentioned, a *recursive relationship* is one in which a relationship can exist between occurrences of the same entity set. (Naturally, such a condition is found within a unary relationship.)

For example, a 1:M unary relationship can be expressed by "an EMPLOYEE may manage many EMPLOYEEs, and each EMPLOYEE is managed by one EMPLOYEE." And, as long as polygamy is not legal, a 1:1 unary relationship may be expressed by "an EMPLOYEE may be married to one and only one other EMPLOYEE." Finally, the M:N unary relationship may be expressed by "a COURSE may be a prerequisite to many other COURSEs, and each COURSE may have many other COURSEs as prerequisites." These relationships are shown in Figure 4.18.

FIGURE 4.18 **AN ER REPRESENTATION OF RECURSIVE RELATIONSHIPS**

The 1:1 relationship shown in Figure 4.18 can be implemented in the single table shown in Figure 4.19. Note that we can determine that James Ramirez is married to Louise Ramirez, who is married to James Ramirez. And Anne Jones is married to Anton Shapiro, who is married to Anne Jones.

FIGURE 4.19 **THE 1:1 RECURSIVE RELATIONSHIP "EMPLOYEE IS MARRIED TO EMPLOYEE"**

Table name: EMPLOYEE_V1 Database name: Ch04_PartCo

EMP_NUM	EMP_LNAME	EMP_FNAME	EMP_SPOUSE
345	Ramirez	James	347
346	Jones	Anne	349
347	Ramirez	Louise	345
348	Delaney	Robert	
349	Shapiro	Anton	346

Unary relationships are common in manufacturing industries. For example, Figure 4.20 illustrates that a rotor assembly (C-130) is composed of many parts, but each part is used to create only one rotor assembly. Figure 4.20 indicates that a rotor assembly is composed of four 2.5-cm washers, two cotter pins, one 2.5-cm steel shank,

four 10.25-cm rotor blades, and two 2.5-cm hex nuts. The relationship implemented in Figure 4.20 thus enables us to track each part within each rotor assembly we manufacture.

FIGURE 4.20 ANOTHER UNARY RELATIONSHIP: "PART CONTAINS PART"

Table name: PART_V1 Database name: Ch04_PartCo

PART_CODE	PART_DESCRIPTION	PART_IN_STOCK	PART_UNITS_NEEDED	PART_OF_PART
AA21-6	2.5 cm. washer, 1.0 mm. rim	432	4	C-130
AB-121	Cotter pin, copper	1034	2	C-130
C-130	Rotor assembly	36		
E129	2.5 cm. steel shank	128	1	C-130
X10	10.25 cm. rotor blade	345	4	C-130
X34AW	2.5 cm. hex nut	879	2	C-130

If a part can be used to assemble several different kinds of other parts and is itself composed of many parts, two tables are required to implement the "PART contains PART" relationship. Figure 4.21 illustrates such an environment. Parts tracking has become increasingly important as managers become more aware of the legal ramifications of producing more complex output. In fact, in many industries, especially those involving aviation, full parts tracking is mandatory.

FIGURE 4.21 IMPLEMENTATION OF THE M:N RECURSIVE "PART CONTAINS PART" RELATIONSHIP

Table name: COMPONENT Database name: Ch04_PartCo

COMP_CODE	PART_CODE	COMP_PARTS_NEEDED
C-130	AA21-6	4
C-130	AB-121	2
C-130	E129	1
C-131A2	E129	1
C-130	X10	4
C-131A2	X10	1
C-130	X34AW	2
C-131A2	X34AW	2

Table name: PART

PART_CODE	PART_DESCRIPTION	PART_IN_STOCK
AA21-6	2.5 cm. washer, 1.0 mm. rim	432
AB-121	Cotter pin, copper	1,034
C-130	Rotor assembly	36
E129	2.5 cm. steel shank	128
X10	10.25 cm. rotor blade	345
X34AW	2.5 cm. hex nut	879

The M:N recursive relationship might be more familiar in a school environment. For instance, note how the M:N "COURSE requires COURSE" relationship illustrated in Figure 4.18 is implemented in Figure 4.22. In this example, MATH-243 is a prerequisite to QM-261 and QM-362, while both MATH-243 and QM-261 are prerequisites to QM-362.

FIGURE 4.22 IMPLEMENTATION OF THE M:N "COURSE REQUIRES COURSE" RECURSIVE RELATIONSHIP

Table name: COURSE Database name: Ch04_TinyCollege

	CRS_CODE	DEPT_CODE	CRS_DESCRIPTION	CRS_CREDIT
▶ ⊞	ACCT-211	ACCT	Accounting I	3
⊞	ACCT-212	ACCT	Accounting II	3
⊞	CIS-220	CIS	Intro. to Microcomputing	3
⊞	CIS-420	CIS	Database Design and Implementation	4
⊞	MATH-243	MATH	Mathematics for Managers	3
⊞	QM-261	CIS	Intro. to Statistics	3
⊞	QM-362	CIS	Statistical Applications	4

Table name: PREREQ

	CRS_CODE	PRE_TAKE
▶	CIS-420	CIS-220
	QM-261	MATH-243
	QM-362	MATH-243
	QM-362	QM-261

Finally, the 1:M recursive relationship "EMPLOYEE manages EMPLOYEE," shown in Figure 4.18, is implemented in Figure 4.23.

FIGURE 4.23 IMPLEMENTATION OF THE 1:M "EMPLOYEE MANAGES EMPLOYEE" RECURSIVE RELATIONSHIP

Table name: EMPLOYEE_V2 Database name: Ch04_PartCo

	EMP_CODE	EMP_LNAME	EMP_MANAGER
▶	101	Waddell	102
	102	Orincona	
	103	Jones	102
	104	Reballoh	102
	105	Robertson	102
	106	Deltona	102

4.1.9 COMPOSITE ENTITIES

In the original ER model described by Chen, relationships do not contain attributes. You should recall from Chapter 3 that the relational model generally requires the use of 1:M relationships. (You also should recall that the 1:1 relationship has its place, but it should be used with caution and proper justification.) If M:N relationships are encountered, we must create a bridge between the entities that display such relationships. Recall that the *bridge entity* (also known as a *composite* entity) is composed of the primary keys of each of the entities to be connected. (An example of such a bridge is shown in Figure 4.24.) A composite entity in the Chen model is represented by a diamond shape within a rectangle. Because the Crow's Foot model is implementation oriented, it does not identify the composite entity as such. Instead, the composite entity is identified by the solid relationship line between the parent and child entities, thereby indicating the presence of a strong (identifying) relationship.

As you examine Figure 4.24, note that the composite ENROLL entity is existence-dependent on the other two entities; its composition is based on the primary keys of the entities that are connected by the composite entity. The composite entity may also contain additional attributes that play no role in the connective process. For example, although the entity must be composed of at least the STUDENT and CLASS primary keys, it may also include such additional attributes as grades, absences, and other data uniquely identified by the student's performance in a specific class.

FIGURE 4.24 CONVERTING THE M:N RELATIONSHIP INTO TWO 1:M RELATIONSHIPS

Table name: **STUDENT** Database name: **Ch04_CollegeTry**

		STU_NUM	STU_LNAME
▶	+	321452	Bowser
	+	324257	Smithson

Table name: **ENROLL**

	CLASS_CODE	STU_NUM	ENROLL_GRADE
▶	10014	321452	C
	10014	324257	B
	10018	321452	A
	10018	324257	B
	10021	321452	C
	10021	324257	C

Table name: **CLASS**

		CLASS_CODE	CRS_CODE	CLASS_SECTION	CLASS_TIME	CLASS_ROOM	PROF_NUM
▶	+	10014	ACCT-211	3	TTh 2:30-3:45 p.m.	BUS252	342
	+	10018	CIS-220	2	MWF 9:00-9:50 a.m.	KLR211	114
	+	10021	QM-261	1	MWF 8:00-8:50 a.m.	KLR200	114

Finally, keep in mind that the ENROLL table's key (CLASS_CODE and STU_NUM) is entirely composed of the primary keys of the CLASS and STUDENT tables. Therefore, no null entries are possible in the ENROLL table's key attributes.

Implementing the small database shown in Figure 4.24 requires that we define the relationships clearly. Specifically, we must know the 1 and the M sides of each of the relationships, and we must know whether the relationships are mandatory or optional. For example, it is worth noting the following points:

- A class may exist (at least at the start of registration) even though it contains no students at all. Therefore, if we examine Figure 4.25, an optional symbol should appear on the STUDENT side of the M:N relationship between STUDENT and CLASS.

 You might argue that to be classified as a STUDENT, a person must be enrolled in at least one CLASS. Therefore, CLASS is mandatory to STUDENT from a purely conceptual point of view. However, when a student is admitted to college, that student has not (yet) signed up for any classes. Therefore, at *least initially*, CLASS is optional to STUDENT. Note that the practical considerations in the data environment help dictate the use of optionalities. If CLASS is *not* optional to STUDENT—from a database point of view—a class assignment must be made when the student is admitted. And that's *not* how the process actually works. In short, the optionality reflects practice.

 Because the M:N relationship between STUDENT and CLASS is decomposed into two 1:M relationships through ENROLL, the optionalities must be transferred to ENROLL (see Figure 4.26). In other words, it now becomes possible for a class not to occur in ENROLL if no student has yet signed up for that class. Because a class need not occur in ENROLL, the ENROLL entity becomes optional to CLASS. And, because the ENROLL entity is created before any students have signed up for a class, the ENROLL entity is optional to STUDENT, too. (At least initially!)

FIGURE 4.25 THE M:N RELATIONSHIP BETWEEN STUDENT AND CLASS

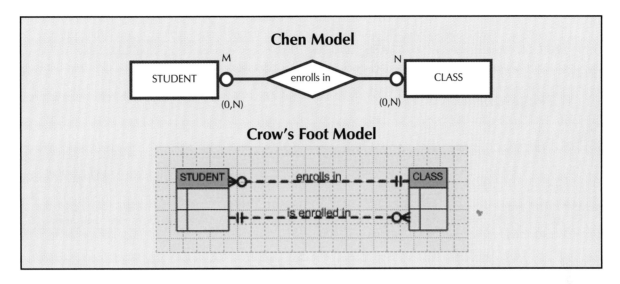

FIGURE 4.26 A COMPOSITE ENTITY IN AN ERD

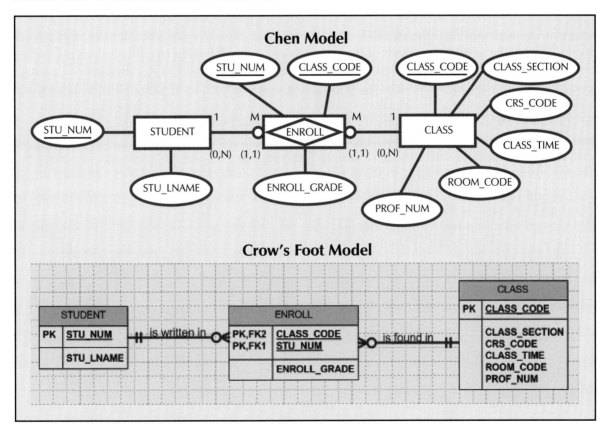

- As students begin to sign up for their classes, they will be entered into the ENROLL entity. Naturally, if the student takes more than one class, that student will occur more than once in ENROLL. For example, note that STU_NUM = 321452 occurs three times in the ENROLL table in Figure 4.24. On the other hand, each student occurs only once in the STUDENT entity. (Note that the STUDENT table in Figure 4.24 has

only one STU_NUM = 321452 entry in it.) Therefore, the relationship between STUDENT and ENROLL is shown to be 1:M in Figure 4.26, with the M on the ENROLL side.

- If you examine the tables shown in Figure 4.24, you will see that a class can occur more than once in the ENROLL table. For example, CLASS_CODE = 10014 occurs twice in the ENROLL table. However, CLASS_CODE = 10014 occurs only once in the CLASS table, to reflect that the relationship between CLASS and ENROLL is 1:M. Note that, in Figure 4.26, the M is located on the ENROLL side, while the 1 is located on the CLASS side.

N O T E

The ERD in Figure 4.26 contains all the information required to model the simplified student enrollment process. However, the inclusion of the attributes tends to clutter a diagram, especially when an ER model contains many entities, and each of those entities contains many attributes. Therefore, although the ER format shown in Figure 4.26 serves a useful teaching role, database designers generally use ER models that do not include the attribute ellipses. Most current Chen ERDs contain entity boxes that list the attributes just as the Crow's Foot entity boxes do.

4.1.10 ENTITY SUPERTYPES AND SUBTYPES

Most businesses employ people with a wide range of skills and special qualifications. For example, an aviation business might employ pilots and many other types of employees. Describing the different types of employees within a single entity would be awkward at best. For one thing, pilots (for example) must meet special requirements such as minimum flight hours, training flight checks, and so on, that are not required of other employees. So, if all employee characteristics and special qualifications were described by a single EMPLOYEE entity and stored as shown in Figure 4.27, you would either have a lot of nulls or you would have to make a lot of needless dummy entries. For example, if employees Lewis, Lange, Williams, Duzak, and Travis are pilots, Figure 4.27 illustrates that their special pilot characteristics, EMP_LICENSE, EMP_RATINGS, and EMP_MED_TYPE, would cause a large number of nulls for other employees who are not pilots. A pilot must have a periodically renewed medical certification type (EMP_MED_TYPE) to fly legally. For example, a Type I medical certificate is required to exercise the privilege of an airline transport pilot's (ATP) license, a Type II is required for a commercial license, and a Type III for a private pilot's license.

FIGURE 4.27 NULLS CREATED BY UNIQUE ATTRIBUTES

Table name: EMPLOYEE_V1 Database name: Ch04_AirCo

EMP_NUM	EMP_LNAME	EMP_LICENSE	EMP_RATINGS	EMP_MED_TYPE	EMP_HIRE_DATE
100	Kolmycz				15-Mar-88
101	Lewis	ATP	SEL/MEL/Instr/CFII	1	25-Apr-89
102	Vandam				20-Dec-93
103	Jones				28-Aug-03
104	Lange	ATP	SEL/MEL/Instr	1	20-Oct-97
105	Williams	COM	SEL/MEL/Instr/CFI	2	08-Nov-97
106	Duzak	COM	SEL/MEL/Instr	2	05-Jan-04
107	Diante				02-Jul-97
108	Wiesenbach				18-Nov-95
109	Travis	COM	SEL/MEL/SES/Instr/CFII	1	14-Apr-01
110	Genkazi				01-Dec-03

Clearly, pilots share certain characteristics with other employees, such as a last name (EMP_LNAME) and hire date (EMP_HIRE_DATE). On the other hand, there are many characteristics that are not shared. It is these unshared characteristics that create problems when you try to store all employee attributes in the same table. Fortunately, these problems are solved easily by using a generalization hierarchy to represent entity sets that share common characteristics.

Basically, the generalization hierarchy depicts the parent-child relationship we encountered in the discussion of the hierarchical database model in Chapter 2. In a relational database context, the **generalization hierarchy** depicts a relationship between a higher-level supertype entity and a lower-level subtype entity (see Figure 4.28). In other words, the **supertype** entity contains the shared attributes, while the **subtype** entity contains the unique attributes.

FIGURE 4.28 **A GENERALIZATION HIERARCHY**

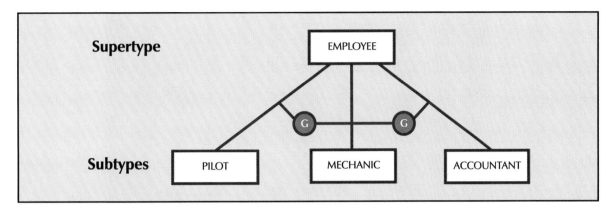

NOTE

Although it might be tempting to argue that the nulls in the EMPLOYEE table in Figure 4.27 could be eliminated by using flags, such a solution is not feasible when the table is expanded to include pilot-specific numeric values such as total flight hours, year-to-date flight hours, and so on. If you use a flag of -999 or 0 (or any other numeric flag) for non-pilots, then computed values such as average flight hours, average year-to-date flight hours, and minimum and maximum flight hours become meaningless. A possible solution to this problem might be to create a job classification attribute, and then perform all computations subject to this attribute limitation. This solution would yield correctly computed numeric values, but it would also lead to more complex queries and a likely proliferation of job classification attribute values and different types of classification attributes. Ultimately, the designer must use professional judgment about the approach to take. We have found the supertype/subtype approach useful in our design work.

As you examine Figure 4.28, note that its relationship is one of inheritance. That is, a subtype entity inherits its attributes and its relationships from the supertype entity. For example, pilots, mechanics, and accountants all have names, home addresses, and so on. However, as you discovered earlier, employees can also have attributes that are unique to their craft. In other words, the supertype entity set is related to several unique subtype entity sets known as disjoint subtypes. **Disjoint subtypes**, also known as **non-overlapping subtypes**, are subtypes that contain a subset of the supertype entity set, but each entity instance (row) of the supertype can appear in only one of the disjoint subtypes. For example, in Figure 4.28, an employee (supertype) who is a pilot (subtype) can only appear in the PILOT subtype, not in any of the other disjoint subtypes. Such disjoint subtypes are indicated by the symbol "G".

The supertype and its subtype(s) maintain a 1:1 relationship. For example, the generalization hierarchy lets us replace the undesirable EMPLOYEE table structure in Figure 4.27 with two tables, one representing the supertype EMPLOYEE and the other representing the subtype PILOT. (See Figure 4.29.)

FIGURE 4.29 THE EMPLOYEE/PILOT SUPERTYPE/SUBTYPE RELATIONSHIP

Table name: EMPLOYEE (the supertype) **Database name: Ch04_AirCo**

	EMP_NUM	EMP_LNAME	EMP_HIRE_DATE
	100	Kolmycz	15-Mar-88
	101	Lewis	25-Apr-89
	102	Vandam	20-Dec-93
	103	Jones	28-Aug-03
	104	Lange	20-Oct-97
	105	Williams	08-Nov-97
	106	Duzak	05-Jan-04
	107	Diante	02-Jul-97
	108	Wiesenbach	18-Nov-95
	109	Travis	14-Apr-01
	110	Genkazi	01-Dec-03

Table name: PILOT (the subtype)

	EMP_NUM	PIL_LICENSE	PIL_RATINGS	PIL_MED_TYPE
	101	ATP	SEL/MEL/Instr/CFII	1
	104	ATP	SEL/MEL/Instr	1
	105	COM	SEL/MEL/Instr/CFI	2
	106	COM	SEL/MEL/Instr	2
	109	COM	SEL/MEL/SES/Instr/CFII	1

Some supertypes contain overlapping subtypes. Overlapping subtypes are subtypes that contain a subset of the supertype entity set, but each entity instance (row) of the supertype can appear in more than one subtype. For example, an employee may be a professor as well as an administrator, because a department chair may also teach courses. Similarly, a person may be an employee and a student. Because an employee also may be a student, STUDENT and EMPLOYEE are overlapping subtypes of the supertype PERSON, just as PROFESSOR and ADMINISTRATOR are overlapping subtypes of the supertype EMPLOYEE. For example, in Figure 4.30, an employee (supertype) could be both a professor and an administrator. Such overlapping subtypes are labeled with "Gs" in Figure 4.30.

N O T E

We have adopted the supertype/subtype symbols G and Gs used by Toby J. Teorey, *Database Modeling and Design: The Entity Relationship Approach*, Morgan Kaufmann, 1990.

As you examine Figures 4.28 and 4.30, note that the classifications G and Gs are usually determined by business rules. For example, if a business rule specifies that an employee can be a pilot and a mechanic at the same time, Figure 4.28 would use Gs to label the PILOT and MECHANIC subtypes.

FIGURE 4.30 A GENERALIZATION HIERARCHY WITH OVERLAPPING SUBTYPES

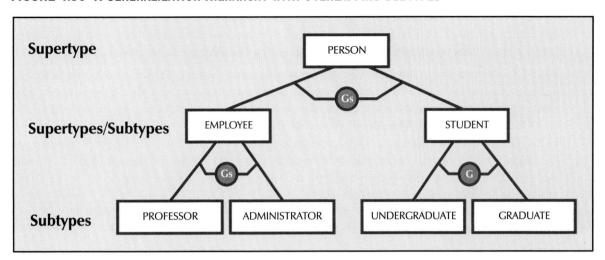

4.2 A COMPARISON OF ER MODELING SYMBOLS

The ER connectivity and cardinality symbols we have used thus far are typical of the Chen and Crow's Foot models. However, you should be aware of other ER styles, including the Rein85 and the IDEF1X. Although these models are all based on the same modeling concepts, the alternate styles were developed because they fit computer-based modeling tools more easily than the original Chen model. It is quite likely that you will convert even the finest Chen model into these (or very similar) models when you are using **computer-assisted systems engineering (CASE)** tools in the database environment. The following list summarizes the major characteristics of each model.

- The *Chen model* is based on Peter Chen's landmark work "The Entity-Relationship Model: Toward a Unified View of Data," *ACM Transactions on Database Systems* 1(1), March, 1976. The Chen model moved conceptual modeling into the practical database design arena by establishing the basic building blocks: entities and relationships. Chen's work was enhanced by T. J. Teorey, D. Yang, and J. P. Fry when they published "A Logical Design Methodology for Relational Databases Using the Extended Entity Relationship Model," *ACM Computing Surveys*, June, 1986, pp. 197-222. The Chen model's basic structure, with the enhancements made by Teorey, Yang, and Fry, became a dominant player in the CASE tool market during the late 1980s and early 1990s. (See especially Mike Ricciuti's "Database Vendors Make Their CASE," *Datamation* 38(5), March 1992.) Excelerator, a CASE tool of choice for many database designers in the early 1990s, is perhaps the best "pure" Chen modeling tool. Although the Chen model is no longer the dominant ERD generator, all current ERD tools find their conceptual origins in the Chen model.

- The *Crow's Foot model*, developed by C. W. Bachman, was made popular by the Knowledgeware modeling tool. Rather than using the Chen-style connectivity designations 1 and M and cardinality designations such as (0,1), (0,N), (1,1), and (1,N), the Crow's Foot approach uses the sticklike symbols illustrated in Figure 4.31. (The name "Crow's Foot" is a reflection of the symbol used for the connectivity M, which resembles a three-toed bird's foot.) The Crow's Foot model combines connectivity and cardinality information in a single symbol set. Unlike the Chen methodology, the Crow's Foot model cannot detail cardinalities other than 0, 1, or N. For example, the cardinality (5,25) cannot be shown in a Crow's Foot model. However, the commercial modeling tools that use the Crow's Foot—such as Microsoft Visio Professional—allow you to define such cardinalities in a data dictionary.

- The **Rein85 model** was developed by D. Reiner in 1985. Although it is based on the same modeling conventions as the Crow's Foot, its symbols are quite different, and it operates at a higher level of abstraction than the Crow's Foot. The Rein85 methodology does not recognize cardinalities explicitly, relying on connectivities to lead to logical cardinality conclusions. The Rein85 symbols are displayed in Figure 4.31.

- **IDEF1X** is a derivative of the integrated computer-aided manufacturing (ICAM) studies that were conducted in the late 1970s. ICAM became the source of graphical methods for defining the functions, data structures, and dynamics of manufacturing businesses. The integration of these methods became known as IDEF (ICAM DEFinition). The original version of IDEF, developed by Hughes Aircraft, became known as IDEF1. The extended version of IDEF1, known as IDEF1X, became the U.S. Air Force standard and has gained acceptance as a general manufacturing data-modeling tool. As you examine Figure 4.31, note that IDEF1X uses fewer symbols than the other modeling methods, thus providing fewer explicit details of the type and extent of the relationships being modeled.

FIGURE 4.31 A COMPARISON OF ER MODELING SYMBOLS

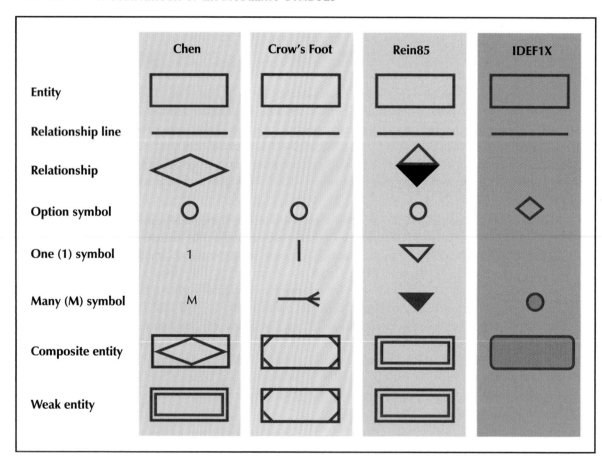

> **NOTE**
>
> Figure 4.31 shows Crow's Foot composite and weak entities to reflect the early implementations of this model. However, modern modeling tools such as Microsoft Visio Professional do not depict the composite and weak entities. Instead, the existence of weak and composite entities is inferred from the relationship lines, which are solid when the relationship between parent and child entities is strong or identifying. In addition, the nature of the entity can be established by examining the PK/FK depictions. Therefore, the special weak/composite depictions are redundant in the Crow's Foot model.

To illustrate the use of the four methods, let us examine the invoicing model we discussed in Chapter 3, Section 3.7. Make sure you pay particular attention to the contents of the tables shown in Figure 3.28 and its relational schema presented in Figure 3.29. We will also use the following business rules:

- In the following invoicing model, we will assume that the CUSTOMER list contains potential as well as actual customers. Therefore, a customer on this list may not (yet) have made any purchase that would generate an invoice. In other words, INVOICE is optional to CUSTOMER.

- Some products kept in inventory are never sold and may, therefore, never show up in an invoice. Therefore, INVOICE is optional to PRODUCT. However, keep in mind that the M:N relationship between INVOICE and PRODUCT is implemented through LINE, in order to decompose the M:N relationship into two 1:M relationships. Thus LINE becomes optional to PRODUCT, because an unsold product will never appear in an invoice line.

Using the information in Section 3.7 of Chapter 3, plus the preceding business rules, the four modeling methods are illustrated in Figures 4.32 through 4.35.

FIGURE 4.32 **THE CHEN REPRESENTATION OF THE INVOICING PROBLEM**

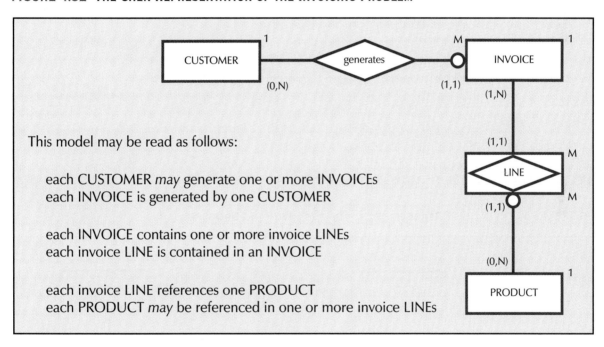

FIGURE 4.33 THE CROW'S FOOT REPRESENTATION OF THE INVOICING PROBLEM

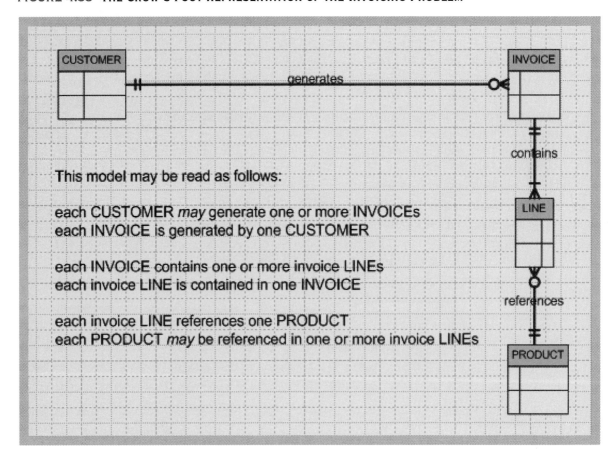

This model may be read as follows:

each CUSTOMER *may* generate one or more INVOICEs
each INVOICE is generated by one CUSTOMER

each INVOICE contains one or more invoice LINEs
each invoice LINE is contained in one INVOICE

each invoice LINE references one PRODUCT
each PRODUCT *may* be referenced in one or more invoice LINEs

FIGURE 4.34 THE REin85 REPRESENTATION OF THE INVOICING PROBLEM

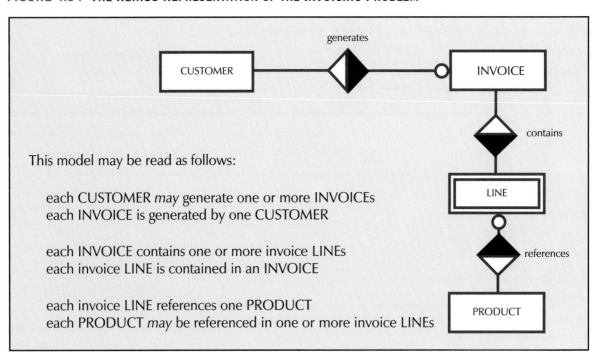

This model may be read as follows:

each CUSTOMER *may* generate one or more INVOICEs
each INVOICE is generated by one CUSTOMER

each INVOICE contains one or more invoice LINEs
each invoice LINE is contained in an INVOICE

each invoice LINE references one PRODUCT
each PRODUCT *may* be referenced in one or more invoice LINEs

FIGURE 4.35 THE IDEF1X REPRESENTATION OF THE INVOICING PROBLEM

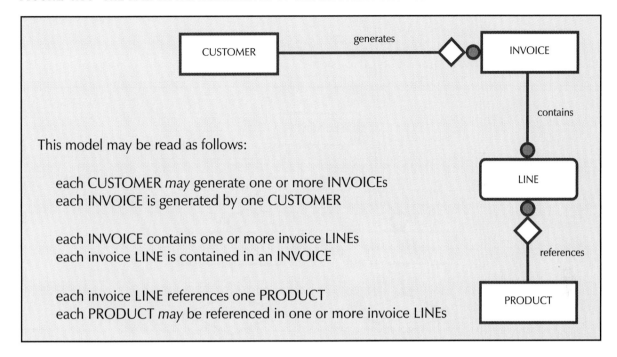

This model may be read as follows:

each CUSTOMER *may* generate one or more INVOICEs
each INVOICE is generated by one CUSTOMER

each INVOICE contains one or more invoice LINEs
each invoice LINE is contained in an INVOICE

each invoice LINE references one PRODUCT
each PRODUCT *may* be referenced in one or more invoice LINEs

4.3 DEVELOPING AN ER DIAGRAM

The process of database design is an iterative rather than a linear or sequential process. The verb *iterate* literally means "to do again or repeatedly." An **iterative process** is thus one based on repetition of processes and procedures. For example, building an ER diagram usually begins with a general narrative of the organization's operations and procedures. Next, the basic ER model is graphically depicted and is presented for review. During the review process it is likely that additional objects, attributes, and relationships will be uncovered. Therefore, the basic ER model will be modified to incorporate the newly discovered ER components. Subsequently, another round of reviews may yield additional components or the clarification of the existing diagram. The process is repeated until the end users and designers agree that the ER diagram is a fair representation of the organization's activities and functions.

During the design process, the database designer does not simply depend on interviews to help define entities, attributes, and relationships. A surprising amount of information can be gathered by examining the business forms and reports that an organization uses in its daily operations.

To illustrate the use of the iterative process that ultimately yields a workable ER diagram, suppose we start with an initial interview with the Tiny College administrators. The interview process yields a basic description of the college's activities.

1. Tiny College (TC) is divided into several schools: a school of business, a school of arts and sciences, a school of education, and a school of applied sciences. Each school is administered by a dean. Note that the cardinality can be expressed by (1,1) for the entity DEAN and by (1,1) for the entity SCHOOL. (The smallest number of deans per school is one, as is the largest number, and each dean is assigned to only one school.) Therefore, a 1:1 relationship exists between DEAN and SCHOOL.

 There are, of course, several ways to evaluate the relationship between DEAN and SCHOOL. Each dean is a member of a larger group known as administrators, so we may include deans in an ADMINISTRATOR entity. However, deans also hold professorial rank and may teach a class. Therefore, deans may also be included within the entity PROFESSOR. In addition, administrators and professors are also

employees. So, depending on the information requirements and the database designer's view of those requirements, it might be useful to evaluate the feasibility of creating the supertype/subtype relationship shown in the ER segment in Figure 4.36.

FIGURE 4.36 A SUPERTYPE/SUBTYPE RELATIONSHIP

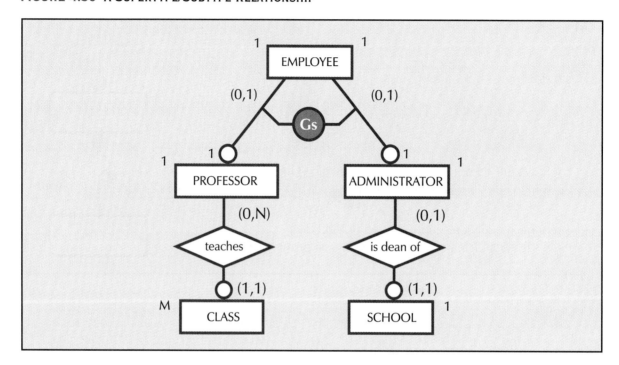

Always keep in mind that good database design requires the designer to consider the information requirements and the organization's business rules. Given Tiny College's operational environment, the professional characteristics (academic degrees, academic rank, etc.) of its administrators and professors are identical. And, because none of the other employees (secretaries, custodians, drivers, mechanics, etc.) is required to have such academic attributes, the relevant supertype/subtype relationship is reduced to professor/employee. This supertype/subtype relationship may be represented at the implementation level by the ER segment shown in Figure 4.37.

Because most DBMSs do not support the supertype/subtype relationship directly, designers convert it into a 1:1 relationship, as shown in Figure 4.37, in which a PROFESSOR is an EMPLOYEE. In this case, not all employees are professors, so PROFESSOR is optional to EMPLOYEE. (In other words, an employee is not required to be a professor.)

FIGURE 4.37 A SUPERTYPE/SUBTYPE RELATIONSHIP IN AN ERD

Chen Model

Crow's Foot Model

NOTE

The existence of 1:1 relationships often indicates that attributes are misidentified as entities, so take particular care in evaluating the need for such 1:1 relationships. In this case, the 1:1 relationship is *properly* defined by the nature of the supertype/subtype relationship.

2. Each school is composed of several departments. For example, the school of business has an accounting department, a management/marketing department, an economics/finance department, and a computer information systems department. Note again the cardinality rules: The smallest number of departments operated by a school is one, and the largest number of departments is indeterminate (N). On the other hand, each department belongs to only a single school; thus, the cardinality is expressed by (1,1). That is, the minimum number of schools that a department belongs to is one, as is the maximum number.

Given these basic descriptions of the college's activities, we can draw the ER diagram segment shown in Figure 4.38. (Remember from Figure 4.37 that PROFESSOR is optional to EMPLOYEE—not all employees are professors. Note also that the PROFESSOR entity is existence-dependent on the EMPLOYEE entity *and* it inherits its PK from the EMPLOYEE entity. Therefore, the relationship between EMPLOYEE and PROFESSOR is strong, while PROFESSOR is a weak entity.)

FIGURE 4.38 THE FIRST TINY COLLEGE ERD SEGMENT

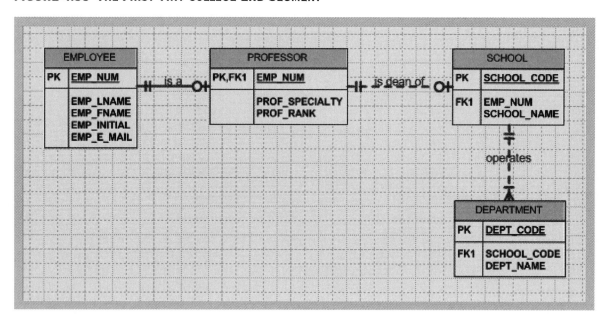

> **NOTE**
>
> It is (again) appropriate to evaluate the reason for maintaining the 1:1 relationship between PROFESSOR and SCHOOL in the "is dean of" relationship. It is worth repeating that the existence of 1:1 relationships often indicates a misidentification of attributes as entities. In this case, we could easily eliminate the 1:1 relationship by storing the dean's attributes in the SCHOOL entity. This solution also would make it easier to answer the query, "Who is the school's dean, and what are that dean's credentials?" The downside of this solution is that it requires the duplication of data that are already stored in the PROFESSOR table, thus setting the stage for anomalies. However, because each school is run by a single dean, the problem of data duplication is rather minor. The selection of one approach over another often depends on information requirements, transaction speed, and the database designer's professional judgment. In short, do not use 1:1 relationships lightly, and make sure that each 1:1 relationship within the database design is defensible.

3. Each department offers several courses. For example, the management/marketing department offers courses such as Introduction to Management, Principles of Marketing, and Production Management. The ER diagram segment for this condition is shown in Figure 4.39. Note that this relationship is based on the way Tiny College operates. If, for example, Tiny College had some departments that were classified as "research only," such departments would not offer courses and, therefore, the COURSE entity would be optional to the DEPARTMENT entity.

4. We illustrated the relationship between COURSE and CLASS in Figure 4.14. Nevertheless, it is worth repeating that a CLASS is a section of a COURSE. That is, a department may offer several sections (classes) of the same database course. Each of those classes is taught by a professor at a given time in a given place. In short, there exists a 1:M relationship between COURSE and CLASS. However, because a course may exist in Tiny College's course catalog even when it is not offered as a class in a current class schedule, CLASS is optional to COURSE. Therefore, the relationship between COURSE and CLASS looks like Figure 4.40.

FIGURE 4.39 THE SECOND TINY COLLEGE ERD SEGMENT

FIGURE 4.40 THE THIRD TINY COLLEGE ERD SEGMENT

5. Each department may have many professors assigned to it. One of those professors chairs the department. Only one of the professors can chair the department to which (s)he is assigned, and no professor is required to accept the chair position. Therefore, DEPARTMENT is optional to PROFESSOR in the "chairs" relationship. These relationships are summarized in the ER segment shown in Figure 4.41.

FIGURE 4.41 THE FOURTH TINY COLLEGE ERD SEGMENT

6. Each professor may teach up to four classes; each class is a section of a course. A professor may also be on a research contract and teach no classes at all. The ER diagram segment in Figure 4.42 depicts these conditions.

FIGURE 4.42 THE FIFTH TINY COLLEGE ERD SEGMENT

7. A student may enroll in several classes, but (s)he takes each class only once during any given enrollment period. For example, during the current enrollment period a student may decide to take five classes— Statistics, Accounting, English, Database, and History—but that student would not be enrolled in the same Statistics class five times during the enrollment period! Each student may enroll in up to six classes, and each class may have up to 35 students in it, thus creating an M:N relationship between STUDENT and CLASS. Because a CLASS can initially exist (at the start of the enrollment period) even though no students have enrolled in it, STUDENT is optional to CLASS in the M:N relationship. This M:N relationship must be divided into two 1:M relationships through the use of the ENROLL entity shown in the ERD segment in Figure 4.43. But note that the optional symbol is shown next to ENROLL: if a class exists that has no students enrolled in it, that class never occurs in the ENROLL table. Note also that the ENROLL entity is weak: it is existence-dependent, and its (composite) PK is composed of the PKs of the STUDENT and CLASS entities.

FIGURE 4.43 THE SIXTH TINY COLLEGE ERD SEGMENT

8. Each department has several (hopefully many!) students whose major is offered by that department. However, each student has only a single major and is, therefore, associated with a single department. (See Figure 4.44.) However, in the Tiny College environment, it is possible—at least for a while—for a student not to declare a major field of study. Such a student would not be associated with a department, and, therefore, DEPARTMENT is optional to STUDENT. It is worth repeating that the relationships between entities and the entities themselves reflect the organization's operating environment. That is, the business rules define the ERD components!

FIGURE 4.44 THE SEVENTH TINY COLLEGE ERD SEGMENT

9. Each student has an advisor in his or her department; each advisor counsels several students. An advisor is also a professor, but not all professors advise students. Therefore, STUDENT is optional to PROFESSOR in the "PROFESSOR advises STUDENT" relationship. (See Figure 4.45.)

FIGURE 4.45 THE EIGHTH TINY COLLEGE ERD SEGMENT

10. When you examine the CLASS entity in Figure 4.46, you'll note that this entity contains a ROOM_CODE attribute. Given our naming conventions, it is clear that ROOM_CODE is a FK to another entity. Clearly, because a class is taught in a room, it is reasonable to assume that the ROOM_CODE in CLASS is the FK to an entity named ROOM. In turn, each room is located in a building, so we'll create the last Tiny College ERD by observing that a BUILDING can contain many ROOMs, but each ROOM is found in a single BUILDING. (See Figure 4.46.) In this ERD segment, it is clear that some buildings do not contain (class) rooms. For example, a storage building might not contain any named rooms at all.

FIGURE 4.46 THE NINTH TINY COLLEGE ERD SEGMENT

Using the preceding summary, we can identify the following entities:

SCHOOL COURSE

DEPARTMENT CLASS

EMPLOYEE ENROLL (the bridge entity between STUDENT and CLASS)

PROFESSOR STUDENT

BUILDING ROOM

Once we have discovered the relevant entities, we can then define the initial set of relations among these enti-
ties. Next, we describe the entity attributes. Identifying the attributes of the entities helps us better understand the
relationships among entities. Table 4.2 summarizes the ER model's components and names the entities and their
relations.

TABLE 4.2 COMPONENTS OF THE ER MODEL

ENTITY	RELATIONSHIP	CONNECTIVITY	ENTITY
SCHOOL	operates	1:M	DEPARTMENT
DEPARTMENT	has	1:M	STUDENT
DEPARTMENT	employs	1:M	PROFESSOR
DEPARTMENT	offers	1:M	COURSE
COURSE	generates	1:M	CLASS
PROFESSOR	is an	1:1	EMPLOYEE
PROFESSOR	is dean of	1:1	SCHOOL
PROFESSOR	chairs	1:1	DEPARTMENT
PROFESSOR	teaches	1:M	CLASS
PROFESSOR	advises	1:M	STUDENT
STUDENT	enrolls in	1:M	CLASS
BUILDING	contains	1:M	ROOM
ROOM	is used for	1:M	CLASS

Note: ENROLL is the composite entity that implements the relationship "STUDENT enrolls in CLASS."

We must also define the connectivity and cardinality for the just-discovered relations by querying the end user
extensively. Having defined the ER model's components, we can now draw the ER diagram, or conceptual
schema, depicted in Figure 4.47. Actually, the entity attributes and their domains should also be displayed in the
ER diagram. However, to avoid crowding the diagram, the entity attributes can be depicted separately.

FIGURE 4.47 THE COMPLETED TINY COLLEGE ERD

4.4 THE CHALLENGE OF DATABASE DESIGN: CONFLICTING GOALS

Database designers often must make design compromises that are triggered by conflicting goals, such as adherence to design standards (design elegance), processing speed, and information requirements.

- The database design must conform to design standards. Such standards have guided us to develop logical structures that minimize data redundancies, thereby minimizing the likelihood of destructive data anomalies. You have also learned how standards prescribed avoiding nulls to the greatest extent possible, thus triggering the development of the supertype/subtype relationships. In fact, you have learned that design standards govern the presentation of all the components within the database design. In short, design standards allow us to work with well-defined components and to evaluate the interaction of those components with some precision. Without design standards, it is nearly impossible to formulate a proper design process, to evaluate an existing design, or to trace the likely logical impact of changes in design.

- In many organizations, particularly those generating large numbers of transactions, high processing speeds are often a top priority in database design. High processing speed means minimal access time, which may be achieved by minimizing the number and complexity of logically desirable relationships. For example, a "perfect" design might use a 1:1 supertype/subtype relationship to avoid nulls, while a higher-transaction-speed design might combine the two tables to avoid the use of an additional relationship, using dummy entries to avoid the nulls. If the focus is on data-retrieval speed, we might also be forced to include derived attributes in the design.

- The quest for timely information might be the focus of database design. Complex information requirements may dictate data transformations, and they may expand the number of entities and attributes within the design. Therefore, the database may have to sacrifice some of its "clean" design structures and/or some of its high transaction speed, to ensure maximum information generation. For example, suppose that a detailed sales report must be generated periodically. The sales report includes all invoice subtotals, taxes, and totals, and even the invoice lines include subtotals. If the sales report includes hundreds of thousands (or even millions) of invoices, then computing the totals, taxes, and subtotals is likely to take some time. If those computations had been made and the results had been stored as derived attributes in the INVOICE and LINE tables at the time of the transaction, the real-time transaction speed might have declined, but that loss of speed would only be noticeable if there were many simultaneous transactions. The cost of a slight loss of transaction speed at the front end and the addition of multiple derived attributes is likely to pay off when the sales reports are generated. (Not to mention the fact that it would be simpler to generate the queries.)

A design that meets all the logical requirements and design conventions is an important goal. However, if this perfect design fails to meet the customer's transaction speed and/or information requirements, the designer will not have done a proper job from the end user's point of view. Compromises are a fact of life in the real world of database design.

Even as we focus on the entities, attributes, relationships, and constraints, we should begin thinking about end-user requirements such as performance, security, shared access, and data integrity. The designer must consider processing requirements and verify that all the update, retrieval, and deletion options are available. Finally, a design is of little value unless the end product is capable of delivering all specified query and reporting requirements.

You are quite likely to discover that even the best ER design process produces an ERD that will require further changes mandated by operational requirements. Such changes should not discourage you from using the ER process. ER modeling is essential in the development of a sound design capable of meeting the demands of adjustment and growth. Using ER diagrams yields perhaps the richest bonus of all: a thorough understanding of how the organization really functions.

There are occasional design and implementation problems that do not yield "clean" implementation solutions. To get a sense of the design and implementation choices a database designer faces, let's revisit the 1:1 recursive relationship "EMPLOYEE is married to EMPLOYEE" we first examined in Figure 4.19. Figure 4.48 shows three different ways of implementing such a relationship.

FIGURE 4.48 VARIOUS IMPLEMENTATIONS OF A 1:1 RECURSIVE RELATIONSHIP

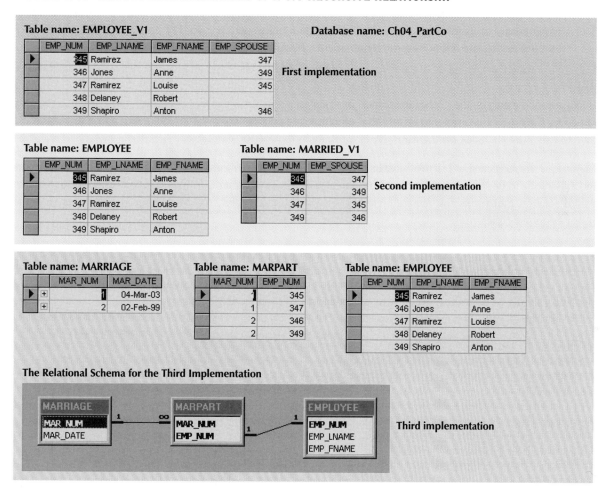

As you examine the EMPLOYEE_V1 table in Figure 4.48, note that this table is likely to yield data anomalies. For example, if Anne Jones divorces Anton Shapiro, two records must be updated—by setting the respective EMP_SPOUSE values to null—to properly reflect this change. If only one record is updated, inconsistent data occur. The problem becomes even worse if several of the divorced employees then marry each other. In addition, this implementation also produces undesirable nulls for employees who are *not* married to other employees in the company.

Another approach would be to create a new entity shown as MARRIED_V1 in a 1:M relationship with EMPLOYEE (see Figure 4.48, second implementation). This second implementation does eliminate the nulls for employees who are not married to somebody working for the same company. (Such employees would not be entered into the MARRIED_V1 table.) However, this approach still yields possible duplicate values. For example, the marriage between employees 345 and 347 may still appear twice, once as 345,347 and once as 347,345. (Since each of these permutations is unique the first time it appears, the creation of a unique index will not solve the problem.)

As you can see, the first two implementations yield several problems:

- Both solutions use synonyms. The EMPLOYEE_V1 uses EMP_NUM and SPOUSE_NUM to refer to an employee. The MARRIED_V1 table uses the same synonyms.

- Both solutions are likely to produce inconsistent data. For example, it is possible to enter employee 345 married to employee 347 and to also enter employee 348 as married to employee 345.

- Both solutions make it possible to make data entries show one employee married to several other employees. For example, it is possible to have data pairs such as 345,347 and 348,347 and 349,347 that will not violate entity integrity requirements, because they are all unique.

A third approach would be to have two new entities MARRIAGE and MARPART in a 1:M relationship. MARPART contains the EMP_NUM foreign key to EMPLOYEE (see the relational schema in Figure 4.48). This third approach would be the preferred solution in a relational environment. But even this third approach requires some fine-tuning. For example, to ensure that an employee occurs only once in any given marriage, you would have to use a unique index on the EMP_NUM attribute in the MARPART table.

As you can see, a recursive 1:1 relationship yields many different solutions with varying degrees of effectiveness and adherence to basic design principles. Your job as a database designer is to use your professional judgment to yield a solution that meets the requirements imposed by business rules, the processing requirements, and the basic design principles.

Finally, document, document, and document! Put all the activities in writing. Then review what you've written. Documentation not only helps you keep on track during the design process, but also enables you (or those following you) to pick up the design thread when the time comes to modify the design. Although the need for documentation should be obvious, one of the most vexing problems in database and systems analysis work is that the "put it in writing" rule often is not observed in all of the design and implementation stages. The development of organizational documentation standards is a very important aspect of ensuring data compatibility and coherence.

SUMMARY

The entity relationship (ER) model uses ER diagrams to represent the conceptual database as viewed by the end user. The ER model's three main components are entities, relationships, and attributes. The ER diagram also includes connectivity and cardinality notations.

Connectivity describes the relationship classification (1:1, 1:M, M:N). Cardinality expresses the specific number of entity occurrences associated with an occurrence of a related entity. Connectivities and cardinalities are usually based on business rules. In turn, the business rules are derived from detailed descriptions of an organization's data environment, transactions, and information requirements. If these descriptions are inadequate or inaccurate, the business rules that are derived from such descriptions are likely to yield inadequate or inaccurate data models.

In a relational database model, 1:M is the preferred relationship. The 1:1 relationship is relatively rare and must be used with caution. Because the M:N relationship yields data redundancies, it must not be implemented in a transaction-focused database. If a M:N relationship is identified, it must be modeled as a set of 1:M relationships through a composite entity.

ER symbols are used to graphically depict the ER model's components and relationships. Such graphic representations are the database equivalent of a building architect's blueprint. Just as it is inadvisable to build a building before carefully planning its structures and functions and defining them in a blueprint, it is inadvisable to design databases without first constructing carefully considered data models that are represented by ERDs.

ERDs may be based on many different ER models. The Chen model's use of numeric connectivities and cardinalities enables the database designer to model the nature of entity occurrences and constraints in the ERD. The

Crow's Foot, Rein85, and IDEF1X models rely on internal data dictionaries to track specific cardinality values. Therefore, the Chen-based models tend to make better communication tools from a *conceptual* perspective. However, the Crow's Foot, Rein85, and IDEF1X models, with their more abstract symbol-based connectivities and cardinalities and their greater focus on the implementation aspects of a database design, are more easily adapted to CASE tools than the basic Chen model. Because the Crow's Foot model is readily available in a wide variety of database modeling tools, it is our preferred model in this and subsequent chapters. However, regardless of which methodology is selected, the modeling logic remains the same. Because no ER model can accurately portray all the real-world data and action constraints, application software must be used to augment the implementation of at least some of the business rules.

To avoid data redundancies and the proliferation of nulls, entities can also be classified as supertypes and subtypes within a generalization hierarchy. The supertype contains the shared set of entity characteristics, while the subtype contains the unique characteristics. In effect, the generalization hierarchy depicts the supertype entity as a parent of each of the subtype entities; that is, the subtype inherits all the general characteristics and relationships from its parent supertype. The relationship between supertype and subtype is 1:1; that is, each subtype row corresponds to only one row in the supertype, and vice versa.

Database designers, no matter how well they are able to produce designs that conform to all applicable modeling conventions, often are forced to make design compromises. These compromises are required because end users may have vital transaction speed and/or information requirements that prevent the use of "perfect" modeling logic and adherence to all modeling conventions. Therefore, database designers must use professional judgment to determine how and to what extent the modeling conventions are subject to modification. To ensure that professional judgments are sound, database designers must have detailed and in-depth knowledge of the data-modeling conventions. It is also important to document the design process from beginning to end; this can help keep the design process on track, and make modifications much easier down the road.

K E Y T E R M S

binary relationship	IDEF1X	Rein85 model
cardinality	identifying relationship	simple attribute
composite attribute	iterative process	single-valued attribute
computer-assisted systems engineering (CASE)	mandatory	strong relationship
degree	multivalued attribute	subtype
derived attribute	non-identifying relationship	supertype
disjoint subtypes	non-overlapping subtypes	ternary relationship
existence-dependent	optionality	unary relationship
existence-independent	overlapping subtypes	weak entity
generalization hierarchy	participant	weak relationship
	recursive relationship	

R E V I E W Q U E S T I O N S

1. What two conditions must be met before an entity can be classified as a weak entity? Give an example of a weak entity.
2. What is a strong (or identifying) relationship, and how is it depicted in a Crow's Foot ERD?
3. Given the business rule "an employee may have many degrees," discuss its effect on attributes, entities, and relationships. (*Hint*: Remember what a multivalued attribute is and how it might be implemented.)

4. What is a composite entity, and when is it used?

5. Suppose you are working within the framework of the conceptual model in Figure Q4.5.

FIGURE Q4.5 THE CONCEPTUAL MODEL FOR QUESTION 5

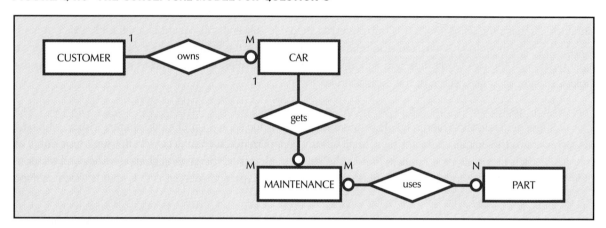

Given the conceptual model in Figure Q4.5,
 a. Write the business rules that are reflected in it.
 b. Identify all the cardinalities.
 c. Explain how you would modify the design to remove the M:N relationship between MAINTENANCE and PART.

6. When might you want to use entity supertypes and subtypes and what must be the relationship between them? Give an example.

7. How would you (graphically) identify each of the following ER model components in a Crow's Foot model?
 a. an entity
 b. the cardinality (0,N)
 c. a weak relationship
 d. a strong relationship

8. The Hudson Engineering Group (HEG) has contacted you to create a conceptual model whose application will meet the expected database requirements for its training program. The HEG administrator gives you the following description of the training group's operating environment:

 The HEG has 12 instructors and can handle up to 30 trainees per class. HEG offers five "advanced technology" courses, each of which may generate several classes. If a class has fewer than 10 trainees in it, it will be canceled. It is, therefore, possible for a course not to generate any classes. Each class is taught by one instructor. Each instructor may teach up to two classes or may be assigned to do research only. Each trainee may take up to two classes per year.

 Given this information, do the following:
 a. Define all the entities and relationships. (Use Table 4.2 as your guide.)
 b. Describe the relationship between instructor and class in terms of connectivity, cardinality, and existence-dependence.

9. Discuss the difference between a composite key and a composite attribute. How would each be indicated in an ER diagram?

10. What two courses of action are available to a designer when a multivalued attribute is encountered?

11. What is a derived attribute? Give an example.

12. How is a relationship between entities indicated in an ER diagram? Give an example, using the Chen and Crow's Foot data models.

13. Discuss two ways in which the 1:M relationship between COURSE and CLASS can be implemented. (*Hint*: Think about relationship strength.)

14. How is a composite entity represented in an ER diagram, and what is its function? (Illustrate both the Chen and the Crow's Foot models. (Note that Visio Professional uses a relationship to let the designer depict a composite entity in a Crow's Foot ERD—the entity box is not identifiable as a composite entity.)

15. Given the following business rules, create the appropriate Chen and Crow's Foot ER diagrams for each of the specified relationships:
 a. A company operates four departments.
 b. Each department employs employees.
 c. Each of the employees may or may not have one or more dependents.
 d. Each employee may or may not have an employment history.

16. Using the ER diagram components developed in Question 15, create Chen and Crow's Foot ER diagrams that include all the components.

17. What three (often conflicting) database requirements must be addressed in database design?

18. Briefly, but precisely, explain the difference between single-valued attributes and simple attributes. Give an example of each.

19. What are multivalued attributes, and how can they be handled within the database design?

 The final four questions are based on the Chen ER diagram in Figure Q4.20.

FIGURE Q4.20 THE ERD FOR QUESTIONS 20–23

20. Write the proper cardinalities for (a,b) _____ (c,d) _____ (e,f) _____
 (g,h) _____ (i,j) _____ (k,l) _____ (m,n) _____ and (o,p) _____.

21. Write the proper connectivities for: W _____ X _____ Y _____ and Z _____.

22. What two attributes must be contained in the composite entity? Use proper terminology in your answer.

23. Describe precisely the composition of the weak entity's primary key. Use proper terminology in your answer.

24. Convert the Chen ERD in Figure Q4.20 to a Crow's Foot ERD.

P R O B L E M S

The first three problems are based on the Chen ER model in Figure P4.1.

FIGURE P4.1 THE ERD FOR PROBLEMS 1–3

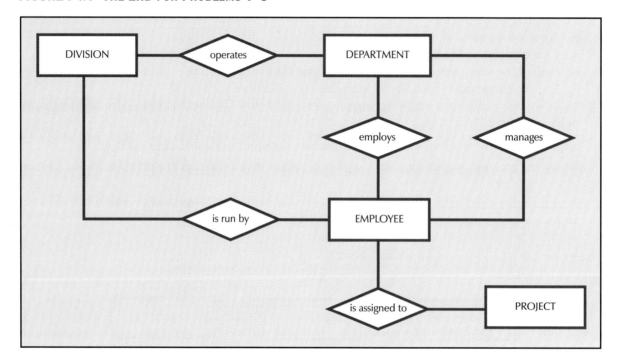

1. Use the following business rules to write all appropriate connectivities in the ER diagram:
 a. A department employs many employees, but each employee is employed by one department.
 b. Some employees, known as "rovers," are not assigned to any department.
 c. A division operates many departments, but each department is operated by one division.
 d. An employee may be assigned to many projects, and a project may have many employees assigned to it.
 e. A project must have at least one employee assigned to it.
 f. One of the employees manages each department, and each department is managed by only one employee.
 g. One of the employees runs each division, and each division is run by only one employee.
2. Write all the cardinalities into the model.
3. Modify the ER model by splitting the M:N relationship into two 1:M relationships that are connected through a composite entity. Then rewrite the connectivities and cardinalities to match the changes you have made.
4. Convert the Chen model you have developed in Problems 1–3 to a Crow's Foot model. Include at least the minimum number of attributes required to implement the model.
5. Temporary Employment Corporation (TEC) places temporary workers in companies during peak periods. TEC's manager gives you the following description of the business:

 • TEC has a file of candidates who are willing to work.

 • If the candidate has worked before, that candidate has a specific job history. (Naturally, no job history exists if the candidate has never worked.) Each time the candidate worked, one additional job history record was created.

- Each candidate has earned several qualifications. Each qualification may be earned by more than one candidate. (For example, it is possible for more than one candidate to have earned a BBA degree or a Microsoft Network Certification. And clearly a candidate may have earned both a BBA and a Microsoft Network Certification.)

- TEC also has a list of companies that request temporaries.

- Each time a company requests a temporary employee, TEC makes an entry in the Openings folder. This folder contains an opening number, company name, required qualifications, starting date, anticipated ending date, and hourly pay.

- Each opening requires only one specific or main qualification.

- When a candidate matches the qualification, (s)he is given the job, and an entry is made in the Placement Record folder. This folder contains an opening number, candidate number, total hours worked, and so on. In addition, an entry is made in the job history for the candidate.

- TEC uses special codes to describe a candidate's qualifications for an opening. The list of codes is shown in Table P4.5.

TABLE P4.5 **TEC QUALIFICATION CODES**

CODE	DESCRIPTION
SEC-45	Secretarial work, at least 45 words per minute
SEC-60	Secretarial work, at least 60 words per minute
CLERK	General clerking work
PRG-VB	Programmer, Visual Basic
PRG-C++	Programmer, C++
DBA-ORA	Database Administrator, Oracle
DBA-DB2	Database Administrator, IBM DB2
DBA-SQLSERV	Database Administrator, MS SQL Server
SYS-1	Systems Analyst, level 1
SYS-2	Systems Analyst, level 2
NW-NOV	Network Administrator, Novell experience
WD-CF	Web Developer, ColdFusion

TEC's management wants to keep track of the following entities:

COMPANY

OPENING

QUALIFICATION

CANDIDATE

JOB_HISTORY

PLACEMENT

Given this information, do the following:

 a. Draw the Crow's Foot ER diagrams for this enterprise.

 b. Identify all possible relationships.

 c. Identify the connectivity for each relationship.

 d. Identify the mandatory/optional dependencies for the relationships.

 e. Resolve all M:N relationships.

6. The Jonesburgh County Basketball Conference (JCBC) is an amateur basketball association. Each city in the county has one team that represents it. Each team has a maximum of 12 players and a minimum of nine players. Each team also has up to three coaches (offensive, defensive, and PT coaches). Each team plays two games (home and visitor) against each of the other teams during the season. Given these conditions, do the following:

 a. Identify the connectivity of each relationship.

 b. Identify the type of dependency that exists between CITY and TEAM.

 c. Identify the cardinality between teams and players, and between teams and city.

 d. Identify the dependency between coach and team, and between team and player.

 e. Draw the Chen and Crow's Foot ER diagrams to represent the JCBC database.

7. Automata Inc. produces specialty vehicles by contract. The company operates several departments, each of which builds a particular vehicle, such as a limousine, a truck, a van, or an RV.

When a new vehicle is built, the department places an order with the purchasing department to request specific components. Automata's purchasing department is interested in creating a database to keep track of orders and to accelerate the process of delivering materials.

The order received by the purchasing department can contain several different items. An inventory is maintained so that the most frequently requested items are delivered almost immediately. When an order comes in, it is checked to determine whether the requested item is in inventory. If an item is not in inventory, it must be ordered from a supplier. Each item may have several suppliers.

Given this functional description of the processes encountered at Automata's purchasing department, do the following:

 a. Identify all the main entities.

 b. Identify all the relations and connectivities among entities.

 c. Identify the type of existence dependency in all relations.

 d. Give some examples of the types of reports that can be obtained from the database.

8. Create an ERD based on the Crow's Foot model, using the following requirements:

- An INVOICE is written by a SALESREP. Each sales representative can write many invoices, but each invoice is written by a single sales representative.

- The INVOICE is written for a single CUSTOMER. However, each customer can have many invoices.

- An INVOICE can include many detail lines (LINE), which describe the products bought by the customer.

- The product information is stored in a PRODUCT entity.

- The product's vendor information is found in a VENDOR entity.

> **N O T E**
>
> The ERD must reflect business rules that you are (within reason) free to define. Make sure that your ERD reflects the conditions you require. Finally, make sure that you include the attributes that would permit the model to be successfully implemented.

9. Given the following brief summary of business rules for the ROBCOR catering service, and using the Crow's Foot ER methodology, draw the fully labeled ERD. Make sure to include all appropriate entities, relationships, connectivities, and cardinalities.

NOTE

Limit your ERD to entities and relationships based on the business rules shown here. In other words, do NOT add "realism" to your design by expanding or refining the business rules! However, make sure that you include the attributes that would permit the model to be successfully implemented.

Each dinner is based on a single entree, but each entree can be served at many dinners. A guest can attend many dinners, and each dinner can be attended by many guests. Each dinner invitation can be mailed to many guests, and each guest can receive many invitations.

10. Using the Crow's Foot methodology, create an ERD that can be implemented for a medical clinic, using at least the following business rules:

- A patient can make many appointments with one or more doctors in the clinic, and a doctor can accept appointments with many patients. However, each appointment is made with only one doctor, and each appointment references a single patient.

- Emergency cases do not require an appointment. However, an emergency is entered into the appointment book as "unscheduled" for appointment management purposes.

- If kept, an appointment yields a visit with the doctor specified in the appointment. The visit yields a diagnosis and, when appropriate, treatment.

- Each visit updates the patient's records to provide a medical history.

- Each patient visit creates a bill. Each patient visit is billed by one doctor, and each doctor can bill many patients.

- Each bill must be paid. However, a bill may be paid in many installments, and a payment may cover more than one bill.

- A patient can pay the bill directly, or the bill may be the basis for a claim submitted to an insurance company.

- If the bill is paid by an insurance company, the deductible is submitted to the patient for payment.

11. Tiny College is so pleased with your design and implementation of its student registration/tracking system that it wants you to expand the design to include its motor pool. A brief description of operations follows:

- Faculty members may use the vehicles owned by Tiny College for officially sanctioned travel. For example, its vehicles may be used by faculty members to travel to off-campus learning centers, to travel to locations at which research papers are presented, to transport students to officially sanctioned locations, and to travel for public service purposes. The vehicles used for such purposes are managed by Tiny College's TFBS (Travel Far But Slowly) Center.

- Using reservation forms, each department may reserve vehicles for its faculty, who are responsible for filling out the appropriate trip completion form at the end of each trip. The reservation form includes the expected departure date, vehicle type required, destination, and the name of the authorized faculty member. When the faculty member arrives to pick up the vehicle, (s)he must sign a check-out form to log out the vehicle and pick up a trip completion form. (The TFBS employee who releases the vehicle for use also signs the check-out form.) The faculty member's trip completion form includes the faculty member's identification code, the vehicle's identification, the odometer readings at the start and end of the trip, maintenance complaints (if any), gallons of fuel purchased (if any), and the Tiny College credit card used to pay for the fuel. If fuel has been purchased, the credit card receipt must be stapled to the trip completion form. Upon receipt of the faculty trip completion form, the faculty member's department is billed at a mileage rate based on the vehicle type (sedan, station wagon, panel truck, minivan,

minibus) used. (*Hint*: Do *not* use more entities than are necessary. Remember the difference between attributes and entities!)

- All vehicle maintenance is performed by TFBS. Each time a vehicle requires maintenance, a maintenance log entry is completed on a prenumbered maintenance log form. The maintenance log form includes the vehicle identification, a brief description of the type of maintenance required, the initial log entry date, the date on which the maintenance was completed, and the identification of the mechanic who released the vehicle back into service. (Only mechanics who have an inspection authorization may release the vehicle back into service.)

- As soon as the log form has been initiated, the log form's number is transferred to a maintenance detail form; the log form's number is also forwarded to the parts department manager, who fills out a parts usage form on which the maintenance log number is recorded. The maintenance detail form contains separate lines for each maintenance item performed, the parts used, and the identification of the mechanic who performed the maintenance item. When all the maintenance items have been completed, the maintenance detail form is stapled to the maintenance log form, the maintenance log form's completion date is filled out, and the mechanic who releases the vehicle back to service signs the form. The stapled forms are then filed, to be used later as the source for various maintenance reports.

- TFBS maintains a parts inventory, including oil, oil filters, air filters, and belts of various types. The parts inventory is monitored daily to monitor parts usage and to reorder parts that reach the "minimum quantity on hand" level. To track parts usage, the parts manager requires each mechanic to sign out the parts that are used to perform each vehicle's maintenance; the parts manager records the maintenance log number under which the part is used.

- Each month, TFBS issues a set of reports. These reports include the mileage driven by vehicle, by department, and by faculty members within the department. In addition, various revenue reports are generated by vehicle and department. A detailed parts usage report is also filed each month. Finally, a vehicle maintenance summary is created each month.

Given this brief summary of operations, draw the appropriate (and fully labeled) ER diagram. Use the Chen methodology to indicate entities, relationships, connectivities, and cardinalities.

12. Given the following information, produce an ERD—based on the Crow's Foot model—that can be implemented. Make sure to include all appropriate entities, relationships, connectivities, and cardinalities.

- EverFail company is in the quick oil change and lube business. Although customers bring in their cars for what is described as "quick oil changes," EverFail also replaces windshield wipers, oil filters, and air filters, subject to customer approval. The invoice contains the charges for the oil used and all parts used, and a standard labor charge. When the invoice is presented, customers pay cash, use a credit card, or write a check. EverFail does not extend credit. EverFail's database is to be designed to keep track of all components in all transactions.

- Given the high parts usage of the business operations, EverFail must maintain careful control of its parts (oil, wipers, oil filters, air filters) inventory. Therefore, if parts reach their minimum on-hand quantity, the parts in low supply must be reordered from an appropriate vendor. EverFail maintains a vendor list, which contains both the vendors actually used and the potential vendors.

- Periodically, EverFail mails updates to customers, based on the date of the car's service. EverFail also tracks each customer's car mileage.

NOTE

Problems 13 and 14 may be used as the basis for class projects. These problems illustrate the challenge of translating a description of operations to a set of business rules that will define the components for an ERD that can be successfully implemented. These problems can also be used as the basis for discussions about the components and contents of a proper description of operations. One of the things you must learn to do if you want to create databases that can be successfully implemented is to separate the "generic" background material from the details that directly affect database design. You must also keep in mind that many constraints cannot be incorporated into the database design; instead, such constraints are handled by the applications software.

Although the description of operations in Problem 13 deals with a Web-based business, the focus should be on the *database* aspects of the design, rather than on its interface and the transaction management details. In fact, the argument can easily be made that the existence of Web-based businesses has made database design more important than ever. (You can possibly get away with a bad database design if you sell only a few items per day, but the problems of poorly designed databases are compounded as the number of transactions increases.)

13. Use the following descriptions of the operations of RC_Models Company to complete this exercise.

RC_Models Company sells its products—plastic models (aircraft, ships, and cars) and "add-on" decals for those models—through its Internet Web site (*www.rc_models.com*). Models and decals are available in scales that vary from 1/144 to 1/32.

Customers use the Web site to select the products and pay by credit card. If a product is not currently available, it is placed on back order at the customer's discretion. (Back orders are not charged to the customer until the back order is shipped.) When the customer transactions are completed, the invoice is printed, and the products listed on the invoice are pulled from inventory for shipment. (The invoice includes a shipping charge.) The printed invoice is enclosed in the shipping container. The customer credit card charges are transmitted to the CC Bank, at which RC_Models Company maintains a commercial account. (*Note:* The CC Bank is *not* part of the RC_Models database!)

RC_Models Company tracks customer purchases and periodically sends out promotional materials. Because the RC_Models Company management requires detailed information to conduct its operations, numerous reports are available. These reports include, but are not limited to, customer purchases by product category and amount, product turnover, and revenues by product and customer. If a product has not recorded a sale within four weeks of being stocked, it is removed from the inventory and scrapped.

Many of the customers on the RC_Models customer list have bought RC_Models products. However, RC_Models Company also has purchased a copy of the *FineScale Modeler* magazine subscription list in order to market its products to customers who have not yet purchased any of the RC_Models Company products. In addition, customer data are recorded when potential customers request product information.

RC_Models Company orders its products directly from the manufacturers. For example, the plastic models are ordered from Tamiya, Academy, Revell/Monogram, and so on. Decals are ordered from Aeromaster, Tauro, WaterMark, and others. (*Note:* Not all manufacturers in the RC_Models Company database have already received orders.) All orders are placed via the manufacturers' Web sites, and the order amounts are automatically handled through RC_Models' commercial bank account with the CC Bank. Orders are automatically placed when the product inventory reaches the specified minimum quantity on hand. (The number of product units ordered depends on the minimum order quantity specified for each of the products.)

a. Given this (brief and incomplete) description of operations for RC_Models Company, write all applicable business rules to establish entities, relationships, optionalities, connectivities, and cardinalities. (*Hint:* Use the following three business rules as examples, and write the remaining business rules in the same format.)

- A customer may generate many invoices.
- Each invoice is generated by only one customer.
- Some customers have not (yet) generated an invoice.

b. Draw the fully labeled and implementable Crow's Foot ERD based on the business rules you wrote in Problem 13, Part a. Include all entities, relationships, optionalities, connectivities, and cardinalities.

14. Use the following description of the operations of the RC_Charter2 Company to complete this exercise.

The RC_Charter2 Company operates a fleet of aircraft under the FAR Part 135 (air taxi or charter) certificate, of the Federal Air Regulations—FARs—that are enforced by the Federal Aviation Administration (FAA). These aircraft are available for air taxi (charter) operations within the United States and Canada.

Charter companies provide so-called "unscheduled" operations—that is, charter flights only take place after a customer reserves the use of an aircraft to fly at a customer-designated date and time to one or more customer-designated destination(s), transporting passengers, cargo, or some combination of passengers and cargo. A customer can, of course, reserve many different charter flights (trips) during any time frame. However, for billing purposes each charter trip is reserved by one and only one customer. Some of RC_Charter2's customers do not use the company's charter operations; instead, they may purchase fuel, use maintenance services, or other RC_Charter2 services. (*Note:* This database design will focus on the charter operations only.)

Each charter trip yields revenue for the RC_Charter2 Company. Such revenue is generated by the charges that are paid by the customer upon the completion of each flight. The charter flight charges are a function of the aircraft model used, the distance flown, waiting time, special customer requirements, and crew expenses. The distance mileage charges are computed by multiplying the round-trip miles by the model's charge per mile. Round-trip miles are based on the actual navigational path flown. The sample route traced in Figure P4.14 illustrates the procedure. Note that the number of round-trip miles is calculated to be 130 + 200 + 180 + 390 = 900.

FIGURE P4.14 ROUND-TRIP MILE DETERMINATION

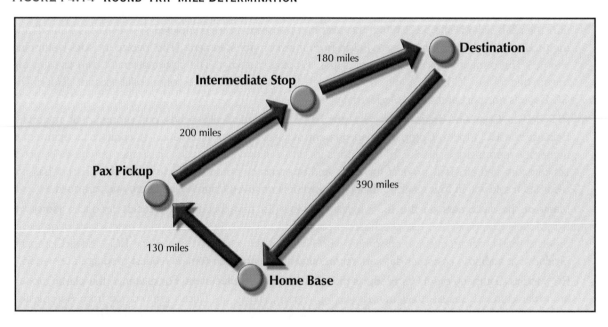

Depending on whether or not the customer has RC_Charter2 credit authorization, a customer may:

• Pay the entire charter bill upon the completion of the charter flight.

• Pay a part of the charter bill and charge the remainder to the account. The charge amount may not exceed the available credit.

• Charge the entire charter bill to the account. The charge amount may not exceed the available credit.

Customers may pay all or part of the existing balance(s) for previous charter trips. Such payments may be made at any time and are not necessarily tied to a specific charter trip.

The charter mileage charge includes the expense of the pilot(s) and other crew that are required by FAR 135. However, if customers require *additional* crew *not* required by FAR 135, those customers are charged for such crew members on an hourly basis. The hourly crew-member charge is based on each crew member's qualifications.

The database must be able to handle crew assignment. Each charter trip requires the use of an aircraft, and a crew flies each aircraft. The smaller piston-engine-powered charter aircraft require a crew consisting of only a single pilot. Larger aircraft—that is, aircraft having a gross take-off weight of 12,500 lbs. or more—and jet-powered aircraft require a pilot and a copilot, while some of the larger aircraft that are used to transport passengers may require flight attendants as part of the crew. Some of the older aircraft require the assignment of a flight engineer, and larger cargo-carrying aircraft require the assignment of a loadmaster. In short, a crew can consist of more than one person, and not all crew members are pilots.

The charter flight's aircraft waiting charges are computed by multiplying the hours waited by the model's hourly waiting charge. Crew expenses are limited to meals and overnight expenses such as hotel/motel charges and transportation to and from such hotels/motels.

The RC_Charter2 database must be designed to generate a monthly summary of all charter trips, expenses, and revenues derived from the charter records. Such records are based on the data that each pilot in command is required to record for each charter trip: the trip date(s) and time(s), destination(s), aircraft number, pilot (and other crew) data, distance flown, fuel usage, and other data pertinent to the charter flight. Such charter data are then used to generate monthly reports that detail revenue and operating cost information for customers, aircraft, and pilots. All pilots and other crew members are RC_Charter2 Company employees, that is, the company does not use contract pilots and crew.

FAR Part 135 operations are conducted under a very strict set of requirements that govern crew-member licensing and training. For example, pilots must have earned either a Commercial license or an Airline Transport Pilot (ATP) license. Both licenses require appropriate ratings. Ratings are specific competency requirements. For example:

- To operate a multiengine aircraft designed for takeoffs and landings on land only, the appropriate rating would be MEL, or Multi-Engine Landplane. If a multiengine aircraft can take off and land on water, the appropriate rating would be MES, or Multi-Engine Seaplane.

- The instrument rating is based on a demonstrated ability to conduct all flight operations with sole reference to cockpit instrumentation. The instrument rating is required to operate an aircraft under Instrument Meteorological Conditions (IMC), and all such operations are governed under FAR-specified Instrument Flight Rules (IFR). In contrast, operations conducted under "good weather" or *visual* flight conditions are based on the FAR Visual Flight Rules (VFR).

- The type rating is required for all aircraft with a take-off weight over 12,500 lbs. or for aircraft that are purely jet-powered. (If an aircraft uses jet engines to drive propellers, such an aircraft is said to be turbo-prop-powered. A "turboprop"—that is, a turbo-propeller-powered aircraft—does not require a type rating, unless it meets the 12,500-lb. weight limitation.

Although pilot licenses and ratings are not time-limited, exercising the privilege of the license and ratings under Part 135 requires both *a current medical certificate and a current Part 135 checkride*. The following distinctions are important:

- The medical certificate may be Class I or Class II. The Class I medical is more stringent than the Class II, and it must be renewed each six months. The Class II medical must be renewed yearly. If the Class I medical is not renewed during the six-month period, it automatically reverts to a Class II certificate. If the Class II medical is not renewed within the specified period, it automatically reverts to a Class III medical, which is not valid for commercial flight operations.

- A Part 135 checkride is a practical flight examination that must be successfully completed every six months. The checkride includes all flight maneuvers and procedures specified in Part 135.

Nonpilot crew members must also have the proper certificates to meet specific job requirements. For example, loadmasters need an appropriate certificate, as do flight attendants. In addition, crew members such as loadmasters and flight attendants, who may be required in operations that involve large aircraft (more than 12,500-lb. takeoff weight and passenger configurations over 19 passengers) are also required to periodically pass a written and practical exam. The RC_Charter2 Company is required to keep a complete record of all test types, dates, and results for each crew member, as well as pilot medical certificate examination dates.

In addition, all flight crew members are required to submit to periodic drug testing, and the results must be tracked, too. (Note that nonpilot crew members are not required to take pilot-specific tests such as Part 135 checkrides. Nor are pilots required to take crew tests such as loadmaster and flight attendant practical exams.) However, many crew members have licenses and/or certifications in several areas. For example, a pilot may have an ATP and a loadmaster certificate. If that pilot is assigned to be a loadmaster on a given charter flight, the loadmaster certificate is required. Similarly, a flight attendant may have earned a commercial pilot's license. Sample data formats are shown in Table P4.14.

TABLE P4.14 SAMPLE DATA FORMATS

Part A Tests

TEST CODE	TEST DESCRIPTION	TEST FREQUENCY
1	Part 135 Flight Check	6 months
2	Medical, Class 1	6 months
3	Medical, Class 2	12 months
4	Loadmaster Practical	12 months
5	Flight Attendant Practical	12 months
6	Drug test	Random
7	Operations, written exam	6 months
...		

Part B Results

EMPLOYEE	TEST CODE	TEST DATE	TEST RESULT
101	1	12-Nov-03	Pass-1
103	6	23-Dec-03	Pass-1
112	4	23-Dec-03	Pass-2
103	7	11-Jan-04	Pass-1
112	7	16-Jan-04	Pass-1
101	7	16-Jan-04	Pass-1
101	6	11-Feb-04	Pass-2
125	2	15-Feb-04	Pass-1
...			

Pilots and other crew members must receive recurrency training appropriate to their work assignments. *Recurrency training* is based on an FAA-approved curriculum that is job-specific. For example, pilot recurrency training includes a review of all applicable Part 135 flight rules and regulations, weather data interpretation, company flight operations requirements, specified flight procedures, and so on. The RC_Charter2 Company is required to keep a complete record of all recurrency training for each crew member subject to such training.

The RC_Charter2 Company is required to maintain a detailed record of all crew credentials and all training mandated by Part 135. *A complete record must be kept of each requirement and of all the compliance data.*

To conduct a charter flight, the company must have a properly maintained aircraft available. A pilot who meets all of the FAA's licensing and currency requirements must fly this aircraft as Pilot in Command (PIC). For those aircraft that are powered by piston engines or turboprops and have a gross takeoff weight under

12,500 lbs., single-pilot operations are permitted under Part 135, as long as a properly maintained autopilot is available. However, even if FAR Part 135 permits single-pilot operations, many customers require the presence of a copilot who is capable of conducting the flight operations under Part 135.

Part C Licenses and Certifications

LICENSE OR CERTIFICATE	LICENSE OR CERTIFICATE DESCRIPTION
ATP	Airline Transport Pilot
Comm	Commercial license
Med-1	Medical certificate, class 1
Med-2	Medical certificate, class 2
Instr	Instrument rating
MEL	Multi-engine Land aircraft rating
LM	Load Master
FA	Flight Attendant
……..	…………..

Part D Licenses and Certificates Held by Employees

EMPLOYEE	LICENSE OR CERTIFICATE	DATE EARNED
101	Comm	12-Nov-93
101	Instr	28-Jun-94
101	MEL	9-Aug-94
103	Comm	21-Dec-95
112	FA	23-Jun-98
103	Instr	18-Jan-96
112	LM	27-Nov-02
…	………….	…………..

The RC_Charter2 operations manager anticipates the lease of turbojet-powered aircraft, and such aircraft are required to have a crew consisting of a pilot and copilot. Both pilot and copilot must meet the same Part 135 licensing, ratings, and training requirements.

The company also leases larger aircraft that exceed the 12,500-lb. gross takeoff weight. Such aircraft can carry the number of passengers that requires the presence of one or more flight attendants. If such aircraft carry cargo weighing over 12,500 lbs., a loadmaster must be assigned as a crew member to supervise loading and securing such cargo. *The database must be designed to meet the anticipated additional charter crew assignment capability.*

a. Given this (incomplete) description of operations, write all applicable business rules to establish entities, relationships, optionalities, connectivities, and cardinalities. (*Hint:* Use the following five business rules as examples, and write the remaining business rules in the same format.)

 • A customer may request many charter trips.
 • Each charter trip is requested by only one customer.
 • Some customers have not (yet) requested a charter trip.
 • An employee may be assigned to serve as a crew member on many charter trips.
 • Each charter trip may have many employees assigned to it to serve as crew members.

b. Draw the fully labeled and implementable Crow's Foot ERD based on the business rules you wrote in Step 14a. Include all entities, relationships, optionalities, connectivities, and cardinalities.

5

NORMALIZATION OF DATABASE TABLES

In this chapter, you will learn:

❖ What normalization is and what role it plays in the database design process

❖ About the normal forms 1NF, 2NF, 3NF, BCNF, and 4NF

❖ How normal forms can be transformed from lower normal forms to higher normal forms

❖ That normalization and ER modeling are used concurrently to produce a good database design

❖ That some situations require denormalization to generate information efficiently

PREVIEW

Good database design must be matched to good table structures. In this chapter, you will learn to evaluate and design good table structures to control data redundancies, thereby avoiding data anomalies. The process that yields such desirable results is known as normalization.

In order to recognize and appreciate the characteristics of a good table structure, it is useful to examine a poor one. Therefore, we will begin by examining the characteristics of a poor table structure and the problems created by it. You will then learn how to correct a poor table structure. This methodology will yield important dividends: You will know how to design a good table structure and how to repair an existing poor one.

You will discover not only that data anomalies can be eliminated through normalization, but also that a properly normalized set of table structures is actually less complicated to use than an unnormalized set. In addition, you will learn that the normalized set of table structures more faithfully reflects an organization's real operations.

FIVE

FIVE

5.1 DATABASE TABLES AND NORMALIZATION

Having good relational database software is not enough to avoid the data redundancy we discussed in Chapter 1, "Database Systems." If the database tables are treated as though they were files in a file system, the RDBMS never has a chance to demonstrate its superior data-handling capabilities.

The table is a basic building block in the database design process. Consequently, the table's structure is of great interest to us. Ideally, the database design process we explored in Chapter 4, "Entity Relationship (ER) Modeling," yields good table structures. Yet it is possible to create poor table structures even in a good database design. So how do we recognize a poor table structure, and how do we produce a good table? The answer to both questions is based on normalization. **Normalization** is a process for evaluating and correcting table structures to minimize data redundancies, thereby helping to eliminate data anomalies.

Normalization works through a series of stages called normal forms. The first three stages are described as first normal form (1NF), second normal form (2NF), and third normal form (3NF). From a structural point of view, 2NF is better than 1NF, and 3NF is better than 2NF. For most business database design purposes, 3NF is as high as we need to go in the normalization process. (Actually, you will discover in Section 5.3 that properly designed 3NF structures will also meet the requirements of 4NF.) And some very specialized applications may require normalization beyond the 4NF. Such applications are usually driven by statistical research requirements that fall outside the scope of most business operations. Therefore, we will not cover normal forms beyond 4NF.

Although normalization is a very important database design ingredient, you should not assume that the highest level of normalization is always the most desirable. Generally, the higher the normal form, the more joins are required to produce a specified output and the more slowly the database system responds to end-user demands. A successful design must also consider end-user demand for fast performance. Therefore, you will occasionally be expected to *denormalize* some portions of a database design in order to meet performance requirements. (**Denormalization** produces a lower normal form; that is, a 3NF will be converted to a 2NF through denormalization.) However, *the price you pay for increased performance through denormalization is a larger amount of data redundancy.*

5.1.1 THE NEED FOR NORMALIZATION

To illustrate the normalization process, we will examine a simple business application. In this case, we will explore the simplified database activities of a construction company that manages several building projects. Each project has its own project number, name, employees assigned to it, and so on. Each employee has an employee number, name, and job classification such as engineer or computer technician.

The company charges its clients by billing the hours spent on each contract. The hourly billing rate is dependent on the employee's position. (For example, one hour of computer technician time is billed at a different rate than one hour of engineer time.) Periodically, a report is generated that contains the information displayed in Table 5.1.

TABLE 5.1 A SAMPLE REPORT LAYOUT

PROJ. NUM.	PROJECT NAME	EMPLOYEE NUMBER	EMPLOYEE NAME	JOB CLASS.	CHG/HOUR	HOURS BILLED	TOTAL CHARGE
15	Evergreen	103	June E. Arbough	Elec. Engineer	$84.50	23.8	$2,011.10
		101	John G. News	Database Designer	$105.00	19.4	$2,037.00
		105	Alice K. Johnson *	Database Designer	$105.00	35.7	$3,748.50
		106	William Smithfield	Programmer	$35.75	12.6	$450.45
		102	David H. Senior	Systems Analyst	$96.75	23.8	$2,302.65
				Subtotal			**$10,549.70**
18	Amber Wave	114	Annelise Jones	Applications Designer	$48.10	24.6	$1,183.26
		118	James J. Frommer	General Support	$18.36	45.3	$831.71
		104	Anne K. Ramoras *	Systems Analyst	$96.75	32.4	$3,135.70
		112	Darlene M. Smithson	DSS Analyst	$45.95	44.0	$2,021.80
				Subtotal			**$7,171.47**
22	Rolling Tide	105	Alice K. Johnson	Database Designer	$105.00	64.7	$6,793.50
		104	Anne K. Ramoras	Systems Analyst	$96.75	48.4	$4,682.70
		113	Delbert K. Joenbrood *	Applications Designer	$48.10	23.6	$1,135.16
		111	Geoff B. Wabash	Clerical Support	$26.87	22.0	$591.14
		106	William Smithfield	Programmer	$35.75	12.8	$457.60
				Subtotal			**$13,660.10**
25	Starflight	107	Maria D. Alonzo	Programmer	$35.75	24.6	$879.45
		115	Travis B. Bawangi	Systems Analyst	$96.75	45.8	$4,431.15
		101	John G. News *	Database Designer	$105.00	56.3	$5,911.50
		114	Annelise Jones	Applications Designer	$48.10	33.1	$1,592.11
		108	Ralph B. Washington	Systems Analyst	$96.75	23.6	$2,283.30
		118	James J. Frommer	General Support	$18.36	30.5	$559.98
		112	Darlene M. Smithson	DSS Analyst	$45.95	41.4	$1,902.33
				Subtotal			**$17,559.82**
				Total			**$48,941.09**

* Indicates project leader

The Total Charge in Table 5.1 is a *derived* attribute and, at this point, is not stored in this table.

The easiest short-term way to generate the required report might seem to be a table whose contents correspond to the reporting requirements. (See Figure 5.1.)

FIGURE 5.1 A TABLE IN THE REPORT FORMAT

	PROJ_NUM	PROJ_NAME	EMP_NUM	EMP_NAME	JOB_CLASS	CHG_HOUR	HOURS
▶	15	Evergreen	103	June E. Arbough	Elect. Engineer	$84.50	23.8
			101	John G. News	Database Designer	$105.00	19.4
			105	Alice K. Johnson *	Database Designer	$105.00	35.7
			106	William Smithfield	Programmer	$35.75	12.6
			102	David H. Senior	Systems Analyst	$96.75	23.8
	18	Amber Wave	114	Annelise Jones	Applications Designer	$48.10	24.6
			118	James J. Frommer	General Support	$18.36	45.3
			104	Anne K. Ramoras *	Systems Analyst	$96.75	32.4
			112	Darlene M. Smithson	DSS Analyst	$45.95	44.0
	22	Rolling Tide	105	Alice K. Johnson	Database Designer	$105.00	64.7
			104	Anne K. Ramoras	Systems Analyst	$96.75	48.4
			113	Delbert K. Joenbrood *	Applications Designer	$48.10	23.6
			111	Geoff B. Wabash	Clerical Support	$26.87	22.0
			106	William Smithfield	Programmer	$35.75	12.8
	25	Starflight	107	Maria D. Alonzo	Programmer	$35.75	24.6
			115	Travis B. Bawangi	Systems Analyst	$96.75	45.8
			101	John G. News *	Database Designer	$105.00	56.3
			114	Annelise Jones	Applications Designer	$48.10	33.1
			108	Ralph B. Washington	Systems Analyst	$96.75	23.6
			118	James J. Frommer	General Support	$18.36	30.5
			112	Darlene M. Smithson	DSS Analyst	$45.95	41.4

Table name: RPT_FORMAT Database name: Ch05_ConstructCo

ADDITIONAL CD CONTENT

The databases used to illustrate the material in this chapter are found on the CD bundled with this book in the \Databases\Student\Ch05 folder.

As you examine the data in Figure 5.1, note that it reflects the assignment of employees to projects. Apparently an employee can be assigned to more than one project. For example, Darlene Smithson (EMP_NUM = 112) has been assigned to two projects: Amber Wave and Starflight. Given the structure of the data set, each project includes only a single occurrence of any one employee. Therefore, knowing the PROJ_NUM and EMP_NUM value will let you find the job classification and its hourly charge. In addition, you will know the total number of hours worked by each employee on each project. (The total charge—a derived attribute whose value can be computed by multiplying the hours billed and the charge per hour—has not been included in Figure 5.1. No structural harm is done if this derived attribute is included.)

Unfortunately, the structure of the data set in Figure 5.1 does not conform to the requirements we discussed in Chapter 3, "The Relational Database Model," nor does it handle data very well:

1. The project number (PROJ_NUM) is apparently intended to be a primary key, or at least a part of a primary key, but it contains nulls. (Given the preceding discussion, you know that PROJ_NUM + EMP_NUM will define each row.)

2. The table entries invite data inconsistencies. For example, the JOB_CLASS value "Elect. Engineer" might be entered as "Elect.Eng." in some cases, "El. Eng." in others, and "EE" in still others.

3. The table displays data redundancies. These data redundancies yield the following anomalies:

 a. *Update anomalies.* Modifying the JOB_CLASS for employee number 105 requires (potentially) many alterations, one for each EMP_NUM = 105.

 b. *Insertion anomalies.* Just to complete a row definition, a new employee must be assigned to a project. If the employee is not yet assigned, a phantom project must be created to complete the employee data entry.

 c. *Deletion anomalies.* If employee 103 quits, deletions must be made for every entry in which EMP_NUM = 103. As such deletions are made, other vital data are lost, too.

In spite of these structural deficiencies, the table structure *appears* to work; the report is generated with ease. Unfortunately, the report may yield different results, depending on what data anomaly has occurred. For example, if you want to print a report to show the total "hours worked" value by the job classification "Database Designer," that report will not include data for "DB Design" and "Database Design" data entries. (We have actually witnessed such reporting anomalies, which drive managers up the proverbial wall—and which cannot be fixed through applications programming.)

Even if very careful data entry auditing can eliminate most of the reporting problems (at high cost!) it is easy to demonstrate that even a simple data entry becomes inefficient. Given the existence of update anomalies, suppose Darlene M. Smithson is assigned to work on the Evergreen project. The data entry clerk must update the PROJECT file with the entry:

15	Evergreen	112	Darlene M. Smithson	DSS Analyst	$45.95	0.0

to match the attributes PROJ_NUM, PROJ_NAME, EMP_NUM, EMP_NAME, JOB_CLASS, CHG_HOUR, and HOURS. (When Ms. Smithson has just been assigned to the project, she has not yet worked, so the total number of hours worked is 0.0.)

> **N O T E**
>
> Note again that the naming convention makes it easy to see what each attribute stands for and what its likely origin is. For example, PROJ_NAME uses the prefix PROJ to indicate that the attribute is associated with the PROJECT table, while the NAME component is self-documenting, too. However, do keep in mind that name length is also an issue, especially in the prefix designation. Therefore, we elected to use the prefix CHG, rather than CHARGE. (Given this database's context, it is not likely that this prefix can be misunderstood.)

Each time another employee is assigned to a project, some data entries (such as PROJ_NAME, EMP_NAME, CHG_HOUR) are unnecessarily repeated. Imagine the data entry chore when 200 or 300 table entries must be made. Note that the entry of the employee number should be sufficient to identify Darlene M. Smithson, her job description, and her hourly charge. Because there is only one person identified by the number 112, that person's characteristics (name, job classification, and so on) should not have to be typed in each time the main file is updated. Unfortunately, the structure displayed in Figure 5.1 does not make allowances for such a possibility.

The data redundancy evident in Figure 5.1 leads to wasted disk space. What's more, data redundancy produces data anomalies. For example, suppose the data entry clerk had entered the data as:

15	Evergeen	112	Darla Smithson	DCS Analyst	$45.95	0.0

At first glance, the data entry appears to be correct. But is Evergeen the same project as Evergreen? And is DCS Analyst supposed to be DSS Analyst? Is Darla Smithson the same person as Darlene M. Smithson? Such confusion is a data integrity problem, caused because the data entry failed to conform to the rule that all copies of redundant data must be identical.

The possibility of introducing data integrity problems caused by data redundancy must be considered when a database is designed. The relational database environment is especially well-suited to help the designer overcome these problems.

CONVERSION TO FIRST NORMAL FORM

Because the relational model views the data as part of a table or collection of tables in which all key values must be identified, the data depicted in Figure 5.1 may not be stored as shown. Note that Figure 5.1 contains what is known as repeating groups. A **repeating group** derives its name from the fact that a group of multiple (related) entries can exist for any *single* key attribute occurrence. If you examine Figure 5.1, note that each single project number (PROJ_NUM) occurrence can reference a group of related data entries. For example, the Evergreen project (PROJ_NUM = 15) in Figure 5.1 shows five entries at this point—and these entries are related because they each share the PROJ_NUM = 15 characteristic. Each time a new record is entered for the Evergreen project, the number of entries in the group grows by one.

A relational table must not contain repeating groups. The existence of repeating groups provides evidence that the RPT_FORMAT table in Figure 5.1 fails to meet even the lowest normal form requirements, thus reflecting data redundancies.

Normalizing the table structure will reduce these data redundancies. If repeating groups do exist, they must be eliminated by making sure that each row defines a single entity. In addition, the dependencies must be identified to diagnose the normal form. Identification of the normal form will let you know where you are in the normal-ization process. The normalization process starts with a simple three-step procedure.

Step 1: Eliminate the Repeating Groups

Start by presenting the data in a tabular format, where each cell has a single value and there are no repeating groups. To eliminate the repeating groups, eliminate the nulls by making sure that each repeating group attribute contains an appropriate data value. This change converts the RPT_FORMAT table in Figure 5.1 to the DATA_ORG_1NF table in Figure 5.2.

FIGURE 5.2 DATA ORGANIZATION: FIRST NORMAL FORM

Table name: DATA_ORG_1NF Database name: Ch05_ConstructCo

PROJ_NUM	PROJ_NAME	EMP_NUM	EMP_NAME	JOB_CLASS	CHG_HOUR	HOURS
15	Evergreen	103	June E. Arbough	Elect. Engineer	$84.50	23.8
15	Evergreen	101	John G. News	Database Designer	$105.00	19.4
15	Evergreen	105	Alice K. Johnson *	Database Designer	$105.00	35.7
15	Evergreen	106	William Smithfield	Programmer	$35.75	12.6
15	Evergreen	102	David H. Senior	Systems Analyst	$96.75	23.8
18	Amber Wave	114	Annelise Jones	Applications Designer	$48.10	24.6
18	Amber Wave	118	James J. Frommer	General Support	$18.36	45.3
18	Amber Wave	104	Anne K. Ramoras *	Systems Analyst	$96.75	32.4
18	Amber Wave	112	Darlene M. Smithson	DSS Analyst	$45.95	44.0
22	Rolling Tide	105	Alice K. Johnson	Database Designer	$105.00	64.7
22	Rolling Tide	104	Anne K. Ramoras	Systems Analyst	$96.75	48.4
22	Rolling Tide	113	Delbert K. Joenbrood *	Applications Designer	$48.10	23.6
22	Rolling Tide	111	Geoff B. Wabash	Clerical Support	$26.87	22.0
22	Rolling Tide	106	William Smithfield	Programmer	$35.75	12.8
25	Starflight	107	Maria D. Alonzo	Programmer	$35.75	24.6
25	Starflight	115	Travis B. Bawangi	Systems Analyst	$96.75	45.8
25	Starflight	101	John G. News *	Database Designer	$105.00	56.3
25	Starflight	114	Annelise Jones	Applications Designer	$48.10	33.1
25	Starflight	108	Ralph B. Washington	Systems Analyst	$96.75	23.6
25	Starflight	118	James J. Frommer	General Support	$18.36	30.5
25	Starflight	112	Darlene M. Smithson	DSS Analyst	$45.95	41.4

Step 2: Identify the Primary Key

The layout in Figure 5.2 represents much more than a mere cosmetic change. Even a casual observer will note that PROJ_NUM is not an adequate (primary) key because the project number does not uniquely identify all the remaining entity (row) attributes. For example, the PROJ_NUM value 15 can identify any one of five employees. To maintain a proper primary key that will *uniquely* identify any attribute value, the new key must be composed of a *combination* of PROJ_NUM and EMP_NUM. For example, using the data shown in Figure 5.2, if you know that PROJ_NUM = 15 and EMP_NUM = 103, the entries for the attributes PROJ_NAME, EMP_NAME, JOB_CLASS, CHG_HOUR, and HOURS must be Evergreen, June E. Arbough, Elect. Engineer, $85.50, and 23.8, respectively.

Step 3: Identify all Dependencies

The identification of the PK in Step 2 means that you have already identified the following dependency:
PROJ_NUM, EMP_NUM → PROJ_NAME, EMP_NAME, JOB_CLASS, CHG_HOUR, HOURS

That is, the PROJ_NAME, EMP_NAME, JOB_CLASS, CHG_HOUR, and HOURS values are all dependent on—that is, they are determined by—the combination of PROJ_NUM and EMP_NUM. There are additional dependencies. For example, the project number identifies (determines) the project name. In other words, the project name is dependent on the project number. You can write this dependency as:

PROJ_NUM → PROJ_NAME

Also, if you know an employee number, you also know that employee's name, that employee's job classification, and that employee's charge per hour. Therefore, you can identify the dependency shown next:

EMP_NUM → EMP_NAME, JOB_CLASS, CHG_HOUR

However, given the previous dependency components, you can see that knowing the job classification means knowing the charge per hour for that job classification. In other words, you can identify one last dependency:

JOB_CLASS → CHG_HOUR

The dependencies we have just examined can also be depicted with the help of the diagram shown in Figure 5.3. Because such a diagram depicts all the dependencies found within a given table structure, it is known as a **dependency diagram**. Dependency diagrams are very helpful in getting a bird's-eye view of all the relationships among a table's attributes, and their use makes it much less likely that you might overlook an important dependency.

As you examine Figure 5.3, note the following dependency diagram features:
1. The primary key attributes are bold, underlined, and shaded in a different color.
2. The arrows above the attributes indicate all desirable dependencies, that is, dependencies that are based on the primary key. In this case, note that the entity's attributes are dependent on the *combination* of PROJ_NUM and EMP_NUM.

3. The arrows below the dependency diagram indicate less-desirable dependencies. Two types of such dependencies exist:

a. *Partial dependencies.* You need to know only the PROJ_NUM to determine the PROJ_NAME; that is, the PROJ_NAME is dependent on only part of the primary key. And you need to know only the EMP_NUM to find the EMP_NAME, the JOB_CLASS, and the CHG_HOUR. Dependencies based on only a part of a composite primary key are called **partial dependencies**.

b. *Transitive dependencies.* As you examine Figure 5.4, note that CHG_HOUR is dependent on JOB_CLASS. Because neither CHG_HOUR nor JOB_CLASS is a prime attribute—that is, neither attribute is at least part of a key—we have a condition known as a transitive dependency. (In other words, a **transitive dependency** is a dependency of one nonprime attribute on another nonprime attribute.) The problem with transitive dependencies is that they still yield data anomalies.

FIGURE 5.3 A DEPENDENCY DIAGRAM: FIRST NORMAL FORM (1NF)

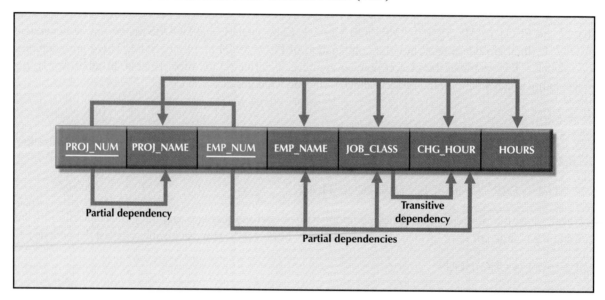

NOTE

The term **first normal form (1NF)** describes the tabular format in which:

- All the key attributes are defined.
- There are no repeating groups in the table. In other words, each row/column intersection can contain one and only one value, not a set of values.
- All attributes are dependent on the primary key.

All relational tables satisfy the 1NF requirements. The problem with the 1NF table structure shown in Figure 5.3 is that it contains partial dependencies—that is, dependencies based on only a part of the primary key.

While partial dependencies are sometimes used for performance reasons, they should be used with great caution. (If the information requirements seem to dictate the use of partial dependencies, it is time to evaluate the need for a Data Warehouse design, discussed in Chapter 12, "The Data Warehouse.") Such caution is warranted because a table that contains partial dependencies is still subject to data redundancies and, therefore, to various anomalies. The data redundancies are caused because every row entry requires duplication of data. For example, if the user makes 20 table entries for the condition EMP_NUM = 105 during the course of a day, the

EMP_NAME, JOB_CLASS, and CHG_HOUR must be entered each time, even though the attribute values are identical for EMP_NUM = 105. Such duplication of effort is very inefficient. What's more, the duplication of effort helps create data anomalies; nothing prevents the user from typing slightly different versions of the employee name, the position, or the hourly pay. For instance, the employee name for EMP_NUM = 102 might be entered as Dave Senior or D. Senior. The project name also might be entered correctly as Evergreen or misspelled as Evergeen. Such data anomalies violate the relational database's integrity and consistency rules.

5.1.3 CONVERSION TO SECOND NORMAL FORM

Fortunately, the relational database design can be improved easily by converting the database into a format known as the second normal form (2NF). The 1NF-to-2NF conversion is simple: Starting with the 1NF format displayed in Figure 5.3, you do the following:

Step 1: Identify All Key Components

Write each key component on a separate line, and then write the original (composite) key on the last line:

PROJ_NUM

EMP_NUM

PROJ_NUM EMP_NUM

Each component will become the key in a new table. In other words, the original table is now divided into three tables. We'll call these tables PROJECT, EMPLOYEE, and ASSIGN, respectively.

Step 2: Identify the Dependent Attributes

Use Figure 5.3 to determine which attributes are dependent on which other attributes. The dependencies for the original key components are found by examining the arrows below the dependency diagram shown in Figure 5.3. In other words, the three new tables, PROJECT, EMPLOYEE, and ASSIGN, are described by:

PROJECT (**PROJ_NUM**, PROJ_NAME)

EMPLOYEE (**EMP_NUM**, EMP_NAME, JOB_CLASS, CHG_HOUR)

ASSIGN (**PROJ_NUM**, **EMP_NUM**, ASSIGN_HOURS)

Because the number of hours spent on each project by each employee is dependent on both PROJ_NUM and EMP_NUM in the ASSIGN table, we place these hours in the ASSIGN table as ASSIGN_HOURS.

> **NOTE**
>
> The ASSIGN table contains a composite primary key composed of the attributes PROJ_NUM and EMP_NUM. Any attribute that is at least part of a key is known as a **prime attribute** or a **key attribute**. Therefore, both PROJ_NUM and EMP_NUM are prime (or key) attributes. Conversely, a **nonprime attribute**, or a **nonkey attribute**, is not even part of a key.

The results of steps 1 and 2 are displayed in Figure 5.4. At this point, most of the anomalies we discussed earlier have been eliminated. For example, if you now want to add/change/delete a PROJECT record, you only need to go to the PROJECT table and add/change/delete only one row.

FIGURE 5.4 SECOND NORMAL FORM (2NF) CONVERSION RESULTS

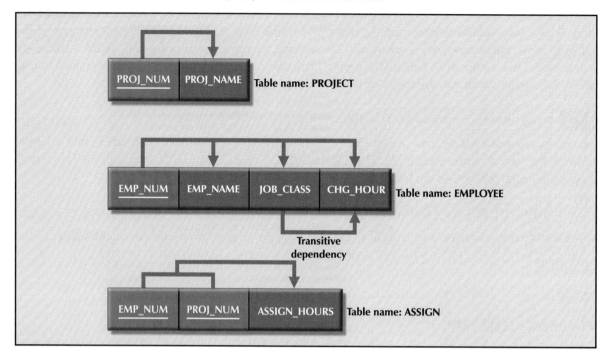

NOTE

A table is in **second normal form (2NF)** if:

- It is in 1NF.

 and

- It includes no partial dependencies; that is, no attribute is dependent on only a portion of the primary key. (But it is still possible for a table in 2NF to exhibit transitive dependency; that is, one or more attributes may be functionally dependent on nonkey attributes.)

Because a partial dependency can exist only if a table's primary key is composed of several attributes, a table whose primary key consists of only a single attribute is automatically in 2NF if it is in 1NF.

Figure 5.4 still shows a transitive dependency, which can generate anomalies. For example, if the charge per hour changes for a job classification which is held by many employees, that change must be made for *each* of those employees. If you forget to update some of the employee records that are affected by the charge per hour change, different employees with the same job description will generate different hourly charges.

5.1.4 CONVERSION TO THIRD NORMAL FORM

The data anomalies created by the database organization shown in Figure 5.4 are easily eliminated by completing the following three steps:

Step 1: Identify Each New Determinant

For every transitive dependency, write its determinant as a PK for a new table. (A **determinant** is any attribute whose value determines other values within a row.) If you have three different transitive dependencies, you will

have three different determinants. Figure 5.4 shows only one table that contains a transitive dependency. Therefore, write the determinant for this transitive dependency:

JOB_CLASS.

Step 2: Identify the Dependent Attributes

Identify the attributes that are dependent on each determinant identified in Step 1 and identify the dependency. In this case, you write:

JOB_CLASS → CHG_HOUR

Name the table to reflect its contents and function. In this case, JOB seems appropriate.

Step 3: Remove the Dependent Attributes from Transitive Dependencies

Eliminate all the dependent attributes in the transitive relationship(s) from each of the tables that have such a transitive relationship. In this example, eliminate CHG_HOUR from the EMPLOYEE table shown in Figure 5.4 to leave the EMPLOYEE table dependency definition as:

EMP_NUM → EMP_NAME, JOB_CLASS

Note that the JOB_CLASS remains in the EMPLOYEE table to serve as the FK.

Draw a new dependency diagram to show all the tables you have defined in Steps 1–3. Check the new tables as well as the tables you modified in Step 3 to make sure that each table has a determinant and that no table contains inappropriate dependencies.

When you have completed Steps 1–3, you will see the results in Figure 5.5. (The usual procedure is to complete Steps 1–3 by simply drawing the revisions as you make them.)

FIGURE 5.5 THIRD NORMAL FORM (3NF) CONVERSION RESULTS

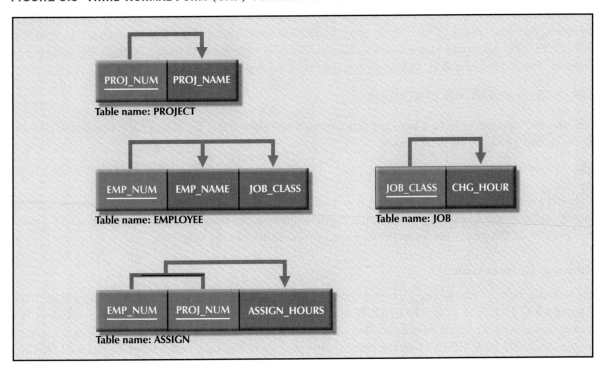

In other words, after the conversion has been completed, our database contains four tables:

PROJECT (**PROJ_NUM**, PROJ_NAME)

ASSIGN (**PROJ_NUM**, **EMP_NUM**, ASSIGN_HOURS)

EMPLOYEE (**EMP_NUM**, EMP_NAME, JOB_CLASS)

JOB (**JOB_CLASS**, CHG_HOUR)

Note that this conversion has eliminated the original EMPLOYEE table's transitive dependency; the tables are now said to be in third normal form (3NF).

> **N O T E**
>
> A table is in **third normal form (3NF)** if:
>
> - It is in 2NF.
>
> *and*
>
> - It contains no transitive dependencies.

5.1.5 IMPROVING THE DESIGN

The table structures are cleaned up to eliminate the troublesome initial partial and transitive dependencies. So we can now focus on improving the database's ability to provide information and on enhancing its operational characteristics. Remember that normalization cannot, by itself, be relied on to make good designs. Instead, it is valuable because its use helps eliminate data redundancies. Therefore, we have made the following changes:

PK Assignment

As the number of employees grows, a JOB_CLASS value must be entered each time a new employee is entered into the EMPLOYEE table. Unfortunately, it is too easy to make data-entry errors that lead to referential integrity violations. For example, a JOB_CLASS entry of DB Designer, rather than Database Designer, into the EMPLOYEE table will trigger such a violation, because the JOB table does not have a JOB_CODE entry of DB Designer. The addition of a JOB_CODE attribute produces the dependency:

JOB_CODE → JOB_CLASS, CHG_HOUR

This new attribute does produce a transitive dependency if you assume that the JOB_CODE is a proper primary key, because it produces the dependency:

JOB_CLASS → CHG_HOUR

A transitive dependency exists because a nonkey attribute—the JOB_CLASS—determines the value of another nonkey attribute—the CHARGE. However, this transitive dependency is an easy price to pay, because the presence of JOB_CODE greatly decreases the likelihood of referential integrity violations.

Naming Conventions

It is best to adhere to the naming conventions we outlined in Chapter 2, "Data Models." Therefore, we have changed CHG_HOUR to JOB_CHG_HOUR to indicate its association with the JOB table. In addition, the attribute name JOB_CLASS does not quite describe entries such as Systems Analyst, Database Designer, and so on; the label JOB_DESCRIPTION fits the entries better. Also, you may have noticed that we changed HOURS

to ASSIGN_HOURS when we converted the normal form from 1NF to 2NF. This change lets us associate the hours worked with the ASSIGN table.

Attribute Atomicity

It generally is good practice to pay attention to the *atomicity* requirement. (An **atomic** attribute is one that cannot be further subdivided. Such an attribute is said to display **atomicity**.) Clearly, the use of the EMP_NAME in the EMPLOYEE table is not atomic, because EMP_NAME can be decomposed into a last name, a first name, and an initial. By improving the degree of atomicity, we also gain querying flexibility. For example, if we use EMP_LNAME, EMP_FNAME, and EMP_INITIAL, we can easily generate phone lists by sorting last names, first names, and initials. Such a task would be very difficult if we kept the name components within a single attribute.

Adding Attributes

If the EMPLOYEE table were used in a real-world environment, quite a few other attributes would have to be added. For example, year-to-date gross salary payments, Social Security payments, or Medicare payments, are desirable. In our example, we have added an employee hire date. The EMP_HIREDATE can then be used to track employee "job longevity" and serve as a basis for awarding bonuses to long-term employees and for other morale-enhancing measures.

Adding Relationships

The system's ability to supply detailed information about each project's manager is ensured by using the EMP_NUM as a foreign key in PROJECT. This action ensures that we can access the details of each PROJECT's manager data without producing unnecessary and undesirable data duplication.

Refining PKs

Although the combination of EMP_NUM and PROJ_NUM may seem to be an acceptable primary key, using a transaction code such as ASSIGN_NUM is more precise and yields greater flexibility. For example, if the EMP_NUM and PROJ_NUM combination is used as the primary key and an employee makes two "hours worked" entries in the ASSIGN table, this action violates the entity integrity requirement. Even if we add the ASSIGN_DATE as part of a composite PK, we will still generate an entity integrity violation if any employee makes two or more entries for the same project on the same day. (The employee might have worked on the project during a few hours in the morning, then worked again on that project later that day.) The same data entry yields no problems if ASSIGN_NUM is used as the primary key, especially if the application software is designed to generate a new ASSIGN_NUM value automatically each time an entry is made. However, retaining EMP_NUM and PROJ_NUM as foreign keys is essential, because these attributes implement the relationships between ASSIGN, EMPLOYEE, and PROJECT. Thus we can track the details of all work performed by summing the ASSIGN_HOUR values by project, by employee, by date, and—through the relationship between EMPLOYEE and JOB—by job description.

Maintaining Historical Accuracy

Writing the job charge per hour into the ASSIGN table is crucial to maintain the historical accuracy of the data in the ASSIGN table. It would be appropriate to name this attribute ASSIGN_CHG_HOUR. Although this attribute would appear to have the same value as JOB_CHG_HOUR, that is true *only* if the JOB_CHG_HOUR value remains forever the same. However, it is reasonable to assume that the job charge per hour will change over time. But suppose that we relied on billing the charges to each project by multiplying the hours worked on the project in the ASSIGN table and the charge per hour found in the JOB table. Those charges would always show the current charge per hour stored in the JOB table, rather than the charge per hour that was in effect at

the time of the assignment. (See Chapter 3, Section 3.7 for a more detailed discussion on how historical data accuracy is maintained within the database.)

Using Derived Attributes

Finally, we can use a derived attribute in the ASSIGN table to store the actual charge made to a project. This derived attribute, to be named ASSIGN_CHARGE, is the result of multiplying the ASSIGN_HOURS by the ASSIGN_CHG_HOUR. From a strictly database point of view, such derived attribute values can be calculated when they are needed to write reports or invoices. However, storing the derived attribute in the table makes it easier to write the application software to produce the desired results. Also, if there are many transactions to be reported and/or summarized, the availability of the derived attribute will save reporting time. (If the calculation is done at the time of data entry, it will be completed when the end user presses the Enter key, thus avoiding a time penalty.)

We have included these enhancements in the tables and in the dependency diagrams shown in Figure 5.6.

FIGURE 5.6 THE COMPLETED DATABASE

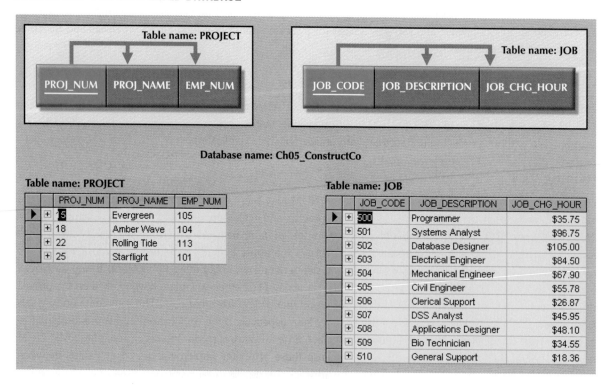

Database name: Ch05_ConstructCo

Table name: PROJECT

		PROJ_NUM	PROJ_NAME	EMP_NUM
▶	+	15	Evergreen	105
	+	18	Amber Wave	104
	+	22	Rolling Tide	113
	+	25	Starflight	101

Table name: JOB

		JOB_CODE	JOB_DESCRIPTION	JOB_CHG_HOUR
▶	+	500	Programmer	$35.75
	+	501	Systems Analyst	$96.75
	+	502	Database Designer	$105.00
	+	503	Electrical Engineer	$84.50
	+	504	Mechanical Engineer	$67.90
	+	505	Civil Engineer	$55.78
	+	506	Clerical Support	$26.87
	+	507	DSS Analyst	$45.95
	+	508	Applications Designer	$48.10
	+	509	Bio Technician	$34.55
	+	510	General Support	$18.36

FIGURE 5.6 THE COMPLETED DATABASE (CONTINUED)

Table name: ASSIGN

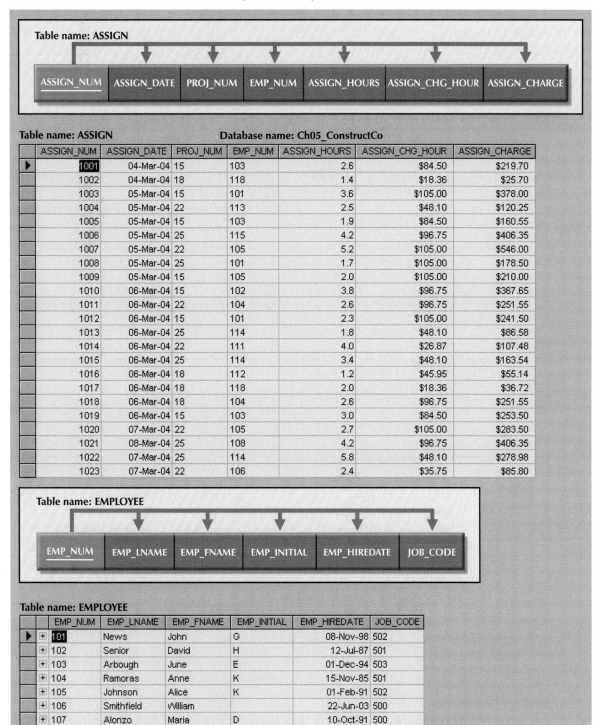

| ASSIGN_NUM | ASSIGN_DATE | PROJ_NUM | EMP_NUM | ASSIGN_HOURS | ASSIGN_CHG_HOUR | ASSIGN_CHARGE |

Table name: ASSIGN Database name: Ch05_ConstructCo

ASSIGN_NUM	ASSIGN_DATE	PROJ_NUM	EMP_NUM	ASSIGN_HOURS	ASSIGN_CHG_HOUR	ASSIGN_CHARGE
1001	04-Mar-04	15	103	2.6	$84.50	$219.70
1002	04-Mar-04	18	118	1.4	$18.36	$25.70
1003	05-Mar-04	15	101	3.6	$105.00	$378.00
1004	05-Mar-04	22	113	2.5	$48.10	$120.25
1005	05-Mar-04	15	103	1.9	$84.50	$160.55
1006	05-Mar-04	25	115	4.2	$96.75	$406.35
1007	05-Mar-04	22	105	5.2	$105.00	$546.00
1008	05-Mar-04	25	101	1.7	$105.00	$178.50
1009	05-Mar-04	15	105	2.0	$105.00	$210.00
1010	06-Mar-04	15	102	3.8	$96.75	$367.65
1011	06-Mar-04	22	104	2.6	$96.75	$251.55
1012	06-Mar-04	15	101	2.3	$105.00	$241.50
1013	06-Mar-04	25	114	1.8	$48.10	$86.58
1014	06-Mar-04	22	111	4.0	$26.87	$107.48
1015	06-Mar-04	25	114	3.4	$48.10	$163.54
1016	06-Mar-04	18	112	1.2	$45.95	$55.14
1017	06-Mar-04	18	118	2.0	$18.36	$36.72
1018	06-Mar-04	18	104	2.6	$96.75	$251.55
1019	06-Mar-04	15	103	3.0	$84.50	$253.50
1020	07-Mar-04	22	105	2.7	$105.00	$283.50
1021	08-Mar-04	25	108	4.2	$96.75	$406.35
1022	07-Mar-04	25	114	5.8	$48.10	$278.98
1023	07-Mar-04	22	106	2.4	$35.75	$85.80

Table name: EMPLOYEE

| EMP_NUM | EMP_LNAME | EMP_FNAME | EMP_INITIAL | EMP_HIREDATE | JOB_CODE |

Table name: EMPLOYEE

EMP_NUM	EMP_LNAME	EMP_FNAME	EMP_INITIAL	EMP_HIREDATE	JOB_CODE
101	News	John	G	08-Nov-98	502
102	Senior	David	H	12-Jul-87	501
103	Arbough	June	E	01-Dec-94	503
104	Ramoras	Anne	K	15-Nov-85	501
105	Johnson	Alice	K	01-Feb-91	502
106	Smithfield	William		22-Jun-03	500
107	Alonzo	Maria	D	10-Oct-91	500
108	Washington	Ralph	B	22-Aug-89	501
109	Smith	Larry	W	18-Jul-95	501
110	Olenko	Gerald	A	11-Dec-93	505
111	Wabash	Geoff	B	04-Apr-89	506
112	Smithson	Darlene	M	23-Oct-92	507
113	Joenbrood	Delbert	K	15-Nov-94	508
114	Jones	Annelise		20-Aug-91	508
115	Bawangi	Travis	B	25-Jan-90	501
116	Pratt	Gerald	L	05-Mar-95	510
117	Williamson	Angie	H	19-Jun-94	509
118	Frommer	James	J	04-Jan-04	510

Figure 5.6 is a vast improvement over the original database design. If the application software is designed properly, the most active table (ASSIGN) requires the entry of only the PROJ_NUM, EMP_NUM, and ASSIGN_HOURS values. The values for the attributes ASSIGN_NUM and ASSIGN_DATE can be generated by the application. For example, the ASSIGN_NUM can be created by using a counter, and the ASSIGN_DATE can be the system date read by the application and automatically entered into the ASSIGN table. In addition, the application software can automatically insert the correct ASSIGN_CHG_HOUR value by writing the appropriate JOB table's JOB_CHG_HOUR value into the ASSIGN table. (The JOB and ASSIGN tables are related though the JOB_CODE!) If the JOB table's JOB_CHG_HOUR value changes, the next insertion of that value into the ASSIGN table will reflect the change automatically. The table structure thus minimizes the need for human intervention. In fact, if the system requires the employees to enter their own work hours, the EMP_NUM can be scanned into the ASSIGN table by using a magnetic card reader that enters the employee's identity. The ASSIGN table's structure thus can set the stage for maintaining some desired level of security.

5.1.6 LIMITATIONS ON SYSTEM-ASSIGNED KEYS

Although our design meets the vital entity and referential requirements, there are still some concerns the designer must address. You should recall that we created the JOB_CODE attribute and designated it to be the JOB table's primary key (PK) to ensure entity integrity in the JOB table. The DBMS may then be used to have the system assign the PK values. However, it is useful to remember that the JOB_CODE does not prevent us from making the entries in the JOB table shown in Table 5.2.

TABLE 5.2 DUPLICATE ENTRIES IN THE JOB TABLE

JOB_CODE	JOB_DESCRIPTION	JOB_CHG_HOUR
511	Programmer	$35.75
512	Programmer	$35.75

Clearly, the data entries in Table 5.2 are inappropriate because they duplicate existing records—yet there has been no violation of either entity integrity or referential integrity. This "multiple duplicate records" problem was created when we added the JOB_CODE to become the PK. (When the JOB_DESCRIPTION was initially designated to be the PK, the DBMS would ensure unique values for all job description entries if it were asked to enforce entity integrity. But that option created the problems that caused us to use the JOB_CODE attribute in the first place!) In any case, if JOB_CODE is to be the PK, we still must ensure the existence of unique values in the JOB_DESCRIPTION through the use of a unique index.

Note that all the remaining tables (PROJECT, ASSIGN, and EMPLOYEE) are subject to the same limitations. For example, if we use the EMP_NUM attribute in the EMPLOYEE table as the PK, we can make multiple entries for the same employee. To avoid that problem, we might create a unique index for EMP_LNAME, EMP_FNAME, and EMP_INITIAL. But how would we then deal with two employees named Joe B. Smith? In this case, we might use another (preferably externally defined) attribute such as Social Security number to serve as the basis for a unique index.

It is worth repeating that database design often involves trade-offs and the exercise of professional judgment. In a real-world environment, we must strike a balance between design integrity and flexibility. For example, we might design the ASSIGN table to use a unique index on PROJ_NUM, EMP_NUM and ASSIGN_DATE if we want to limit an employee to only one ASSIGN_HOURS entry per date. This limitation would ensure that employees couldn't enter the same hours multiple times for any given date. Unfortunately, this limitation likely is undesirable from a managerial point of view. After all, if an employee works several different times on a given project during any given day, it must be possible to make multiple entries for that same employee and the same

project during that day. In this case, the best solution might be to add a new externally defined attribute—such as a stub, voucher, or ticket number—to ensure uniqueness. In any case, frequent data audits would be appropriate.

5.1.7 THE BOYCE-CODD NORMAL FORM (BCNF)

A table is in **Boyce-Codd normal form** (**BCNF**) if every determinant in the table is a candidate key. (Recall from Chapter 3, "The Relational Database Model," that a candidate key has the same characteristics as a primary key, but for some reason it was not chosen to be the primary key.) Clearly, if a table contains only one candidate key, the 3NF and the BCNF are equivalent. Putting this proposition another way, BCNF can be violated only if the table contains more than one candidate key.

Most designers consider the Boyce-Codd normal form (BCNF) as a special case of the 3NF. In fact, if you use the techniques we have shown, most tables conform to the BCNF requirements once the 3NF is reached. So how can a table be in 3NF and not be in BCNF? To answer this question you must keep in mind that a transitive dependency exists when one nonprime attribute is dependent on another nonprime attribute.

In other words, a table is in 3NF if it is in 2NF and there are no transitive dependencies. But what about a case in which a nonkey attribute is the determinant of a key attribute? That condition does not violate 3NF, yet it fails to meet the BCNF requirements, because BCNF requires that *every* determinant in the table be a candidate key.

The situation we have just described (a 3NF table that fails to meet BCNF requirements) is most easily understood by examining Figure 5.7.

FIGURE 5.7 A TABLE THAT IS IN 3NF BUT NOT IN BCNF

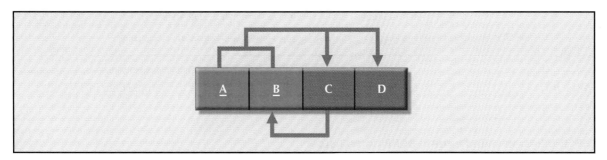

As you examine Figure 5.7, note these functional dependencies:

$A + B \rightarrow C, D$

$C \rightarrow B$

The table structure shown in Figure 5.7 has no partial dependencies, nor does it contain transitive dependencies. (The condition $C \rightarrow B$ indicates that *a nonkey attribute determines part of the primary key*—and *this* dependency is *not* transitive!) Thus, the table structure in Figure 5.6 meets the 3NF requirements. Yet the condition $C \rightarrow B$ causes the table to fail to meet the BCNF requirements.

To convert the table structure in Figure 5.7 into table structures that are in 3NF and in BCNF, first change the primary key to A + C. This is an appropriate action, because the dependency $C \rightarrow B$ means that C is, in effect, a superset of B. At this point, the table is in 1NF, because it contains a partial dependency $C \rightarrow B$. Next, follow our standard decomposition procedures to produce the results shown in Figure 5.8.

FIGURE 5.8 DECOMPOSITION TO BCNF

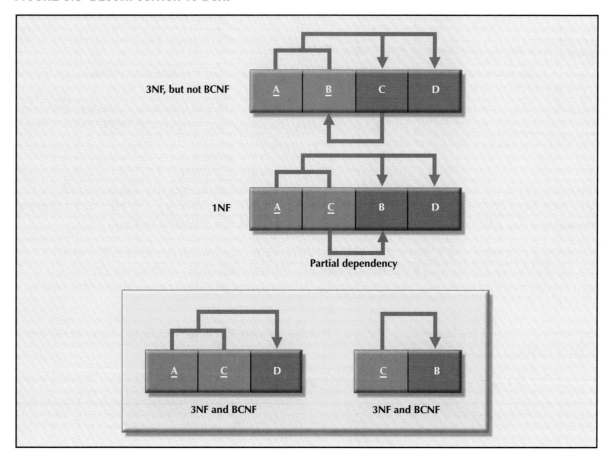

To see how this procedure can be applied to an actual problem, let's examine the sample data in Table 5.3.

TABLE 5.3 SAMPLE DATA FOR A BCNF CONVERSION

STU_ID	STAFF_ID	CLASS_CODE	ENROLL_GRADE
125	25	21334	A
125	20	32456	C
135	20	28458	B
144	25	27563	C
144	20	32456	B

Table 5.3 reflects the following conditions:

- Each CLASS_CODE identifies a class uniquely. This condition illustrates the examples we used in Chapters 2 and 3, in which a course might generate many classes. For example, a course labeled INFS 420 might be taught in two classes (sections), each identified by a unique code to facilitate registration. Thus the CLASS_CODE 32456 might identify INFS 420, class section 1, while the CLASS_CODE 32457 might identify INFS 420, class section 2. Or the CLASS_CODE 28458 might identify QM 362, class section 5.

- A student can take many classes. Note, for example, that student 125 has taken both 21334 and 32456, earning the grades A and C, respectively.

- A staff member can teach many classes, but each class is taught by only one staff member. Note that staff member 20 teaches the classes identified as 32456 and 28458.

The structure shown in Table 5.3 is reflected in Panel A of Figure 5.9:

STU_ID + STAFF_ID → CLASS_CODE, ENROLL_GRADE

CLASS_CODE → STAFF_ID

FIGURE 5.9 ANOTHER BCNF DECOMPOSITION

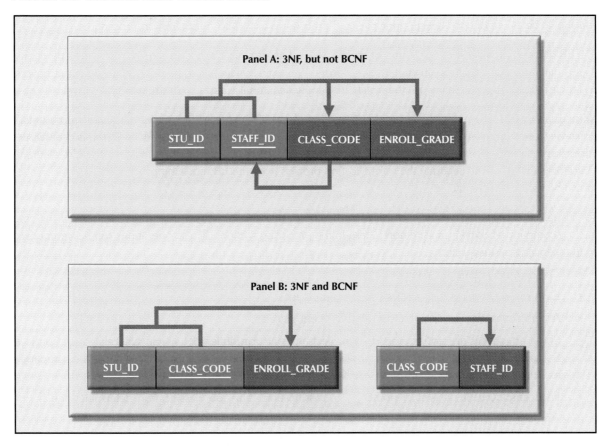

Panel A of Figure 5.9 shows a structure that is clearly in 3NF, but the table represented by this structure has a major problem; because it is trying to describe two things, it causes anomalies. For example, if a different staff member is assigned to teach class 32456, two rows will require updates, thus producing an update anomaly. And, if student 135 drops class 28458, we lose information about who taught that class, thus producing a deletion anomaly. The solution to this problem is to decompose the table structure, following the procedure we outlined earlier. Note that the decomposition of Panel B shown in Figure 5.9 yields two table structures that conform to both 3NF and BCNF requirements.

Remember that a table is in BCNF if *every* determinant in that table is a candidate key. Therefore, if a table contains only one candidate key, 3NF and BCNF are equivalent.

5.2 NORMALIZATION AND DATABASE DESIGN

The tables shown in Figure 5.6 were created to illustrate how normalization procedures can produce good tables from poor ones. We suspect that you will have ample opportunity to put this skill into practice when you begin to work with real-world databases. *Normalization should be part of the design process.* Therefore, make sure that proposed entities meet the required normal form *before* the table structures are created. Keep in mind that, if you follow the design procedures we have discussed in Chapters 3 and 4, the likelihood of data anomalies will be small. (But even the best of us are known to make occasional design mistakes that come to light during normalization checks!) However, many of the real-world databases you encounter will either have been improperly designed or burdened with anomalies if they were improperly modified during the course of time. And that means you may be asked to redesign and modify existing databases that are, in effect, anomaly traps. Therefore, you should be aware of good design principles and procedures as well as normalization procedures.

The ER diagram provides the big picture, or macro view, of an organization's data requirements and operations. An ER diagram is created through an iterative process. We begin by identifying relevant entities, their attributes, and their relationships. Then we use the results to identify additional entities and attributes.

Normalization procedures focus on the characteristics of specific entities; that is, normalization represents a micro view of the entities within the ER diagram. And, as you learned in the previous sections of this chapter, the normalization process may yield additional entities and attributes to be incorporated into the ER diagram. Therefore, it is difficult to separate the normalization process from the ER modeling process; the two techniques should be used concurrently.

To illustrate the proper role of normalization in the design process, let's reexamine the operations of the contracting company whose tables were normalized in the preceding sections. These operations can be summarized by the following business rules:

- The company manages many projects.

- Each project requires the services of many employees.

- An employee may be assigned to several different projects.

- Some employees are not assigned to a project and perform duties not specifically related to a project. Some employees are part of a labor pool, to be shared by all project teams. For example, the company's executive secretary would not be assigned to any one particular project.

- Each employee has a (single) primary job classification. This job classification determines the hourly billing rate.

- Many employees can have the same job classification. For example, the company employs more than one electrical engineer.

Given this simple description of the company's operations, two entities and their attributes are initially defined:

- PROJECT (**PROJ_NUM**, PROJ_NAME)

- EMPLOYEE (**EMP_NUM**, EMP_LNAME, EMP_FNAME, EMP_INITIAL, JOB_DESCRIPTION, JOB_CHG_HOUR)

These two entities constitute the initial ER diagram shown in Figure 5.10.

After creating the initial ER diagram shown in Figure 5.10, the normal forms are defined:

- PROJECT is in 3NF and needs no modification at this point.

- EMPLOYEE requires additional scrutiny. The JOB_DESCRIPTION defines job classifications such as Systems Analyst, Database Designer, and Programmer. In turn, these classifications determine the billing rate, JOB_CHG_HOUR. Therefore, EMPLOYEE contains a transitive dependency.

FIGURE 5.10 **THE INITIAL ERD FOR A CONTRACTING COMPANY**

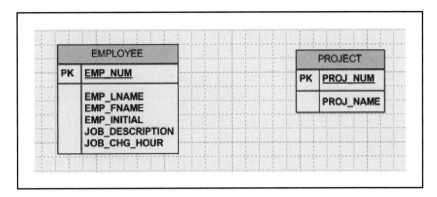

The removal of EMPLOYEE's transitive dependency yields three entities:

- PROJECT (**PROJ_NUM**, PROJ_NAME)
- EMPLOYEE (**EMP_NUM**, EMP_LNAME, EMP_FNAME, EMP_INITIAL, JOB_CODE)
- JOB (**JOB_CODE**, JOB_DESCRIPTION, JOB_CHG_HOUR)

Because the normalization process yields an additional entity (JOB), we modify the initial ER diagram as shown in Figure 5.11.

FIGURE 5.11 **THE MODIFIED ERD FOR A CONTRACTING COMPANY**

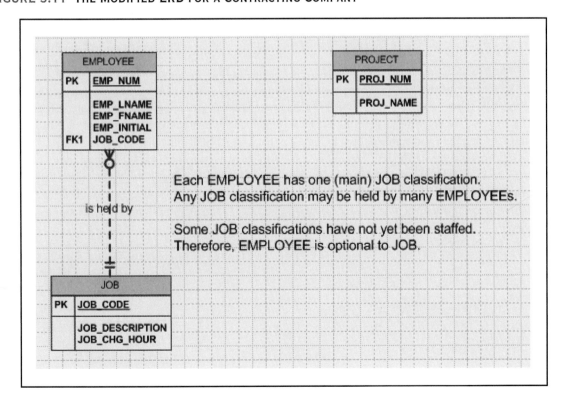

To represent the M:N relationship between EMPLOYEE and PRODUCT, you might think that two 1:M relationships could be used—an employee can be assigned to many projects and each project can have many employees assigned to it (see Figure 5.12). Unfortunately, this representation yields a design that can not be correctly implemented.

FIGURE 5.12 THE INCORRECT REPRESENTATION OF A M:N RELATIONSHIP

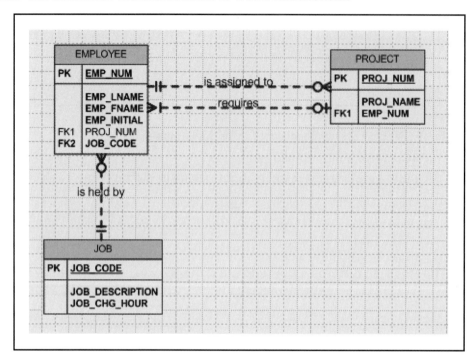

Because the M:N relationship between EMPLOYEE and PROJECT cannot be implemented, the ERD in Figure 5.11 must be modified to include the ASSIGNMENT entity to track the assignment of employees to projects, thus yielding the ER diagram shown in Figure 5.13. The ASSIGNMENT entity in Figure 5.13 uses the primary keys from the entities PROJECT and EMPLOYEE to serve as its foreign keys. However, note that in this implementation ASSIGNMENT is not a composite entity, because its PK is ASSIGN_NUM, which is not inherited from either the PROJECT or the EMPLOYEE. Therefore, the *enters* relationship between EMPLOYEE and ASSIGNMENT and the *requires* relationship between PROJECT and ASSIGNMENT are shown as weak or nonidentifying.

FIGURE 5.13 THE FINAL (IMPLEMENTABLE) ERD FOR A CONTRACTING COMPANY

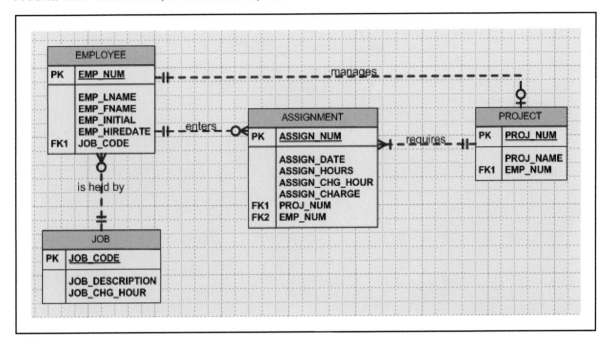

As you examine Figure 5.13, note that the ASSIGN_HOURS attribute is assigned to the composite entity named ASSIGN. Because we will likely need detailed information about each project's manager, the creation of a "manages" relationship is useful. The "manages" relationship is implemented through the foreign key in PROJECT. Finally, some additional attributes may be created to improve the system's ability to generate additional information. For example, you may want to include the date on which the employee was hired (EMP_HIREDATE) in order to keep track of worker longevity. Based on this last modification, the model should include four entities and their attributes:

PROJECT (**PROJ_NUM**, PROJ_NAME, EMP_NUM)

EMPLOYEE (**EMP_NUM**, EMP_LNAME, EMP_FNAME, EMP_INITIAL, EMP_HIREDATE,

JOB_CODE)

JOB (**JOB_CODE**, JOB_DESCRIPTION, JOB_CHG_HOUR)

ASSIGNMENT (**ASSIGN_NUM**, ASSIGN_DATE, ASSIGN_HOURS, ASSIGN_CHG_HOUR,

ASSIGN_CHARGE, EMP_NUM, PROJ_NUM)

The design process is now on the right track. The ER diagram represents the operations accurately, and the entities now reflect their conformance to 3NF. The combination of normalization and ER modeling yields a useful ER diagram, whose entities may now be translated into appropriate table structures. As you examine Figure 5.13, note that PROJECT is optional to EMPLOYEE in the "manages" relationship. This optionality exists because not all employees manage projects. The final database contents are shown in Figure 5.14.

FIGURE 5.14 THE IMPLEMENTED DATABASE FOR THE CONTRACTING COMPANY

Table name: EMPLOYEE

Database name: Ch05_ConstructCo

EMP_NUM	EMP_LNAME	EMP_FNAME	EMP_INITIAL	EMP_HIREDATE	JOB_CODE
101	News	John	G	08-Nov-98	502
102	Senior	David	H	12-Jul-87	501
103	Arbough	June	E	01-Dec-94	503
104	Ramoras	Anne	K	15-Nov-85	501
105	Johnson	Alice	K	01-Feb-91	502
106	Smithfield	William		22-Jun-03	500
107	Alonzo	Maria	D	10-Oct-91	500
108	Washington	Ralph	B	22-Aug-89	501
109	Smith	Larry	W	18-Jul-95	501
110	Olenko	Gerald	A	11-Dec-93	505
111	Wabash	Geoff	B	04-Apr-89	506
112	Smithson	Darlene	M	23-Oct-92	507
113	Joenbrood	Delbert	K	15-Nov-94	508
114	Jones	Annelise		20-Aug-91	508
115	Bawangi	Travis	B	25-Jan-90	501
116	Pratt	Gerald	L	05-Mar-95	510
117	Williamson	Angie	H	19-Jun-94	509
118	Frommer	James	J	04-Jan-04	510

Table name: JOB

JOB_CODE	JOB_DESCRIPTION	JOB_CHG_HOUR
500	Programmer	$35.75
501	Systems Analyst	$96.75
502	Database Designer	$105.00
503	Electrical Engineer	$84.50
504	Mechanical Engineer	$67.90
505	Civil Engineer	$55.78
506	Clerical Support	$26.87
507	DSS Analyst	$45.95
508	Applications Designer	$48.10
509	Bio Technician	$34.55
510	General Support	$18.36

Table name: PROJECT

PROJ_NUM	PROJ_NAME	EMP_NUM
15	Evergreen	105
18	Amber Wave	104
22	Rolling Tide	113
25	Starflight	101

Table name: ASSIGNMENT

ASSIGN_NUM	ASSIGN_DATE	PROJ_NUM	EMP_NUM	ASSIGN_HOURS	ASSIGN_CHG_HOUR	ASSIGN_CHARGE
1001	04-Mar-04	15	103	2.6	$84.50	$219.70
1002	04-Mar-04	18	118	1.4	$18.36	$25.70
1003	05-Mar-04	15	101	3.6	$105.00	$378.00
1004	05-Mar-04	22	113	2.5	$48.10	$120.25
1005	05-Mar-04	15	103	1.9	$84.50	$160.55
1006	05-Mar-04	25	115	4.2	$96.75	$406.35
1007	05-Mar-04	22	105	5.2	$105.00	$546.00
1008	05-Mar-04	25	101	1.7	$105.00	$178.50
1009	05-Mar-04	15	105	2.0	$105.00	$210.00
1010	06-Mar-04	15	102	3.8	$96.75	$367.65
1011	06-Mar-04	22	104	2.6	$96.75	$251.55
1012	06-Mar-04	15	101	2.3	$105.00	$241.50
1013	06-Mar-04	25	114	1.8	$48.10	$86.58
1014	06-Mar-04	22	111	4.0	$26.87	$107.48
1015	06-Mar-04	25	114	3.4	$48.10	$163.54
1016	06-Mar-04	18	112	1.2	$45.95	$55.14
1017	06-Mar-04	18	118	2.0	$18.36	$36.72
1018	06-Mar-04	18	104	2.6	$96.75	$251.55
1019	06-Mar-04	15	103	3.0	$84.50	$253.50
1020	07-Mar-04	22	105	2.7	$105.00	$283.50
1021	08-Mar-04	25	108	4.2	$96.75	$406.35
1022	07-Mar-04	25	114	5.8	$48.10	$278.98
1023	07-Mar-04	22	106	2.4	$35.75	$85.80

5.3 HIGHER-LEVEL NORMAL FORMS

You may encounter poorly designed databases or you may be asked to convert spreadsheets into a database format in which multiple multivalued attributes exist. For example, let's consider the possibility that an employee can have multiple assignments and can also be involved in multiple service organizations.

Suppose employee 10123 does volunteer work for the Red Cross and United Way. In addition, the same employee might be assigned to work on three projects: 1, 3, and 4. Figure 5.15 illustrates how this set of facts can be recorded in very different ways.

FIGURE 5.15 TABLES WITH MULTIVALUED DEPENDENCIES

Database name: Ch05_Service

Table name: VOLUNTEER_V1

EMP_NUM	ORG_CODE	ASSIGN_NUM
10123	RC	1
10123	UW	3
10123		4

Table name: VOLUNTEER_V2

EMP_NUM	ORG_CODE	ASSIGN_NUM
10123	RC	
10123	UW	
10123		1
10123		3
10223		4

Table name: VOLUNTEER_V3

EMP_NUM	ORG_CODE	ASSIGN_NUM
10123	RC	1
10123	RC	3
10123	UW	4

If you examine the tables in Figure 5.14, we seem to have a problem. The attributes ORG_CODE and ASSIGN_NUM each may have many different values. That is, the tables contain two sets of independent multivalued dependencies. (One employee can have many service entries and many assignment entries.) The presence of multiple sets of independent multivalued dependencies means that, if versions 1 and 2 are implemented, the tables are likely to contain quite a few null values and, in fact, these tables do not even have a viable candidate key. (The EMP_NUM values are not unique, so they cannot be PKs. No combination of the attributes in table versions 1 and 2 can be used to create a PK, because some of them contain nulls.) Such a condition is not desirable, especially if there are thousands of employees, many of whom may have multiple job assignments and many service activities. Version 3 at least has a PK, but it is composed of all the attributes in the table. In fact, version 3 meets 3NF requirements, yet it contains many redundancies that are clearly undesirable.

The solution is to eliminate the problems caused by independent multivalued dependencies. You do this by creating the ASSIGNMENT and SERVICE_V1 tables depicted in Figure 5.16. As you examine Figure 5.16, note that neither the ASSIGNMENT nor the SERVICE_V1 table contains independent multivalued dependencies. Such tables are said to be in fourth normal form (4NF). We can define 4NF this way:

> **NOTE**
>
> A table is in **fourth normal form (4NF)** if it is in 3NF and has no multiple sets of multivalued dependencies.

If you follow the proper design procedures we have illustrated in this book, you should never encounter the previously described problem. Specifically, the discussion of 4NF is largely academic if you make sure that your tables always conform to the following two rules:

1. All attributes must be dependent on the primary key, but they must be independent of each other.

2. No row may contain two or more multivalued facts about an entity.

Even higher-level normal forms exist. However, normal forms such as the 5NF and domain-key normal form (DKNF) are not likely to be encountered in a business environment and are mainly of theoretical interest. Because we are focusing on practical applications of database techniques, we will not cover the higher-level normal forms in this text.

FIGURE 5.16 A SET OF TABLES IN 4NF

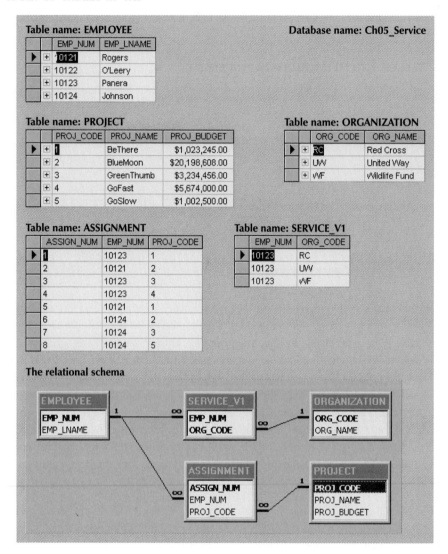

5.4 DENORMALIZATION

Although the creation of normalized relations is an important database design goal, it is only one of many such goals. Good database design also considers processing requirements. As tables are decomposed to conform to normalization requirements, the number of database tables expands. Joining the larger number of tables takes additional disk input/output (I/O) operations and processing logic, thereby reducing system speed. Consequently, there may be very occasional circumstances that allow some degree of denormalization in order to increase processing speed.

Keep in mind that the advantage of higher processing speed must be carefully weighed against the disadvantage of data anomalies. On the other hand, some anomalies are of only theoretical interest. For example, should

people in a real-world database environment worry that a ZIP_CODE determines CITY in a CUSTOMER table whose primary key is the customer number? Is it really practical to produce a separate table:

ZIP (**ZIP_CODE**, CITY)

to eliminate a transitive dependency from the CUSTOMER table? (Perhaps your answer to this question changes if you are in the business of producing mailing lists.) Our advice is simple: use some common sense during the normalization process.

Normalization purity is often difficult to sustain in the modern database environment. The conflicts between design efficiency, information requirements, and processing speed are often resolved through compromises that may include denormalization. You will also learn (in Chapter 12) that lower normalization forms occur (and are even required) in specialized databases known as *data warehouses*. Such specialized databases reflect the ever-growing demand for greater scope and depth in the data on which decision support systems increasingly rely. You will discover that the data warehouse routinely uses 2NF structures in its complex, multilevel, multisource data environment. In short, although normalization is very important, especially in the so-called production database environment we have examined thus far, 2NF is no longer disregarded as it once was.

Although 2NF tables cannot always be avoided, we do not mean to minimize the problem of working with tables that contain partial and/or transitive dependencies in a production database environment. Aside from possibly creating troublesome data anomalies, unnormalized tables in a production database tend to suffer from these defects:

- Data updates are less efficient because programs that read and update tables must deal with larger tables.

- Indexing is much more cumbersome. It simply is not practical to build all the indexes required for the many attributes that may be located in a single unnormalized table.

- Unnormalized tables yield no simple strategies for creating virtual tables known as *views*. (You will learn how to create and use views in Chapter 6, "Introduction to Structured Query Language (SQL).")

Remember this: good design cannot be created in the application programs that use a database. Also keep in mind that unnormalized database tables often lead to various data redundancy disasters in production databases such as the ones we have examined thus far. In other words, use denormalization cautiously and make sure that you can explain why—under some circumstances—the unnormalized tables are a better choice than their normalized counterparts.

SUMMARY

Normalization is a technique used to design tables in which data redundancies are minimized. The first three normal forms (1NF, 2NF, and 3NF) are most commonly encountered. From a structural point of view, higher normal forms are better than lower normal forms because higher normal forms yield relatively fewer data redundancies in the database. In other words, 3NF is better than 2NF, which is better than 1NF. Almost all business designs use the 3NF as the ideal normal form. (A special, more restricted 3NF is known as Boyce-Codd normal form, or BCNF.)

A table is in 1NF when all the key attributes are defined and when all remaining attributes are dependent on the primary key. However, a table in 1NF can still contain both partial and transitive dependencies. (A partial dependency is one in which an attribute is functionally dependent on only a part of a multi-attribute primary key. A transitive dependency is one in which one attribute is functionally dependent on another nonkey attribute.) Naturally, a table with a single-attribute primary key cannot exhibit partial dependencies.

A table is in 2NF when it is in 1NF and contains no partial dependencies. Therefore, a 1NF table is automatically in 2NF if its primary key is based on only a single attribute. A table in 2NF may still contain transitive dependencies.

A table is in 3NF if it is in 2NF and contains no transitive dependencies. Given this definition of 3NF, the Boyce-Codd normal form (BCNF) is merely a special 3NF case, in which all the determinant keys are candidate keys. Naturally, if a table has only a single candidate key, a 3NF table is automatically in BCNF.

A table that is not in 3NF may be split into new tables until all the tables meet the 3NF requirements. The process is illustrated in Figures 5.17 through 5.19.

FIGURE 5.17 THE INITIAL 1NF STRUCTURE

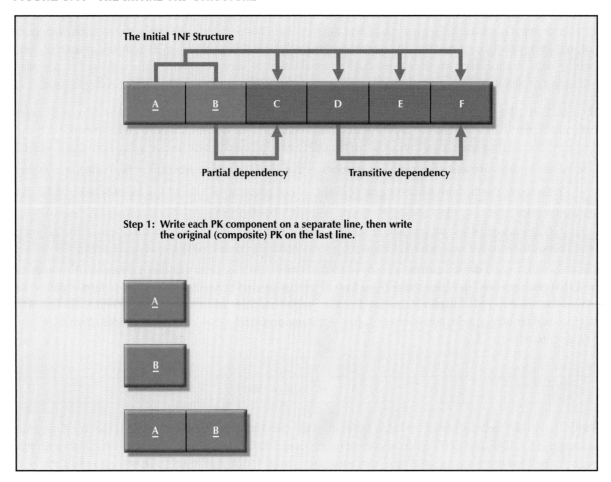

FIGURE 5.18 IDENTIFYING POSSIBLE PK ATTRIBUTES

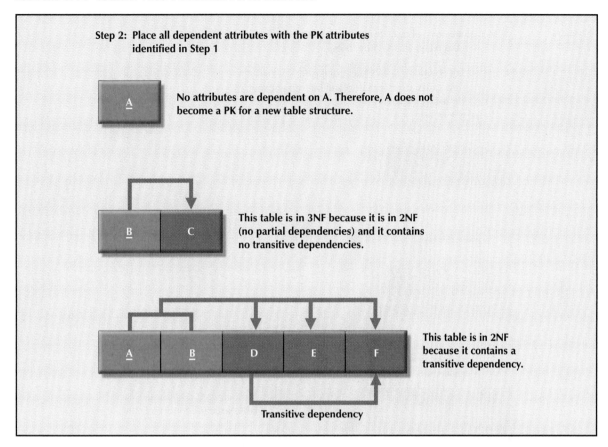

Normalization is an important part—but only a part—of the design process. As entities and attributes are defined during the ER modeling process, subject each entity (set) to normalization checks and form new entity (sets) as required. Incorporate the normalized entities into the ER diagram and continue the iterative ER process until all entities and their attributes are defined and all equivalent tables are in 3NF.

A table in 3NF may contain multivalued dependencies that produce either numerous null values or redundant data. Therefore, it may be necessary to convert a 3NF table to the fourth normal form (4NF) by splitting such a table to remove the multivalued dependencies. Thus, a table is in 4NF if it is in 3NF and contains no multivalued dependencies.

The larger the number of tables, the more additional disk I/O is required to join them, and the greater the amount of processing logic. Therefore, tables are sometimes denormalized to yield less I/O in order to increase processing speed. Unfortunately, you pay for the increased processing speed by making the (larger table) data updates less efficient, by making indexing more cumbersome, and by introducing data redundancies that are likely to yield data anomalies. In the design of production databases, use denormalization sparingly and cautiously.

FIGURE 5.19 TABLE STRUCTURES BASED ON THE SELECTED PKs

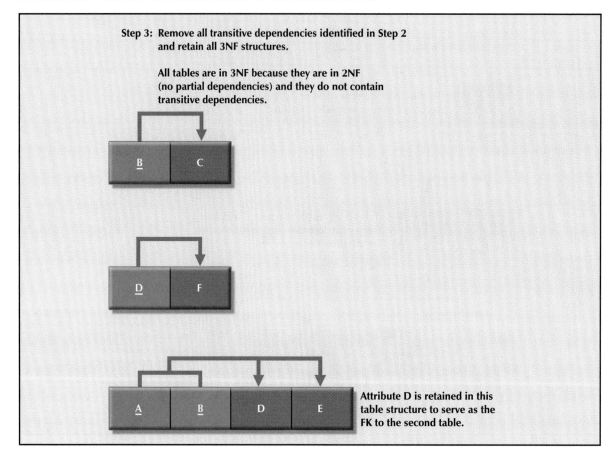

Step 3: Remove all transitive dependencies identified in Step 2
 and retain all 3NF structures.

 All tables are in 3NF because they are in 2NF
 (no partial dependencies) and they do not contain
 transitive dependencies.

Attribute D is retained in this
table structure to serve as the
FK to the second table.

KEY TERMS

atomic	first normal form (1NF)	partial dependency
atomicity	fourth normal form (4NF)	prime attribute
Boyce-Codd normal form (BCNF)	key attribute	repeating group
denormalization	nonkey attribute	second normal form (2NF)
dependency diagram	nonprime attribute	third normal form (3NF)
determinant	normalization	transitive dependency

REVIEW QUESTIONS

1. What is normalization?
2. When is a table in 1NF?
3. When is a table in 2NF?
4. When is a table in 3NF?
5. When is a table in BCNF?

6. Given the dependency diagram shown in Figure Q5.6, answer Questions 6a through 6c:

FIGURE Q5.6 THE DEPENDENCY DIAGRAM FOR QUESTION 6

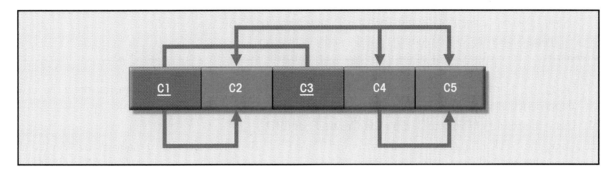

a. Identify and discuss each of the indicated dependencies.
b. Create a database whose tables are at least in 2NF, showing the dependency diagrams for each table.
c. Create a database whose tables are at least in 3NF, showing the dependency diagrams for each table.

7. What is a partial dependency? With what normal form is it associated?

8. What three data anomalies are likely to be the result of data redundancy? How can such anomalies be eliminated?

9. Define and discuss the concept of transitive dependency.

PROBLEMS

1. Using the INVOICE table structure shown in Table P5.1, draw its dependency diagram and identify all dependencies (including all partial and transitive dependencies). You can assume that the table does not contain repeating groups and that any invoice number may reference more than one product. (*Hint*: This table uses a composite primary key.)

TABLE P5.1 SAMPLE INVOICE RECORDS

ATTRIBUTE NAME	SAMPLE VALUE	SAMPLE VALUE	SAMPLE VALUE	SAMPLE VALUE	SAMPLE VALUE
INV_NUM	211347	211347	211347	211348	211349
PROD_NUM	AA-E3422QW	QD-300932X	RU-995748G	AA-E3422QW	GH-778345P
SALE_DATE	15-Jan-2004	15-Jan-2004	15-Jan-2004	15-Jan-2004	16-Jan-2004
PROD_LABEL	Rotary sander	0.25-in. drill bit	Band saw	Rotary sander	Power drill
VEND_CODE	211	211	309	211	157
VEND_NAME	NeverFail, Inc.	NeverFail, Inc.	BeGood, Inc.	NeverFail, Inc.	ToughGo, Inc.
QUANT_SOLD	1	8	1	2	1
PROD_PRICE	$49.95	$3.45	$39.99	$49.95	$87.75

2. Using the initial dependency diagram drawn in Problem 1, remove all partial dependencies, draw the new dependency diagrams, and identify the normal forms for each table structure you created.

N O T E

You can assume that any given product is supplied by a single vendor, but a vendor can supply many products. Therefore, it is proper to conclude that the following dependency exists:

PROD_NUM → PROD_DESCRIPTION, PROD_PRICE, VEND_CODE, VEND_NAME

(*Hint*: Your actions should produce three dependency diagrams.)

3. Using the table structures you created in Problem 2, remove all transitive dependencies, draw the new dependency diagrams, and identify the normal forms for each table structure you created.

4. Using the results of Problem 3, draw the Crow's Foot ERD.

5. Using the STUDENT table structure shown in Table P5.5, draw its dependency diagram and identify all dependencies (including all transitive dependencies).

TABLE P5.5 SAMPLE STUDENT RECORDS

ATTRIBUTE NAME	SAMPLE VALUE	SAMPLE VALUE	SAMPLE VALUE	SAMPLE VALUE
STU_NUM	211343	200128	199876	223456
STU_LNAME	Stephanos	Smith	Jones	McKulski
STU_MAJOR	Accounting	Accounting	Marketing	Statistics
DEPT_CODE	ACCT	ACCT	MKTG	MATH
DEPT_NAME	Accounting	Accounting	Marketing	Mathematics
DEPT_PHONE	4356	4356	4378	3420
COLLEGE_NAME	Business Admin	Business Admin	Business Admin	Arts & Sciences
ADVISOR_LNAME	Grastrand	Grastrand	Gentry	Chen
ADVISOR_OFFICE	T201	T201	T228	J331
ADVISOR_BLDG	Torre Building	Torre Building	Torre Building	Jones Building
ADVISOR_PHONE	2115	2115	2123	3209
STU_GPA	3.87	2.78	2.31	3.58
STU_HOURS	75	45	117	87
STU_CLASS	Junior	Sophomore	Senior	Junior

6. Using the initial dependency diagram drawn in Problem 5, normalize the structure to meet the 3NF requirements to the greatest practical extent. If you believe that practical considerations dictate using a 2NF structure, explain why your decision to retain 2NF is appropriate. If necessary, add or modify attributes to create appropriate determinants and to adhere to the naming conventions.

NOTE

Although the completed student hours (STU_HOURS) do determine the student classification (STU_CLASS), this dependency is not as obvious as you might initially assume it to be. For example, a student is considered a Junior if that student has completed between 61 and 90 credit hours. Therefore, a student who is classified as a Junior may have completed 66, 72, 87, or any number of hours within the specified range of 61–90 hours. In short, any hour value within a specified range will define the classification.

7. Using the results of Problem 6, draw the Crow's Foot ERD.

NOTE

This ER diagram constitutes a small segment of a university's full-blown design. For example, this segment might be combined with the Tiny College presentation in Chapter 5.

8. To keep track of office furniture, computers, printers, and so on, the FOUNDIT company uses the table structure shown in Table P5.8:

TABLE P5.8 SAMPLE ITEM RECORDS

ATTRIBUTE NAME	SAMPLE VALUE	SAMPLE VALUE	SAMPLE VALUE
ITEM_ID	231134-678	342245-225	254668-449
ITEM_LABEL	HP DeskJet 895Cse	HP Toner	DT Scanner
ROOM_NUMBER	325	325	123
BLDG_CODE	NTC	NTC	CSF
BLDG_NAME	Nottooclear	Nottooclear	Canseefar
BLDG_MANAGER	I. B. Rightonit	I. B. Rightonit	May B. Next

Given this information, draw the dependency diagram. Make sure you label the transitive and/or partial dependencies.

9. Starting with the dependency diagram drawn for Problem 8, create a set of dependency diagrams that meet 3NF requirements. Rename attributes to meet the naming conventions, and create new entities and attributes as necessary.

10. Using the results of Problem 9, draw the Crow's Foot ERD.

NOTE

Problems 11–14 may be combined to create a case or a miniproject.

11. The table structure shown in Table P5.11 contains many unsatisfactory components and characteristics. (For example, there are several multivalued attributes, naming conventions are violated, and some attributes are not atomic.)

TABLE P5.11 SAMPLE EMPLOYEE RECORDS

ATTRIBUTE NAME	SAMPLE VALUE	SAMPLE VALUE	SAMPLE VALUE	SAMPLE VALUE
EMP_NUM	1003	1018	1019	1023
EMP_LNAME	Willaker	Smith	McGuire	McGuire
EMP_EDUCATION	BBA, MBA	BBA		BS, MS, Ph.D.
JOB_CLASS	SLS	SLS	JNT	DBA
EMP_DEPENDENTS	Gerald (spouse), Mary (daughter), John (son)		JoAnne (spouse)	George (spouse) Jill (daughter)
DEPT_CODE	MKTG	MKTG	SVC	INFS
DEPT_NAME	Marketing	Marketing	General Service	Info. Systems
DEPT_MANAGER	Jill H. Martin	Jill H. Martin	Hank B. Jones	Carlos G. Ortez
EMP_TITLE	Sales Agent	Sales Agent	Janitor	DB Admin
EMP_DOB	23-Dec-1968	28-Mar-1979	18-May-1982	20-Jul-1959
EMP_HIRE_DATE	14-Oct-1997	15-Jan-2004	21-Apr-2003	15-Jul-1999
EMP_TRAINING	L1, L2		L1	L1, L3, L8, L15
EMP_BASE_SALARY	$38,255.00	$30,500.00	$19.750.00	$127,900.00
EMP_COMMISSION_RATE	0.015	0.010		

Given this structure, draw its dependency diagram. Label all transitive and/or partial dependencies.

12. Break up the dependency diagram you drew in Problem 11 to produce dependency diagrams that are in 3NF. (*Hint:* You might have to create a few new attributes. Also, make sure that the new dependency diagrams contain attributes that meet proper design criteria; that is, make sure there are no multivalued attributes, that the naming conventions are met, and so on.)

13. Using the results of Problem 12, draw the Crow's Foot ERD.

14. Using the results of Problem 13, draw the relational schema.

NOTE

Problems 15–17 may be combined to serve as a case or a miniproject.

15. Suppose you are given the following business rules to form the basis for a database design. The database must enable the manager of a company dinner club to mail invitations to the club's members, to plan the meals, to keep track of who attends the dinners, and so on.

• Each dinner serves many members, and each member may attend many dinners.

• A member receives many invitations, and each invitation is mailed to many members.

• A dinner is based on a single entree, but an entree may be used as the basis for many dinners. For example, a dinner might be composed of a fish entree, rice, and corn. Or the dinner might be composed of a fish entree, baked potato, and string beans.

• A member may attend many dinners, and each dinner may be attended by many members.

Because the manager is not a database expert, the first attempt at creating the database uses the following structure:

TABLE P5.15 SAMPLE RESERVATION RECORDS

ATTRIBUTE NAME	SAMPLE VALUE	SAMPLE VALUE	SAMPLE VALUE
MEMBER_NUM	214	235	214
MEMBER_NAME	Alice B. VanderVoort	Gerald M. Gallega	Alice B. VanderVoort
MEMBER_ADDRESS	325 Meadow Park	123 Rose Court	325 Meadow Park
MEMBER_CITY	Murkywater	Highlight	Murkywater
MEMBER_ZIPCODE	12345	12349	12345
INVITE_NUM	8	9	10
INVITE_DATE	23-Feb-2004	12-Mar-2004	23-Feb-2004
ACCEPT_DATE	27-Feb-2004	15-Mar-2004	27-Feb-2004
DINNER_DATE	15-Mar-2004	17-Mar-2004	15-Mar-2004
DINNER_ATTENDED	Yes	Yes	No
DINNER_CODE	DI5	DI5	DI2
DINNER_DESCRIPTION	Glowing Sea Delight	Glowing Sea Delight	Ranch Superb
ENTREE_CODE	EN3	EN3	EN5
ENTREE_DESCRIPTION	Stuffed crab	Stuffed crab	Marinated steak
DESSERT_CODE	DE8	DE5	DE2
DESSERT_DESCRIPTION	Chocolate mousse with raspberry sauce	Cherries jubilee	Apple pie with honey crust

Given this structure, draw its dependency diagram. Label all transitive and/or partial dependencies. (*Hint:* This structure uses a composite primary key.)

16. Break up the dependency diagram you drew in Problem 15 to produce dependency diagrams that are in 3NF. (*Hint:* You might have to create a few new attributes. Also, make sure that the new dependency diagrams contain attributes that meet proper design criteria; that is, make sure there are no multivalued attributes, that the naming conventions are met, and so on.)

17. Using the results of Problem 16, draw the Crow's Foot ERD.

NOTE

Problems 18–20 may be combined to serve as a case or a miniproject.

18. The manager of a consulting firm has asked you to evaluate a database that contains the table structure shown in Table P5.18.

TABLE P5.18 SAMPLE CLIENT RECORDS

ATTRIBUTE NAME	SAMPLE VALUE	SAMPLE VALUE	SAMPLE VALUE
CLIENT_NUM	298	289	289
CLIENT_NAME	Marianne R. Brown	James D. Smith	James D. Smith
CLIENT_REGION	Midwest	Southeast	Southeast
CONTRACT_DATE	10-Feb-2004	15-Feb-2004	12-Mar-2004
CONTRACT_NUMBER	5841	5842	5843
CONTRACT_AMOUNT	$2,985,00.00	$670,300.00	$1,250,000.00
CONSULT_CLASS_1	Database Administration	Internet Services	Database Design
CONSULT_CLASS_2	Web Applications		Database Administration
CONSULT_CLASS_3			Network Installation
CONSULT_CLASS_4			
CONSULTANT_NUM_1	29	34	25
CONSULTANT_NAME_1	Rachel G. Carson	Gerald K. Ricardo	Angela M. Jamison
CONSULTANT_REGION_1	Midwest	Southeast	Southeast
CONSULTANT_NUM_2	56	38	34
CONSULTANT_NAME_2	Karl M. Spenser	Anne T. Dimarco	Gerald K. Ricardo
CONSULTANT_REGION_2	Midwest	Southeast	Southeast
CONSULTANT_NUM_3	22	45	
CONSULTANT_NAME_3	Julian H. Donatello	Geraldo J. Rivera	
CONSULTANT_REGION_3	Midwest	Southeast	
CONSULTANT_NUM_4		18	
CONSULTANT_NAME_4		Donald Chen	
CONSULTANT_REGION_4		West	

The table was created to enable the manager to match clients with consultants. The objective is to match a client within a given region with a consultant in that region, and to make sure that the client's need for specific consulting services is properly matched to the consultant's expertise. For example, if the client needs help with database design and is located in the Southeast, the objective is to make a match with a consultant who is located in the Southeast and whose expertise is in database design. (Although the consulting company manager tries to match consultant and client locations to minimize travel expense, it is not always possible to do so.) The following basic business rules are maintained:

- Each client is located in one region.

- A region can contain many clients.

- Each consultant can work on many contracts.

- Each contract may require the services of many consultants.

- A client can sign more than one contract, but each contract is signed by one client.

- Each contract may cover multiple consulting classifications. (For example, a contract may list consulting services in database design and networking.)

- Each consultant is located in one region.

- A region can contain many consultants.

- Each consultant has one or more areas of expertise (class). For example, a consultant might be classified as an expert in both database design and networking.

- Each area of expertise (class) can have many consultants in it. For example, the consulting company may employ many consultants who are networking experts.

Given this brief description of the requirements and the business rules, draw the dependency diagram for the preceding (very poor!) table structure. Label all transitive and/or partial dependencies.

19. Break up the dependency diagram you drew in Problem 18 to produce dependency diagrams that are in 3NF. (*Hint*: You might have to create a few new attributes. Also, make sure that the new dependency diagrams contain attributes that meet proper design criteria; that is, make sure there are no multivalued attributes, that the naming conventions are met, and so on.)

20. Using the results of Problem 19, draw the Crow's Foot ERD.

21. Given the sample records in the CHARTER table shown in Figure P5.21, draw the dependency diagram for this table structure. Make sure you label all dependencies. CHAR_PAX indicates the number of passengers carried. The CHAR_MILES entry is based on round-trip miles, including pickup points. (*Hint*: Look at the data values to determine the nature of the relationships. For example, note that employee Melton has flown two charter trips as pilot and one as copilot.)

TABLE P5.21 SAMPLE CHARTER RECORDS

ATTRIBUTE NAME	SAMPLE VALUE	SAMPLE VALUE	SAMPLE VALUE	SAMPLE VALUE
CHAR_TRIP	10232	10233	10234	10235
CHAR_DATE	15-Jan-2004	15-Jan-2004	16-Jan-2004	17-Jan-2004
CHAR_CITY	STL	MIA	TYS	ATL
CHAR_MILES	580	1,290	524	768
CUST_NUM	784	231	544	784
CUST_LNAME	Brown	Hanson	Bryana	Brown
CHAR_PAX	5	12	2	5
CHAR_CARGO	235 lbs.	18,940 lbs.	348 lbs.	155 lbs.
PILOT	Melton	Chen	Henderson	Melton
COPILOT		Henderson	Melton	
FLT_ENGINEER		O'Shaski		
LOAD_MASTER		Benkasi		
AC_NUMBER	1234Q	3456Y	1234Q	2256W
MODEL_CODE	PA31-350	CV-580	PA31-350	PA31-350
MODEL_SEATS	10	38	10	10
MODEL_CHG_MILE	$2.79	$23.36	$2.79	$2.79

22. Decompose the dependency diagram in Problem 21 to create table structures that are all in 3NF. Make sure you label all dependencies.

23. Draw the Crow's Foot ERD to reflect the properly decomposed dependency diagrams you created in Problem 22. Make sure that the ERD yields a database that can track all the data shown in Problem 21. Show all entities, relationships, connectivities, optionalities, and cardinalities.

N O T E

Use the dependency diagram shown in Figure P5.24 to work Problems 24–26.

FIGURE P5.24 THE INITIAL DEPENDENCY DIAGRAM FOR PROBLEMS 24–26

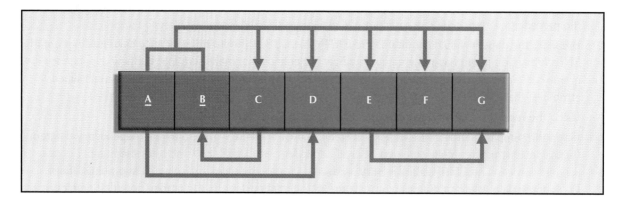

24. Break up the dependency diagram to create two new dependency diagrams, one in 3NF and one in 2NF.

25. Modify the dependency diagrams you produced in Problem 24 to produce a set of dependency diagrams that are all in 3NF. To keep the entire collection of attributes together, copy the 3NF dependency diagram from Problem 24 and then show the new dependency diagrams that are also in 3NF. (*Hint*: One of your dependency diagrams will be in 3NF, but not in BCNF.)

26. Modify the dependency diagrams in Problem 25 to produce a collection of dependency diagrams that are all in 3NF and BCNF. To ensure that all attributes are accounted for, copy the 3NF dependency diagrams you showed in Problem 25 and then show the new 3NF and BCNF dependency diagrams.

27. Suppose that you have been given the table structure and data shown in Table P5.27, which was imported from an Excel spreadsheet. The data reflect that a professor can have multiple advisees, can serve on multiple committees, and can edit more than one journal.

TABLE P5.27 SAMPLE PROFESSOR RECORDS

ATTRIBUTE NAME	SAMPLE VALUE	SAMPLE VALUE	SAMPLE VALUE	SAMPLE VALUE
EMP_NUM	123	104	118	
PROF_RANK	Professor	Asst. Professor	Assoc. Professor	Assoc. Professor
EMP_NAME	Ghee	Rankin	Ortega	Smith
DEPT_CODE	CIS	CHEM	CIS	ENG
DEPT_NAME	Computer Info. Systems	Chemistry	Computer Info. Systems	English
PROF_OFFICE	KDD-567	BLF-119	KDD-562	PRT-345
ADVISEE	1215, 2312, 3233, 2218, 2098	3102, 2782, 3311, 2008, 2876, 2222, 3745, 1783, 2378	2134, 2789, 3456, 2002, 2046, 2018, 2764	2873, 2765, 2238, 2901, 2308
COMMITTEE_CODE	PROMO, TRAF APPL, DEV	DEV	SPR, TRAF	PROMO, SPR DEV
JOURNAL_CODE	JMIS, QED, JMGT		JCIS, JMGT	

Given the information in Table P5.27:

a. Draw the dependency diagram.

b. Identify the multivalued dependencies.

c. Create the dependency diagrams to yield a set of table structures in 3NF.

d. Eliminate the multivalued dependencies by converting the affected table structures to 4NF.

e. Draw the Crow's Foot ERD to reflect the dependency diagrams you drew in (c). (Note: You may have to create additional attributes to define the proper PKs and FKs. Make sure that all your attributes conform to the naming conventions.)

PART

III

ADVANCED DESIGN
AND IMPLEMENTATION

6

INTRODUCTION TO STRUCTURED QUERY LANGUAGE (SQL)

In this chapter, you will learn:

❖ The basic commands and functions of SQL

❖ How to use SQL for data administration (to create tables, indexes, and views)

❖ How to use SQL for data manipulation (to add, modify, delete, and retrieve data)

❖ How to use SQL to query a database to extract useful information

In this chapter, you will learn the basics of Structured Query Language (SQL). SQL, pronounced S-Q-L or "sequel," is composed of commands that enable users to create database and table structures, perform various types of data manipulation and data administration, and query the database to extract useful information. All relational DBMS software supports SQL, and many software vendors have developed extensions to the basic SQL command set.

Because SQL's vocabulary is simple, SQL is relatively easy to learn. Its simplicity is enhanced by the fact that much of its work takes place behind the scenes. For example, a single command creates the complex table structures required to store and manipulate data successfully. Furthermore, SQL is a nonprocedural language; that is, the user specifies what must be done, but not how it is to be done. To issue SQL commands, end users and programmers do not need to know the physical data storage format nor the complex activities that take place when any SQL command is executed.

Although quite useful and powerful, SQL is not meant to stand alone in the applications arena. Data entry is possible but awkward, as are data corrections and additions. SQL itself does not create menus, special report forms, overlays, pop-ups, or any of the other utilities and screen devices that end users usually expect. Instead, those features are available as vendor-supplied enhancements. SQL focuses on data definition (creating tables, indexes, and views) and data manipulation (adding, modifying, deleting, and retrieving data), the basic functions you will learn about in this chapter. In spite of its limitations, you will discover that SQL is a powerful information extraction and data management tool.

6.1 INTRODUCTION TO SQL

Ideally, a database language must enable us to create database and table structures, to perform basic data management chores (add, delete, and modify), and to perform complex queries designed to transform the raw data into useful information. Moreover, it must perform such basic functions with minimal user effort, and its command structure and syntax must be easy to learn. Finally, it must be portable; that is, it must conform to some basic standard so that we do not have to relearn the basics when we move from one RDBMS to another.

SQL meets these ideal database language requirements well. First, SQL functions fit into two broad categories:

- It is a *data definition language:* SQL includes commands to create database objects such as tables, indexes, and views as well as commands to define access rights to those database objects. The data definition commands you will learn in this chapter are listed in Table 6.1.

- It is a *data manipulation language:* It includes commands to insert, update, delete, and retrieve data within the database tables. The data manipulation commands you will learn in this chapter are listed in Table 6.2.

Second, SQL is relatively easy to learn. Its basic command set has a vocabulary of less than 100 words. Better yet, SQL is a nonprocedural language: you merely have to command *what* is to be done; you don't have to worry about *how* it is to be done.

Finally, the American National Standards Institute (ANSI) prescribes a standard SQL—the most recent version is known as SQL-99 or SQL3.[1] The ANSI SQL standards are also accepted by the International Organization for Standardization (ISO).[2] Although adherence to the ANSI/ISO SQL standard is usually required in commercial and government contract database specifications, many RDBMS vendors add their own special enhancements. Consequently, it is seldom possible to move a SQL-based application from one RDBMS to another without making some changes.

Don't become discouraged by the fact that several SQL dialects exist, because the differences among the various dialects are minor. Whether you use Oracle, Microsoft SQL Server, IBM's DB2, Microsoft Access, or any other well-established RDBMS, our experience is that a software manual will be sufficient to get you up to SQL speed if you know the material presented in this chapter. In short, the knowledge you gain in this chapter is portable.

There are some very good reasons for studying SQL basics:

- The ANSI standardization effort has led to a *de facto* query standard for relational databases. In fact, many relational database experts are inclined to argue that "If it's not SQL, it's not relational."

- SQL has become the basis for present and expected future DBMS integration efforts, allowing us to link hierarchical, network, relational, and object-oriented databases.

- SQL has become the catalyst in the development of distributed databases and client/server database architectures. We will examine these developments in Chapter 10, "Distributed Database Management Systems," and Appendix F, "Multi-tier Client/Server Systems."

In Chapter 1, "Database Systems," you learned that a query is a spur of the moment question. Actually, in the SQL environment the word "query" covers both questions and actions. For the RDBMS, a query is basically a SQL statement that must be executed. Given the broader definition of the word "query" in the relational database environment, a SQL statement—which may include multiple SQL commands and functions—may be written to solicit an answer to a question or to perform (command) an action.

[1] "Information Technology – Database Languages – SQL – Part 2: Foundation (SQL/Foundation)," Document Number: ANSI/ISO/IEC 9075-2-1999, from the American National Standards Institute, *webstore.ansi.org.*

[2] Note that ISO is a Greek word meaning "equal." ISO is also an acronym for International Standards Organization. This organization is composed of national standards bodies of over 75 countries.

Most SQL queries are used to answer questions such as "what products currently held in inventory are priced over $100 and what is the quantity on hand for each of those products?" or "how many employees have been hired since November 1, 2003 by each of the company's departments?" However, many SQL queries are used to perform actions such as adding or deleting table rows or changing attribute values within tables. Still other SQL queries create new tables or indexes.

Tables 6.1 and 6.2 list the SQL commands you will learn in this chapter. These SQL commands are grouped in two categories: *data definition* and *data manipulation*. Keep in mind that the table contents do not define the sequence in which you will learn these SQL commands. Instead, the table contents are designed to give you a bird's-eye view of the SQL command structure.

TABLE 6.1 **SQL Data Definition Commands**

COMMAND OR OPTION	DESCRIPTION
CREATE SCHEMA AUTHORIZATION	Creates a database schema
CREATE TABLE	Creates a new table in the user's database schema
NOT NULL	Constraint that ensures that a column will not have null values
UNIQUE	Constraint that ensures that a column will not have duplicate values
PRIMARY KEY	Defines a primary key for a table
FOREIGN KEY	Defines a foreign key for a table
DEFAULT	Defines a default value for a column (when no value is given)
CHECK	Constraint used to validate data in a column
CREATE INDEX	Creates an index for a table
CREATE VIEW	Creates a dynamic subset of rows/columns from one or more tables
ALTER TABLE	Modifies a table's definition (adds, modifies, or deletes attributes or constraints)
CREATE TABLE AS	Creates a new table based on a query in the user's database schema
DROP TABLE	Permanently deletes a table (and thus its data)
DROP INDEX	Permanently deletes an index
DROP VIEW	Permanently deletes a view

TABLE 6.2 **Data Manipulation Commands**

COMMAND OR OPTION	DESCRIPTION
INSERT	Inserts row(s) into a table
SELECT	Selects attributes from rows in one or more tables or views
WHERE	Restricts the selection of rows based on a conditional expression
GROUP BY	Groups the selected rows based on one or more attributes
HAVING	Restricts the selection of grouped rows based on a condition
ORDER BY	Orders the selected rows

TABLE 6.2 DATA MANIPULATION COMMANDS (CONTINUED)

COMMAND OR OPTION	DESCRIPTION
UPDATE	Modifies attribute's values in one or more of table's rows
DELETE	Deletes one or more rows from a table
COMMIT	Permanently saves data changes
ROLLBACK	Restores data to their original values
Comparison operators	
=, <, >, <=, >=, <>	Used in conditional expressions
Logical operators	
AND/OR/NOT	Used in conditional expressions
Special operators	Used in conditional expressions
BETWEEN	Checks whether an attribute value is within a range
IS NULL	Checks whether an attribute value is null
LIKE	Checks whether an attribute value matches a given string pattern
IN	Checks whether an attribute value matches any value within a value list
EXISTS	Checks if a subquery returns any rows or not
DISTINCT	Limits values to unique values
Aggregate functions	Used with SELECT to return mathematical summaries on columns
COUNT	Returns the number of rows with non-null values for a given column
MIN	Returns the minimum attribute value found in a given column
MAX	Returns the maximum attribute value found in a given column
SUM	Returns the sum of all values for a given column
AVG	Returns the average of all values for a given column

6.2 DATA DEFINITION COMMANDS

Before we explore the SQL syntax, let's first examine the simple database model and the database tables that will form the basis for the many SQL examples we'll explore in this chapter. In this way, you'll start with a clear picture of the data environment, thus making it easier for you to understand what we are trying to do and how SQL is used to help us do it.

6.2.1 THE DATABASE MODEL

To illustrate the SQL commands you will learn in this chapter, we will use a simple database composed of the following tables: CUSTOMER, INVOICE, LINE, PRODUCT, and VENDOR. This database model is shown in Figure 6.1.

FIGURE 6.1 THE DATABASE MODEL

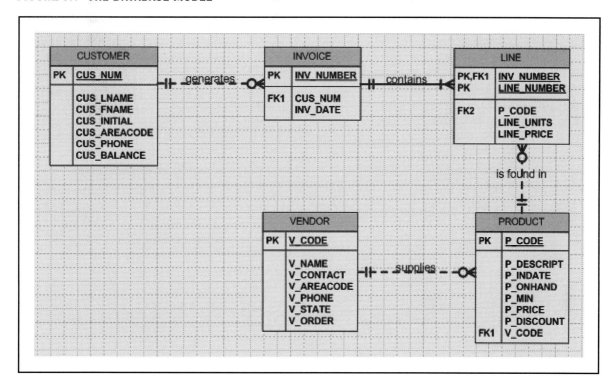

The database model in Figure 6.1 reflects the following business rules:

- A customer may generate many invoices. Each invoice is generated by one customer.

- An invoice contains one or more invoice lines. Each invoice line is associated with one invoice.

- Each invoice line references one product. A product may be found in many invoice lines. (You can sell more than one hammer to more than one customer!)

- A vendor *may* supply many products. Some vendors do not (yet?) supply products. (For example, a vendor list may include *potential* vendors.)

- If a product is vendor-supplied, that product is supplied by only a single vendor.

- Some products are not supplied by a vendor. (For example, some products may be produced "in-house" or may have been bought on the open market.)

As you can see in Figure 6.1 the database model contains many tables. However, to illustrate the initial set of data definition commands, we will focus our attention on the PRODUCT and VENDOR tables. You will have the opportunity to use the remaining tables later in this chapter or in the problem section.

ADDITIONAL CD CONTENT

The database model in Figure 6.1 is implemented in the Microsoft Access **Ch06_SaleCo** database located on the student CD in the **\Databases\Student\Ch06** folder. (This database contains a few additional tables that are not reflected in Figure 6.1. These tables are used for discussion purposes only.)

 If you use Microsoft Access, you can use the database we have supplied. However, we suggest that you create your own database structures so you can practice the SQL commands illustrated in this chapter.

NOTE

NOTE TO ORACLE USERS

If you are using Oracle DBMS, we have included SQL script files to create the tables and load the data in the database. The script files are located on the CD in the \SQL\Ch06 folder. How you connect to the Oracle database depends on how the Oracle software was installed on your server and on the access paths and methods defined and managed by the database administrator. Please follow the instructions provided by your instructor or your school's technology department.

To make sure that you have a point of reference when we demonstrate the effect of the SQL queries, we have listed the contents of the PRODUCT and VENDOR tables in Figure 6.2.

FIGURE 6.2 THE PRODUCT AND VENDOR TABLES

Table name: PRODUCT Database name: Ch06_SaleCo

P_CODE	P_DESCRIPT	P_INDATE	P_ONHAND	P_MIN	P_PRICE	P_DISCOUNT	V_CODE
11QER/31	Power painter, 15 psi., 3-nozzle	03-Nov-03	8	5	$109.99	0.00	25595
13-Q2/P2	7.25-in. pwr. saw blade	13-Dec-03	32	15	$14.99	0.05	21344
14-Q1/L3	9.00-in. pwr. saw blade	13-Nov-03	18	12	$17.49	0.00	21344
1546-QQ2	Hrd. cloth, 1/4-in., 2x50	15-Jan-04	15	8	$39.95	0.00	23119
1558-QW1	Hrd. cloth, 1/2-in., 3x50	15-Jan-04	23	5	$43.99	0.00	23119
2232/QTY	B&D jigsaw, 12-in. blade	30-Dec-03	8	5	$109.92	0.05	24288
2232/QWE	B&D jigsaw, 8-in. blade	24-Dec-03	6	5	$99.87	0.05	24288
2238/QPD	B&D cordless drill, 1/2-in.	20-Jan-04	12	5	$38.95	0.05	25595
23109-HB	Claw hammer	20-Jan-04	23	10	$9.95	0.10	21225
23114-AA	Sledge hammer, 12 lb.	02-Jan-04	8	5	$14.40	0.05	
54778-2T	Rat-tail file, 1/8-in. fine	15-Dec-03	43	20	$4.99	0.00	21344
89-WRE-Q	Hicut chain saw, 16 in.	07-Feb-04	11	5	$256.99	0.05	24288
PVC23DRT	PVC pipe, 3.5-in., 8-ft	20-Feb-04	188	75	$5.87	0.00	
SM-18277	1.25-in. metal screw, 25	01-Mar-04	172	75	$6.99	0.00	21225
SW-23116	2.5-in. wd. screw, 50	24-Feb-04	237	100	$8.45	0.00	21231
WR3/TT3	Steel matting, 4'x8'x1/6", .5" mesh	17-Jan-04	18	5	$119.95	0.10	25595

Table name: VENDOR

V_CODE	V_NAME	V_CONTACT	V_AREACODE	V_PHONE	V_STATE	V_ORDER
21225	Bryson, Inc.	Smithson	615	223-3234	TN	Y
21226	SuperLoo, Inc.	Flushing	904	215-8995	FL	N
21231	D&E Supply	Singh	615	228-3245	TN	Y
21344	Gomez Bros.	Ortega	615	889-2546	KY	N
22567	Dome Supply	Smith	901	678-1419	GA	N
23119	Randsets Ltd.	Anderson	901	678-3998	GA	Y
24004	Brackman Bros.	Browning	615	228-1410	TN	N
24288	ORDVA, Inc.	Hakford	615	898-1234	TN	Y
25443	B&K, Inc.	Smith	904	227-0093	FL	N
25501	Damal Supplies	Smythe	615	890-3529	TN	N
25595	Rubicon Systems	Orton	904	456-0092	FL	Y

As you examine the contents of the PRODUCT and VENDOR tables, note the following features. (These features correspond to the business rules reflected in the ERD shown in Figure 6.1.)

- The VENDOR table contains vendors who are not referenced in the PRODUCT table. Apparently, some vendors in the VENDOR table are there on a "standby" basis. That is, some vendors have never supplied a product, but they are kept in the VENDOR table as *potential* suppliers.

- Database designers note this possibility by saying that PRODUCT is *optional* to VENDOR, because a vendor may exist without a reference to a product. We examined such optional relationships in detail in Chapter 4, "Entity Relationship (ER) Modeling."

- Existing V_CODE values in the PRODUCT table must (and do) have a match in the VENDOR table to ensure referential integrity.

- A few products are supplied factory-direct, a few are made in-house, and a few may have been bought in a special warehouse sale. In other words, a product is not necessarily supplied by a vendor. Therefore, VENDOR is optional to PRODUCT.

A few of the conditions we just described were made for the sake of illustrating specific SQL features. For example, we have elected to use null V_CODE values in the PRODUCT table to (later) illustrate how you can track such nulls using SQL. (We examined the null concept in detail in Chapter 3, "The Relational Database Model." Such nulls also reflect the last business rule mentioned in the preceding bulleted list.)

6.2.2 CREATING THE DATABASE

In Chapter 1 you learned that a DBMS is a collection of programs that manage a database structure. The database is a shared, integrated computer structure that houses a collection of metadata and end-user data. The metadata and end-user data are stored in tables. Before you can use a new RDBMS, two tasks must be completed: first, create the database structure and second, create the tables that will hold the end-user data.

To complete the first task, the RDBMS creates the physical files that will hold the database. When you create a new database, the RDBMS automatically creates the data dictionary tables—to store the metadata—and creates a default database administrator. Creating the physical files that will hold the database means interacting with the operating system and the file systems supported by the operating system. Therefore, creating the database structure is the one feature that tends to differ substantially from one RDBMS to another. The good news is that it is relatively easy to create a database structure, regardless of which RDBMS you use. (We will show you how to create an Oracle database structure and how to manage the database environment within that structure in Chapter 15, "Database Administration.")

If you use Microsoft Access, creating the database is simple: start Access, select **File/New/Blank Database**, specify the folder in which you want to store the database, and name the database. However, if you work in the database environment typically used by larger organizations, you will probably use an enterprise RDBMS such as Oracle, SQL Server, or DB2. Given their security requirements and their greater complexity, these database products require a more elaborate database creation process. Typically, professional database administrators use the DBMS interface to help them create and manage databases.

You will be relieved to discover that, *with the exception of the database creation process*, most RDBMS vendors use SQL that deviates little from the ANSI standard SQL. Nevertheless, you might occasionally encounter minor syntax differences. For example, most RDBMSs require that any SQL command be ended with a semicolon. However, some SQL implementations do not use a semicolon. We will alert you about important differences among implementations through Note boxes.

6.2.3 THE DATABASE SCHEMA

If you are using an enterprise RDBMS such as Oracle, DB2, or SQL Server, you must first be authenticated by the RDBMS before you can start creating tables. **Authentication** is the process through which the DBMS verifies that only registered users are able to access the database. To be authenticated, you must log on to the RDBMS using a user ID and a password created by the database administrator. In an enterprise RDBMS, every user ID is associated with a database schema.

In a SQL environment, a **schema** is a group of database objects—such as tables and indexes—that are related to each other. Usually, the schema belongs to a single user or application. A single database can hold multiple schemas belonging to different users or applications. Think of a schema as a logical grouping of database objects

such as tables, indexes, and views. In short, the schema is a *logical* database. Schemas are useful in that they group tables by owner (or function) and enforce a first level of security by allowing the user to see only the tables that belong to him or her.

ANSI SQL standards define a command to create a database schema:

CREATE SCHEMA AUTHORIZATION {creator};

Therefore, if the creator is JONES, use the command:

CREATE SCHEMA AUTHORIZATION JONES;

Most enterprise RDBMS support this command at various levels of compliance. However, this command is seldom used directly—that is, at the command line level. (By creating a user, the DBMS automatically assigns a schema to that user.) When it *is* used, the CREATE SCHEMA AUTHORIZATION command must be issued by the user who owns the schema. That is, if you log on as JONES, you can only use CREATE SCHEMA AUTHORIZATION JONES.

For most RDBMSs, the CREATE SCHEMA AUTHORIZATION is optional. That is why we will focus on the ANSI SQL commands required to create and manipulate tables.

6.2.4 DATA TYPES

After the database structure has been created, you are ready to define the PRODUCT and VENDOR table structures within the database. The table-creating SQL commands will be based on the data dictionary shown in Table 6.3.

As you examine the data dictionary in Table 6.3, note particularly the data types we have selected. Keep in mind that data type selection is usually dictated by the nature of the data and by the intended use. For example:

- P_PRICE clearly requires some kind of numeric data type; defining it as a character field is not acceptable.

- Just as clearly, a vendor name is an obvious candidate for a character data type. For example, VARCHAR(35) fits well because vendor names are "variable-length" character strings, and we have, in this case, allowed such strings to be up to 35 characters long.

- U.S. state abbreviations are always two characters, so CHAR(2) is a logical choice.

- Selecting P_INDATE to be a (Julian) DATE field, rather than a character field, is desirable because the Julian dates allow us to make simpler date comparisons and to perform date arithmetic. For instance, if we want to know how many days are between March 1, 1999, and April 15, 2004, we can use 15-APR-2004 – 01-MAR-1999, *if we have used DATE fields.*

 If we use DATE fields, we can also determine what the date will be 60 days from February 15, 2004, by using 15-FEB-2004, + 60. Or we can use the RDBMS' system date—SYSDATE in Oracle, Date() in MS Access—to determine the answer to questions such as "what will be the date 60 days from today's date?" For example: we might use SYSDATE + 60 (in Oracle) or Date() + 60 (in Access).

Date arithmetic capability is particularly useful in billing. Perhaps we want our system to start charging interest on a customer balance 60 days after the invoice is generated. Such simple date arithmetic would be impossible if we used a character data type.

TABLE 6.3 DATA DICTIONARY FOR THE CH06_SALECO DATABASE

TABLE NAME	ATTRIBUTE NAME	CONTENTS	TYPE	FORMAT	RANGE	REQUIRED	PK OR FK	FK REFERENCED TABLE
PRODUCT	P_CODE	Product code	CHAR(10)	XXXXXXXXX	NA	Y	PK	
	P_DESCRIPT	Product description	VARCHAR(35)	Xxxxxxxxxxx	NA	Y		
	P_INDATE	Stocking date	DATE	DD-MON-YYYY	NA	Y		
	P_ONHAND	Units available	SMALLINT	####	0-9999	Y		
	P_MIN	Minimum units	SMALLINT	####	0-9999	Y		
	P_PRICE	Product price	NUMBER(8,2)	####.##	0.00-9999.00	Y		
	P_DISCOUNT	Discount rate	NUMBER(4,2)	0.##	0.00-0.20	Y		
	V_CODE	Vendor code	INTEGER	###	100-999		FK	VENDOR
VENDOR	V_CODE	Vendor code	INTEGER	####	1000-9999	Y	PK	
	V_NAME	Vendor name	CHAR(35)	Xxxxxxxxxxx	NA	Y		
	V_CONTACT	Contact person	CHAR(15)	Xxxxxxxxxxx	NA	Y		
	V_AREACODE	Area code	CHAR(3)	999	NA	Y		
	V_PHONE	Phone number	CHAR(8)	999-9999	NA	Y		
	V_STATE	State	CHAR(2)	XX	NA	Y		
	V_ORDER	Previous order	CHAR(1)	X	Y or N	Y		

FK = Foreign Key
PK = Primary Key

CHAR = Fixed character length data, 1 to 255 characters.
VARCHAR = Variable character length data, 1 to 2,000 characters. May also be labeled VARCHAR2 in Oracle.
NUMBER = Numeric data. NUMBER(9,2) is used to specify numbers with two decimal places and up to nine digits long, including the decimal places. Some RDBMSs permit the use of a MONEY or a CURRENCY data type.
INT = Integer values only.
SMALLINT = Small integer values only.
DATE formats may vary. Commonly accepted formats are: 'DD-MON-YYYY', 'DD-MON-YY', 'MM/DD/YYYY', or 'MM/DD/YY'.

* Not all the ranges shown here will be illustrated in this chapter. However, you can use these constraints to practice writing your own constraints.

On the other hand, data type selection can require professional judgment. For example, you must make a decision about the V_CODE's data type as follows:

- If you want the computer to generate new vendor codes by adding 1 to the largest recorded vendor code, you must classify V_CODE as a numeric attribute. (You cannot perform mathematical procedures on character data!) The designation INTEGER will ensure that only the counting numbers (integers) can be used. Most SQL implementations also permit the use of SMALLINT for integer values up to six digits.

- If you do not want to perform mathematical procedures based on V_CODE, you should classify it as a character attribute, even though it is composed entirely of numbers. Character data are "quicker" to process in queries. Therefore, if there is no need to perform mathematical procedures on the attribute, store it as a character attribute.

We will use the first option to demonstrate the SQL procedures in this chapter.

When you define the attribute's data type, you must pay close attention to the expected use of the attributes for sorting and data retrieval purposes. For example, in a real estate application, an attribute that represents the numbers of bathrooms in a home (H_BATH_NUM) could be assigned the CHAR(3) data type because it is highly unlikely the application will do any addition, multiplication, or division with the number of bathrooms. Based on the CHAR(3) data type definition, valid H_BATH_NUM values would be '2','1','2.5','10'. However, this data type decision creates potential problems. For example, if an application sorts the homes by number of bathrooms, a query would "see" the value '10' as less than '2', which is clearly incorrect. (That's because the ANSI code for the first character in "10," that is the "1," is less than the ANSI code for the character "2.") For the same reason, if you want to create a query that lists the homes with three or more bathrooms, the home with 10 bathrooms will not even make the list. And if you want to calculate the average number of bathrooms for the houses located in a given community, the CHAR(3) data type will not be appropriate. Clearly, you must give some thought to the expected use of the data in order to properly define the attribute data type.

The data dictionary in Table 6.3 contains only a few of the data types supported by SQL. We have limited the selection of data types to ensure that almost any RDBMS can be used to implement our examples. If your RDBMS is fully ANSI SQL-compliant, it will support many more data types than the ones shown in Table 6.4. And many RDBMSs support data types beyond the ones specified in ANSI SQL.

TABLE 6.4 SOME COMMON SQL DATA TYPES

DATA TYPE	FORMAT	COMMENTS
Numeric	NUMBER(L,D)	Used to store numbers (fixed and floating point) in the range 10^{-130} to 10^{126}, with up to 38 decimal places. Allows you to specify numbers in terms of storage length and the number of decimal places. The declaration NUMBER(7,2) indicates numbers that will be stored with two decimal places and may be up to six digits long, including the sign and the decimal place. Examples: 12.32, -134.99. If you specify only a single parameter by writing NUMBER(7), only the number length is specified, allowing you to store numbers that are up to six digits long, without specifying the number of decimal places. Therefore, both 0.023 and -243562 are acceptable when you specify NUMBER(7). The designation NUMBER, without parameters, is also allowed. However, this "no parameter" designation uses the system default specifications for any numbers you may store.

TABLE 6.4 SOME COMMON SQL DATA TYPES (CONTINUED)

DATA TYPE	FORMAT	COMMENTS
	INTEGER	May be abbreviated as INT. Integers are (whole) counting numbers, so they cannot be used if you want to store numbers that require decimal places. Up to 11 digits. Examples: 12, 36576, 9989765.
	SMALLINT	Like INTEGER, but limited to integer values up to six digits. If your integer values are relatively small, use SMALLINT instead of INT to produce better storage optimization.
	DECIMAL(L,D)	Like the NUMBER specification, but the storage length is a *minimum* specification. That is, greater lengths are acceptable, but smaller ones are not. DECIMAL(9,2), DECIMAL(9), and DECIMAL are all acceptable.
Character	CHAR(L)	Fixed length character data for up to 255 characters. If you store strings that are not as long as the CHAR parameter value, the remaining spaces are filled with blanks. Therefore, if you specify CHAR(25), strings like "Smith" and "Katzenjammer" are stored as Smithbbbbbbbbbbbbbbbbbbbbb and Katzenjammerbbbbbbbbbbbbb. (The letter "b" indicates a blank, not the letter "b".) In short, if you must store variable-length strings, do not use CHAR, because this designation wastes storage space. However, a U. S. phone area code is always three digits long, so CHAR(3) would be appropriate if you want to store such codes.
	VARCHAR(L)	Variable-length character data. The designation VARCHAR(25) will let you store characters that are up to 25 characters long. Unlike the CHAR designation, VARCHAR will not store trailing blanks. Therefore, if you specify VARCHAR(25), strings like "Smith" and "Katzenjammer" are stored without trailing blanks. VARCHAR is more efficient than CHAR if variable-length character data are to be stored. Oracle users may use VARCHAR2, as well as VARCHAR.
Date	DATE	Stores dates in the Julian date format. The year value is four digits long, 0001 through 9999, the month values require two digits 01 through 12, and the day values require two digits, 01 through 31.

In addition to the data types shown in Table 6.4, SQL supports several other data types, including: TIME, TIMESTAMP, REAL, DOUBLE, FLOAT, and intervals such as INTERVAL DAY TO HOUR. Many RDBMSs also have expanded the data type list to OLE (object linking and embedding), LOGICAL, CURRENCY, auto-number (Access), and sequence (Oracle). We mention these data types only to let you know that a wealth of data specifications exists, making SQL a very powerful tool in the database environment. However, because this chapter is designed to introduce the SQL basics, we will limit our discussion to the data types summarized in Table 6.4.

6.2.5 CREATING TABLE STRUCTURES

Given the PRODUCT and VENDOR table characteristics, we are ready to implement these table structures with the help of SQL, using the **CREATE TABLE** syntax shown next:

CREATE TABLE *tablename* (
 column1 *data type* *[constraint]* [,
 column2 *data type* *[constraint]*] [,
 PRIMARY KEY (*column1* [, *column2*])] [,
 FOREIGN KEY (*column1* [, *column2*]) REFERENCES *tablename*] [,
 CONSTRAINT *constraint*]);

To make the SQL code more readable, most SQL programmers use one line per column (attribute) definition. In addition, spaces are used to line up the attribute characteristics and constraints. Finally, both table and attribute

names are fully capitalized. These conventions are used in the following examples that create VENDOR and PRODUCT tables, and throughout the book.

> **N O T E**
>
> **NOTES ABOUT SQL SYNTAX NOTATION**
> Syntax notation for SQL commands used in this book:
>
> | CAPITALS | Required SQL command keywords |
> | *italics* | An end-user provided parameter (generally required) |
> | {a \| b \| ..} | A mandatory parameter, use one option from "\|" separated option list |
> | [......] | An optional parameter—anything inside square brackets is optional |
> | *tablename* | The name of a table |
> | *column* | The name of an attribute in a table |
> | *data type* | A valid data type definition |
> | *constraint* | A valid constraint definition |
> | *condition* | A valid conditional expression (evaluates to true or false) |
> | *columnlist* | One or more column names or expressions separated by commas |
> | *tablelist* | One or more table names separated by commas |
> | *conditionlist* | One or more conditional expressions separated by logical operators |
> | *expression* | A simple value (that is, 76 or 'Married') or a formula (that is, P_PRICE – 10) |

CREATE TABLE VENDOR (

				Sample Data
V_CODE	INTEGER	NOT NULL	UNIQUE,	21266
V_NAME	VARCHAR(35)	NOT NULL,		SuperLoo, Inc.
V_CONTACT	VARCHAR(15)	NOT NULL,		Flushing
V_AREACODE	CHAR(3)	NOT NULL,		904
V_PHONE	CHAR(8)	NOT NULL,		215-8995
V_STATE	CHAR(2)	NOT NULL,		FL
V_ORDER	CHAR(1)	NOT NULL,		N

PRIMARY KEY (V_CODE));

> **N O T E**
>
> Neither the Sample Data header nor the sample data values are part of the SQL code.
>
> - Because the PRODUCT table contains a foreign key that references VENDOR, create the VENDOR table first. (In fact, the "M" side of a relationship always references the "1" side, so you must *always* create the table for the "1" side first in a 1:M relationship.)
>
> - If your RDBMS does not support the VARCHAR and FCHAR format, use CHAR.
>
> - If your RDBMS does not support SINT or SMALLINT, use INTEGER or INT. If INTEGER is not supported, use NUMBER.
>
> - If you use MS Access, you can use the NUMBER data type, but you cannot use the number delimiters at the SQL level. For example, NUMBER(8,2) to indicate numbers with up to eight characters and two decimal places is fine in Oracle, but you cannot use it in Access—use NUMBER without the delimiters.
>
> - If your RDBMS does not support primary and foreign key designations or the UNIQUE specification, delete them from the SQL code shown here.
>
> - If you use the PRIMARY KEY designation in Oracle, you do not need the NOT NULL and UNIQUE specifications. In fact, in Oracle 9i you will get an error message if you use these specifications with PRIMARY KEY.
>
> - The ON UPDATE CASCADE clause is part of the ANSI standard, but it may not be supported by your RDBMS. In that case, delete the ON UPDATE CASCADE clause.

```
CREATE TABLE PRODUCT (                                        Sample Data
    P_CODE          VARCHAR(10)     NOT NULL UNIQUE,     2238/QPD
    P_DESCRIPT      VARCHAR(35)     NOT NULL,            Rat-tail file, 1/8-in., fine
    P_INDATE        DATE            NOT NULL,            1-Nov-03
    P_ONHAND        SMALLINT        NOT NULL,            43
    P_MIN           SMALLINT        NOT NULL,            20
    P_PRICE         NUMBER(8,2)     NOT NULL,            4.99
    P_DISCOUNT      NUMBER(4,2)     NOT NULL,            0.05
    V_CODE          INTEGER,                             21344
PRIMARY KEY (P_CODE),
FOREIGN KEY (V_CODE) REFERENCES VENDOR
ON UPDATE CASCADE);
```

As you examine the preceding SQL table-creating command sequences, note the following features:

- The NOT NULL specifications for the attributes ensure that a data entry will be made. If it is crucial to have the data available, the NOT NULL specification will not allow the end user to leave the attribute empty (with no data entry at all). Because this specification is made at the table level and stored in the data dictionary, application programs can use this information to create the data dictionary validation automatically.

- The UNIQUE specification creates a unique index in the respective attribute. Use it to avoid duplicated values in a column.

- The primary key attributes contain both a NOT NULL and a UNIQUE specification. These specifications enforce the entity integrity requirements. If the NOT NULL and UNIQUE specifications are not supported, use PRIMARY KEY without the specifications. (For example, if you designate the PK in MS Access, the NOT NULL and UNIQUE specifications are automatically assumed and they are not spelled out.)

N O T E

If you are working with a composite primary key, the primary key's attributes are all contained within the parentheses and are separated with commas. For example, the LINE table in Figure 6.1 has a primary key that consists of the two attributes INV_NUMBER and LINE_NUMBER. Therefore, you would define the primary key by typing:

PRIMARY KEY (INV_NUMBER, LINE_NUMBER),

The order of the primary key components is important, because the indexing starts with the first-mentioned attribute, then proceeds with the next attribute, and so on. In this example, the line numbers would be ordered within each of the invoice numbers:

INV_NUMBER	LINE_NUMBER
1001	1
1001	2
1002	1
1003	1
1003	2

- The entire table definition is enclosed in parentheses. A comma is used to separate each table element (attributes, primary key, and foreign key) definition.

- The ON UPDATE CASCADE specification ensures that if you make a change in any VENDOR's V_CODE, this change is automatically applied to all foreign key references throughout the system (cascade) to ensure that referential integrity is maintained. (Although the ON UPDATE CASCADE clause is part of the ANSI standard, some RDBMSs, such as Oracle, do not support ON UPDATE CASCADE. If your RDBMS does not support the clause, delete it from the code shown here.)

- A RDBMS will automatically enforce referential integrity for foreign keys. That is, you cannot have an invalid entry in the foreign key column and, at the same time, you cannot delete a vendor row as long as there is a product row that references that vendor.

- The command sequence ends with a semicolon. (Remember, your RDBMS may require that you omit the semicolon.)

N O T E

NOTE ABOUT COLUMN NAMES

Do *not* use mathematical symbols such as +, -, /, and so on. For example, PER-NUM may generate an error message, but PER_NUM is acceptable. Also, do *not* use reserved words. **Reserved words** are words used by SQL to perform specific functions. For example, in some RDBMSs, the column name INITIAL will generate the message, "invalid column name."

N O T E

NOTE TO ORACLE USERS

Pressing the ENTER key after each line has been typed automatically generates a "line number" if no semicolon is encountered before the ENTER key is pressed. For example, Oracle's execution of the CREATE TABLE command will look like this:

```
    CREATE TABLE PRODUCT (
2           P_CODE              VARCHAR2(10)
3           CONSTRAINT PRODUCT_P_CODE_PK PRIMARY KEY,
4           P_DESCRIPT          VARCHAR2(35)    NOT NULL,
5           P_INDATE            DATE            NOT NULL,
6           P_ONHAND            NUMBER          NOT NULL,
7           P_MIN               NUMBER          NOT NULL,
8           P_PRICE             NUMBER(8,2)     NOT NULL,
9           P_DISCOUNT          NUMBER(5,2)     NOT NULL,
10          V_CODE              NUMBER,
11          CONSTRAINT PRODUCT_V_CODE_FK
12          FOREIGN KEY (V_CODE) REFERENCES VENDOR);
13
```

As you examine the preceding SQL command sequence, note the following:

- The attribute definition for P_CODE starts in line 2 and ends with a comma at the end of line 3.

- The CONSTRAINT clause (line 3) allows you to define (and name) a constraint in ORACLE. The constraint named PRODUCT_P_CODE_PK is a PRIMARY KEY constraint for the P_CODE attribute.

- Examples of constraints are: NOT NULL, UNIQUE, PRIMARY KEY, FOREIGN KEY, and CHECK. For additional details about constraints, see Section 6.2.6, "SQL Constraints."

- To define a PRIMARY KEY constraint, you could also use the following syntax:

 P_CODE VARCHAR2(10) PRIMARY KEY,

 In this case, Oracle would automatically name the constraint.

- Lines 11 and 12 define a FOREIGN KEY constraint name PRODUCT_V_CODE_FK for the attribute V_CODE. The CONSTRAINT clause is generally used at the end of the CREATE TABLE command sequence.

- *If you do not name the constraints yourself, Oracle will automatically assign a name. Unfortunately, the Oracle-assigned name makes sense only to Oracle, so you will have a difficult time deciphering it later. We strongly urge you to assign a name that makes sense to human beings!*

6.2.6 SQL CONSTRAINTS

In Chapter 2, you learned that adherence to entity integrity and referential integrity rules is crucial in a relational database environment. Fortunately, most SQL implementations support both integrity rules. Entity integrity is enforced automatically if the primary key is specified in the CREATE TABLE command sequence. For example, we can create the VENDOR table structure and set the stage for the enforcement of entity integrity rules by using:

PRIMARY KEY (V_CODE)

As you reexamine the PRODUCT table's CREATE TABLE sequence, note that referential integrity has been enforced by specifying in the PRODUCT table:

FOREIGN KEY (V_CODE) REFERENCES VENDOR ON UPDATE CASCADE

This foreign key constraint definition ensures that:

- We cannot delete a vendor from the VENDOR table if at least one product row references this vendor. This is the default behavior for the treatment of foreign keys.

- On the other hand, if a change is made in an existing VENDOR table's V_CODE, this change must be reflected automatically in any PRODUCT table V_CODE reference (ON UPDATE CASCADE). This restriction makes it impossible for a V_CODE value to exist in the PRODUCT table pointing to a nonexisting VENDOR table V_CODE value. In other words, the ON UPDATE CASCADE specification ensures the preservation of referential integrity. (Oracle does not support ON UPDATE CASCADE.)

In general, ANSI SQL permits the use of ON DELETE and ON UPDATE clauses to cover any of the following actions: CASCADE, SET NULL, or SET DEFAULT. (See Section 6.9.1, "General Rules Governing Relationships among Tables," in this chapter for a more detailed discussion of these options.)

> **NOTE**
>
> **NOTE ABOUT REFERENTIAL CONSTRAINT ACTIONS**
> The support for the referential constraints actions varies from product to product. For example:
>
> - MS Access, SQL Server, and Oracle support ON DELETE CASCADE.
> - MS Access and SQL Server support ON UPDATE CASCADE.
> - Oracle does not support ON UPDATE CASCADE.
> - Oracle supports SET NULL.
> - MS Access and SQL Server do not support SET NULL.
>
> Please refer to your product manuals for additional information on referential constraints.
>
> While MS Access does not support ON DELETE CASCADE or ON UPDATE CASCADE at the SQL line level, it does support it through the relationship window interface. In fact, whenever you try to establish a relationship between two tables, Access will automatically pop up the relationship window interface.

Besides the PRIMARY KEY and FOREIGN KEY constraints, ANSI SQL standard also defines the following constraints:

- The NOT NULL constraint is used to ensure that a column does not accept nulls.
- The UNIQUE constraint is used to ensure that all values in a column are unique.
- The DEFAULT constraint is used to assign a value to an attribute when a new row is added to a table. The end user may, of course, enter a value other than the default value.

- The CHECK constraint is used to validate data when an attribute value is entered. The check constraint does precisely what its name suggests: checks to see if a specified condition exists. Examples of such constraints might be:

 The minimum order value must be at least 10.

 The date must be after April 15, 2003.

If the CHECK constraint is met for the specified attribute, that is, the condition is true, the data are accepted for that attribute. If the condition is found to be false, an error message is generated and the data are not accepted.

Note that the CREATE TABLE command lets you define constraints in two different places:

1. When you create the column definition (known as a *column constraint.*)
2. When you use the CONSTRAINT keyword (known as a *table constraint.*)

A column constraint applies just to one column; a table constraint can apply to many columns. These constraints are supported at varying levels of compliance by enterprise RDBMSs.

We will use Oracle to illustrate the use of SQL constraints. For example, note that the following SQL command sequence uses the DEFAULT and CHECK constraints to define the table named CUSTOMER.

```
CREATE TABLE CUSTOMER (
CUS_CODE          NUMBER              PRIMARY KEY,
CUS_LNAME         VARCHAR(15)         NOT NULL,
CUS_FNAME         VARCHAR(15)         NOT NULL,
CUS_INITIAL       CHAR(1),
CUS_AREACODE      CHAR(3)             DEFAULT '615' NOT NULL
                                      CHECK(CUS_AREACODE IN ('615','713','931')),
CUS_PHONE         CHAR(8)             NOT NULL,
CUS_BALANCE       NUMBER(9,2)         DEFAULT 0.00,
CONSTRAINT CUS_UI1 UNIQUE (CUS_LNAME, CUS_FNAME));
```

In this case, the CUS_AREACODE attribute is assigned a default value of '615'. Therefore, if a new CUSTOMER table row is added and the end user makes no entry for the area code, the '615' value will be recorded. Also, note that the CHECK condition restricts the values for the customer's area code to 615, 713, and 931; any other values will be rejected.

It is important to note that the DEFAULT value only applies when new rows are added to a table and then only if no value is entered for the customer's area code. (The default value is not used when the table is modified.) In contrast, the CHECK condition is always validated, whether a customer row is added *or modified*. However, while the CHECK condition may include any valid expression, it applies only to the attributes in the table being checked. If you want to check for conditions that include attributes in other tables, you must use triggers (see Chapter 7, "Advanced SQL.") Finally, the last line of the CREATE TABLE command sequence creates a unique index constraint (named CUS_UI1) on the customer's last name and first name. This index will prevent the entry of two customers with the same last name and first name. (We have used this index merely to illustrate the process. Clearly, it should be possible to have more than one person named 'John Smith' in the CUSTOMER table.)

N O T E

NOTE TO MS ACCESS USERS
MS Access will not accept the DEFAULT or CHECK constraints. However, MS Access will accept the CONSTRAINT CUS_UI1 UNIQUE (CUS_LNAME, CUS_FNAME) line and create the unique index.

In the following SQL command to create the INVOICE table, we use the DEFAULT constraint to assign a default date to a new invoice, and the CHECK constraint to validate that the invoice date is greater than January 1, 2002.

```
CREATE TABLE INVOICE (
INV_NUMBER          NUMBER      PRIMARY KEY,
CUS_CODE            NUMBER      NOT NULL REFERENCES CUSTOMER(CUS_CODE),
INV_DATE            DATE        DEFAULT SYSDATE NOT NULL,
CONSTRAINT INV_CK1 CHECK (INV_DATE > TO_DATE('01-JAN-2002','DD-MON-YYYY')));
```

In this case, notice the following:

- The CUS_CODE attribute definition contains REFERENCES CUSTOMER (CUS_CODE) to indicate that the CUS_CODE is a foreign key. This is another way to define a foreign key.

- The DEFAULT constraint uses the SYSDATE special function. This function always returns today's date.

- The invoice date (INV_DATE) attribute is automatically given today's date (returned by SYSDATE) when adding a new row and no value is given for the attribute.

- A CHECK constraint is used to validate that the invoice date is greater than 'January 1, 2002'. When comparing a date to a manually entered date in a CHECK clause, Oracle requires the use of the TO_DATE function. The TO_DATE function takes two parameters, the literal date and the date format used.

Our final SQL command sequence creates the LINE table. The LINE table has a composite primary key (INV_NUMBER, LINE_NUMBER) and uses a unique constraint in INV_NUMBER and P_CODE to ensure that the same product is not ordered twice in the same invoice.

```
CREATE TABLE LINE (
INV_NUMBER          NUMBER          NOT NULL,
LINE_NUMBER         NUMBER(2,0)     NOT NULL,
P_CODE              VARCHAR(10)     NOT NULL,
LINE_UNITS          NUMBER(9,2)     DEFAULT 0.00 NOT NULL,
LINE_PRICE          NUMBER(9,2)     DEFAULT 0.00 NOT NULL,
PRIMARY KEY (INV_NUMBER,LINE_NUMBER),
FOREIGN KEY (INV_NUMBER) REFERENCES INVOICE ON DELETE CASCADE,
FOREIGN KEY (P_CODE) REFERENCES PRODUCT(P_CODE),
CONSTRAINT LINE_UI1 UNIQUE(INV_NUMBER, P_CODE));
```

In the creation of the LINE table, note that we added a UNIQUE constraint to prevent the duplication of an invoice line. A UNIQUE constraint is enforced through the creation of a unique index. Also, note that we have used the ON DELETE CASCADE foreign key action to enforce referential integrity. The use of ON DELETE CASCADE is recommended for weak entities to ensure that the deletion of a row in the strong entity automatically triggers the deletion of the corresponding rows in the dependent weak entity. In this case, the deletion of an INVOICE row will automatically delete all the LINE rows related to that invoice. In the following section you will learn more about indexes and how to use SQL commands to create them.

6.2.7 SQL INDEXES

You learned in Chapter 3 what indexes are and how they may be used to improve the efficiency of searches and to avoid duplicate column values. In the previous section you saw how to declare unique indexes on selected attributes when the table is created. In fact, when you declare a primary key, the DBMS automatically creates a unique index. Even with this feature, you will often need additional indexes. Therefore, the ability to create

indexes quickly and efficiently at any time is important. Using the **CREATE INDEX** command, SQL indexes can be created on the basis of any selected attribute. The syntax is:

CREATE [UNIQUE] INDEX *indexname* ON *tablename(column1 [, column2])*

For example, the following command sequence creates an index named P_CODEX, based on the attribute P_CODE stored in the PRODUCT table:

CREATE INDEX P_CODEX ON PRODUCT(P_CODE);

SQL does not let you write over an existing index without first warning you, thus preserving the index structure within the data dictionary. Using the UNIQUE index qualifier, you can even create an index that prevents you from using a value that has been used before. Such a feature is especially useful when the index field (attribute) is a primary key (PK) whose values must not be duplicated:

CREATE UNIQUE INDEX P_CODEX ON PRODUCT(P_CODE);

If you now try to enter a duplicate P_CODE value, SQL produces an error message: "duplicate value in index." Many RDBMSs, including Access, automatically create a unique index on the PK attribute(s) when you declare the PK.

If the index is based on two or more attributes, it is known as a **composite index**. Unique composite indexes are often used to prevent data duplication. For example, consider the case illustrated in Table 6.5, in which required employee test scores are stored. (Perhaps the employees require periodic testing to establish their competency.) Given the structure of Table 6.5, the PK is EMP_NUM + TEST_NUM. The third test entry for employee 111 meets entity integrity requirements—the combination 111,3 is unique—yet the WEA test entry is clearly duplicated.

TABLE 6.5 A DUPLICATED TEST RECORD

EMP_NUM	TEST_NUM	TEST_CODE	TEST_DATE	TEST_SCORE
110	1	WEA	15-May-2003	93
110	2	WEA	12-May-2004	87
111	1	HAZ	14-Dec-2002	91
111	2	WEA	18-Feb-2004	95
111	3	WEA	18-Feb-2004	95
112	1	CHEM	17-Aug-2003	91

Such duplication could have been avoided through the use of a unique composite index, using the attributes EMP_NUM, TEST_CODE, and TEST_DATE:

CREATE UNIQUE INDEX EMP_TESTDEX ON TEST(EMP_NUM, TEST_CODE, TEST_DATE);

A common practice is to create an index on any field that is used as a search key or for comparison purposes in a conditional expression. For example, if you want to create a report of all products by vendor, it would be useful to create an index on the V_CODE attribute in the PRODUCT table. Remember that a vendor can supply many products. Therefore, you should *not* create a UNIQUE index in this case. Better yet, to make this search as efficient as possible, a composite index is recommended. For example, you might create the composite index named VENPRODX by typing:

CREATE INDEX VENPRODX ON PRODUCT(V_CODE, P_CODE);

By default all indexes produce results that are listed in ascending order, but you can also specify the creation of an index that yields output in descending order. For example, if you routinely print a report that lists all products ordered by price from the highest to lowest, you could create an index named PROD_PRICEX by typing:

CREATE INDEX PROD_PRICEX ON PRODUCT(P_PRICE DESC);

To delete an index, use the DROP INDEX command:

DROP INDEX *indexname*

For example, if you want to eliminate the PROD_PRICEX index, type:

DROP INDEX PROD_PRICEX;

After creating the tables and some indexes, we are ready to start entering data. In the following sections we will use two tables (VENDOR and PRODUCT) to demonstrate most of the data manipulation commands. You will have the opportunity to use the remaining tables later in this chapter or in the problem section.

6.3 DATA MANIPULATION COMMANDS

In this section you will learn how to use the basic SQL data manipulation commands shown in Table 6.6.

TABLE 6.6 COMMON SQL DATA MANIPULATION COMMANDS

COMMAND	DESCRIPTION
INSERT	Lets you insert data into a table, one row at a time. Used to make the initial data entries into a new table structure or to add data to a table that already contains data.
SELECT	Lists the table contents.
COMMIT	Lets you permanently save your work to disk.
UPDATE	Enables you to make changes to column values in one or more data rows.
ROLLBACK	Restores the database table contents to their original condition (since the last COMMIT).
DELETE	Enables you to delete one or more data rows.

6.3.1 ADDING TABLE ROWS

SQL requires the use of the **INSERT** command to enter data into a table. The INSERT command's basic syntax looks like this:

INSERT INTO *tablename* VALUES (*value1, value2, ... , valuen*)

Because the PRODUCT table uses its V_CODE to reference the VENDOR table's V_CODE, it is clear that an integrity violation will occur if those VENDOR table V_CODE values don't yet exist. Therefore, we need to enter the VENDOR rows before we enter the PRODUCT rows. Given the VENDOR table structure we defined earlier and the sample VENDOR data shown in Figure 6.2, you would enter the first two data rows as follows:

INSERT INTO VENDOR
 VALUES (21225,'Bryson, Inc.','Smithson','615','223-3234','TN','Y');
INSERT INTO VENDOR
 VALUES (21226,'Superloo, Inc.','Flushing','904','215-8995','FL','N');

and so on, until all the VENDOR table records have been entered.

(To see the contents of the VENDOR table, type SELECT * FROM VENDOR;)

The PRODUCT table rows would be entered in the same fashion, using the PRODUCT data shown in Figure 6.2. For example, the first two data rows would be entered as follows, pressing ENTER at the end of each line:

INSERT INTO PRODUCT
 VALUES ('11QER/31','Power painter, 15 psi., 3-nozzle','03-Nov-03',8,5,109.99,0.00,25595);
INSERT INTO PRODUCT
 VALUES ('13-Q2/P2','7.25-in. pwr. saw blade','13-Dec-03',32,15,14.99, 0.05, 21344);

(To see the contents of the PRODUCT table, type SELECT * FROM PRODUCT;)

N O T E

Date entry is a function of the date format expected by the DBMS. For example, March 25, 2004, might be shown as '25-Mar-2004' (in Microsoft Access and Oracle) or in other presentation formats, depending on your RDBMS. MS Access requires the use of # delimiters when performing any computations or comparisons based on date attributes, as in P_INDATE >= #25-Mar-04#.

As you examine the preceding data entry lines, observe that:

- The row contents are entered between parentheses. Note that the first character after VALUES is a parenthesis and that the last character in the command sequence is also a parenthesis.

- Character (string) and date values must be entered between apostrophes (').

- Numerical entries are *not* enclosed in apostrophes.

- Attribute entries are separated by commas.

- A value is required for each column in the table.

This version of the INSERT commands adds one table row at a time.

As you can see, data entry is rather cumbersome in SQL. Remember that end-user applications are best created with utilities that generate attractive (and easy to use!) input screens. For example, we used Microsoft Access to create a form-based data view and entry screen like the one shown in Figure 6.3.

Inserting Rows with Null Attributes

Thus far, we have shown you how to enter rows in which all the attribute values are specified. But what do you do if a product does not have a vendor, or if you don't (yet) know the vendor code? In that case, you would want to leave the vendor code null. To enter a null, we can use the following syntax:

INSERT INTO PRODUCT
 VALUES ('BRT-345','Titanium drill bit', '18-Oct-02', 75, 10, 4.50, 0.06, NULL);

Incidentally, note that the NULL entry is accepted only because the V_CODE attribute is optional—we did not use the NOT NULL declaration in the CREATE TABLE statement for this attribute.

FIGURE 6.3 **A DATA VIEW AND ENTRY FORM**

Inserting Rows with Optional Attributes

There may be occasions on which more than one attribute is optional. Rather than declaring each attribute as NULL in the INSERT command, we can indicate just the attributes that have required values. We do this by listing the attribute names for which we are providing values inside parentheses after the table name. For the purpose of this example, assume that the only required attributes for the PRODUCT table are P_CODE and P_DESCRIPT:

INSERT INTO PRODUCT(P_CODE, P_DESCRIPT)
 VALUES ('BRT-345','Titanium drill bit');

6.3.2 SAVING TABLE CHANGES

Any changes made to the table contents are not physically saved on disk until you close the database, close the program you are using, or use the **COMMIT** command. If you are using the database and a power outage or some other interruption occurs before you issue the COMMIT command, your changes will be lost and only the original table contents will be retained. The syntax for the COMMIT command is:

COMMIT [WORK]

The COMMIT command will permanently save *any* changes—such as rows added, attributes modified, or rows deleted—made to any table in the database. Therefore, if you intend to make your changes to the PRODUCT table permanent, it is a good idea to save those changes by using

COMMIT;

N O T E

NOTE TO MS ACCESS USERS

MS Access doesn't support the COMMIT command. Access automatically saves changes after the execution of each SQL command.

We do not want to leave you with the impression that the COMMIT command's sole purpose is to save changes. In fact, the ultimate purpose of the COMMIT and ROLLBACK commands (see Section 6.3.5) is to ensure database update integrity in transaction management. (You will see how such issues are addressed in Chapter 9, "Transaction Management and Concurrency Control.")

6.3.3 LISTING TABLE ROWS

The **SELECT** command is used to list the contents of a table. The syntax of the SELECT command is as follows:

SELECT *columnlist*
 FROM *tablename*

The *columnlist* represents one or more attributes, separated by commas. You could use the * (asterisk) as a wildcard character to list all attributes. (A **wildcard character** is a symbol that can be used as a general substitute for other characters or commands.) For example, to list all attributes and all rows of the PRODUCT table, use:

SELECT * FROM PRODUCT;

Figure 6.4 shows the output generated by this command.

N O T E

Figure 6.4 shows all the rows in the PRODUCT table to serve as the basis for subsequent discussions. If you entered only the PRODUCT table's first two records as shown in the preceding section, the output of the preceding SELECT command will show only the rows you have entered. Don't worry about the difference between your SELECT output and the output we show in Figure 6.4. When you have completed the work in this section, your VENDOR and PRODUCT tables will have been created and populated with the correct rows that will be used in future sections.

FIGURE 6.4 THE CONTENTS OF THE PRODUCT TABLE

P_CODE	P_DESCRIPT	P_INDATE	P_ONHAND	P_MIN	P_PRICE	P_DISCOUNT	V_CODE
11QER/31	Power painter, 15 psi., 3-nozzle	03-Nov-03	8	5	$109.99	0.00	25595
13-Q2/P2	7.25-in. pwr. saw blade	13-Dec-03	32	15	$14.99	0.05	21344
14-Q1/L3	9.00-in. pwr. saw blade	13-Nov-03	18	12	$17.49	0.00	21344
1546-QQ2	Hrd. cloth, 1/4-in., 2x50	15-Jan-04	15	8	$39.95	0.00	23119
1558-QW1	Hrd. cloth, 1/2-in., 3x50	15-Jan-04	23	5	$43.99	0.00	23119
2232/QTY	B&D jigsaw, 12-in. blade	30-Dec-03	8	5	$109.92	0.05	24288
2232/QWE	B&D jigsaw, 8-in. blade	24-Dec-03	6	5	$99.87	0.05	24288
2238/QPD	B&D cordless drill, 1/2-in.	20-Jan-04	12	5	$38.95	0.05	25595
23109-HB	Claw hammer	20-Jan-04	23	10	$9.95	0.10	21225
23114-AA	Sledge hammer, 12 lb.	02-Jan-04	8	5	$14.40	0.05	
54778-2T	Rat-tail file, 1/8-in. fine	15-Dec-03	43	20	$4.99	0.00	21344
89-WRE-Q	Hicut chain saw, 16 in.	07-Feb-04	11	5	$256.99	0.05	24288
PVC23DRT	PVC pipe, 3.5-in., 8-ft	20-Feb-04	188	75	$5.87	0.00	
SM-18277	1.25-in. metal screw, 25	01-Mar-04	172	75	$6.99	0.00	21225
SW-23116	2.5-in. wd. screw, 50	24-Feb-04	237	100	$8.45	0.00	21231
WR3/TT3	Steel matting, 4'x8'x1/6", .5" mesh	17-Jan-04	18	5	$119.95	0.10	25595

> **NOTE**
>
> Your listing may not be in the order shown in Figure 6.4. The listings shown here are the result of system-controlled primary-key-based index operations. You will later learn how to control the output so that it will always conform to the order you have specified. For the time being, just check your listing to make sure that all of the records are accounted for, and don't worry about the order in which the records appear.

> **NOTE**
>
> **NOTE TO ORACLE USERS**
> Some SQL implementations (such as Oracle's) *cut* the attribute labels to fit the width of the column. However, Oracle's SQL*Plus lets you set the display column width to display the complete attribute name. You can also change the display format, regardless of how the data are stored in the table. For example, if you want to display dollar symbols and commas in the P_PRICE output, you can declare:
> COLUMN P_PRICE FORMAT $99,999.99
> to change the output 12346.67 to $12,346.67.
> In the same manner, to display only the first 12 characters of the P_DESCRIPT attribute, use:
> COLUMN P_DESCRIPT FORMAT A12 TRUNCATE

Although the SQL commands may be issued on a single line, command sequences are best structured when the SQL command's components are shown on separate (and indented) lines. Therefore, we will write the preceding command this way:

SELECT *
 FROM PRODUCT;

Or, instead of using a wildcard character, each of the table's attributes may be listed:

SELECT P_CODE, P_DESCRIPT, P_INDATE, P_ONHAND, P_MIN, P_PRICE, P_DISCOUNT, V_CODE
 FROM PRODUCT;

Using the just-described formatting convention makes it much easier to see the components of the SQL statements, thus making it easier to trace the SQL logic and, if necessary, to make corrections. The number of spaces used in the indentation is up to you.

You should remember from Chapter 3 that *the output of every relational operator is another relation*. You also learned that relation is another word for the relational table. Therefore, when you run a SELECT command on a table, the RDBMS returns a set of one or more rows that have the same characteristics as a relational table. In addition, note that the SELECT command lists all rows from the table you specified in the FROM clause. This is a very important characteristic of SQL commands. By default, most SQL data manipulation commands will operate over an entire table (or relation). That is why SQL commands are said to be *set-oriented* commands. A SQL set-oriented command works over a set of rows. The set may include one or more columns and zero or more rows from one or more tables.

6.3.4 UPDATING TABLE ROWS

Use the **UPDATE** command to modify data in a table. The syntax for this command is:

UPDATE *tablename*
SET *columnname = expression* [, *columnname = expression*]
[WHERE *conditionlist*];

For example, if you want to change P_INDATE from December 13, 2003 to January 18, 2004 in the second row of the PRODUCT table (see Figure 6.4), use the primary key (13-Q2/P2) to locate the correct (second) row. Therefore, type:

UPDATE PRODUCT
 SET P_INDATE = '18-JAN-2004'
 WHERE P_CODE = '13-Q2/P2';

If more than one attribute is to be updated in the row, separate the corrections with commas:

UPDATE PRODUCT
 SET P_INDATE = '18-JAN-2004', P_PRICE = 16.99, P_MIN = 10
 WHERE P_CODE = '13-Q2/P2';

By the way, what would have happen if the previous UPDATE command had not included the WHERE condition? Answer: the P_INDATE, P_PRICE and P_MIN values would have been changed in *all* rows of the PRODUCT table. Remember, the UPDATE command is a set-oriented operator. Therefore, if you don't specify a WHERE condition, the UPDATE command will apply the changes to *all* rows in the specified table.

Confirm the correction(s) by using this command to check the PRODUCT table's listing:

SELECT * FROM PRODUCT;

6.3.5 RESTORING TABLE CONTENTS

If you have not yet used the COMMIT command to permanently store the changes in the database, you may restore the database to its previous condition with the **ROLLBACK** command. ROLLBACK undoes any changes made and brings the data back to the values that existed before the changes were made. To restore the data to their "pre-change" condition, type

ROLLBACK;

and press ENTER. Use the SELECT statement again to see that the ROLLBACK did, in fact, restore the data to their original values. We will examine COMMIT and ROLLBACK in greater detail in Chapter 9.

COMMIT and ROLLBACK only work with data manipulation commands that are used to add, modify, or delete table rows. For example, let's assume that you perform these actions:

1. CREATE a table called SALES.
2. INSERT 10 rows in the SALES table.
3. UPDATE two rows in the SALES table.
4. Execute the ROLLBACK command.

Will the SALES table be removed by the ROLLBACK command? No, the ROLLBACK command will undo *only* the results of the INSERT and UPDATE commands. All data definition commands (CREATE TABLE) are automatically committed to the data dictionary and cannot be rolled back.

> **N O T E**
>
> **NOTE TO MS ACCESS USERS**
> MS Access doesn't support the ROLLBACK command.

Some RDBMSs (like Oracle) will automatically COMMIT data changes when issuing data definition commands. For example, if you had used the CREATE INDEX command after updating the two rows in the previous example, all previous changes would have been committed automatically and doing a ROLLBACK afterward wouldn't undo anything. *Please check your RDBMS manual to understand these subtle differences.*

6.3.6 DELETING TABLE ROWS

It is easy to delete a table row using the **DELETE** statement; the syntax is:

DELETE FROM *tablename*
 [WHERE *conditionlist*];

For example, if you want to delete from the PRODUCT table the product that you added earlier whose product code (P_CODE) is 'BRT-345', use:

DELETE FROM PRODUCT
 WHERE P_CODE = 'BRT-345';

Note that in this example the primary key value lets SQL find the exact record to be deleted. However, deletions are not limited to a primary key match; any attribute may be used. For example, if you examine your PRODUCT table, note that there are several products for which the P_MIN attribute is equal to 5. The following command will delete all rows from the PRODUCT table for which the P_MIN is equal to 5:

DELETE FROM PRODUCT
 WHERE P_MIN = 5;

Check the PRODUCT table's contents again to verify that all products with P_MIN equal to 5 have been deleted.

Finally, remember that DELETE is a set-oriented command. And keep in mind that the WHERE condition is optional. Therefore, if you do not specify a WHERE condition, *all* rows from the specified table will be deleted!

6.3.7 INSERTING TABLE ROWS WITH A SELECT SUBQUERY

You learned in Section 6.3.1 how to use the INSERT statement to add rows to a table. In that section, you added rows one row at a time. In this section you will learn how to add multiple rows to a table using another table as the source of the data (multiple rows at a time). The syntax for the INSERT statement is:

INSERT INTO *tablename* SELECT *columnlist* FROM *tablename* ;

In this case, the INSERT statement uses a SELECT subquery. A **subquery**, also known as a **nested query** or an **inner query**, is a query that is embedded (or nested) inside another query. The inner query is always executed first by the RDBMS. Given the previous SQL statement, the INSERT portion represents the outer query and the SELECT portion represents the inner query or subquery. You can nest queries (place queries inside queries) many levels deep; in every case the output of the inner (lower level) query is used as the input for the outer (higher level) query. In Chapter 7, "Advanced SQL," you will learn more about the different types of subqueries.

The values returned by the SELECT subquery should match the attributes and data types of the table in the INSERT statement. If the table into which you are inserting rows has one date attribute, one number attribute, and one character attribute, the SELECT subquery should return one or more rows in which the first column has date values, the second has number values, and the third has character values.

Populating the VENDOR and PRODUCT Tables

The following steps will guide you through the process of populating the VENDOR and PRODUCT tables with the data to be used for the forthcoming sections. To accomplish this task, we'll use two tables named V and P as the data source. V and P have the same table structure (attributes) as our VENDOR and PRODUCT tables.

ADDITIONAL CD CONTENT

MS ACCESS USERS

The V and P tables are located in the **Ch06_SaleCo** database located on the student CD in the **\Databases\Student\Ch06** folder. To learn how to work within the SQL environment, we recommend that you try the process described next. However, you can also simply copy the **Ch06_SaleCo** database and its contents from your CD.

ADDITIONAL CD CONTENT

ORACLE USERS

The Oracle version of the commands assumes the instructor has created the TEACHER schema and has run the scripts to create the V and P tables in that schema. (A copy of the scripts is located in the **\SQL\Ch06** folder on your CD.) Otherwise, please follow the instructions specific to your school setup given by your instructor.

The following steps will be used to populate your VENDOR and PRODUCT tables. (We're assuming that you have already created and used the PRODUCT and VENDOR tables to practice the SQL commands in the previous sections.)

- Delete all rows from the PRODUCT and VENDOR tables.
 - DELETE FROM PRODUCT;
 - DELETE FROM VENDOR;
- Add the rows to VENDOR by copying all rows from V.
 - If you are using MS Access type:

 INSERT INTO VENDOR SELECT * FROM V;
 - If you are using Oracle type:

 INSERT INTO VENDOR SELECT * FROM TEACHER.V;
- Add the rows to PRODUCT by copying all rows from P.
 - If you are using MS Access type:

 INSERT INTO PRODUCT SELECT * FROM P;
 - If you are using Oracle type:

 INSERT INTO PRODUCT SELECT * FROM TEACHER.P;
- Oracle users must permanently save the changes: COMMIT;

If you followed the steps above, you now have the VENDOR and PRODUCT tables populated with the data that will be used in the remaining section of the chapter.

ADDITIONAL CD CONTENT

ORACLE USERS

If you are using Oracle, please run the **ch06dbinit.sql** script file to create all tables and load the data in the database. This script file populates the remaining tables (CUSTOMER, INVOICE, LINE, EMP, and EMPLOYEE). The script files are located on the CD in the **\SQL\Ch06** folder. To connect to the database, please follow the instructions specific to your school setup given by your instructor.

6.4 SELECT QUERIES

In this section you will learn how to fine-tune the SELECT command by adding restrictions to the search criteria. SELECT, coupled with appropriate search conditions, is an incredibly powerful tool that enables you to transform data into information. For example, in the following sections you'll learn how to create queries that can be used to answer questions such as "What products were supplied by a particular vendor?", "Which products are priced below $10.00?", and "How many products supplied by a given vendor were sold between January 5, 2002 and March 20, 2002?"

6.4.1 SELECTING ROWS WITH CONDITIONAL RESTRICTIONS

You can select partial table contents by placing restrictions on the rows to be included in the output. This is done by adding conditional restrictions to the SELECT statement, using the WHERE clause. The following syntax enables you to specify which rows to select:

SELECT *columnlist*
 FROM *tablelist*
 [WHERE *conditionlist*] ;

The SELECT statement will retrieve all rows that match the specified condition(s)—also known as the *conditional criteria*—you specified in the WHERE clause. The *conditionlist* in the WHERE clause of the SELECT statement is represented by one or more conditional expressions, separated by logical operators. Keep in mind that the WHERE clause is optional. If no rows match the specified criteria in the WHERE clause, you may see a blank screen or a message that tells you that no rows were retrieved. For example, the query:

SELECT P_DESCRIPT, P_INDATE, P_PRICE, V_CODE
 FROM PRODUCT
 WHERE V_CODE = 21344;

returns the description, date, and price of products with a vendor code of 21344, as shown in Figure 6.5.

FIGURE 6.5 **SELECTED PRODUCT TABLE ATTRIBUTES FOR VENDOR CODE 21344**

P_DESCRIPT	P_INDATE	P_PRICE	V_CODE
7.25-in. pwr. saw blade	13-Dec-03	$14.99	21344
9.00-in. pwr. saw blade	13-Nov-03	$17.49	21344
Rat-tail file, 1/8-in. fine	15-Dec-03	$4.99	21344

Microsoft Access users can use the Access QBE (query by example) query generator. Although Access QBE generates its own "native" version of SQL, you can also elect to type standard SQL in the Access SQL window, as

shown in the third screen segment of Figure 6.6. Figure 6.6 shows the Access QBE screen, the SQL window's QBE-generated SQL, and the listing of the modified SQL.

FIGURE 6.6 THE MICROSOFT ACCESS QBE AND ITS SQL

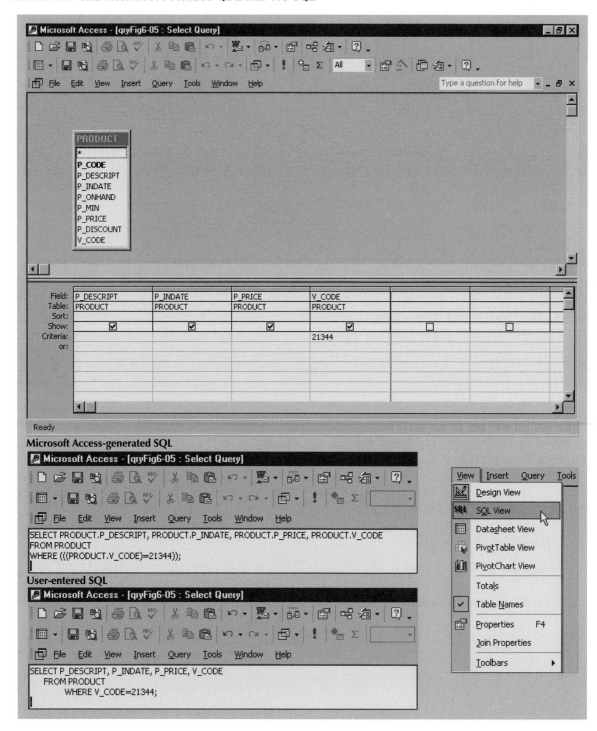

NOTE

NOTE TO MS ACCESS USERS

The MS Access QBE interface automatically designates the data source by using the table name as a prefix. You will discover later that the table name prefix is used to avoid ambiguity when the same column name appears in multiple tables. For example, both the VENDOR and the PRODUCT tables contain the V_CODE attribute. Therefore, if both tables are used—as they would be in a join—the source of the V_CODE attribute must be specified.

The SQL command structure thus provides almost limitless query flexibility. Numerous conditional restrictions may be placed on the selected table contents. For example, the comparison operators shown in Table 6.7 may be used to restrict output.

TABLE 6.7 **COMPARISON OPERATORS**

SYMBOL	MEANING
=	Equal to
<	Less than
<=	Less than or equal to
>	Greater than
>=	Greater than or equal to
<> or !=	Not equal to

The following example uses the "not equal to" operator:

SELECT P_DESCRIPT, P_INDATE, P_PRICE, V_CODE
 FROM PRODUCT
 WHERE V_CODE <> 21344;

The output, shown in Figure 6.7, lists all the rows for which the vendor code is *not* 21344.

FIGURE 6.7 **SELECTED PRODUCT TABLE ATTRIBUTES FOR VENDOR CODES OTHER THAN 21344**

P_DESCRIPT	P_INDATE	P_PRICE	V_CODE
Power painter, 15 psi., 3-nozzle	03-Nov-03	$109.99	25595
Hrd. cloth, 1/4-in., 2x50	15-Jan-04	$39.95	23119
Hrd. cloth, 1/2-in., 3x50	15-Jan-04	$43.99	23119
B&D jigsaw, 12-in. blade	30-Dec-03	$109.92	24288
B&D jigsaw, 8-in. blade	24-Dec-03	$99.87	24288
B&D cordless drill, 1/2-in.	20-Jan-04	$38.95	25595
Claw hammer	20-Jan-04	$9.95	21225
Hicut chain saw, 16 in.	07-Feb-04	$256.99	24288
1.25-in. metal screw, 25	01-Mar-04	$6.99	21225
2.5-in. wd. screw, 50	24-Feb-04	$8.45	21231
Steel matting, 4'x8'x1/6", .5" mesh	17-Jan-04	$119.95	25595

As you examine Figure 6.7, note that rows with nulls in the V_CODE column (see Figure 6.4) are not included in the SELECT command's output.

The command sequence:

SELECT P_DESCRIPT, P_ONHAND, P_MIN, P_PRICE
 FROM PRODUCT
 WHERE P_PRICE <= 10;

will yield the output shown in Figure 6.8.

FIGURE 6.8 SELECTED PRODUCT TABLE ATTRIBUTES WITH A P_PRICE RESTRICTION

P_DESCRIPT	P_ONHAND	P_MIN	P_PRICE
Claw hammer	23	10	$9.95
Rat-tail file, 1/8-in. fine	43	20	$4.99
PVC pipe, 3.5-in., 8-ft	188	75	$5.87
1.25-in. metal screw, 25	172	75	$6.99
2.5-in. wd. screw, 50	237	100	$8.45

Using Comparison Operators on Character Attributes

Because computers identify all characters by their (numeric) American Standard Code for Information Interchange (ASCII) codes, comparison operators may even be used to place restrictions on character-based attributes. Therefore, the command:

SELECT P_CODE, P_DESCRIPT, P_ONHAND, P_MIN, P_PRICE
 FROM PRODUCT
 WHERE P_CODE < '1558-QW1';

would be correct and would yield a list of all rows in which the P_CODE is alphabetically less than 1558-QW1. (Because the ASCII code value for the letter B is greater than the value of the letter A, it follows that A is less than B.) Therefore, the output will be generated as shown in Figure 6.9.

FIGURE 6.9 SELECTED PRODUCT TABLE ATTRIBUTES: THE ASCII CODE EFFECT

P_CODE	P_DESCRIPT	P_ONHAND	P_MIN	P_PRICE
11QER/31	Power painter, 15 psi., 3-nozzle	8	5	$109.99
13-Q2/P2	7.25-in. pwr. saw blade	32	15	$14.99
14-Q1/L3	9.00-in. pwr. saw blade	18	12	$17.49
1546-QQ2	Hrd. cloth, 1/4-in., 2x50	15	8	$39.95

NOTE

ABOUT ASCII CODES

All characters (such as the letter T, the number 8, or the symbol +) are assigned a code by the computer. For example, Table 6.8 shows just a few ASCII codes:

TABLE 6.8 SAMPLE CHARACTERS AND THEIR ASCII CODES

CHARACTER	ASCII CODE	CHARACTER	ASCII CODE
A	65	*	42
a	97	&	38
B	66	1	49
b	98	2	50
[	91	3	51

If you are curious about the full 256-code ASCII character set, check your computer's operating manual.

String (character) comparisons are made from left to right. This left-to-right comparison is especially useful when attributes such as names are to be compared. For example, the string "Ardmore" would be judged *greater than* the string "Aarenson" but *less than* the string "Brown"; such results may be used to generate alphabetic listings like those found in a phone directory.

If the characters 0–9 are stored as strings, the same left-to-right string comparisons may lead to apparent anomalies. For example, the ASCII code for the character "5" is, as expected, *greater than* the ASCII code for the character "4." Yet the same "5" will also be judged *greater than* the string "44" because the *first* character in the string "44" is less than the string "5." For this reason you may get some unexpected results from date comparisons if the dates are stored in character format. For example, the left-to-right ASCII character comparison would force the conclusion that the date "01/01/2004" occurred *before* "12/31/2003." Since the leftmost character "0" in "01/01/2004" is *less than* the leftmost character "1" in "12/31/2003," "01/01/2004" is *less than* "12/31/2003." Naturally, if data strings are stored in a yyyy/mm/dd format, the comparisons will yield appropriate results, but few of us would expect such a nonstandard date presentation. That's why all current RDBMSs support "date" data types, and that's why you should use them. In addition, using "date" data types gives you the benefit of date arithmetic.

Using Comparison Operators on Dates

Date procedures are often more software-specific than most other SQL procedures. For example, the query to list all the rows in which the inventory stock dates occur on or after January 20, 2004 will look like this:

SELECT P_DESCRIPT, P_ONHAND, P_MIN, P_PRICE, P_INDATE
 FROM PRODUCT
 WHERE P_INDATE >= '20-Jan-2004';

(Remember that MS Access users must use the # delimiters for dates. For example, you would use #20-Jan-04# in the above WHERE clause.) The date-restricted output is shown in Figure 6.10.

FIGURE 6.10 **SELECTED PRODUCT TABLE ATTRIBUTES: DATE RESTRICTION**

	P_DESCRIPT	P_ONHAND	P_MIN	P_PRICE	P_INDATE
▶	B&D cordless drill, 1/2-in.	12	5	$38.95	20-Jan-04
	Claw hammer	23	10	$9.95	20-Jan-04
	Hicut chain saw, 16 in.	11	5	$256.99	07-Feb-04
	PVC pipe, 3.5-in., 8-ft	188	75	$5.87	20-Feb-04
	1.25-in. metal screw, 25	172	75	$6.99	01-Mar-04
	2.5-in. wd. screw, 50	237	100	$8.45	24-Feb-04

Using Computed Columns and Column Aliases

Suppose you want to determine the total value of each of the products currently held in inventory. Logically, this determination requires the multiplication of each product's quantity on hand by its current price. You can accomplish this task with the following command:

SELECT P_DESCRIPT, P_ONHAND, P_PRICE, P_ONHAND*P_PRICE
 FROM PRODUCT;

Entering this SQL command in Access generates the output shown in Figure 6.11.

FIGURE 6.11 **SELECT STATEMENT WITH A COMPUTED COLUMN**

	P_DESCRIPT	P_ONHAND	P_PRICE	Expr1
▶	Power painter, 15 psi., 3-nozzle	8	$109.99	$879.92
	7.25-in. pwr. saw blade	32	$14.99	$479.68
	9.00-in. pwr. saw blade	18	$17.49	$314.82
	Hrd. cloth, 1/4-in., 2x50	15	$39.95	$599.25
	Hrd. cloth, 1/2-in., 3x50	23	$43.99	$1,011.77
	B&D jigsaw, 12-in. blade	8	$109.92	$879.36
	B&D jigsaw, 8-in. blade	6	$99.87	$599.22
	B&D cordless drill, 1/2-in.	12	$38.95	$467.40
	Claw hammer	23	$9.95	$228.85
	Sledge hammer, 12 lb.	8	$14.40	$115.20
	Rat-tail file, 1/8-in. fine	43	$4.99	$214.57
	Hicut chain saw, 16 in.	11	$256.99	$2,826.89
	PVC pipe, 3.5-in., 8-ft	188	$5.87	$1,103.56
	1.25-in. metal screw, 25	172	$6.99	$1,202.28
	2.5-in. wd. screw, 50	237	$8.45	$2,002.65
	Steel matting, 4'x8'x1/6", .5" mesh	18	$119.95	$2,159.10

SQL will accept any valid expressions (or formulas) in the computed columns. Such formulas can contain any valid mathematical operators and functions that are applied to attributes in any of the tables specified in the FROM clause of the SELECT statement. Note also that Access will automatically add an Expr label to all computed columns. (The first computed column would be labeled Expr1, the second Expr2, and so on.) If you are using Oracle, note that Oracle will use the actual formula text as the label for the computed column.

To make the output more readable, the SQL standard permits the use of aliases for any column in a SELECT statement. An **alias** is an alternate name given to a column or table in any SQL statement. For example, you can rewrite our previous SQL statement as:

SELECT P_DESCRIPT, P_ONHAND, P_PRICE, P_ONHAND*P_PRICE AS TOTVALUE
 FROM PRODUCT;

The output of this command is shown in Figure 6.12.

FIGURE 6.12 **SELECT STATEMENT WITH A COMPUTED COLUMN AND AN ALIAS**

P_DESCRIPT	P_ONHAND	P_PRICE	TOTVALUE
Power painter, 15 psi., 3-nozzle	8	$109.99	$879.92
7.25-in. pwr. saw blade	32	$14.99	$479.68
9.00-in. pwr. saw blade	18	$17.49	$314.82
Hrd. cloth, 1/4-in., 2x50	15	$39.95	$599.25
Hrd. cloth, 1/2-in., 3x50	23	$43.99	$1,011.77
B&D jigsaw, 12-in. blade	8	$109.92	$879.36
B&D jigsaw, 8-in. blade	6	$99.87	$599.22
B&D cordless drill, 1/2-in.	12	$38.95	$467.40
Claw hammer	23	$9.95	$228.85
Sledge hammer, 12 lb.	8	$14.40	$115.20
Rat-tail file, 1/8-in. fine	43	$4.99	$214.57
Hicut chain saw, 16 in.	11	$256.99	$2,826.89
PVC pipe, 3.5-in., 8-ft	188	$5.87	$1,103.56
1.25-in. metal screw, 25	172	$6.99	$1,202.28
2.5-in. wd. screw, 50	237	$8.45	$2,002.65
Steel matting, 4'x8'x1/6", .5" mesh	18	$119.95	$2,159.10

You could also use a computed column, an alias, and date arithmetic in a single query. For example, assume that you want to get a list of "out-of-warranty" products. In this case, we have arbitrarily defined out-of-warranty products as those products that have been stored more than 90 days. Therefore, the P_INDATE is at least 90 days less than the current (system) date. The MS Access version of this query is shown next:

SELECT P_CODE, P_INDATE, DATE() - 90 AS CUTDATE
 FROM PRODUCT
 WHERE P_INDATE <= DATE() - 90;

The Oracle version of the same query is shown below:

SELECT P_CODE, P_INDATE, SYSDATE - 90 AS CUTDATE
 FROM PRODUCT
 WHERE P_INDATE <= SYSDATE - 90;

Note that DATE() and SYSDATE are special functions that return today's date in MS Access and Oracle, respectively. You could use the DATE() and SYSDATE functions anywhere a date literal is expected, such as in the value list of an INSERT statement, in an UPDATE statement when changing the value of a date attribute, or in a SELECT statement as shown here. Of course, the previous query output would change based on today's date.

Suppose that a manager wants to have a list of all the products, the dates they were received, and the warranty expiration date (90 days from when the product was received). To generate that list, type:

SELECT P_CODE, P_INDATE, P_INDATE + 90 AS EXPDATE
 FROM PRODUCT;

Note that you can use all arithmetic operators with date attributes as well as with numeric attributes.

6.4.2 ARITHMETIC OPERATORS: THE RULE OF PRECEDENCE

As you saw in the previous example, you can use arithmetic operators with table attributes in a column list or in a conditional expression. In fact, SQL commands are often used in conjunction with the arithmetic operators shown in Table 6.9.

TABLE 6.9 **THE ARITHMETIC OPERATORS**

ARITHMETIC OPERATOR	DESCRIPTION
+	Add
-	Subtract
*	Multiply
/	Divide
^	Raise to the power of (Some applications use ** instead of ^.)

Do not confuse the multiplication symbol (*) with the wildcard symbol used by some SQL implementations such as Microsoft Access; the latter is used only in string comparisons, while the former is used in conjunction with mathematical procedures.

As you perform mathematical operations on attributes, remember the rules of precedence. As the name suggests, the **rules of precedence** are the rules that establish the order in which computations are completed. For example, note the order of the following computational sequence:

1. Perform operations within parentheses.

2. Perform power operations.

3. Perform multiplications and divisions.

4. Perform additions and subtractions.

The application of the rules of precedence will tell you that $8 + 2 * 5 = 8 + 10 = 18$, but $(8 + 2) * 5 = 10 * 5 = 50$. Similarly, $4 + 5^2 * 3 = 4 + 25 * 3 = 79$, but $(4 + 5)^2 * 3 = 81 * 3 = 243$, while the operation expressed by $(4 + 5^2) * 3$ yields the answer $(4 + 25) * 3 = 29 * 3 = 87$.

6.4.3 LOGICAL OPERATORS: AND, OR, AND NOT

In the real world, a search of data normally involves multiple conditions. For example, when you are buying a new house, you look for a certain area, 3 bedrooms, 2.5 bathrooms, two stories, etc. In the same way, SQL allows you to have multiple conditions in a query through the use of logical operators. The logical operators are AND, OR, and NOT. For example, if you want a list of the table contents for either the V_CODE = 21344 **OR** the V_CODE = 24288, you can use the following command sequence:

SELECT P_DESCRIPT, P_INDATE, P_PRICE, V_CODE
 FROM PRODUCT
 WHERE V_CODE = 21344 OR V_CODE = 24288;

This command generates the six rows, shown in Figure 6.13, that match the logical restriction.

FIGURE 6.13 **SELECTED PRODUCT TABLE ATTRIBUTES: THE LOGICAL OR**

P_DESCRIPT	P_INDATE	P_PRICE	V_CODE
7.25-in. pwr. saw blade	13-Dec-03	$14.99	21344
9.00-in. pwr. saw blade	13-Nov-03	$17.49	21344
B&D jigsaw, 12-in. blade	30-Dec-03	$109.92	24288
B&D jigsaw, 8-in. blade	24-Dec-03	$99.87	24288
Rat-tail file, 1/8-in. fine	15-Dec-03	$4.99	21344
Hicut chain saw, 16 in.	07-Feb-04	$256.99	24288

The logical **AND** has the same SQL syntax requirement. The following command generates a list of all rows for which P_PRICE is less than $50.00 AND for which P_INDATE is a date occurring after January 15, 2004:

SELECT P_DESCRIPT, P_INDATE, P_PRICE, V_CODE
 FROM PRODUCT
 WHERE P_PRICE < 50
 AND P_INDATE > '15-Jan-2004';

This command will produce the output shown in Figure 6.14.

FIGURE 6.14 SELECTED PRODUCT TABLE ATTRIBUTES: THE LOGICAL AND

P_DESCRIPT	P_INDATE	P_PRICE	V_CODE
B&D cordless drill, 1/2-in.	20-Jan-04	$38.95	25595
Claw hammer	20-Jan-04	$9.95	21225
PVC pipe, 3.5-in., 8-ft	20-Feb-04	$5.87	
1.25-in. metal screw, 25	01-Mar-04	$6.99	21225
2.5-in. wd. screw, 50	24-Feb-04	$8.45	21231

You can combine the logical OR with the logical AND to place further restrictions on the output. For example, suppose you want a table listing for the following conditions:

- The P_INDATE is after January 15, 2004, and the P_PRICE is less than $50.00.
- Or, the V_CODE is 24288.

The required listing can be produced by using:

SELECT P_DESCRIPT, P_INDATE, P_PRICE, V_CODE
 FROM PRODUCT
 WHERE (P_PRICE < 50 AND P_INDATE > '15-Jan-2004')
 OR V_CODE = 24288;

Note the use of parentheses to combine logical restrictions. Where you place the parentheses depends on how you want the logical restrictions to be executed. Conditions listed within parentheses are always executed first. The preceding query yields the output shown in Figure 6.15.

FIGURE 6.15 SELECTED PRODUCT TABLE ATTRIBUTES: THE LOGICAL AND AND OR

P_DESCRIPT	P_INDATE	P_PRICE	V_CODE
B&D jigsaw, 12-in. blade	30-Dec-03	$109.92	24288
B&D jigsaw, 8-in. blade	24-Dec-03	$99.87	24288
B&D cordless drill, 1/2-in.	20-Jan-04	$38.95	25595
Claw hammer	20-Jan-04	$9.95	21225
Hicut chain saw, 16 in.	07-Feb-04	$256.99	24288
PVC pipe, 3.5-in., 8-ft	20-Feb-04	$5.87	
1.25-in. metal screw, 25	01-Mar-04	$6.99	21225
2.5-in. wd. screw, 50	24-Feb-04	$8.45	21231

Note that the three rows with the V_CODE = 24288 are included regardless of the P_INDATE and P_PRICE entries for those rows.

The use of the logical operators OR and AND can become quite complex when numerous restrictions are placed on the query. In fact, a specialty field in mathematics known as **Boolean algebra** is dedicated to the use of logical operators.

The logical operator **NOT** is used to negate the result of a conditional expression. That is, in SQL all conditional expressions evaluate to true or false. If an expression is true, the row is selected; if it is false, the row is not selected. The NOT logical operator is typically used to find what rows *do not* match a certain condition. For example, if you want to see a listing of all the rows for which the vendor code is not 21344, use the command sequence:

SELECT *
 FROM PRODUCT
 WHERE NOT (V_CODE = 21344);

Note that we have enclosed the condition in parentheses; this practice is optional but it is highly recommended for clarity. The logical NOT can be combined with AND and OR.

> **N O T E**
>
> If your SQL version does not support the logical NOT, you can generate the required output by using the condition:
> WHERE V_CODE < > 21344
> If your version of SQL does not support < >, use:
> WHERE V_CODE != 21344

6.4.4 SPECIAL OPERATORS

ANSI-standard SQL allows the use of special operators in conjunction with the WHERE clause. These special operators include:

- **BETWEEN**—Used to check whether an attribute value is within a range.
- **IS NULL**—Used to check whether an attribute value is null.
- **LIKE**—Used to check whether an attribute value matches a given string pattern.
- **IN**—Used to check whether an attribute value matches any value within a value list.
- **EXISTS**—Used to check if a subquery returns any rows or not.

The BETWEEN Special Operator

If you use software that implements a standard SQL, the operator BETWEEN may be used to check whether an attribute value is within a range of values. For example, if you want to see a listing for all products whose prices are between $50.00 and $100.00, use the following command sequence:

SELECT *
 FROM PRODUCT
 WHERE P_PRICE BETWEEN 50.00 AND 100.00;

If your DBMS does not support BETWEEN, you can use:

SELECT *
 FROM PRODUCT
 WHERE P_PRICE > 50.00 AND P_PRICE < 100.00;

The IS NULL Special Operator

Standard SQL allows the use of IS NULL to check for a null attribute value. For example, suppose that you want to list all products that do not have a vendor assigned (V_CODE is null). Such a null entry could be found by using the command sequence:

SELECT P_CODE, P_DESCRIPT, V_CODE
 FROM PRODUCT
 WHERE V_CODE IS NULL;

Similarly, if you want to check a null date entry, the command sequence is:

SELECT P_CODE, P_DESCRIPT, P_INDATE
 FROM PRODUCT
 WHERE P_INDATE IS NULL;

Note that SQL uses a special operator to test for nulls. Why? Couldn't you just enter a condition like "V_CODE = NULL"? No. The reason is that **NULL** is technically not a "value" (such as the number 0—zero—or the blank space), but a special property of an attribute that represents precisely the absence of any value at all.

The LIKE Special Operator

The LIKE special operator is used in conjunction with the wildcards to find patterns within string attributes. Standard SQL allow you to use the percent sign (%) and underscore (_) wildcard characters to make matches when the entire string is not known:

- % means any and all *following* characters are eligible. For example,

 'J%' includes Johnson, Jones, Jernigan, July, and J-231Q

 'Jo%' includes Johnson and Jones

- _ means any *one* character may be substituted for the underscore. For example,

 '_23-456-6789' includes 123-456-6789, 223-456-6789, and 323-456-6789

 '_23-_56-678_' includes 123-156-6781, 123-256-6782, and 823-956-6788

 '_o_es' includes Jones, Cones, Cokes, totes, and roles

NOTE

Some RDBMSs, such as Microsoft Access, use the wildcard characters * and ? instead of % and _.

For example, the following query would find all the VENDOR rows for contacts whose last names begin with Smith.

SELECT V_NAME, V_CONTACT, V_AREACODE, V_PHONE
 FROM VENDOR
 WHERE V_CONTACT LIKE 'Smith%';

If you check the original VENDOR data in Figure 6.2 again, you'll see that this SQL query will yield three records: two Smiths and one Smithson.

Keep in mind that most SQL implementations yield case-sensitive searches. For example, Oracle will not yield a return that includes 'Jones' if you use the wildcard search delimiter 'jo%' in a search for last names, because 'Jones' begins with a capital and your wildcard search starts with a lowercase 'j'. On the other hand, MS Access searches are not case sensitive.

For example, suppose you typed the following query in Oracle:

```
SELECT V_NAME, V_CONTACT, V_AREACODE, V_PHONE
    FROM VENDOR
        WHERE V_CONTACT LIKE 'SMITH%';
```

No rows are printed. What happened?

What happened is that character-based queries may be case sensitive. That is, an uppercase character has a different ASCII code than a lowercase character, thus causing 'SMITH', 'Smith', and 'smith' to be evaluated as different (unequal) entries. Because the table contains no vendor whose last name begins with (uppercase) 'SMITH', the (uppercase) 'SMITH%' used in the query cannot make a match. Matches can be made only if the query entry is written exactly like the table entry.

Some RDBMSs, such as Microsoft Access, automatically make the necessary conversions to eliminate case sensitivity. Others, such as Oracle, provide a special UPPER function to convert both table and query character entries to uppercase. (The conversion is done in the computer's memory only; it has no effect on how the value is actually stored in the table.) So, if you want to avoid a no-match result based on case sensitivity, and if your RDBMS allows the use of the UPPER function, you can generate the same results by using the query:

```
SELECT V_NAME, V_CONTACT, V_AREACODE, V_PHONE
    FROM VENDOR
        WHERE UPPER(V_CONTACT) LIKE 'SMITH%';
```

The preceding query would produce a list including all rows that contain a last name that begins with 'Smith', regardless of uppercase or lowercase letter combinations such as 'Smith', 'smith', 'SMITH', and so on.

The logical operators may be used in conjunction with the special operators. For instance, the query:

```
SELECT V_NAME, V_CONTACT, V_AREACODE, V_PHONE
    FROM VENDOR
        WHERE V_CONTACT NOT LIKE 'Smith%';
```

will yield an output of all vendors whose names do not start with Smith.

Suppose that you do not know whether a person's name is spelled Johnson or Johnsen. The wildcard character _ lets you find a match for either spelling. The proper search would be instituted by the query:

```
SELECT *
    FROM VENDOR
        WHERE V_CONTACT LIKE 'Johns_n'
```

The wildcards thus allow us to make matches when only approximate spellings are known. Wildcard characters may be used in combinations. For example, the wildcard search based on the string '_l%' can yield the strings Al, Alton, Elgin, Blakeston, blank, bloated, and eligible.

The IN Special Operator

Many queries that would require the use of the logical OR can be more easily handled with the help of the special operator IN. For example, the query:

SELECT *
 FROM PRODUCT
 WHERE V_CODE = 21344
 OR V_CODE = 24288;

can be handled more efficiently with:

SELECT *
 FROM PRODUCT
 WHERE V_CODE IN (21344, 24288);

Note that the IN operator uses a value list. The values in the list must all be of the same data type. Each of the values in the value list is compared to the attribute, in this case V_CODE. If the V_CODE value matches any of the values in the list, the row is selected. In this example, the rows selected will be only those in which the V_CODE is either 21344 or 24288.

If the attribute used is of a character data type, then the list values must be enclosed in single quotation marks. For instance, if the V_CODE had been defined as CHAR(5) during the table-creation process, the preceding query would have read:

SELECT *
 FROM PRODUCT
 WHERE V_CODE IN ('21344', '24288');

The IN operator is especially valuable when it is used in conjunction with sub-queries. For example, suppose you want to list the V_CODE and V_NAME of only those vendors that provide products. In this case, you could use a subquery within the IN operator to automatically generate the value list. The query would be:

SELECT V_CODE, V_NAME
 FROM VENDOR
 WHERE V_CODE IN (SELECT V_CODE FROM PRODUCT);

The preceding query will be executed in two steps:

1. The inner query or subquery will generate a list of V_CODE values from the PRODUCT tables. These V_CODE values represent the vendors that supply products.

2. The IN operator will compare the values generated by the subquery to the V_CODE values in the VENDOR table and will select only the rows with matching values—that is, the vendors that provide products.

The IN special operator will receive additional attention in Chapter 7 where you will learn more about subqueries.

The EXISTS Special Operator

EXISTS can be used whenever there is a requirement to execute a command based on the result of another query. That is, if a subquery returns any rows, then run the main query, otherwise don't. For example, the following query will list all vendors but only if there are products to order:

SELECT * FROM VENDOR
 WHERE EXISTS (SELECT * FROM PRODUCT WHERE P_ONHAND <= P_MIN);

The EXISTS special operator is used in the following example to list all vendors but only if there are products with the quantity on hand, less than double the minimum quantity:

SELECT * FROM VENDOR
 WHERE EXISTS (SELECT * FROM PRODUCT
 WHERE P_ONHAND < P_MIN*2);

The EXISTS special operator will receive additional attention in the next chapter where you will learn more about subqueries.

6.5 ADVANCED DATA DEFINITION COMMANDS

In Section 6.2 you learned how to create a table structure and to enter, modify, delete, and list data. In this section you will learn how to change (alter) the table structure by changing attribute characteristics and by adding columns. You will then learn how to enter data into the new or modified columns. Finally, you will learn how to copy tables or parts of tables, and how to delete tables.

All changes in the table structure are made by using the **ALTER** command, followed by a keyword that produces the specific change you wish to make. Three options are available: ADD, MODIFY, and DROP. ADD enables you to add a column and MODIFY enables you to change column characteristics. DROP allows you delete a column from a table. Most RDBMSs do not allow you to delete a column (unless the column does not contain any values), because such an action may delete crucial data that are used by other tables.

The basic syntax to add or modify columns is:

ALTER TABLE *tablename*
{ADD | MODIFY} (*columnname datatype* [{ADD | MODIFY} *columnname datatype*]) ;

The ALTER TABLE command can also be used to add table constraints. In that case the syntax would be:

ALTER TABLE *tablename*
ADD *constraint* [ADD *constraint*] ;

where *constraint* refers to a constraint definition like the ones you learned in Section 6.2.6.

You could also use the ALTER TABLE command to remove a column or a table constraint. The syntax would be as follows:

ALTER TABLE *tablename*
 DROP {PRIMARY KEY | COLUMN *columnname* | CONSTRAINT *constraintname* };

Notice that when removing a constraint, you need to specify the name given to the constraint. This is one reason why it is always advisable to name your constraints in your CREATE TABLE or ALTER TABLE statement.

6.5.1 CHANGING A COLUMN'S DATA TYPE

Using the just-discussed ALTER syntax, the (integer) V_CODE in the PRODUCT table can be changed to a character V_CODE by using:

ALTER TABLE PRODUCT
 MODIFY (V_CODE CHAR(5));

Some RDBMSs (such as Oracle) do not let you change data types unless the column to be changed is empty. For example, if you want to change the V_CODE field from the current number definition to a character

definition, the above command will yield an error message because the V_CODE column already contains data. The error message is easily explained. If you change the V_CODE data type, remember that the V_CODE in PRODUCT references the V_CODE in VENDOR. If the data types don't match, there is a referential integrity violation, thus triggering the error message. If the V_CODE column does not contain data, the preceding command sequence will produce the expected table structure alteration if the foreign key reference was not specified during the PRODUCT table's creation.

6.5.2 CHANGING A COLUMN'S DATA CHARACTERISTICS

If the column to be changed already contains data, you can make any changes in the column's characteristics if those changes do not alter the data *type*. For example, if you want to increase the width of the P_PRICE column to nine digits, use the command:

ALTER TABLE PRODUCT
 MODIFY (P_PRICE DECIMAL(9,2));

If you now list the table contents again, you see that the column width of P_PRICE has increased by one digit. (Your SQL version may require that the parentheses be dropped.)

> **NOTE**
>
> Some DBMS vendors may impose limitations on when it's possible to change attribute characteristics. For example, Oracle will let you increase (but not decrease) the size of a column. The reason for this restriction is that an attribute modification will affect the integrity of the data in the database. In fact, some attribute changes can be done only if there are no data in any rows for the affected attribute.

6.5.3 ADDING A COLUMN

You can alter an existing table by adding one or more columns. In the following example, we will add the column named P_SALECODE to the PRODUCT table. (We will use this column later to determine whether goods that have been in inventory for a certain length of time should be placed on special sale.)

Suppose that we expect the P_SALECODE entries to be 1, 2, or 3. Because we are not going to perform arithmetic with the P_SALECODE, the P_SALECODE will be classified as a single-character attribute. Note the inclusion of all the required information in the following ALTER command:

ALTER TABLE PRODUCT
 ADD (P_SALECODE CHAR(1));

> **ADDITIONAL CD CONTENT**
>
> To enable you to see the individual effects of each of the updates in the following sections, we made several copies of the PRODUCT table in the Ch06_SaleCo database, one named PRODUCT_2 and one named PRODUCT_3. Each of the two copies includes the new P_SALECODE column. We preserved the original PRODUCT table to match the query results in the preceding sections.
>
> Because many of our students assured us that they found it much easier to understand the update queries when their effects were examined separately, we have illustrated the update results in Section 6.6.5 with the help of the PRODUCT_2 and PRODUCT_3 tables. However, if you want to see the *cumulative* effect of all the UPDATE commands, you can continue using the PRODUCT table with the P_SALECODE modification and all of the changes you will make in the following sections. (You may even want to use both options, first to examine the individual effects of the update queries and then to examine the cumulative effects. The selection of one option over the other is ultimately a judgment call.)

When adding a column, be careful not to include the NOT NULL clause for the new column. Doing so will cause an error message because if you add a new column to a table that already has rows, the existing rows will default to a value of null for the new column. Therefore, it is not possible to add the NOT NULL clause for this new column. (You can, of course, add the NOT NULL clause to the table structure after all the data for the new column have been entered and the column no longer contains nulls.)

6.5.4 DROPPING A COLUMN

Occasionally, you may want to modify a table by deleting a column. Suppose you want to delete the V_ORDER attribute from the VENDOR table. To accomplish this task, you would use the following command:

```
ALTER TABLE VENDOR
    DROP COLUMN V_ORDER;
```

Again, some RDBMSs impose restrictions on the deletion of an attribute. For example, you may not drop attributes that are involved in foreign key relationships, nor may you delete an attribute of a table that contains only that one attribute.

6.5.5 ADVANCED DATA UPDATES

To make data entries in an existing row's columns, SQL employs the UPDATE command. The UPDATE command only updates data in existing rows. For example, to enter the P_SALECODE value '2' in the fourth row, use the UPDATE command together with the primary key P_CODE '1546-QQ2'. Enter the value by using the command sequence:

```
UPDATE PRODUCT
    SET P_SALECODE = '2'
        WHERE P_CODE = '1546-QQ2';
```

The effect of this command yields the output shown in Figure 6.16. (We have used the PRODUCT_2 table in the MS Access database, so substitute PRODUCT_2 in the above query if you have used that table.)

FIGURE 6.16 THE EFFECT OF DATA ENTRY INTO THE NEW P_SALECODE COLUMN

P_CODE	P_DESCRIPT	P_INDATE	P_ONHAND	P_MIN	P_PRICE	P_DISCOUNT	V_CODE	P_SALECODE
1546-QQ2	Hrd. cloth, 1/4-in., 2x50	15-Jan-04	15	8	$39.95	0.00	23119	2

Subsequent data may be entered the same way, defining each entry location by its primary key (P_CODE) and its column location (P_SALECODE). For example, if we want to enter the P_SALECODE value '1' for the P_CODE values '2232/QWE' and '2232/QTY', we use:

```
UPDATE PRODUCT
    SET P_SALECODE = '1'
        WHERE P_CODE IN ('2232/QWE', '2232/QTY');
```

If your RDBMS does not support IN, use the following command:

```
UPDATE PRODUCT
    SET P_SALECODE = '1'
        WHERE P_CODE = '2232/QWE' OR P_CODE = '2232/QTY';
```

The results of your efforts can be checked by using:

SELECT P_CODE, P_DESCRIPT, P_INDATE, P_PRICE, P_SALECODE
 FROM PRODUCT;

to yield the output shown in Figure 6.17. (If you used the PRODUCT_2 table, substitute PRODUCT_2 in the above queries.)

FIGURE 6.17 **UPDATE OF THE P_SALECODE COLUMN IN MULTIPLE DATA ROWS**

P_CODE	P_DESCRIPT	P_INDATE	P_PRICE	P_SALECODE
1QER/31	Power painter, 15 psi., 3-	03-Nov-03	$109.99	
13-Q2/P2	7.25-in. pwr. saw blade	13-Dec-03	$14.99	
14-Q1/L3	9.00-in. pwr. saw blade	13-Nov-03	$17.49	
1546-QQ2	Hrd. cloth, 1/4-in., 2x50	15-Jan-04	$39.95	2
1558-QW1	Hrd. cloth, 1/2-in., 3x50	15-Jan-04	$43.99	
2232/QTY	B&D jigsaw, 12-in. blade	30-Dec-03	$109.92	1
2232/QWE	B&D jigsaw, 8-in. blade	24-Dec-03	$99.87	1
2238/QPD	B&D cordless drill, 1/2-in.	20-Jan-04	$38.95	
23109-HB	Claw hammer	20-Jan-04	$9.95	
23114-AA	Sledge hammer, 12 lb.	02-Jan-04	$14.40	
54778-2T	Rat-tail file, 1/8-in. fine	15-Dec-03	$4.99	
89-WRE-Q	Hicut chain saw, 16 in.	07-Feb-04	$256.99	
PVC23DRT	PVC pipe, 3.5-in., 8-ft	20-Feb-04	$5.87	
SM-18277	1.25-in. metal screw, 25	01-Mar-04	$6.99	
SW-23116	2.5-in. wd. screw, 50	24-Feb-04	$8.45	
WR3/TT3	Steel matting, 4'x8'x1/6",	17-Jan-04	$119.95	

Although the UPDATE sequences just shown allow you to enter values into specified table cells, the process is very cumbersome. Fortunately, if a relationship can be established between the entries and the existing columns, that relationship may be used to assign values to their appropriate slots. For example, suppose that we want to place sales codes based on the P_INDATE into the table, using the following schedule:

P_INDATE	P_SALECODE
before December 25, 2003	2
between January 16 and February10, 2004	1

We will use the PRODUCT table to demonstrate how to make these changes. The following two command sequences will make the appropriate assignments:

UPDATE PRODUCT
 SET P_SALECODE = '2'
 WHERE P_INDATE < '25-Dec-2003';

UPDATE PRODUCT
 SET P_SALECODE = '1'
 WHERE P_INDATE >= '16-Jan-2004'
 AND P_INDATE <='10-Feb-2004';

(If you used the PRODUCT_3 table, substitute PRODUCT_3 in the above query.)

To check the results of these two command sequences, use:

SELECT P_CODE, P_DESCRIPT, P_INDATE, P_PRICE, P_SALECODE
 FROM PRODUCT;

to produce the output shown in Figure 6.18. (We have shown only the results of these latter two update queries to avoid mixing the results of these two updates and the updates that were shown in Figure 6.17. To preserve the separate update results, we actually used the PRODUCT_3 table as the recipient of the preceding two UPDATE commands. (The SELECT query was actually executed on this PRODUCT_3 table to produce the results shown in Figure 6.18.)

FIGURE 6.18 THE EFFECT OF MULTIPLE DATA UPDATES IN THE PRODUCT TABLE (MS ACCESS)

P_CODE	P_DESCRIPT	P_INDATE	P_PRICE	P_SALECODE
11QER/31	Power painter, 15 psi., 3-nozzle	03-Nov-2003	$109.99	2
13-Q2/P2	7.25-in. pwr. saw blade	13-Dec-2003	$14.99	2
14-Q1/L3	9.00-in. pwr. saw blade	13-Nov-2003	$17.49	2
1546-QQ2	Hrd. cloth, 1/4-in., 2x50	15-Jan-2004	$39.95	
1558-QW1	Hrd. cloth, 1/2-in., 3x50	15-Jan-2004	$43.99	
2232/QTY	B&D jigsaw, 12-in. blade	30-Dec-2003	$109.92	
2232/QWE	B&D jigsaw, 8-in. blade	24-Dec-2003	$99.87	2
2238/QPD	B&D cordless drill, 1/2-in.	20-Jan-2004	$38.95	1
23109-HB	Claw hammer	20-Jan-2004	$9.95	1
23114-AA	Sledge hammer, 12 lb.	02-Jan-2004	$14.40	
54778-2T	Rat-tail file, 1/8-in. fine	15-Dec-2003	$4.99	2
89-WRE-Q	Hicut chain saw, 16 in.	07-Feb-2004	$256.99	1
PVC23DRT	PVC pipe, 3.5-in., 8-ft	20-Feb-2004	$5.87	
SM-18277	1.25-in. metal screw, 25	01-Mar-2004	$6.99	
SW-23116	2.5-in. wd. screw, 50	24-Feb-2004	$8.45	
WR3/TT3	Steel matting, 4'x8'x1/6", .5" mesh	17-Jan-2004	$119.95	1

If you have made *all* the updates shown in this section using your Oracle account, your PRODUCT table will look like Figure 6-18A. (If you have made *all* the updates in this section using MS Access, your Access table will yield the same results—but keep in mind that the order of the rows in the Access table may not be the same as in the Oracle table.)

The arithmetic operators are particularly useful in data updates. For example, if the quantity on hand in your PRODUCT table has dropped below the minimum desirable value, you'll order more of the product. Suppose, for example, that you have ordered 20 units of product 2232/QWE. When the 20 units arrive, you'll want to add them to inventory using:

UPDATE PRODUCT
 SET P_ONHAND = P_ONHAND + 20
 WHERE P_CODE = '2232/QWE';

If you want to add 10 percent to the price for all products that have current prices below $50, you may use:

UPDATE PRODUCT
 SET P_PRICE = P_PRICE*1.10
 WHERE P_PRICE < 50.00;

FIGURE 6.18A THE CUMULATIVE EFFECT OF THE MULTIPLE UPDATES IN THE PRODUCT TABLE (ORACLE)

```
Oracle SQL*Plus
File  Edit  Search  Options  Help
SQL> SELECT P_CODE, P_DESCRIPT, P_INDATE, P_PRICE, P_SALECODE FROM PRODUCT;

P_CODE      P_DESCRIPT                       P_INDATE    P_PRICE  P_SALECODE
----------  -------------------------------  ----------  -------  ----------
11QER/31    Power painter, 15 psi., 3-nozzle 03-NOV-03   109.99 2
13-Q2/P2    7.25-in. pwr. saw blade          13-DEC-03    14.99 2
14-Q1/L3    9.00-in. pwr. saw blade          13-NOV-03    17.49 2
1546-QQ2    Hrd. cloth, 1/4-in., 2x50        15-JAN-04    39.95 2
1558-QW1    Hrd. cloth, 1/2-in., 3x50        15-JAN-04    43.99
2232/QTY    B&D jigsaw, 12-in. blade         30-DEC-03   109.92 1
2232/QWE    B&D jigsaw, 8-in. blade          24-DEC-03    99.87 2
2238/QPD    B&D cordless drill, 1/2-in.      20-JAN-04    38.95 1
23109-HB    Claw hammer                      20-JAN-04     9.95 1
23114-AA    Sledge hammer, 12 lb.            02-JAN-04    14.4
54778-2T    Rat-tail file, 1/8-in. fine      15-DEC-03     4.99 2
89-WRE-Q    Hicut chain saw, 16 in.          07-FEB-04   256.99 1
PVC23DRT    PVC pipe, 3.5-in., 8-ft          20-FEB-04     5.87
SM-18277    1.25-in. metal screw, 25         01-MAR-04     6.99
SW-23116    2.5-in. wd. screw, 50            24-FEB-04     8.45
WR3/TT3     Steel matting, 4'x8'x1/6", .5" mesh 17-JAN-04 119.95 1

16 rows selected.

SQL>
```

> **NOTE**
>
> The output of the preceding UPDATE queries may not match the output of the same queries when executed against your database (you may have not added all rows, may have skipped some queries, and so on). In any case, check the effect of your query to verify it does what it is supposed to do with your own data set. Use a SELECT statement on the tables and attributes you are modifying before you run the query and after you run the query. The objective is for you to understand what the SQL query you are executing does, regardless of the data set. Once you master SQL you should be able to apply it to any database and any tables. Of course, you can always reload the original tables from the student CD.

6.5.6 COPYING PARTS OF TABLES

As you will discover in later chapters on database design, sometimes it may be necessary to break up a table structure into several component parts (or smaller tables). Fortunately, SQL allows us to copy the contents of selected table columns so that the data need not be reentered manually into the newly created table(s). For example, if you want to copy P_CODE, P_DESCRIPT, and P_PRICE from the PRODUCT table to a new table named PART, first create the PART table structure, as shown next.

```
CREATE TABLE PART(
PART_CODE            CHAR(8)       NOT NULL     UNIQUE,
PART_DESCRIPT        CHAR(35),
PART_PRICE           DECIMAL(8,2),
PRIMARY KEY (PART_CODE));
```

Note that the PART column names need not be identical to those of the original table, and that the new table need not have the same number of columns as the original table. In this case, the first column in the PART table is PART_CODE, rather than the original P_CODE found in the PRODUCT table. And the PART table contains

only three columns, rather than the seven columns found in the PRODUCT table. However, column characteristics must match; you cannot copy a character-based attribute into a numeric structure, and vice versa.

Next you need to add the rows to the new PART table using the PRODUCT table rows. To do this you need to use the INSERT command you learned in Section 6.3.7. The syntax is:

INSERT INTO *target_tablename*[(*target_columnlist*)]
 SELECT *source_columnlist*
 FROM *source_tablename*;

Note that the target column list is required if the source column list doesn't match all the attribute names and characteristics of the target table (including the order of the columns). Otherwise, you do not need to specify the target column list. In our example, we need to specify the target column list in the INSERT command below because the column names of the target table are different:

INSERT INTO PART (PART_CODE, PART_DESCRIPT, PART_PRICE)
 SELECT P_CODE, P_DESCRIPT, P_PRICE FROM PRODUCT;

The contents of the PART table can now be examined by using the query:

SELECT * FROM PART;

to generate the new PART table's contents, shown in Figure 6.19.

FIGURE 6.19 **PART ATTRIBUTES COPIED FROM THE PRODUCT TABLE**

PART_CODE	PART_DESCRIPT	PART_PRICE
11QER/31	Power painter, 15 psi., 3-nozzle	$109.99
13-Q2/P2	7.25-in. pwr. saw blade	$14.99
14-Q1/L3	9.00-in. pwr. saw blade	$17.49
1546-QQ2	Hrd. cloth, 1/4-in., 2x50	$39.95
1558-QW1	Hrd. cloth, 1/2-in., 3x50	$43.99
2232/QTY	B&D jigsaw, 12-in. blade	$109.92
2232/QWE	B&D jigsaw, 8-in. blade	$99.87
2238/QPD	B&D cordless drill, 1/2-in.	$38.95
23109-HB	Claw hammer	$9.95
23114-AA	Sledge hammer, 12 lb.	$14.40
54778-2T	Rat-tail file, 1/8-in. fine	$4.99
89-WRE-Q	Hicut chain saw, 16 in.	$256.99
PVC23DRT	PVC pipe, 3.5-in., 8-ft	$5.87
SM-18277	1.25-in. metal screw, 25	$6.99
SW-23116	2.5-in. wd. screw, 50	$8.45
WR3/TT3	Steel matting, 4'x8'x1/6", .5" mesh	$119.95

SQL also provides another way to rapidly create a new table based on selected columns and rows of an existing table. In this case, the new table will copy the attribute names, data characteristics, and rows of the original table. The Oracle version of this command is:

CREATE TABLE PART AS
 SELECT P_CODE AS PART_CODE, P_DESCRIPT AS PART_DESCRIPT, P_PRICE AS PART_PRICE
 FROM PRODUCT;

> **NOTE**
>
> **NOTE TO ORACLE USERS**
>
> If the PART table already exists, Oracle will not let you overwrite the existing table. To run this command, you must first delete the existing PART table (see Section 6.5.8).

> **NOTE**
>
> **NOTE TO MS ACCESS USERS**
>
> The MS Access version of this command is:
> SELECT P_CODE AS PART_CODE, P_DESCRIPT AS PART_DESCRIPT, P_PRICE AS PART_PRICE INTO PART
> FROM PRODUCT;
> If the PART table exists already, MS Access will ask you if you want to delete the existing table and continue with the creation of the new PART table.

The SQL command we have just shown you will create a new PART table with PART_CODE, PART_DESCRIPT, and PART_PRICE columns. In addition, all the data rows (for the selected columns) will be automatically copied. *But note that no entity integrity (primary key) or referential integrity (foreign key) rules are automatically applied to the new table.* In the next section you will learn how to define the PK and FK to enforce entity and referential integrity, respectively.

6.5.7 ADDING PRIMARY AND FOREIGN KEY DESIGNATIONS

In the previous section you learned how to create a new table based on another table. Unfortunately, the process we illustrated in the previous section created a new PART table without the inclusion of the integrity rules. (In particular, there is no primary key.) To define the primary key for the new PART table use the following command:

ALTER TABLE PART
 ADD PRIMARY KEY (PART_CODE);

In Section 6.2.2, you declared the primary and foreign keys when we created the PRODUCT and VENDOR table structures to make sure that the RDBMS would enforce the integrity rules. But as you just learned, the integrity rules are not transferred when the table derives its data from one or more other tables. (You should recall from Chapter 5, "Normalization of Database Tables," that the normalization process often requires you to break up tables into sets of related tables.)

Quite aside from the fact that the integrity rules are not automatically transferred to a new table that derives its data from one or more other tables, there are several other scenarios that would leave you without entity and referential integrity enforcement. For example, you might have forgotten to define the primary and foreign keys when you created the tables. Or, if you had imported tables from a different database, you would have discovered that the importing procedure will not transfer the integrity rules. In any case, you can reestablish the integrity rules via the ALTER command. For example, if the PRODUCT table's primary key had not yet been created, you would use:

ALTER TABLE PRODUCT
 ADD PRIMARY KEY (P_CODE);

If the PRODUCT table's foreign key had not yet been designated, it could be designated by:

ALTER TABLE PRODUCT
 ADD FOREIGN KEY (V_CODE) REFERENCES VENDOR;

If neither the PRODUCT table's primary key nor its foreign key had been designated, you could incorporate both changes at once, using:

ALTER TABLE PRODUCT
 ADD PRIMARY KEY (P_CODE)
 ADD FOREIGN KEY (V_CODE) REFERENCES VENDOR;

Even composite primary keys and multiple foreign keys can be designated in a single SQL command. For example, if you wanted to enforce the integrity rules for the LINE table shown in Figure 6.1, you could use:

ALTER TABLE LINE
 ADD PRIMARY KEY (INV_NUMBER, LINE_NUMBER)
 ADD FOREIGN KEY (INV_NUMBER) REFERENCES INVOICE
 ADD FOREIGN KEY (PROD_CODE) REFERENCES PRODUCT;

6.5.8 DELETING A TABLE FROM THE DATABASE

A table can be deleted from the database through the **DROP** command. For example, we can delete the just-created PART table with:

DROP TABLE PART;

You can only drop a table if that table is not participating as the "one" side of any relationship. If you try to drop such a table, the RDBMS will generate an error message to indicate that a foreign key integrity violation has occurred.

6.6 ADVANCED SELECT QUERIES

One of the most important advantages of SQL is its ability to produce complex free-form queries. The logical operators that were introduced earlier to update table contents work just as well in the query environment. In addition, SQL provides useful functions that count, find minimum and maximum values, calculate averages, and so on. Better yet, SQL allows the user to limit queries to only those entries having no duplicates or entries whose duplicates may be grouped.

6.6.1 ORDERING A LISTING

The **ORDER BY** clause is especially useful if the listing order is important to you. The syntax is:

SELECT *columnlist*
 FROM *tablelist*
 [WHERE *conditionlist*]
 [ORDER BY *columnlist* [ASC | DESC]] ;

Although you have the option of declaring the ordering type—ascending or descending—note that the default order is ascending. For example, if you want the contents of the PRODUCT table listed by P_PRICE in ascending order, use:

SELECT P_CODE, P_DESCRIPT, P_INDATE, P_PRICE
 FROM PRODUCT
 ORDER BY P_PRICE;

The output is shown in Figure 6.20. Note that ORDER BY yields an ascending price listing.

FIGURE 6.20 **SELECTED PRODUCT TABLE ATTRIBUTES: ORDERED BY (ASCENDING) P_PRICE**

P_CODE	P_DESCRIPT	P_INDATE	P_PRICE
54778-2T	Rat-tail file, 1/8-in. fine	15-Dec-03	$4.99
PVC23DRT	PVC pipe, 3.5-in., 8-ft	20-Feb-04	$5.87
SM-18277	1.25-in. metal screw, 25	01-Mar-04	$6.99
SW-23116	2.5-in. wd. screw, 50	24-Feb-04	$8.45
23109-HB	Claw hammer	20-Jan-04	$9.95
23114-AA	Sledge hammer, 12 lb.	02-Jan-04	$14.40
13-Q2/P2	7.25-in. pwr. saw blade	13-Dec-03	$14.99
14-Q1/L3	9.00-in. pwr. saw blade	13-Nov-03	$17.49
2238/QPD	B&D cordless drill, 1/2-in.	20-Jan-04	$38.95
1546-QQ2	Hrd. cloth, 1/4-in., 2x50	15-Jan-04	$39.95
1558-QW1	Hrd. cloth, 1/2-in., 3x50	15-Jan-04	$43.99
2232/QWE	B&D jigsaw, 8-in. blade	24-Dec-03	$99.87
2232/QTY	B&D jigsaw, 12-in. blade	30-Dec-03	$109.92
11QER/31	Power painter, 15 psi., 3-nozzle	03-Nov-03	$109.99
WR3/TT3	Steel matting, 4'x8'x1/6", .5" mesh	17-Jan-04	$119.95
89-WRE-Q	Hicut chain saw, 16 in.	07-Feb-04	$256.99

Comparing the listing in Figure 6.20 to the actual table contents shown in Figure 6.2, you will see that in Figure 6.20, the lowest-priced product is listed first, followed by the next lowest-priced product, and so on. However, although ORDER BY produces a sorted output, the actual table contents are unaffected by the ORDER command.

To produce the list in descending order, you would enter:

SELECT P_CODE, P_DESCRIPT, P_INDATE, P_PRICE
 FROM PRODUCT
 ORDER BY P_PRICE DESC;

Ordered listings are used frequently. For example, suppose you want to create a phone directory. It would be helpful if you could produce an ordered sequence (last name, first name, initial) in three stages:

1. ORDER BY last name.
2. Within the last names, ORDER BY first name.
3. Within the order created in Step 2, ORDER BY middle initial.

Such a multilevel ordered sequence is known as a **cascading order sequence**, and it can be created easily by listing several attributes, separated by commas, after the ORDER BY clause. The cascading order sequence is the basis for any telephone directory. To illustrate a cascading order sequence, let's use the EMPLOYEE data shown in Figure 6.21.

FIGURE 6.21 **PARTIAL LISTING OF EMPLOYEE TABLE CONTENTS**

EMP_NUM	EMP_TITLE	EMP_LNAME	EMP_FNAME	EMP_INITIAL	EMP_AREACODE	EMP_PHONE
100	Mr.	Kolmycz	George	D	615	324-5456
101	Ms.	Lewis	Rhonda	G	615	324-4472
102	Mr.	Vandam	Rhett		901	675-8993
103	Ms.	Jones	Anne	M	615	898-3456
104	Mr.	Lange	John	P	901	504-4430
105	Mr.	Williams	Robert	D	615	890-3220
106	Mrs.	Smith	Jeanine	K	615	324-7883
107	Mr.	Diante	Jorge	D	615	890-4567
108	Mr.	Wiesenbach	Paul	R	615	897-4358
109	Mr.	Smith	George	K	901	504-3339
110	Mrs.	Genkazi	Leighla	W	901	569-0093
111	Mr.	Washington	Rupert	E	615	890-4925
112	Mr.	Johnson	Edward	E	615	898-4387
113	Ms.	Smythe	Melanie	P	615	324-9006
114	Ms.	Brandon	Marie	G	901	882-0845
115	Mrs.	Saranda	Hermine	R	615	324-5505
116	Mr.	Smith	George	A	615	890-2984

Given the contents of the EMPLOYEE table, the SQL command:

SELECT EMP_LNAME, EMP_FNAME, EMP_INITIAL, EMP_AREACODE, EMP_PHONE
 FROM EMPLOYEE
 ORDER BY EMP_LNAME, EMP_FNAME, EMP_INITIAL;

will yield the results shown in Figure 6.22.

FIGURE 6.22 **TELEPHONE LIST QUERY RESULTS**

EMP_LNAME	EMP_FNAME	EMP_INITIAL	EMP_AREACODE	EMP_PHONE
Brandon	Marie	G	901	882-0845
Diante	Jorge	D	615	890-4567
Genkazi	Leighla	W	901	569-0093
Johnson	Edward	E	615	898-4387
Jones	Anne	M	615	898-3456
Kolmycz	George	D	615	324-5456
Lange	John	P	901	504-4430
Lewis	Rhonda	G	615	324-4472
Saranda	Hermine	R	615	324-5505
Smith	George	A	615	890-2984
Smith	George	K	901	504-3339
Smith	Jeanine	K	615	324-7883
Smythe	Melanie	P	615	324-9006
Vandam	Rhett		901	675-8993
Washington	Rupert	E	615	890-4925
Wiesenbach	Paul	R	615	897-4358
Williams	Robert	D	615	890-3220

The ORDER BY clause is useful in many applications, especially because the DESC qualifier may be invoked. For example, getting the most recent items listed first is a standard procedure. Typically, invoice due dates are listed in descending order. Or, if you want to examine budgets, it's probably useful to start by looking at the largest budget line items.

You can use the ORDER BY clause in conjunction with other SQL commands, too. For example, note the use of restrictions on date and price in the following command sequence:

SELECT P_DESCRIPT, V_CODE, P_INDATE, P_PRICE
 FROM PRODUCT
 WHERE P_INDATE < '21-Jan-2004' AND
 P_PRICE <= 50.00
 ORDER BY V_CODE, P_PRICE DESC;

The output is shown in Figure 6.23.

FIGURE 6.23 **A QUERY BASED ON MULTIPLE RESTRICTIONS**

P_DESCRIPT	V_CODE	P_INDATE	P_PRICE
Sledge hammer, 12 lb.		02-Jan-04	$14.40
Claw hammer	21225	20-Jan-04	$9.95
9.00-in. pwr. saw blade	21344	13-Nov-03	$17.49
7.25-in. pwr. saw blade	21344	13-Dec-03	$14.99
Rat-tail file, 1/8-in. fine	21344	15-Dec-03	$4.99
Hrd. cloth, 1/2-in., 3x50	23119	15-Jan-04	$43.99
Hrd. cloth, 1/4-in., 2x50	23119	15-Jan-04	$39.95
B&D cordless drill, 1/2-in.	25595	20-Jan-04	$38.95

As you examine the output in Figure 6.23, note that, for each V_CODE, the P_PRICE values are descending. For example, the P_PRICE values for the V_CODE = 23119 are shown in descending order from $43.99 to $39.95, while the P_PRICE values for V_CODE = 21344 are shown in descending order from $17.49 to $4.99.

N O T E

- If the ordering column has nulls, they are listed either first or last (depending on the RDBMS).
- The ORDER BY clause must always be listed last in the SELECT command sequence.

6.6.2 LISTING UNIQUE VALUES

How many *different* vendors are currently represented in the PRODUCT table? A simple listing (SELECT) is not very useful if the table contains several thousand rows and we have to sift out the vendor codes manually. Fortunately, SQL's **DISTINCT** clause is designed to produce a list of only those values that are different from one another. For example, the command:

SELECT DISTINCT V_CODE
 FROM PRODUCT;

will yield only the different (distinct) vendor codes (V_CODE) that are encountered in the PRODUCT table, as shown in Figure 6.24. (Note that the first output row shows the null.)

FIGURE 6.24 A LISTING OF DISTINCT (DIFFERENT) V_CODE VALUES IN THE PRODUCT TABLE

V_CODE
21225
21231
21344
23119
24288
25595

N O T E

Oracle will place the null V_CODE at the bottom of the list, while Access places it at the top of the list. The placement of the null does not affect the list contents. In Oracle, you could use ORDER BY V_CODE NULLS FIRST to place nulls at the top of the list. The Oracle default is to place the nulls at the bottom of the list.

6.6.3 AGGREGATE FUNCTIONS

SQL will perform various mathematical summaries for you, such as counting the number of rows that contain a specified condition, finding the minimum or maximum values for some specified attribute, summing the values in a specified column, and averaging the values in a specified column. These aggregate functions are shown in Table 6.10.

TABLE 6.10 SOME BASIC SQL AGGREGATE FUNCTIONS

FUNCTION	OUTPUT
COUNT	The number of rows containing "non null" values
MIN	The minimum attribute value encountered in a given column
MAX	The maximum attribute value encountered in a given column
SUM	The sum of all values for a given column
AVG	The arithmetic mean (average) for the specified column

Because we want to illustrate the standard SQL command formats, we will present most of the remaining input and output sequences with the help of Oracle.

COUNT

Because **COUNT** is designed to tally the number of specific "values" of an attribute, it can be used in conjunction with the DISTINCT clause. For example, suppose you want to find out how many different vendors are in the PRODUCT table. The answer, generated by the first SQL code set shown in Figure 6.25, is 6. This answer indicates that six different VENDOR codes were found in the PRODUCT table. (As you examine the contents of the PRODUCT table, note that the nulls were not counted as V_CODE values.)

FIGURE 6.25 **COUNT FUNCTION OUTPUT EXAMPLES**

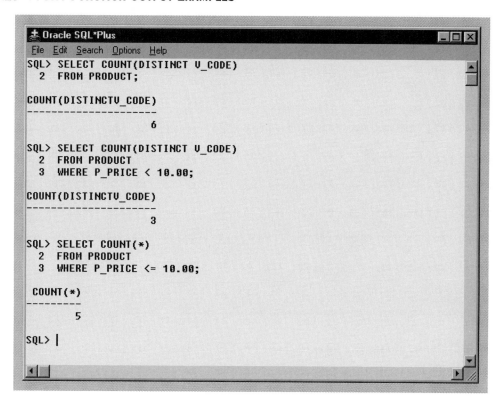

The aggregate functions can be combined with the SQL commands we explored earlier. For example, the second SQL command set in Figure 6.25 supplies the answer to the question "How many vendors referenced in the PRODUCT table have supplied products with prices that are less than, or equal to, $10.00?" The answer is 3, indicating that three vendors referenced in the PRODUCT table have supplied products that meet this price specification.

Note that the third SQL command set in Figure 6.25 uses the COUNT(*) command to answer the question "How many rows in the PRODUCT table have a P_PRICE value less than, or equal to, $10.00?" The answer, 5, indicates that five products have a listed price that meets this price specification. The asterisk (*) is only allowed inside the COUNT aggregate function.

The COUNT aggregate function uses one parameter within the parentheses, generally a column name such as COUNT(V_CODE) or COUNT(P_CODE). The parameter may also be an expression, such as COUNT(DISTINCT V_CODE) or COUNT(P_PRICE+10). Using this syntax, COUNT always returns the number of non-null values in the given column. (Whether the column values are computed or show stored table row values is immaterial). In contrast, the syntax COUNT(*) returns the number of total rows returned by the query, including the rows that contain nulls. In the example in Figure 6.25, SELECT COUNT(P_CODE) FROM PRODUCT and SELECT COUNT(*) FROM PRODUCT will yield the same answer, because there are no null values in the P_CODE primary key column.

ADDITIONAL CD CONTENT

NOTE TO MS ACCESS USERS

MS Access does not support the use of COUNT with the DISTINCT clause. If you want to use such queries in MS Access, you must create subqueries with DISTINCT and NOT NULL clauses. For example, the equivalent MS Access queries for the first two queries shown in Figure 6.25 are:

```
SELECT COUNT(*)
    FROM
        (SELECT DISTINCT V_CODE FROM PRODUCT WHERE V_CODE IS NOT NULL)
```

and

```
SELECT COUNT(*) FROM
    (SELECT DISTINCT(V_CODE) FROM
        (SELECT V_CODE, P_PRICE FROM PRODUCT
            WHERE V_CODE IS NOT NULL AND P_PRICE < 10)
    )
```

We have saved these two queries as **qry6-25a** and **qry6-25b** in your **Ch06_SaleCo** (Access) database— check your CD. MS Access does add a "trailer" at the end of the query after you have executed it, but you can delete that trailer the next time you use the query.

As you can see, adherence to SQL standards is likely to simplify the queries. You will learn more about subqueries in Chapter 7, "Advanced SQL."

MAX and MIN

The **MAX** and **MIN** functions help us find answers to problems such as the:

- Highest (maximum) price in the PRODUCT table.
- Lowest (minimum) price in the PRODUCT table.

The correct answer to the first problem, $256.99, is supplied by the first SQL command set in Figure 6.26. The second SQL command set shown in Figure 6.26 yields the minimum price of $4.99.

FIGURE 6.26 MAX AND MIN FUNCTION OUTPUT EXAMPLES

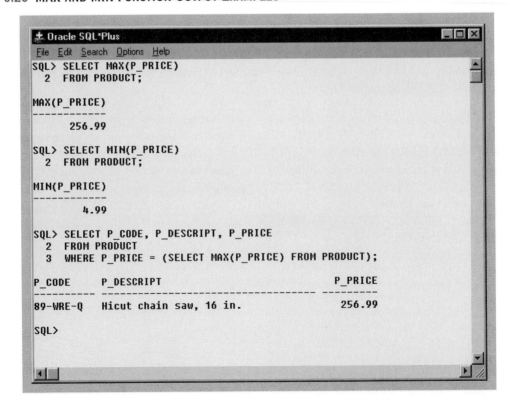

The third SQL command set in Figure 6.26 demonstrates that the numeric functions may also be used in conjunction with more complex queries. However, you must remember that *the numeric functions yield only one value* based on all the values found in the table: a single maximum value, a single minimum value, a single count, or a single average value. *It is easy to overlook this warning.* For example, let's examine the question, "Which product has the highest price?"

Although this query seems simple enough, the SQL command sequence:

SELECT P_CODE, P_DESCRIPT, P_PRICE
　　FROM PRODUCT
　　　　WHERE P_PRICE = MAX(P_PRICE);

will not yield the expected results because the use of MAX(P_PRICE) to the right side of a comparison operator is incorrect, thus producing an error message. The aggregate function MAX(*columnname*) can only be used in the column list of a SELECT statement. Also, when you are using a comparison that uses an equality symbol, you can only use a single value to the right of the equal sign.

To answer the question, therefore, we must compute the maximum price first, then compare it to each of the prices returned by the query. To do this you need a nested query. In this case, the nested query is composed of two parts:

* The *inner query*, which is executed first.
* The *outer query*, which is executed last. (The outer query is always the first SQL command you will encounter—in this case, SELECT—in the command sequence.)

Using the following command sequence as an example, note that the inner query first finds the maximum price value, which is stored in memory. Since the outer query now has a value to compare each P_PRICE value to, the query executes properly.

SELECT P_CODE, P_DESCRIPT, P_PRICE
　　FROM PRODUCT
　　　　WHERE P_PRICE = (SELECT MAX(PRICE) FROM PRODUCT);

The execution of this nested query yields the correct answer shown below the third (nested) SQL command set in Figure 6.26.

The MAX and MIN aggregate functions can also be used with date columns. For example, to find out the product that has the oldest date, we would use MIN(P_INDATE). In the same manner, to find out the most recent product, we would use MAX(P_INDATE).

SUM

The **SUM** function computes the total sum for any specified attribute, using whatever condition(s) you have imposed. For example, if you want to compute the total amount owned by your customers, you could use the following command:

SELECT SUM(CUS_BALANCE) AS TOTBALANCE FROM CUSTOMER;

You could also compute the sum total of an expression. For example, if you want to find the total value of all the items carried in inventory, you may use:

SELECT SUM(P_ONHAND*P_PRICE) AS TOTVALUE
 FROM PRODUCT;

because the total value is the sum of the product of the quantity on hand and price for all items. (See Figure 6.27.)

FIGURE 6.27 THE TOTAL VALUE OF ALL ITEMS IN THE PRODUCT TABLE

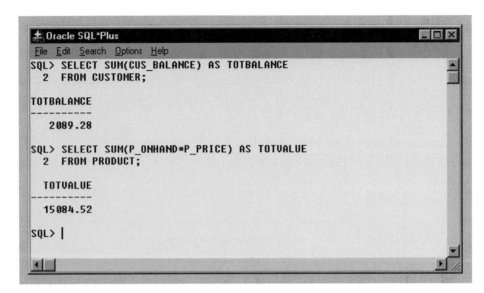

AVG

The **AVG** function format conforms to that of MIN and MAX and is, therefore, subject to the same operating restrictions. The first SQL command set shown in Figure 6.28 shows how a simple average P_PRICE value can be generated, to yield the computed average price of 56.17125. The second SQL command set in Figure 6.30 produces the five output lines that describe products whose prices exceed the average product price. Note that the second query uses nested SQL commands and the ORDER BY clause we examined earlier.

6.6.4 GROUPING DATA

Frequency distributions can be created quickly and easily by using the **GROUP BY** clause within the SELECT statement. The syntax is:

SELECT *columnlist*
 FROM *tablelist*
 [WHERE *conditionlist*]
 [GROUP BY *columnlist*]
 [HAVING *conditionlist*]
 [ORDER BY *columnlist* [ASC | DESC]] ;

FIGURE 6.28 **AVG FUNCTION OUTPUT EXAMPLES**

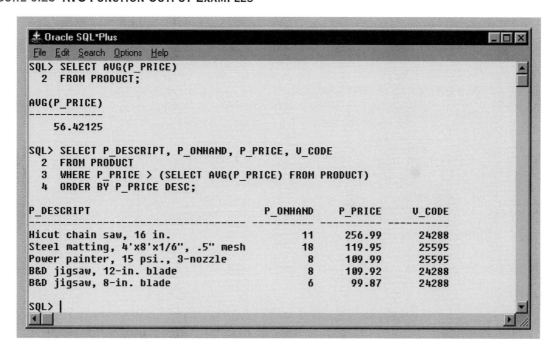

The GROUP BY clause is generally used when you have attribute columns combined with aggregate functions in the SELECT statement. For example, to determine the minimum price for each sales code, use the first SQL command set shown in Figure 6.29:

SELECT P_SALECODE, MIN(P_PRICE)
 FROM PRODUCT
 GROUP BY P_SALECODE;

The second SQL command set in Figure 6.29 generates the average price within each sales code. Note that the P_SALECODE nulls are included within the grouping.

The GROUP BY clause is valid only when used in conjunction with one of the SQL aggregate functions, such as COUNT, MIN, MAX, AVG, and SUM. For example, if you try to group the output by using:

SELECT V_CODE, P_CODE, P_DESCRIPT, P_PRICE
 FROM PRODUCT
 GROUP BY V_CODE;

you will generate a "not a GROUP BY expression" error. However, if you write the preceding SQL command sequence in conjunction with some aggregate function, the GROUP BY clause works properly. Note that Figure 6.30 shows an error message for the first SQL command sequence. The second SQL command sequence answers the question "How many products are supplied by each vendor?" properly, because it uses a COUNT aggregate function.

FIGURE 6.29 **GROUP BY CLAUSE OUTPUT EXAMPLES**

```
Oracle SQL*Plus                                            _ □ ×
File  Edit  Search  Options  Help
SQL> SELECT P_SALECODE, MIN(P_PRICE)
  2  FROM PRODUCT
  3  GROUP BY P_SALECODE;

P MIN(P_PRICE)
- ------------
1        9.95
2        4.99
         5.87

SQL> SELECT P_SALECODE, AVG(P_PRICE)
  2  FROM PRODUCT
  3  GROUP BY P_SALECODE;

P AVG(P_PRICE)
- ------------
1      107.152
2       47.88
        15.94

SQL> |
```

FIGURE 6.30 **INCORRECT AND CORRECT USE OF THE GROUP BY CLAUSE**

```
Oracle SQL*Plus                                            _ □ ×
File  Edit  Search  Options  Help
SQL> SELECT V_CODE, P_CODE, P_DESCRIPT, P_PRICE
  2  FROM PRODUCT
  3  GROUP BY V_CODE;
SELECT V_CODE, P_CODE, P_DESCRIPT, P_PRICE
               *
ERROR at line 1:
ORA-00979: not a GROUP BY expression

SQL> SELECT V_CODE, COUNT(DISTINCT (P_CODE))
  2  FROM PRODUCT
  3  GROUP BY V_CODE;

   V_CODE COUNT(DISTINCT(P_CODE))
--------- -----------------------
    21225                       2
    21231                       1
    21344                       3
    23119                       2
    24288                       3
    25595                       3
                                2

7 rows selected.

SQL> |
```

As you examine the output listing in Figure 6.30, note that the last output line shows a null for the V_CODE, indicating that two products were not supplied by a vendor. Perhaps these products were produced "in house," or they may have been bought via a "non-vendor" channel. Or the person making the data entry merely forgot to enter a vendor code. (Remember that nulls can mean many things.)

When using the GROUP BY clause with a SELECT statement:

- The SELECT's *columnlist* must include a combination of column names and aggregate functions.

- The GROUP BY clause's *columnlist* must include all non-aggregate function columns specified in the SELECT's *columnlist*. If required, you could also group by any aggregate function columns that appear in the SELECT's *columnlist*.

- The GROUP BY clause *columnlist* can include any columns from the tables in the FROM clause of the SELECT statement, even if they do not appear in the SELECT's *columnlist*.

The GROUP BY Feature's HAVING Clause

A particularly useful extension of the GROUP BY feature is the **HAVING** clause. Basically, HAVING operates like the WHERE clause in the SELECT statement. However, the WHERE clause applies to columns and expressions for individual rows, while the HAVING clause is applied to the output of a GROUP BY operation. For example, suppose you want to generate a listing of the number of products in the inventory supplied by each vendor. But this time you want to limit the listing to products whose prices average below $10.00. The first part of this requirement is satisfied with the help of the GROUP BY clause, as illustrated in the first SQL command set in Figure 6.31. Note that we have used the HAVING clause in conjunction with the GROUP BY clause in the second SQL command set in Figure 6.31, to generate the desired result.

FIGURE 6.31 AN APPLICATION OF THE HAVING CLAUSE

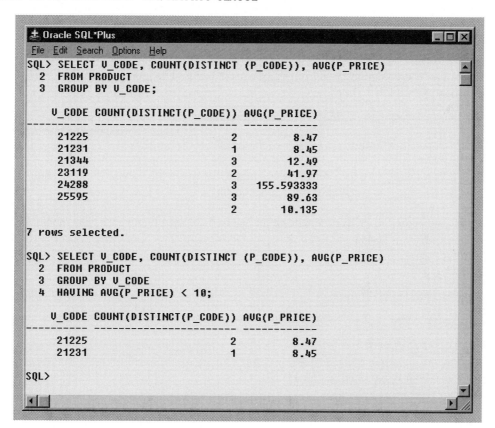

Using the WHERE clause—instead of the HAVING clause—in the second SQL command set in Figure 6.31 will produce an error message.

To demonstrate the flexibility of SQL, we can combine multiple clauses and aggregate functions. For example, the following SQL statement will:

- Aggregate the total cost of products grouped by V_CODE.

- Select only the rows having totals that exceed $500.

- List the results in descending order by the total cost.

```
SELECT V_CODE, SUM(P_ONHAND * P_PRICE) AS TOTCOST
    FROM PRODUCT
        GROUP BY V_CODE
            HAVING (SUM(P_ONHAND*P_PRICE)>500)
                ORDER BY SUM(P_ONHAND*P_PRICE) DESC;
```

Note the syntax used in the HAVING and ORDER BY clauses; in both cases you must specify the column expression (formula) used in the SELECT's column list, rather than the column alias (TOTCOST). Some RDBMS vendors will allow you to substitute the column expression with the column alias, while others will not.

6.7 VIRTUAL TABLES: CREATING A VIEW

As you learned in Section 6.3.3, the output of a relational operator (like SELECT) is another relation (or table). Suppose that at the end of every day, you would like to get a list of all products to reorder, that is, products with a quantity on hand that is less than or equal to the minimum quantity. Instead of typing the same query at the end of every day, wouldn't it be better to permanently save that query in the database? That's the function of a relational view. A **view** is a virtual table based on a SELECT query. The query can contain columns, computed columns, aliases, and aggregate functions from one or more tables. The tables on which the view is based are called **base tables**.

You can create a view by using the **CREATE VIEW** command.

CREATE VIEW *viewname* AS SELECT *query*

The CREATE VIEW statement is a data definition command that stores the subquery specification—the SELECT statement used to generate the virtual table—in the data dictionary.

Note the syntax used in the first SQL command set in Figure 6.32, which creates a view named PRODUCT_3. This view contains only the designated three attributes (P_DESCRIPT, P_ONHAND, and P_PRICE), and only rows in which the price is over $50.00. The second SQL command sequence in Figure 6.32 shows the rows that make up the view.

N O T E

NOTE TO MS ACCESS USERS
The CREATE VIEW command is not directly supported in MS Access. To create a view in MS Access, you just need to create an SQL query and then save it.

FIGURE 6.32 CREATING A VIRTUAL TABLE WITH THE **CREATE VIEW** COMMAND

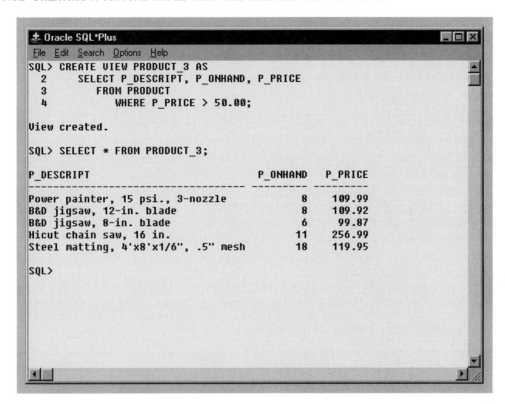

A relational view has several special characteristics:

- You can use the name of a view anywhere a table name is expected in a SQL statement.

- Views are dynamically updated. That is, the view is re-created on demand each time it is invoked. Therefore, if new products are added (or deleted) to meet the criterion P_PRICE > 50.00, those new products will automatically appear (or disappear) in the PRODUCT_3 view the next time it is invoked.

- Views provide a level of security in the database because the view can restrict users to only specified columns and specified rows in a table. For example, if you have a company with hundreds of employees in several departments, you could give the secretary of each department a view of only certain attributes and only for the employees that belong to that department.

- Views may also be used as the basis for reports. For example, if you need a report that shows a summary of total product cost and quantity on hand statistics grouped by vendor, you could create a view as:

CREATE VIEW SUMPRDXVEN AS
SELECT V_CODE, SUM(P_ONHAND*P_PRICE) AS TOTCOST,
 MAX(P_ONHAND) AS MAXQTY, MIN(P_ONHAND) AS MINQTY,
 AVG(P_ONHAND) AS AVGQTY
 FROM PRODUCT
 GROUP BY V_CODE;

In Chapter 7 you will learn more about views and, in particular, about updating data in base tables through views.

6.8 JOINING DATABASE TABLES

The ability to combine (join) tables on common attributes is perhaps the most important distinction between a relational database and other databases. A join is performed when data are retrieved from more than one table at a time. (If necessary, review the natural join and equijoin definitions and examples in Chapter 3. Remember that SQL is based on relational algebra.)

To join tables you simply enumerate the tables in the FROM clause of the SELECT statement. The DBMS will create the Cartesian product of every table in the FROM clause. However, to get the correct result—that is, a natural join—you must select only the rows in which the common attribute values match. (See Chapter 3 for a review of the join relational operation.) This is done with the WHERE clause. Use the WHERE clause to indicate the common attributes that are used to link the tables (sometimes referred to as the *join condition*).

The join condition is generally composed of an equality comparison between the foreign key and the primary key of related tables. For example, suppose we want to join our two tables, VENDOR and PRODUCT. Because V_CODE is the foreign key in the PRODUCT table and the primary key in the VENDOR table, the link is established on V_CODE. (See Table 6.11.)

TABLE 6.11 CREATING LINKS THROUGH FOREIGN KEYS

TABLE	ATTRIBUTES TO BE SHOWN	LINKING ATTRIBUTE
PRODUCT	P_DESCRIPT, P_PRICE	V_CODE
VENDOR	V_COMPANY, V_PHONE	V_CODE

When the same attribute name appears in more than one of the joined tables, the source table of the attributes listed in the SELECT command sequence must be defined. To join the PRODUCT and VENDOR tables, you would use:

```
SELECT PRODUCT.P_DESCRIPT, PRODUCT.P_PRICE, VENDOR.V_NAME,
     VENDOR.V_CONTACT, VENDOR.V_AREACODE, VENDOR.V_PHONE
       FROM PRODUCT, VENDOR
         WHERE PRODUCT.V_CODE = VENDOR.V_CODE;
```

The preceding command sequence produces the output shown in Figure 6.33.

FIGURE 6.33 THE RESULTS OF A JOIN

P_DESCRIPT	P_PRICE	V_NAME	V_CONTACT	V_AREACODE	V_PHONE
Claw hammer	$9.95	Bryson, Inc.	Smithson	615	223-3234
1.25-in. metal screw, 25	$6.99	Bryson, Inc.	Smithson	615	223-3234
2.5-in. wd. screw, 50	$8.45	D&E Supply	Singh	615	228-3245
7.25-in. pwr. saw blade	$14.99	Gomez Bros.	Ortega	615	889-2546
9.00-in. pwr. saw blade	$17.49	Gomez Bros.	Ortega	615	889-2546
Rat-tail file, 1/8-in. fine	$4.99	Gomez Bros.	Ortega	615	889-2546
Hrd. cloth, 1/4-in., 2x50	$39.95	Randsets Ltd.	Anderson	901	678-3998
Hrd. cloth, 1/2-in., 3x50	$43.99	Randsets Ltd.	Anderson	901	678-3998
B&D jigsaw, 12-in. blade	$109.92	ORDVA, Inc.	Hakford	615	898-1234
B&D jigsaw, 8-in. blade	$99.87	ORDVA, Inc.	Hakford	615	898-1234
Hicut chain saw, 16 in.	$256.99	ORDVA, Inc.	Hakford	615	898-1234
Power painter, 15 psi., 3-nozzle	$109.99	Rubicon Syste	Orton	904	456-0092
B&D cordless drill, 1/2-in.	$38.95	Rubicon Syste	Orton	904	456-0092
Steel matting, 4'x8'x1/6", .5" mesh	$119.95	Rubicon Syste	Orton	904	456-0092

Your output might be presented in a different order because the SQL command produces a listing in which the order of the columns is not relevant. In fact, you are likely to get a different order of the same listing the next time you execute the command. However, you can generate a more predictable list by using an ORDER BY clause:

SELECT P_DESCRIPT, P_PRICE, V_NAME, V_CONTACT, V_AREACODE, V_PHONE
 FROM PRODUCT, VENDOR
 WHERE PRODUCT.V_CODE = VENDOR.V_CODE;
 ORDER BY P_PRICE;

In this case, your listing will always be arranged from the lowest price to the highest.

> **N O T E**
>
> We used table names as prefixes in the preceding SQL command sequence. For example, we used PRODUCT.P_PRICE, rather than P_PRICE. Most current-generation RDBMSs do not require the table names to be used as prefixes, unless the same attribute name occurs in several of the tables being joined. In this case, V_CODE is used as a foreign key in PRODUCT and as a primary key in VENDOR, so we must use the table names as prefixes in the WHERE clause. In other words, we may write the previous query as:
> SELECT P_DESCRIPT, P_PRICE, V_NAME,
> V_CONTACT, V_AREACODE, V_PHONE
> FROM PRODUCT, VENDOR
> WHERE PRODUCT.V_CODE = VENDOR.V_CODE;
> Naturally, if an attribute name occurs in several places, its origin (table) must be specified. If you fail to provide such a specification, SQL will generate an error message to indicate that you have been ambiguous about the attribute's origin.

Note that the preceding SQL command sequence joins a row in the PRODUCT table with a row in the VENDOR table in which the V_CODE values are the same, as indicated in the WHERE clause's condition. Because any vendor can deliver any number of ordered products, the PRODUCT table may contain multiple V_CODE entries for each V_CODE entry in the VENDOR table. In other words, each V_CODE in VENDOR can be matched with many V_CODE rows in PRODUCT.

If you do not specify the WHERE clause, the result will be the Cartesian product of PRODUCT and VENDOR. (Review Chapter 3 to revisit these terms, if necessary.) Because the PRODUCT table contains 16 rows and the VENDOR table contains 11 rows, the Cartesian product would produce a listing of (16 x 11) = 176 rows. (That's because each row in PRODUCT would be joined to each row in the VENDOR table.)

All the SQL commands can be used on the joined tables. For example, the following command sequence is quite acceptable in SQL:

SELECT P_DESCRIPT, P_PRICE, V_NAME, V_CONTACT, V_AREACODE, V_PHONE
 FROM PRODUCT, VENDOR
 WHERE PRODUCT.V_CODE = VENDOR.V_CODE
 AND P_INDATE > '15-Jan-2004';

This command sequence will produce the output shown in Figure 6.34.

FIGURE 6.34 **AN ORDERED AND LIMITED LISTING AFTER A JOIN**

P_DESCRIPT	P_PRICE	V_NAME	V_CONTACT	V_AREACODE	V_PHONE
1.25-in. metal screw, 25	$6.99	Bryson, Inc.	Smithson	615	223-3234
2.5-in. wd. screw, 50	$8.45	D&E Supply	Singh	615	228-3245
Claw hammer	$9.95	Bryson, Inc.	Smithson	615	223-3234
B&D cordless drill, 1/2-in.	$38.95	Rubicon Systems	Orton	904	456-0092
Steel matting, 4'x8'x1/6", .5" mesh	$119.95	Rubicon Systems	Orton	904	456-0092
Hicut chain saw, 16 in.	$256.99	ORDVA, Inc.	Hakford	615	898-1234

When joining three or more tables, you need to specify a join condition for each pair of tables. The number of join conditions will always be N-1, where N represents the number of tables listed in the FROM clause. For example, if you have three tables, you must have two join conditions. If you have five tables, you must have four join conditions. And so on.

Remember, the join condition will match the foreign key of a table to the primary key of the related table. For example, using Figure 6.1, if you want to list the customer last name, invoice number, invoice date, and products descriptions for all the invoices for customer 10014, you must type the following:

SELECT CUS_LNAME, INV_NUMBER, INV_DATE, P_DESCRIPT
 FROM CUSTOMER, INVOICE, LINE, PRODUCT
 WHERE CUSTOMER.CUS_NUM = INVOICE.CUS_NUM AND
 INVOICE.INV_NUMBER = LINE.INV_NUMBER AND
 LINE.P_CODE = PRODUCT.P_CODE AND
 CUSTOMER.CUS_NUM = 10014
 ORDER BY INV_NUMBER;

Finally, be careful not to create circular join conditions. For example, if table A is related to table B, table B is related to table C, and table C is also related to table A, create only two join conditions: join A with B and B with C. Do not join C with A!

6.8.1 JOINING TABLES WITH AN ALIAS

An alias may be used to identify the source table from which the data are taken. Note that we have used the aliases P and V to label the PRODUCT and VENDOR tables in the next command sequence. Any legal table name may be used as an alias. (We have also dropped the table name prefixes, because the attribute listing contains no duplicate names in the SELECT statement.)

SELECT P_DESCRIPT, P_PRICE, V_NAME, V_CONTACT, V_AREACODE, V_PHONE
 FROM PRODUCT P, VENDOR V
 WHERE P.V_CODE = V.V_CODE
 ORDER BY P_PRICE;

6.8.2 RECURSIVE JOINS

An alias is especially useful when a table must be joined to itself in a **recursive query**. For example, suppose that we are working with the EMP table shown in Figure 6.35.

FIGURE 6.35 **THE CONTENTS OF THE EMP TABLE**

EMP_NUM	EMP_TITLE	EMP_LNAME	EMP_FNAME	EMP_INITIAL	EMP_DOB	EMP_HIRE_DATE	EMP_AREACODE	EMP_PHONE	EMP_MGR
100	Mr.	Kolmycz	George	D	15-Jun-42	15-Mar-85	615	324-5456	
101	Ms.	Lewis	Rhonda	G	19-Mar-65	25-Apr-86	615	324-4472	100
102	Mr.	Vandam	Rhett		14-Nov-58	20-Dec-90	901	675-8993	100
103	Ms.	Jones	Anne	M	16-Oct-74	28-Aug-94	615	898-3456	100
104	Mr.	Lange	John	P	08-Nov-71	20-Oct-94	901	504-4430	105
105	Mr.	Williams	Robert	D	14-Mar-75	08-Nov-98	615	890-3220	
106	Mrs.	Smith	Jeanine	K	12-Feb-68	05-Jan-89	615	324-7883	105
107	Mr.	Diante	Jorge	D	21-Aug-74	02-Jul-94	615	890-4567	105
108	Mr.	Wiesenbach	Paul	R	14-Feb-66	18-Nov-92	615	897-4358	
109	Mr.	Smith	George	K	18-Jun-61	14-Apr-89	901	504-3339	108
110	Mrs.	Genkazi	Leighla	W	19-May-70	01-Dec-90	901	569-0093	108
111	Mr.	Washington	Rupert	E	03-Jan-66	21-Jun-93	615	890-4925	105
112	Mr.	Johnson	Edward	E	14-May-61	01-Dec-83	615	898-4387	100
113	Ms.	Smythe	Melanie	P	15-Sep-70	11-May-99	615	324-9006	105
114	Ms.	Brandon	Marie	G	02-Nov-56	15-Nov-79	901	882-0845	108
115	Mrs.	Saranda	Hermine	R	25-Jul-72	23-Apr-93	615	324-5505	105
116	Mr.	Smith	George	A	08-Nov-65	10-Dec-88	615	890-2984	108

Using the data in the EMP table shown in Figure 6.35, how would we generate a list of all employees with their managers' names? We can accomplish this task by joining the EMP table to itself. In this case, we'll also use aliases to differentiate the tables. The SQL command sequence would look like this:

SELECT E.EMP_MGR, M.EMP_LNAME, E.EMP_NUM, E.EMP_LNAME
 FROM EMP E, EMP M
 WHERE E.EMP_MGR=M.EMP_NUM
 ORDER BY E.EMP_MGR;

The output generated by this command sequence is shown in Figure 6.36.

FIGURE 6.36 **USING AN ALIAS TO JOIN A TABLE TO ITSELF**

EMP_NUM	A.EMP_LNAME	EMP_MGR	B.EMP_LNAME
112	Johnson	100	Kolmycz
103	Jones	100	Kolmycz
102	Vandam	100	Kolmycz
101	Lewis	100	Kolmycz
115	Saranda	105	Williams
113	Smythe	105	Williams
111	Washington	105	Williams
107	Diante	105	Williams
106	Smith	105	Williams
104	Lange	105	Williams
116	Smith	108	Wiesenbach
114	Brandon	108	Wiesenbach
110	Genkazi	108	Wiesenbach
109	Smith	108	Wiesenbach

N O T E

In MS Access, add AS to the previous SQL command sequence, making it read:
 SELECT E.EMP_MGR, M.EMP_LNAME, E.EMP_NUM, E.EMP_LNAME
 FROM EMP AS E, EMP AS M
 WHERE E.EMP_MGR=M.EMP_NUM
 ORDER BY E.EMP_MGR;

6.8.3 OUTER JOINS

Figure 6.33 showed the results of joining the PRODUCT and VENDOR tables. If you examine the output, note that 14 product rows have been listed. If you compare the output to the PRODUCT table in Figure 6.2, you will note that there are two missing products. Why? The reason is that there are two products with nulls in the V_CODE attribute. Because there is no matching null "value" in the VENDOR table's V_CODE attribute, they do not show up in the final output based on the join. Also, if you examine the VENDOR table in Figure 6.2, you will notice that there are several vendors that have no matching V_CODE in the PRODUCT table. To include these rows in the final join output, we must use an outer join.

There are two types of outer joins: left and right (see Chapter 3). Given the contents of the PRODUCT and VENDOR tables, the following left outer join will show all VENDOR rows and all matching PRODUCT rows.

```
SELECT P_CODE, VENDOR.V_CODE, V_NAME
    FROM VENDOR
        LEFT JOIN PRODUCT ON VENDOR.V_CODE = PRODUCT.V_CODE;
```

The preceding command will work fine in MS Access and Oracle 9i. Older versions of Oracle require you to use the following left outer join command syntax:

```
SELECT P_CODE, VENDOR.V_CODE, V_NAME
    FROM VENDOR, PRODUCT
        WHERE VENDOR.V_CODE = PRODUCT.V_CODE (+)
```

Figure 6.37 shows the output generated by the left outer join command in MS Access. Oracle yields the same result, but shows the output in a different order.

FIGURE 6.37 THE LEFT OUTER JOIN RESULTS

P_CODE	V_CODE	V_NAME
23109-HB	21225	Bryson, Inc.
SM-18277	21225	Bryson, Inc.
	21226	SuperLoo, Inc.
SW-23116	21231	D&E Supply
13-Q2/P2	21344	Gomez Bros.
14-Q1/L3	21344	Gomez Bros.
54778-2T	21344	Gomez Bros.
	22567	Dome Supply
1546-QQ2	23119	Randsets Ltd.
1558-QW1	23119	Randsets Ltd.
	24004	Brackman Bros.
2232/QTY	24288	ORDVA, Inc.
2232/QWE	24288	ORDVA, Inc.
89-WRE-Q	24288	ORDVA, Inc.
	25443	B&K, Inc.
	25501	Damal Supplies
11QER/31	25595	Rubicon Systems
2238/QPD	25595	Rubicon Systems
WR3/TT3	25595	Rubicon Systems

The right outer join will join both tables and show all product rows with all matching vendor rows. The SQL command for the right outer join is:

```
SELECT PRODUCT.P_CODE, VENDOR.V_CODE, V_NAME
    FROM VENDOR
        RIGHT JOIN PRODUCT ON VENDOR.V_CODE = PRODUCT.V_CODE;
```

The older Oracle version of the right outer join SQL command is:

```
SELECT P_CODE, VENDOR.V_CODE, V_NAME
    FROM VENDOR, PRODUCT
        WHERE VENDOR.V_CODE (+) = PRODUCT.V_CODE;
```

Figure 6.38 shows the output generated by the right outer join command sequence in MS Access. Again, Oracle yields the same result, but shows the output in a different order.

FIGURE 6.38 THE RIGHT OUTER JOIN RESULTS

P_CODE	V_CODE	V_NAME
23114-AA		
PVC23DRT		
23109-HB	21225	Bryson, Inc.
SM-18277	21225	Bryson, Inc.
SW-23116	21231	D&E Supply
13-Q2/P2	21344	Gomez Bros.
14-Q1/L3	21344	Gomez Bros.
54778-2T	21344	Gomez Bros.
1546-QQ2	23119	Randsets Ltd.
1558-QW1	23119	Randsets Ltd.
2232/QTY	24288	ORDVA, Inc.
2232/QWE	24288	ORDVA, Inc.
89-WRE-Q	24288	ORDVA, Inc.
11QER/31	25595	Rubicon Systems
2238/QPD	25595	Rubicon Systems
WR3/TT3	25595	Rubicon Systems

In Chapter 7 you will learn more about joins and you will learn how to use the latest ANSI SQL standard syntax.

6.9 CONVERTING AN ER MODEL INTO A DATABASE STRUCTURE

Converting any ER model to a set of tables in a database requires following specific rules that govern such a conversion. The application of these rules requires an understanding of the effects of updates and deletions on the tables in the database. Before we discuss these rules in detail, let's briefly review a simple model, its schema, and the SQL commands used to generate the tables. The model, implemented in the Ch06_Artist database, conforms to the following conditions:

- A painter might paint many paintings. To be considered a painter in the ARTIST database, the painter must have painted at least one painting. This business rule decrees that the cardinality is (1,N) in the relationship between PAINTER and PAINTING.

- Each painting is painted by one (and only one) painter.

- A painting might (or might not) be exhibited in a gallery; that is, GALLERY is an optional entity to the PAINTING entity.

Given this description, we create a simple ER model and some matching tables for the ARTIST database shown in Figure 6.39.

FIGURE 6.39 THE CH06_ARTIST DATABASE ERD AND TABLES

The Ch06_Artist Database ERD

	PAINTER			PAINTING			GALLERY
PK	PTR_NUM		PK	PTNG_NUM		PK	GAL_NUM
	PTR_LASTNAME	paints	FK2	PTR_NUM	accepts		GAL_OWNER
	PTR_FIRSTNAME		FK1	GAL_NUM			GAL_AREACODE
	PTR_AREACODE			PTNG_TITLE			GAL_PHONE
	PTR_PHONE			PTNG_PRICE			GAL_RATE

Table name: PAINTER

	PTR_NUM	PTR_LASTNAME	PTR_FIRSTNAME	PTR_AREACODE	PTR_PHONE
+	123	Ross	Georgette	901	885-4567
+	126	Itero	Julio	901	346-1112
+	127	Geoff	George	615	221-4456

Table name: PAINTING **Database name: Ch06_Artist**

PTNG_NUM	PTNG_TITLE	PTNG_PRICE	PTR_NUM	GAL_NUM
1338	Dawn Thunder	$245.50	123	5
1339	A Faded Rose	$6,723.00	123	
1340	The Founders	$567.99	126	6
1341	Hasty Pudding Exit	$145.50	123	
1342	Plastic Paradise	$8,328.99	126	6
1343	Roamin'	$785.00	127	6
1344	Wild Waters	$999.00	127	5
1345	Stuff 'n Such 'n Some	$9,800.00	123	5

Table name: GALLERY

	GAL_NUM	GAL_OWNER	GAL_AREACODE	GAL_PHONE	GAL_RATE
+	5	L. R. Gilliam	901	123-4456	0.37
+	6	G. G. Waters	405	353-2243	0.45

The tables and attributes that correspond to this ER model are summarized in the data dictionary in Table 6.12.

Given the information in Figure 6.39 and Table 6.12, note that:

- PTR_NUM in the PAINTING table is the foreign key that references the PAINTER table. Because the relationship between PAINTER and PAINTING is mandatory, the PTR_NUM foreign key must be classified as NOT NULL.

- GAL_NUM in PAINTING is the foreign key that references the GALLERY table. Because the relationship between PAINTING and GALLERY is optional, the GAL_NUM foreign key may be null.

Given these Ch06_Artist database table structures, let us now examine the effect of the following data manipulation events of the foreign key constraint actions:

1. *Adding a painter (row) to the PAINTER table.* Adding a painter does not cause any problems, because the PAINTER table does not have any dependencies in other tables.

2. *Updating the PAINTER table's primary key.* Changing a PAINTER key causes problems in the database because some paintings in the PAINTING table may make reference to this key. The option is to use the UPDATE CASCADE. This option makes sure that a change in the PAINTER's PTR_NUM automatically triggers the required changes in the PTR_NUM foreign key found in other tables. This is the recommended option. This behavior (UPDATE CASCADE) is not supported by some RDBMS products such as Oracle.

TABLE 6.12 A DATA DICTIONARY FOR THE CH06_ARTIST DATABASE

TABLE NAME	ATTRIBUTE NAME	CONTENTS	TYPE	FORMAT	RANGE	REQUIRED	PK OR FK	FK REFERENCED TABLE
PAINTER	PTR_NUM	Painter number	CHAR(4)	9999	1000-9999	Y	PK	
	PTR_LASTNAME	Painter last name	VARCHAR(15)	Xxxxxxxxxxx		Y		
	PTR_FIRSTNAME	Painter first name	VARCHAR(15)	Xxxxxxxxxxx		Y		
	PTR_AREACODE	Painter area code	CHAR(3)	999				
	PTR_PHONE	Painter phone	CHAR(8)	999-9999				
GALLERY	GAL_NUM	Gallery number	CHAR(4)	9999	1000-9999	Y	PK	
	GAL_OWNER	Gallery owner	VARCHAR(35)	Xxxxxxxxxxx				
	GAL_AREACODE	Gallery area code	CHAR(3)	999		Y		
	GAL_PHONE	Gallery phone	CHAR(8)	999-9999		Y		
	GAL_RATE	Gallery commission rate (pct.)	NUMBER(4,2)	99.99	0.00-60.00	Y		
PAINTING	PTNG_NUM	Painting number	CHAR(4)	9999	1000-9999	Y	PK	
	PTNG_TITLE	Painting title	VARCHAR(35)	Xxxxxxxxxxx				
	PTNG_PRICE	Painting Price	NUMBER(9,2)	99,999.99	10.00-99,999.99	Y		
	PTR_NUM	Painter number	CHAR(4)	9999	1000-9999	Y	FK	PAINTER
	GAL_NUM	Gallery number	CHAR(4)	9999	1000-9999	Y	FK	GALLERY

FK = Foreign Key

PK = Primary key

CHAR = Fixed character length data, 1 to 255 characters.

VARCHAR = Variable character length data, 1 to 2,000 characters. May also be labeled VARCHAR2.

NUMBER = Numeric data. NUMBER(9,2) is used to specify numbers with two decimal places and up to nine digits long, including the decimal places. Some RDBMSes permit the use of a MONEY or a CURRENCY data type.

3. *Deleting a painter (row) from the PAINTER table.* If we delete a row (painter) from the PAINTER table, the PAINTING table may contain references to a painter who no longer exists, thereby creating a deletion anomaly. (A painting does not cease to exist just because the painter does.) Given this situation, it is wise to restrict the ability to delete a row from a table if there is a foreign key in another table that references that row. The restriction means that we can delete a painter from the PAINTER table only if there is no foreign key in another table related to this painter row. This behavior is enforced automatically by the RDBMS when using the FOREIGN KEY clause.

4. *Adding a gallery (row) to the GALLERY table.* Adding a new row does not affect the database because the GALLERY does not have dependencies in other tables.

5. *Updating the GALLERY table's primary key.* Changing a primary key value in a GALLERY row requires that all foreign keys making reference to it be updated, as well. Therefore, we must use an UPDATE CASCADE clause. This option makes sure that a change in the GALLERY's GAL_NUM automatically triggers the required changes in the GAL_NUM foreign key found in other tables. This is the recommended option. This behavior (UPDATE CASCADE) is not supported by some RDBMS products such as Oracle.

6. *Deleting a gallery (row) from the GALLERY table.* Deleting a GALLERY row creates problems in the database if there are rows in the PAINTING table that make reference to that GALLERY row's primary key. Because GALLERY is optional to PAINTING, we may set all the deleted gallery GAL_NUM values to null (DELETE SET NULL). Or we may want the database user to be alerted to the problem by specifying that the deletion of a GALLERY row is permitted only if there is no foreign key (GAL_NUM) in another table that requires the GALLERY row's existence. This behavior is enforced automatically by the RDBMS when using the FOREIGN KEY clause.

Armed with these results, we can now transform the ER model into a set of tables by using the following SQL commands:

1. Create the PAINTER table:

   ```
   CREATE TABLE PAINTER (
   PTR_NUM              CHAR(4)        NOT NULL     UNIQUE,
   PTR_LASTNAME         CHAR(15)       NOT NULL,
   PTR_FIRSTNAME        CHAR(15),      NOT NULL
   PTR_AREACODE         CHAR(3),
   PTR_PHONE            CHAR(8),
   PRIMARY KEY (PTR_NUM));
   ```

2. Create the GALLERY table:

   ```
   CREATE TABLE GALLERY (
   GAL_NUM              CHAR(4)        NOT NULL     UNIQUE,
   GAL_OWNER            CHAR(35),
   GAL_AREACODE         CHAR(3)        NOT NULL,
   GAL_PHONE            CHAR(8)        NOT NULL,
   GAL_RATE             NUMBER(4,2),
   PRIMARY KEY (GAL_NUM));
   ```

3. Create the PAINTING table:

```
CREATE TABLE PAINTING (
PTNG_NUM              CHAR(4)        NOT NULL    UNIQUE,
PTNG_TITLE            CHAR(35),
PTNG_PRICE            NUMBER(9,2),
PTR_NUM              CHAR(4)        NOT NULL,
GAL_NUM              CHAR(4),
PRIMARY KEY (PNTG_NUM),
FOREIGN KEY (PTR_NUM) REFERENCES PAINTER
     ON UPDATE CASCADE,
FOREIGN KEY (GAL_NUM) REFERENCES GALLERY
     ON UPDATE CASCADE);
```

After creating the database tables and entering their contents, you are now ready for data entry, queries, and reports. Note that the decisions made by the designer to govern data integrity are reflected in the foreign key rules. Implementation decisions vary according to the problem being addressed.

6.9.1 GENERAL RULES GOVERNING RELATIONSHIPS AMONG TABLES

Given our experience with the simple Ch06_Artist database in the previous section, let us next formulate a general set of rules to help us create any database table structure that will meet the required integrity constraints.

1. All primary keys must be defined as NOT NULL and UNIQUE. If your applications software does not support the NOT NULL option, you should enforce such a condition by using programming techniques. This is another argument for using DBMS software that meets ANSI SQL standards!

2. Define all foreign keys to conform to the following requirements for binary relationships.

1:M Relationships

Create the foreign key by putting the primary key of the "one" in the table of the "many." The "one" side is referred to as the **parent table**, and the "many" side is referred to as a **dependent table**. Observe the following foreign key rules:

If both sides are MANDATORY:

Column constraint:	NOT NULL
FK constraint:	Default behavior (on delete restrict)
	ON UPDATE CASCADE

If both sides are OPTIONAL:

Column constraint:	NULL ALLOWED
FK constraint:	ON DELETE SET NULL
	ON UPDATE CASCADE

If one side is OPTIONAL and one side is MANDATORY:

a. If the "many" and the mandatory components of the relationship are on the same side in the ER diagram, a NULL ALLOWED condition must be defined for the dependent table's foreign key. The foreign key rules should be:

Column constraint: NULL ALLOWED

FK constraint: ON DELETE SET NULL or

 Default behavior (on delete restrict)

 ON UPDATE CASCADE

b. If the "many" and the mandatory components of the relationship are not on the same side in the ER diagram, a NOT NULL condition must be defined for the dependent table's foreign key. Deletion and update in the parent table of the foreign key should be subject to default behavior (on delete restrict) and UPDATE CASCADE restrictions.

Weak Entity

a. Put the key of the parent table (the strong entity) in the weak entity. The key of the weak entity will be a composite key composed of the parent table key and the weak entity candidate key, if any. (The designer may decide to create a new unique ID for the entity.)

b. The weak entity relationship conforms to the same rules as the 1:M relationship, except for the following foreign key restrictions:

Column constraint: NOT NULL

FK constraint: ON DELETE CASCADE

 ON UPDATE CASCADE

M:N Relationships

Convert the M:N relationship to a composite (bridge) entity consisting of (at least) the parent tables' primary keys. The composite entity primary key is thus a composite key, subject to the NOT NULL restriction.

1:1 Relationships

If both entities are in a mandatory participation in the relationship and they do not participate in other relationships, it is most likely that the two entities should be part of the same entity.

Table 6.13 summarizes the ramifications of the foreign key actions that could be used to represent multiple cases of 1:1, 1:M, and M:N relationships.

TABLE 6.13 A SUMMARY OF FOREIGN KEY RULES

RELATIONSHIP	FOREIGN KEY LOCATION	THE ENTITIES PARTICIPATING IN THE RELATIONSHIPS ARE …	KEY ATTRIBUTE CONSTRAINT	FOREIGN KEY ACTIONS	
				DELETE	UPDATE
M:N	New entity: composite key	Both mandatory	NN	R	C
		Both optional	NN	C	C
		One mandatory, one optional			
		Operations on mandatory side	NN	R	C
		Operations on optional side	NN	C	C
1:M	Many side	Both mandatory	NN	R	C
		Both optional	NA	SN or R	C
		One mandatory, one optional			
		Operations on mandatory side	NA	SN or R	C
		Operations on optional side	NN	R	C
1:1	Foreign key placement is a matter of informed choice. Put the FK in the ERD's optional side, the strong entity, the most frequent accessed side, or the side dictated by the case semantics. *Do not put the FK in both sides.*	Both mandatory	NN	R	C
		Both optional	NA	SN	C
		One mandatory, one optional			
		Operations on mandatory side	NA	SN	C
		Operations on optional side	NN	R	C
Weak	Weak entity		NN*	C	C
Multi-valued Attributes	Create a set of new tables in 1:M relationships. Conform to the weak entity rules.		NN	C	C

NN = Not Null NA = Null Allowed R = Restrict

SN = Set to Null C = Cascade * = Inherited from parent entity

S U M M A R Y

In this chapter you learned the basics of the Structured Query Language (SQL) command set. The SQL commands can be divided into two overall categories: data definition language (DDL) commands and data manipulation language (DML) commands.

Before creating tables, you need to get familiar with the data types available in your RDBMS. The ANSI standard data types are supported by all RDBMS vendors in different ways. The basic data types are: NUMBER, INTEGER, CHAR, VARCHAR, and DATE.

The basic data definition commands allow you to create tables, indexes, and views. You also learned about the many SQL constraints that can be used with columns. The commands you learned are:

CREATE TABLE, CREATE INDEX, CREATE VIEW, ALTER TABLE, DROP TABLE, DROP VIEW, and DROP INDEX.

You also learned about the DML commands that allow you to add, modify, and delete rows from tables. The basic DML commands are: SELECT, INSERT, UPDATE, DELETE, COMMIT, and ROLLBACK.

The INSERT command is used to add new rows to tables. The UPDATE command is used to modify data values in existing rows of a table. The DELETE command is used to delete rows from tables. The COMMIT and ROLLBACK commands are used to permanently save or roll back changes made to the rows. Once you COMMIT the changes, you cannot undo the changes with a ROLLBACK command.

The SELECT statement is the main data retrieval command in SQL. A SELECT statement has the following syntax:

SELECT *columnlist*
 FROM *tablelist*
 [WHERE *conditionlist*]
 [GROUP BY *columnlist*]
 [HAVING *conditionlist*]
 [ORDER BY *columnlist* [ASC | DESC]] ;

The column list represents one or more column names separated by commas. The column list could also include computed columns, aliases, and aggregate functions. A computed column is represented by an expression or formula, for example, P_PRICE * P_ONHAND. The FROM clause contains a list of table names or view names.

The WHERE clause can be used with the SELECT, UPDATE, and DELETE statements to restrict the rows affected by the DDL command. The condition list represents one or more conditional expressions separated by logical operators (AND/OR/NOT). The conditional expression can contain any comparison operators (=, >, <, >=, <=,<>) as well as special operators (BETWEEN, IS NULL, LIKE, IN, and EXISTS).

Aggregate functions (COUNT, MIN, MAX, AVG) are special functions that perform arithmetic computations over a set of rows. The aggregate functions are usually used in conjunction with the GROUP BY clause to group the output of aggregate computations by one or more attributes. The HAVING clause is used to restrict the output of the GROUP BY clause by selecting only the aggregate rows that match a given condition.

The ORDER BY clause is used to sort the output of a SELECT statement. The ORDER BY clause can sort by one or more columns and use either an ascending or descending order.

You can join the output of multiple tables with the SELECT statement. The join operation is performed every time you specify two or more tables in the FROM clause and use a join condition in the WHERE clause to match the foreign key of one table to the primary key of the related table. If you do not specify a join condition, the DBMS will automatically perform a Cartesian product of the tables you specify in the FROM clause.

The natural join uses the join condition to match only rows with equal values in the specified columns. You could also do a right outer join and left outer join to select the rows that have no matching values in the other related table.

KEY TERMS

alias	dependent table	OR
ALTER	DISTINCT	ORDER BY
AND	DROP	parent table
authentication	EXISTS	recursive query
AVG	GROUP BY	reserved words
base table	HAVING	ROLLBACK
BETWEEN	IN	rules of precedence
Boolean algebra	inner query	schema
cascading order sequence	INSERT	SELECT
COMMIT	IS NULL	subquery
composite index	LIKE	SUM
COUNT	MAX	UPDATE
CREATE INDEX	MIN	view
CREATE TABLE	nested query	wildcard character
CREATE VIEW	NOT	
DELETE	null	

REVIEW QUESTIONS

ADDITIONAL CD CONTENT

All Review Questions are based on the **Ch06_Review** database located in the **Databases\Student\Ch06** folder on the student CD. This database is stored in the Microsoft Access format. If you use another DBMS such as Oracle, SQL Server, or DB2, use its import utilities to move the Access database contents to the DBMS of your choice.

The **Ch06_Review** database stores data for a consulting company that tracks all charges to all projects. The charges are based on the hours worked by each employee on each project. The structure and contents of the Ch06_Review database are shown in Figure Q6.1.

FIGURE Q6.1 STRUCTURE AND CONTENTS OF THE CH06_REVIEW DATABASE

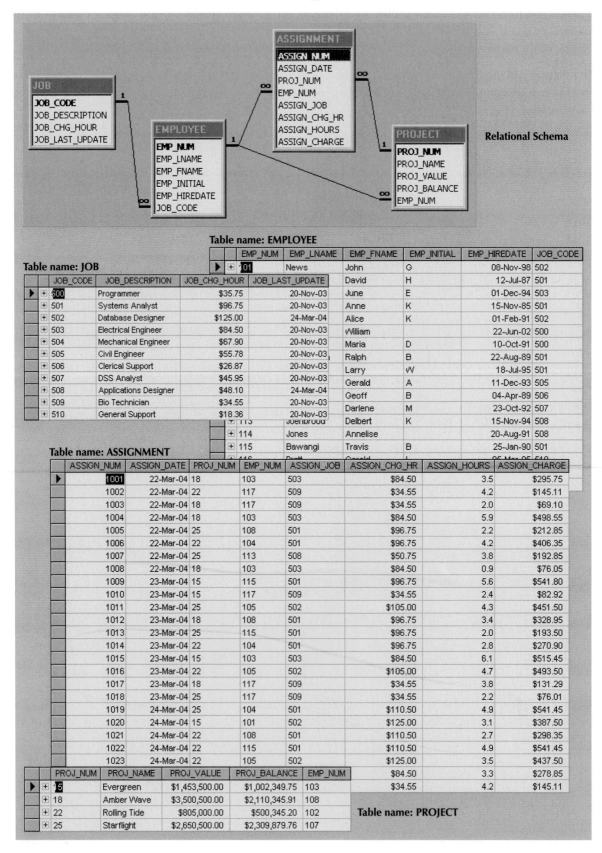

Relational Schema

Table name: EMPLOYEE

		EMP_NUM	EMP_LNAME	EMP_FNAME	EMP_INITIAL	EMP_HIREDATE	JOB_CODE
▶	+	101	News	John	G	08-Nov-98	502
				David	H	12-Jul-87	501
				June	E	01-Dec-94	503
				Anne	K	15-Nov-85	501
				Alice	K	01-Feb-91	502
				William		22-Jun-02	500
				Maria	D	10-Oct-91	500
				Ralph	B	22-Aug-89	501
				Larry	W	18-Jul-95	501
				Gerald	A	11-Dec-93	505
				Geoff	B	04-Apr-89	506
				Darlene	M	23-Oct-92	507
	+	113	Joenbrood	Delbert	K	15-Nov-94	508
	+	114	Jones	Annelise		20-Aug-91	508
	+	115	Bawangi	Travis	B	25-Jan-90	501

Table name: JOB

		JOB_CODE	JOB_DESCRIPTION	JOB_CHG_HOUR	JOB_LAST_UPDATE
▶	+	500	Programmer	$35.75	20-Nov-03
	+	501	Systems Analyst	$96.75	20-Nov-03
	+	502	Database Designer	$125.00	24-Mar-04
	+	503	Electrical Engineer	$84.50	20-Nov-03
	+	504	Mechanical Engineer	$67.90	20-Nov-03
	+	505	Civil Engineer	$55.78	20-Nov-03
	+	506	Clerical Support	$26.87	20-Nov-03
	+	507	DSS Analyst	$45.95	20-Nov-03
	+	508	Applications Designer	$48.10	24-Mar-04
	+	509	Bio Technician	$34.55	20-Nov-03
	+	510	General Support	$18.36	20-Nov-03

Table name: ASSIGNMENT

	ASSIGN_NUM	ASSIGN_DATE	PROJ_NUM	EMP_NUM	ASSIGN_JOB	ASSIGN_CHG_HR	ASSIGN_HOURS	ASSIGN_CHARGE
▶	1001	22-Mar-04	18	103	503	$84.50	3.5	$295.75
	1002	22-Mar-04	22	117	509	$34.55	4.2	$145.11
	1003	22-Mar-04	22	117	509	$34.55	2.0	$69.10
	1004	22-Mar-04	18	103	503	$84.50	5.9	$498.55
	1005	22-Mar-04	25	108	501	$96.75	2.2	$212.85
	1006	22-Mar-04	22	104	501	$96.75	4.2	$406.35
	1007	22-Mar-04	25	113	508	$50.75	3.8	$192.85
	1008	22-Mar-04	18	103	503	$84.50	0.9	$76.05
	1009	23-Mar-04	15	115	501	$96.75	5.6	$541.80
	1010	23-Mar-04	15	117	509	$34.55	2.4	$82.92
	1011	23-Mar-04	25	105	502	$105.00	4.3	$451.50
	1012	23-Mar-04	18	108	501	$96.75	3.4	$328.95
	1013	23-Mar-04	25	115	501	$96.75	2.0	$193.50
	1014	23-Mar-04	22	104	501	$96.75	2.8	$270.90
	1015	23-Mar-04	15	103	503	$84.50	6.1	$515.45
	1016	23-Mar-04	22	105	502	$105.00	4.7	$493.50
	1017	23-Mar-04	18	117	509	$34.55	3.8	$131.29
	1018	23-Mar-04	25	117	509	$34.55	2.2	$76.01
	1019	24-Mar-04	25	104	501	$110.50	4.9	$541.45
	1020	24-Mar-04	15	101	502	$125.00	3.1	$387.50
	1021	24-Mar-04	22	108	501	$110.50	2.7	$298.35
	1022	24-Mar-04	22	115	501	$110.50	4.9	$541.45
	1023	24-Mar-04	22	105	502	$125.00	3.5	$437.50
						$84.50	3.3	$278.85
						$34.55	4.2	$145.11

	PROJ_NUM	PROJ_NAME	PROJ_VALUE	PROJ_BALANCE	EMP_NUM	
▶	+	15	Evergreen	$1,453,500.00	$1,002,349.75	103
	+	18	Amber Wave	$3,500,500.00	$2,110,345.91	108
	+	22	Rolling Tide	$805,000.00	$500,345.20	102
	+	25	Starflight	$2,650,500.00	$2,309,879.76	107

Table name: PROJECT

As you examine Figure Q6.1, note that the ASSIGNMENT table stores the JOB_CHG_HOUR values as a "native" attribute (ASSIGN_CHG_HR) to maintain historical accuracy of the data. The JOB_CHG_HOUR values are likely to change over time. In fact, we have made sure that there is a JOB_CHG_HOUR change that will be reflected in the ASSIGNMENT table. And, naturally, the employee primary job assignment may change, so the ASSIGN_JOB is also stored. Because these attributes are required to maintain the historical accuracy of the data, they are *not* redundant. For example, note that the JOB table was updated on 24-Mar-2004 for JOB_CODE 502 (Database Designer) and shows a JOB_CHG_HOUR of $125.00. The ASSIGNMENT table shows this new charge in its ASSIGN_CHG_HR, starting on 24-Mar-2004. (Look at records 1020 and 1023 in the ASSIGNMENT table. The "old" ASSIGN_CHG_HR was $105.00 per hour. See records 1011 and 1016 for 23-Mar-2004.)

Given the Ch06_Review database's structure and contents shown in Figure Q6.1, use SQL commands to answer questions 1–25.

1. Write the SQL code that will create the table structure for a table named EMP_1. This table is a subset of the EMPLOYEE table. The basic EMP_1 table structure is summarized in Table Q6.1. (Note that the JOB_CODE is the FK to JOB.)

TABLE Q6.1 THE **EMP_1** TABLE STRUCTURE

ATTRIBUTE (FIELD) NAME	DATA DECLARATION
EMP_NUM	CHAR(3)
EMP_LNAME	VARCHAR(15)
EMP_FNAME	VARCHAR(15)
EMP_INITIAL	CHAR(1)
EMP_HIREDATE	DATE
JOB_CODE	CHAR(3)

2. Having created the table structure in Question 1, write the SQL code to enter the first two rows for the table shown in Figure Q6.2.

FIGURE Q6.2 THE **CONTENTS OF THE EMP_1** TABLE

EMP_NUM	EMP_LNAME	EMP_FNAME	EMP_INITIAL	EMP_HIREDATE	JOB_CODE
101	News	John	G	08-Nov-98	502
102	Senior	David	H	12-Jul-87	501
103	Arbough	June	E	01-Dec-94	500
104	Ramoras	Anne	K	15-Nov-85	501
105	Johnson	Alice	K	01-Feb-91	502
106	Smithfield	William		22-Jun-02	500
107	Alonzo	Maria	D	10-Oct-91	500
108	Washington	Ralph	B	22-Aug-89	501
109	Smith	Larry	W	18-Jul-95	501

3. Assuming that the data shown in the EMP_1 table have been entered, write the SQL code that will list all attributes for a job code of 502.

4. Write the SQL code that will save the changes made to the EMP_1 table.

5. Write the SQL code to change the job code to 501 for the person whose employee number is 106. After you have completed the task, examine the results, and then reset the job code to its original value.

6. Write the SQL code to delete the row for the person named William Smithfield, who was hired on June 22, 2002 and whose job code classification is 500. (*Hint*: Use logical operators to include all the information given in this problem.)

7. Write the SQL code that will restore the data to its original status; that is, the table should contain the data that existed before you made the changes in Questions 5 and 6.

8. Write the SQL code to create a copy of EMP_1, naming the copy EMP_2. Then write the SQL code that will add the attributes EMP_PCT and PROJ_NUM to its structure. The EMP_PCT is the bonus percentage to be paid to each employee. The new attribute characteristics are shown next:

EMP_PCT NUMBER(4,2)

PROJ_NUM CHAR(3)

(*Note*: If your SQL implementation allows it, you may use DECIMAL(4,2), rather than NUMBER(4,2).)

9. Write the SQL code to enter an EMP_PCT value of 3.85 for the person whose employee number (EMP_NUM) is 103. Next, enter the remaining EMP_PCT values shown in Figure Q6.9:

FIGURE Q6.9 THE CONTENTS OF THE EMP_2 TABLE

EMP_NUM	EMP_LNAME	EMP_FNAME	EMP_INITIAL	EMP_HIREDATE	JOB_CODE	EMP_PCT	PROJ_NUM
101	News	John	G	08-Nov-98	502	5.00	
102	Senior	David	H	12-Jul-87	501	8.00	
103	Arbough	June	E	01-Dec-94	500	3.85	
104	Ramoras	Anne	K	15-Nov-85	501	10.00	
105	Johnson	Alice	K	01-Feb-91	502	5.00	
106	Smithfield	William		22-Jun-02	500	6.20	
107	Alonzo	Maria	D	10-Oct-91	500	5.15	
108	Washington	Ralph	B	22-Aug-89	501	10.00	
109	Smith	Larry	W	18-Jul-95	501	2.00	

10. Using a single command sequence, write the SQL code that will enter the project number (PROJ_NUM) = 18 for all employees whose job classification (JOB_CODE) is 500.

11. Using a single command sequence, write the SQL code that will enter the project number (PROJ_NUM) = 25 for all employees whose job classification (JOB_CODE) is 502 or higher. When you finish Questions 10 and 11, the EMP_2 table will contain the data shown in Figure Q6.11:

FIGURE Q6.11 THE EMP_2 TABLE CONTENTS AFTER THE MODIFICATIONS

EMP_NUM	EMP_LNAME	EMP_FNAME	EMP_INITIAL	EMP_HIREDATE	JOB_CODE	EMP_PCT	PROJ_NUM
101	News	John	G	08-Nov-98	502	5.00	25
102	Senior	David	H	12-Jul-87	501	8.00	
103	Arbough	June	E	01-Dec-94	500	3.85	18
104	Ramoras	Anne	K	15-Nov-85	501	10.00	
105	Johnson	Alice	K	01-Feb-91	502	5.00	25
106	Smithfield	William		22-Jun-02	500	6.20	18
107	Alonzo	Maria	D	10-Oct-91	500	5.15	18
108	Washington	Ralph	B	22-Aug-89	501	10.00	
109	Smith	Larry	W	18-Jul-95	501	2.00	

(You may assume that the table has been saved again at this point!)

12. Write the SQL code that will enter a PROJ_NUM of 14 for those employees who were hired before January 1, 1992 and whose job code is at least 501. (You may assume that the table will be restored to its condition preceding this question.)

13. Write the two SQL command sequences required to:
 a. Create a temporary table named TEMP_1, whose structure is composed of the EMP_2 attributes EMP_NUM and EMP_PCT.
 b. Copy the matching EMP_2 values into the TEMP_1 table.

14. Write the SQL command that will delete the newly created TEMP_1 table from the database.

15. Write the SQL code required to list all employees whose last names start with 'Smith'. In other words, the rows for both Smith and Smithfield should be included in the listing.

16. Using the EMPLOYEE, JOB, and PROJECT tables in the Ch06_Review database (see Figure Q6.1), write the SQL code that will produce the results shown in Figure Q6.16.

FIGURE Q6.16 **THE QUERY RESULTS FOR QUESTION 16**

PROJ_NAME	PROJ_VALUE	PROJ_BALANCE	EMP_LNAME	EMP_FNAME	EMP_INITIAL	JOB_CODE	JOB_DESCRIPTION	JOB_CHG_HOUR
Rolling Tide	$805,000.00	$500,345.20	Senior	David	H	501	Systems Analyst	$96.75
Evergreen	$1,453,500.00	$1,002,349.75	Arbough	June	E	500	Programmer	$35.75
Starflight	$2,650,500.00	$2,309,879.76	Alonzo	Maria	D	500	Programmer	$35.75
Amber Wave	$3,500,500.00	$2,110,345.91	Washington	Ralph	B	501	Systems Analyst	$96.75

17. Write the SQL code that will produce a virtual table named REP_1, containing the same information that was shown in Question 16.

18. Write the SQL code to find the average bonus percentage in the EMP_2 table you created in Question 11.

19. Write the SQL code that will produce a listing for the data in the EMP_2 table in ascending order by the bonus percentage.

20. Write the SQL code that will list only the different project numbers found in the EMP_2 table.

21. Write the SQL code to calculate the ASSIGN_CHARGE values in the ASSIGNMENT table in the Ch06_Review database. (See Figure Q6.1.) Note that ASSIGN_CHARGE is a derived attribute that is calculated by multiplying ASSIGN_CHG_HR by ASSIGN_HOURS.

22. Using the data in the ASSIGNMENT table, write the SQL code that will yield, for each employee, the total number of hours worked and the total charges stemming from those hours worked. The results of running this query are shown in Figure Q6.22.

FIGURE Q6.22 **TOTAL HOURS AND CHARGES BY EMPLOYEE**

EMP_NUM	EMP_LNAME	SumOfASSIGN_HOURS	SumOfASSIGN_CHARGE
101	News	3.1	$387.50
103	Arbough	19.7	$1,664.65
104	Ramoras	11.9	$1,218.70
105	Johnson	12.5	$1,382.50
108	Washington	8.3	$840.15
113	Joenbrood	3.8	$192.85
115	Bawangi	12.5	$1,276.75
117	Williamson	18.8	$649.54

23. Write a query to produce the total number of hours and charges for each of the projects represented in the ASSIGNMENT table. The output is shown in Figure Q6.23.

FIGURE Q6.23 **TOTAL HOURS AND CHARGES BY PROJECT**

PROJ_NUM	SumOfASSIGN_HOURS	SumOfASSIGN_CHARGE
15	20.5	$1,806.52
18	23.7	$1,544.80
22	27.0	$2,593.16
25	19.4	$1,668.16

24. Write the SQL code to generate the total hours worked and the total charges made by all employees. The results are shown in Figure Q6.24. (*Hint:* This is a nested query. If you use Microsoft Access, you can generate this result by using the query output in Figure Q6.22 as the basis for the query that will produce the output in Figure Q6.24.)

FIGURE Q6.24 **TOTAL HOURS AND CHARGES, ALL EMPLOYEES**

SumOfSumOfASSIGN_HOURS	SumOfSumOfASSIGN_CHARGE
90.6	$7,612.64

25. Write the SQL code to generate the total hours worked and the total charges made to all projects. The results should be the same as those shown in Figure Q6.24. (*Hint:* This is a nested query. If you use Microsoft Access, you can generate this result by using the query output in Figure Q6.23 as the basis for the query.)

P R O B L E M S

A D D I T I O N A L C D C O N T E N T

Problems 1–15 are based on the **Ch06_AviaCo** database located in the **Databases\Student\Ch06** folder on the student CD. This database is stored in the Microsoft Access format. If you use another DBMS such as Oracle, SQL Server, or DB2, use its import utilities to move the Access database contents to the DBMS of your choice.

Before you attempt to write any SQL queries, familiarize yourself with the **Ch06_AviaCo** database structure and contents shown in Figure P6.1. Although the relational schema does not show optionalities, keep in mind that all pilots are employees, but not all employees are flight crew members. (Although in this database the crew member assignments all involve pilots and copilots, the design is sufficiently flexible to accommodate crew member assignments—such a load masters and flight attendants—of people who are not pilots. That's why the relationship between CHARTER and EMPLOYEE is implemented through CREW.) Note also that this design implementation does not include multivalued attributes. For example, multiple ratings such as Instrument and Certified Flight Instructor ratings are stored in the (composite) EARNEDRATINGS table. Nor does the CHARTER table include multiple crew assignments, which are properly stored in the CREW table.

We have not identified the tables you see in Figure P6.1. Given the relational schema and the use of the recommended attribute naming conventions, you will be able to identify all the tables and attributes you will need to write the SQL queries in this problem set. But do note that only a few of the CHARTER table attributes are shown. For example, the relational schema shows that the CHARTER table includes the customer number (CUS_CODE) as the FK to CUSTOMER.

FIGURE P6.1 STRUCTURE AND CONTENTS OF THE CH06_AVIACO DATABASE

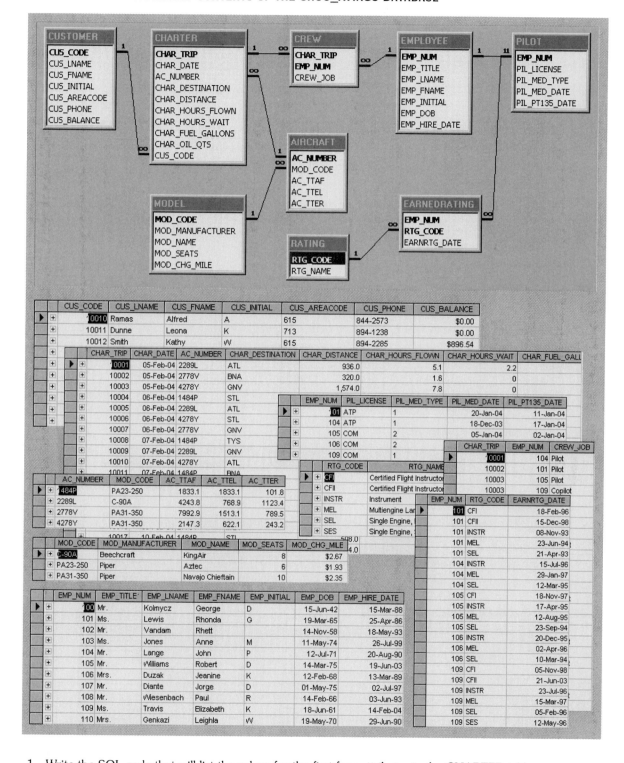

1. Write the SQL code that will list the values for the first four attributes in the CHARTER table.

2. Using the contents of the CHARTER table, write the SQL query that will produce the output shown in Figure P6.2. Note that the output is limited to selected attributes for aircraft number 2778V.

FIGURE P6.2 **PROBLEM 2 QUERY RESULTS**

CHAR_DATE	AC_NUMBER	CHAR_DESTINATION	CHAR_DISTANCE	CHAR_HOURS_FLOWN
05-Feb-04	2778V	BNA	320	1.6
06-Feb-04	2778V	GNV	1,574	7.9
08-Feb-04	2778V	MOB	884	4.8
09-Feb-04	2778V	MQY	312	1.5

3. Create a virtual table (named AC2778V) containing the output presented in Problem 2.

4. Produce the output shown in Figure P6.4 for aircraft 2778V. Note that this output includes data from the CHARTER and CUSTOMER tables. (*Hint*: Use a JOIN in this query.)

FIGURE P6.4 **PROBLEM 4 QUERY RESULTS**

CHAR_DATE	AC_NUMBER	CHAR_DESTINATION	CUS_LNAME	CUS_AREACODE	CUS_PHONE
08-Feb-04	2778V	MOB	Ramas	615	844-2573
09-Feb-04	2778V	MQY	Dunne	713	894-1238
06-Feb-04	2778V	GNV	Smith	615	894-2285
05-Feb-04	2778V	BNA	Brown	615	297-1228

5. Produce the output shown in Figure P6.5. The output, derived from the CHARTER and MODEL tables, is limited to February 6, 2004. (*Hint*: The join passes through another table. Note that the "connection" between CHARTER and MODEL requires the existence of AIRCRAFT, because the CHARTER table does not contain a foreign key to MODEL. However, CHARTER does contain AC_NUMBER, a foreign key to AIRCRAFT, which contains a foreign key to MODEL.)

FIGURE P6.5 **PROBLEM 5 QUERY RESULTS**

CHAR_DATE	CHAR_DESTINATION	AC_NUMBER	MOD_NAME	MOD_CHG_MILE
06-Feb-04	STL	1484P	Aztec	$1.93
06-Feb-04	ATL	2289L	KingAir	$2.67
06-Feb-04	STL	4278Y	Navajo Chieftain	$2.35
06-Feb-04	GNV	2778V	Navajo Chieftain	$2.35

6. Modify the query in Problem 5 to include data from the CUSTOMER table. This time, the output is limited to charter records generated since February 9, 2004. (The query results are shown in Figure P6.6.

FIGURE P6.6 **PROBLEM 6 QUERY RESULTS**

CHAR_DATE	CHAR_DESTINATION	AC_NUMBER	MOD_NAME	MOD_CHG_MILE	CUS_LNAME
09-Feb-04	ATL	4278Y	Navajo Chieftain	$2.35	Williams
09-Feb-04	MQY	2778V	Navajo Chieftain	$2.35	Dunne
09-Feb-04	GNV	2289L	KingAir	$2.67	Brown
10-Feb-04	TYS	4278Y	Navajo Chieftain	$2.35	Williams
10-Feb-04	STL	1484P	Aztec	$1.93	Orlando

7. Modify the query in Problem 6 to produce the output shown in Figure P6.7. The date limitation in Problem 6 applies to this problem, too. Note that this query includes data from the CREW and EMPLOYEE tables. Note: You may wonder why the date restriction seems to generate more records than it did in Problem 6. Actually, the number of (CHARTER) records is the same, but there are several records that are listed twice

to reflect a crew of two: a pilot and a copilot. For example, the record for the 09-Feb-2004 flight to GNV, using aircraft 2289L, required a crew consisting of a pilot (Lange) and a copilot (Lewis).

FIGURE P6.7 PROBLEM 7 QUERY RESULTS

CHAR_DATE	CHAR_DESTINATION	AC_NUMBER	MOD_CHG_MILE	CHAR_DISTANCE	EMP_NUM	CREW_JOB	EMP_LNAME
09-Feb-04	GNV	2289L	$2.67	1,645	104	Pilot	Lange
09-Feb-04	GNV	2289L	$2.67	1,645	101	Copilot	Lewis
09-Feb-04	MQY	2778V	$2.35	312	109	Pilot	Travis
09-Feb-04	MQY	2778V	$2.35	312	105	Copilot	Williams
09-Feb-04	ATL	4278Y	$2.35	936	106	Pilot	Duzak
10-Feb-04	STL	1484P	$1.93	508	101	Pilot	Lewis
10-Feb-04	TYS	4278Y	$2.35	644	105	Pilot	Williams
10-Feb-04	TYS	4278Y	$2.35	644	104	Copilot	Lange

8. Modify the query in Problem 5 to include the computed (derived) attribute "fuel per hour." (*Hint*: It is possible to use SQL to produce computed "attributes" that are not stored in any table. For example, the following SQL query is perfectly acceptable:

SELECT CHAR_DISTANCE, CHAR_FUEL_GALLONS/CHAR_DISTANCE

 FROM CHARTER;

(The above query produces the "gallons per mile flown" value.) Use a similar technique on joined tables to produce the "gallons per hour" output shown in Figure P6.8. (Note that 67.2 gallons/1.5 hours produces 44.8 gallons per hour.)

FIGURE P6.8 PROBLEM 8 QUERY RESULTS

CHAR_DATE	AC_NUMBER	MOD_NAME	CHAR_HOURS_FLOWN	CHAR_FUEL_GALLONS	Expr1
09-Feb-04	2778V	Navajo Chieftain	1.5	67.2	44.8
09-Feb-04	2289L	KingAir	6.7	459.5	68.5820895522388
09-Feb-04	4278Y	Navajo Chieftain	6.1	302.6	49.6065573770492
10-Feb-04	4278Y	Navajo Chieftain	3.8	167.4	44.0526315789474
10-Feb-04	1484P	Aztec	3.1	105.5	34.0322580645161

Query output such as the "gallons per hour" result shown in Figure P6.8 provide managers with very important information. In this case, why is the fuel burn for the Navajo Chieftain 4278Y flown on 9-Feb-04 so much higher than the fuel burn for that aircraft on 10-Feb-04? Such a query result may lead to additional queries to find out who flew the aircraft or what special circumstances might have prevailed. Is the fuel burn difference due to poor fuel management by the pilot, does it reflect an engine fuel metering problem, or was there an error in the fuel recording? The ability to generate useful query output is an important management asset.

NOTE

The output format is determined by the RDBMS you use. In this example, the Access software defaulted to an output heading labeled "Expr1" to indicate the expression resulting from the division

[CHARTER]![CHAR_FUEL_GALLONS]/[CHARTER]![CHAR_HOURS]

created by its expression builder. Oracle defaults to the full division label. You should learn to control the output format with the help of your RDBMS's utility software. For instance, Access allows you to use its properties box to redefine the output format to produce the output shown in Figure P6.8A.

FIGURE P6.8A USING THE SYSTEM SOFTWARE TO FORMAT THE OUTPUT

CHAR_DATE	AC_NUMBER	MOD_NAME	CHAR_HOURS_FLOWN	CHAR_FUEL_GALLONS	Gallons/Hour
09-Feb-04	2778V	Navajo Chieftain	1.5	67.2	44.80
09-Feb-04	2289L	KingAir	6.7	459.5	68.58
09-Feb-04	4278Y	Navajo Chieftain	6.1	302.6	49.61
10-Feb-04	4278Y	Navajo Chieftain	3.8	167.4	44.05
10-Feb-04	1484P	Aztec	3.1	105.5	34.03

9. Create a query to produce the output shown in Figure P6.9. Note that, in this case, the computed attribute requires data found in two different tables. (*Hint*: The MODEL table contains the charge per mile, and the CHARTER table contains the total miles flown.) Note also that the output is limited to charter records generated since February 9, 2004. In addition, the output is ordered by date and, within the date, by the customer's last name.

FIGURE P6.9 PROBLEM 9 QUERY RESULTS

CHAR_DATE	CUS_LNAME	CHAR_DISTANCE	MOD_CHG_MILE	Mileage Charge
09-Feb-04	Brown	1,645	$2.67	$4,392.15
09-Feb-04	Dunne	312	$2.35	$733.20
09-Feb-04	Williams	936	$2.35	$2,199.60
10-Feb-04	Orlando	508	$1.93	$980.44
10-Feb-04	Williams	644	$2.35	$1,513.40

10. Use the techniques that produced the output in Problem 9 to produce the charges shown in Figure P6.10. The total charge to the customer is computed by:

Miles flown * charge per mile

Hours waited * $50 per hour

The miles flown (CHAR_DISTANCE) value is found in the CHARTER table, the charge per mile (MOD_CHG_MILE) is found in the MODEL table, and the hours waited (CHAR_HOURS_WAIT) are found in the CHARTER table.

FIGURE P6.10 PROBLEM 10 QUERY RESULTS

CHAR_DATE	CUS_LNAME	Mileage Charge	Waiting Charge	Total Charge
09-Feb-04	Brown	$4,392.15	$0.00	$4,392.15
09-Feb-04	Dunne	$733.20	$0.00	$733.20
09-Feb-04	Williams	$2,199.60	$105.00	$2,304.60
10-Feb-04	Orlando	$980.44	$0.00	$980.44
10-Feb-04	Williams	$1,513.40	$225.00	$1,738.40

11. Create the SQL query that will produce a list of customers who have an unpaid balance. The required output is shown in Figure P6.11. Note that the balances are listed in descending order.

FIGURE P6.11 A LIST OF CUSTOMERS WITH UNPAID BALANCES

CUS_LNAME	CUS_FNAME	CUS_INITIAL	CUS_BALANCE
Olowski	Paul	F	$1,285.19
O'Brian	Amy	B	$1,014.56
Smith	Kathy	W	$896.54
Orlando	Myron		$673.21
Smith	Olette	K	$453.98

12. Find the average customer balance, the minimum balance, the maximum balance, and the total of the unpaid balances. The resulting values are shown in Figure P6.12.

FIGURE P6.12 CUSTOMER BALANCE SUMMARY

Average Balance	Minimum Balance	Maximum Balance	Total Unpaid Bills
$432.35	$0.00	$1,285.19	$4,323.48

13. Using the CHARTER table as the source, group the aircraft data, then use the SQL functions to produce the output shown in Figure P6.13. (We used utility software to modify the headers, so your headers may have a different "look.")

FIGURE P6.13 THE AIRCRAFT DATA SUMMARY STATEMENT

AC_NUMBER	Number of Trips	Total Distance	Average Distance	Total Hours	Average Hours
1484P	4	1,976	494.0	12.0	3.0
2289L	4	5,178	1,294.5	24.1	6.0
2778V	4	3,090	772.5	15.8	4.0
4278Y	6	5,268	878.0	30.4	5.1

14. Write the SQL code to generate the output shown in Figure P6.14. Note that the listing includes all CHARTER flights that did not include a copilot crew assignment. (*Hint:* The crew assignments are listed in the CREW table. Also, note that the pilot's last name requires access to the EMPLOYEE table, while the MOD_CODE requires access to the MODEL table.)

FIGURE P6.14 A LISTING OF ALL CHARTER FLIGHTS THAT DID NOT USE A COPILOT

CHAR_TRIP	CHAR_DATE	AC_NUMBER	MOD_NAME	CHAR_HOURS_FLOWN	EMP_LNAME	CREW_JOB
10001	05-Feb-2004	2289L	KingAir	5.1	Lange	Pilot
10002	05-Feb-2004	2778V	Navajo Chieftain	1.6	Lewis	Pilot
10004	06-Feb-2004	1484P	Aztec	2.9	Duzak	Pilot
10005	06-Feb-2004	2289L	KingAir	5.7	Lewis	Pilot
10006	06-Feb-2004	4278Y	Navajo Chieftain	2.6	Travis	Pilot
10008	07-Feb-2004	1484P	Aztec	4.1	Duzak	Pilot
10009	07-Feb-2004	2289L	KingAir	6.6	Williams	Pilot
10010	07-Feb-2004	4278Y	Navajo Chieftain	6.2	Wiesenbach	Pilot
10012	08-Feb-2004	2778V	Navajo Chieftain	4.8	Lewis	Pilot
10013	08-Feb-2004	4278Y	Navajo Chieftain	3.9	Williams	Pilot
10014	09-Feb-2004	4278Y	Navajo Chieftain	6.1	Duzak	Pilot
10017	10-Feb-2004	1484P	Aztec	3.1	Lewis	Pilot

15. Write a query that will list the employee ages and the date the query was run. The required output is shown in Figure P6.15. (As you can tell, the query was run on April 7, 2003, so the employee ages are current as of that date.)

FIGURE 6.15 EMPLOYEE AGES AND DATE OF QUERY

EMP_NUM	EMP_LNAME	EMP_FNAME	EMP_HIRE_DATE	EMP_DOB	Age	Query Date
100	Kolmycz	George	15-Mar-88	15-Jun-42	60	07-Apr-03
101	Lewis	Rhonda	25-Apr-86	19-Mar-65	38	07-Apr-03
102	Vandam	Rhett	18-May-93	14-Nov-58	44	07-Apr-03
103	Jones	Anne	26-Jul-99	11-May-74	28	07-Apr-03
104	Lange	John	20-Aug-90	12-Jul-71	31	07-Apr-03
105	Williams	Robert	19-Jun-03	14-Mar-75	28	07-Apr-03
106	Duzak	Jeanine	13-Mar-89	12-Feb-68	35	07-Apr-03
107	Diante	Jorge	02-Jul-97	01-May-75	27	07-Apr-03
108	Wiesenbach	Paul	03-Jun-93	14-Feb-66	37	07-Apr-03
109	Travis	Elizabeth	14-Feb-04	18-Jun-61	41	07-Apr-03
110	Genkazi	Leighla	29-Jun-90	19-May-70	32	07-Apr-03

ADDITIONAL CD CONTENT

Problems 16–33 are based on the Ch06_SaleCo database located in the Databases\Student\Ch06 folder on the student CD. This database is stored in the Microsoft Access format. If you use another DBMS such as Oracle, SQL Server, or DB2, use its import utilities to move the Access database contents to the DBMS of your choice.

The Ch06_SaleCo database's structure and contents are shown in Figure P6.16.

16. Build the Ch06_SaleCo database and its table structures. Use the data listings as your guide to define the data characteristics. For example, the PROD_CODE is always composed of eight characters, so it may be defined as FCHAR(8). (Naturally, if your RDBMS does not support FCHAR, use CHAR(8), instead.)

17. Enter the data into the table structures you defined in Problem 16.

FIGURE P6.16 THE CH06_SALECO DATABASE STRUCTURE AND CONTENTS

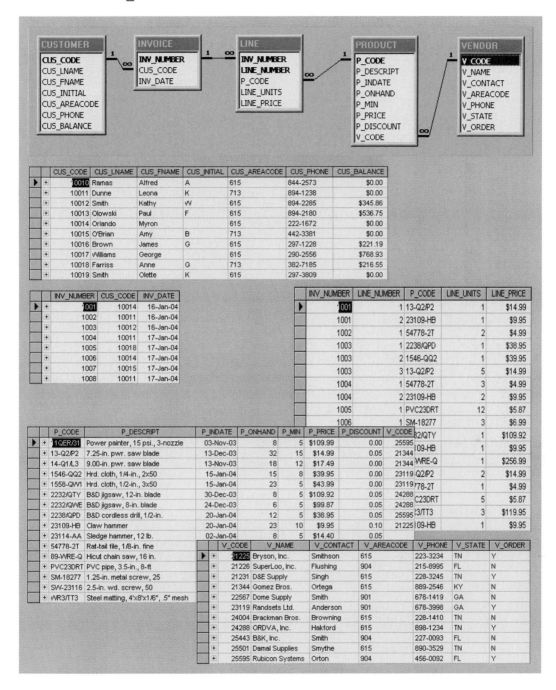

18. Generate a listing of all purchases made by the customers, using the output shown in Figure P6.18 as your guide. (*Hint*: ORDER BY CUS_CODE in the INVOICE table.)

FIGURE P6.18 **LIST OF CUSTOMER PURCHASES**

CUS_CODE	INV_NUMBER	INV_DATE	P_DESCRIPT	LINE_UNITS	LINE_PRICE
10011	1002	16-Jan-04	Rat-tail file, 1/8-in. fine	2	$4.99
10011	1004	17-Jan-04	Claw hammer	2	$9.95
10011	1004	17-Jan-04	Rat-tail file, 1/8-in. fine	3	$4.99
10011	1008	17-Jan-04	Claw hammer	1	$9.95
10011	1008	17-Jan-04	PVC pipe, 3.5-in., 8-ft	5	$5.87
10011	1008	17-Jan-04	Steel matting, 4'x8'x1/6", .5" mesh	3	$119.95
10012	1003	16-Jan-04	7.25-in. pwr. saw blade	5	$14.99
10012	1003	16-Jan-04	B&D cordless drill, 1/2-in.	1	$38.95
10012	1003	16-Jan-04	Hrd. cloth, 1/4-in., 2x50	1	$39.95
10014	1001	16-Jan-04	7.25-in. pwr. saw blade	1	$14.99
10014	1001	16-Jan-04	Claw hammer	1	$9.95
10014	1006	17-Jan-04	1.25-in. metal screw, 25	3	$6.99
10014	1006	17-Jan-04	B&D jigsaw, 12-in. blade	1	$109.92
10014	1006	17-Jan-04	Claw hammer	1	$9.95
10014	1006	17-Jan-04	Hicut chain saw, 16 in.	1	$256.99
10015	1007	17-Jan-04	7.25-in. pwr. saw blade	2	$14.99
10015	1007	17-Jan-04	Rat-tail file, 1/8-in. fine	1	$4.99
10018	1005	17-Jan-04	PVC pipe, 3.5-in., 8-ft	12	$5.87

19. Using the output shown in Figure P6.19 as your guide, generate the listing of customer purchases, including the subtotals for each of the invoice line numbers. (*Hint*: Modify the query format used to produce the listing of customer purchases in Problem 18, delete the INV_DATE column, then add the derived (computed) attribute LINE_UNITS * LINE_PRICE to calculate the subtotals.)

FIGURE P6.19 **SUMMARY OF CUSTOMER PURCHASES WITH SUBTOTALS**

CUS_CODE	INV_NUMBER	P_DESCRIPT	Units Bought	Unit Price	Subtotal
10011	1002	Rat-tail file, 1/8-in. fine	2	$4.99	$9.98
10011	1004	Claw hammer	2	$9.95	$19.90
10011	1004	Rat-tail file, 1/8-in. fine	3	$4.99	$14.97
10011	1008	Claw hammer	1	$9.95	$9.95
10011	1008	PVC pipe, 3.5-in., 8-ft	5	$5.87	$29.35
10011	1008	Steel matting, 4'x8'x1/6", .5" mesh	3	$119.95	$359.85
10012	1003	7.25-in. pwr. saw blade	5	$14.99	$74.95
10012	1003	B&D cordless drill, 1/2-in.	1	$38.95	$38.95
10012	1003	Hrd. cloth, 1/4-in., 2x50	1	$39.95	$39.95
10014	1001	7.25-in. pwr. saw blade	1	$14.99	$14.99
10014	1001	Claw hammer	1	$9.95	$9.95
10014	1006	1.25-in. metal screw, 25	3	$6.99	$20.97
10014	1006	B&D jigsaw, 12-in. blade	1	$109.92	$109.92
10014	1006	Claw hammer	1	$9.95	$9.95
10014	1006	Hicut chain saw, 16 in.	1	$256.99	$256.99
10015	1007	7.25-in. pwr. saw blade	2	$14.99	$29.98
10015	1007	Rat-tail file, 1/8-in. fine	1	$4.99	$4.99
10018	1005	PVC pipe, 3.5-in., 8-ft	12	$5.87	$70.44

20. Now modify the query used in Problem 19 to produce the summary shown in Figure P6.20.

FIGURE P6.20 **CUSTOMER PURCHASE SUMMARY**

CUS_CODE	CUS_BALANCE	Total Purchases
▶ 10011	$0.00	$444.00
10012	$345.86	$153.85
10014	$0.00	$422.77
10015	$0.00	$34.97
10018	$216.55	$70.44

21. Modify the query in Problem 20 to include the number of individual product purchases made by each customer. (In other words, if the customer's invoice is based on three products, one per LINE_NUMBER, you are counting three product purchases. If you examine the original invoice data, you will note that customer 10011 generated three invoices, which contained a total of six lines, each representing a product purchase.) Your output values must match those shown in Figure P6.21.

FIGURE P6.21 **CUSTOMER TOTAL PURCHASE AMOUNTS AND NUMBER OF PURCHASES**

CUS_CODE	CUS_BALANCE	Total Purchases	Number of Purchases
▶ 10011	$0.00	$444.00	6
10012	$345.86	$153.85	3
10014	$0.00	$422.77	6
10015	$0.00	$34.97	2
10018	$216.55	$70.44	1

22. Use a query to compute the average purchase amount per product made by each customer. (Hint: Use the results of Problem 21 as the basis for this query.) Your output values must match those shown in Figure P6.22. Note that the Average Purchase Amount is equal to the Total Purchases divided by the Number of Purchases.

FIGURE P6.22 **AVERAGE PURCHASE AMOUNT BY CUSTOMER**

CUS_CODE	CUS_BALANCE	Total Purchases	Number of Purchases	Average Purchase Amount
▶ 10011	$0.00	$444.00	6	$74.00
10012	$345.86	$153.85	3	$51.28
10014	$0.00	$422.77	6	$70.46
10015	$0.00	$34.97	2	$17.49
10018	$216.55	$70.44	1	$70.44

23. Create a query to produce the total purchase per invoice, to generate the results shown in Figure P6.23. The Invoice Total is the sum of the product purchases in the LINE that corresponds to the INVOICE.

FIGURE P6.23 **INVOICE TOTALS**

INV_NUMBER	Invoice Total
1001	$24.94
1002	$9.98
1003	$153.85
1004	$34.87
1005	$70.44
1006	$397.83
1007	$34.97
1008	$399.15

24. Use a query to show the invoices and invoice totals, as shown in Figure P6.24. (*Hint*: Group by the CUS_CODE.)

FIGURE P6.24 **INVOICE TOTALS BY CUSTOMER**

CUS_CODE	INV_NUMBER	Invoice Total
10011	1002	$9.98
10011	1004	$34.87
10011	1008	$399.15
10012	1003	$153.85
10014	1001	$24.94
10014	1006	$397.83
10015	1007	$34.97
10018	1005	$70.44

25. Write a query to produce the number of invoices and the total purchase amounts by customer, using the output shown in Figure P6.25 as your guide. (Compare this summary to the results shown in Problem 24.)

FIGURE P6.25 **NUMBER OF INVOICES AND TOTAL PURCHASE AMOUNTS BY CUSTOMER**

CUS_CODE	Number of Invoices	Total Customer Purchases
10011	3	$444.00
10012	1	$153.85
10014	2	$422.77
10015	1	$34.97
10018	1	$70.44

26. Using the query results in Problem 25 as your basis, write a query to generate the total number of invoices, the invoice total for all of these invoices, the smallest invoice amount, the largest invoice amount, and the average of all the invoices. (*Hint*: Check the figure output in Problem 25.) Your output must match Figure P6.26.

FIGURE P6.26 **NUMBER OF INVOICES, INVOICE TOTALS, MINIMUM, MAXIMUM, AND AVERAGE SALES**

	Total Invoices	Total Sales	Min Sale	Max Sale	Avg Sale
▶	5	$1,126.03	$34.97	$444.00	$225.21

27. List the balance characteristics of the customers who have made purchases during the current invoice cycle—that is, for the customers who appear in the INVOICE table. The results of this query are shown in Figure P6.27.

FIGURE P6.27 **BALANCES FOR CUSTOMERS WHO MADE PURCHASES**

	CUS_CODE	CUS_BALANCE
▶	10011	$0.00
	10012	$345.86
	10014	$0.00
	10015	$0.00
	10018	$216.55

28. Using the results of the query created for Problem 27, provide a summary of customer balance characteristics, as shown in Figure P6.28.

FIGURE P6.28 **BALANCE SUMMARY FOR CUSTOMERS WHO MADE PURCHASES**

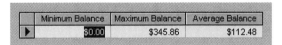

	Minimum Balance	Maximum Balance	Average Balance
▶	$0.00	$345.86	$112.48

29. Create a query to find the customer balance characteristics for all customers, including the total of the outstanding balances. The results of this query are shown in Figure P6.29.

FIGURE P6.29 **CUSTOMER BALANCE SUMMARY FOR ALL CUSTOMERS**

	Total Balances	Minimum Balance	Maximum Balance	Average Balance
▶	$2,089.28	$0.00	$768.93	$208.93

30. Find the listing of customers who did not make purchases during the invoicing period. Your output must match the output shown in Figure P6.30.

FIGURE P6.30 CUSTOMER BALANCES FOR CUSTOMERS WHO DID NOT MAKE PURCHASES

```
CUS_CODE CUS_BALANCE
-------- -----------
   10010           0
   10013      536.75
   10016      221.19
   10017      768.93
   10019           0

SQL> |
```

31. Now find the customer balance summary for all customers who have not made purchases during the current invoicing period. The results are shown in Figure P6.31.

FIGURE P6.31 SUMMARY OF CUSTOMER BALANCES FOR CUSTOMERS WHO DID NOT MAKE PURCHASES

```
SUM(CUS_BALANCE) MIN(CUS_BALANCE) MAX(CUS_BALANCE) AVG(CUS_BALANCE)
---------------- ---------------- ---------------- ----------------
         1526.87                0           768.93          305.374

SQL> |
```

32. Create a query to produce the summary of the value of products currently in inventory. Note that the value of each product is produced by the multiplication of the units currently in inventory and the unit price. The results of this query are shown in Figure P6.32.

FIGURE P6.32 VALUE OF PRODUCTS CURRENTLY IN INVENTORY

P_DESCRIPT	P_ONHAND	P_PRICE	Subtotal
Power painter, 15 psi., 3-nozzle	8	$109.99	$879.92
7.25-in. pwr. saw blade	32	$14.99	$479.68
9.00-in. pwr. saw blade	18	$17.49	$314.82
Hrd. cloth, 1/4-in., 2x50	15	$39.95	$599.25
Hrd. cloth, 1/2-in., 3x50	23	$43.99	$1,011.77
B&D jigsaw, 12-in. blade	8	$109.92	$879.36
B&D jigsaw, 8-in. blade	6	$99.87	$599.22
B&D cordless drill, 1/2-in.	12	$38.95	$467.40
Claw hammer	23	$9.95	$228.85
Sledge hammer, 12 lb.	8	$14.40	$115.20
Rat-tail file, 1/8-in. fine	43	$4.99	$214.57
Hicut chain saw, 16 in.	11	$256.99	$2,826.89
PVC pipe, 3.5-in., 8-ft	188	$5.87	$1,103.56
1.25-in. metal screw, 25	172	$6.99	$1,202.28
2.5-in. wd. screw, 50	237	$8.45	$2,002.65
Steel matting, 4'x8'x1/6", .5" mesh	18	$119.95	$2,159.10

33. Using the results of the query created in Problem 32, find the total value of the product inventory. The results are shown in Figure P6.33.

FIGURE P6.33 TOTAL VALUE OF ALL PRODUCTS IN INVENTORY

7

ADVANCED SQL

In this chapter, you will learn:

❖ About the relational set operators UNION, UNION ALL, INTERSECT, and MINUS

❖ How to use the advanced SQL JOIN operator syntax

❖ About the different types of subqueries and correlated queries

❖ How to use SQL functions to manipulate dates, strings, and other data

❖ How to create and use updatable views

❖ How to create and use triggers and stored procedures

❖ How to create embedded SQL

P R E V I E W

In Chapter 6, "An Introduction to Structured Query Language," you learned the basic SQL data definition and data manipulation commands used to create and manipulate relational data. In this chapter, you will build on what you learned in Chapter 6 and learn how to use more advanced SQL features.

In this chapter, you will learn about the SQL relational set operators (UNION, INTERSECT and MINUS) and how those operators are used to merge the results of multiple queries. Joins are at the heart of SQL. Therefore, you must learn how to use the SQL JOIN statement to extract information from multiple tables. In the previous chapter you learned how cascading queries inside other queries can be useful in certain circumstances. In this chapter, you also will learn about the different styles of subqueries that can be implemented in a SELECT statement. Finally, you'll learn more of SQL's many functions to extract information from data, including manipulation of dates and strings, as well as computations based on stored or even derived data.

In the real world, business procedures require the execution of clearly defined actions when a specific event occurs, such as the addition of a new invoice or a student's enrollment in a class. Such procedures can be applied within the DBMS through the use of triggers and stored procedures. In addition, SQL facilitates the application of business procedures when it is embedded in a programming language such as COBOL or C++.

S E V E N

A D D I T I O N A L C D C O N T E N T

Most of the examples used in this chapter are based on Oracle9*i*. If you want to see the examples in action, you must load the required database tables. We have included Oracle SQL script files to create the tables and load the data in the database. The script files are located on the CD in the **\SQL\Ch07** folder. How you connect to the Oracle database depends on how the Oracle software is installed on your server and on the access paths and methods defined and managed by the database administrator. Please follow the instructions provided by your instructor or your school's technology department.

7.1 RELATIONAL SET OPERATORS

In Chapter 3, "The Relational Database Model," you learned about the eight general relational operators. In this section you will learn how to use SQL commands (UNION, INTERSECT, and MINUS) to implement the union, intersection, and difference relational operators.

In previous chapters you learned that SQL data manipulation commands are set oriented, that is, they operate over entire sets of rows and columns (tables) at once. Using sets, you can combine two or more sets to create new sets (or relations). That's precisely what the UNION, INTERSECT, and MINUS statements do. Note that in relational database terms, you can use the words sets, relations, and tables interchangeably because they all provide a conceptual view of the data set as it is presented to the relational database user.

N O T E

Keep in mind that the SQL-99 standard defines the operations that all DBMSs must perform on data, but it leaves the implementation details to the DBMS vendors. Therefore, some advanced SQL features may not work on all DBMS implementations. Also, some DBMS vendors may implement additional features not found in the SQL standard.

UNION, INTERSECT, and MINUS are the names of the SQL statements as implemented in Oracle. The SQL standard uses the keyword EXCEPT to refer to the difference (MINUS) relational operator. Other RDBMS vendors may use a different command name or might not implement a given command at all.

To learn more about the ANSI/ISO SQL standards, check the ANSI Web site (*www.ansi.org*) to find out how to obtain the latest standard documents in electronic form.

UNION, INTERSECT, and MINUS work properly only if relations are *union-compatible*. In SQL terms, union-compatible means that the names of the relation attributes must be the same and their data types must be identical. In practice, some RDBMS vendors require the data types to be "compatible," but not necessarily "exactly the same." For example, compatible data types are VARCHAR (35) and CHAR (15)—in this case, both attributes store character (string) values; the only difference is the string size. Another example of compatible data types would be NUMBER and SMALLINT. Both data types are used to store numeric values.

N O T E

Some DBMS products may require union-compatible tables to have *identical* data types.

7.1.1 UNION

Suppose that SaleCo has bought another company. SaleCo's management wants to make sure that the acquired company's customer list is properly merged with SaleCo's customer list. Because it is quite possible that some customers have purchased goods from both companies, the two lists may contain common customers. SaleCo's management wants to make sure that customer records are not duplicated when the two customer lists are merged. The UNION query is a perfect tool to generate a combined listing of customers—one that excludes duplicate records.

ADDITIONAL CD CONTENT

The **Ch07_SaleCo** database on the CD that accompanies this book illustrates the UNION commands and functions in this section. Look in the **\Databases\Student\Ch07** folder. The file **CH07-SQL-Text** is the Oracle script accompanying the **Ch07_SaleCo** database.

The UNION statement combines rows from two or more queries *without including duplicate rows*. The syntax of the UNION statement is:

query UNION *query*

In other words, the UNION statement combines the output of two SELECT queries. (Remember that the SELECT statements must be union-compatible. That is, they must return the same attribute names and similar data types.)

To demonstrate the use of the UNION statement in SQL, we'll use the CUSTOMER and CUSTOMER_2 tables in the **Ch07_SaleCo** database. To show the combined CUSTOMER and CUSTOMER_2 records without the duplicates, the UNION query is written as follows:

SELECT CUS_LNAME, CUS_FNAME, CUS_INITIAL, CUS_AREACODE, CUS_PHONE FROM CUSTOMER
UNION
SELECT CUS_LNAME, CUS_FNAME, CUS_INITIAL, CUS_AREACODE, CUS_PHONE FROM CUSTOMER_2;

Figure 7.1 shows the contents of the CUSTOMER and CUSTOMER_2 tables and the result of the UNION query. Although we have used MS Access to show you the results, similar results can also be obtained with Oracle9*i*.

FIGURE 7.1 **UNION QUERY RESULT**

As you examine Figure 7.1, note the following:

- The CUSTOMER table contains 10 rows, while the CUSTOMER_2 table contains seven rows.

- Customers Dunne and Olowski are included in the CUSTOMER table as well as in the CUSTOMER_2 table.

- The UNION query yields 15 records, because the duplicate records of customers Dunne and Olowski are not included. In short, the UNION query yields a unique set of records.

NOTE

The SQL standard calls for the elimination of duplicate rows when using the UNION SQL statement. However, some DBMS vendors may not adhere to this standard. Check your DBMS manual to see if the UNION statement is supported and, if so, *how* it is supported.

The UNION statement can be used to unite more than just two queries. For example, let's assume that you have four union-compatible queries named T1, T2, T3, and T4. With the UNION statement you can combine the output of all four queries into a single result set. The SQL statement will be similar to this:

SELECT column-list FROM T1
UNION
SELECT column-list FROM T2
UNION
SELECT column-list FROM T3
UNION
SELECT column-list FROM T4;

7.1.2 UNION ALL

If SaleCo's management wants to know how many customers are on *both* the Customer and Customer_2 lists, a UNION ALL query can be used to produce a relation that retains the duplicate rows. Therefore, the following query will keep all rows from both queries (including the duplicate rows) and return 17 rows.

SELECT CUS_LNAME, CUS_FNAME, CUS_INITIAL, CUS_AREACODE, CUS_PHONE FROM CUSTOMER
UNION ALL
SELECT CUS_LNAME, CUS_FNAME, CUS_INITIAL, CUS_AREACODE, CUS_PHONE FROM CUSTOMER_2;

Running the preceding UNION ALL query produces the result shown in Figure 7.2.

FIGURE 7.2 UNION ALL QUERY RESULT

Like the UNION statement, the UNION ALL statement can be used to unite more than just two queries.

7.1.3 INTERSECT

If SaleCo's management wants to know which customer records are duplicated in the CUSTOMER and CUSTOMER_2 tables, the INTERSECT statement can be used to combine rows from two queries, returning only the rows that appear in both sets. The syntax for the INTERSECT statement is:

query INTERSECT *query*

To generate the list of duplicate customer records, you can use:

SELECT CUS_LNAME, CUS_FNAME, CUS_INITIAL, CUS_AREACODE, CUS_PHONE FROM CUSTOMER
INTERSECT
SELECT CUS_LNAME, CUS_FNAME, CUS_INITIAL, CUS_AREACODE, CUS_PHONE FROM CUSTOMER_2;

The INTERSECT statement can be used to generate additional useful customer information. For example, the following query returns the customer codes for all customers who are located in area code 615 and who have made purchases. (If a customer has made a purchase, there must be an invoice record for that customer.)

SELECT CUS_CODE FROM CUSTOMER WHERE CUS_AREACODE = '615'
INTERSECT
SELECT DISTINCT CUS_CODE FROM INVOICE;

Figure 7.3 shows both sets of SQL statements and their output.

FIGURE 7.3 **INTERSECT QUERY RESULT**

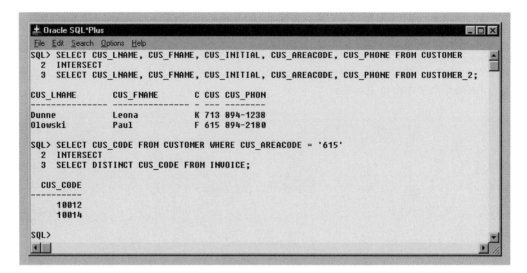

N O T E

MS Access does not support the INTERSECT query, nor does it support other complex queries we'll explore in this chapter. At least in some cases, Access might be able to give you the desired results if you use an alternate query format or procedure. (For example, although Access does not support SQL triggers and stored procedures, you can use Visual Basic code to perform similar actions.) However, our objective is to show you how some important standard SQL features may be used.

7.1.4 MINUS

The MINUS statement in SQL combines rows from two queries and returns only the rows that appear in the first set but not in the second. The syntax for the MINUS statement is:

query MINUS *query*

For example, if the SaleCo managers want to know what customers in the CUSTOMER table are not found in the CUSTOMER_2 table, they can use:

SELECT CUS_LNAME, CUS_FNAME, CUS_INITIAL, CUS_AREACODE, CUS_PHONE FROM CUSTOMER
MINUS
SELECT CUS_LNAME, CUS_FNAME, CUS_INITIAL, CUS_AREACODE, CUS_PHONE FROM CUSTOMER_2;

If the managers want to know what customers in the CUSTOMER_2 table are not found in the CUSTOMER table, they merely switch the table designations:

SELECT CUS_LNAME, CUS_FNAME, CUS_INITIAL, CUS_AREACODE, CUS_PHONE FROM CUSTOMER_2
MINUS
SELECT CUS_LNAME, CUS_FNAME, CUS_INITIAL, CUS_AREACODE, CUS_PHONE FROM CUSTOMER;

You can extract much useful information by combining MINUS with various clauses such as WHERE. For example, the following query returns the customer codes for all customers located in area code 615 minus the ones who have made purchases, leaving the ones who have not made purchases.

SELECT CUS_CODE FROM CUSTOMER WHERE CUS_AREACODE = '615'
MINUS
SELECT DISTINCT CUS_CODE FROM INVOICE;

Figure 7.4 shows the preceding three SQL statements and their output.

FIGURE 7.4 MINUS QUERY RESULTS

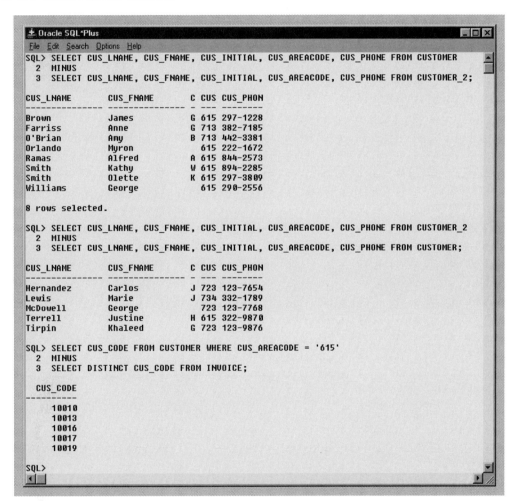

NOTE

Some DBMS products do not support the INTERSECT or MINUS statements, while others may implement the difference relational operator in SQL as EXCEPT. Please consult your DBMS manual to see if the statements illustrated here are supported by your DBMS.

7.1.5 SYNTAX ALTERNATIVES

If your DBMS doesn't support the INTERSECT and MINUS statements, you could use IN and NOT IN sub-queries to obtain similar results. For example, the query below will produce the same results as the INTERSECT query shown in Section 7.1.3.

SELECT CUS_CODE FROM CUSTOMER
 WHERE CUS_AREACODE = '615' AND
 CUS_CODE IN (SELECT DISTINCT CUS_CODE FROM INVOICE);

Figure 7.5 shows the use of the INTERSECT alternative.

FIGURE 7.5 **INTERSECT** ALTERNATIVE

NOTE

MS Access will generate an input request for the CUS_AREACODE if you used apostrophes around the area code. (If you supply the 615 area code, the query will execute properly.) You can eliminate that problem by using standard double quotation marks, writing the WHERE clause in the second line of the preceding SQL statement as

WHERE CUS_AREACODE = "615" AND

MS Access will also accept single quotation marks.

Using the same alternative to the MINUS statement, you can generate the output for the third MINUS query shown in Section 7.1.4 by using:

SELECT CUS_CODE FROM CUSTOMER
 WHERE CUS_AREACODE = '615' AND
 CUS_CODE NOT IN (SELECT DISTINCT CUS_CODE FROM INVOICE);

The results of this query are shown in Figure 7.6. Note that the query output includes only the customers who have not made any purchases and who, therefore, have not generated invoices.

FIGURE 7.6 MINUS ALTERNATIVE

7.2 SQL JOIN OPERATORS

The relational join operation merges rows from two tables and returns the rows that:

- Have common values in common columns (natural join) or,

- Meet a given join condition (equality or inequality) or,

- Have common values in common columns or have no matching values (outer join).

In Chapter 6 you learned how to use the SELECT statement in conjunction with the WHERE clause to join two or more tables. For example, you can join the PRODUCT and VENDOR tables through their common V_CODE by writing:

SELECT P_CODE, P_DESCRIPT, P_PRICE, V_NAME
 FROM PRODUCT, VENDOR
 WHERE PRODUCT.V_CODE = VENDOR.V_CODE;

The preceding SQL join syntax is sometimes referred as an "old-style" join. Note that the FROM clause contains the tables being joined and that the WHERE clause contains the join condition(s) used to join the tables.

As you examine the preceding query, note the following points:

- The FROM clause indicates which tables are to be joined. If three or more tables are included, the join operation takes place two tables at a time, starting from left to right. For example, if you are joining tables T1, T2, and T3, first table T1 is joined to T2; the results of that join are then joined to table T3.

- The join condition in the WHERE clause tells the SELECT statement which rows will be returned. In this case the SELECT statement returns all rows for which the V_CODE values in the PRODUCT and VENDOR tables are equal.

- The number of join conditions is always equal to the number of tables being joined minus one. For example, if you join three tables (T1, T2, and T3), you will have two join conditions (j1 and j2). All join conditions are connected through an AND logical operator. The first join condition (j1) defines the join criteria for T1 and T2. The second join condition (j2) defines the join criteria for the output of the first join and table T3.

- Generally, the join condition will be an equality comparison of the primary key in one table and the related foreign key in the second table.

Join operations can be classified as: inner joins and outer joins. The **inner join** is the traditional join in which only rows that meet a given criteria are selected. The join criteria can be an equality condition (natural join or equijoin) or an inequality condition (theta join). An **outer join** returns not only the matching rows but also the rows with unmatched attribute values for one or both tables to be joined. The SQL standard also introduces a special type of join that returns the same result as the Cartesian product of two sets or tables.

In this section you will learn different ways to express join operations that meet the ANSI SQL standard. (See Table 7.1.) It is useful to remember that not all DBMS vendors provide the same level of SQL support and some do not support the join styles we show in this section. We have used Oracle9i, Release 2, to demonstrate the use of the following queries. Please refer to your DBMS manual if you are using a different DBMS.

TABLE 7.1 SQL JOIN EXPRESSION STYLES

JOIN CLASSIFICATION	JOIN TYPE	SQL SYNTAX EXAMPLE	DESCRIPTION
Cross	CROSS JOIN	SELECT * FROM T1, T2	Returns the Cartesian product of T1 and T2—old style
		SELECT * FROM T1 CROSS JOIN T2	Returns the Cartesian product of T1 and T2
Inner	Old-Style JOIN	SELECT * FROM T1, T2 WHERE T1.C1=T2.C1	Returns only the rows that meet the join condition in the WHERE clause—old style. Only rows with matching values are selected.
	NATURAL JOIN	SELECT * FROM T1 NATURAL JOIN T2	Returns only the rows with matching values in the matching columns. The matching columns must have the same names and similar data types.
	JOIN USING	SELECT * FROM T1 JOIN T2 USING (C1)	Returns only the rows with matching values in the columns indicated in the USING clause
	JOIN ON	SELECT * FROM T1 JOIN T2 ON T1.C1=T2.C1	Returns only the rows that meet the join condition indicated in the ON clause
Outer	LEFT JOIN	SELECT * FROM T1 LEFT OUTER JOIN T2 ON T1.C1=T2.C1	Returns rows with matching values and includes all rows from the left table (T1) with unmatched values
	RIGHT JOIN	SELECT * FROM T1 RIGHT OUTER JOIN T2 ON T1.C1=T2.C1	Returns rows with matching values and includes all rows from the right table (T2) with unmatched values
	FULL JOIN	SELECT * FROM T1 FULL OUTER JOIN T2 ON T1.C1=T2.C1	Returns rows with matching values and includes all rows from both tables (T1 and T2) with unmatched values

7.2.1 CROSS JOIN

A **cross join** performs a relational product (also known as the Cartesian product) of two tables. The cross join syntax is:

SELECT *column-list* FROM *table1* CROSS JOIN *table2*

For example,

SELECT * FROM INVOICE CROSS JOIN LINE;

will perform a cross join of the INVOICE and LINE tables. This CROSS JOIN query will generate 144 rows. (There were 8 invoice rows and 18 line rows, thus yielding 8 × 18 = 144 rows.)

You can also perform a cross join that yields only specified attributes. For example, you can specify:

SELECT INVOICE.INV_NUMBER, CUS_CODE, INV_DATE, P_CODE
 FROM INVOICE CROSS JOIN LINE;

The results generated through this SQL statement can also be generated by using the following syntax:

SELECT INVOICE.INV_NUMBER, CUS_CODE, INV_DATE, P_CODE
 FROM INVOICE, LINE;

7.2.2 NATURAL JOIN

Recall from Chapter 3 that a natural join returns all rows with matching values in the matching columns and eliminates duplicate columns. This style of query is used when the tables share one or more common attributes with common names. The natural join syntax is:

SELECT *column-list* FROM *table1* NATURAL JOIN *table2*

The natural join will perform the following tasks:

- Determine the common attribute(s) by looking for attributes with identical names and compatible data types.
- Select only the rows with common values in the common attribute(s).
- If there are no common attributes, return the relational product of the two tables.

The following example performs a natural join of the CUSTOMER and INVOICE tables and returns only selected attributes:

SELECT CUS_CODE, CUS_LNAME, INV_NUMBER, INV_DATE
 FROM CUSTOMER NATURAL JOIN INVOICE;

The SQL code and its results are shown at the top of Figure 7.7.

You are not limited to two tables. For example, you can perform a natural join of the INVOICE, LINE, and PRODUCT tables and project only selected attributes by writing:

SELECT INV_NUMBER, P_CODE, P_DESCRIPT, LINE_UNITS, LINE_PRICE
 FROM INVOICE NATURAL JOIN LINE NATURAL JOIN PRODUCT;

The SQL code and its results are shown at the bottom of Figure 7.7.

FIGURE 7.7 **NATURAL JOIN RESULT**

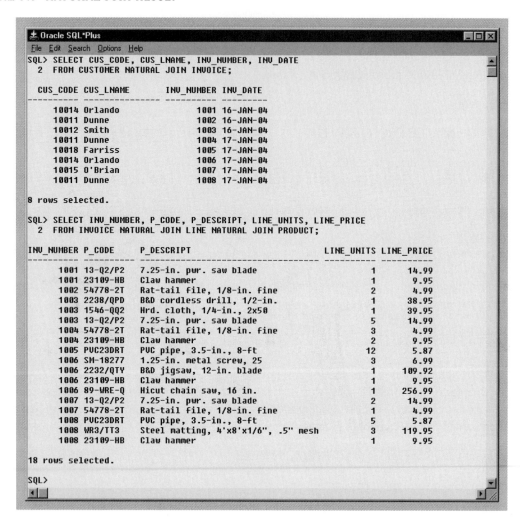

One important difference between the natural join and the "old-style" join syntax is that the natural join does not require the use of a table qualifier for the common attributes. Note that in the first natural join example, you projected CUS_CODE, yet the projection did not require any table qualifier, even though the CUS_CODE attribute appears in both CUSTOMER and INVOICE tables. The same can be said of the INV_NUMBER attribute in the second natural join example.

7.2.3 JOIN WITH USING CLAUSE

A second way to express a join is through the USING keyword. This query will return only the rows with matching values in the column indicated in the USING clause—and this column must exist in both tables. The syntax is:

SELECT *column-list* FROM *table1* JOIN *table2* USING (*common-column*)

To see the JOIN USING query in action, let's perform a join of the INVOICE and LINE tables by writing:

SELECT INV_NUMBER, P_CODE, P_DESCRIPT, LINE_UNITS, LINE_PRICE
 FROM INVOICE JOIN LINE USING (INV_NUMBER)
 JOIN PRODUCT USING (P_CODE);

The SQL statement will produce the results shown in Figure 7.8.

FIGURE 7.8 JOIN USING RESULT

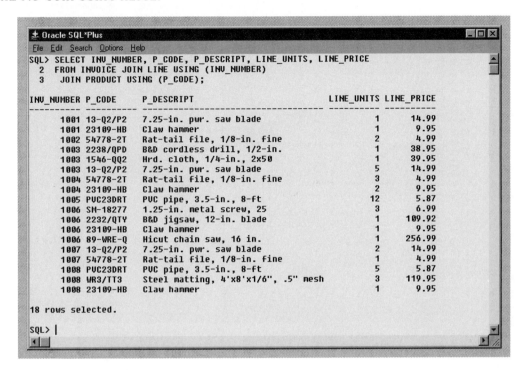

```
SQL> SELECT INV_NUMBER, P_CODE, P_DESCRIPT, LINE_UNITS, LINE_PRICE
  2  FROM INVOICE JOIN LINE USING (INV_NUMBER)
  3    JOIN PRODUCT USING (P_CODE);

INV_NUMBER P_CODE     P_DESCRIPT                          LINE_UNITS LINE_PRICE
---------- ---------- ---------------------------------- ---------- ----------
      1001 13-Q2/P2   7.25-in. pwr. saw blade                     1      14.99
      1001 23109-HB   Claw hammer                                 1       9.95
      1002 54778-2T   Rat-tail file, 1/8-in. fine                 2       4.99
      1003 2238/QPD   B&D cordless drill, 1/2-in.                 1      38.95
      1003 1546-QQ2   Hrd. cloth, 1/4-in., 2x50                  1      39.95
      1003 13-Q2/P2   7.25-in. pwr. saw blade                     5      14.99
      1004 54778-2T   Rat-tail file, 1/8-in. fine                 3       4.99
      1004 23109-HB   Claw hammer                                 2       9.95
      1005 PVC23DRT   PVC pipe, 3.5-in., 8-ft                    12       5.87
      1006 SM-18277   1.25-in. metal screw, 25                    3       6.99
      1006 2232/QTY   B&D jigsaw, 12-in. blade                    1     109.92
      1006 23109-HB   Claw hammer                                 1       9.95
      1006 89-WRE-Q   Hicut chain saw, 16 in.                     1     256.99
      1007 13-Q2/P2   7.25-in. pwr. saw blade                     2      14.99
      1007 54778-2T   Rat-tail file, 1/8-in. fine                 1       4.99
      1008 PVC23DRT   PVC pipe, 3.5-in., 8-ft                     5       5.87
      1008 WR3/TT3    Steel matting, 4'x8'x1/6", .5" mesh         3     119.95
      1008 23109-HB   Claw hammer                                 1       9.95

18 rows selected.

SQL>
```

As was the case with the NATURAL JOIN command, the JOIN USING operand does not require the use of table qualifiers. As a matter of fact, Oracle9i will return an error if you specify the table name in the USING clause.

7.2.4 JOIN ON Clause

The previous two join styles used common attribute names in the joining tables. Another way to express a join when the tables have no common attribute names is to use the JOIN ON operand. This query will return only the rows that meet the indicated join condition. The join condition will typically include an equality comparison expression of two columns. (The columns may or may not share the same name, but must obviously have comparable data types.) The syntax is:

SELECT *column-list* FROM *table1* JOIN *table2* ON *join-condition*

The following example will perform a join of the INVOICE and LINE tables using the ON clause. The result is shown in Figure 7.9.

SELECT INVOICE.INV_NUMBER, P_CODE, P_DESCRIPT, LINE_UNITS, LINE_PRICE
 FROM INVOICE JOIN LINE ON INVOICE.INV_NUMBER = LINE.INV_NUMBER
 JOIN PRODUCT ON LINE.P_CODE = PRODUCT.P_CODE;

FIGURE 7.9 JOIN ON RESULT

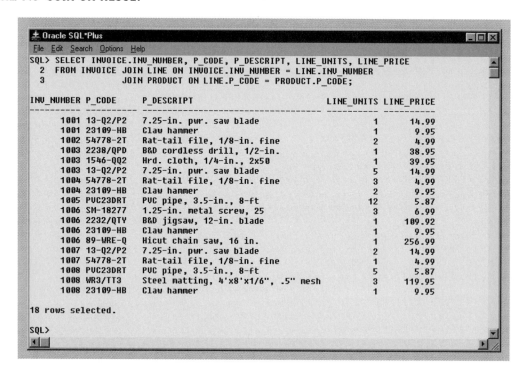

As you examine the preceding query, note that unlike the NATURAL JOIN and the JOIN USING operands, the JOIN ON clause requires a table qualifier for the common attributes. If you do not specify the table qualifier, you will get a "column ambiguously defined" error message.

Keep in mind that the JOIN ON syntax lets you perform a join even when the tables do not share a common attribute name. For example, to generate a list of all employees with the manager's names you can use the following (recursive) query:

SELECT E.EMP_MGR, M.EMP_LNAME, E.EMP_NUM, E.EMP_LNAME
 FROM EMP E JOIN EMP M ON E.EMP_MGR = M.EMP_NUM
 ORDER BY E.EMP_MGR;

7.2.5 OUTER JOINS

An outer join returns not only the rows matching the join condition (that is, rows with matching values in the common columns) but also the rows with unmatched values. The ANSI standard defines three types of outer joins: left, right, and full. The left and right designations reflect the order in which the tables are processed by the DBMS. Remember that join operations take place two tables at a time. The first table named in the FROM clause will be the left side, and the second table named will be the right side. If joining three or more tables, the result of joining the first two tables becomes the left side; the third table becomes the right side.

The left outer join returns not only the rows matching the join condition (that is, rows with matching values in the common column), but also the rows in the left side table with unmatched values in the right side table. The syntax is:

SELECT column-list
 FROM table1 LEFT [OUTER] JOIN table2
 ON join-condition

For example, the following query will list the product code, vendor code, and vendor name for all products and will also include those vendors with no matching products:

SELECT P_CODE, VENDOR.V_CODE, V_NAME
 FROM VENDOR LEFT JOIN PRODUCT ON VENDOR.V_CODE = PRODUCT.V_CODE;

The preceding SQL code and its result is shown in Figure 7.10.

FIGURE 7.10 **LEFT JOIN RESULT**

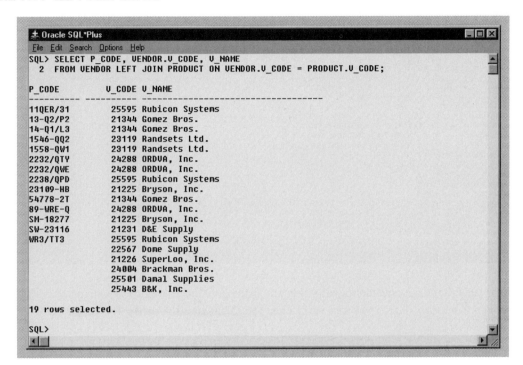

The right outer join returns not only the rows matching the join condition (that is, rows with matching values in the common column), but also the rows in the right side table with unmatched values in the left side table. The syntax is:

SELECT column-list
 FROM *table1* RIGHT [OUTER] JOIN *table2*
 ON *join-condition*

For example, the following query will list the product code, vendor code, and vendor name for all products and will also include those products that do not have a matching vendor code:

SELECT P_CODE, VENDOR.V_CODE, V_NAME
 FROM VENDOR RIGHT JOIN PRODUCT ON VENDOR.V_CODE = PRODUCT.V_CODE;

The SQL code and its output are shown in Figure 7.11.

FIGURE 7.11 **RIGHT JOIN RESULT**

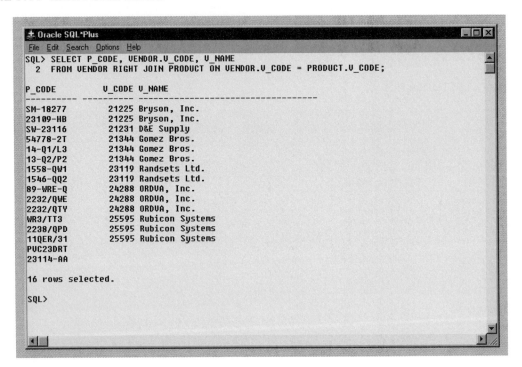

The full outer join returns not only the rows matching the join condition (that is, rows with matching values in the common column), but also all the rows with unmatched values in either side table. The syntax is:

SELECT column-list
 FROM *table1* FULL [OUTER] JOIN *table2*
 ON *join-condition*

For example, the following query will list the product code, vendor code, and vendor name for all products and will include all product rows (products without matching vendors) and also all vendor rows (vendors without matching products).

SELECT P_CODE, VENDOR.V_CODE, V_NAME
 FROM VENDOR FULL JOIN PRODUCT ON VENDOR.V_CODE = PRODUCT.V_CODE;

The SQL code and its result is shown in Figure 7.12.

FIGURE 7.12 **FULL JOIN RESULT**

```
Oracle SQL*Plus                                                    _ □ ×
File  Edit  Search  Options  Help
SQL> SELECT P_CODE, VENDOR.V_CODE, V_NAME
  2  FROM VENDOR FULL JOIN PRODUCT ON VENDOR.V_CODE = PRODUCT.V_CODE;

P_CODE        V_CODE V_NAME
----------    ---------- --------------------------------
11QER/31       25595 Rubicon Systems
13-Q2/P2       21344 Gomez Bros.
14-Q1/L3       21344 Gomez Bros.
1546-QQ2       23119 Randsets Ltd.
1558-QW1       23119 Randsets Ltd.
2232/QTY       24288 ORDVA, Inc.
2232/QWE       24288 ORDVA, Inc.
2238/QPD       25595 Rubicon Systems
23109-HB       21225 Bryson, Inc.
54778-2T       21344 Gomez Bros.
89-WRE-Q       24288 ORDVA, Inc.
SM-18277       21225 Bryson, Inc.
SW-23116       21231 D&E Supply
WR3/TT3        25595 Rubicon Systems
               22567 Dome Supply
               21226 SuperLoo, Inc.
               24004 Brackman Bros.
               25501 Damal Supplies
               25443 B&K, Inc.
23114-AA
PVC23DRT

21 rows selected.

SQL>
```

7.3 SUBQUERIES AND CORRELATED QUERIES

The use of joins allows a relational database to get information from two or more tables. For example, the following query would allow you to get the customers' data with their respective invoices by joining the CUSTOMER and INVOICE tables.

SELECT INV_NUMBER, INVOICE.CUS_CODE, CUS_LNAME, CUS_FNAME
 FROM CUSTOMER, INVOICE
 WHERE CUSTOMER.CUS_CODE = INVOICE.CUS_CODE;

In the previous query, the data from both tables (CUSTOMER and INVOICE) are processed at once, matching rows with shared CUS_CODE values.

However, it is often necessary to process data based on *other* processed data. Suppose, for example, that you want to generate a list of vendors who provide products. (Recall that not all vendors in the VENDOR table have provided products—some of them were only *potential* vendors.) In Chapter 6 you learned that you could generate such a list by writing the following query:

SELECT V_CODE, V_NAME FROM VENDOR
 WHERE V_CODE NOT IN (SELECT V_CODE FROM PRODUCT);

Similarly, to generate a list of all products with a price greater than or equal to the average product price, you can write the following query:

SELECT P_CODE, P_PRICE FROM PRODUCT
 WHERE P_PRICE >= (SELECT AVG(P_PRICE) FROM PRODUCT);

In both of these cases, you needed to get information that was not previously known:

1. What vendors provide products?
2. What is the average price of all products?

In both cases, you used a subquery to generate the required information that could then be used as input for the originating query.

Although you learned how to use subqueries in Chapter 6, let's review the basic characteristics of a subquery:

* A subquery is a query (SELECT statement) inside a query.
* A subquery is normally expressed inside parentheses.
* The first query in the SQL statement is known as the outer query.
* The query inside the SQL statement is known as the inner query.
* The inner query is executed first.
* The output of an inner query is used as the input for the outer query.
* The entire SQL statement is sometimes referred to as a nested query.

In this section you will learn more about the practical use of subqueries. You already know that a subquery is based on the use of the SELECT statement to return one or more values to another query. But subqueries have a wide range of uses. For example, you can use a subquery within a SQL data manipulation language (DML) statement (INSERT, UPDATE, or DELETE) where a value or a list of values (such as multiple vendor codes or a table) is expected. Table 7.2 uses simple examples to summarize the use of SELECT subqueries in DML statements.

TABLE 7.2 SELECT SUBQUERY EXAMPLES

SELECT SUBQUERY EXAMPLES	EXPLANATION
INSERT INTO PRODUCT SELECT * FROM P;	Inserts all rows from the table P into the PRODUCT Table. Both tables must have the same attributes. The subquery returns all rows from table P.
UPDATE PRODUCT SET P_PRICE = (SELECT AVG(P_PRICE) FROM PRODUCT) WHERE V_CODE IN (SELECT V_CODE FROM VENDOR WHERE V_AREACODE = '615');	Updates the product price to the average product price, but only for the products that are provided by vendors who have an area code equal to 615. The first subquery returns the average price; the second subquery returns the list of vendors with an area code equal to 615.
DELETE FROM PRODUCT WHERE V_CODE IN (SELECT V_CODE FROM VENDOR WHERE V_AREACODE = '615');	Deletes the PRODUCT table rows that are provided by vendors with an area code equal to '615'. The subquery returns the list of vendors' codes with area code equal to 615.

Using the examples shown in Table 7.2, note that the subquery is always at the right side of a comparison or assigning expression. Also, a subquery returns either one value or multiple values. To be precise, the subquery can return:

* One single value (one column and one row). This subquery is used anywhere a single value is expected, like in the right side of a comparison expression (such as in the UPDATE example above when you assign the average price to the product's price). Obviously, when you assign a value to an attribute, that value is a single value, not a list of values. Therefore, the subquery must return only one value (one column, one row). If the query returns multiple values, the DBMS will generate an error.

- A list of values (one column and multiple rows). This type of subquery is used anywhere a list of values is expected; such as when using the IN clause (that is, when comparing the vendor code to a list of vendors). Again, in this case, there is only one column of data with multiple value instances. This type of subquery is used frequently in combination with the IN operator in a WHERE conditional expression.

- A virtual table (multicolumn, multirow set of values). This type of subquery can be used anywhere a table is expected, such as when using the FROM clause. We will see this type of query later in this chapter.

It's important to note that a subquery can return no values at all; it is a NULL. In such cases, the output of the outer query may result in an error or a null empty set, depending where the subquery is used (in a comparison, an expression or a table set).

In the following sections you will learn how to write subqueries within the SELECT statement to retrieve data from the database.

7.3.1 WHERE SUBQUERIES

The most common type of subquery uses an inner SELECT subquery on the right side of a WHERE comparison expression. For example, to find all products with a price greater than or equal to the average product price, you write the following query:

```
SELECT P_CODE, P_PRICE FROM PRODUCT
    WHERE P_PRICE >= (SELECT AVG(P_PRICE) FROM PRODUCT);
```

The output of the preceding query is shown in Figure 7.13. Note that this type of query, when used in a >, <, =, >=, or <= conditional expression, requires a subquery that returns only one single value (one column, one row). The value generated by the subquery must be of a "comparable" data type; if the attribute to the left of the comparison symbol is a character type, then the subquery must return a character string. Also, if the query returns more than a single value, the DBMS generates an error.

Subqueries can also be used in combination with joins. For example, the following query lists all the customers that ordered the product "Claw hammer":

```
SELECT DISTINCT CUS_CODE, CUS_LNAME, CUS_FNAME
    FROM CUSTOMER   JOIN INVOICE USING (CUS_CODE)
            JOIN LINE USING (INV_NUMBER)
            JOIN PRODUCT USING (P_CODE)
        WHERE P_CODE = (SELECT P_CODE FROM PRODUCT WHERE P_DESCRIPT = 'Claw hammer');
```

The result of this query is shown in Figure 7.13.

FIGURE 7.13 WHERE Subquery Examples

```
Oracle SQL*Plus
File  Edit  Search  Options  Help
SQL> SELECT P_CODE, P_PRICE FROM PRODUCT
  2  WHERE P_PRICE >= (SELECT AVG(P_PRICE) FROM PRODUCT);

P_CODE         P_PRICE
----------  ----------
11QER/31       109.99
2232/QTY       109.92
2232/QWE        99.87
89-WRE-Q       256.99
WR3/TT3        119.95

SQL> SELECT DISTINCT CUS_CODE, CUS_LNAME, CUS_FNAME
  2  FROM CUSTOMER JOIN INVOICE USING (CUS_CODE)
  3              JOIN LINE USING (INV_NUMBER)
  4              JOIN PRODUCT USING (P_CODE)
  5  WHERE P_CODE = (SELECT P_CODE FROM PRODUCT WHERE P_DESCRIPT = 'Claw hammer');

  CUS_CODE CUS_LNAME       CUS_FNAME
---------- ---------------- ----------------
     10011 Dunne           Leona
     10014 Orlando         Myron

SQL> |
```

In the preceding example, the inner query finds the P_CODE for the product "Claw hammer." The P_CODE is then used to restrict the selected rows to only those where the P_CODE in the LINE table matches the P_CODE for "Claw hammer." Note that the previous query could have been written this way:

SELECT DISTINCT CUS_CODE, CUS_LNAME, CUS_FNAME
 FROM CUSTOMER JOIN INVOICE USING (CUS_CODE)
 JOIN LINE USING (INV_NUMBER)
 JOIN PRODUCT USING (P_CODE)
 WHERE P_DESCRIPT = 'Claw hammer';

But what happens if this query encounters the "Claw hammer" string in more than one product description? You would get an error message. In order to compare one value to a list of values, you must use an IN operand as shown in the next section.

7.3.2 IN Subqueries

What would you do if you wanted to find all customers who have purchased a "hammer" or any kind of saw or saw blade? Note that the product table has two different types of hammers: "Claw hammer" and "Sledge hammer." Also, note that there are multiple occurrences of products that contain "saw" in their product description. For example, there are saw blades, jigsaws, and so on. In such cases, you would need to compare the P_CODE not to one product code (single value), but to a list of product code values. When you want to compare a single attribute to a list of values, you use the IN operator. If the P_CODE values are not known beforehand, but they can be derived using a query, then you must use an IN subquery. The following example lists all customers that have purchased hammers or saws or saw blades.

SELECT DISTINCT CUS_CODE, CUS_LNAME, CUS_FNAME
 FROM CUSTOMER JOIN INVOICE USING (CUS_CODE)
 JOIN LINE USING (INV_NUMBER)
 JOIN PRODUCT USING (P_CODE)
 WHERE P_CODE IN (SELECT P_CODE FROM PRODUCT
 WHERE P_DESCRIPT LIKE '%hammer%'
 OR P_DESCRIPT LIKE '%saw%');

The result of this query is shown in Figure 7.14.

FIGURE 7.14 IN SUBQUERY EXAMPLE

7.3.3 HAVING SUBQUERIES

Just as you can use subqueries with the WHERE clause, you can use a subquery with a HAVING clause. Remember from Chapter 6 that the HAVING clause is used to restrict the output of a GROUP BY query by applying a conditional criteria to the grouped rows. For example, to list all products with the total quantity sold greater than the average quantity sold, you will write the following query:

SELECT P_CODE, SUM(LINE_UNITS)
 FROM LINE
 GROUP BY P_CODE
 HAVING SUM(LINE_UNITS) > (SELECT AVG(LINE_UNITS) FROM LINE);

The result of this query is shown in Figure 7.15.

FIGURE 7.15 HAVING SUBQUERY EXAMPLE

7.3.4 MULTIROW SUBQUERY OPERATORS: ANY AND ALL

So far, you have learned that you must use an IN subquery when you need to compare a value to a list of values. But the IN subquery uses an equality operator; that is, it only selects those rows that match (are equal to) at least one of the values in the list. What happens if you need to do an inequality comparison (> or <) of one value to a list of values?

For example, suppose you want to know what products have a product cost that is greater than all individual product costs for products provided by vendors from Florida.

```
SELECT P_CODE, P_ONHAND*P_PRICE
    FROM PRODUCT
        WHERE P_ONHAND*P_PRICE > ALL (SELECT P_ONHAND*P_PRICE
                                        FROM PRODUCT
                                            WHERE V_CODE IN (SELECT V_CODE
                                                FROM VENDOR
                                                    WHERE V_STATE = 'FL'));
```

The result of this query is shown in Figure 7.16.

FIGURE 7.16 MULTIROW SUBQUERY OPERATOR EXAMPLE

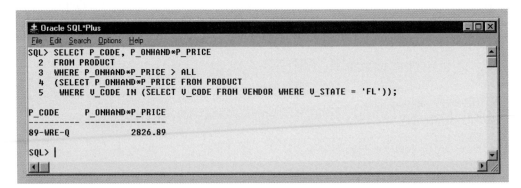

As you examine the query and its output in Figure 7.16, it's important to note the following points:

- The query is a typical example of a nested query.
- The query has one outer SELECT statement with a SELECT subquery (we will call it sqA) containing a second SELECT subquery (we will call it sqB).
- The last SELECT subquery (sqB) is executed first and returns a list of all vendors from Florida.
- The first SELECT subquery (sqA) uses the output of the SELECT subquery (sqB). The sqA subquery returns the list of product costs for all products provided by vendors from Florida.
- The use of the ALL operator allows you to compare a single value (P_ONHAND*P_PRICE) with a list of values returned by the first subquery (sqA) using a comparison operator other than equals.
- For a row to appear in the result set, it has to meet the criterion P_ONHAND*P_PRICE > ALL the individual values returned by the subquery sqA. The values returned by sqA are a list of product costs. In fact, "greater than ALL" is equivalent to greater than the highest product cost of the list. In the same way, a condition of "less than ALL" is equivalent to less than the lowest product cost of the list.

Another powerful operator is the ANY multirow operator (near-cousin of the ALL multirow operator). The ANY operator will allow you to compare a single value to a list of values selecting only the rows for which the inventory cost is greater than any value of the list, or less than any value of the list. You could use the equal to ANY operator, which would be the equivalent of the IN operator.

7.3.5 FROM SUBQUERIES

So far we have seen how the SELECT statement uses subqueries within WHERE, HAVING, and IN statements, and how the ANY or ALL operators are used for multirow subqueries. In all those cases, the subquery was part of a conditional expression and it always appeared at the right side of the expression. In this section you will learn how to use subqueries in the FROM clause.

As you already know, the FROM clause specifies the table(s) from which the data will be drawn. Because the output of a SELECT statement is another table (or more precisely a "virtual" table), you could use a SELECT subquery in the FROM clause. For example, assume that you want to know all customers who have purchased products 13-Q2/P2 *and* 23109-HB. All product purchases are stored in the LINE table. It is easy to find out who purchased any given product just by searching the P_CODE attribute in the LINE table. But, in this case, you want to know all customers who purchased both products, not just one. You could write the following query:

SELECT DISTINCT CUSTOMER.CUS_CODE, CUSTOMER.CUS_LNAME
 FROM CUSTOMER,
 (SELECT INVOICE.CUS_CODE
 FROM INVOICE NATURAL JOIN LINE WHERE P_CODE = '13-Q2/P2') CP1,
 (SELECT INVOICE.CUS_CODE
 FROM INVOICE NATURAL JOIN LINE WHERE P_CODE = '23109-HB') CP2
 WHERE CUSTOMER.CUS_CODE = CP1.CUS_CODE
 AND CP1.CUS_CODE = CP2.CUS_CODE;

The result of this query is shown in Figure 7.17.

FIGURE 7.17 FROM SUBQUERY EXAMPLE

As you examine Figure 7.17, note that the first subquery returns all customers who purchased product 13-Q2/P2, while the second subquery returns all customers who purchased product 23109-HB. So, in this FROM subquery you are joining the CUSTOMER table with two virtual tables. The join condition selects only the rows with matching CUS_CODE values in each table (base or virtual).

In the previous chapter you learned that a view is also a virtual table, therefore, you can use a view name anywhere a table is expected. So you could create two views: one listing all customers who purchased product 13-Q2/P2 and another one listing all customers who purchased product 23109-HB. Doing so, you could write the query as:

CREATE VIEW CP1 AS
 SELECT INVOICE.CUS_CODE FROM INVOICE NATURAL JOIN LINE WHERE P_CODE = '13-Q2/P2';

CREATE VIEW CP2 AS
 SELECT INVOICE.CUS_CODE FROM INVOICE NATURAL JOIN LINE WHERE P_CODE = '23109-HB';
SELECT DISTINCT CUS_CODE, CUS_LNAME
 FROM CUSTOMER NATURAL JOIN CP1 NATURAL JOIN CP2;

You may be tempted to speculate that the above query could also be written using the following syntax:

SELECT CUS_CODE, CUS_LNAME
 FROM CUSTOMER NATURAL JOIN INVOICE NATURAL JOIN LINE
 WHERE P_CODE = '13-Q2/P2' AND P_CODE = '23109-HB';

But, if you examine this query carefully, you will note that a P_CODE cannot be equal to two different values at the same time! Therefore, this query will not return any rows.

7.3.6 ATTRIBUTE LIST SUBQUERIES

The SELECT statement uses the attribute list to indicate what columns to project in the resulting set. Those columns can be attributes of base tables, computed attributes, or the result of an aggregate function. The attribute list can also include a subquery expression, also known as an inline subquery. A subquery in the attribute list must return one single value, otherwise an error code is raised. For example, a simple inline query can be used to list the difference between each product's price and the average product price:

SELECT P_CODE, P_PRICE, (SELECT AVG(P_PRICE) FROM PRODUCT) AS AVGPRICE,
 P_PRICE – (SELECT AVG(P_PRICE) FROM PRODUCT) AS DIFF
 FROM PRODUCT;

Figure 7.18 shows the result of this query.

FIGURE 7.18 INLINE SUBQUERY EXAMPLE

```
SQL> SELECT P_CODE, P_PRICE, (SELECT AVG(P_PRICE) FROM PRODUCT) AS AVGPRICE,
  2          P_PRICE-(SELECT AVG(P_PRICE) FROM PRODUCT) AS DIFF
  3  FROM PRODUCT;

P_CODE        P_PRICE    AVGPRICE        DIFF
----------  ----------  ----------  ----------
11QER/31      109.99    56.42125    53.56875
13-Q2/P2       14.99    56.42125   -41.43125
14-Q1/L3       17.49    56.42125   -38.93125
1546-QQ2       39.95    56.42125   -16.47125
1558-QW1       43.99    56.42125   -12.43125
2232/QTY      109.92    56.42125    53.49875
2232/QWE       99.87    56.42125    43.44875
2238/QPD       38.95    56.42125   -17.47125
23109-HB        9.95    56.42125   -46.47125
23114-AA       14.4     56.42125   -42.02125
54778-2T        4.99    56.42125   -51.43125
89-WRE-Q      256.99    56.42125   200.56875
PVC23DRT        5.87    56.42125   -50.55125
SM-18277        6.99    56.42125   -49.43125
SW-23116        8.45    56.42125   -47.97125
WR3/TT3       119.95    56.42125    63.52875

16 rows selected.

SQL>
```

As you examine Figure 7.18, note that the inline query output returns one single value (the average product's price) and that value is the same in every row. Note also that the query used the full expression instead of the column aliases when computing the difference. In fact, if you try to use the alias in the difference expression, you will get an error message. The column alias cannot be used in computations in the attribute list if the alias is defined in the same attribute list. This DBMS requirement is due to the way the DBMS parses and executes queries.

Another example will help you understand the use of attribute list subqueries and column aliases. For example, suppose you want to know the product code, the total sales by product, and the contribution by employee of each product sales. To get the sales by product you only need to use the LINE table. To compute the contribution by employee you need to know the number of employees (from the EMPLOYEE table). As you study the tables' structure you can see that the LINE and EMPLOYEE tables do not share a common attribute. In fact, you don't need a common attribute. You only need to know the total number of employees, not the total employees related to each product. So, to answer this query you would write the following code:

```
SELECT P_CODE, SUM(LINE_UNITS*LINE_PRICE) AS SALES,
       (SELECT COUNT(*) FROM EMPLOYEE) AS ECOUNT,
         SUM(LINE_UNITS*LINE_PRICE)/(SELECT COUNT(*) FROM EMPLOYEE) AS CONTRIB
  FROM LINE
       GROUP BY P_CODE;
```

The result of this query is shown in Figure 7.19.

FIGURE 7.19 **ANOTHER EXAMPLE OF AN INLINE SUBQUERY**

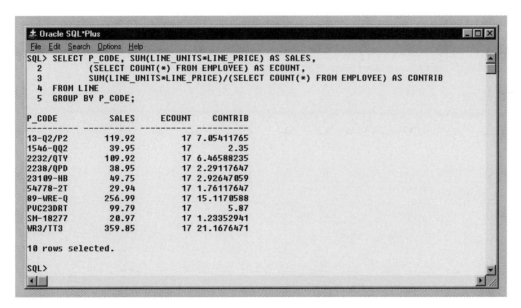

As you can see in Figure 7.19, the number of employees remains the same for each row in the result set. The use of this type of subquery is limited to certain instances where you need to include data from other tables that are not directly related to main table(s) in the query. The value will remain the same for each row, like a constant in a programming language (although you will learn another use of inline subqueries later in the "Correlated Subqueries" section). Note that you cannot use an alias in the attribute list to write the expression that computes the contribution per employee.

Another way to write this same query using column aliases requires the use of a subquery in the FROM clause, as shown next:

```
SELECT P_CODE, SALES, ECOUNT, SALES/ECOUNT AS CONTRIB
    FROM (SELECT P_CODE, SUM(LINE_UNITS*LINE_PRICE) AS SALES,
            (SELECT COUNT(*) FROM EMPLOYEE) AS ECOUNT
                FROM LINE
                    GROUP BY P_CODE);
```

In this case, you are actually using two subqueries. The subquery in the FROM clause executes first and returns a virtual table with three columns: P_CODE, SALES, and ECOUNT. The FROM subquery contains an inline subquery that returns the number of employees as ECOUNT. Because the outer query is receiving the output of the inner query, you could now refer to the columns in the outer subquery using the column aliases.

> **N O T E**
>
> The previous query would run fine in Oracle9*i* but not in Oracle8*i*.

7.3.7 CORRELATED SUBQUERIES

Until now, all subqueries you have learned execute independently. That is, each subquery in a command sequence executes in a serial fashion, one after another. The inner subquery executes first; its output is used by the outer query which then executes, until the last outer query executes (the first SQL statement in the code).

In contrast, a **correlated subquery** is a subquery that executes once for each row in the outer query. This process is similar to the typical nested loop in a programming language. For example:

```
FOR X = 1 TO 2
  FOR Y = 1 TO 3
    PRINT "X = "X, "Y = "Y
  END
END
```

will yield the output

```
X = 1      Y = 1
X = 1      Y = 2
X = 1      Y = 3
X = 2      Y = 1
X = 2      Y = 2
X = 2      Y = 3
```

Note that the outer loop X = 1 TO 2 begins the process by setting X = 1, then the inner loop Y = 1 TO 3 is completed for each X outer loop value. The relational DBMS uses the same sequence to produce correlated subquery results:

1. It initiates the outer query.

2. For each row of the outer query result set, it executes the inner query, by passing the outer row to the inner query.

This process is the opposite of the subqueries we have seen so far. This query is called a *correlated* subquery because the inner query is *related* to the outer query because the inner query references a column of the outer subquery.

To see the correlated subquery in action, suppose you want to know all product sales in which the units sold value is greater than the average units sold value for *that* product (as opposed to the average for *all* products). In this case, the following procedure must be completed:

1. Compute the average units sold value for a product.

2. Compare the average computed in Step 1 to the units sold in each sale row, and then select only the rows in which the number of units sold is greater.

Note that the following correlated query completes the preceding two-step process:

```
SELECT INV_NUMBER, P_CODE, LINE_UNITS
    FROM LINE LS
        WHERE LS.LINE_UNITS > (SELECT AVG(LINE_UNITS) FROM LINE LA
                                WHERE LA.P_CODE = LS.P_CODE);
```

The first example in Figure 7.20 shows the result of this query.

FIGURE 7.20 CORRELATED SUBQUERY EXAMPLES

As you examine the top query and its result in Figure 7.20, note that we are using the LINE table more than once. Therefore, we must use table aliases. In this case, the inner query computes the average units sold of the product that matches the P_CODE of the outer query P_CODE. That is, the inner query runs once using the first product code found in the (outer) LINE table and returns the average sale for that product. If the units sold in that (outer) LINE row is greater than the average computed, then the row is added to the output. Then the inner query runs again, this time using the second product code found in the (outer) LINE table, and the process repeats until the inner query has run for all rows in the (outer) LINE table. In this case, the inner query will be repeated as many times as there are rows in the outer query.

To verify the results and to give you an example of how you can combine subqueries, you can add a correlated inline subquery to the previous query. This correlated inline subquery will show the average units sold column for each product. (See the second query and its results in Figure 7.20.) As you can see, this new query contains a correlated inline subquery that computes the average units sold per each product. This way, not only do you get an answer, but you can verify that the answer is correct.

Correlated subqueries can also be used with the EXISTS special operator. For example, suppose you want to know all customers who have placed an order lately. In this case, you could use a correlated subquery like the first one shown in Figure 7.21:

SELECT CUS_CODE, CUS_LNAME, CUS_FNAME
 FROM CUSTOMER
 WHERE EXISTS (SELECT CUS_CODE FROM INVOICE
 WHERE INVOICE.CUS_CODE = CUSTOMER.CUS_CODE);

FIGURE 7.21 EXISTS CORRELATED SUBQUERY EXAMPLES

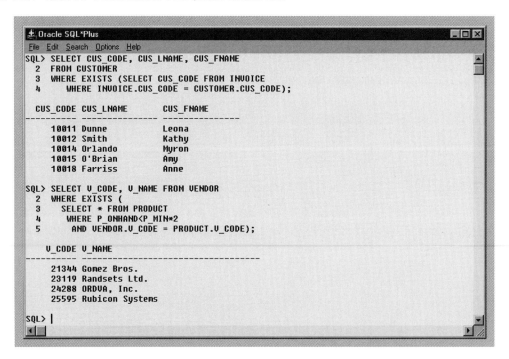

The second example of an EXISTS correlated subquery in Figure 7.21 will help you understand the usage of correlated queries. For example, suppose that you want to know what vendors you must contact to start ordering products that are approaching the minimum quantity on hand value. In particular, you want to know the vendor code and name for vendors of products having a quantity on hand that is less than double the minimum quantity. The query that answers that question is shown next:

SELECT V_CODE, V_NAME FROM VENDOR
 WHERE EXISTS (
 SELECT * FROM PRODUCT
 WHERE P_ONHAND < P_MIN * 2
 AND VENDOR.V_CODE = PRODUCT.V_CODE);

As you examine the second query in Figure 7.21, note that:

1. The inner correlated subquery runs using the first vendor.

2. If there are any products that match the condition (quantity on hand is less than double the minimum quantity), the vendor code and name are listed in the output.

3. Next, the correlated subquery runs using the second vendor, and the process repeats itself until all vendors are used.

7.4 SQL FUNCTIONS

The data in databases are the basis of critical business information. Generating information from data often requires many data manipulation procedures. Sometimes, such data manipulation involves the decomposition of data elements. For example, an employee's date of birth can be subdivided into a day, a month, and a year. A product manufacturing code, for example SE-05-2-09-1234-1-3/12/04-19:26:48, may be designed to record the manufacturing region, plant, shift, production line, employee number, date, and time. For years, conventional programming languages have had special functions that enabled programmers to perform data transformations like these data decompositions. If you know a modern programming language, it's very likely that the SQL functions we'll introduce in this section will look familiar.

SQL functions are very useful tools. You'll need to use functions when you want to list all employees ordered by year of birth, or when your marketing department wants you to generate a list of all customers ordered by zip code and the first three digits of their telephone number. In both of these cases, you'll need to use data elements that are not present as such in the database, but by using a SQL function they can be derived from an existing attribute. Functions always use a numerical, date, or string value. The value may be part of the command itself (a constant or literal) or it may be an attribute located in a table. Therefore, a function may appear anywhere in a SQL statement where a value or an attribute can be used.

There are many types of SQL functions, such as arithmetic, trigonometric, string, date, and time functions. This section will not explain all these types of functions in detail, but instead gives you a brief overview of the most useful ones.

> **N O T E**
>
> Although the main DBMS vendors support the SQL functions covered here, the syntax or degree of support may differ. In fact, DBMS vendors invariably add their own functions to their products to lure new customers. The functions covered in this section represent just a small portion of functions supported by your DBMS. Please read your DBMS SQL reference manual for a complete list of available functions.

7.4.1 DATE AND TIME FUNCTIONS

All SQL-standard DBMSs support date and time functions. All date functions take one parameter (of a date or character data type) and return a value (character, numeric, or date type). Unfortunately, date/time data types are implemented differently by different DBMS vendors. This problem occurs because the ANSI SQL standard defines the support of date data types but does not say how those data types are to be stored; instead, it lets the vendor deal with that issue. This is in keeping with the separation of the logical nature of the data from the physical details of the implementation.

Because date/time functions differ from vendor to vendor, this section will cover basic date/time functions for MS Access/SQL Server and for Oracle. Table 7.3 shows a list of selected MS Access/SQL Server date/time functions.

TABLE 7.3 SELECTED MS ACCESS/SQL SERVER DATE/TIME FUNCTIONS

FUNCTION	EXAMPLE(S)
YEAR Returns a four-digit year. Syntax: YEAR(date_value)	Lists all employees born in 1962: SELECT EMP_LNAME, EMP_FNAME, EMP_DOB, YEAR(EMP_DOB) AS YEAR FROM EMPLOYEE WHERE YEAR(EMP_DOB) = 1966;
MONTH Returns a two-digit month code. Syntax: MONTH(date_value)	Lists all employees born in November: SELECT EMP_LNAME, EMP_FNAME, EMP_DOB, MONTH(EMP_DOB) AS MONTH FROM EMPLOYEE WHERE MONTH(EMP_DOB) = 11;
DAY Returns the number of the day Syntax: DAY(date_value)	List all employees born on the 14th day of the month: SELECT EMP_LNAME, EMP_FNAME, EMP_DOB, DAY(EMP_DOB) AS DAY FROM EMPLOYEE WHERE DAY(EMP_DOB) = 14;
DATE() Returns today's date	List how many days are left until Christmas: SELECT #25-Dec-2004# - DATE(); Note two features: • There is no FROM clause (this is acceptable in MS Access). • The Christmas date is enclosed in # signs because you are doing date arithmetic.

Table 7.4 show the equivalent date/time functions as used in Oracle. Note that Oracle uses the same function (TO_CHAR) to extract the different parts of a date. Also, there is another function (TO_DATE) that is used to convert character strings to a valid Oracle date format that can be used in date arithmetic.

TABLE 7.4 SELECTED ORACLE DATE/TIME FUNCTIONS

FUNCTION	EXAMPLE(S)
TO_CHAR Returns a character string or a formatted string from a date value. Syntax: TO_CHAR(date_value, fmt) fmt = format used, can be: MONTH: name of month MON: three-letter month name MM: two-digit month D: number for day of week DD: number day of the month DAY: name of day of week YYYY: four-digit year value YY: two-digit year value	List all employees born in 1962: SELECT EMP_LNAME, EMP_FNAME, EMP_DOB, TO_CHAR(EMP_DOB,'YYYY') AS YEAR FROM EMPLOYEE WHERE TO_CHAR(EMP_DOB,'YYYY') = '1966'; Lists all employees born in November: SELECT EMP_LNAME, EMP_FNAME, EMP_DOB, TO_CHAR(EMP_DOB,'MM') AS MONTH FROM EMPLOYEE WHERE TO_CHAR(EMP_DOB,'MM') = '11'; List all employees born on the 14th day of the month: SELECT EMP_LNAME, EMP_FNAME, EMP_DOB, TO_CHAR(EMP_DOB,'DD') AS DAY FROM EMPLOYEE WHERE TO_CHAR(EMP_DOB,'DD') = '14';

TABLE 7.4 SELECTED ORACLE DATE/TIME FUNCTIONS (CONTINUED)

FUNCTION	EXAMPLE(S)
TO_DATE Returns a date value using a character string and a date format mask. Also used to translate a date between formats. Syntax: TO_DATE(char_value, fmt) fmt = format used, can be: MONTH: name of month MON: three-letter month name MM: two-digit month D: number for day of week DD: number day of the month DAY: name of day of week YYYY: four-digit year value YY: two-digit year value	List the approximate ages of the employees on the company's 10th anniversary date (11/25/2004): SELECT EMP_LNAME, EMP_FNAME, EMP_DOB, '11/25/2004'AS ANIV_DATE, (TO_DATE('11/25/1994','MM/DD/YYYY') – EMP_DOB)/365 AS YEARS FROM EMPLOYEE ORDER BY YEARS; Note the following: • '11/25/2004' is just a text string, not a date. • The TO_DATE function translates the text string to a valid Oracle date used in date arithmetic. How many days between Thanksgiving and Christmas 2004? SELECT TO_DATE('2004/12/25','YYYY/MM/DD') - TO_DATE('NOVEMBER 25, 2004','MONTH DD, YYYY') FROM DUAL; Note the following: • The TO_DATE function translates the text string to a valid Oracle date used in date arithmetic. • DUAL is an Oracle's pseudo table used only for cases where a table is not really needed.
SYSDATE Returns today's date.	List how many days are left until Christmas: SELECT TO_DATE('25-Dec-2004','DD-MON-YYYY') – SYSDATE FROM DUAL; Notice two things: • DUAL is an Oracle's pseudo table used only for cases where a table is not really needed. • The Christmas date is enclosed in a TO_DATE function to translate the date to a valid date format.
ADD_MONTHS Adds a number of months to a date. Useful to add months or years to a date. Syntax: ADD_MONTHS(date_value, n) n = number of months	List all products with their expiration date (two years from the purchase date): SELECT P_CODE, P_INDATE, ADD_MONTHS(P_INDATE,24) FROM PRODUCT ORDER BY ADD_MONTHS(P_INDATE,24);
LAST_DAY Returns the date of the last day of the month given in a date. Syntax: LAST_DAY(date_value)	List all employees that were hired within the last seven days of a month: SELECT EMP_LNAME, EMP_FNAME, EMP_HIRE_DATE FROM EMPLOYEE WHERE EMP_HIRE_DATE >= LAST_DAY(EMP_HIRE_DATE)-7;

7.4.2 NUMERIC FUNCTIONS

Numeric functions can be grouped in many different ways, such as algebraic, trigonometric, logarithmic, and so on. In this section you will learn two very useful functions. Do not confuse the SQL aggregate functions you saw in the previous chapter with the numeric functions in this section. The first group operates over a set of values (multiple rows, hence the name of aggregate functions), while the numeric functions covered here operate over a single row value. Numeric functions take one numeric parameter and return one value. Table 7.5 shows a selected group of numeric functions available in an Oracle DBMS.

TABLE 7.5 SELECTED ORACLE NUMERIC FUNCTIONS

FUNCTION	EXAMPLE(S)
ABS Returns the absolute value of a number. Syntax: ABS(numeric_value)	List absolute values: SELECT 1.95, -1.93, ABS(1.95), ABS(-1.93) FROM DUAL;
ROUND Rounds a value to a specified precision (number of digits). Syntax: ROUND(numeric_value, p) p = precision	List the product prices rounded to one and zero decimal places: SELECT P_CODE, P_PRICE, ROUND(P_PRICE,1) AS PRICE1, ROUND(P_PRICE,0) AS PRICE0 FROM PRODUCT;
TRUNC Truncates a value to a specified precision (number of digits). Syntax: TRUNC(numeric_value, p) p = precision	List the product price rounded to one and zero decimal places and truncated: SELECT P_CODE, P_PRICE, ROUND(P_PRICE,1) AS PRICE1, ROUND(P_PRICE,0) AS PRICE0, TRUNC(P_PRICE,0) AS PRICEX FROM PRODUCT;
CEIL / FLOOR Returns the smallest integer greater than or equal to a number, or returns the largest integer equal to or less than a number, respectively. Syntax: CEIL(numeric_value) FLOOR(numeric_value)	List the product price, smallest integer greater than or equal to the product price, and the largest integer equal to or less than the product price: SELECT P_PRICE, CEIL(P_PRICE), FLOOR(P_PRICE) FROM PRODUCT;

7.4.3 STRING FUNCTIONS

String manipulations are among the most-used functions in programming. If you have ever created a report using any programming language, you know the importance of properly concatenating strings of characters, printing names in uppercase, or knowing the length of a given attribute. Table 7.6 shows a subset of useful string manipulation functions in Oracle.

TABLE 7.6 SELECTED ORACLE STRING FUNCTIONS

FUNCTION	EXAMPLE(S)
\|\| Concatenates data from two different character columns and returns a single column. Syntax: strg_value \|\| strg_value	List all employee names (concatenated): SELECT EMP_LNAME \|\| ', ' \|\| EMP_FNAME AS NAME 　　FROM EMPLOYEE;
UPPER / LOWER Returns a string in all capitals or all lowercase. Syntax: UPPER(strg_value) LOWER(strg_value)	List all employee names in all capitals (concatenated): SELECT UPPER(EMP_LNAME) \|\| ', ' \|\| UPPER(EMP_FNAME) AS NAME 　　FROM EMPLOYEE; List all employee names in all lowercase (concatenated): SELECT LOWER(EMP_LNAME) \|\| ', ' \|\| LOWER(EMP_FNAME) AS NAME 　　FROM EMPLOYEE;
SUBSTR Returns a substring or part of a given string parameter. Syntax: SUBSTR(strg_value, p, l) p = start position l = length of characters	List the first three characters of all employees' phone numbers: SELECT EMP_PHONE, SUBSTR(EMP_PHONE,1,3) 　　FROM EMPLOYEE; Generate a list of employee user IDs using the first character of first name and first 7 characters of last name: SELECT EMP_FNAME, EMP_LNAME, 　　　　SUBSTR(EMP_FNAME,1,1) \|\| SUBSTR(EMP_LNAME,1,7) 　　FROM EMPLOYEE;
LENGTH Returns the number of characters in a string value. Syntax: LENGTH(strg_value)	List all employees' last names and the length of their names, ordered descended by last name length: SELECT EMP_LNAME, LENGTH(EMP_LNAME) AS NAMESIZE 　　FROM EMPLOYEE 　　　　ORDER BY NAMESIZE DESC;

7.4.4 CONVERSION FUNCTIONS

Conversion functions allow you to take a value of a given data type and convert it to the equivalent value in another data type. In Section 7.4.1, you learned about two of the basic conversion functions: TO_CHAR and TO_DATE. Note that the TO_CHAR function takes a date value and return a character string representing a day, a month, or a year. In the same way, the TO_DATE function takes a character string representing a date and returns an actual date in Oracle format.

In this section you will see how the TO_CHAR function is used to convert numbers to a formatted character string and the TO_NUMBER function is used to convert text strings to numeric values. A summary of the selected functions is shown in Table 7.7.

TABLE 7.7 SELECTED ORACLE CONVERSION FUNCTIONS

FUNCTION	EXAMPLE(S)
TO_CHAR (numeric) Returns a character string or a formatted string from a numeric value. Very useful to format numeric columns in reports. Syntax: TO_CHAR(numeric_value, fmt) fmt = format used, can be: 9 = displays a digit 0 = displays a leading zero , = displays the comma . = displays the decimal point $ = displays the dollar sign	List all product prices, quantity on hand, percent discount, and total inventory cost using formatted values: SELECT P_CODE, TO_CHAR(P_PRICE,'$999.99') AS PRICE, TO_CHAR(P_ONHAND,'9,999.99') AS QUANTITY, TO_CHAR(P_DISCOUNT, '0.99') AS DISC, TO_CHAR(P_PRICE*P_ONHAND, '$99,999.99') AS TOTAL_COST FROM PRODUCT;
TO_CHAR (date) Returns a character string or a formatted character string from a date value. Syntax: TO_CHAR(date_value, fmt) fmt = format used, can be: MONTH: name of month MON: three-letter month name MM: two-digit month D: number for day of week DD: number day of the month DAY: name of day of week YYYY: four-digit year value YY: two-digit year value	List all employees' dates of birth using different date formats: SELECT EMP_LNAME, EMP_DOB, TO_CHAR(EMP_DOB, 'DAY, MONTH DD, YYYY') AS "DATE OF BIRTH" FROM EMPLOYEE; SELECT EMP_LNAME, EMP_DOB, TO_CHAR(EMP_DOB, 'YYYY/MM/DD') AS "DATE OF BIRTH" FROM EMPLOYEE;
TO_NUMBER Returns a formatted number from a character string using a given format. Syntax: TO_NUMBER(char_value, fmt) fmt = format used, can be: 9 = displays a digit 0 = displays a leading zero , = displays the comma . = displays the decimal point $ = displays the dollar sign B = leading blank S = leading sign MI = trailing minus sign	This function is useful to convert text strings to numeric values when importing data to a table from another source in text format. For example, the query shown below uses the TO_NUMBER function to convert text formatted to Oracle default numeric values using the format masks given: SELECT TO_NUMBER('-123.99', 'S999.99'), TO_NUMBER(' 99.78-','B999.99MI') FROM DUAL;
NVL Replaces a null with a string in the results of a query. Syntax: NVL(x, y) x = attribute or expression. y= value to return if x is null.	If x is null, then NVL returns y. If x is not null, then NVL returns x. The data type of the return value is always the same as the data type of x. Useful to avoid errors caused by incorrect calculation when one of the arguments is null. For example, assuming the P_DISCOUNT attribute can have null values, you would use the following expression: SELECT P_CODE, P_PRICE, P_PRICE*NVL(P_DISCOUNT,0) FROM PRODUCT;

TABLE 7.7 SELECTED ORACLE CONVERSION FUNCTIONS (CONTINUED)

FUNCTION	EXAMPLE(S)
DECODE Compares an attribute or expression with a series of values and returns an associated value or a default value if no match is found. Syntax: DECODE(e, x, y, d) e = attribute or expression x = value to compare e with. y = value to return if e = x d = default value to return if e is not equal to x.	The following example, will: • Compare V_STATE to 'CA', if the values match, it returns .08. • Compare V_STATE to 'FL', if the values match, it returns .05. • Compare V_STATE to 'TN', if the values match, it returns .085. • If there is no match, it returns 0.00 (the default value). SELECT V_CODE, V_STATE, DECODE(V_STATE,'CA', .08, 'FL', .05, 'TN', .085, 0.00) AS TAX FROM VENDOR

7.5 ORACLE SEQUENCES

If you use MS Access, you might be familiar with the Auto-number data type. In MS Access, you can use the Auto-number data type to define a column in your table that will be automatically populated with unique numeric values. In fact, if you create a table in MS Access and forget to define a primary key, MS Access will offer to create a primary key column; if you accept, you will notice that MS Access created a column named ID with an Auto-number data type. After you define a column as an Auto-number type, every time you insert a row into the table, MS Access automatically adds a value to that column; starting with 1 and increasing the value by 1 in every new row you add. Also, you cannot include that column in your INSERT statements—Access will not let you edit that value at all!

Similarly, in Oracle, you can use a "sequence" to assign values to a column on a table. But an Oracle sequence is very different from the Access Auto-number data type and deserves close scrutiny:

- Oracle sequences are an independent object in the database. (Sequences are not a data type.)
- Oracle sequences have a name and can be used anywhere a value is expected.
- Oracle sequences are not tied to a table or a column.
- Oracle sequences generate a numeric value that can be assigned to any column in any table.
- The table attribute to which you assigned a value based on a sequence can be edited and modified.
- An Oracle sequence can be created and deleted anytime.

The basic syntax to create a sequence in Oracle is:

CREATE SEQUENCE *name* [START WITH *n*] [INCREMENT BY *n*] [CACHE | NOCACHE]

where:

- *name* is the name of the sequence.
- *n* is an integer value that could be positive or negative.
- *START WITH* specifies the initial sequence value. (The default value is 1.)
- *INCREMENT BY* determines the value by which the sequence is incremented. (The default increment value is 1.) The sequence increment can be a positive or negative to enable you to create ascending or descending sequences.)
- The *CACHE* or *NOCACHE* clause indicates whether or not Oracle will preallocate sequence numbers in memory. (Oracle preallocates 20 values by default.)

For example, you could create a sequence to automatically assign values to the customer code each time a new customer is added and another sequence to automatically assign values to the invoice number each time a new invoice is added. The SQL code to accomplish these tasks is shown next:

CREATE SEQUENCE CUS_CODE_SEQ START WITH 20010 NOCACHE;
CREATE SEQUENCE INV_NUMBER_SEQ START WITH 4010 NOCACHE;

You can always check all the sequences you have created by using the following SQL command, illustrated in Figure 7.22:

SELECT * FROM USER_SEQUENCES;

FIGURE 7.22 ORACLE SEQUENCE

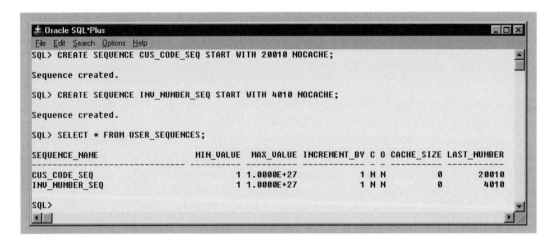

To use sequences during data entry you must use two special pseudo columns: NEXTVAL and CURRVAL. NEXTVAL retrieves the next available value from a sequence. CURRVAL retrieves the current value of a sequence. For example, you can use the following code to enter a new customer:

INSERT INTO CUSTOMER
 VALUES (CUS_CODE_SEQ.NEXTVAL, 'Connery', 'Sean', NULL, '615', '898-2007', 0.00);

The preceding SQL statement adds a new customer to the CUSTOMER table and assigns the value 20010 to the CUS_CODE attribute. Let's examine some important sequence characteristics:

- CUS_CODE_SEQ.NEXTVAL always retrieves the next available value from the sequence.

- Each time you use NEXTVAL, the sequence is incremented.

- Once a sequence value is used (through NEXTVAL), it cannot be used again. If, for some reason, your SQL statement rolls back, *the sequence value does not roll back*. If you issue another SQL statement (with another NEXTVAL), the next available sequence value will be returned to the user—it will look like the sequence skips a number.

- You can always issue an INSERT statement without using the sequence.

CURRVAL retrieves the current value of a sequence, that is, the last sequence number used, which was generated with a NEXTVAL. You cannot use CURRVAL unless a NEXTVAL has been previously issued in the same session. The main use for CURRVAL is to enter rows in dependent tables. For example, the INVOICE and LINE tables in the **Ch07_SaleCo** database are related in a one-to-many relationship through the INV_NUMBER attribute. You can use the INV_NUMBER_SEQ sequence to automatically generate invoice numbers. Then,

using CURRVAL, you can get the latest INV_NUMBER used and assign it to the related INV_NUMBER foreign key attribute in the LINE table. For example:

```
INSERT INTO INVOICE
    VALUES (INV_NUMBER_SEQ.NEXTVAL, 20010, SYSDATE);
INSERT INTO LINE
    VALUES (INV_NUMBER_SEQ.CURRVAL, 1,'13-Q2/P2', 1, 14.99);
INSERT INTO LINE
    VALUES (INV_NUMBER_SEQ.CURRVAL, 2,'23109-HB', 1, 9.95);
COMMIT;
```

The results are shown in Figure 7.23.

FIGURE 7.23 ORACLE SEQUENCE EXAMPLES

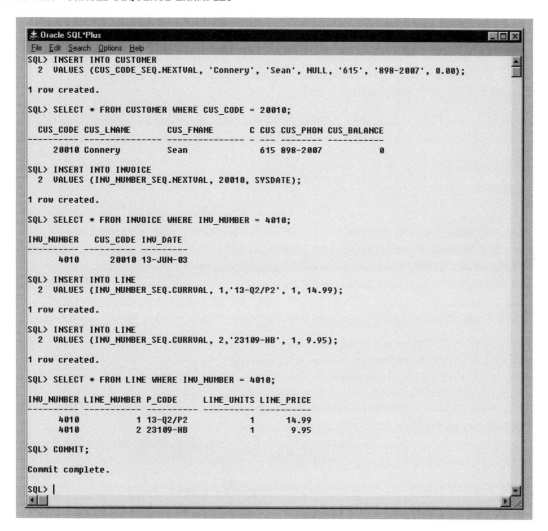

In the example shown in Figure 7.23, INV_NUMBER_SEQ.NEXTVAL retrieves the next available sequence number (4010) and assigns it to the INV_NUMBER column in the INVOICE table. Note also the use of the SYSDATE attribute to automatically insert the current date in the INV_DATE attribute. Next, the following two INSERT statements add the products being sold to the LINE table. In this case, INV_NUMBER_SEQ.CURRVAL refers to the last-used INV_NUMBER_SEQ sequence number (4010). In this way, the relationship between INVOICE and LINE is established automatically. The COMMIT statement at the end of the command sequence

makes the changes permanent. Of course, you could also issue a ROLLBACK, in which case the rows you inserted in INVOICE and LINE tables would be rolled back (but the sequence number would not be rolled back). Once you use a sequence number (with NEXTVAL) there is no way to re-use it! This "no re-use" characteristic is designed to guarantee that the sequence will always generate unique values.

Remember these points when you think about sequences:

- The use of sequences is optional. You can always enter the values manually.

- A sequence is not associated with a table. If you recall the examples in Figure 7.23, we created two distinct sequences, one for customer code values and one for invoice number values, but you could have created just one sequence and used it to generate unique values for both tables.

Finally, you can always drop a sequence from a database with a DROP SEQUENCE command. For example, to drop the sequences we created earlier, you would type:

DROP SEQUENCE CUS_CODE_SEQ;
DROP SEQUENCE INV_NUMBER_SEQ;

Dropping a sequence does not delete the values you assigned to table attributes (CUS_CODE and INV_NUMBER); it only deletes the sequence object from the database. The *values* you assigned to the table columns (CUS_CODE and INV_NUMBER) remain in the database.

Because we will use the CUSTOMER and INVOICE tables in subsequent examples, we want to keep the original data set. Therefore, we'll delete the customer, invoice, and line rows we just added by using the following commands:

DELETE FROM INVOICE WHERE INV_NUMBER = 4010;
DELETE FROM CUSTOMER WHERE CUS_CODE = 20010;
COMMIT;

These commands delete the recently added invoice with all the invoice line rows associated with the invoice (the LINE table's INV_NUMBER foreign key was defined with the ON DELETE CASCADE option) and the recently added customer. The COMMIT statement saves all changes to permanent storage.

N O T E

At this point, you'll need to re-create the CUS_CODE_SEQ and INV_NUMBER_SEQ sequences, as they will be used again later in the chapter. Enter:
CREATE SEQUENCE CUS_CODE_SEQ START WITH 20010 NOCACHE;
CREATE SEQUENCE INV_NUMBER_SEQ START WITH 4010 NOCACHE;

7.6 UPDATABLE VIEWS

In Chapter 6, you learned how to create a view and you learned why and how views were used. We will now take a look at how views can be made to serve common data management tasks executed by database administrators.

One of the most common operations in production database environments is using batch update routines to update a master table attribute (field) with transaction data. As the name implies, a **batch update routine** pools multiple transactions into a single batch to update a master table field *in a single operation*. For example, a batch update routine is commonly used to update a product's quantity on hand based on summary sales transactions. Such routines are typically run as overnight batch jobs to update the quantity on hand of products in inventory by entering sales transactions performed in remote areas, during periods when the system was offline, by traveling salespersons, and so on.

To demonstrate a batch update routine, let's begin by defining the master product table (PRODMASTER) and the product monthly sales totals table (PRODSALES) shown in Figure 7.24. As you examine the tables in Figure 7.24, note that there is a 1:1 relationship between the two tables.

ADDITIONAL CD CONTENT

For MS Access users, the PRODMASTER and PRODSALES tables are located in the **Ch07_UV** database, which is located on the student CD in the **\Databases\Student\Ch07** folder.

FIGURE 7.24 **THE PRODMASTER AND PRODSALES TABLES**

ADDITIONAL CD CONTENT

For Oracle users, all SQL commands you see in this section are located in script files on the student CD in the **\SQL\Ch07\updatableviews** folder. After you locate the script files (**uv-01.sql** through **uv-04.sql**), you can copy and paste the command sequences into your SQL*Plus program.

Using the tables in Figure 7.24, let's update the PRODMASTER table by subtracting the PRODSALES table's product monthly sales quantity (PS_QTY) from the PRODMASTER table's PROD_QOH. To produce the required update, the update query would be written like this:

UPDATE PRODMASTER, PRODSALES
 SET PRODMASTER.PROD_QOH = PROD_QOH - PS_QTY
 WHERE PRODMASTER.PROD_ID = PRODSALES.PROD_ID;

Note that this update statement reflects the following sequence of events:

- Join the PRODMASTER and PRODSALES tables.
- Update the PROD_QOH attribute in each row of the PRODMASTER table with the matching PROD_ID in the PRODSALES table.

To be used in a batch update, the PRODSALES data must be stored in a base table, rather than in a view. This query will work fine in Microsoft Access, but Oracle will return the error message shown in Figure 7.25.

Oracle produced the error message because Oracle expects to find a single table name in the UPDATE statement. In fact, if you are using Oracle, you cannot join tables in the UPDATE statement. To solve this problem, you will have to create an *updatable* view. As its name suggests, an **updatable view** is a view that can be used to update attributes in the base table(s) that are used in the view. You must realize that *not all views are updatable*. Actually, there are several restrictions that govern updatable views, and some of them are vendor-specific.

FIGURE 7.25 THE ORACLE UPDATE ERROR MESSAGE

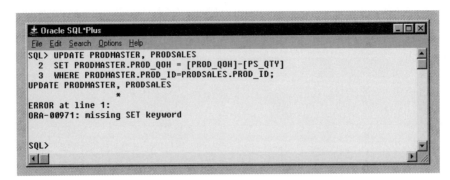

> **NOTE**
>
> Keep in mind that our examples are generated in Oracle9i. To see what restrictions are placed on updatable views by the DBMS you are using, we recommend that you check the appropriate DBMS manuals.

The most common updatable view restrictions are:

- GROUP BY expressions or aggregate functions cannot be used in the updatable views.
- You cannot use set operators such as UNION, INTERSECT, MINUS, and so on.
- Most restrictions are based on the use of JOINs or group operators in views.

To meet the Oracle limitations, we have created an updatable view named PSVUPD, shown in Figure 7.26.

FIGURE 7.26 CREATING AN UPDATABLE VIEW IN ORACLE

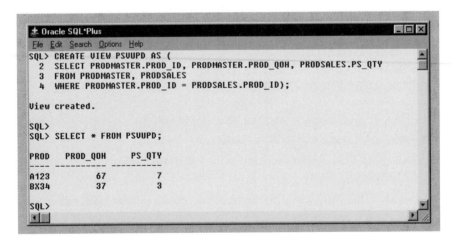

One easy way to determine if a view can be used to update a base table is to examine the view's output. *If the primary key columns of the base table you want to update still have unique values in the view, the base table is updatable.* For example, if the PROD_ID column of the view returns the 'A123' or 'BX34' values more than once, the PRODMASTER table cannot be updated through the view.

After creating the updatable view shown in Figure 7.26, you can use the UPDATE command to update the view, thereby updating the PRODMASTER table. Figure 7.27 shows how the UPDATE command is used and what the final contents of the PRODMASTER table are after the UPDATE has been executed.

FIGURE 7.27 **PRODMASTER TABLE UPDATE, USING AN UPDATABLE VIEW**

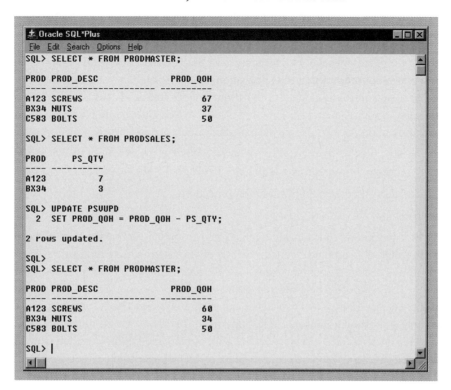

Although the batch update procedure we have just illustrated meets the goal of updating a master table with data from a transaction table, the real-world preferred solution to the update problem is to use procedural SQL. We will explore procedural SQL in the next section.

7.7 PROCEDURAL SQL

Thus far, you have learned to use SQL to read, write, and delete data in the database. For example, you learned to update values in a record, to add records, and to delete records. Unfortunately, SQL does not support the *conditional* execution of procedures that are typically supported by a programming language, using the general format:

IF <condition>
 THEN <perform procedure>
 ELSE <perform alternate procedure>
ENDIF

SQL also fails to support the looping operations in programming languages that permit the execution of repetitive actions typically encountered in a programming environment. The typical format is:

DO WHILE
 <perform procedure>
END DO

Traditionally, if you wanted to perform conditional (IF-THEN-ELSE) or looping (DO-WHILE) types of operations, that is, a procedural type of programming, you used a programming language such as COBOL, FORTRAN, C, Pascal, or Visual Basic. That's why many older (so-called legacy) business applications are based on enormous numbers of COBOL program lines. Although this approach is still common, it usually involves the duplication of

application code in many programs. Therefore, when procedural changes are required, program modifications must be made in many different programs. An environment characterized by such redundancies often creates data management and data administration problems.

A better approach is to isolate critical code and then have all application programs call this shared code. The advantage of this modular approach is that the application code is isolated in a single program, thus yielding better maintenance and logic control. In any case, the rise of distributed databases (see Chapter 10, "Distributed Database Management Systems") and object-oriented databases (see Chapter 11, "Object Oriented Databases") required that more application code be stored and executed within the database. To meet this requirement, most RDBMS vendors created numerous programming language extensions. These extensions include:

- Flow control procedural programming structures (IF-THEN-ELSE, DO-WHILE) for logic representation.

- Variable declaration and designation within the procedures.

- Error management.

To remedy the lack of procedural functionality in SQL, and to provide some standardization within the many vendor offerings, the SQL 99 standard defined the use of persistent stored modules. A **persistent stored module (PSM)** is a block of code (containing standard SQL statements and procedural extensions) that is stored and executed at the DBMS server. The PSM represents business logic that can be encapsulated, stored, and shared among multiple database users. A PSM lets an administrator assign specific access rights to a stored module to ensure that only authorized users can use it. Support for persistent stored modules is left for each vendor to implement. In fact, some RDBMSs (such as Oracle, SQL Server, and DB2) supported stored procedure modules within the database for many years before the official standard was promulgated.

Oracle implements PSMs through its procedural SQL language. **Procedural SQL (PL/SQL)** is a language that makes it possible to use and store procedural code and SQL statements within the database. Procedural SQL makes it possible to merge SQL and traditional programming constructs such as variables, conditional processing (IF-THEN-ELSE), basic loops (FOR and WHILE loops,) and error trapping. The procedural code is executed as a unit by the DBMS when it is invoked (directly or indirectly) by the end user. End users can use PL/SQL to create:

- Anonymous PL/SQL blocks

- Triggers (covered in Section 7.7.1)

- Stored procedures (covered in Section 7.7.2 and Section 7.7.3)

- PL/SQL functions (covered in Section 7.7.4)

Please do not confuse PL/SQL functions with SQL's built-in aggregate functions such as MIN and MAX that you learned about in Chapter 6. SQL built-in functions can be used only within SQL statements, while PL/SQL functions are mainly invoked within PL/SQL programs such as triggers or stored procedures. Functions can also be called within SQL statements, provided they conform to very specific rules that are dependent on your DBMS environment.

N O T E

We will demonstrate the use of PL/SQL, triggers, and stored procedures within the context of an Oracle DBMS. All examples in the following sections assume the use of an Oracle9i RDBMS.

Using Oracle SQL*Plus, you can write a PL/SQL code block by enclosing the commands inside BEGIN and END clauses. For example, the following PL/SQL block inserts a new row in the VENDOR table (see Figure 7.28):

```
BEGIN
   INSERT INTO VENDOR
   VALUES (25678,'Microsoft Corp.', 'Bill Gates','765','546-8484','WA','N');
END;
/
```

FIGURE 7.28 ANONYMOUS PL/SQL BLOCK EXAMPLES

```
Oracle SQL*Plus                                                         _ □ ×
File  Edit  Search  Options  Help
SQL> BEGIN
  2    INSERT INTO VENDOR
  3    VALUES (25678,'Microsoft Corp.', 'Bill Gates','765','546-8484','WA','N');
  4    END;
  5    /

PL/SQL procedure successfully completed.

SQL> SET SERVEROUTPUT ON
SQL>
SQL> BEGIN
  2    INSERT INTO VENDOR
  3    VALUES (25772,'Clue Store','Issac Hayes','456','323-2009','VA','N');
  4    DBMS_OUTPUT.PUT_LINE('New Vendor Added!');
  5    END;
  6    /
New Vendor Added!

PL/SQL procedure successfully completed.

SQL> SELECT * FROM VENDOR;

    V_CODE V_NAME                        V_CONTACT        V_A V_PHONE  V_ V
---------- ----------------------------- ---------------- --- -------- -- -
     21225 Bryson, Inc.                  Smithson         615 223-3234 TN Y
     21226 SuperLoo, Inc.                Flushing         904 215-8995 FL N
     21231 D&E Supply                    Singh            615 228-3245 TN Y
     21344 Gomez Bros.                   Ortega           615 889-2546 KY N
     22567 Dome Supply                   Smith            901 678-1419 GA N
     23119 Randsets Ltd.                 Anderson         901 678-3998 GA Y
     24004 Brackman Bros.                Browning         615 228-1410 TN N
     24288 ORDVA, Inc.                   Hakford          615 898-1234 TN Y
     25443 B&K, Inc.                     Smith            904 227-0093 FL N
     25501 Damal Supplies                Smythe           615 890-3529 TN N
     25595 Rubicon Systems               Orton            904 456-0092 FL Y
     25678 Microsoft Corp.               Bill Gates       765 546-8484 WA N
     25772 Clue Store                    Issac Hayes      456 323-2009 VA N

13 rows selected.

SQL>
```

The PL/SQL block shown in Figure 7.28 is known as an **anonymous PL/SQL block** because it has not been given a specific name. (Incidentally, note that the block's last line uses a forward slash ("/") to indicate the end of the command-line entry.) This type of PL/SQL block executes as soon as you press the Enter key after typing the forward slash. Following the PL/SQL block's execution, you will see the message "PL/SQL procedure successfully completed."

But suppose you want a more specific message displayed on the SQL*Plus screen after a procedure completes, for example, "New Vendor Added." To produce a more specific message, you must do two things:

1. At the SQL > prompt type: SET SERVEROUTPUT ON. This SQL*Plus command enables the client console (SQL*Plus) to receive messages from the server side (Oracle DBMS). Remember, just like standard SQL, the PL/SQL code (anonymous blocks, triggers, and procedures) are executed at the server side, not at the client side. (To stop receiving messages from the server, you would enter SET SERVEROUT OFF.)

2. To send messages from the PL/SQL block to the SQL*Plus console, use the DBMS_OUTPUT.PUT_LINE function.

The following anonymous PL/SQL block inserts a row in the VENDOR table and displays the message "New Vendor Added!" (See Figure 7.28.)

```
BEGIN
   INSERT INTO VENDOR
   VALUES (25772,'Clue Store', 'Issac Hayes', '456','323-2009','VA', 'N');
   DBMS_OUTPUT.PUT_LINE('New Vendor Added!');
END;
/
```

In Oracle, you can use the SQL*Plus command SHOW ERRORS to help you diagnose errors found in PL/SQL blocks. The SHOW ERRORS command yields additional debugging information whenever you generate an error after creating or executing a PL/SQL block.

The following example of an anonymous PL/SQL block demonstrates several of the constructs supported by the procedural language. Please remember that the exact syntax of the language is vendor-dependent; in fact, many vendors enhance their products with proprietary features.

```
DECLARE
W_P1 NUMBER(3) := 0;
W_P2 NUMBER(3) := 10;
W_NUM NUMBER(2) := 0;
BEGIN
WHILE W_P2 < 300 LOOP
   SELECT COUNT(P_CODE) INTO W_NUM FROM PRODUCT
   WHERE P_PRICE BETWEEN W_P1 AND W_P2;
   DBMS_OUTPUT.PUT_LINE('There are ' | | W_NUM | | ' Products with price between ' | | W_P1 | | ' and '
                        | | W_P2);
   W_P1 := W_P2 + 1;
   W_P2 := W_P2 + 50;
END LOOP;
END;
/
```

The block's code and execution are shown in Figure 7.29.

FIGURE 7.29 ANONYMOUS PL/SQL BLOCK WITH VARIABLES AND LOOPS

```
Oracle SQL*Plus
File  Edit  Search  Options  Help
SQL> DECLARE
  2   W_P1   NUMBER(3)  := 0;
  3   W_P2   NUMBER(3)  := 10;
  4   W_NUM  NUMBER(2)  := 0;
  5   BEGIN
  6   WHILE W_P2 < 300 LOOP
  7       SELECT COUNT(P_CODE) INTO W_NUM FROM PRODUCT
  8       WHERE P_PRICE BETWEEN W_P1 AND W_P2;
  9       DBMS_OUTPUT.PUT_LINE('There are ' || W_NUM || ' Products with price between ' || W_P1 || ' and ' || W_P2);
 10       W_P1 := W_P2 + 1;
 11       W_P2 := W_P2 + 50;
 12   END LOOP;
 13   END;
 14   /
There are 5 Products with price between 0 and 10
There are 6 Products with price between 11 and 60
There are 3 Products with price between 61 and 110
There are 1 Products with price between 111 and 160
There are 0 Products with price between 161 and 210
There are 1 Products with price between 211 and 260

PL/SQL procedure successfully completed.

SQL>
```

The PL/SQL block shown in Figure 7.29 has the following characteristics:

- The PL/SQL block starts with the DECLARE section in which you declare the variable names, data types, and if desired, an initial value. Supported data types are shown in Table 7.8.

TABLE 7.8 PL/SQL BASIC DATA TYPES

DATA TYPE	DESCRIPTION
CHAR	Character values of a fixed length. For example: W_ZIP CHAR(3)
VARCHAR2	Variable length character values. For example: W_FNAME VARCHAR2(15)
NUMBER	Numeric values. For example: W_PRICE NUMBER(6,2)
DATE	Date values. For example: W_EMP_DOB DATE
%TYPE	Inherits the data type from a variable that you have declared previously or from an attribute of a database table. For example: W_PRICE PRODUCT.P_PRICE%TYPE Assigns W_PRICE the same data type as the P_PRICE column in the PRODUCT table.

- A WHILE loop is used. Note the syntax:

 WHILE *condition* LOOP

 PL/SQL statements;

 END LOOP

- The SELECT statement uses the INTO keyword to assign the output of the query to a PL/SQL variable. You can only use the INTO keyword inside a PL/SQL block of code. If the SELECT statement returns more than one value, you will get an error.

- Note the use of the string concatenation symbol " | | " to display the output.

- Each statement inside the PL/SQL code must end with a semi-colon ";".

> **N O T E**
>
> PL/SQL blocks can contain only standard SQL data manipulation language (DML) commands such as SELECT, INSERT, UPDATE, and DELETE. The use of data definition language (DDL) commands is not directly supported in a PL/SQL block.

The most useful feature of PL/SQL blocks is that they let you create code that can be named, stored, and executed—either implicitly or explicitly—by the DBMS. This capability is especially desirable when you need to use triggers and stored procedures. We will explore database triggers and stored procedures in the next sections.

7.7.1 TRIGGERS

Automation of business procedures and automatically maintaining data integrity and consistency are critical in a modern business environment. One of the most critical business procedures is proper inventory management. For example, you want to make sure that current product sales can be supported with sufficient product availability. Therefore, it is necessary to ensure that a product order be written to a vendor when that product's inventory drops below its minimum allowable quantity on hand. Better yet, how about ensuring that this task is completed automatically?

To accomplish automatic product ordering, you first must make sure that the product's quantity on hand reflects an up-to-date and consistent value. After the appropriate product availability requirements have been set, two key issues must be addressed:

1. Business logic requires an update of the product quantity on hand each time there is a sale of that product.

2. If the product's quantity of hand falls below its minimum allowable inventory (quantity on hand) level, the product must be reordered.

To accomplish these two tasks, you could write multiple SQL statements: one to update the product quantity on hand and another to update the product reorder flag. Next, you would have to run each statement in the correct order each time there is a new sale. Such a multistage process is inefficient, because it requires that a series of SQL statements be written and executed each time a product is sold. Worse, this SQL environment requires that somebody must remember to perform these SQL tasks.

A **trigger** is procedural SQL code that is *automatically* invoked by the RDBMS upon the occurrence of a given data manipulation event. It is useful to remember that:

- A trigger is always invoked before or after a data row is inserted, updated, or deleted.

- A trigger is always associated with a database table.

- Each database table may have one or more triggers.

- A trigger is executed as part of the transaction that triggered it.

Triggers are critical to proper database operation and management. For example,

- Triggers can be used to enforce constraints that cannot be enforced at the DBMS design and implementation levels.

- Triggers add functionality by automating critical actions and providing appropriate warnings and suggestions for remedial action. In fact, one of the most common uses for triggers is to facilitate referential integrity enforcement.

- Triggers can be used to update table values, insert records in tables, and call other stored procedures.

Triggers play a critical role in making the database truly useful, and add processing power to the RDBMS and to the database system as a whole. Oracle recommends triggers for:

- Auditing purposes (creating audit logs)
- Automatic generation of derived column values
- Enforcement of business or security constraints
- Creating replica tables for backup purposes

To see how a trigger is created and used, let's examine a simple inventory management problem. For example, if a product's quantity on hand is updated when the product is sold, the system should (automatically) check whether the quantity on hand falls below its minimum allowable quantity. To demonstrate this process, let's use the PRODUCT table in Figure 7.30. Note the use of the minimum order quantity (P_MIN_ORDER) and the product reorder flag (P_REORDER) columns. The P_MIN_ORDER indicates the minimum quantity for a restocking order. The P_REORDER column is a numeric field to indicate whether the product needs to be reordered or not (1 = Yes, 0 = No). The initial P_REORDER values will be set to 0 (No) to serve as the basis for the initial trigger development.

FIGURE 7.30 THE PRODUCT TABLE

```
🛢 Oracle SQL*Plus                                                        _ □ ×
File  Edit  Search  Options  Help
SQL> SELECT * FROM PRODUCT;

P_CODE    P_DESCRIPT      P_INDATE   P_ONHAND P_MIN P_PRICE P_DISCOUNT V_CODE P_MIN_ORDER P_REORDER
--------  --------------  ---------  -------- ----- ------- ---------- ------ ----------- ---------
11QER/31  Power painter,  03-NOV-03         8     5  109.99       0.00  25595          25         0
13-Q2/P2  7.25-in. pwr.   13-DEC-03        32    15   14.99       0.05  21344          50         0
14-Q1/L3  9.00-in. pwr.   13-NOV-03        18    12   17.49       0.00  21344          50         0
1546-QQ2  Hrd. cloth, 1/  15-JAN-04        15     8   39.95       0.00  23119          35         0
1558-QW1  Hrd. cloth, 1/  15-JAN-04        23     5   43.99       0.00  23119          25         0
2232/QTY  B&D jigsaw, 12  30-DEC-03         8     5  109.92       0.05  24288          15         0
2232/QWE  B&D jigsaw, 8-  24-DEC-03         6     5   99.87       0.05  24288          15         0
2238/QPD  B&D cordless d  20-JAN-04        12     5   38.95       0.05  25595          12         0
23109-HB  Claw hammer     20-JAN-04        23    10    5.95       0.10  21225          25         0
23114-AA  Sledge hammer,  02-JAN-04         8     5   14.40       0.05                 12         0
54778-2T  Rat-tail file,  15-DEC-03        43    20    4.99       0.00  21344          25         0
89-WRE-Q  Hicut chain sa  07-FEB-04        11     5  256.99       0.05  24288          10         0
PVC23DRT  PVC pipe, 3.5-  20-FEB-04       188    75    5.87       0.00                 50         0
SM-18277  1.25-in. metal  01-MAR-04       172    75    6.99       0.00  21225          50         0
SW-23116  2.5-in. wd. sc  24-FEB-04       237   100    8.45       0.00  21231         100         0
WR3/TT3   Steel matting,  17-JAN-04        18     5  119.95       0.10  25595          10         0

16 rows selected.

SQL>
```

ADDITIONAL CD CONTENT

Oracle users can run the **PRODLIST.SQL** script file to format the output of the PRODUCT table shown in Figure 7.30. The script file is located in the **\SQL\CH07\sqlplus** folder. (The PRODUCT table is also shown in the **Ch07_SaleCo** database that is stored in MS Access format in the **\Databases\Student\Ch07** folder.)

Given the PRODUCT table listing shown in Figure 7.30, we will create a trigger to evaluate the product's quantity on hand, P_ONHAND. If the quantity on hand is below the minimum quantity shown in P_MIN, the trigger will set the P_REORDER column to 1. (Remember that the number 1 in the P_REORDER column represents "Yes.") The syntax to create a trigger in Oracle is:

CREATE OR REPLACE TRIGGER trigger_name
 [BEFORE / AFTER] [DELETE / INSERT / UPDATE OF column_name] ON table_name
 [FOR EACH ROW]

```
[DECLARE]
[variable_name          data type[:= initial_value] ]
BEGIN
   PL/SQL instructions;
   ..........
END;
```

As you can see, a trigger definition contains the following parts:

- The triggering timing: BEFORE or AFTER. This timing indicates when the trigger's PL/SQL code executes; in this case, before or after the triggering statement is completed.

- The triggering event: the statement that causes the trigger to execute (INSERT, UPDATE, or DELETE).

- The triggering level: There are two types of triggers: statement-level triggers and row-level triggers.

 - A **statement-level trigger** is assumed if you omit the FOR EACH ROW keywords. This type of trigger is executed once, before or after the triggering statement completes. This is the default case.

 - A **row-level trigger** requires the use of the FOR EACH ROW keywords. This type of trigger is executed once for each row affected by the triggering statement. (In other words, if you update 10 rows, the trigger executes 10 times.)

- The triggering action: The PL/SQL code enclosed between the BEGIN and END keywords. Each statement inside the PL/SQL code must end with a semicolon ";".

In the PRODUCT table's case, you will create a statement-level trigger that is implicitly executed AFTER an UPDATE of the P_ONHAND attribute for an existing row or AFTER an INSERT of a new row in the PRODUCT table. The trigger action executes an UPDATE statement that compares the P_ONHAND with the P_MIN column. If the value of P_ONHAND is equal to or less than P_MIN, the trigger updates the P_REORDER to 1. To create the trigger, we will use Oracle's SQL*Plus. The trigger code is shown in Figure 7.31.

FIGURE 7.31 CREATING THE TRG_PRODUCT_REORDER TRIGGER

ADDITIONAL CD CONTENT

The **Sql\Ch07\triggers** folder on the Student CD contains the source code for all the triggers shown in this section.

To test the TRG_PRODUCT_REORDER trigger, let's update the quantity on hand of product '11QER/31' to 4. After the UPDATE completes, the trigger is automatically fired and the UPDATE statement (inside the trigger code) sets the P_REORDER to 1 for all products that are below the minimum. (See Figure 7.32.)

FIGURE 7.32 VERIFYING THE **TRG_PRODUCT_REORDER** TRIGGER EXECUTION

The trigger shown in Figure 7.32 seems to work fine, but what happens if you reduce the minimum quantity of product '2232/QWE'? Figure 7.33 shows that when you update the minimum quantity, the quantity on hand of the product '2232/QWE' falls below the new minimum, but the reorder flag is still 0. Why?

FIGURE 7.33 THE **P_REORDER** VALUE MISMATCH AFTER UPDATE OF THE **P_MIN** ATTRIBUTE

The answer is simple: you updated the P_MIN column, but the trigger is never executed. TRG_PRODUCT_ REORDER only executes *after* an update of the P_ONHAND column! To avoid this inconsistency, you must modify the trigger event to execute after an update of the P_MIN field, too. The updated trigger code is shown in Figure 7.34.

FIGURE 7.34 SECOND VERSION OF THE **TRG_PRODUCT_REORDER** TRIGGER

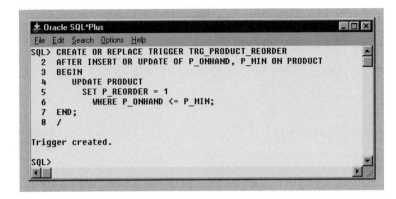

To test this new trigger version, let's change the minimum quantity for product '23114-AA' to 8. After this update, the trigger makes sure that the reorder flag is properly set for all the products in our PRODUCT table. (See Figure 7.35.)

FIGURE 7.35 SUCCESSFUL TRIGGER EXECUTION AFTER THE **P_MIN** VALUE IS UPDATED

```
Oracle SQL*Plus
File Edit Search Options Help
SQL> SELECT * FROM PRODUCT WHERE P_CODE = '23114-AA';

P_CODE   P_DESCRIPT      P_INDATE  P_ONHAND P_MIN P_PRICE P_DISCOUNT V_CODE P_MIN_ORDER P_REORDER
-------- --------------- --------- -------- ----- ------- ---------- ------ ----------- ---------
23114-AA Sledge hammer,  02-JAN-04        8     5   14.40       0.05                 12         0

SQL> UPDATE PRODUCT
  2     SET P_MIN = 10
  3         WHERE P_CODE = '23114-AA';

1 row updated.

SQL> SELECT * FROM PRODUCT WHERE P_CODE = '23114-AA';

P_CODE   P_DESCRIPT      P_INDATE  P_ONHAND P_MIN P_PRICE P_DISCOUNT V_CODE P_MIN_ORDER P_REORDER
-------- --------------- --------- -------- ----- ------- ---------- ------ ----------- ---------
23114-AA Sledge hammer,  02-JAN-04        8    10   14.40       0.05                 12         1

SQL>
```

This second version of the trigger seems to work well, but what happens if you change the P_ONHAND value for product '11QER/31', as shown in Figure 7.36? Nothing! (Note that the reorder flag is *still* set to 1.) Why didn't the trigger change the reorder flag to 0?

FIGURE 7.36 THE **P_REORDER** VALUE MISMATCH AFTER INCREASING THE **P_ONHAND** VALUE

```
Oracle SQL*Plus                                                              _ □ ×
File  Edit  Search  Options  Help
SQL> SELECT * FROM PRODUCT WHERE P_CODE = '11QER/31';

P_CODE     P_DESCRIPT     P_INDATE  P_ONHAND P_MIN P_PRICE P_DISCOUNT V_CODE P_MIN_ORDER P_REORDER
--------   --------------  ---------  -------- ----- ------- ---------- ------ ----------- ---------
11QER/31 Power painter, 03-NOV-03        4     5  109.99       0.00  25595         25          1

SQL> UPDATE PRODUCT
  2      SET P_ONHAND = P_ONHAND + P_MIN_ORDER
  3          WHERE P_CODE = '11QER/31';

1 row updated.

SQL> SELECT * FROM PRODUCT WHERE P_CODE = '11QER/31';

P_CODE     P_DESCRIPT     P_INDATE  P_ONHAND P_MIN P_PRICE P_DISCOUNT V_CODE P_MIN_ORDER P_REORDER
--------   --------------  ---------  -------- ----- ------- ---------- ------ ----------- ---------
11QER/31 Power painter, 03-NOV-03       29     5  109.99       0.00  25595         25          1

SQL>
```

The answer is that our trigger does not consider all possible cases. Let's examine the second version of the TRG_PRODUCT_REORDER trigger code (Figure 7.34) in more detail:

- The trigger fires after the triggering statement completes. Therefore, the DBMS always executes two statements (INSERT plus UPDATE or UPDATE plus UPDATE). That is, after you do an update of P_MIN or P_ONHAND or you insert a new row in the PRODUCT table, the trigger executes another UPDATE statement automatically.

- The triggering action performs an UPDATE that updates *all* the rows in the PRODUCT table, *even if the triggering statement updates just one row!* This can affect the performance of our database. Imagine what will happen if you have a PRODUCT table with 519,128 rows and you insert just one product. The trigger will update all 519,129 rows (519,128 original rows plus the one you inserted), including the rows that do not need an update!

- The trigger only sets the P_REORDER value to 1; it never resets the value to 0 when such an action is clearly required when the inventory level is back to a value greater than the minimum value.

In short, the second version of the TRG_PRODUCT_REORDER trigger still does not complete all of the necessary steps. Now let's modify the trigger to handle all update scenarios, as shown in Figure 7.37.

FIGURE 7.37 THE THIRD VERSION OF THE **TRG_PRODUCT_REORDER** TRIGGER

```
Oracle SQL*Plus                                                              _ □ ×
File  Edit  Search  Options  Help
SQL> CREATE OR REPLACE TRIGGER TRG_PRODUCT_REORDER
  2      BEFORE INSERT OR UPDATE OF P_ONHAND, P_MIN ON PRODUCT
  3      FOR EACH ROW
  4      BEGIN
  5         IF :NEW.P_ONHAND <= :NEW.P_MIN THEN
  6      :NEW.P_REORDER := 1;
  7         ELSE
  8             :NEW.P_REORDER := 0;
  9         END IF;
 10      END;
 11      /

Trigger created.

SQL>
```

The trigger in Figure 7.37 sports several new features:

- First, the trigger is executed *before* the actual triggering statement completes. In Figure 7.37, the triggering timing is defined in line 2, BEFORE INSERT OR UPDATE. This clearly indicates that the triggering statement is executed before the INSERT or UPDATE completes, unlike the previous trigger examples.

- Second, the trigger is a row-level trigger instead of a statement-level trigger. The FOR EACH ROW keywords make the trigger a row-level trigger. Therefore, this trigger executes once for each row affected by the triggering statement.

- Third, the trigger action uses the :NEW attribute reference to change the value of the P_REORDER attribute.

The use of the :NEW attribute references deserves a more detailed explanation. To understand its use, you must first consider a basic computing tenet: *all changes are done first in primary memory, then to permanent memory*. In other words, the computer cannot change anything directly in permanent storage (disk); it must first read the data from permanent storage to primary memory, then it makes the change in primary memory, and finally, it writes the changed data back to permanent memory (disk).

The DBMS does exactly the same thing, plus something additional. Because ensuring data integrity is critical, the DBMS makes two copies of every row being changed by a DML (INSERT, UPDATE, or DELETE) statement. (You will learn more about this in Chapter 9, "Transaction Management and Concurrency Control.") The first copy contains the original ("old") values of the attributes before the changes. The second copy contains the changed ("new") values of the attributes that will be permanently saved to the database (after any changes made by an INSERT, UPDATE, or DELETE). You can use :OLD to refer to the original values; you can use :NEW to refer to the changed values (the values that will be stored in the table). You can use :NEW and :OLD attribute references only within the PL/SQL code of a database trigger action. For example:

- IF :NEW.P_ONHAND < = :NEW.P_MIN compares the quantity on hand with the minimum quantity of a product. Remember that this is a row-level trigger. Therefore, this comparison is done for each row that is updated by the triggering statement.

- Although the trigger is a BEFORE trigger, this does not mean that the triggering statement hasn't executed yet. To the contrary, the triggering statement has already taken place—otherwise the trigger would not have fired and the :NEW values would not exist. Remember, BEFORE means *before* the changes are permanently saved to disk, but *after* the changes are made in memory.

- The trigger uses the :NEW reference to assign a value to the P_REORDER column before the UPDATE or INSERT results are permanently stored in the table. The assignment is always done to the :NEW value, never to the :OLD value, and the assignment always uses the " := " assignment operator. The :OLD values are *read only* values, you cannot change them. Note that :NEW.P_REORDER := 1; assigns the value 1 to the P_REORDER column and :NEW.P_REORDER := 0; assigns the value 0 to the P_REORDER column.

- This new trigger version does not use any DML statement!

Before testing the new trigger, note that product '11QER/31' currently has a quantity on hand that is above the minimum quantity, yet the reorder flag is set to 1. Given this condition, the reorder flag must be zero. After creating the new trigger, we can execute an UPDATE statement to fire it, as shown in Figure 7.38.

FIGURE 7.38 EXECUTION OF THE THIRD TRIGGER VERSION

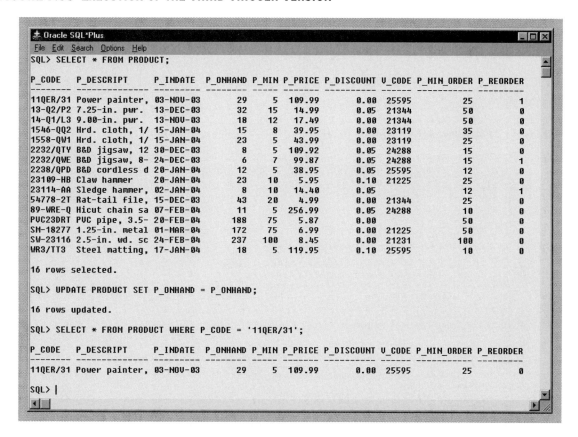

As you examine Figure 7.38, note the following important features:

- The trigger is automatically invoked for each affected row—in this case, all rows of the PRODUCT table. If your triggering statement would have affected only three rows, not all PRODUCT rows will have the correct P_REORDER value set. That's the reason the triggering statement was set up as shown in Figure 7.38.

- The trigger will run only if you insert a new product row or you update P_ONHAND or P_MIN. If you update any other attribute, the trigger doesn't run.

You can also use a trigger to update an attribute in a table other than the one being modified. For example, suppose that you would like to create a trigger that automatically reduces the quantity on hand of a product with every sale. To accomplish this task, you must create a trigger for the LINE table that updates a row in the PRODUCT table. The sample code for this trigger is shown in Figure 7.39.

FIGURE 7.39 TRG_LINE_PROD TRIGGER TO UPDATE THE PRODUCT QUANTITY ON HAND

Note that the TRG_LINE_PROD row-level trigger executes after inserting a new invoice's LINE, and reduces the quantity on hand of the recently sold product by the number of units sold. This row-level trigger updates a row in a different table (PRODUCT) using the :NEW values of the recently added LINE row.

A third trigger example shows the use of variables within a trigger. In this case, you want to update the customer balance (CUS_BALANCE) in the CUSTOMER table each time after inserting a new LINE row. The trigger code is shown in Figure 7.40.

FIGURE 7.40 TRG_LINE_CUS TRIGGER TO UPDATE THE CUSTOMER BALANCE

```
Oracle SQL*Plus
File  Edit  Search  Options  Help
SQL> CREATE OR REPLACE TRIGGER TRG_LINE_CUS
  2  AFTER INSERT ON LINE
  3  FOR EACH ROW
  4  DECLARE
  5  W_CUS CHAR(5);
  6  W_TOT NUMBER:= 0;    -- to compute total cost
  7  BEGIN
  8     -- this trigger fires up after an INSERT of a LINE
  9     -- it will update the CUS_BALANCE in CUSTOMER
 10
 11     -- 1) get the CUS_CODE
 12     SELECT CUS_CODE INTO W_CUS
 13       FROM INVOICE
 14        WHERE INVOICE.INV_NUMBER = :NEW.INV_NUMBER;
 15
 16     -- 2) compute the total of the current line
 17     W_TOT := :NEW.LINE_PRICE * :NEW.LINE_UNITS;
 18
 19     -- 3) Update the CUS_BALANCE in CUSTOMER
 20     UPDATE CUSTOMER
 21       SET CUS_BALANCE = CUS_BALANCE + W_TOT
 22        WHERE CUS_CODE = W_CUS;
 23
 24     DBMS_OUTPUT.PUT_LINE(' * * * Balance updated for customer: ' || W_CUS);
 25
 26  END;
 27  /

Trigger created.

SQL>
```

Let's examine the trigger in Figure 7.40 carefully:

- The trigger is a row-level trigger that executes after each new LINE row is inserted.

- The DECLARE section in the trigger is used to declare any variables used inside the trigger code.

- You can declare a variable by assigning a name, a data type, and optionally, an initial value; as in the case of the W_TOT variable.

- The first step in the trigger code is to get the customer code (CUS_CODE) from the related INVOICE table. Note that the SELECT statement returns only one attribute (CUS_CODE) from the INVOICE table. And note that that attribute returns only one value as specified by the use of the WHERE clause *to restrict the query output to a single value.*

- Note also the use of the INTO clause within the SELECT statement. You use the INTO clause to assign a value from a SELECT statement to a variable (W_CUS) used within a trigger.

- The second step in the trigger code computes the total of the line by multiplying the :NEW.LINE_UNITS times :NEW.LINE_PRICE and assigning the result to the W_TOT variable.

- The final step updates the customer balance by using an UPDATE statement and the W_TOT and W_CUS trigger variables.

- Double dashes (--) are used to indicate comments within the PL/SQL block.

Let's summarize the triggers created in this section:

- The TRG_PROD_REORDER is a row-level trigger that updates P_REORDER in PRODUCT when a new product is added or the P_ONHAND or P_MIN columns are updated.

- The TRG_LINE_PROD is a row-level trigger that automatically reduces the P_ONHAND in PRODUCT when a new row is added to the LINE table.

- TRG_LINE_CUS is a row-level trigger that automatically increases the CUS_BALANCE in CUSTOMER when a new row is added in the LINE table.

The use of triggers facilitates the automation of multiple data management tasks. Although triggers are independent objects, they are associated with database tables. When you delete a table, all trigger objects are deleted with it. However, if you need to delete a trigger without deleting the table, you could use the following command:

DROP TRIGGER *trigger_name*

7.7.2 STORED PROCEDURES

A **stored procedure** is a named collection of procedural and SQL statements. Just like database triggers, stored procedures are stored in the database. One of the major advantages of stored procedures is that they can be used to encapsulate and represent business transactions. For example, we can create a stored procedure to represent a product sale, a credit update, or the addition of a new customer. By doing this, we can encapsulate SQL statements within a single stored procedure and execute them as a single transaction. There are two clear advantages to the use of stored procedures:

1. Stored procedures substantially reduce network traffic and increase performance. Because the stored procedure is stored at the server, there is no transmission of individual SQL statements over the network. The use of stored procedures improves system performance, because all transactions are executed locally on the RDBMS, and each SQL statement does not have to travel over the network.

2. Stored procedures help reduce code duplication by means of code isolation and code sharing (creating unique PL/SQL modules that are called by application programs), thereby minimizing the chance of errors and the cost of application development and maintenance.

To create a stored procedure, you use the following syntax:

CREATE OR REPLACE PROCEDURE *procedure_name* [(*argument* [IN/OUT] *data-type*, ...)] [IS/AS]
 [*variable_name* *data type*[:= *initial_value*]]
BEGIN
 PL/SQL or SQL statements;
 …
END;

Note the following important points about stored procedures and their syntax:

- *argument* specifies the parameters that are passed to the stored procedure. A stored procedure could have zero or more arguments or parameters.

- *IN/OUT* indicates whether the parameter is for input or for output or both.

- *data-type* is one of the procedural SQL data types used in the RDBMS. The data types normally match the data types used in the RDBMS table-creation statement.

- Variables can be declared between the keywords IS and BEGIN. You must specify the variable name, its data type, and optionally, an initial value.

To illustrate stored procedures, let's assume that you want to create a procedure (PRC_PROD_DISCOUNT) to assign an additional 5% discount for all products when the quantity on hand is more than or equal to twice the minimum quantity. Figure 7.41 shows how the stored procedure is created.

FIGURE 7.41 CREATING THE **PRC_PROD_DISCOUNT** STORED PROCEDURE

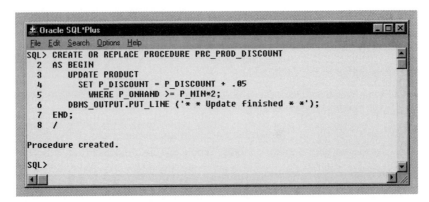

ADDITIONAL CD CONTENT

The **Sql\Ch07\procedures** folder on your student CD contains the source code for all the stored procedures shown in this section.

As you examine Figure 7.41, note that the PRC_PROD_DISCOUNT stored procedure uses the DBMS_OUTPUT.PUT_LINE function to display a message when the procedure executes. (This action assumes you previously ran SET SERVEROUTPUT ON.)

To execute the stored procedure you must use the following syntax:

EXEC *procedure_name*[(*parameter_list*)];

For example, to see the results of running the PRC_PROD_DISCOUNT stored procedure, you can use the EXEC PRC_PROD_DISCOUNT command shown in Figure 7.42.

Using Figure 7.42 as your guide, you can see how the product discount attribute for all products with the quantity on hand more than or equal to twice the minimum quantity was increased by 5 percent. (Compare the first PRODUCT table listing to the second PRODUCT table listing.)

One of the main advantages of procedures is that you can pass values to them. For example, our previous PRC_PRODUCT_DISCOUNT procedure worked fine, but what if you wanted to make the percentage increase an input variable? In this case, we can pass an argument to represent the rate of increase to the procedure. Figure 7.43 shows the code for this procedure.

FIGURE 7.42 RESULTS OF THE PRC_PROD_DISCOUNT STORED PROCEDURE

```
± Oracle SQL*Plus                                                          _ □ ×
File  Edit  Search  Options  Help

P_CODE    P_DESCRIPT    P_INDATE   P_ONHAND P_MIN P_PRICE P_DISCOUNT V_CODE P_MIN_ORDER P_REORDER
--------  ------------  ---------  -------- ----- ------- ---------- ------ ----------- ---------
11QER/31  Power painter, 03-NOV-03      29     5  109.99       0.00  25595          25         0
13-Q2/P2  7.25-in. pwr.  13-DEC-03      32    15   14.99       0.05  21344          50         0
14-Q1/L3  9.00-in. pwr.  13-NOV-03      18    12   17.49       0.00  21344          50         0
1546-QQ2  Hrd. cloth, 1/ 15-JAN-04      15     8   39.95       0.00  23119          35         0
1558-QW1  Hrd. cloth, 1/ 15-JAN-04      23     5   43.99       0.00  23119          25         0
2232/QTY  B&D jigsaw, 12 30-DEC-03       8     5  109.92       0.05  24288          15         0
2232/QWE  B&D jigsaw, 8-  24-DEC-03       6     7   99.87       0.05  24288          15         1
2238/QPD  B&D cordless d 20-JAN-04      12     5   38.95       0.05  25595          12         0
23109-HB  Claw hammer    20-JAN-04      23    10    5.95       0.10  21225          25         0
23114-AA  Sledge hammer, 02-JAN-04       8    10   14.40       0.05                 12         1
54778-2T  Rat-tail file, 15-DEC-03      43    20    4.99       0.00  21344          25         0
89-WRE-Q  Hicut chain sa 07-FEB-04      11     5  256.99       0.05  24288          10         0
PVC23DRT  PVC pipe, 3.5- 20-FEB-04     188    75    5.87       0.00                 50         0
SM-18277  1.25-in. metal 01-MAR-04     172    75    6.99       0.00  21225          50         0
SW-23116  2.5-in. wd. sc 24-FEB-04     237   100    8.45       0.00  21231         100         0
WR3/TT3   Steel matting, 17-JAN-04      18     5  119.95       0.10  25595          10         0

16 rows selected.

SQL> EXEC PRC_PROD_DISCOUNT;
* * Update finished * *

PL/SQL procedure successfully completed.

SQL> SELECT * FROM PRODUCT;

P_CODE    P_DESCRIPT    P_INDATE   P_ONHAND P_MIN P_PRICE P_DISCOUNT V_CODE P_MIN_ORDER P_REORDER
--------  ------------  ---------  -------- ----- ------- ---------- ------ ----------- ---------
11QER/31  Power painter, 03-NOV-03      29     5  109.99       0.05  25595          25         0
13-Q2/P2  7.25-in. pwr.  13-DEC-03      32    15   14.99       0.10  21344          50         0
14-Q1/L3  9.00-in. pwr.  13-NOV-03      18    12   17.49       0.00  21344          50         0
1546-QQ2  Hrd. cloth, 1/ 15-JAN-04      15     8   39.95       0.00  23119          35         0
1558-QW1  Hrd. cloth, 1/ 15-JAN-04      23     5   43.99       0.05  23119          25         0
2232/QTY  B&D jigsaw, 12 30-DEC-03       8     5  109.92       0.05  24288          15         0
2232/QWE  B&D jigsaw, 8-  24-DEC-03       6     7   99.87       0.05  24288          15         1
2238/QPD  B&D cordless d 20-JAN-04      12     5   38.95       0.10  25595          12         0
23109-HB  Claw hammer    20-JAN-04      23    10    5.95       0.15  21225          25         0
23114-AA  Sledge hammer, 02-JAN-04       8    10   14.40       0.05                 12         1
54778-2T  Rat-tail file, 15-DEC-03      43    20    4.99       0.05  21344          25         0
89-WRE-Q  Hicut chain sa 07-FEB-04      11     5  256.99       0.10  24288          10         0
PVC23DRT  PVC pipe, 3.5- 20-FEB-04     188    75    5.87       0.05                 50         0
SM-18277  1.25-in. metal 01-MAR-04     172    75    6.99       0.05  21225          50         0
SW-23116  2.5-in. wd. sc 24-FEB-04     237   100    8.45       0.05  21231         100         0
WR3/TT3   Steel matting, 17-JAN-04      18     5  119.95       0.15  25595          10         0

16 rows selected.

SQL>
```

FIGURE 7.43 SECOND VERSION OF THE PRC_PROD_DISCOUNT STORED PROCEDURE

```
± Oracle SQL*Plus                                                          _ □ ×
File  Edit  Search  Options  Help
SQL> CREATE OR REPLACE PROCEDURE PRC_PROD_DISCOUNT(WPI IN NUMBER) AS
  2  BEGIN
  3    IF ((WPI <= 0) OR (WPI >= 1)) THEN -- validate WPI parameter
  4      DBMS_OUTPUT.PUT_LINE('Error: Value must be greater than 0 and less than 1');
  5    ELSE        -- if value is greater than 0 and less than 1
  6  UPDATE PRODUCT
  7    SET P_DISCOUNT = P_DISCOUNT + WPI
  8      WHERE P_ONHAND >= P_MIN*2;
  9  DBMS_OUTPUT.PUT_LINE ('* * Update finished * *');
 10    END IF;
 11  END;
 12  /

Procedure created.

SQL>
```

Figure 7.44 shows the execution of the second version of the PRC_PROD_DISCOUNT stored procedure. Note that if the procedure requires arguments, those arguments must be enclosed in parentheses and they must be separated by commas.

FIGURE 7.44 RESULTS OF THE SECOND VERSION OF THE PRC_PROD_DISCOUNT STORED PROCEDURE

Stored procedures are also useful to encapsulate shared code to represent business transactions. For example, you can create a simple stored procedure to add a new customer. By using a stored procedure, all programs can call the stored procedure by name each time a new customer is to be added. Naturally, if new customer attributes are added later, you would need to modify the stored procedure. However, the programs that use the stored procedure would not need to know the name of the newly added attribute and would only need to add a new parameter to the procedure call. (Take a look at the PRC_CUS_ADD stored procedure shown in Figure 7.45.)

FIGURE 7.45 THE PRC_CUS_ADD STORED PROCEDURE

```
Oracle SQL*Plus
File  Edit  Search  Options  Help
SQL> CREATE OR REPLACE PROCEDURE PRC_CUS_ADD
  2  (W_LN IN VARCHAR, W_FN IN VARCHAR, W_INIT IN VARCHAR, W_AC IN VARCHAR, W_PH IN VARCHAR)
  3  AS
  4  BEGIN
  5  -- note that the procedure uses the CUS_CODE_SEQ sequence created earlier
  6  -- attribute names are required when not giving values for all table attributes
  7     INSERT INTO CUSTOMER(CUS_CODE,CUS_LNAME, CUS_FNAME, CUS_INITIAL, CUS_AREACODE, CUS_PHONE)
  8            VALUES (CUS_CODE_SEQ.NEXTVAL, W_LN, W_FN, W_INIT, W_AC, W_PH);
  9     DBMS_OUTPUT.PUT_LINE ('Customer ' || W_LN || ', ' || W_FN || ' added.');
 10  END;
 11  /

Procedure created.

SQL> EXEC PRC_CUS_ADD('Walker','Johnie',NULL,'615','84-DRUNK');
Customer Walker, Johnie added.

PL/SQL procedure successfully completed.

SQL> SELECT * FROM CUSTOMER WHERE CUS_LNAME = 'Walker';

  CUS_CODE CUS_LNAME       CUS_FNAME       C CUS CUS_PHON CUS_BALANCE
---------- --------------- --------------- - --- -------- -----------
     20010 Walker          Johnie            615 84-DRUNK           0

SQL> EXEC PRC_CUS_ADD('Lowery', 'Denisee', NULL, NULL, NULL);
BEGIN PRC_CUS_ADD('Lowery', 'Denisee', NULL, NULL, NULL); END;

*
ERROR at line 1:
ORA-01400: cannot insert NULL into ("TEACHER"."CUSTOMER"."CUS_AREACODE")
ORA-06512: at "TEACHER.PRC_CUS_ADD", line 7
ORA-06512: at line 1

SQL>
```

As you examine Figure 7.45, note these features:

- The PRC_CUS_ADD procedure uses several parameters, one for each required attribute in the CUSTOMER table.

- The stored procedure uses the CUS_CODE_SEQ sequence to generate a new customer code.

- The required parameters—those specified in the table definition—must always be included and can be null *only* if the table specifications permit nulls for that parameter. For example, note that the second customer addition was unsuccessful, because the CUS_AREA_CODE is a required attribute and cannot be null.

- The procedure displays a message in the SQL*Plus console to let the user know the customer was added.

The next two examples further illustrate the use of sequences within stored procedures. In this case, let's create two stored procedures:

1. The PRC_INV_ADD procedure adds a new invoice.

2. The PRC_LINE_ADD procedure adds a new product line row for a given invoice.

Both procedures are shown in Figure 7.46. (Note the use of a variable in the PRC_LINE_ADD procedure to get the product price from the PRODUCT table.)

FIGURE 7.46 THE PRC_INV_ADD AND PRC_LINE_ADD STORED PROCEDURES

```
Oracle SQL*Plus
File  Edit  Search  Options  Help
SQL> CREATE OR REPLACE PROCEDURE PRC_INV_ADD (W_CUS_CODE IN VARCHAR2, W_DATE IN DATE)
  2   AS BEGIN
  3      INSERT INTO INVOICE
  4             VALUES(INV_NUMBER_SEQ.NEXTVAL, W_CUS_CODE, W_DATE);
  5      DBMS_OUTPUT.PUT_LINE('Invoice added');
  6   END;
  7  /

Procedure created.

SQL> CREATE OR REPLACE PROCEDURE PRC_LINE_ADD (W_LN IN NUMBER, W_P_CODE IN VARCHAR2, W_LU NUMBER)
  2   AS
  3    W_LP NUMBER := 0.00;
  4   BEGIN
  5      -- GET THE PRODUCT PRICE
  6    SELECT P_PRICE INTO W_LP
  7           FROM PRODUCT
  8                WHERE P_CODE = W_P_CODE;
  9
 10      -- ADDS THE NEW LINE ROW
 11    INSERT INTO LINE
 12           VALUES(INV_NUMBER_SEQ.CURRVAL, W_LN, W_P_CODE, W_LU, W_LP);
 13
 14    DBMS_OUTPUT.PUT_LINE('Invoice line ' || W_LN || ' added');
 15   END;
 16  /

Procedure created.

SQL>
```

To test the procedures shown in Figure 7.46:

1. Call the PRC_INV_ADD procedure with the new invoice data as arguments.

2. Call the INV_LINE_ADD procedure and pass the product line arguments.

This process is illustrated in Figure 7.47.

FIGURE 7.47 TESTING THE **PRC_INV_ADD** AND **PRC_LINE_ADD** PROCEDURES

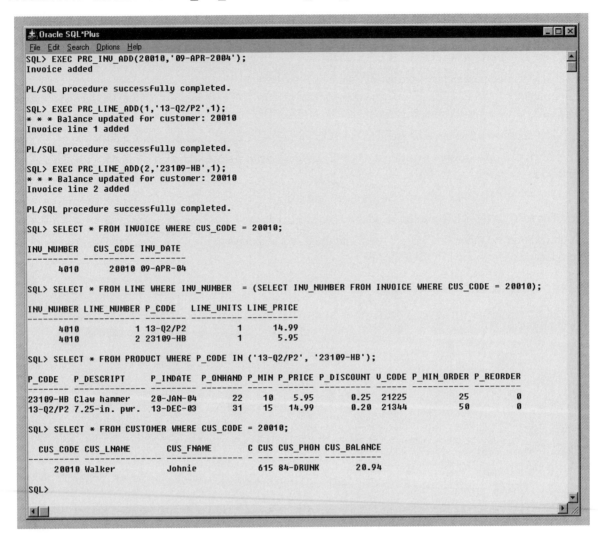

7.7.3 PL/SQL PROCESSING WITH CURSORS

Until now, all the SQL statements you have used inside a PL/SQL block (trigger or stored procedure) have returned a single value. If the SQL statement returns more than one value, you will generate an error. If you want to use a SQL statement that returns more than one value inside your PL/SQL code, you need to use a cursor. A **cursor** is a special construct used in procedural SQL to hold the data rows returned by a SQL query. You can think of a cursor as a reserved area of memory in which the output of the query is stored, like an array holding columns and rows. Cursors are held in a reserved memory area in the DBMS server, not in the client computer.

There are two types of cursors: implicit and explicit. An **implicit cursor** is automatically created in procedural SQL when the SQL statement returns only one value. Up to this point, all the examples we have used created an implicit cursor. An **explicit cursor** is created to hold the output of a SQL statement that may return two or

more rows (but could return zero or only one row!). To create an explicit cursor, you use the following syntax inside a PL/SQL DECLARE section:

CURSOR *cursor_name* IS *select-query*;

Once you have declared a cursor, you can use specific PL/SQL cursor processing commands (OPEN, FETCH, and CLOSE) anywhere between the BEGIN and END keywords of the PL/SQL block. Table 7.9 summarizes the main use of each of these commands:

TABLE 7.9 **CURSOR PROCESSING COMMANDS**

CURSOR COMMAND	EXPLANATION
OPEN	Opening the cursor executes the SQL command and populates the cursor with data, getting the cursor ready for processing. The cursor declaration command only reserves a named memory area for the cursor; it doesn't populate the cursor with the data. Before you can use a cursor, you need to open it. For example: OPEN *cursor_name*
FETCH	Once the cursor is opened, you can use the FETCH command to retrieve data from the cursor and copy it to the PL/SQL variables for processing. The syntax is: FETCH cursor_name INTO variable1 [, variable2, ...] The PL/SQL variables used to hold the data must be declared in the DECLARE section and must have data types compatible with the columns retrieved by the SQL command. If the cursor's SQL statement returns five columns, then there must be five PL/SQL variables to receive the data from the cursor. This type of processing resembles the "one-record-at-a-time" processing used in previous database models. The first time you fetch a row from the cursor, the first row of data from the cursor is copied to the PL/SQL variables; the second time you fetch a row from the cursor, the second row of data is placed in the PL/SQL variables, and so on.
CLOSE	The CLOSE command closes the cursor for processing.

Cursor-style processing involves retrieving data from the cursor, one row at a time. Once you open a cursor, it becomes an active data set. This data set contains a "current" row pointer. Therefore, after opening a cursor, the current row is the first row of the cursor.

When you fetch a row from the cursor, the data from the "current" row in the cursor is copied to the PL/SQL variables. After the fetch, the "current" row pointer moves to the next row in the set and continues until it reaches the end of the cursor.

How do you know how many rows are in the cursor? Or how do you know if you have reached the end of the cursor data set? You know because cursors have special attributes that convey important information. Table 7.10 summarizes the cursor attributes.

TABLE 7.10 CURSOR ATTRIBUTES

ATTRIBUTE	DESCRIPTION
%ROWCOUNT	Returns the number of rows fetched so far. If the cursor is not OPEN, it returns an error. If no FETCH has been done, but the cursor is OPEN, it returns 0.
%FOUND	Returns TRUE if the last FETCH returned a row. Returns FALSE if the last FETCH did not return any row. If the cursor is not OPEN, it returns an error. If no FETCH has been done, it contains NULL.
%NOTFOUND	Returns TRUE if the last FETCH did not return any row. Returns FALSE if the last FETCH returned a row. If the cursor is not OPEN returns an error. If no FETCH has been done, if contains NULL.
%ISOPEN	Returns TRUE if the cursor is open (ready for processing) or FALSE if the cursor is closed. Remember, before you can use a cursor you must open it.

To illustrate the use of cursors, let's use a simple stored procedure example that lists all products that have a quantity on hand greater than the average quantity on hand for all products. The code is shown in Figure 7.48.

FIGURE 7.48 A SIMPLE PRC_CURSOR_EXAMPLE

```
SQL> CREATE OR REPLACE PROCEDURE PRC_CURSOR_EXAMPLE IS
  2    W_P_CODE PRODUCT.P_CODE%TYPE;
  3    W_P_DESCRIPT PRODUCT.P_DESCRIPT%TYPE;
  4    W_TOT  NUMBER(3);
  5    CURSOR PROD_CURSOR IS
  6      SELECT P_CODE, P_DESCRIPT
  7        FROM PRODUCT
  8        WHERE P_ONHAND > (SELECT AVG(P_ONHAND) FROM PRODUCT);
  9  BEGIN
 10    DBMS_OUTPUT.PUT_LINE('PRODUCTS WITH P_ONHAND > AVG(P_ONHAND)');
 11    DBMS_OUTPUT.PUT_LINE('=====================================');
 12    OPEN PROD_CURSOR;
 13    LOOP
 14      FETCH PROD_CURSOR INTO W_P_CODE, W_P_DESCRIPT;
 15      EXIT WHEN PROD_CURSOR%NOTFOUND;
 16      DBMS_OUTPUT.PUT_LINE(W_P_CODE ||' -> ' || W_P_DESCRIPT );
 17    END LOOP;
 18    DBMS_OUTPUT.PUT_LINE('=====================================');
 19    DBMS_OUTPUT.PUT_LINE('TOTAL PRODUCT PROCESSED ' || PROD_CURSOR%ROWCOUNT);
 20    DBMS_OUTPUT.PUT_LINE('--- END OF REPORT ----');
 21    CLOSE PROD_CURSOR;
 22  END;
 23  /

Procedure created.

SQL> EXEC PRC_CURSOR_EXAMPLE;
PRODUCTS WITH P_ONHAND > AVG(P_ONHAND)
=====================================
PVC23DRT -> PVC pipe, 3.5-in., 8-ft
SW-18277 -> 1.25-in. metal screw, 25
SW-23116 -> 2.5-in. wd. screw, 50
=====================================
TOTAL PRODUCT PROCESSED 3
--- END OF REPORT ----

PL/SQL procedure successfully completed.

SQL>
```

As you examine the stored procedure code shown in Figure 7.48, note the following important characteristics:

- Lines 2 and 3 use the %TYPE data type in the variable definition section. As indicated in Table 7.8, the %TYPE data type is used to indicate that the given variable inherits the data type from a variable previously declared or from an attribute of a database table. In this case, you are using the %TYPE to indicate that the W_P_CODE and W_P_DESCRIPT will have the same data type as the respective columns in the PRODUCT table. This way, you ensure that the PL/SQL variable will have a compatible data type.

- Line 5 declares the PROD_CURSOR cursor.

- Line 12 opens the PROD_CURSOR cursor and populates it.

- Line 13 uses the LOOP statement to loop through the data in the cursor, fetching one row at a time.

- Line 14 uses the FETCH command to retrieve a row from the cursor and place it in the respective PL/SQL variables.

- Line 15 uses the EXIT command to evaluate when there are no more rows in the cursor (using the %NOTFOUND cursor attribute) and exit the loop.

- Line 19 uses the %ROWCOUNT cursor attribute to get the total number of rows processed.

- Line 21 issues the CLOSE PROD_CURSOR command to close the cursor.

The use of cursors, combined with standard SQL, makes relational databases very desirable because they enable programmers to work in the best of both worlds: set-oriented processing and record-oriented processing. Any experienced programmer knows it is always better to use the tool that best fits the job. Sometimes you may be better off manipulating data in a set-oriented environment, while at other times it might be better to use a record-oriented environment. Procedural SQL lets you have your proverbial cake and eat it, too. Procedural SQL provides functionality that enhances the capabilities of the DBMS while maintaining a high degree of manageability.

7.7.4 PL/SQL STORED FUNCTIONS

Using programmable or procedural SQL, you can also create your own stored functions. Stored procedures and functions are very similar. A **stored function** is basically a named group of procedural and SQL statements that returns a value (indicated by a RETURN statement in its program code). To create a function, you use the following syntax:

```
CREATE FUNCTION function_name (argument IN data-type, ... ) RETURN data-type [IS]
BEGIN
    PL/SQL statements;
    ...
    RETURN (value or expression);
END;
```

Stored functions can only be invoked from within stored procedures or triggers and cannot be invoked from SQL statements (unless the function follows some very specific compliance rules). Remember not to confuse built-in SQL functions (such as MIN, MAX, and AVG) with stored functions.

7.8 EMBEDDED SQL

There is little doubt that SQL's popularity as a data manipulation language is in part due to its ease of use and its powerful data-retrieval capabilities. But in the real world, database systems are related to other systems and programs, and you still need a conventional programming language such as COBOL, C++, or Visual Basic to integrate database systems with other programs and systems. If you are developing Web applications, you are very likely to be familiar with Java, ASP, ColdFusion, and so on. Yet, almost regardless of the programming tools you

use, if your Web application or Windows-based GUI system requires access to a database such as MS Access, SQL Server, Oracle, or DB2, you will likely need to use SQL to manipulate the data in the database.

Embedded SQL is a term used to refer to SQL statements that are contained within an application programming language such as COBOL, C++, ASP, Java, or ColdFusion. The program being developed may be a standard binary executable in Windows or Linux, or it may be a Web application designed to run over the Internet. No matter what language you use, if it contains embedded SQL statements it is called the **host language**. Embedded SQL is still the most common approach to maintaining procedural capabilities in DBMS-based applications. However, mixing SQL with procedural languages requires that you understand some key differences between SQL and procedural languages:

- Run-time mismatch: Remember that SQL is a nonprocedural, interpreted language; that is, each instruction is parsed, the syntax checked, and executed one instruction at a time.[1] All the processing takes place at the server side. Meanwhile, the host language is generally a binary-executable program (also known as a compiled program). The host program typically runs at the client side in its own memory space (different from the DBMS environment).

- Processing mismatch: Conventional programming languages (COBOL, ADA, FORTRAN, PASCAL, C++, and PL/I) process one data element at a time. Although you can use arrays to hold data, you still process the array elements one row at a time. This is especially true for file manipulation, where the host language typically will manipulate data one record at a time.

- Data type mismatch. SQL provides several data types, but some of those data types may not match data types used in different host languages (for example date and varchar2 data types).

To bridge these differences, the Embedded SQL standard defines a framework to integrate SQL within several programming languages. The Embedded SQL framework defines the following:

- A standard syntax to identify embedded SQL code within the host language (EXEC SQL/END-EXEC).

- A standard syntax to identify host variables. Host variables are variables in the host language that will receive data from the database (through the embedded SQL code) and process the data in the host language. All host variables are preceded by a colon (":").

- A communication area used to exchange status and error information between SQL and the host language. This communications area contains two variables—SQLCODE and SQLSTATE.

Another way to interface host languages and SQL is through the use of a call level interface (CLI), in which the programmer writes to an application programming interface (API). A common CLI in Windows is the one provided by the Open Database Connectivity (ODBC) interface. (We have provided additional coverage of these topics in Appendix F, "Client/Server Systems," and in Chapter 14, "Web Database Development.")

Before continuing, let's explore the process required to create and run an executable program with embedded SQL statements. If you have ever programmed in COBOL or C++, you are familiar with the multiple steps

[1] We are particularly grateful for the thoughtful comments provided by Emil T. Cipolla, who teaches at Mt. St. Mary College and whose IBM experience is the basis for his considerable and practical expertise.

required to generate the final executable program. Although the specific details vary among language and DBMS vendors, the following general steps are standard:

1. The programmer writes embedded SQL code within the host language instructions. The code follows the standard syntax required for the host language and embedded SQL.

2. A preprocessor is used to transform the embedded SQL into specialized procedure calls that are DBMS- and language-specific. The preprocessor is provided by the DBMS vendor and is specific to the host language.

3. The program is compiled using the host language compiler. The compiler creates an object code module for the program containing the DBMS procedure calls.

4. The object code is linked to the respective library modules and generates the executable program. This process binds the DBMS procedure calls to the DBMS run-time libraries. Additionally, the binding process typically creates an "access plan" module that contains instructions to run the embedded code at run time.

5. The executable is run and the embedded SQL statement retrieves data from the database.

Note that you can embed individual SQL statements or even an entire PL/SQL block. Up to this point in the book, you have used a DBMS-provided application (SQL*Plus) to write SQL statements and PL/SQL blocks in an interpretive mode to address one-time or ad hoc data requests. However, it is extremely difficult and awkward to use ad hoc queries to process transactions inside a host language. Programmers typically embed SQL statements within a host language which it is compiled once and executed as often as needed. To embed SQL into a host language follow the syntax shown next:

```
EXEC SQL
    SQL statement;
END-EXEC.
```

The preceding syntax will work for SELECT, INSERT, UPDATE, or DELETE statements. For example, the following embedded SQL code will delete employee 109, George Smith, from the EMPLOYEE table:

```
EXEC SQL
    DELETE FROM EMPLOYEE WHERE EMP_NUM = 109;
END-EXEC.
```

Remember, the preceding embedded SQL statement is compiled to generate an executable statement. Therefore, the statement is fixed permanently and can never change—unless, of course, the programmer changes it. Each time the program runs, it deletes the same row. In short, the preceding code is good only for the first run; all subsequent runs will more than likely generate an error. Clearly, this code would be more useful if we could specify a variable to indicate the employee number to be deleted.

In embedded SQL, all host variables are preceded by a colon (":"). The host variables may be used to send data from the host language to the embedded SQL, or they can be used to receive the data from the embedded SQL. To use a host variable, you must first declare it in the host language. Common practice is to use similar host variable names as the SQL source attributes. For example, if you are using COBOL, you would define the host variables in the Working Storage section. Then you would refer to them in the embedded SQL section by preceding them with the colon (":"). For example, to delete an employee whose employee number is represented by the host variable "W_EMP_NUM", you would write the following code:

```
EXEC SQL
    DELETE FROM EMPLOYEE WHERE EMP_NUM = :W_EMP_NUM;
END-EXEC.
```

At run time, the host variable value will be used to execute the embedded SQL statement. What happens if the employee you are trying to delete doesn't exist in the database? How do you know that the statement has been completed without errors? As we mentioned previously, the Embedded SQL standard defines a SQL communication area to hold status and error information. In COBOL, such an area is known as the SQLCA area and is defined in the Data Division, as shown next.

```
EXEC SQL
    INCLUDE SQLCA
END-EXEC.
```

The SQLCA area contains two variables for status and error reporting. Table 7.11 shows some of the main values returned by the variables and their meaning.

TABLE 7.11 **SQL STATUS AND ERROR REPORTING VARIABLES**

VARIABLE NAME	VALUE	EXPLANATION
SQLCODE		Old-style error reporting supported for backward compatibility only. Returns an integer value (positive or negative).
	0	Successful completion of command.
	100	No data. The SQL statement did not return any rows, or did not select, update, or delete any rows.
	-999	Any negative value indicates an error occurred.
SQLSTATE		Added by SQL-92 standard to provide predefined error codes. Defined as a character string (5 characters long).
	"00000"	Successful completion of command.
		Multiple values in the format "XXYYY" where: XX-> represents the class code. YYY-> represents the subclass code.

The following embedded SQL code illustrates the use of the SQLCODE within a COBOL program.

```
EXEC SQL
    SELECT EMP_LNAME, EMP_LNAME INTO :W_EMP_FNAME, :W_EMP_LNAME
        WHERE EMP_NUM = :W_EMP_NUM;
END-EXEC.
IF SQLCODE = 0 THEN
    PERFORM DATA_ROUTINE
ELSE
    PERFORM ERROR_ROUTINE
END-IF.
```

In this example, the SQLCODE host variable is checked to determine if the query completed successfully. If that is the case, the DATA_ROUTINE is performed. Otherwise, the ERROR_ROUTINE is performed.

Just as with PL/SQL, embedded SQL requires the use of cursors to hold data from a query that returns more than one value. If COBOL is used, the cursor can be declared either in the Working Storage Section or in the

Procedure Division. The cursor must be declared and processed as you learned earlier in this chapter. To declare a cursor you use the syntax shown in the following example:

```
EXEC SQL
     DECLARE PROD_CURSOR FOR
          SELECT P_CODE, P_DESCRIPT FROM PRODUCT
               WHERE P_ONHAND > (SELECT AVG(P_ONHAND) FROM PRODUCT);
END-EXEC.
```

Next, you must open the cursor to make the cursor ready for processing:

```
EXEC SQL
     OPEN PROD_CURSOR;
END-EXEC.
```

To process the data rows in the cursor you use the FETCH command to retrieve one row of data at a time and place the values in the host variables. The SQLCODE must be checked to ensure that the FETCH command completed successfully. This section of code typically constitutes part of a routine in the COBOL program. Such a routine is executed with the PERFORM command. For example:

```
EXEC SQL
     FETCH PROD_CURSOR INTO :W_P_CODE, :W_P_DESCRIPT;
END-EXEC.
IF SQLCODE = 0 THEN
     PERFORM DATA_ROUTINE
ELSE
     PERFORM ERROR_ROUTINE
END-IF.
```

When all rows have been processed, you close the cursor, as shown next:

```
EXEC SQL
     CLOSE PROD_CURSOR;
END-EXEC.
```

Thus far, you have seen examples of embedded SQL in which the programmer used predefined SQL statements and parameters. Therefore, the end users of the programs are limited to the actions that were specified in the application programs. This style of embedded SQL is known as **static SQL**. This means that, the SQL statements will not change while the application is running. For example, the SQL statement may read like this:

```
SELECT P_CODE, P_DESCRIPT, P_ONHAND, P_PRICE
     FROM PRODUCT
          WHERE P_PRICE > 100;
```

Note that the attributes, tables, and conditions are known in the preceding SQL statement. Unfortunately, end users seldom work in a static environment. They are more much likely to require the flexibility of defining their data access requirements on the fly. Therefore, the end user requires that SQL be as dynamic as their data access requirements.

Dynamic SQL is a term used to describe an environment in which the SQL statement is not known in advance, but instead is generated at run time. In a dynamic SQL environment, a program can generate the SQL statements at run time that are required to respond to ad hoc queries. In such an environment, neither the programmer

nor the end user is likely to know precisely what kind of queries are to be generated or how those queries are to be structured. For example, a dynamic SQL equivalent of the preceding example could be:

```
SELECT :W_ATTRIBUTE_LIST
    FROM :W_TABLE
        WHERE :W_CONDITION;
```

Note that the attribute list and the condition are not known until the end user specifies them. W_TABLE, W_ATRIBUTE_LIST, and W_CONDITION are text variables that contain the end-user input values used in the query generation. Because the program uses the end-user input to build the text variables, the end user can run the same program multiple times to generate different outputs. For example, in one instance the end user might want to know what products have a price less than $100; in another case, the end user might want to know how many units of a given product are available for sale at any given moment.

Although dynamic SQL is clearly flexible, such flexibility carries a price. Dynamic SQL tends to be much slower that static SQL and it requires more computer resources (overhead). In addition, you are more likely to find different levels of support and incompatibilities between DBMS vendors.

S U M M A R Y

SQL provides relational set operators to combine the output of two queries to generate a new relation. The UNION and UNION ALL set operators combine the output of two (or more) queries and produce a new relation with all unique (UNION) or duplicate (UNION ALL) rows from both queries. The INTERSECT relational set operator selects only the common rows. The MINUS set operator selects only the rows that are different. UNION, INTERSECT, and MINUS require union-compatible relations.

If the DBMS does not support the INTERSECT and MINUS relational set operators, you can use alternate SQL queries.

Operations that join tables can be classified as inner joins and outer joins. An inner join is the traditional join in which only rows that meet a given criteria are selected. The join criteria can be an equality condition (natural join or equijoin) or an inequality condition (theta join). An outer join returns the matching rows as well as the rows with unmatched attribute values for one or both tables to be joined.

The SQL standard defines the CROSS, INNER, and OUTER join specifications.

A natural join returns all rows with matching values in the matching columns and eliminates duplicate columns. This style of query is used when the tables share a common attribute with a common name. One important difference between the natural join and the "old-style" join syntax is that the natural join does not require the use of a table qualifier for the common attributes.

Joins may use keywords such as USING and ON. If the USING clause is used, the query will return only the rows with matching values in the column indicated in the USING clause; this column must exist in both tables. If the ON clause is used, the query will return only the rows that meet the specified join condition. The join condition will typically include an equality comparison expression of two columns. (The columns may or may not share the same name, but must have comparable data types.) OUTER JOINs may be classified as LEFT, RIGHT, and FULL.

Subqueries and correlated queries are used when it is necessary to process data based on *other* processed data. That is, the query uses results that were previously unknown and that are generated by another query. Subqueries may be used with the FROM, WHERE, IN, and HAVING clauses in a SELECT statement. A subquery may return a single row or multiple rows. If the subquery returns multiple rows, you need to use the ANY or ALL

multirow operators. Subqueries can also be used within a SELECT statement's attribute list; in this case, the subquery is called an inline subquery. However, an inline subquery can only return a single value.

Most subqueries are executed in a serial fashion. That is, the outer query initiates the data request and then the inner subquery is executed. The inner query's output is then used by the outer query, which then executes the remaining query action. This process continues until the last outer query executes. In contrast, a correlated subquery is a subquery that is executed once for each row in the outer query. This process is similar to the typical nested loop in a programming language. A correlated subquery is so named because the inner query is related to the outer query because the inner query references a column of the outer subquery.

SQL functions are used to extract or transform data. The most frequently used functions are date and time functions that generate current dates and times. The results of the function output can be used to store values in a database table, to serve as the basis for the computation of derived variables, or to serve as a basis for data comparisons. For example, a date function may be used to record the current date when a record is generated, it may be used to compute the number of days that have elapsed since an invoice was generated, or it may be used to see if a payment was made more than 90 days after the invoice date to determine whether to assign an interest penalty. The date and time function format is vendor-specific. For example, MS Access uses DATE() to return the current date and Oracle uses SYSDATE. Aside from time and date functions, there are numerous other numeric and string functions. Finally, there are conversion functions that are used to convert one data format to another.

Oracle sequences may be used to generate values to be assigned to a record. For example, a sequence may be used to number invoices automatically. MS Access uses an autonumber data type to generate numeric sequences.

Procedural SQL (PL/SQL) may be used to create triggers, stored procedures, and PL/SQL functions. A trigger is procedural SQL code that is automatically invoked by the DBMS upon the occurrence of a specified data manipulation event (UPDATE, INSERT, or DELETE). Triggers are critical to proper database operation and management. They help automate various transaction and data management processes and they can be used to enforce constraints that cannot be enforced at the DBMS design and implementation levels. Although triggers are independent objects, they are associated with database tables. Each database table may have one or more triggers. When you delete a table, all associated trigger objects are deleted.

A stored procedure is a named collection of SQL statements. Just like database triggers, stored procedures are stored in the database. One of the major advantages of stored procedures is that they can be used to encapsulate and represent complete business transactions. For example, a stored procedure may be created to represent a product sale, a credit update, or the addition of a new customer. Because many SQL statements can be written within a single stored procedure, which is then executed as a single transaction, the use of stored procedures substantially reduces network traffic and increases system performance. Stored procedures help reduce code duplication by creating unique PL/SQL modules that are called by the application programs, thereby minimizing the chance of errors and the cost of application development and maintenance.

If SQL statements are designed to return more than one value inside the PL/SQL code, a cursor is needed. A cursor is a special construct used in procedural SQL to hold the data rows returned by a SQL query. You can think of a cursor as a reserved area of memory in which the output of the query is stored, like an array holding columns and rows. Cursors are held in a reserved memory area in the DBMS server, rather than in the client computer.

There are two types of cursors: implicit and explicit. An implicit cursor is automatically created in procedural SQL when the SQL statement returns only one value. An explicit cursor holds the output of a SQL statement that may return two or more rows.

Embedded SQL refers to the use of SQL statements within an application programming language such as COBOL, C++, ASP, Java, or ColdFusion. The language in which the SQL statements are embedded is called the host language. Embedded SQL is still the most common approach to maintaining procedural capabilities in DBMS-based applications.

anonymous PL/SQL block

batch update routine

correlated subquery

cross join

cursor

dynamic SQL

embedded SQL

explicit cursor

host language

implicit cursor

inner join

outer join

persistent stored module (PSM)

procedural SQL (PL/SQL)

row-level trigger

statement-level trigger

static SQL

stored function

stored procedure

trigger

updatable view

R E V I E W Q U E S T I O N S

1. The relational set operators UNION, INTERSECT, and MINUS work properly only if the relations are union-compatible. What does *union-compatible* mean and how would you check for this condition?

2. What is the difference between UNION and UNION ALL? Write the syntax for each.

3. Suppose that you have two tables: EMPLOYEE and EMPLOYEE_1. The EMPLOYEE table contains the records for three employees: Alice Cordoza, John Cretchakov, and Anne McDonald. The EMPLOYEE_1 table contains the records for employees John Cretchakov and Mary Chen. Given this information, what is the query output for the UNION query? (List the query output.)

4. Given the employee information in Question 3, what is the query output for the UNION ALL query? (List the query output.)

5. Given the employee information in Question 3, what is the query output for the INTERSECT query? (List the query output.)

6. Given the employee information in Question 3, what is the query output for the MINUS query? (List the query output.)

7. What is a CROSS JOIN? Give an example of its syntax.

8. What three join types are included in the OUTER JOIN classification?

9. Using tables named T1 and T2, write a query example for each of the three join types you described in Question 8. Assume that T1 and T2 share a common column named C1.

10. What is a subquery and what are its basic characteristics?

11. What is a correlated subquery? Give an example.

12. What MS Access/SQL Server function should you use to calculate the number of days between the current date and January 25, 1999?

13. What Oracle function should you use to calculate the number of days between the current date and January 25, 1999?

14. Suppose that a PRODUCT table contains two attributes, PROD_CODE and VEND_CODE. These two attributes have values of ABC, 125, DEF, 124, GHI, 124, and JKL, 123, respectively. The VENDOR table contains a single attribute, VEND_CODE, with values 123, 124, 125, and 126, respectively. (The VEND_CODE attribute in the PRODUCT table is a foreign key to the VEND_CODE in the VENDOR table.) Given this information, what would be the query output for:
 a. A UNION query based on these two tables?
 b. A UNION ALL query based on these two tables?
 c. An INTERSECT query based on these two tables?
 d. A MINUS query based on these two tables?

15. What Oracle string function should you use to list the first three characters of a company's EMP_LNAME values? Give an example, using a table named EMPLOYEE.

16. What is an Oracle sequence? Write its syntax.

17. What is a trigger and what is its purpose? Give an example.

18. What is a stored procedure and why is it particularly useful? Give an example.

19. What is embedded SQL and how is it used?

20. What is dynamic SQL and how does it differ from static SQL?

P R O B L E M S

Use the database tables in Figure P7.1 as the basis for problems 1–18.

A D D I T I O N A L C D C O N T E N T

The **Ch07_SimpleCo** database is located on the student CD in the **\Databases\Student\Ch07** folder. To duplicate this data set in Oracle, we have included scripts files in the **\SQL\Ch07\Ch07_SimpleCo** folder.

FIGURE P7.1 CH07_SIMPLECO DATABASE TABLES

1. Create the tables. (Use the MS Access example shown in Figure P7.1 to see what table names and attributes to use.)

2. Insert the data into the tables you created in Problem 1.

3. Write the query that will generate a combined list of customers (from the tables CUSTOMER and CUSTOMER_2) that do not include the duplicate customer records. (Note that only the customer named Juan Ortega shows up in both customer tables.)

4. Write the query that will generate a combined list of customers to include the duplicate customer records.

5. Write the query that will show only the duplicate customer records.

6. Write the query that will generate only the records that are unique to the CUSTOMER_2 table.

7. Write the query to show the invoice number, the customer number, the customer name, the invoice date, and the invoice amount for all the customers with a customer balance of $1,000 or more.

8. Write the query that will show the invoice number, the average invoice amount, and the difference between the average invoice amount and the actual invoice amount.

9. Write the query that will write Oracle sequences to produce automatic customer number and invoice number values. Start the customer numbers at 1000 and the invoice numbers at 5000.

10. Modify the CUSTOMER table to included two new attributes: CUST_DOB and CUST_AGE. Customer 1000 was born on March 15, 1969 and customer 1001 was born on December 22, 1977.

11. Assuming you completed Problem 10, write the query that would list the names and ages of your customers.

12. Assuming that the CUSTOMER table contains a CUST_AGE attribute, write the query to update the values in this attribute. (*Hint*: use the results of the previous query.)

13. Write the query that would list the average age of your customers. (Assume that the CUSTOMER table has been modified to include the CUST_DOB and the derived CUST_AGE attribute.)

14. Write the trigger to update the CUST_BALANCE in the CUSTOMER table when a new invoice record is entered. (Assume that the sale is a credit sale.) Test the trigger using the following new INVOICE record:

 8005, 1001, '27-APR-04', 225.40

 Name the trigger **trg_updatecustbalance**.

15. Write a procedure to add a new customer to the CUSTOMER table. Use the following values in the new record:

 1002, 'Rauthor', 'Peter', 0.00

 Name the procedure **prc_cust_add**. Run a query to see if the record has been added.

16. Write a procedure to add a new invoice record to the INVOICE table. Use the following values in the new record:

 8006, 1000, '30-APR-04', 301.72

 Name the procedure **prc_invoice_add**. Run a query to see if the record has been added.

17. Write a trigger to update the customer balance when an invoice is deleted. Name the trigger **trg_updatecustbalance2**.

18. Write a procedure to delete an invoice giving the invoice number as a parameter. Name the procedure **prc_inv_delete**. Test the procedure by deleting invoices 8005 and 8006.

N O T E

The following problem sets can serve as the basis for a class project or case.

Use the **Ch07_SaleCo** database to work Problems 19–22.

A D D I T I O N A L C D C O N T E N T

The **Ch07_SaleCo2** database used in Problems 19–22 is located on the student CD in the **\Databases\Student\Ch07** folder. To duplicate this data set in Oracle, we have included script files in the **\SQL\Ch07\Ch07_SaleCo2** folder.

19. Create a trigger named **trg_line_total** to write the LINE_TOTAL value into the LINE table every time you add a new LINE row. (The LINE_TOTAL value is the product of the LINE_UNITS and the LINE_PRICE values.)

20. Create a trigger named **trg_line_prod** that will automatically update the product quantity on hand for each product sold after a new LINE row is added.

21. Create a stored procedure named **prc_inv_amounts** to update the INV_SUBTOTAL, INV_TAX, and INV_TOTAL. The procedure takes the invoice number as a parameter. The INV_SUBTOTAL is the sum of the LINE_TOTAL amounts for the invoice, the INV_TAX is the product of the INV_SUBTOTAL and the tax rate (8%), and the INV_TOTAL is the sum of the INV_SUBTOTAL and the INV_TAX.

FIGURE P7.19 CH07_SALECO2 DATABASE TABLES

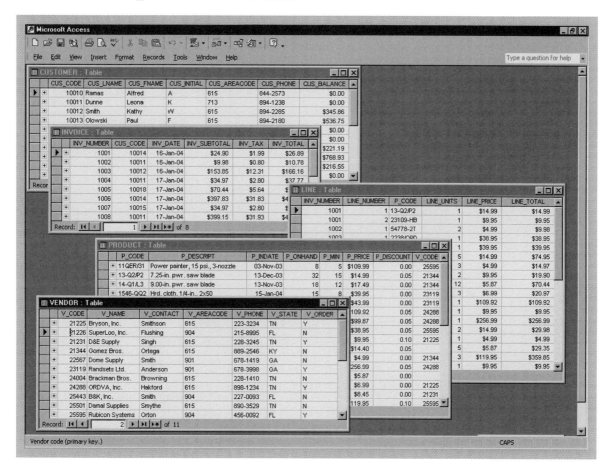

22. Create a procedure named **prc_cus_balance_update** that will take the invoice number as a parameter and update the customer balance. (*Hint*: You can use the DECLARE section to define a TOTINV numeric variable that holds the computed invoice total.)

Use the **Ch07_AviaCo** database to work Problems 23–34.

ADDITIONAL CD CONTENT

The **Ch07_AviaCo** database used for Problems 23–34 is located on the student CD in the \Databases\Student\Ch07 folder. To duplicate this data set in Oracle, we have included script files in the \SQL\Ch07\Ch07_AviaCo folder.

FIGURE P7.23 CH07_AVIACO DATABASE TABLES

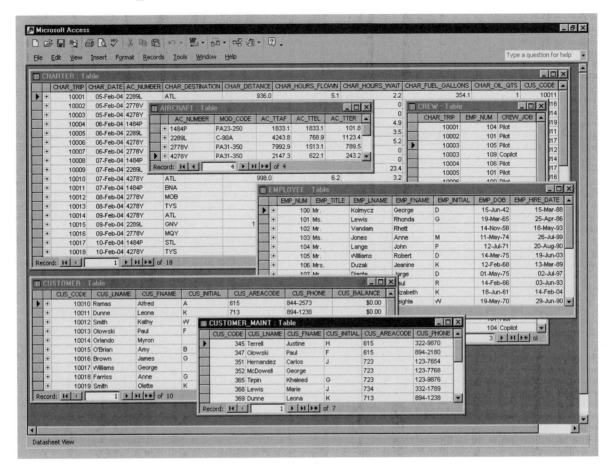

23. Modify the MODEL table to add the following attribute and insert the values shown in Table P7.23.

TABLE P7.23 THE NEW ATTRIBUTE FOR THE MODEL TABLE

ATTRIBUTE NAME	ATTRIBUTE DESCRIPTION	ATTRIBUTE TYPE	ATTRIBUTE VALUES
MOD_WAIT_CHG	Waiting charge per hour for each model:	Numeric	$100 for C-90A $50 for PA23-250 $75 for PA31-350

24. Write the queries to update the MOD_WAIT_CHG attribute values based on Problem 23.
25. Modify the CHARTER table to add the attributes shown in Table P7.25.

TABLE P7.25 **THE NEW ATTRIBUTES FOR THE CHARTER TABLE**

ATTRIBUTE NAME	ATTRIBUTE DESCRIPTION	ATTRIBUTE TYPE
CHAR_WAIT_CHG	Waiting charge for each model. (Copied from the MODEL table.)	Numeric
CHAR_FLT_CHG_HR	Flight charge per mile for each model. (Copied from the MODEL table using the MOD_CHG_MILE attribute.)	Numeric
CHAR_FLT_CHG	Flight charge. (Calculated by CHAR_HOURS_FLOWN x CHAR_FLT_CHG_HR.)	Numeric
CHAR_TAX_CHG	CHAR_FLT_CHG x tax rate (8%).	Numeric
CHAR_TOT_CHG	CHAR_FLT_CHG + CHAR_TAX_CHG.	Numeric
CHAR_PYMT	Amount paid by customer.	Numeric
CHAR_BALANCE	Balance remaining after payment.	Numeric

26. Write the sequence of commands required to update the CHAR_WAIT_CHG attribute values in the CHARTER table. (*Hint*: Use either an updatable view or a stored procedure.)

27. Write the sequence of commands required to update the CHAR_FLT_CHG_HR attribute values in the CHARTER table. (*Hint*: Use either an updatable view or a stored procedure.)

28. Write the command required to update the CHAR_FLT_CHG attribute values in the CHARTER table.

29. Write the command required to update the CHAR_TAX_CHG attribute values in the CHARTER table.

30. Write the command required to update the CHAR_TOT_CHG attribute values in the CHARTER table.

31. Modify the PILOT table to add the attribute shown in Table P7.31.

TABLE P7.31 **THE NEW ATTRIBUTE FOR THE PILOT TABLE**

ATTRIBUTE NAME	ATTRIBUTE DESCRIPTION	ATTRIBUTE TYPE
PIL_PIC_HRS	Pilot in command (PIC) hours. Updated by adding the CHARTER table's CHAR_HOURS_FLOWN to the PIL_PIC_HRS when the CREW table shows the CREW_JOB to be pilot.	Numeric

32. Create a trigger named **trg_char_hours** that will automatically update the AIRCRAFT table when a new CHARTER row is added. Use the CHARTER table's CHAR_HOURS_FLOWN to update the AIRCRAFT table's AC_TTAF, AC_TTEL, and AC_TTER values.

33. Create a trigger named **trg_pic_hours** that will automatically update the PILOT table when a new CREW row is added and the CREW table uses a 'pilot' CREW_JOB entry. Use the CHARTER table's CHAR_HOURS_FLOWN to update the PILOT table's PIL_PIC_HRS only when the CREW table uses a 'pilot' CREW_JOB entry.

34. Create a trigger named **trg_cust_balance** that will automatically update the CUSTOMER table's CUST_BALANCE when a new CHARTER row is added. Use the CHARTER table's CHAR_TOT_CHG as the update source. (Assume that all charter charges are charged to the customer balance.)

8

DATABASE DESIGN

In this chapter, you will learn:

❖ That successful database design must reflect the information system of which the database is a part

❖ That successful information systems are developed within a framework known as the Systems Development Life Cycle (SDLC)

❖ That, within the information system, the most successful databases are subject to frequent evaluation and revision within a framework known as the Database Life Cycle (DBLC)

❖ How to conduct evaluation and revision within the SDLC and DBLC frameworks

❖ What database design strategies exist: top-down vs. bottom-up design, and centralized vs. decentralized design

Database design is a part of a larger picture called an information system. Within the information system we not only collect, store, and retrieve data, but also transform that raw data into useful information. Given timely and relevant information, decision makers are able to make the decisions that allow organizations to prosper and grow. Database designs that fail to recognize that the database is part of this larger whole are not likely to be successful. That is, database designers must recognize that the database is a critical means to an end, rather than an end in itself. (Managers want the database to serve their management needs, but too many databases seem to require that managers alter their routines to fit the database requirements.) To use an analogy, you don't own a car just for the sake of car ownership. Rather, ownership of a car serves the purpose of providing you with reliable transportation.

Information systems don't just happen; they are the product of a carefully staged development process. To determine the need for an information system and to establish its limits, we use a process known as systems analysis. Within systems analysis, we create the actual information system through a process known as systems development.

An organization does not exist within a static environment. Even its own growth creates a demand for new and expanded information. Therefore, as soon as an information system is completed, it is subject to a continuous evaluation. Do the system's applications still perform the intended functions? How can they be enhanced to meet changing requirements? Does the database still serve its intended functions as the data repository from which the applications produce the appropriate information?

Sooner or later even the finest information system becomes obsolete. If it no longer fits the business requirements, how do you create a system that does? To paraphrase a timeworn notion: the only constant in the systems development arena is change. This continuous process of creation, maintenance, enhancement, and replacement constitutes the Systems Development Life Cycle.

A similar cycle applies to databases. The database is created, maintained, and enhanced. When even enhancement can no longer stretch the database's usefulness, and the database can no longer perform its functions adequately, it may have to be replaced. This Database Life Cycle is carefully traced in this chapter.

8.1 CHANGING DATA INTO INFORMATION

As important as data are, raw facts are seldom immediately useful to a decision maker. (After all, staring at a stack of computer printouts that list page after page of collected facts is not likely to lead to sudden insight.) You learned in Chapter 1, "Database Systems," that the decision maker needs information, which is defined as data processed and presented in a meaningful form.

To the decision maker, information is as different from data as a brick building is from bricks. And, just as a collection of bricks can be used to create buildings of any size, shape, and function, data must be put together or transformed to produce information that can create insight. Such transformations can be as simple as tabulating the data, thereby making certain data patterns more obvious. For example, if managers believe that the age and gender distribution of clients might help define appropriate marketing strategies, a simple data organization technique known as *cross-classification* is likely to be helpful. The results of such cross-classifications are displayed in cross-classification tables. An example of a cross-classification table, showing the age and gender distribution for 9,040 clients, is shown in Table 8.1. Such tables are probably the most frequently used data organization devices, and—certainly in our experience—they are used in conjunction with a wide variety of organizational structures and managerial styles.

TABLE 8.1 A SIMPLE CROSS-CLASSIFICATION TABLE: TRANSFORMING DATA INTO INFORMATION

	UNDER 25	25 – 45	46 – 60	61 AND OVER	TOTAL
Male	119	1,892	2,641	876	5,528
Female	48	1,117	1,805	542	3,512
Total	167	3,009	4,446	1,418	9,040

The word **transformation** is used to describe any process that changes data into information. A data transformation can be based on the simple tabulation that produced the summary displayed in Table 8.1; it can be based on a detailed report, or it can be based on a complex set of statistical procedures. Whatever transformation method is used, decision making is usually based on information produced by some form of data transformation. Without data we cannot make transformations; without transformations we cannot create the information on which decisions are based.

8.2 THE INFORMATION SYSTEM

Basically, a *database* is a carefully designed and constructed repository of facts. The fact repository is a part of a larger whole known as an information system. An **information system** provides for data collection, storage, and retrieval. It also facilitates the transformation of data into information, and the management of both data and information. Thus, a complete information system is composed of people, hardware, software, the database(s), application programs, and procedures. **Systems analysis** is the process that establishes the need for and the extent of an information system. The process of creating an information system is known as **systems development**.

> **N O T E**
>
> This chapter is not meant to cover all the aspects of systems analysis and development—these usually are covered in a separate course or book. However, this chapter is designed to help you develop a better understanding of database design, implementation, and management issues that are affected by the information system in which the database is a critical component.

Within the framework of systems development, applications transform data into the information that forms the basis for decision making. Applications usually produce formal reports, tabulations, and graphic displays designed to produce insight. Figure 8.1 illustrates that every application is composed of two parts: the data and the code (program instructions) by which the data are transformed into information. Data and code work together to represent real-world business functions and activities. At any given moment, physically stored data represent, in effect, a snapshot of the business. But the picture is not complete until we understand the business activities that are represented by the code.

FIGURE 8.1 GENERATING INFORMATION FOR DECISION MAKING

The performance of an information system depends on a triad of factors:

- Database design and implementation
- Application design and implementation
- Administrative procedures

Although this book emphasizes the (arguably most important!) database design and implementation segment of the triad, failure to address the other two segments will likely yield a poorly functioning information system. Creating a sound information system is hard work: systems analysis and development require much planning to ensure that all the activities will interface with each other, that they will complement each other, and that they will be completed on time.

In a broad sense, the term **database development** is used to describe the process of database design and implementation. The primary objective in database design is to create complete, normalized, nonredundant (to the extent possible), and fully integrated conceptual, logical, and physical database models. The implementation phase includes creating the database storage structure, loading data into the database, and providing for data management.

In order to make the procedures discussed in this chapter broadly applicable, we will focus on the elements that are common to all information systems. Consequently, most of the processes and procedures described in this chapter do not depend on the size, type, or complexity of the database being implemented. For example, regardless of the size, type, and complexity of the database, you must plan, analyze, and design. However, the procedures that would be used to design a small database—such as a database for a neighborhood shoe store—do not precisely scale up to the procedures that would be needed to design a database for a large corporation or even a segment of such a corporation. To use an analogy, building a small house requires a blueprint, just as building the Golden Gate Bridge does, but the bridge requires much more complex and further-ranging planning, analysis, and design than the house. For example, the house designer won't have to consider the complex effects of wind on the structure or on the surrounding structures.

Although the processes and procedures will be shown to conform to well-defined standards, particular circumstances might dictate different general approaches to database design. Once you are familiar with the standards, we will address top-down versus bottom-up, and centralized versus decentralized design issues in the last section of this chapter.

8.3 THE SYSTEMS DEVELOPMENT LIFE CYCLE (SDLC)

The **Systems Development Life Cycle** (**SDLC**) traces the history (life cycle) of an information system. Perhaps more important to the system designer, the SDLC provides the "big picture" within which the database design and application development can be mapped out and evaluated.

You might ask yourself why we even examine the SDLC. Isn't this book supposed to be about database design and implementation? The answer to this question is simple: database design takes place within an information system. In short, it is difficult to divorce database design from the SDLC, and vice versa.

As illustrated in Figure 8.2, the traditional SDLC is divided into five phases: planning, analysis, detailed systems design, implementation, and maintenance. The SDLC is an iterative rather than a sequential process. For example, the details of the feasibility study might help refine the initial assessment, and the details discovered during the user requirements portion of the SDLC might help refine the feasibility study.

Because the Database Life Cycle (DBLC) fits into and resembles the Systems Development Life Cycle (SDLC), a brief description of the SDLC is in order.

8.3.1 PLANNING

The SDLC's planning phase yields a general overview of the company and its objectives. An initial assessment of the information-flow-and-extent requirements must be made during this discovery portion of the SDLC. Such an assessment should answer some important questions:

- *Should the existing system be continued?* If the information generator does its job well, there is no point in modifying or replacing it. To quote an old saying, "If it ain't broke, don't fix it."

- *Should the existing system be modified?* If the initial assessment indicates deficiencies in the extent and flow of the information, minor (or even major) modifications may be in order. The participants in the initial assessment must keep in mind the distinction between wishes and needs, if modifications are considered.

- *Should the existing system be replaced?* The initial assessment might indicate that the current system's flaws are beyond fixing. Given the effort required to create a new system, a careful distinction between wishes and needs is perhaps even more important in this case than it is in modifying the system.

FIGURE 8.2 THE SYSTEMS DEVELOPMENT LIFE CYCLE (SDLC)

Phase		Action(s)	Section
Planning	○ ○	**Initial assessment** **Feasibility study**	**8.3.1**
Analysis	● ● ●	**User requirements** **Existing system evaluation** **Logical system design**	**8.3.2**
Detailed systems design	●	**Detailed system specification**	**8.3.3**
Implementation	● ●	**Coding, testing, and debugging** **Installation, fine-tuning**	**8.3.4**
Maintenance	● ● ●	**Evaluation** **Maintenance** **Enhancement**	**8.3.5**

Participants in the SDLC's initial assessment must begin to study and evaluate alternate solutions. If it is decided that a new system is necessary, the next question is, "Is it feasible?" The feasibility study must address the following:

- *The technical aspects of hardware and software requirements.* The decisions might not (yet) be vendor-specific, but they must address the nature of the hardware (PC, minicomputer, or mainframe) and the software requirements (single- or multiuser operating systems, database type and software, programming languages to be used by the applications, and so on).

- *The system cost.* The admittedly mundane question "Can we afford it?" is crucial. (And the answer to this question might force a careful review of the initial assessment!) It bears repeating that a million-dollar solution to a thousand-dollar problem is not defensible.

8.3.2 ANALYSIS

Problems defined during the planning phase are examined in greater detail during the analysis phase. We must make a macroanalysis of both individual needs and organizational needs, addressing questions such as:

- What are the precise requirements of the current system's end users?
- Do those requirements fit into the overall information requirements?

The analysis phase of the SDLC is, in effect, a thorough *audit* of user requirements.

The existing hardware and software systems are also studied during the analysis phase. The result of analysis should be a better understanding of the system's functional areas, actual and potential problems, and opportunities.

End users and system designer(s) must work together to identify processes and to uncover potential problem areas. Such cooperation is vital to defining the appropriate performance objectives by which the new system can be judged.

Along with studying user requirements and the existing systems, the analysis phase includes the creation of a logical systems design. The logical design must specify the appropriate conceptual data model, inputs, processes, and expected output requirements.

In creating a logical design, the designer might use tools such as data flow diagrams (DFD), hierarchical input process output (HIPO) diagrams, or entity relationship (ER) diagrams. The database design's data-modeling activities take place at this point to discover and describe all entities and their attributes, and the relationships among the entities within the database.

Defining the logical system also yields functional descriptions of the system's components (modules) for each process within the database environment. All data transformations (processes) are described and documented, using such systems analysis tools as DFDs. The conceptual data model is validated against these processes.

8.3.3 DETAILED SYSTEMS DESIGN

In the detailed systems design phase, the designer completes the design of the system's processes. This includes all the necessary technical specifications for the screens, menus, reports, and other devices that might be used to help make the system a more efficient information generator. The steps are laid out for conversion from the old to the new system. Training principles and methodologies are also planned and must be submitted for management approval.

> **NOTE**
>
> Because we have focused our attention on the details of the systems design process, we have not—until this point—*explicitly* recognized the fact that management approval is needed at all stages of the process. Such approval is needed because a GO decision requires funding. There are many "GO/NO GO" decision points along the way to a completed systems design!

8.3.4 IMPLEMENTATION

During the implementation phase, the hardware, the DBMS software, and application programs are installed, and the database design is implemented. During the initial stages of the implementation phase, the system enters into a cycle of coding, testing, and debugging, until it is ready to be delivered. The actual database is created, and the system is customized by the creation of tables and views, user authorizations, and so on.

The database contents may be loaded interactively or in batch mode, using a variety of methods and devices:

- Customized user programs
- Database interface programs
- Conversion programs that import the data from a different file structure, using batch programs, a database utility, or both

The system is subjected to exhaustive testing until it is ready for use. Traditionally, the implementation and testing of a new system took 50 to 60 percent of the total development time. However, the advent of sophisticated application generators and debugging tools has substantially decreased coding and testing time. After testing is concluded, the final documentation is reviewed and printed, and end users are trained. The system is in full operation at the end of this phase but will be continuously evaluated and fine-tuned.

8.3.5 MAINTENANCE

Almost as soon as the system is operational, end users begin to request changes in it. Those changes generate system maintenance activities, which can be grouped into three types:[1]

1. *Corrective maintenance* in response to systems errors.

2. *Adaptive maintenance* due to changes in the business environment.

3. *Perfective maintenance* to enhance the system.

Because every request for structural change requires retracing the SDLC steps, the system is, in a sense, always at some stage of the SDLC!

Each system has a predetermined operational life span. The actual operational life span of a system depends on its perceived utility. There are several reasons for reducing the operational life of certain systems. Rapid technological change has been one reason, especially for systems based on processing speed and expandability. Another common reason is the cost of maintaining a system.

If the system's maintenance cost is high, its value becomes suspect. **Computer-assisted systems engineering** (**CASE**) technology—for example, System Architect or Visio Professional software—helps make it possible to produce better systems within a reasonable amount of time and at a reasonable cost. In addition, the more structured, better documented, and especially *standardized* implementation of CASE-produced applications tends to prolong the operational life of systems by making them easier and cheaper to update and maintain.[2]

8.4 THE DATABASE LIFE CYCLE (DBLC)

Within the larger information system, the database, too, is subject to a life cycle. The **Database Life Cycle** (**DBLC**) contains six phases (Figure 8.3): database initial study, database design, implementation and loading, testing and evaluation, operation, and maintenance and evolution.

8.4.1 THE DATABASE INITIAL STUDY

If a designer has been called in, chances are that the current system has failed to perform functions deemed vital by the company. (You don't call the plumber unless the pipes leak!) So, in addition to examining the current system's operation within the company, the designer must determine how and why the current system fails. This means spending a lot of time talking with (mostly listening to!) end users. Although database design is a very technical business, it is also very people-oriented. Database designers must be excellent communicators, and they must have finely tuned interpersonal skills.

Depending on the complexity and scope of the database environment, the database designer might be a lone operator or part of a systems development team composed of a project leader, one or more senior systems analysts, and one or more junior systems analysts. We will use the word *designer* generically to cover a wide range of design team compositions.

[1] See "Software Maintenance Revisited: A Product Life Cycle Perspective," E. Reed Doke and Neil E. Swanson, Information Executive 4(1), Winter 1991, pp. 8–11. The date on this reference may cause you to consider it outdated, but it remains relevant today. Although the software environment changes with dizzying frequency, especially with respect to its interface, most of the underlying principles of software design, implementation, and management have enjoyed remarkable longevity.

[2] For example, if you have used Microsoft's Visio Professional to develop your database design, you already know that Visio Professional tests the internal consistency of your ERDs when you create the relationships. Visio Professional will implement the FKs according to the design's entity types (weak, strong) and the nature of the relationship (identifying, non-identifying) between those entities. When you see the implementation's result, you will immediately know whether the results are what you wanted. In addition, if there are circular arguments in the design, Visio Professional will make that clear. Therefore, you will be able to spot design problems before they become cemented into the implementation.

FIGURE 8.3 THE DATABASE LIFE CYCLE (DBLC)

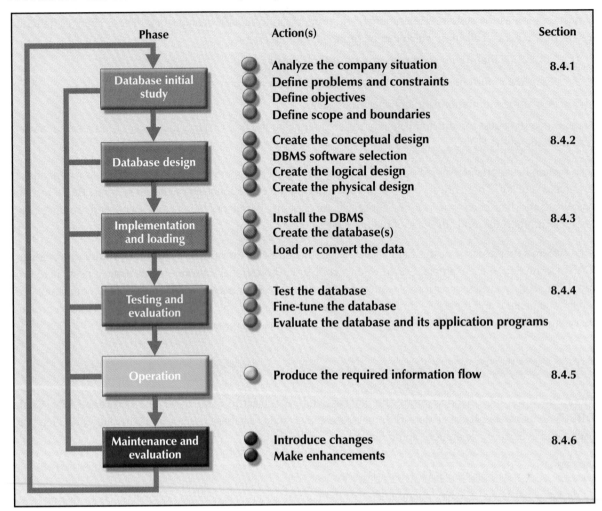

The overall purpose of the database initial study is to:

- Analyze the company situation.
- Define problems and constraints.
- Define objectives.
- Define scope and boundaries.

Figure 8.4 depicts the interactive and iterative processes required to complete the first phase of the DBLC successfully. As you examine Figure 8.4, note that the database initial study phase leads to the development of the database system objectives. Using Figure 8.4 as our discussion template, let us now examine each of its components in greater detail.

Analyze the Company Situation

The word *analysis* literally means "to break up any whole into its parts so as to find out their nature, function, and so on." The *company situation* describes the general conditions in which a company operates, its organizational structure, and its mission. To analyze the company situation, the database designer must discover what the company's operational components are, how they function, and how they interact.

FIGURE 8.4 SUMMARY OF ACTIVITIES IN THE DATABASE INITIAL STUDY

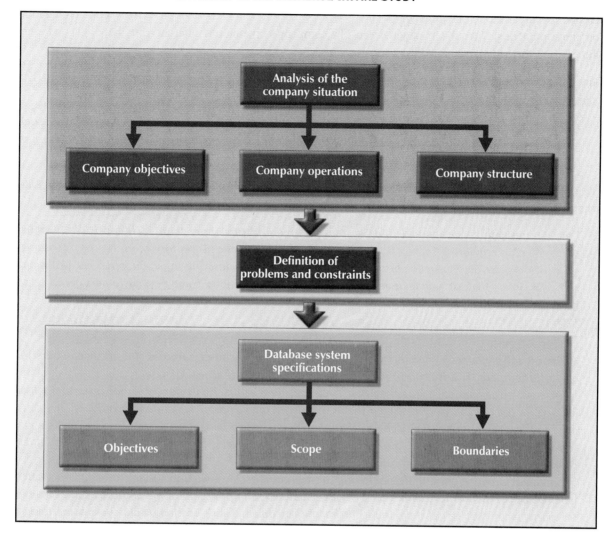

These issues must be resolved:

- *What is the organization's general operating environment, and what is its mission within that environment?* The design must satisfy the operational demands created by the organization's mission. For example, a mail-order business is likely to have (database) operational requirements that are quite different from those of a manufacturing concern.

- *What is the organization's structure?* Knowing who controls what and who reports to whom is quite useful when you are trying to define required information flows, specific report and query formats, and so on.

Define Problems and Constraints

The designer has both formal and informal sources of information. If the company has existed for any length of time, it already has some kind of system in place (either manual or computer-based). How does the existing system function? What input does the system require? What documents does the system generate? How is the system output used? By whom? Studying the paper trail can be very informative. Quite aside from the official version of the system's operation, there is also the more informal real version; the designer must be shrewd enough to see how these differ.

The problem definition process might initially appear to be unstructured. Company end users are often unable to describe the larger scope of company operations precisely or to identify the real problems encountered during company operations. Often the managerial view of a company's operation is different from that of the end users who perform the actual routine work.

During the initial problem definition process, the designer is likely to collect very broad problem descriptions. For example, note these concerns expressed by the president of a fast-growing manufacturing company:

> *Although the rapid growth is gratifying, members of the management team are concerned that such growth is beginning to undermine the ability to maintain a high customer-services standard and, perhaps worse, to diminish manufacturing standards control.*

The problem definition process quickly leads to a host of general problem descriptions. For example, the marketing manager comments:

> *I'm working with an antiquated filing system. We manufacture more than 1,700 specialty machine parts. When a regular customer calls in, we can't get a very quick inventory scan. If a new customer calls in, we can't do a current parts search by using a simple description, so we often do a machine setup for a part that we have in inventory. That's wasteful. Besides, some new customers get irritated if we can't give a quick response.*

The production manager comments:

> *At best it takes hours to generate the reports I need for scheduling purposes. I don't have hours for quick turnarounds. It's difficult to manage what I don't have information about.*
>
> *I don't get quick product request routing. Take machine setup. Right now, I've got operators either waiting for the right stock or getting it themselves when a new part is scheduled for production. I can't afford to have an operator doing chores that a much lower-paid worker ought to be doing. There's just too much waiting around with the current scheduling. I'm losing too much time, and my schedules back up. Our overtime bill is ridiculous.*
>
> *I sometimes produce parts that are already in inventory, because we don't seem to be able to match what we've got in inventory with what we have scheduled. Shipping yells at me because I can't turn out the parts, when they've got them in inventory one bay down. That's costing us big bucks sometimes.*
>
> *New reports can take days or even weeks to get to my office. And I need a ton of reports to schedule personnel, downtime, training, et cetera. I can't get new reports that I need NOW. What I need is the ability to get quick updates on percent defectives, percent rework, the effectiveness of training, you name it. I need such reports by shift, by date, by any characteristic I can think of, to help me manage scheduling, training, you name it.*

A machine operator comments:

> *It takes a long time to set my stuff up. If I get my schedule banged up because John doesn't get the paperwork on time, I wind up looking for setup specs, startup material, bin assignments, and other stuff. Sometimes, I spend two or three hours just setting up. Now you know why I can't meet schedules. I try to be productive, but I'm spending too much time getting ready to do my job.*

After the initial declarations, the database designer must continue to carefully probe in order to generate additional information that will help define the problems within the larger framework of company operations. How does the marketing manager's customer problem fit within the broader set of marketing department activities? How does the solution to the customer problem help meet marketing department and other company objectives? How do the marketing department's activities relate to those of the other departments? This last question is especially important. Note that there are common threads in the problems described by the marketing and production department managers. If the inventory query capability can be improved, both departments are likely to find some simple solutions to at least some of the problems.

Finding precise answers is important, especially about the (operational) relationships among business units. If a proposed system will solve the marketing department's problems but exacerbate those of the production department, we have not made much progress. Using an analogy, suppose your home water bill is too high. You have determined the problem: the faucets leak. The solution? You step outside and cut off the water supply to the house. Is that an adequate solution? Or, would the replacement of faucet washers do a better job of solving the problem? You may find the leaking faucet scenario simplistic, yet almost any experienced database designer can find (admittedly more complicated and less obvious) similar instances of so-called database problem solving.

Even the most complete and accurate problem definition does not always lead to the perfect solution. The real world usually intrudes to limit the design of even the most elegant database by imposing constraints. Such constraints include time, budget, personnel, and a host of others. If you must have a solution within a month and within a $12,000 budget, a solution that takes two years to develop at a cost of $100,000 is not a solution. The designer must learn to distinguish between what's perfect and what's possible.

Define Objectives

A proposed database system must be designed to help solve at least the major problems identified during the problem discovery process. As the list of problems unfolds, several common sources are likely to be discovered. For example, both the marketing manager and the production manager in our example seem to be plagued by inventory inefficiencies. If the designer can create a database that sets the stage for more efficient parts management, both departments gain. The initial objective, therefore, might be to create an efficient inventory query and management system.

> **NOTE**
>
> In trying to develop solutions, the database designer must look for the *source* of the problems. There are (too) many cases of database systems that failed to satisfy the end users because they were designed to treat the *symptoms* of the problems rather than their source.

Note that the initial study phase also yields proposed problem solutions. The designer's job is to make sure that the database system objectives, as seen by the designer, correspond to those envisioned by the end user(s). In any case, the database designer must begin to address the following questions:

- What is the proposed system's initial objective?
- Will the system interface with other existing or future systems in the company?
- Will the system share the data with other systems or users?

Define Scope and Boundaries

The designer must recognize the existence of two sets of limits: scope and boundaries. The system's **scope** defines the extent of the design, according to operational requirements. Will the database design encompass the entire organization, one or more departments within that organization, or one or more functions of a single department? The designer must know the "size of the ballpark." Knowing the database design scope helps define the required data structures, the type and number of entities, the physical size of the database, and so on.

The proposed system is also subject to limits known as **boundaries**, which are external to the system. Has any designer ever been told, "We have all the time in the world. Use an unlimited budget and use as many people as are needed to make the design come together?" Boundaries are also imposed by existing hardware and software. Ideally, the designer can choose the hardware and software that will best accomplish the system goals. In fact, software selection is an important aspect of the Systems Development Life Cycle. Unfortunately, in the real world a system often must be designed around existing hardware. The scope and boundaries thus become the factors

that force the design into a specific mold, and it is the designer's job to design the best system possible within those constraints. (Note that problem definitions and the objectives sometimes must be reshaped to meet the system scope and boundaries.)

8.4.2 DATABASE DESIGN

The second phase focuses on the design of the database model that will support company operations and objectives. We are now ready to begin what is arguably the most critical DBLC phase: making sure that the final product meets user and system requirements. In the process of database design, we must concentrate on the data characteristics required to build the database model. At this point, we have two views of the data within the system: the business view of data as a source of information, and the designer's view of the data structure, its access, and the activities required to transform the data into information. Figure 8.5 contrasts these views. Note that the different views can be summarized by looking at the terms *what* and *how*. Defining data is an integral part of the DBLC's second phase.

FIGURE 8.5 TWO VIEWS OF DATA: BUSINESS MANAGER AND DESIGNER

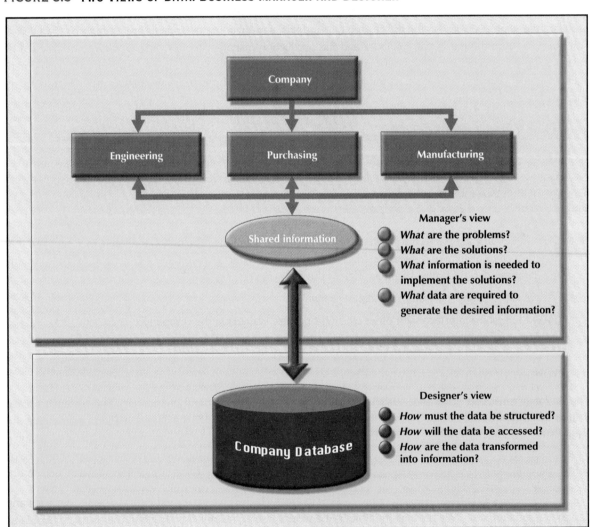

As you begin to examine the procedures required to complete the design phase in the DBLC, remember these points:

- The process of database design is loosely related to the analysis and design of a larger system. The data component is only one element of a larger information system.

- The systems analysts or systems programmers are in charge of designing the other system components. Their activities create the procedures that will help transform the data within the database into useful information.

- The database design does not constitute a sequential process. Rather it is an iterative process that provides continuous feedback designed to trace previous steps.

Within phase 2 (database design) of the Database Life Cycle, we can trace the database design process depicted in Figure 8.6. Look at the procedure flow in the figure.

ADDITIONAL CD CONTENT

In **Appendixes D and E** on the student CD, "The University Lab: Conceptual Design" and "The University Lab: Conceptual Design Verification, Logical Design, and Implementation," you will learn exactly what happens during each of these stages when we develop real databases.

FIGURE 8.6 PROCEDURE FLOW IN THE DATABASE DESIGN

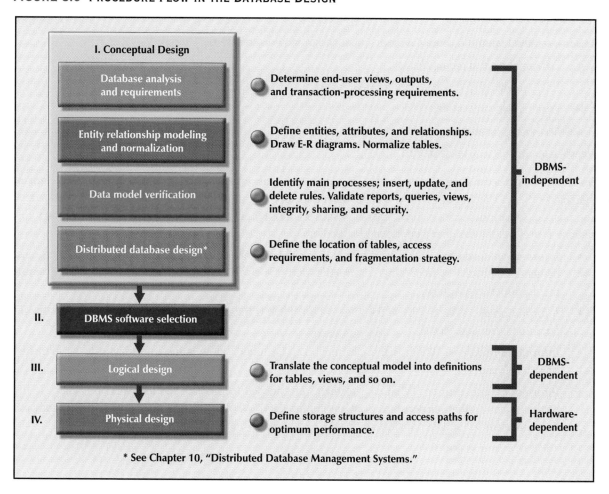

We are now ready to discuss in detail each of the components in Figure 8.6. Knowing those details will help you successfully design and implement databases in a real-world setting.

I. Conceptual Design

In the **conceptual design** stage, data modeling is used to create an abstract database structure that represents real-world objects in the most realistic way possible. The conceptual model must embody a clear understanding of the business and its functional areas. At this level of abstraction, the type of hardware and/or database model to be used might not yet have been identified. Therefore, the design must be software- and hardware-independent, so that the system can be set up within any hardware and software platform chosen later.

Keep in mind the following **minimal data rule**:

> ### *All that is needed is there, and all that is there is needed.*

In other words, make sure that all data needed are in the model, and that all data in the model are needed. All data elements required by the database transactions must be defined in the model, and all data elements defined in the model must be used by at least one database transaction.

However, as you apply the minimal data rule, avoid an excessive short-term bias. Focus not only on the immediate data needs of the business, but also on the future data needs. Thus, the database design must leave room for future modifications and additions, thereby ensuring that the business's investment in information resources will endure.

As you examine Figure 8.6, note that conceptual design requires four steps:

1. Data analysis and requirements
2. Entity relationship modeling and normalization
3. Data model verification
4. Distributed database design

We will examine each of these in turn.

Data Analysis and Requirements The first step in conceptual design is to discover the data element characteristics. An effective database is an information factory that produces key ingredients for successful decision making. Appropriate data element characteristics are those that can be transformed into appropriate information. Therefore, the designer's efforts are focused on:

- *Information needs.* What kind of information is needed—that is, what output (reports and queries) must be generated by the system? What information does the current system generate, and to what extent is that information adequate?

- *Information users.* Who will use the information? How is the information to be used? What are the different end-user data views?

- *Information sources.* Where is the information to be found? How is the information to be extracted once it is found?

- *Information constitution.* What data elements are needed to produce the information? What are the data attributes? What relationships exist among the data? What is the data volume? How frequently are the data used? What data transformations are to be used to generate the required information?

The designer obtains the answers to these questions from different sources, to compile the necessary information. Note these sources:

- *Developing and gathering end-user data views.* The database designer and the end user(s) interact to jointly develop a precise description of end-user data views. In turn, the end-user data views will be used to help identify the database's main data elements.

- *Direct observation of the current system: existing and desired output.* The end user usually has an existing system in place (manual or computer-based). The designer reviews the existing system to identify the data and their characteristics. The designer examines the input forms and files (tables) in order to discover the data type and volume. If the end user already has an automated system in place, the designer carefully examines the current and desired reports in order to describe the data required to support the reports.

- *Interfacing with the systems design group.* As we noted earlier in this chapter, the database design process is part of the Systems Development Life Cycle (SDLC). In some cases, the systems analyst in charge of designing the new system will also develop the conceptual database model. (This is usually true in a microcomputer environment.) In other cases, the database design is considered to be part of the database administrator's job. The presence of a database administrator (DBA) usually implies the existence of a formal data-processing department. The DBA designs the database according to the specifications created by the systems analyst.

In order to develop an accurate data model, the designer must have a thorough understanding of the company's data types, extent, and uses. But data do not, by themselves, yield the required understanding of the total business. From a database point of view, the collection of data becomes meaningful only when business rules are defined. Remember from Chapter 2 ("Data Models") that a *business rule* is a brief and precise narrative description of a policy, procedure, or principle within a specific organization's environment. Business rules, derived from a detailed description of an organization's operations, help to create and enforce actions within that organization's environment. If business rules are written properly, they define entities, attributes, relationships, connectivities, cardinalities, and constraints.

To be effective, business rules must be easy to understand, and they must be widely disseminated to ensure that every person in the organization shares a common interpretation of such rules. Business rules describe, in simple language, the main and distinguishing characteristics of the data *as viewed by the company.* Examples of business rules are:

- A customer may make many payments on account.
- Each payment on account is credited to only one customer.
- A customer may generate many invoices.
- Each invoice is generated by only one customer.

Given their critical role in database design, business rules must not be established casually. Poorly defined or inaccurate business rules lead to database designs and implementations that fail to meet the needs of the organization's end users.

Ideally, business rules are derived from a formal *description of operations.* As its name implies, a **description of operations** is a document that provides a precise, detailed, up-to-date, and thoroughly reviewed description of the activities that define an organization's operating environment. (To the database designer, the operating environment is both the data sources and the data users.) Naturally, the organization's operating environment is dependent

on the organization's mission. For example, the operating environment of a university would be quite different from that of a steel manufacturer, an airline, or a nursing home. Yet, no matter how different the organizations may be, the *data analysis and requirements* component of the database design process is enhanced when the data environment and data use are described accurately and precisely within a description of operations.

In a business environment, the main sources of information for the description of operations—and, therefore, of business rules—are company managers, policy makers, department managers, and written documentation such as company procedures, standards, or operations manuals. A faster and more direct source of business rules is direct interviews with end users. Unfortunately, because perceptions differ, the end user can be a relatively less reliable source when it comes to specifying business rules. For example, a maintenance department mechanic may believe that any mechanic may initiate a maintenance procedure, when actually only mechanics with inspection authorization may perform such a task. Such a distinction may seem trivial, but it has major legal consequences. Although end users are crucial contributors to the development of business rules, it pays to verify end-user perceptions. We have discovered too often that interviews with several people who perform the same job may yield very different perceptions of what their job components are. While such a discovery may point to "management problems," that general diagnosis does not help the database designer. Given the discovery of such problems, the database designer's job is to reconcile such differences and verify the results of the reconciliation in order to ensure that the business rules are appropriate and accurate.

Knowing the business rules enables the designer to understand fully how the business works and what role the data plays within company operations. Consequently, the designer must identify the company's business rules and analyze their impact on the nature, role, and scope of data.

Business rules yield several important benefits in the design of new systems:

- They help standardize the company's view of data.
- They constitute a communications tool between users and designers.
- They allow the designer to understand the nature, role, and scope of the data.
- They allow the designer to understand business processes.
- They allow the designer to develop appropriate relationship participation rules and foreign key constraints. (See Chapter 4, "Entity Relationship Modeling.")

The last point is especially noteworthy: whether a given relationship is mandatory or optional is usually a function of the applicable business rule!

Entity Relationship (ER) Modeling and Normalization Before creating the ER model, the designer must communicate and enforce appropriate standards to be used in the documentation of the design. The standards include the use of diagrams and symbols, documentation writing style, layout, and any other conventions to be followed during documentation. Designers often overlook this very important requirement, especially when they are working as members of a design team. Failure to standardize documentation often means a failure to communicate later. And communications failures often lead to poor design work. In contrast, well-defined and enforced standards make design easier and promise (but do not guarantee) a smooth integration of all system components.

Because the business rules usually define the nature of the relationship(s) among the entities, the designer must incorporate them into the conceptual model. The process of defining business rules and developing the conceptual model using ER diagrams can be described using the steps shown in Table 8.2.[3]

[3] See "Linking Rules to Models," Alice Sandifer and Barbara von Halle, *Database Programming and Design*, 4(3), March 1991, pp. 13–16. Although the source seems dated, this remains the current standard. The technology has changed substantially, but the process has not.

TABLE 8.2 **DEVELOPING THE CONCEPTUAL MODEL USING ER DIAGRAMS**

STEP	ACTIVITY
1	Identify, analyze, and refine the business rules.
2	Identify the main entities, using the results of Step 1.
3	Define the relationships among the entities, using the results of Steps 1 and 2.
4	Define the attributes, primary keys, and foreign keys for each of the entities.
5	Normalize the entities.
6	Complete the initial ER diagram.
7	Have the main end users verify the model in Step 6 against the data, information, and processing requirements.
8	Modify the ER diagram, using the results of Step 7.

Some of the steps listed in Table 8.2 take place concurrently. And some, such as the normalization process, can generate the demand for additional entities and/or attributes, thereby causing you to revise the ER model. For example, while identifying two main entities, the designer might also identify the composite bridge entity that represents the many-to-many relationship between those two main entities.

To review, suppose that you are creating a conceptual model for the JollyGood Movie Rental Corporation, whose end users want to track customer movie rental activities. The simple ER diagram presented in Figure 8.7 shows a composite entity that helps track customers and their video rentals. Business rules define the optional nature of the relationships between the entities VIDEO and CUSTOMER depicted in Figure 8.7. (For example, customers are not required to check out a video. A video need not be checked out in order to exist on the shelf. A customer may rent many videos, and a video may be rented by many customers.) Especially note the composite RENTAL entity that connects the two main entities.

FIGURE 8.7 **A COMPOSITE ENTITY**

As you are very likely to discover, the initial ER model may be subjected to several revisions before it meets the system's requirements. Such a revision process is quite natural. You should remember that the ER model is a communications tool as well as a design blueprint. Therefore, the initial ER model should give rise to such questions as "Is this really what you meant?" when you continue to meet with the proposed system users. For example, the ERD shown in Figure 8.7 is far from complete. Clearly, many more attributes must be defined and the dependencies must be checked before the design can be implemented. In addition, the design cannot yet support the typical video rental transactions environment. For example, each video is likely to have many copies available for rental purposes. However, if the VIDEO entity shown in Figure 8.7 is used to store the titles as well as the copies, the design triggers the data redundancies shown in Table 8.3.

TABLE 8.3 DATA REDUNDANCIES IN THE **VIDEO** TABLE

VIDEO_ID	VIDEO_TITLE	VIDEO_COPY	VIDEO_CHG	VIDEO_DAYS
SF-12345FT-1	Adventures on Planet III	1	$2.45	1
SF-12345FT-2	Adventures on Planet III	2	$2.45	1
SF-12345FT-3	Adventures on Planet III	3	$2.45	1
WE-5432GR-1	TipToe Canu and Tyler 2: A Journey	1	$1.99	2
WE-5432GR-2	TipToe Canu and Tyler 2: A Journey	2	$1.99	2

The initial ERD shown in Figure 8.7 must be modified to reflect the answer to the question: "Is more than one copy available for each title?" Also, payment transactions must be supported. (You will have an opportunity to modify this initial design in Problem 5 at the end of the chapter.)

We should not leave the impression that the ER modeling activities (entity/attribute definition, normalization, verification) take place in a precise sequence. In fact, once you have completed the initial ER model, chances are that you will move back and forth among the activities until you are satisfied that the ER model accurately represents a database design that is capable of meeting the required system demands. (The activities often take place in parallel, and the process is iterative.) Figure 8.8 summarizes the ER modeling process interactions. Figure 8.9 summarizes the array of design tools and information sources that the designer can use to produce the conceptual model.

FIGURE 8.8 ER MODELING IS AN ITERATIVE PROCESS BASED ON MANY ACTIVITIES

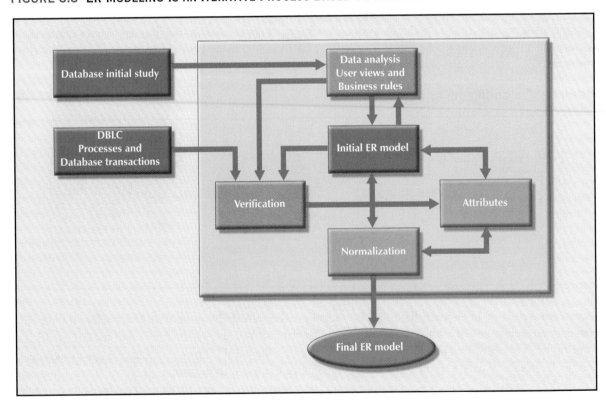

FIGURE 8.9 **CONCEPTUAL DESIGN TOOLS AND INFORMATION SOURCES**

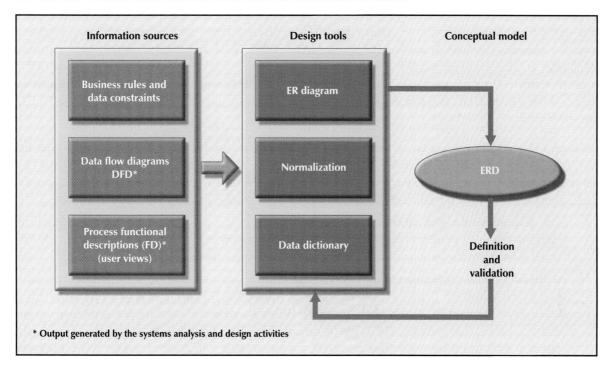

All objects (entities, attributes, relations, views, and so on) are defined in a data dictionary, which is used in tandem with the normalization process to help eliminate data anomalies and redundancy problems. During this ER modeling process, the designer must:

- Define entities, attributes, primary keys, and foreign keys. (The foreign keys serve as the basis for the relationships among the entities.)

- Make decisions about adding new primary key attributes in order to satisfy end-user and/or processing requirements.

- Make decisions about the treatment of multivalued attributes.

- Make decisions about adding derived attributes to satisfy processing requirements.

- Make decisions about the placement of foreign keys in 1:1 relationships.

- Avoid unnecessary ternary relationships.

- Draw the corresponding ER diagram.

- Normalize the data model.

- Include all the data element definitions in the data dictionary.

- Make decisions about standard naming conventions.

The naming conventions requirement is important, yet it is frequently ignored, at the designer's peril. Real database design is generally accomplished by teams. Therefore, it is important to ensure that the team members work in an environment in which naming standards are defined and enforced. Proper documentation is crucial to the successful completion of the design. Therefore, it is very useful to establish procedures that are, in effect, self-documenting.

Although we established some useful entity and attribute naming conventions in Chapter 4, we will revisit them in greater detail here. However, keep in mind that such conventions are sometimes subject to constraints imposed by the DBMS software. In an enterprise-wide database environment, the lowest common denominator rules. For example, Microsoft Access finds the attribute name LINE_ITEM_NUMBER to be perfectly acceptable. Many older DBMSs, however, are likely to truncate such long names when they are exported from one DBMS to another, thus making documentation more difficult. Therefore, table-exporting requirements may dictate the use of shorter names. (The same is true for data types. For example, many older DBMSs cannot handle OLE or memo formats.)

Although it is impossible for us to know what DBMS you will be using, we will use naming conventions that are likely to be acceptable across a reasonably broad range of DBMSs. These naming conventions also will meet the self-documentation requirements to the greatest extent possible. As the older DBMSs fade from the scene, the naming conventions will be more broadly applicable. So, let's try to adhere to the following conventions:

- Use descriptive entity and attribute names wherever possible. For example, in the University Computer Lab database in **Appendix D**, the USER entity contains data about the lab's users, and the LOCATION entity is related to the location of the ITEMs that the lab director wants to track.

- Composite entities usually are assigned a name that describes the relationship they represent. For example, in the University Computer Lab database in **Appendix D**, an ITEM may be stored in many LOCATIONs, and a LOCATION may have many ITEMs stored in it. Therefore, the composite (bridge) entity that links ITEM and LOCATION will be named STORAGE. Occasionally, the designer finds it necessary to show which entities are being linked by the composite entity. In such cases the composite entity name may borrow segments of those entity names. For example, STU_CLASS may be the composite entity that links STUDENT and CLASS. However, this latter naming convention might make the next one more cumbersome, so use it sparingly. (We prefer the composite entity name ENROLL, to indicate that the STUDENT enrolls in a CLASS.)

- An attribute name should be descriptive, and it should contain a prefix that helps identify the table in which it is found. For our purposes, the maximum prefix length will be five characters. For example, the VENDOR table might contain attributes such as VEND_ID and VEND_PHONE. Similarly, the ITEM table might contain attribute names such as ITEM_ID and ITEM_DESCRIPTION. The advantage of this naming convention is that it immediately identifies a table's foreign key(s). For example, if the EMPLOYEE table contains such attributes as EMP_ID, EMP_LNAME, and DEPT_CODE, it is immediately obvious that DEPT_CODE is the foreign key that probably links EMPLOYEE to DEPARTMENT. Naturally, the existence of relationships and table names that start with the same characters might dictate that you bend this naming convention occasionally, as you can see in the next bulleted item.

- If one table is named ORDER and its weak counterpart is named ORDER_ITEM, the prefix ORD will be used to indicate an attribute originating in the ORDER table. The ITEM prefix will identify an attribute originating in the ITEM table. Clearly, we cannot use ORD as a prefix to the attributes originating in the ORDER_ITEM table, so we use a combination of characters such as OI as the prefix to the ORDER_ITEM attribute names. In spite of this limitation, note that it is generally possible to assign prefixes that identify an attribute's origin. (Keep in mind that some RDBMSs use a "reserved word" list. For example, ORDER may be interpreted as a reserved word in a SELECT statement. In that case, use a table name other than ORDER!)

As you can tell, we are not always able to adhere strictly to the naming conventions. Sometimes, the requirement to limit name lengths makes the attribute or entity names less descriptive than we would like. Also, with a large number of entities and attributes in a complex design, we may have to become rather inventive about using proper attribute name prefixes. Then, those prefixes are less helpful in identifying the precise source of the attribute. Nevertheless, the consistent use of prefixes will reduce sourcing doubts significantly. For example, while the prefix CO does not obviously relate to the CHECK_OUT table, it is just as obvious that it does not originate in WITHDRAW, ITEM, or USER.

Adherence to the naming conventions we have just described has served us well. In fact, a common refrain seems to be "I didn't know why you made such a fuss over naming conventions, but now that I'm doing this stuff for real, I've become a true believer."

Data Model Verification The ER model must be verified against the proposed system processes in order to corroborate that the intended processes can be supported by the database model. Verification requires that the model be run through a series of tests against:

- End-user data views and their required transactions: SELECT, INSERT, UPDATE, and DELETE operations and queries and reports.

- Access paths, security, and concurrency control.

> **N O T E**
>
> **Concurrency control** is a feature that allows simultaneous access to a database, while preserving data integrity. A detailed discussion of the need for concurrency control accompanies Table 8.6.

- Business-imposed data requirements and constraints.

Revision of the original database design starts with a careful reevaluation of the entities, followed by a detailed examination of the attributes that describe these entities. This process serves several important purposes:

- The emergence of the attribute details may lead to a revision of the entities themselves. Perhaps some of the components we first believed to be entities will, instead, turn out to be attributes within other entities. Or, what we originally considered to be an attribute may turn out to contain a sufficient number of subcomponents to warrant the introduction of one or more new entities.

- The focus on attribute details can provide clues about the nature of relationships as they are defined by the primary and foreign keys. Improperly defined relationships lead to implementation problems first and to application development problems later.

- To satisfy processing and/or end-user requirements, it might be useful to create a new primary key to replace an existing primary key. For example, in the invoicing example illustrated in Figure 3.30 in Chapter 3, "The Relational Database Model," we created a primary key composed of INV_NUMBER and LINE_NUMBER to replace the original primary key composed of INV_NUMBER and PROD_CODE. We did this to ensure that the items in the invoice would always appear in the same order as they were entered. Sometimes we might create a single-attribute primary key to replace an existing multiple-attribute primary key, to simplify queries and to increase processing speed.

- Unless the entity details—that is, the attributes and their characteristics—are precisely defined, it is difficult to evaluate the extent of the design's normalization. Knowledge of the normalization levels helps guard against undesirable redundancies.

- Because we already have at least a rough database design blueprint, the careful review of its details is likely to lead to revisions. These revisions will help ensure that the design is capable of meeting the end-user requirements.

Because real-world database design is generally done by teams, you should strive to define the design's major components as modules. (A **module** is an information system component that handles a specific function, such as inventory, orders, payroll, and so on.) The creation and use of modules accomplish several important ends:

- The modules, and even the segments within them, can be delegated to design groups within teams, greatly speeding up the development work.

- The modules simplify the design work. The large number of entities within a complex design can be daunting. The modules each contain a much more manageable number of entities.

- The modules can be prototyped quickly. Implementation and applications programming trouble spots can be identified more readily. (Quick prototyping is also a great confidence builder.)

- Even if the entire system can't be brought online quickly, the implementation of one or more modules will demonstrate that progress is being made and that at least part of the system is ready to begin serving the end users.

As useful as modules are, they represent ER model fragments. Fragmentation creates a potential problem: the fragments may not include all of the ER model's components and may, therefore, not be able to support all of the required processes. To avoid this problem, the modules must be verified against the complete ER model. This verification process is detailed in Table 8.4.

TABLE 8.4 **THE ER MODEL VERIFICATION PROCESS**

STEP	ACTIVITY
1	Identify the ER model's central entity.
2	Identify each module and its components.
3	Identify each module's transaction requirements: Internal: Updates/Inserts/Deletes/Queries/Reports External: Module interfaces
4	Verify all processes against the ER model.
5	Make all necessary changes suggested in Step 4.
6	Repeat Steps 2 through 5 for all modules.

Keep in mind that the verification process requires the continuous verification of business transactions as well as system and user requirements. The verification sequence must be repeated for each of the system's modules. Figure 8.10 illustrates the iterative nature of the process.

FIGURE 8.10 **ITERATIVE ER MODEL VERIFICATION PROCESS**

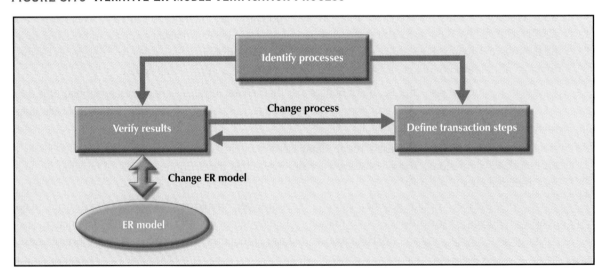

We start the verification process by selecting the central (most important) entity. The central entity is defined in terms of its participation in most of the model's relationships, and it is the focus for most of the system's operations. In other words, to identify the central entity, the designer selects the entity involved in the greatest number of relationships. (In the ER diagram, it is the one that has more lines connected to it than any other.)

The next step is to identify the module or subsystem to which the central entity belongs, and to define that module's boundaries and scope. The entity belongs to the module that uses it most frequently. Once each module is identified, the central entity is placed within the module's framework, to let us focus our attention on the module's details.

Within the central entity/module framework, we must:

- Ensure the module's cohesivity. The term **cohesivity** describes the strength of the relationships found among the module's entities. A module must display *high cohesivity*—that is, the entities must be strongly related, and the module must be complete and self-sufficient.

- Analyze each module's relationships with other modules to address module coupling. **Module coupling** describes the extent to which modules are independent of one another. Modules must display *low coupling*, indicating that they are independent of other modules. Low coupling decreases unnecessary intermodule dependencies, thereby allowing the creation of a truly modular system and eliminating unnecessary relationships among entities.

Processes may be classified according to their:

- Frequency (daily, weekly, monthly, yearly, or exceptions).
- Operational type (INSERT or ADD, UPDATE or CHANGE, DELETE, queries and reports, batches, maintenance, and backups).

All identified processes must be verified against the ER model. If necessary, appropriate changes are implemented. The process verification is repeated for all the model's modules. You can expect that additional entities and attributes will be incorporated into the conceptual model during its validation!

At this point we have defined a conceptual model that is hardware- and software-independent. Such independence ensures the system's portability across platforms. Portability may extend the database's life by making it possible to migrate to another DBMS and/or another hardware platform.

Distributed Database Design Portions of a database may reside in different physical locations. Processes that access the database may also vary from one location to another. For example, a retail process and a warehouse storage process are very likely to be found in different physical locations. If the database process is to be distributed across the system, the designer must also develop the data distribution and allocation strategies for the database. We will examine the design complications introduced by distributed processes in detail in Chapter 10, "Distributed Database Management Systems."

II. DBMS Software Selection

The selection of DBMS software is critical to the information system's smooth operation. Consequently, the proposed DBMS software's advantages and disadvantages should be carefully studied. The end user must be made aware of the limitations of both the DBMS and the database, in order to avoid false expectations.

Although the factors affecting the purchasing decision vary from company to company, some of the most common are:

- *Cost*. Purchase, maintenance, operational, license, installation, training, and conversion costs.

- *DBMS features and tools*. Some database software includes a variety of tools that facilitate the application development task. For example, the availability of query by example (QBE), screen painters, report generators, application generators, data dictionaries, and so on, helps to create a more pleasant work environment for both the end user and the application programmer. Database administrator facilities, query facilities, ease of use, performance, security, concurrency control, transaction processing, and third-party support also influence DBMS software selection.

- *Underlying model*. Hierarchical, network, relational, object/relational, or object.

- *Portability*. Across platforms, systems, and languages.

- *DBMS hardware requirements*. Processor(s), RAM, disk space, etc.

III. Logical Design

Logical design is used to translate the conceptual design into the internal model for a selected database management system (DBMS), such as DB2, SQL Server, Oracle, and Access. Therefore, the logical design is software-dependent.

Logical design requires that all objects in the model be mapped to the specific constructs used by the selected database software. For example, the logical design for a relational DBMS includes the specifications for the tables, indexes, views, transactions, access authorizations, and so on. In the following discussion we will convert a small portion of the simple conceptual model shown in Figure 8.11 into a logical design based on the relational model.

FIGURE 8.11 A SIMPLE CONCEPTUAL MODEL

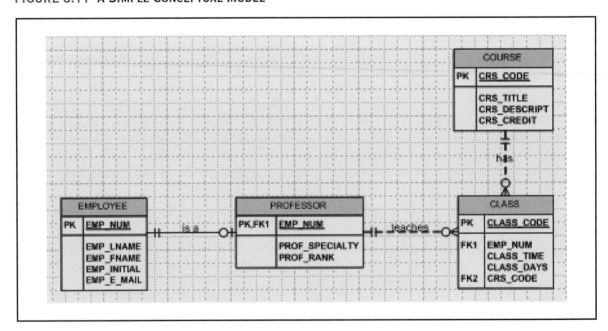

The translation of the conceptual model in Figure 8.11 requires the definition of the attribute domains, the design of the required tables, and appropriate access restriction formats. For example, the domain definitions for the CLASS_CODE, CLASS_DAYS, and CLASS_TIME attributes displayed in the CLASS entity in Figure 8.11 are written this way:

CLASS_CODE is a valid class code.

 Type: numeric

 Range: low value = 1000 high value = 9999

 Display format: 9999

 Length: 4

CLASS_DAYS is a valid day code.

 Type: character

 Display format: XXX

 Valid entries: MWF, TTh, M, T, W, Th, F, S

 Length: 3

CLASS_TIME is a valid time.

 Type: character

 Display format: 99:99 (24-hour clock)

 Display range: 06:00 to 22:00

 Length: 5

The logical design's tables must correspond to the entities (EMPLOYEE, PROFESSOR, COURSE, and CLASS) shown in Figure 8.11's conceptual design, and the table columns must correspond to the attributes specified in the conceptual design. For example, the initial table layout for the COURSE table might look like Table 8.5.

TABLE 8.5 **SAMPLE LAYOUT FOR THE COURSE TABLE**

CRS_CODE	CRS_TITLE	CRS_DESCRIPT	CRS_CREDIT
CIS-4567	Database Systems Design	Design and implementation of database systems. Includes conceptual design, logical design, implementation, and management. Prerequisites: CIS 2040, CIS-2345, and CIS 3680 and upper division standing.	4
QM-3456	Statistics II	Statistical applications. Course requires use of statistical software (MINITAB and SAS) to interpret data. Prerequisites: MATH-2345 and QM-2233.	3

The right to use the database is also specified during the logical design phase. Who will be allowed to use the tables, and what portion(s) of the table(s) will be available to which users? Within a relational framework the answers to such questions require the definition of appropriate access rights and views.

In short, the logical design translates the software-independent conceptual model into a software-dependent model by defining the appropriate domain definitions, the required tables, and the necessary access restrictions. The stage is now set to define the physical requirements that allow the system to function within the selected hardware environment.

IV. Physical Design

Physical design is the process of selecting the data storage and data access characteristics of the database. The storage characteristics are a function of the types of devices supported by the hardware, the type of data access methods supported by the system, and the DBMS. Physical design affects not only the location of the data in the storage device(s) but also the performance of the system.

Physical design is best described as a very technical job, more typical of the client/server and mainframe world than of the PC world. Yet even in the more complex minicomputer and mainframe environments, modern database software has assumed much of the burden of the physical portion of the design and its implementation.

Physical design is particularly important in the older hierarchical and network models described in Chapter 1. Relational databases are more insulated from physical layer details than the older hierarchical and network models. Yet, in spite of the fact that relational models tend to hide the complexities of the computer's physical characteristics, the performance of relational databases is affected by physical-level characteristics. For example, performance can be affected by the characteristics of the storage media, such as seek time, sector and block (page) size, buffer pool size, and number of disk platters and read/write heads. In addition, such factors as the creation of an index can have a considerable performance effect on the relational database's data access speed and efficiency.

Even the type of data request must be analyzed carefully in order to determine the optimum access method for meeting the application requirements, establishing the data volume to be stored, and estimating the performance. Some DBMSs automatically reserve the space required to store the database definition and the user's data in permanent storage devices. This is to ensure that the data are stored in sequentially adjacent locations, thereby reducing data access time and increasing system performance.

Physical design becomes more complex when data are distributed at different locations, because the performance is affected by the communication media's throughput.

Given such complexities as these, it is not surprising that designers favor database software that hides as many of the physical-level activities as possible.

In the preceding sections we have separated the discussions of logical and physical design activities. In fact, logical and physical design can be carried out in parallel, on a table-by-table (or file-by-file) basis. Logical and physical design can also be carried out in parallel when we are working with hierarchical and network models. Such parallel activities require the designer to have a through understanding of the software and hardware in order to take full advantage of both software and hardware characteristics.

8.4.3 IMPLEMENTATION AND LOADING

In most modern relational DBMSs, such as IBM DB2, Microsoft SQL Server, or Oracle, a new database implementation requires the creation of special storage-related constructs to house the end-user tables. These constructs usually include the storage group, the table space, and the tables (see Figure 8.12). Note that a table space may contain more than one table.

FIGURE 8.12 PHYSICAL ORGANIZATION OF A DB2 DATABASE ENVIRONMENT

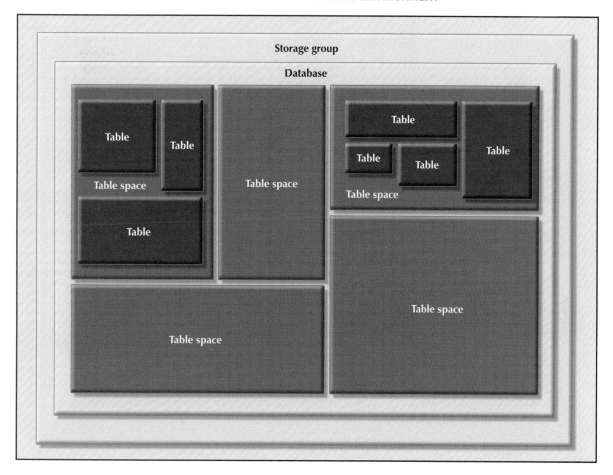

For example, the implementation of the logical design in IBM's DB2 would require that we:

1. Create the database storage group. This step (done by the system administrator, or SYSADM) is mandatory for such mainframe databases as DB2. Other DBMS software may create equivalent storage groups automatically when a database is created (see Step 2). Consult your DBMS manual to see if you must create a storage group and, if so, what the command syntax must be.

2. Create the database within the storage group (also done by the SYSADM).

3. Assign the rights to use the database to a database administrator (DBADM).

4. Create the table space(s) within the database (usually done by a DBADM).

5. Create the table(s) within the table space(s) (also usually done by a DBADM). A generic SQL table creation might look like this:

```
CREATE TABLE COURSE (
        CRS_CODE            CHAR(10)      NOT NULL,
        CRS_TITLE           CHAR(15)      NOT NULL,
        CRS_DESCRIPT        CHAR(8)       NOT NULL
        CRS_CREDIT          NUMBER,
        PRIMARY KEY (CRS_CODE));
```

```
CREATE TABLE CLASS (

    CLASS_CODE      CHAR(4)      NOT NULL,

    CLASS_DAYS      CHAR(3)      NOT NULL,

    CLASS_TIME      CHAR(14)     NOT NULL,

    CLASS_DAY       CHAR(3)      NOT NULL,

    CRS_CODE        CHAR(10)     NOT NULL,

    PRIMARY KEY (CLASS_CODE),

    FOREIGN KEY (CRS_CODE) REFERENCES COURSE;
```

(Note that the COURSE table is created first, because it is referenced by the CLASS table.)

6. Assign access rights to the table spaces and to the tables within specified table spaces (another DBADM duty). Access rights may be limited to views rather than to whole tables. The creation of views is not required for database access in the relational environment, but views are desirable from a security point of view.

Access rights to a table named PROFESSOR may be granted to a person whose identification code is PROB by typing:

```
GRANT USE OF TABLE PROFESSOR

TO PROB;
```

A view, named PROF, may be substituted for the PROFESSOR table:

```
CREATE VIEW PROF

    SELECT EMP_LNAME

        FROM EMPLOYEE

            WHERE PROFESSOR.EMP_NUM = EMPLOYEE.EMP_NUM;
```

After the database has been created, the data must be loaded into the database tables. If the data are currently stored in a format different from that required by the new DBMS, the data must be converted prior to being loaded.

If a hierarchical database is used, the implementation must include the data definition language (DDL) component for each of the databases. In addition, a hierarchical database, such as IMS, requires the creation of program specification blocks (PSBs) that enable programs to access the database.

The use of a network model requires the creation of the database DDL, which must include all record types, location modes, sets, order, insertion modes, set selections, and so on. Network databases also require the use of a device media control language (DMCL) and all necessary views in order to allow programs access to the database.

N O T E

The following summary of database implementation activities assumes the use of a sophisticated DBMS. All new mainframe and client/server DBMSs offer the features discussed next. At the time of this writing, only a few microcomputer DBMSs offer such capability. The more sophisticated microcomputer DBMSs that are capable of supporting these features tend to be derived from the mainframe or minicomputer environment. (For example, DB2, Oracle, and SQL Server are now available for the microcomputer.)

During the implementation and loading phase, we also must address performance, security, backup and recovery, integrity, company standards, and concurrency control. These will be discussed next. (These issues are also addressed in detail in Chapter 9, "Transaction Management and Concurrency Control," and Chapter 10, "Distributed Database Management Systems.")

Performance

Database performance is one of the most important factors in certain database implementations. However, not all DBMSs have performance-monitoring and fine-tuning tools embedded in their software, thus making it difficult to evaluate performance.

Performance evaluation is also rendered more difficult because there is no standard measurement for database performance. Performance varies according to the hardware and software environment used. Naturally, the database's size also affects database performance: a search of 10 tuples will be faster than a search of 100,000 tuples.

Important factors in database performance also include system and database configuration parameters, such as data placement, access path definition, use of indexes, and buffer size.

Although performance is important, it is not the only (nor even the main) factor. The issues discussed next are equally important.

Security

Data stored in the company database must be protected from access by unauthorized users. (It does not take much imagination to predict the likely results when students have access to a student database or when employees have access to payroll data!) Consequently, we must provide for (at least) the following:

- *Physical security* allows physical access to areas by authorized personnel only. Depending on the type of database implementation, however, establishing physical security might not always be practical. For example, a university student research database is not a likely candidate for physical security. The existence of large, multiserver microcomputer networks often makes physical security impractical.

- *Password security* allows the assignment of access rights to specific authorized users. Password security is usually enforced at logon time at the operating system level.

- *Access rights* can be established through the use of database software. The assignment of access rights may restrict operations (CREATE, UPDATE, DELETE, and so on) on predetermined objects such as databases, tables, views, queries, and reports.

- *Audit trails* are usually provided by the DBMS to check for access violations. Although the audit trail is an after-the-fact device, its mere existence can discourage unauthorized use.

- *Data encryption* can be used to render data useless to unauthorized users who might have violated some of the database security layers.

- *Diskless workstations* allow end users to access the database without being able to download the information from their workstations.

Backup and Recovery

Timely data availability is crucial in the information game. Unfortunately, the database can be subject to data loss through unintended data deletion, power outages, and so on. Data backup and recovery procedures create a safety valve, allowing the database administrator to ensure the availability of consistent data.

Integrity

Data integrity is enforced through the proper use of primary and foreign key rules.

Company Standards

Database standards may be partially defined by specific company requirements. The database administrator must implement and enforce such standards.

Concurrency Control

The feature that allows simultaneous access to a database while preserving data integrity is concurrency control. Failure to maintain concurrency control can quickly destroy a database's effectiveness. Consider the scenario depicted in Table 8.6:

1. Initial stock = 500 units.

2. Process A reduces the stock by 150 units; process B reduces the stock by an additional 300 units.

3. The final stock *should be* 500 − 150 − 300 = 50. However, given the improper sequencing illustrated in Table 8.6, the (incorrect) value 200 is stored in the database.

TABLE 8.6 THE NEED FOR CONCURRENCY CONTROL

PROCESS A (MEMORY)	PROCESS B (MEMORY)	DATABASE (STORED STOCK)	TIME LINE
Read 500 ◀━━━━━		500	T1
	Read 500 ◀━━━━━	500	T2
Stock = 500 − 150	Write ━━━━━▶	350	T3
	Stock = 500 − 300—write ━━▶	200	T4
	Value stored in database:	200	

Table 8.6 illustrates what can happen in the database when these transactions take place. Using the sequence of events depicted in Table 8.6, note the following:

1. During time period T1, process A reads the available stock of 500 units.

2. During time period T2, process B reads the available stock before process A reduces the stock by 150 units. Thus, process B "sees" the available stock as 500 units.

3. During time period T3, process A reduces the stock by 150 units and writes the remaining value of 350 units into the database. *But process B still has the originally read 500 stock units in memory.*

4. During time period T4, process B reduces the stock in memory by 300 units, leaving 200 units of stock to be written to the database! Because process B writes to the database after process A, the (incorrect) 200 units will be stored in the database, ready for the next transaction.

This scenario can be prevented by using concurrency-control algorithms. Chapter 9 details the use of locking methods to manage concurrency control.

8.4.4 TESTING AND EVALUATION

Once the data have been loaded into the database, the DBA tests and fine-tunes the database for performance, integrity, concurrent access, and security constraints. The testing and evaluation phase occurs in parallel with applications programming.

Programmers use database tools to *prototype* the applications during the coding of the programs. Tools such as report generators, screen painters, and menu generators are especially useful to the applications programmers during the prototyping phase.

If the database implementation fails to meet some of the system's evaluation criteria, several options may be considered to enhance the system:

- For performance-related issues, the designer must consider fine-tuning specific system and DBMS configuration parameters. The best sources of information are the hardware and software technical reference manuals.

- Modify the physical design. (For example, the proper use of indexes tends to be particularly effective in facilitating pointer movements, thus enhancing performance.)

- Modify the logical design.

- Upgrade or change the DBMS software and/or the hardware platform.

8.4.5 OPERATION

Once the database has passed the evaluation stage, it is considered to be operational. At this point the database, its management, its users, and its application programs constitute a complete information system.

The beginning of the operational phase invariably starts the process of system evolution. As soon as all the targeted end users have entered the operations phase, problems that could not have been foreseen during the testing phase begin to surface. Some of these problems are serious enough to warrant emergency "patchwork," while others are merely minor annoyances. For example, if the database design is implemented to interface with the Web, the sheer volume of transactions may cause even a well-designed system to bog down. In this case, the designers will have to identify the source(s) of the bottleneck(s) and produce alternate solutions, which may include using load-balancing software to distribute the transactions among multiple computers, increasing the available cache for the DBMS, and so on. In any case, the demand for change is the designer's constant, which leads to phase 6, maintenance and evolution.

8.4.6 MAINTENANCE AND EVOLUTION

The database administrator must be prepared to perform routine maintenance activities within the database. Some of the required periodic maintenance activities include:

- Preventive maintenance (backup).

- Corrective maintenance (recovery).

- Adaptive maintenance (enhancing performance, adding entities and attributes, and so on).

- Assignment of access permissions and their maintenance for new and old users.

- Generation of database access statistics to improve the efficiency and usefulness of system audits and to monitor system performance.

- Periodic security audits based on the system-generated statistics.

- Periodic (monthly, quarterly, or yearly) system-usage summaries for internal billing or budgeting purposes.

We noted earlier that the system's only constant is change. Given the likelihood of new information requirements, the demand for additional reports and new query formats will require application changes and, possibly, minor changes in the database components and contents. These changes can be easily implemented only if the database design is flexible and if all the documentation is updated and online. Eventually, even the best-designed database environment will no longer be capable of incorporating such evolutionary changes—then the whole DBLC process begins anew.

You should not be surprised to discover that many of the activities described in the Database Life Cycle (DBLC) remind you of those in the Systems Development Life Cycle (SDLC). After all, the SDLC represents the framework within which the DBLC activities take place. A summary of the parallel activities that take place within the SDLC and the DBLC is shown in Figure 8.13.

FIGURE 8.13 PARALLEL ACTIVITIES IN THE DBLC AND THE SDLC

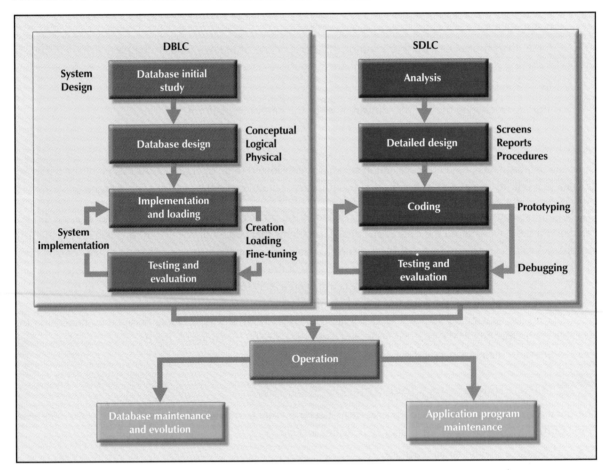

8.5 A SPECIAL NOTE ABOUT DATABASE DESIGN STRATEGIES

There are two classical approaches to database design:

1. **Top-down design** starts by identifying the data sets, and then defines the data elements for each of those sets. This process involves the identification of different entity types and the definition of each entity's attributes.

2. **Bottom-up design** first identifies the data elements (items), and then groups them together in data sets. In other words, first define attributes and then group them to form entities.

The two approaches are illustrated in Figure 8.14. The selection of a primary emphasis on top-down or bottom-up procedures often depends on the scope of the problem or on personal preferences. Although these two methodologies are complementary rather than mutually exclusive, a primary emphasis on a bottom-up approach may be more productive for small databases with few entities, attributes, relations, and transactions. For situations in which the number, variety, and complexity of entities, relations, and transactions is overwhelming, a primarily top-down approach may be more easily managed. Most companies have standards for systems development and database design already in place.

FIGURE 8.14 TOP-DOWN VS. BOTTOM-UP DESIGN SEQUENCING

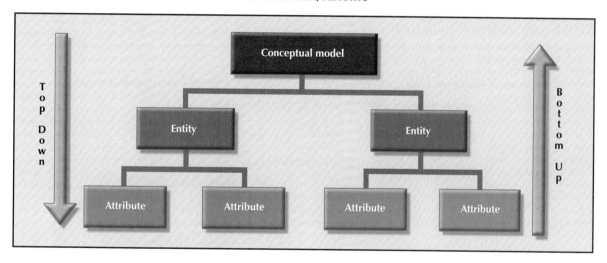

NOTE

Even if a primarily top-down approach is selected, the normalization process that revises existing table structures is (inevitably) a bottom-up technique. ER models constitute a top-down process, even if the selection of attributes and entities can be described as bottom-up. Because both the ER model and normalization techniques form the basis for most designs, the top-down versus bottom-up debate may be based on a distinction, rather than a difference.

8.6 CENTRALIZED vs. DECENTRALIZED DESIGN

The two general approaches (bottom-up and top-down) to database design can be influenced by factors such as the scope and size of the system, the company's management style, or the company's structure (centralized or decentralized). Depending on such factors, the database design may be based on two very different design philosophies:

1. **Centralized design** is productive when the data component is composed of a relatively small number of objects and procedures. The design can be carried out and represented in a fairly simple database.

 Centralized design is typical of relatively simple and/or small databases and can be successfully done by a single person (database administrator) or by a small, informal design team. The company operations and the scope of the problem are sufficiently limited to allow even a single designer to: define the problem(s); create the conceptual design; verify the conceptual design with the user views; define system processes and data constraints to ensure the efficacy of the design; and ensure that the design will comply with all the requirements. (Although centralized design is typical for small companies, do not make the mistake of assuming that centralized design is limited to small companies. Even large companies can operate within a relatively simple database environment!) Figure 8.15 summarizes the centralized design option. Note that a single conceptual design is completed and then validated in the centralized design approach.

FIGURE 8.15 CENTRALIZED DESIGN

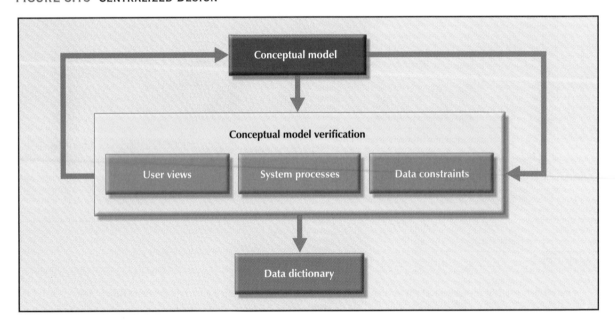

2. Decentralized design might be used when the data component of the system has a considerable number of entities and complex relations on which very complex operations are performed. Decentralized design is also likely to be employed when the problem itself is spread across several operational sites and each element is a subset of the entire data set (see Figure 8.16).

FIGURE 8.16 DECENTRALIZED DESIGN

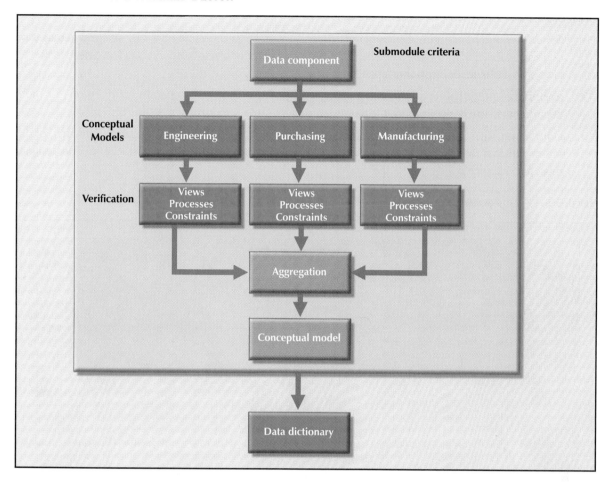

In large and complex projects, the database design typically cannot be done by only one person. Instead, a carefully selected team of database designers is employed to tackle a complex database project. Within the decentralized design framework, the database design task is divided into several modules. Once the design criteria have been established, the lead designer assigns design subsets or modules to design groups within the team.

Because each design group focuses on modeling a subset of the system, the definition of boundaries and the interrelation among data subsets must be very precise. Each design group creates a conceptual data model corresponding to the subset being modeled. Each conceptual model is then verified individually against the user views, processes, and constraints for each of the modules. After the verification process has been completed, all modules are integrated in one conceptual model. Because the data dictionary describes the characteristics of all the objects within the conceptual data model, it plays a vital role in the integration process. Naturally, after the subsets have been aggregated into a larger conceptual model, the lead designer must verify that the combined conceptual model is still able to support all the required transactions.

Keep in mind that the aggregation process requires the designer to create a single model in which various aggregation problems must be addressed (see Figure 8.17):

- *Synonyms and homonyms.* Different departments might know the same object by different names (synonyms), or they might use the same name to address different objects (homonyms). The object can be an entity, an attribute, or a relationship.

- *Entity and entity subtypes.* An entity subtype might be viewed as a separate entity by one or more departments. The designer must integrate such subtypes into a higher-level entity.

- *Conflicting object definitions.* Attributes can be recorded as different types (character, numeric), or different domains can be defined for the same attribute. Constraint definitions, too, can vary. The designer must remove such conflicts from the model.

FIGURE 8.17 SUMMARY OF AGGREGATION PROBLEMS

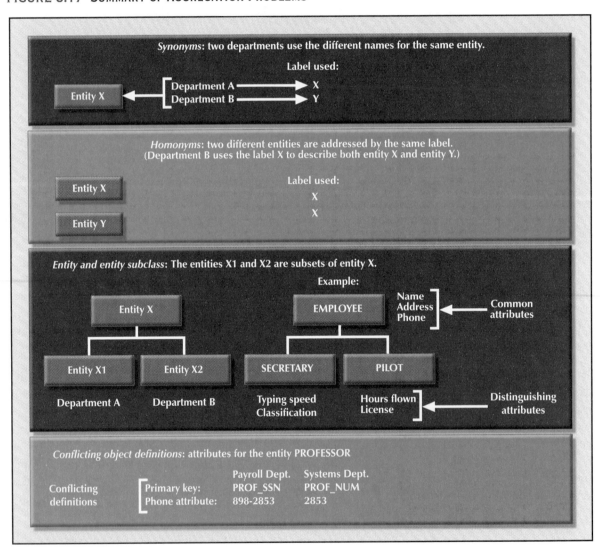

Data constitute the raw material from which information is created. The transformation from data to information is produced when programming code operates on the data, thus producing applications.

An information system is designed to facilitate the transformation of data into information and to manage both data and information. The database is thus a very important part of the information system. Systems analysis is the process that establishes the need for and the extent of an information system. Systems development is the process of creating an information system.

The Systems Development Life Cycle (SDLC) traces the history (life cycle) of an application within the information system. The SDLC can be divided into five phases: planning, analysis, detailed systems design, implementation, and maintenance. The SDLC is an iterative rather than a sequential process.

The Database Life Cycle (DBLC) describes the history of the database within the information system. The DBLC is composed of six phases: the database initial study, the database design, implementation and loading, testing and evaluation, operation, and maintenance and evolution. Like the SDLC, the DBLC is iterative rather than sequential.

Because the database provides the information system's most basic component, the database design and implementation process is of particular interest to us. The database design and implementation process moves through a series of well-defined stages:

1. Data analysis and requirements gathering help determine end-user views and data needs.

2. The end product of Stage 1 yields the definition of relevant entities, attributes, and relationships, which leads in turn to the appropriate entity relationship (ER) model. The ER model's components then give way to a set of normalized tables.

3. The conceptual model is verified through the identification of its main processes and through the definition of the appropriate INSERT, UPDATE, and DELETE rules. In addition, user views and appropriate data constraints are reviewed during the verification process.

4. After verification, a distributed database design is analyzed in order to evaluate table location, database fragmentation strategies, and access requirements.

5. Ideally, the database software selection is made in conjunction with the development of the conceptual model (Stages 1–4). In the less-perfect, practical world, you might discover that the database software is already in place and that the conceptual design is partly dictated by the existence of that software.

6. Given the conceptual model and the database software, the logical design translates the conceptual schema into the DBMS-specific definition of tables, views, and so on. The logical design is thus a function of the selected DBMS characteristics, but it is hardware-independent.

7. The physical design defines specific storage structures and data access paths and is, therefore, hardware-dependent.

8. Implementation constitutes the final stage, in which the data are loaded and the tables and views are actually created. If necessary, attribute data formats are converted into formats that can be accessed efficiently by the selected DBMS. Application coding and testing is completed during this stage.

The conceptual portion of the design may be subject to several variations, based on two design philosophies:

1. *Bottom-up versus top-down.* Bottom-up design first identifies data elements (attributes), and then groups them into sets (entities). Top-down design first identifies the entities, and then defines each entity's attributes. The selection of either approach is a function of the scope of the data problem and personal preference. However, as a general rule, the top-down approach is superior in a complex data environment.

2. *Centralized versus decentralized.* Relatively simple databases are easily created with the centralized approach when the company operations lend themselves to a "bird's-eye" view of the database. A decentralized approach to design may be more appropriate when the company operations are spread across multiple operational sites (each site's data set is a subset of the entire company's operations) or when the database is composed of a very large number of entities that are subject to very complex relationships.

K E Y T E R M S

bottom-up design	Database Life Cycle (DBLC)	physical design
boundaries	decentralized design	scope
centralized design	description of operations	systems analysis
cohesivity	information system	systems development
computer-assisted systems engineering (CASE)	logical design	Systems Development Life Cycle (SDLC)
conceptual design	minimal data rule	top-down design
concurrency control	module	transformation
database development	module coupling	

R E V I E W Q U E S T I O N S

1. What is an information system? What is its purpose?
2. How do systems analysis and systems development fit into a discussion about information systems?
3. Discuss the distinction between data and information.
4. What does the acronym SDLC mean, and what does an SDLC portray?
5. What does the acronym DBLC mean, and what does a DBLC portray?
6. Discuss the distinction between centralized and decentralized conceptual database design.
7. What is the minimal data rule in conceptual design? Why is it important?
8. Discuss the distinction between top-down and bottom-up approaches in database design.
9. What are business rules? Why are they important to a database designer?
10. What is the data dictionary's function in database design?
11. What is concurrency control? Why is it important?

PROBLEMS

1. The ABC Car Service & Repair Centers are owned by the SILENT car dealer; they service and repair only that brand of car. Three ABC Car Service & Repair Centers provide service and repair for the entire state.

 Each of the three centers is independently managed and operated by a shop manager, a receptionist, and at least eight mechanics. Each center maintains a fully stocked parts inventory.

 Each center also maintains a manual file system in which each car's maintenance history is kept: repairs made, parts used, costs, service dates, owner, and so on. Files are also kept to track inventory, purchasing, billing, employees' hours, and payroll.

 You have been contacted by the manager of one of the centers to design and implement a computerized system. Given the preceding information, do the following:

 a. Indicate the most appropriate sequence of activities by labeling each of the following steps in the correct order. (For example, if you think that "Load the database" is the appropriate first step, label it "1.")

 _____ Normalize the conceptual model.
 _____ Obtain a general description of company operations.
 _____ Load the database.
 _____ Create a description of each system process.
 _____ Test the system.
 _____ Draw a data flow diagram and system flow charts.
 _____ Create a conceptual model, using ER diagrams.
 _____ Create the application programs.
 _____ Interview the mechanics.
 _____ Create the file (table) structures.
 _____ Interview the shop manager.

 b. Describe the different modules that you believe the system should include.
 c. How will a data dictionary help you develop the system? Give examples.
 d. What general (system) recommendations might you make to the shop manager? (For example: If the system will be integrated, what modules will be integrated? What benefits would be derived from such an integrated system? Include several general recommendations.)
 e. What do you think is the best approach to conceptual database design? Why?
 f. Name and describe at least four reports that you believe the system should have, and explain their use. Who will use those reports?

2. Suppose you have been asked to create an information system for a manufacturing plant that produces nuts and bolts of many shapes, sizes, and functions. What questions would you ask, and how would the answers to those questions affect the database design?
 a. What do you envision the SDLC to be?
 b. What do you envision the DBLC to be?

3. Suppose you perform the same functions noted in Problem 2 for a larger warehousing operation. How are the two sets of procedures similar? How (and why) are they different?

4. Using the same procedures and concepts employed in Problem 1, how would you create an information system for the Tiny College example in Chapter 4?

5. Write the proper sequence of activities in the design of a video rental database. (The initial ERD was shown in Figure 8.7.) The design must support all rental activities, customer payment tracking, and employee work schedules, as well as track which employees checked out the videos to the customers. After you have finished writing the design activity sequence, complete the ERD to ensure that the database design can be successfully implemented. (Make sure that the design is normalized properly and that it can support the required transactions.)

PART

IV

ADVANCED DATABASE
CONCEPTS

9

TRANSACTION MANAGEMENT AND CONCURRENCY CONTROL

In this chapter, you will learn:

❖ What a database transaction is and what its properties are

❖ How database transactions are managed

❖ What concurrency control is and what role it plays in maintaining the database's integrity

❖ What locking methods are and how they work

❖ How database recovery management is used to maintain database integrity

Database transactions reflect real-world transactions that are triggered by events such as buying a product, registering for a course, or making a deposit in your checking account. Transactions are likely to contain many parts. For example, a sales transaction may require updating the customer's account, adjusting the product inventory, updating the seller's accounts receivable, and so on. All parts of a transaction must be successfully completed to prevent data integrity problems. Therefore, executing and managing transactions are important database system activities.

You will learn that the main database transaction properties are atomicity, durability, serializability, and isolation. After defining these transaction properties, we will demonstrate how SQL can be used to represent transactions. We will also examine the use of transaction logs to ensure the DBMS's ability to recover transactions.

Concurrency control is the management of concurrent transaction execution. Because the number of concurrent transactions tends to grow rapidly when many users are able to access the database, concurrency control is especially important in a multiuser database environment. (Just imagine the number of transactions routinely handled by companies that conduct sales and provide services via the World Wide Web!) We will discuss lost updates, uncommitted data, and inconsistent summaries that can occur during concurrent transaction execution. And you will discover that such problems can be solved when a DBMS scheduler enforces concurrency control. (The scheduler regulates the execution of operations for two or more conflicting transactions.)

You will also learn that the most common algorithms for concurrency control use locks, time stamping, or optimistic methods. Because locks form the basis for the most widely used concurrency control method, we will examine various levels and types of locks. Because locks can also create deadlocks, we will discuss three strategies for deadlock management: detection, prevention, and avoidance. We will then summarize the time stamping and optimistic methods used in concurrency control.

Database contents may be damaged or destroyed by critical operational errors, including transaction management failures. Therefore, you will learn how database recovery management maintains a database's contents by means of various backup procedures. Such backup procedures range from full backups to transaction log backups.

9.1 WHAT IS A TRANSACTION?

To illustrate what transactions are and how they work, let's use the **Ch09_SaleCo database**. The relational schema for this database is shown in Figure 9.1.

A D D I T I O N A L C D C O N T E N T

The database used to illustrate the material in this chapter is found on the CD bundled with this book. Check the **Databases/Student/Ch09** folder to find the database. For example, the source of the relational schema shown in Figure 9.1 is the **Ch09_SaleCo** database located in the above folder.

FIGURE 9.1 THE RELATIONAL SCHEMA FOR THE CH09_SALECO DATABASE

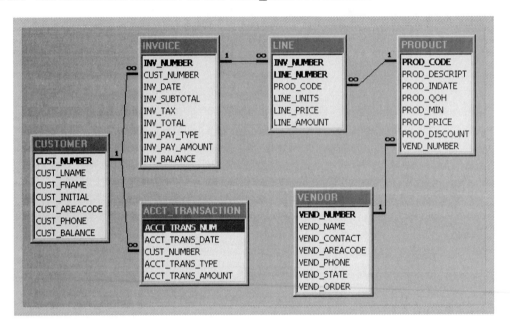

NOTE

Although we use SQL commands to illustrate several transaction and concurrency control issues, you should be able to follow the discussions even if you have not studied Chapter 6, "Introduction to SQL," and Chapter 7, "Advanced SQL." If you don't know SQL, ignore the SQL commands and focus on the discussions. If you do have a working knowledge of SQL, you can use the **Ch09_SaleCo** database to generate your own SELECT and UPDATE examples and to augment the material presented in Chapters 6 and 7 by writing your own triggers and stored procedures.

As you examine the relational schema in Figure 9.1, note the following simplifying features:

- The design stores the customer balance (CUST_BALANCE) value in the CUSTOMER table to indicate the total amount owed by the customer. The CUST_BALANCE attribute is increased whenever the customer makes a purchase on credit, and it is decreased when the customer makes a payment. The inclusion of the current customer account balance in the CUSTOMER table makes it very easy to write a query to determine the current balance for any customer or to generate important summaries such as total, average, minimum, and maximum balances.

- The ACCT_TRANSACTION table records all purchases charged to the customer's credit and all payments to track the details of customer account activity.

Naturally, you can change the database design of the Ch09_SalesCo database to reflect accounting practice more precisely, but this implementation will enable you to track the transactions well enough to serve the purpose of the chapter's discussions.

To understand the concept of a transaction, suppose that you sell a product to a customer. Further suppose that the customer may charge the purchase to the customer's account. Given this scenario, your sales transaction consists of at least the following parts:

- You must write a new customer invoice.

- You must reduce the quantity on hand in the product's inventory.

- You must update the accounts receivable.

- You must update the customer balance.

The preceding sales transaction must be reflected in the database. In database terms, a **transaction** is any action that reads from and/or writes to a database. A transaction may consist of a simple SELECT statement to generate a list of table contents, it may consist of a series of related UPDATE statements to change the values of attributes in various tables, it may consist of a series of INSERT statements to add rows to one or more tables, or it may consist of a combination of SELECT, UPDATE, and INSERT statements. Our sales transaction example will include a combination of INSERT and UPDATE statements.

Given the preceding discussion, you can now augment the definition of a transaction. A *transaction* is a *logical* unit of work that must be either entirely completed or aborted; no intermediate states are acceptable. In other words, a multicomponent transaction, such as the previously mentioned sale, must not be partially completed. Updating only the inventory or only the accounts receivable is not acceptable. All of the SQL statements in the transaction must be completed successfully. If any of the SQL statements fail, the entire transaction is rolled back to the original database state that existed before the transaction started. A successful transaction changes the database from one consistent state to another. A **consistent database state** is one in which all data integrity constraints are satisfied.

To ensure consistency of the database, every transaction must begin with the database in a known consistent state. If the database is not in a consistent state, the transaction will yield an inconsistent database that violates its integrity and business rules. For this reason, subject to limitations we will discuss later, all transactions are controlled and executed by the DBMS to guarantee database integrity.

Most real-world database transactions are formed by two or more database requests. A **database request** is the equivalent of a single SQL statement in an application program or transaction. Therefore, if a transaction is composed of two UPDATE statements and one INSERT statement, the transaction uses three database requests. In turn, each database request generates several input/output (I/O) operations that read from or write to physical storage media.

9.1.1 EVALUATING TRANSACTION RESULTS

Not all transactions update the database. Suppose we want to examine the current balance for customer number 10016 located in the CUSTOMER table. Such a transaction can be completed by using the SQL code:

```
SELECT CUST_NUMBER, CUST_BALANCE
   FROM CUSTOMER
      WHERE CUST_NUMBER = 10016;
```

Although we did not make any changes in the CUSTOMER table, the SQL code represents a transaction because we *accessed* the database. If the database existed in a consistent state before the access, the database remains in a consistent state after the access, because the transaction did not alter the database.

Remember that a transaction may consist of a single SQL statement or a collection of related SQL statements. Let's revisit the previous sales example to illustrate a more complex transaction, using the Ch09_SalesCo database. Suppose that on January 18, 2004 you register the credit sale of one unit of product 89-WRE-Q to customer 10016 in the amount of $277.55. The required transaction affects the INVOICE, LINE, PRODUCT, CUSTOMER, and ACCT_TRANSACTION tables. The SQL statements that represent this transaction are shown next:

```
INSERT INTO INVOICE
   VALUES (1009, 10016,'18-Jan-2004', 256.99, 29.56, 277.55, 'cred', 0.00, 277.55);
INSERT INTO LINE
   VALUES (1009, 1, '89-WRE-Q', 1, 256.99, 256.99);
UPDATE PRODUCT
   SET PROD_QOH = PROD_QOH – 1
     WHERE PROD_CODE = '89-WRE-Q';
UPDATE CUSTOMER
   SET CUST_BALANCE = CUST_BALANCE + 277.55
     WHERE CUST_NUMBER = 10016;
INSERT ACCT_TRANSACTION
   VALUES (10007, '18-Jan-04', 10016, 'charge', 277.55);
COMMIT;
```

The results of the successfully completed transaction are shown in Figure 9.2. (Note that all the records that were involved in the transaction have been highlighted.)

FIGURE 9.2 TRACING THE TRANSACTION IN THE CH09_SALECO DATABASE

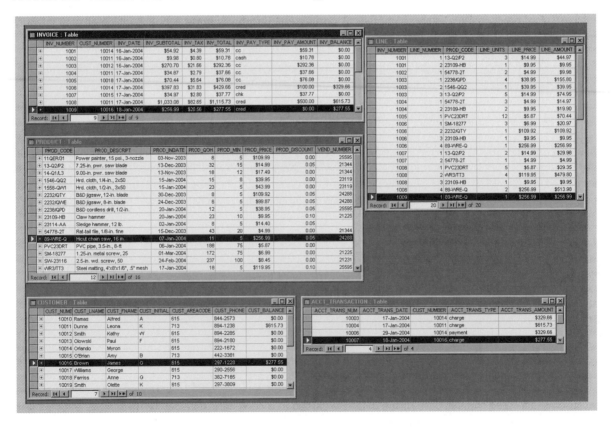

To further your understanding of the transaction results, note the following:

- A new row 1009 was added to the INVOICE table. In this row, we also stored derived attribute values for the invoice subtotal, the tax, the invoice total, and the invoice balance.

- The LINE row for invoice 1009 was added to reflect the purchase of one unit of product '89-WRE-Q' with a price of $256.99. In this row, we have stored the derived attribute values for the line amount.

- The product 89-WRE-Q's quantity on hand (PROD_QOH) in the PRODUCT table was reduced by one (the initial value was 12), thus leaving a quantity on hand of 11.

- The customer balance (CUST_BALANCE) for customer 10016 was updated by adding $277.55 to the existing balance (the initial value was $0.00).

- A new row was added to the ACCT_TRANSACTION table to reflect the new account transaction number 1007.

- The COMMIT statement is used to end a successful transaction (see Section 9.1.3).

Now suppose that the DBMS completes the first three SQL statements. Further suppose that during the execution of the fourth statement (the UPDATE of the CUSTOMER table's CUST_BALANCE value for customer 10016), the computer system experiences a loss of electrical power. If the computer does not have a backup power supply, the transaction cannot be completed. Therefore, the INVOICE and LINE rows were added, the PRODUCT table was updated to represent the sale of product 89-WRE-Q, but customer 10016 was not charged, nor was the required record in the ACCT_TRANSACTION table written. The database is now in an inconsistent state, and it is not usable for subsequent transactions. Assuming that the DBMS supports transaction management, *the DBMS will roll back the database to a previous consistent state.*

> **N O T E**
>
> By default, MS Access does not support transaction management as discussed here. More-sophisticated DBMSs such as Oracle, SQL Server, and DB2 do support the transaction management components discussed in this chapter.

Although the DBMS is designed to recover a database to a previous consistent state when an interruption prevents the completion of a transaction, the transaction itself is defined by the end user or programmer and must be semantically correct. *The DBMS cannot guarantee that the semantic meaning of the transaction truly represents the real-world event.* For example, suppose that, following the sale of 10 units of product 89-WRE-Q, the inventory UPDATE commands were written this way:

```
UPDATE PRODUCT
  SET PROD_QOH = PROD_QOH + 10
    WHERE PROD_CODE = '89-WRE-Q';
```

The sale should have *decreased* the PROD_QOH value for product 89-WRE-Q by 10. Instead, this UPDATE *adds* 10 to product 89-WRE-Q's PROD_QOH value.

Although the UPDATE command's syntax is correct, its use yields incorrect results. Yet the DBMS will execute the transaction anyway. The DBMS cannot evaluate whether the transaction represents the real-world event correctly; that is the end user's responsibility. End users and programmers are capable of introducing many errors in this fashion. Imagine the consequences of reducing the quantity on hand for product 1546-QQ2 instead of product 89-WRE-Q, or of crediting the CUST_BALANCE value for customer 10012 rather than for customer 10016.

Clearly, improper or incomplete transactions can have a devastating effect on database integrity. Some DBMSs, and *especially* the relational variety, provide means by which the user can define enforceable constraints based on business rules. Other integrity rules, such as those governing referential and entity integrity, are enforced automatically by the DBMS when the table structures are properly defined, thereby letting the DBMS validate some transactions. For example, if a transaction inserts a new customer into a customer table and the customer number being inserted already exists, the DBMS will end the transaction with an error code to indicate a violation of the primary key integrity rule.

9.1.2 TRANSACTION PROPERTIES

All transactions must display *atomicity*, *durability*, *serializability*, and *isolation*. We will briefly examine each of these properties.

- **Atomicity** requires that *all* operations (SQL requests) of a transaction be completed; if not, the transaction is aborted. If a transaction T1 has four SQL requests, all four requests must be successfully completed or the entire transaction is aborted. In other words, a transaction is treated as a single, indivisible, logical unit of work.

- **Durability** indicates the permanence of the database's consistent state. When a transaction is completed, the database reaches a consistent state, and that state cannot be lost, even in the event of the system's failure.

- **Serializability** ensures that the concurrent execution of several transactions yields consistent results. More specifically, the concurrent execution of transactions T1, T2, and T3 yields results that appear to have been executed in serial order (one after another). This property is important in multiuser and distributed databases, where multiple transactions are likely to be executed concurrently.

- **Isolation** means that the data used during the execution of a transaction cannot be used by a second transaction until the first one is completed. In other words, if a transaction T1 is being executed and is using the data item X, that data item cannot be accessed by any other transaction (T2 ... T*n*) until T1 ends. This property is particularly useful in multiuser database environments because several different users can access and update the database at the same time.

By its nature, a single-user database system automatically ensures serializability and isolation of the database, because only one transaction is executed at a time. The atomicity and the durability of transactions must be guaranteed by the single-user DBMSs. (Even a single-user DBMS must manage recovery from errors created by operating-system-induced interruptions, power interruptions, and improper application execution.)

Multiuser databases, whether mainframe- or LAN-based, are typically subject to multiple concurrent transactions. Therefore, the multiuser DBMS must implement controls to ensure serializability and isolation of transactions—in addition to atomicity and durability—in order to guard the database's consistency and integrity. For example, if several concurrent transactions are executed over the same data set and the second transaction updates the database before the first transaction is finished, the isolation property is violated, and the database is no longer consistent. The DBMS must manage the transactions by using concurrency control techniques to avoid such undesirable situations.

9.1.3 TRANSACTION MANAGEMENT WITH SQL

The American National Standards Institute (ANSI) has defined standards that govern SQL database transactions. Transaction support is provided by two SQL statements: COMMIT and ROLLBACK. The ANSI standards require that, when a transaction sequence is initiated by a user or an application program, it must continue through all succeeding SQL statements until one of the following four events occurs:

1. A COMMIT statement is reached, in which case all changes are permanently recorded within the database. The COMMIT statement automatically ends the SQL transaction.

2. A ROLLBACK statement is reached, in which case all the changes are aborted and the database is rolled back to its previous consistent state.

3. The end of a program is successfully reached, in which case all changes are permanently recorded within the database. This action is equivalent to COMMIT.

4. The program is abnormally terminated, in which case the changes made in the database are aborted and the database is rolled back to its previous consistent state. This action is equivalent to ROLLBACK.

The use of COMMIT is illustrated in the following simplified sales example, which updates a product's quantity on hand (PROD_QOH) and the customer's balance when the customer buys 2 units of product 1558-QW1 priced at $43.99 per unit, for a total of $87.98, and charges the purchase to their account:

```
UPDATE PRODUCT
  SET PROD_QOH = PROD_QOH - 2
    WHERE PROD_CODE = '1558-QW1';
UPDATE CUSTOMER
  SET CUST_BALANCE = CUST_BALANCE + 87.98
    WHERE CUST_NUMBER = '10011';
COMMIT;
```

(Note that we have simplified the example to make it easier to trace the transaction. In the **Ch09_SaleCo** database, the transaction would involve several additional table updates.)

Actually, the COMMIT statement used in this example is not necessary if the UPDATE statement is the application's last action and the application terminates normally. However, good programming practice dictates that you include the COMMIT statement at the end of a transaction declaration.

A transaction begins implicitly when the first SQL statement is encountered. Not all SQL implementations follow the ANSI standard; some (such as SQL Server) use transaction management statements such as:

BEGIN TRANSACTION;

to indicate the beginning of a new transaction. Other SQL implementations, such as VAX/SQL, allow you to assign characteristics for the transactions as parameters to the BEGIN statement. For example, the Oracle RDBMS uses the SET TRANSACTION statement to declare a new transaction start and its properties.

9.1.4 THE TRANSACTION LOG

A DBMS uses a **transaction log** to keep track of all transactions that update the database. The information stored in this log is used by the DBMS for a recovery requirement triggered by a ROLLBACK statement, a program's abnormal termination, or a system failure such as a network discrepancy or a disk crash. Some RDBMSs use the transaction log to recover a database *forward* to a currently consistent state. After a server failure, for example, Oracle automatically rolls back uncommitted transactions and rolls forward transactions that were committed but not yet written to the physical database.

While the DBMS executes transactions that modify the database, it also automatically updates the transaction log. The transaction log stores:

- A record for the beginning of the transaction.
- For each transaction component (SQL statement) it stores:
 - The type of operation being performed (update, delete, insert).
 - The names of the objects affected by the transaction (the name of the table).
 - The "before" and "after" values for the fields being updated.
 - Pointers to the previous and next transaction log entries for the same transaction.
- The ending (COMMIT) of the transaction.

Although using a transaction log increases the processing overhead of a DBMS, the ability to restore a corrupted database is worth the price.

Table 9.1 illustrates a simplified transaction log that reflects a basic transaction composed of two SQL UPDATE statements. If a system failure occurs, the DBMS will examine the transaction log for all uncommitted or incomplete transactions, and it will restore (ROLLBACK) the database to its previous state on the basis of this information. When the recovery process is completed, the DBMS will write in the log all committed transactions that were not physically written to the database before the failure occurred.

If a ROLLBACK is issued before the termination of a transaction, the DBMS will restore the database only for that particular transaction, rather than for all transactions, in order to maintain the *durability* of the previous transactions. In other words, committed transactions are not rolled back.

The transaction log is itself a database, and it is managed by the DBMS like any other database. The transaction log is subject to common database dangers such as disk-full conditions and disk crashes. Because the transaction log contains some of the most critical data in a DBMS, some implementations support logs on several different disks to reduce the risk of a system failure.

TABLE 9.1 A TRANSACTION LOG

TRL ID	TRX NUM	PREV PTR	NEXT PTR	OPERATION	TABLE	ROW ID	ATTRIBUTE	BEFORE VALUE	AFTER VALUE
341	101	Null	352	START	****Start Transaction				
352	101	341	363	UPDATE	PRODUCT	1558-QW1	PROD_QOH	25	23
363	101	352	365	UPDATE	CUSTOMER	10011	CUST_BALANCE	525.75	615.73
365	101	363	Null	COMMIT	**** End of Transaction				

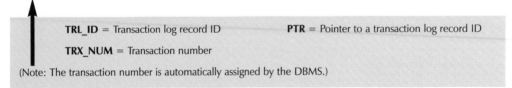

TRL_ID = Transaction log record ID **PTR** = Pointer to a transaction log record ID

TRX_NUM = Transaction number

(Note: The transaction number is automatically assigned by the DBMS.)

9.2 CONCURRENCY CONTROL

The coordination of the simultaneous execution of transactions in a multiprocessing database system is known as concurrency control. The objective of **concurrency control** is to ensure the serializability of transactions in a multiuser database environment. Concurrency control is important because the simultaneous execution of transactions over a shared database can create several data integrity and consistency problems. The three main problems are lost updates, uncommitted data, and inconsistent retrievals.

9.2.1 LOST UPDATES

To illustrate **lost updates**, let's examine a simple PRODUCT table. One of the PRODUCT table's attributes is a product's quantity on hand (PROD_QOH). Assume that you have a product whose current PROD_QOH value

is 35. Now assume that two concurrent transactions occur, T1 and T2, that update the PROD_QOH value for some item in the PRODUCT table. The transactions are:

Transaction	Computation
T1: Purchase 100 units	PROD_QOH = PROD_QOH + 100
T2: Sell 30 units	PROD_QOH = PROD_QOH – 30

Table 9.2 shows the serial execution of these transactions under normal circumstances, yielding the correct answer: PROD_QOH = 105.

TABLE 9.2 **NORMAL EXECUTION OF TWO TRANSACTIONS**

TIME	TRANSACTION	STEP	STORED VALUE
1	T1	Read PROD_QOH	35
2	T1	PROD_QOH = 35 + 100	
3	T1	Write PROD_QOH	135
4	T2	Read PROD_QOH	135
5	T2	PROD_QOH = 135 – 30	
6	T2	Write PROD_QOH	105

But suppose that a transaction is able to read a product's PROD_QOH value from the table *before* a previous transaction (using the *same* product) has been committed. The sequence depicted in Table 9.3 shows how the lost update problem can arise. Note that the first transaction (T1) has not yet been committed when the second transaction (T2) is executed. Therefore, T2 still operates on the value 35, and its subtraction yields 5 in memory. In the meantime, T1 writes the value 135 to disk, which is promptly overwritten by T2. In short, the addition of 100 units is "lost" during the process.

TABLE 9.3 **LOST UPDATES**

TIME	TRANSACTION	STEP	STORED VALUE
1	T1	Read PROD_QOH	35
2	T2	Read PROD_QOH	35
3	T1	PROD_QOH = 35 + 100	
4	T2	PROD_QOH = 35 – 30	
5	T1	Write PROD_QOH (Lost update)	135
6	T2	Write PROD_QOH	5

9.2.2 UNCOMMITTED DATA

The phenomenon of **uncommitted data** occurs when two transactions, T1 and T2, are executed concurrently and the first transaction (T1) is rolled back after the second transaction (T2) has already accessed the uncommitted data—thus violating the isolation property of transactions. To illustrate this possibility, let's use the same transactions described during our lost updates discussion. However, this time the T1 transaction is rolled back to eliminate the addition of the 100 units. Because T2 subtracts 30 from the original 35 units, the correct answer should be 5.

Transaction	Computation	
T1: Purchase 100 units	PROD_QOH = PROD_QOH + 100	**(Rolled back)**
T2: Sell 30 units	PROD_QOH = PROD_QOH − 30	

Table 9.4 shows how, under normal circumstances, the serial execution of these transactions yields the correct answer.

TABLE 9.4 **CORRECT EXECUTION OF TWO TRANSACTIONS**

TIME	TRANSACTION	STEP	STORED VALUE
1	T1	Read PROD_QOH	35
2	T1	PROD_QOH = 35 + 100	
3	T1	Write PROD_QOH	135
4	T1	*****ROLLBACK*****	35
5	T2	Read PROD_QOH	35
6	T2	PROD_QOH = 35 − 30	
7	T2	Write PROD_QOH	5

Table 9.5 shows how the uncommitted data problem can arise when the ROLLBACK is completed after T2 has begun its execution.

TABLE 9.5 **AN UNCOMMITTED DATA PROBLEM**

TIME	TRANSACTION	STEP	STORED VALUE
1	T1	Read PROD_QOH	35
2	T1	PROD_QOH = 35 + 100	
3	T1	Write PROD_QOH	135
4	T2	Read PROD_QOH (Read uncommitted data)	135
5	T2	PROD_QOH = 135 − 30	
6	T1	***** ROLLBACK *****	35
7	T2	Write PROD_QOH	105

9.2.3 INCONSISTENT RETRIEVALS

Inconsistent retrievals occur when a transaction calculates some summary (aggregate) functions over a set of data while other transactions are updating the data. The problem is that the transaction might read some data before they are changed and other data *after* they are changed, thereby yielding inconsistent results.

To illustrate this problem, let's assume the following conditions:

1. T1 calculates the total quantity on hand of the products stored in the PRODUCT table.

2. At the same time, T2 updates the quantity on hand (PROD_QOH) for two of the PRODUCT table's products.

The two transactions are shown in Table 9.6.

TABLE 9.6 RETRIEVAL DURING UPDATE

TRANSACTION T1	TRANSACTION T2
SELECT SUM(PROD_QOH) 　FROM PRODUCT	UPDATE PRODUCT 　SET PROD_QOH = PROD_QOH + 10 　　WHERE PROD_CODE = '1546-QQ2'
	UPDATE PRODUCT 　SET PROD_QOH = PROD_QOH – 10 　　WHERE PROD_CODE = '1558-QW1'
	COMMIT;

While T1 calculates the total quantity on hand (PROD_QOH) for all items, T2 represents the correction of a typing error: the user added 10 units to product 1558-QW1's PROD_QOH, but *meant* to add the 10 units to product 1546-QQ2's PROD_QOH. To correct the problem, the user adds 10 to product 1546-QQ2's PROD_QOH and subtracts 10 from product 1558-QW1's PROD_QOH. (See the two UPDATE statements in Table 9.6.) The initial and final PROD_QOH values are reflected in Table 9.7. (Only a few of the PROD_CODE values for the PRODUCT table are shown. To illustrate the point, the sum for the PROD_QOH values is given for these few products.)

TABLE 9.7 TRANSACTION RESULTS: DATA ENTRY CORRECTION

	BEFORE	AFTER
PROD_CODE	PROD_QOH	PROD_QOH
11QER/31	8	8
13-Q2/P2	32	32
1546-QQ2	15	(15 + 10) ➡ 25
1558-QW1	23	(23 – 10) ➡ 13
2232-QTY	8	8
2232-QWE	6	6
Total	**92**	**92**

Although the final results shown in Table 9.7 are correct after the adjustment, Table 9.8 demonstrates that inconsistent retrievals are possible during the transaction execution, making the result of T1's execution incorrect. The "After" summation shown in Table 9.8 reflects the fact that the value of 25 for product 1546-QQ2 was read *after* the write statement was completed. Therefore, the "After" total is 40 + 25 = 65. The "Before" total reflects the fact that the value of 23 for product 1158-QW1 was read *before* the next write statement was completed to reflect the corrected update of 13. Therefore, the "Before" total is 65 + 23 = 88.

TABLE 9.8 INCONSISTENT RETRIEVALS

TIME	TRANSACTION	ACTION	VALUE	TOTAL
1	T1	Read PROD_QOH for PROD_CODE = '11QER/31'	8	8
2	T1	Read PROD_QOH for PROD_CODE = '13-Q2/P2'	32	40
3	T2	Read PROD_QOH for PROD_CODE = '1546-QQ2'	15	
4	T2	PROD_QOH = 15 + 10		
5	T2	Write PROD_QOH for PROD_CODE = '1546-QQ2'	25	
6	T1	Read PROD_QOH for PROD_CODE = '1546-QQ2'	25	(After) 65
7	T1	Read PROD_QOH for PROD_CODE = '1158-QW1'	23	(Before) 88
8	T2	Read PROD_QOH for PROD_CODE = '1558-QW1'	23	
9	T2	PROD_QOH = 23 – 10		
10	T2	Write PROD_QOH for PROD_CODE = '1558-QW1'	13	
11	T2	***** COMMIT *****		
12	T1	Read PROD_QOH for PROD_CODE = '2232-QTY'	8	96
13	T1	Read PROD_QOH for PROD_CODE = '2232-QWE'	6	102

The computed answer of 102 is obviously wrong, because we know from Table 9.7 that the correct answer is 92. Unless the DBMS exercises concurrency control, a multiuser database environment can create havoc within the information system.

9.2.4 THE SCHEDULER

You now know that *severe* problems can arise when two or more concurrent transactions are executed. You also know that a database transaction involves a series of database I/O operations that take the database from one consistent state to another. Finally, you know that database consistency can be ensured only before and after the execution of transactions. A database always moves through an unavoidable temporary state of inconsistency during a transaction's execution. This temporary inconsistency exists because a computer cannot execute two operations at the same time and must, therefore, execute them serially. During this serial process, the isolation property of transactions prevents them from accessing the data not yet released by other transactions.

In previous examples we executed the operations within a transaction in an arbitrary order. As long as two transactions, T1 and T2, access *unrelated* data, there is no conflict, and the order of execution is irrelevant to the final outcome. But, if the transactions operate on related (or the same) data, conflict is possible among the transaction components, and the selection of one operational order over another may have some undesirable consequences. So how is the correct order determined, and who determines that order? Fortunately, the DBMS handles that tricky assignment for us by using a built-in scheduler.

The **scheduler** is a special DBMS program that establishes the order in which the operations within concurrent transactions are executed. The scheduler *interleaves* the execution of database operations to ensure serializability and isolation of transactions. To determine the appropriate order, the scheduler bases its actions on concurrency control algorithms, such as locking or time stamping methods, which are explained in the next sections.

The scheduler also makes sure that the computer's central processing unit (CPU) is used efficiently. If there were no way to schedule the execution of transactions, all transactions would be executed on a first-come-first-served basis. The problem with this approach is that processing time is wasted when the CPU waits for a READ or WRITE operation to finish, thereby losing several CPU cycles. In short, first-come-first-served scheduling tends to yield unacceptable response times within the multiuser DBMS environment. Therefore, we need some other scheduling method to improve the overall system efficiency by using unused CPU time.

Additionally, the scheduler facilitates data isolation to ensure that two transactions do not update the same data element at the same time. Database operations may require READ and/or WRITE actions that can produce conflicts. For example, Table 9.9 shows the possible conflict scenarios if two transactions, T1 and T2, are executed concurrently over the same data. Using the summary in Table 9.9, note that two operations are in conflict when they access the same data and at least one of them is a WRITE operation.

TABLE 9.9 READ/WRITE CONFLICT SCENARIOS: CONFLICTING DATABASE OPERATIONS MATRIX

	TRANSACTIONS		
	T1	**T2**	**RESULT**
Operations	Read	Read	No conflict
	Read	Write	Conflict
	Write	Read	Conflict
	Write	Write	Conflict

Several methods have been proposed to schedule the execution of conflicting operations in concurrent transactions. These methods have been classified as locking, time stamping, and optimistic. Because locking methods are the most frequently used, we will examine them next.

9.3 CONCURRENCY CONTROL WITH LOCKING METHODS

A **lock** guarantees exclusive use of a data item to a current transaction. In other words, transaction T2 does not have access to a data item that is currently being used by transaction T1. A transaction acquires a lock prior to data access; the lock is released (unlocked) when the transaction is completed, so that another transaction can lock the data item for its exclusive use.

You should recall from our earlier discussion that data consistency cannot be guaranteed *during* a transaction; the database may be in a temporary inconsistent state when several updates are executed. Therefore, locks are required to prevent another transaction from reading inconsistent data.

Most multiuser DBMSs automatically initiate and enforce locking procedures. All lock information is managed by a **lock manager**, which is responsible for assigning and policing the locks used by the transactions.

9.3.1 LOCK GRANULARITY

Lock granularity indicates the level of lock use. Locking can take place at the following levels: database, table, page, row, or even field (attribute).

Database Level

In a **database-level lock**, the entire database is locked, thus preventing the use of any tables in the database by transaction T2 while transaction Tl is being executed. This level of locking is good for batch processes, but it is unsuitable for online multiuser DBMSs. You can imagine how s-l-o-w the data access would be if thousands of transactions had to wait for the previous transaction to be completed before the next one could reserve the entire database. Figure 9.3 illustrates the database-level lock. Note that transactions T1 and T2 cannot access the same database concurrently, *even if they use different tables.*

FIGURE 9.3 A DATABASE-LEVEL LOCKING SEQUENCE

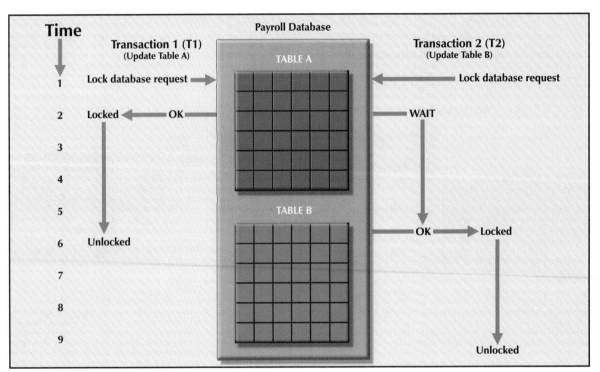

Table Level

In a **table-level lock**, the entire table is locked, preventing access to any row by transaction T2 while transaction T1 is using the table. If a transaction requires access to several tables, each table may be locked. However, two transactions can access the same database as long as they access different tables.

Table-level locks, while less restrictive than database-level locks, cause traffic jams when many transactions are waiting to access the same table. Such a condition is especially irksome if the lock forces a delay when different transactions require access to different parts of the same table; that is, when the transactions would not interfere with each other. Consequently, table-level locks are not suitable for multiuser DBMSs. Figure 9.4 illustrates the effect of a table-level lock. As you examine Figure 9.4, note that transactions T1 and T2 cannot access the same table even if they try to use different rows; T2 must wait until T1 unlocks the table.

FIGURE 9.4 AN EXAMPLE OF A TABLE-LEVEL LOCK

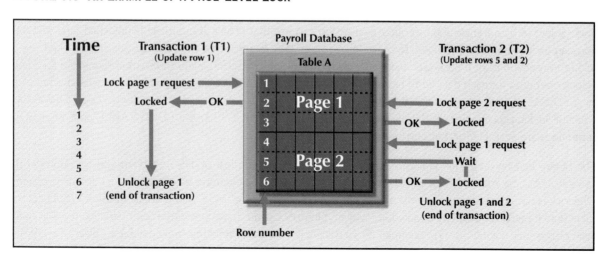

Page Level

In a **page-level lock**, the DBMS will lock an entire diskpage. A **diskpage**, or **page**, is the equivalent of a **diskblock**, which can be described as a (directly addressable) section of a disk. A page has a fixed size, such as 4K, 8K, 16K, and so on. For example, if you want to write only 73 bytes to a 4K page, the entire 4K page must be read from disk, updated in memory, and written back to disk. A table can span several pages, and a page can contain several rows of one or more tables. Page-level locks are currently the most frequently used multiuser DBMS locking method. An example of a page-level lock is shown in Figure 9.5. As you examine Figure 9.5, note that T1 and T2 access the same table while locking different diskpages. If T2 requires the use of a row located on a page that is locked by T1, T2 must wait until the page is unlocked by T1.

FIGURE 9.5 AN EXAMPLE OF A PAGE-LEVEL LOCK

Row Level

The **row-level lock** is much less restrictive than the locks discussed earlier. The DBMS allows concurrent transactions to access different rows of the same table, even if the rows are located on the same page. Although the row-level locking approach improves the availability of data, its management requires high overhead cost. (A lock exists for each row in each table of the database!) Figure 9.6 illustrates the use of a row-level lock.

FIGURE 9.6 AN EXAMPLE OF A ROW-LEVEL LOCK

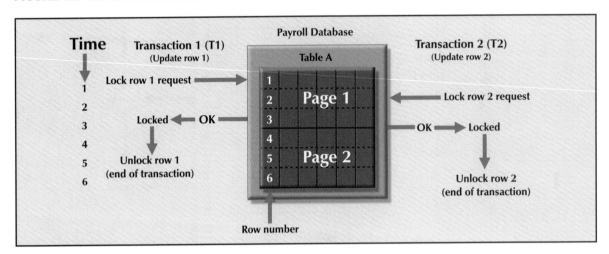

As you examine Figure 9.6, note that both transactions can execute concurrently, even if the requested rows are on the same page. T2 must wait only if it requests the same row as T1.

Field Level

The **field-level lock** allows concurrent transactions to access the same row, as long as they require the use of different fields (attributes) within that row. Although field-level locking clearly yields the most flexible multiuser data access, it is rarely done because it requires an extremely high level of computer overhead.

9.3.2 LOCK TYPES

Regardless of the level of locking, the DBMS may use different lock types: *binary* or *shared/exclusive* locks.

Binary Locks

A **binary lock** has only two states: locked (1) or unlocked (0). If an object—that is, a database, table, page, or row—is locked by a transaction, no other transaction can use that object. If an object is unlocked, any transaction can lock the object for its use. Every database operation requires that the affected object be locked. As a rule, a transaction must unlock the object after its termination. Therefore, every transaction requires a lock and unlock operation for each data item that is accessed. Such operations are automatically managed and scheduled by the DBMS; the user does not need to be concerned about locking or unlocking data items. (Every DBMS has a default locking mechanism. If the end user wants to override the default, the LOCK TABLE and other SQL commands are available for that purpose.)

The binary locking technique is illustrated in Table 9.10, using the lost updates problem you encountered in Table 9.3. As you examine Table 9.10, note that the lock and unlock features eliminate the lost update problem! (The lock is not released until the write statement is completed. Therefore, it is impossible to use a PROD_QOH value until it has been properly updated.) However, binary locks are now considered too restrictive to yield optimal concurrency conditions. For example, if two transactions want to read the same database object, the DBMS

will not allow this to happen, even though neither transaction updates the database (and, therefore, no concurrency problems can occur). Remember from Table 9.9 that concurrency conflicts occur only when two transactions execute concurrently and one of them updates the database.

TABLE 9.10 **AN EXAMPLE OF A BINARY LOCK**

TIME	TRANSACTION	STEP	STORED VALUE
1	T1	Lock PRODUCT	
2	T1	Read PROD_QOH	15
3	T1	PROD_QOH = 15 + 10	
4	T1	Write PROD_QOH	25
5	T1	Unlock PRODUCT	
6	T2	Lock PRODUCT	
7	T2	Read PROD_QOH	23
8	T2	PROD_QOH = 23 – 10	
9	T2	Write PROD_QOH	13
10	T2	Unlock PRODUCT	

Shared/Exclusive Locks

The labels "shared" and "exclusive" indicate the nature of the lock. An **exclusive lock** exists when access is specifically reserved for the transaction that locked the object. The exclusive lock must be used when the potential for conflict exists. (See Table 9.9, Read versus Write.) A **shared lock** exists when concurrent transactions are granted Read access on the basis of a common lock. A shared lock produces no conflict as long as all the concurrent transactions are read-only.

A shared lock is issued when a transaction wants to read data from the database and no exclusive lock is held on that data item. An exclusive lock is issued when a transaction wants to update (write) a data item and no locks are currently held on that data item by any other transaction. Using the shared/exclusive locking concept, a lock can have three states: unlocked, shared (Read), and exclusive (Write).

As you saw in Table 9.9, two transactions conflict only when at least one of them is a Write transaction. Because the two Read transactions can be safely executed at once, shared locks allow several Read transactions to concurrently read the same data item. For example, if transaction T1 has a shared lock on data item X, and transaction T2 wants to read data item X, T2 may also obtain a shared lock on data item X.

If transaction T2 updates data item X, then an exclusive lock is required by T2 over data item X. *The exclusive lock is granted if and only if no other locks are held on the data item.* Therefore, if a shared or exclusive lock is already held on data item X by transaction T1, an exclusive lock cannot be granted to transaction T2, and T2 must wait to begin until T1 commits. This condition is known as the **mutual exclusive rule**—only one transaction at a time can own an exclusive lock on the same object.

Although the possibility of shared locks renders data access more efficient, a shared/exclusive lock schema increases the lock manager's overhead, for several reasons:

- The type of lock held must be known before a lock can be granted.

- Three lock operations exist: READ_LOCK (to check the type of lock), WRITE_LOCK (to issue the lock), and UNLOCK (to release the lock).

- The schema has been enhanced to allow a lock upgrade (from shared to exclusive) and a lock downgrade (from exclusive to shared).

Although locks prevent serious data inconsistencies, their use may lead to two major problems:

- The resulting transaction schedule may not be serializable.
- The schedule may create deadlocks. A database deadlock, which is equivalent to traffic gridlock in a big city, is caused when two transactions wait for each other to unlock data.

Fortunately, both problems can be managed: serializability is guaranteed through a locking protocol known as two-phase locking, and deadlocks can be managed by using deadlock detection and prevention techniques. We will examine these techniques in the next two sections.

9.3.3 TWO-PHASE LOCKING TO ENSURE SERIALIZABILITY

Two-phase locking defines how transactions acquire and relinquish locks. Two-phase locking guarantees serializability, but it does not prevent deadlocks. The two phases are:

1. A growing phase, in which a transaction acquires all the required locks without unlocking any data. Once all locks have been acquired, the transaction is in its locked point.

2. A shrinking phase, in which a transaction releases all locks and cannot obtain any new lock.

The two-phase locking protocol is governed by the following rules:

- Two transactions cannot have conflicting locks.
- No unlock operation can precede a lock operation in the same transaction.
- No data are affected until all locks are obtained—that is, until the transaction is in its locked point.

Figure 9.7 depicts the two-phase locking protocol.

FIGURE 9.7 TWO-PHASE LOCKING PROTOCOL

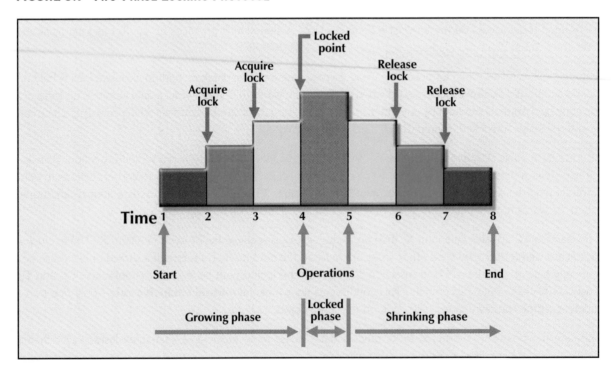

In this example, the transaction acquires all the locks it needs until it reaches its locked point. (In this example, the transaction requires two locks.) When the locked point is reached, the data are modified to conform to the transaction requirements. Finally, the transaction is completed as it releases all of the locks it acquired in the first phase.

9.3.4 DEADLOCKS

A **deadlock** is a condition that occurs when two transactions wait for each other to unlock data. Deadlocks occur when two transactions, T1 and T2, exist in the following mode:

> T1 = access data items X and Y

> T2 = access data items Y and X

If T1 has not unlocked data item Y, T2 cannot begin; if T2 has not unlocked data item X, T1 cannot continue. Consequently, T1 and T2 wait indefinitely, each waiting for the other to unlock the required data item. Such a deadlock is also known as a **deadly embrace**. Table 9.11 demonstrates how a deadlock condition is created.

TABLE 9.11 **HOW A DEADLOCK CONDITION IS CREATED**

TIME	TRANSACTION	REPLY	LOCK STATUS	
0			Data X	Data Y
1	T1:LOCK(X)	OK	Unlocked	Unlocked
2	T2: LOCK(Y)	OK	Locked	Unlocked
3	T1:LOCK(Y)	WAIT	Locked	Locked
4	T2:LOCK(X)	WAIT	Locked **Deadlock**	Locked
5	T1:LOCK(Y)	WAIT	Locked	Locked
6	T2:LOCK(X)	WAIT	Locked	Locked
7	T1:LOCK(Y)	WAIT	Locked	Locked
8	T2:LOCK(X)	WAIT	Locked	Locked
9	T1:LOCK(Y)	WAIT	Locked	Locked
...				
...				
...				
...				

The preceding example used only two concurrent transactions to demonstrate a deadlock condition. In a real-world DBMS, many more transactions can be executed simultaneously, thereby increasing the probability of generating deadlocks. Note that deadlocks are possible only if one of the transactions wants to obtain an exclusive lock on a data item; no deadlock condition can exist among *shared* locks.

Three basic techniques exist to control deadlocks:

- *Deadlock prevention.* A transaction requesting a new lock is aborted if there is the possibility that a deadlock can occur. If the transaction is aborted, all the changes made by this transaction are rolled back, and all locks obtained by the transaction are released. The transaction is then rescheduled for execution. Deadlock prevention works because it avoids the conditions that lead to deadlocking.

- *Deadlock detection.* The DBMS periodically tests the database for deadlocks. If a deadlock is found, one of the transactions (the "victim") is aborted (rolled back and restarted), and the other transaction continues.

- *Deadlock avoidance.* The transaction must obtain all the locks it needs before it can be executed. This technique avoids rollback of conflicting transactions by requiring that locks be obtained in succession. However, the serial lock assignment required in deadlock avoidance increases action response times.

The choice of the best deadlock control method to use depends on the database environment. For example, if the probability of deadlocks is low, deadlock detection is recommended. However, if the probability of deadlocks is high, deadlock prevention is recommended. If response time is not high on the system priority list, deadlock avoidance might be employed.

9.4 CONCURRENCY CONTROL WITH TIME STAMPING METHODS

The **time stamping** approach to scheduling concurrent transactions assigns a global unique time stamp to each transaction. The time stamp value produces an explicit order in which transactions are submitted to the DBMS. Time stamps must have two properties: uniqueness and monotonicity. **Uniqueness** ensures that no equal time stamp values can exist, and **monotonicity**[1] ensures that time stamp values always increase.

All database operations (Read and Write) within the same transaction must have the same time stamp. The DBMS executes conflicting operations in time stamp order, thereby ensuring serializability of the transactions. If two transactions conflict, one is stopped, rolled back, rescheduled, and assigned a new time stamp value.

The disadvantage of the time stamping approach is that each value stored in the database requires two additional time stamp fields: one for the last time the field was read, and one for the last update. Time stamping thus increases the memory needs and the database's processing overhead. Time stamping tends to demand a lot of system resources, because there is a possibility that many transactions may have to be stopped, rescheduled, and restamped.

9.4.1 WAIT/DIE AND WOUND/WAIT SCHEMES

You have learned that time stamping methods are used to manage concurrent transaction execution. In this section, you learn of two schemes used to decide which transaction is rolled back and which continues executing: the wait/die scheme and the wound/wait scheme.[2] We use an example to illustrate the difference. Assume that you have two conflicting transactions T1 and T2, each with a unique time stamp. Let's suppose that T1 has a time stamp of 11548789 and that T2 has a time stamp of 19562545. You can deduce from the time stamps that T1 is the older transaction—the lower time stamp value—and T2 is the newer transaction. Given this scenario, the four possible outcomes are shown in Table 9.12.

[1] The term "monotonicity" is part of the standard concurrency control vocabulary. Our first introduction to this term and to its proper use was an article written by W. H. Kohler, "A Survey of Techniques for Synchronization and Recovery in Decentralized Computer Systems, *Computer Surveys* 3(2), June 1981, pp. 149–283.

[2] The procedure was first described by R. E. Stearnes and P. M. Lewis, II in "System-level Concurrency Control for Distributed Database Systems," *ACM Transactions on Database Systems*, No. 2, June 1978, pp. 178–198.

TABLE 9.12 **WAIT/DIE AND WOUND/WAIT CONCURRENCY CONTROL SCHEMES**

TRANSACTION REQUESTING LOCK	TRANSACTION OWNING LOCK	WAIT/DIE SCHEME	WOUND/WAIT SCHEME
T1 (11548789)	T2 (19562545)	• T1 waits until T2 is completed and T2 releases its locks.	• T1 preempts (rolls back) T2. T2 is rescheduled using the same time stamp.
T2 (19562545)	T1 (11548789)	• T2 Dies (rolls back). • T2 is rescheduled using the same time stamp.	• T2 waits until T1 is completed and T1 releases its locks.

Using the wait/die scheme:

- If the transaction requesting the lock is the older one of the two transactions, it will *wait* until the other transaction is completed and the locks are released.

- If the transaction requesting the lock is the younger of the two transactions, it will *die* (roll back itself) and it is rescheduled using the same time stamp.

In short, in the **wait/die** scheme the older transaction waits and the younger is rolled back and rescheduled.

In the wound/wait scheme:

- If the transaction requesting the lock is the older one of the two transactions, the older transaction will preempt (*wound*) the younger transaction (by rolling it back). T1 preempts T2 when T1 rolls back T2. The younger preempted transaction is rescheduled, using the same time stamp.

- If the transaction requesting the lock is the younger of the two transactions, it waits until the other transaction is completed and the locks are released.

In short, in the **wound/wait** scheme, the older transaction rolls back the younger transaction and reschedules it.

In both schemes, one of the transactions waits for the other transaction to finish and release the locks. However, in many cases a transaction may request multiple locks. How long does a transaction have to wait for each lock request? Obviously, this scenario can cause some transactions to wait indefinitely, causing a deadlock. To prevent this type of deadlock, each lock request has an associated time-out value. If the lock if not granted before the time-out expires, the transaction is rolled back.

9.5 CONCURRENCY CONTROL WITH OPTIMISTIC METHODS

The **optimistic approach** is based on the assumption that the majority of the database operations do not conflict. The optimistic approach does not require locking or time stamping techniques. Instead, a transaction is executed without restrictions until it is committed. Using an optimistic approach, each transaction moves through two or three phases. The phases are *read, validation,* and *write*.[3]

- During the *read phase*, the transaction reads the database, executes the needed computations, and makes the updates to a private copy of the database values. All update operations of the transaction are recorded in a temporary update file, which is not accessed by the remaining transactions.

[3] We first encountered the description of the optimistic approach to concurrency control in an article by H. T. King and J. T. Robinson, "Optimistic Methods for Concurrency Control," *ACM Transactions on Database Systems* 6(2), June 1981, pp. 213–226. Even the most current software is built on conceptual standards that were developed more than two decades ago.

- During the *validation phase*, the transaction is validated to ensure that the changes made will not affect the integrity and consistency of the database. If the validation test is positive, the transaction goes to the write phase. If the validation test is negative, the transaction is restarted, and the changes are discarded.

- During the *write phase*, the changes are permanently applied to the database.

The optimistic approach is acceptable for mostly read or query database systems that require very few update transactions.

In a heavily used DBMS environment, the management of deadlocks—their prevention and detection—constitutes an important DBMS function. The DBMS will use one or more of the techniques we have discussed, as well as variations on those techniques. However, the deadlock cure is sometimes worse than the disease it is supposed to cure. Therefore, it is sometimes necessary to employ database recovery techniques to restore the database to a consistent state.

9.6 DATABASE RECOVERY MANAGEMENT

Database recovery restores a database from a given state, usually inconsistent, to a previously consistent state. Recovery techniques are based on the **atomic transaction property**: All portions of the transaction must be treated as a single logical unit of work, in which all operations must be applied and completed to produce a consistent database. If, for some reason, any transaction operation cannot be completed, the transaction must be aborted, and any changes to the database must be rolled back (undone). In short, transaction recovery reverses all the changes that the transaction made to the database before it was aborted.

Although we have emphasized the recovery of *transactions*, recovery techniques also apply to the *database* or the *system* after some type of critical error has occurred. Backup and recovery functions constitute a very important component of today's DBMSs. Some DBMSs provide functions that allow the database administrator to schedule automatic database backups to permanent secondary storage devices, such as disks or tapes.

The level of backup varies:

- A **full backup** of the database, or dump of the database.

- A **differential backup** of the database, in which only the last modifications to the database (when compared with a previous backup copy) are copied.

- A **transaction log backup** backs up only the transaction log operations that are not reflected in a previous backup copy of the database.

The database backup is stored in a secure place, usually in a different building, and protected against dangers such as fire, theft, flood, and other potential calamities. The backup's existence guarantees database recovery following system (hardware/software) failures.

Failures that plague databases and systems are generally induced by software, hardware, program exemptions, transactions, or external factors.

Software

- Software-induced failures may be traceable to the operating system, the DBMS software, application programs, or viruses.

Hardware

- Hardware-induced failures may include memory chip errors, disk crashes, bad disk sectors, disk full errors, and so on.

Programming Exemption

- Application programs or end users may roll back transactions when certain conditions are defined. For example, a recovery procedure may be initiated if a withdrawal of funds is made when the customer balance is zero, or if an end user has initiated an unintended keyboard error, such as pressing Ctrl+C while running a UNIX application that updates a database at the shell (command prompt) level.

Transaction

- The system detects deadlocks and aborts one of the transactions.

External

- Backups are especially important when a system suffers complete destruction due to fire, earthquake, flood, and so forth.

Depending on the type and the extent of the failure, the recovery process ranges from a minor short-term inconvenience to a major long-term rebuild. Regardless of the extent of the required recovery process, recovery is not possible without backup.

The database recovery process generally follows a predictable scenario. First, determine the type and the extent of the required recovery. If the entire database needs to be recovered to a consistent state, the recovery uses the most recent backup copy of the database in a known consistent state. The backup copy is then rolled forward to restore all subsequent transactions by using the transaction log information. If the database needs to be recovered but the committed portion of the database is still usable, the recovery process uses the transaction log to "undo" all the transactions that were not committed.

9.6.1 TRANSACTION RECOVERY

In Section 9.1.4 you learned about the transaction log structure and how it contains data for database recovery purposes. Database transaction recovery focuses on the different methods used to recover a database from an inconsistent state to a consistent one by using the data in the transaction log.

Before we continue, let's examine four important concepts that affect the recovery process:

- The **write-ahead-log protocol**. This protocol ensures that transaction logs are always written *before* any database data are actually updated. This is to ensure that, in case of a failure, the database can later be recovered to a consistent state using the data in the transaction log.

- **Redundant transaction logs.** Most DBMS keep several copies of the transaction log to ensure that a disk physical failure will not impair the DBMS ability to recover data.

- Database **buffers**. A buffer is a temporary storage area in primary memory used to speed up disk operations. To improve processing time, the DBMS software reads the data from the physical disk and stores a copy of it on a "buffer" in primary memory. When a transaction updates data, it actually updates the copy of the data in the buffer, because it is much faster than accessing the physical disk *every* time. Later on, all buffers that contain updated data are written to a physical disk during a single operation, thereby saving significant processing time.

- Database **checkpoints**. A database checkpoint is an operation in which the DBMS writes all of its updated buffers to disk. While this is happening, the DBMS does not execute any other requests. A checkpoint operation is also registered in the transaction log. By doing this, the physical database and the transaction log will be "in-sync"—remember this is required because update operations update the copy of the data in the buffers and not in the physical database. Checkpoints are automatically scheduled by the DBMS several times per hour. As you will see next, checkpoints also play an important role in transaction recovery.

The database recovery process involves bringing the database to a consistent state after a failure. Transaction recovery procedures generally make use of deferred-write and write-through techniques.

When the recovery procedure uses **deferred write** or **deferred update**, the transaction operations do not immediately update the physical database. Instead, only the transaction log is updated. The database is physically updated only after the transaction reaches its commit point using the transaction log information. If the transaction aborts before it reaches its commit point, no changes (no ROLLBACK or undo) need to be made to the database because the database was never updated. The recovery process for all started and committed transactions (before the failure) will follow these steps:

1. Identify the last checkpoint in the transaction log. This is the last time transaction data was physically saved to disk.

2. For a transaction that started and committed before the last checkpoint, nothing needs to be done, because the data are already saved.

3. For a transaction that performed a commit operation after the last checkpoint, the DBMS uses the transaction log records to redo the transaction and to update the database, using the "after" values in the transaction log. The changes are made in ascending order, from the oldest to the newest.

4. For any transaction with a ROLLBACK operation after the last checkpoint or that was left active (with neither a COMMIT nor a ROLLBACK) before the failure occurred, nothing needs to be done because the database was never updated.

When the recovery procedure uses **write-through** or **immediate update**, the database is immediately updated by transaction operations during the transaction's execution, even before the transaction reaches its commit point. If the transaction aborts before it reaches its commit point, a ROLLBACK or undo operation needs to be done to restore the database to a consistent state. In this case, the ROLLBACK operation will use the transaction log "before" values. The recovery process will follow these steps:

1. Identify the last checkpoint in the transaction log. This is the last time transaction data were physically saved to disk.

2. For a transaction that started and committed before the last checkpoint, nothing needs to be done, because the data are already saved.

3. For a transaction that committed after the last checkpoint, the DBMS redoes the transaction, using the "after" values of the transaction log. Changes are applied in ascending order, from the oldest to the newest.

4. For any transaction with a ROLLBACK operation after the last checkpoint or that was left active (with neither a COMMIT nor a ROLLBACK) before the failure occurred, the DBMS uses the transaction log records to ROLLBACK or undo the operations, using the "before" values in the transaction log. Changes are applied in reverse order, from the newest to the oldest.

You may use the transaction log in Table 9.13 to trace back a simple database recovery process. To make sure that you understand the recovery process, we use a simple transaction log that includes three transactions and one checkpoint. This transaction log includes the transaction components we used earlier in this chapter, so you should already be familiar with the basic process. Given the transaction, the transaction log has the following characteristics:

- Transaction 101 consists of two UPDATE statements that reduce the quantity on hand for product '54778-2T' and increase the customer balance for customer '10011' for a credit sale of two units of product '54778-2T'.

- Transaction 106 is the same credit sales event you saw in section 9.1.1. This transaction represents the credit sale of one unit of product '89-WRE-Q' to customer 10016 in the amount of $277.55. This transaction consists of five SQL DML (three INSERT and two UPDATE) statements.

TABLE 9.13 A TRANSACTION LOG FOR TRANSACTION RECOVERY EXAMPLES

TRL ID	TRX NUM	PREV PTR	NEXT PTR	OPERATION	TABLE	ROW ID	ATTRIBUTE	BEFORE VALUE	AFTER VALUE
341	101	Null	352	START	****Start Transaction				
352	101	341	363	UPDATE	PRODUCT	54778-2T	PROD_QOH	45	43
363	101	352	365	UPDATE	CUSTOMER	10011	CUST_BALANCE	615.73	675.62
365	101	363	Null	COMMIT	**** End of Transaction				
397	106	Null	405	START	****Start Transaction				
405	106	397	415	INSERT	INVOICE	1009			1009,10016, …
415	106	405	419	INSERT	LINE	1009,1			1009,1, 89-WRE-Q,1, ….
419	106	415	427	UPDATE	PRODUCT	89-WRE-Q	PROD_QOH	12	11
423				CHECKPOINT					
427	106	419	431	UPDATE	CUSTOMER	10016	CUST_BALANCE	0.00	277.55
431	106	427	457	INSERT	ACCT_TRANSACTION	10007			1007,18-JAN-2004, …
457	106	431	Null	COMMIT	**** End of Transaction				
521	155	Null	525	START	****Start Transaction				
525	155	521	528	UPDATE	PRODUCT	2232/QWE	PROD_QOH	6	26
528	155	525	Null	COMMIT	**** End of Transaction				

***** C * R * A * S * H * * * * *

- Transaction 155 represents a simple inventory update. This transaction consists of one UPDATE statement that increases the quantity on hand of product '2232/QWE' from 6 units to 26 units.

- A database checkpoint wrote all updated database buffers to disk. The checkpoint event only writes the changes for all previously committed transactions. In this case, the checkpoint applies all changes done by transaction 101 to the database data files.

Using Table 9.13 you can now trace the database recovery process for a DBMS using the deferred update method as follows:

1. Identify the last checkpoint. In this case, the last checkpoint has TRL ID 423. This was the last time database buffers were physically written to disk.

2. Note that transaction 101 started and finished before the last checkpoint. Therefore, all changes were already written to disk and no additional action needs to be taken.

3. For each transaction that committed after the last checkpoint (TRL ID 423), the DBMS will use the transaction log data to write the changes to disk using the after values. For example, for transaction 106:

 - Find COMMIT (TRL ID 457).

 - Use the previous pointer values to locate the start of the transaction (TRL ID 397).

 - Use the next pointer values to locate each DML statement and apply the changes to disk using the after values. (Start with TRL ID 405, then 415, 419, 427 and 431.) Remember that TRL ID 457 was the COMMIT statement for this transaction.

 - Repeat the process for transaction 155.

4. Any other transactions will be ignored. Therefore, for transactions that ended with ROLLBACK or that were left active (do not end with a COMMIT or ROLLBACK) nothing is done because no changes were written to disk.

S U M M A R Y

A transaction is a sequence of database operations that access the database. A transaction represents real-world events. A transaction must be a logical unit of work; that is, no portion of the transaction can exist by itself. Either all parts are executed, or the transaction is aborted. A transaction takes a database from one consistent state to another. A consistent database state is one in which all data integrity constraints are satisfied.

All transactions have four properties:

- *Atomicity*. All parts of the transaction are executed; otherwise, the transaction is aborted.

- *Durability*. The changes made by a transaction cannot be rolled back once the transaction is committed.

- *Serializability*. The result of the concurrent execution of transactions is the same as if they were executed in serial order.

- *Isolation*. Data being used by one transaction cannot be accessed by another transaction until the first transaction is completed.

SQL provides support for transactions through the use of two statements:

- COMMIT saves changes to disk.

- ROLLBACK restores the previous database state.

The SQL transaction is formed by several SQL statements or database requests. Each database request originates several I/O database operations.

The transaction log keeps track of all transactions that modify the database. The information stored in the transaction log is used for recovery (ROLLBACK) purposes.

Concurrency control coordinates the simultaneous execution of transactions. The concurrent execution of transactions can result in three main problems: lost updates, uncommitted data, and inconsistent retrievals.

The scheduler is responsible for establishing the order in which the concurrent transaction operations are executed. The transaction execution order is critical; such an order ensures database integrity in multiuser database systems. Locking, time stamping, and optimistic methods are used by the scheduler to ensure the serializability of transactions.

A lock guarantees unique access to a data item by a transaction. The lock prevents one transaction from using the data item while another transaction is using it. There are several levels of locking: database, table, page, row, and field.

Two types of locks can be used in database systems: binary locks and shared/exclusive locks. A binary lock can have only two states: 1 (locked) or 0 (unlocked). A shared lock is used when a transaction wants to read data from a database and no other transaction is updating the same data. Several shared or "read" locks can exist for a particular item. An exclusive lock is issued when a transaction wants to update (write to) the database and no other locks (shared or exclusive) are held on the data.

Serializability of schedules is guaranteed through the use of a locking protocol known as two-phase locking. The two-phase locking schema has:

1. A growing phase, in which the transaction acquires all the locks that it needs without unlocking any data.
2. A shrinking phase, in which the transaction releases all the locks without acquiring new locks.
3. When two or more transactions wait indefinitely for each other to release a lock, they are in a deadlock, or a deadly embrace. There are three deadlock control techniques: prevention, detection, and avoidance.

Concurrency control with time stamping methods assigns a unique time stamp to each transaction and schedules the execution of conflicting transactions in time stamp order.

Time stamping methods are used to manage concurrent transaction execution. Two schemes are used to decide which transaction is rolled back and which continues executing: the wait/die scheme and the wound/wait scheme.

Concurrency control with optimistic methods assumes that the majority of database transactions do not conflict and that transactions are executed concurrently, using private copies of the data. At commit time the private copies are updated to the database.

Database recovery restores the database from a given state to a previous consistent state. Database backups are permanent copies of the database, stored in a safe place, to be used in case of a critical error in the master database.

KEY TERMS

atomicity	exclusive lock	page
atomic transaction property	full backup	redundant transaction logs
binary lock	immediate update	scheduler
buffers	inconsistent retrievals	serializability
checkpoint	isolation	shared lock
concurrency control	lock	time stamping
consistent database state	lock granularity	transaction
database recovery	database-level lock	transaction log
database request	field-level lock	transaction log backup
deadlock	page-level lock	two-phase locking
deadly embrace	row-level level lock	uncommitted data
deferred update	table-level lock	uniqueness
deferred write	lock manager	wait/die
differential backup	lost update	wound/wait
diskblock	monotonicity	write-ahead-log protocol
diskpage	mutual exclusive rule	write-through
durability	optimistic approach	

REVIEW QUESTIONS

1. Explain the following statement: a transaction is a logical unit of work.
2. What is a consistent database state, and how is it achieved?
3. The DBMS does not guarantee that the semantic meaning of the transaction truly represents the real-world event. What are the possible consequences of this limitation? Give an example.
4. List and discuss the four transaction properties.
5. What is a transaction log, and what is its function?
6. What is a scheduler, what does it do, and why is its activity important to concurrency control?
7. What is a lock, and how does it (generally) work?
8. What is concurrency control, and what is its objective?
9. What is an exclusive lock, and under what circumstances is it granted?
10. What is a deadlock, and how can it be avoided? Discuss several deadlock avoidance strategies.
11. What three levels of backup may be used in database recovery management? Briefly, describe what each of those three backup levels does.

PROBLEMS

1. Suppose that you are a manufacturer of product ABC, which is composed of parts A, B, and C. Each time a new product ABC is created, it must be added to the product inventory, using the PROD_QOH in a table named PRODUCT. And each time the product is created, the parts inventory, using PART_QOH in a table named PART, must be reduced by one each of parts A, B, and C. The sample database contents are shown in Table P9.1

TABLE P9.1 **THE DATABASE FOR PROBLEM 1**

TABLE NAME: PRODUCT	
PROD_CODE	PROD_QOH
ABC	1,1205

TABLE NAME: PART	
PROD_CODE	PROD_QOH
A	567
B	98
C	549

Given this information, answer questions a–e.
 a. How many database requests can you identify for an inventory update for both PRODUCT and PART?
 b. Using SQL, write each database request you have identified in Step a.
 c. Write the complete transaction(s).
 d. Write the transaction log, using Table 9.1 as your template.
 e. Using the transaction log you created in Step d, trace its use in database recovery.

2. Describe the three most common concurrent transaction execution problems. Explain how concurrency control can be used to avoid such problems.

3. What DBMS component is responsible for concurrency control? How is this feature used to resolve conflicts?

4. Using a simple example, explain the use of binary and shared/exclusive locks in a DBMS.

5. Suppose that your database system has failed. Describe the database recovery process and the use of deferred-write and write-through techniques.

ADDITIONAL CD CONTENT

The **Ch09_ABC_Markets** database is located in the **Databases/Student/Ch09** folder on the student CD.

6. ABC Markets sells products to customers. The relational schema shown in Figure P9.6 represents the main entities for ABC's database. Note the following important characteristics:

- A customer can make many purchases, each one represented by an invoice.

 - The CUS_BALANCE is updated with each credit purchase or payment and represents the amount the customer owes.

 - The CUS_BALANCE is increased (+) with every credit purchase and decreased (–) with every customer payment.

 - The date of last purchase is updated with each new purchase made by the customer.

 - The date of last payment is updated with each new payment made by the customer.

- An invoice represents a product purchase by a customer.

 - An INVOICE can have many invoice LINEs, one for each product purchased.

 - The INV_TOTAL represents the total cost of the invoice, including taxes.

 - The INV_TERMS can be "30", "60", or "90" (representing the number of days credit), or "CASH", "CHECK", or "CC".

 - The invoice status can be "OPEN", "PAID", or "CANCEL".

- A product's quantity on hand (P_QTYOH) is updated (decreased) with each product sale.

- A customer can make many payments. The payment type (PMT_TYPE) can be one of the following:

 - "CASH" for cash payments.

 - "CHECK" for check payments.

 - "CC" for credit card payments.

- The payment details (PMT_DETAILS) are used to record data about check or credit card payments:

 - The bank, account #, and check # for check payments.

 - The issuer, credit card #, and expiration date for credit card payments.

Note: Not all entities and attributes are represented in this example. Use only the attributes indicated.

FIGURE P9.6 THE ABC MARKETS RELATIONAL SCHEMA

Using this database, write the SQL code to represent each one of the following transactions. Use BEGIN TRANSACTION and COMMIT to group the SQL statements in logical transactions.

 a. On May 11, 2004, customer '10010' makes a credit purchase (30 days) of one unit of product: '11QER/31' with a unit price of $110.00, and the tax rate is 8%. The invoice number is 10983 and this invoice has only one product line.

 b. On June 3, 2004, customer '10010' makes a payment of $100 in cash. The payment ID is 3428.

 c. Create a simple transaction log (using the format shown in Table 9.13) to represent the actions of the two previous transactions.

10

DISTRIBUTED DATABASE MANAGEMENT SYSTEMS

In this chapter, you will learn:

❖ What a distributed database management system (DDBMS) is and what its components are

❖ How database implementation is affected by different levels of data and process distribution

❖ How transactions are managed in a distributed database environment

❖ How database design is affected by the distributed database environment

T

PREVIEW

In this chapter, you will learn that a single database can be divided into several fragments. The fragments can be stored on different computers within a network. Processing, too, can be dispersed among several different network sites, or nodes. The multisite database forms the core of the distributed database system.

The growth of distributed database systems has been fostered primarily by the dispersion of business operations and the rapid pace of technological change that has made local and wide area networks (LANs and WANs) practical and reliable. The network-based distributed database system is very flexible: It can serve the needs of a small business operating two stores in the same town, or those of a global business.

Although the dispersal of the database within a distributed database system requires a more sophisticated DBMS, the end user should not be burdened by increased operational complexity. That is, the greater complexity of a distributed database system should be transparent to the end user.

The distributed database management system (DDBMS) treats a distributed database as a single *logical* database; therefore, the basic design concepts you learned in earlier chapters remain intact. However, although the end user need not be aware of the distributed database's special characteristics, the distribution of data among different sites in a computer network clearly adds to a system's complexity. For example, the design of a distributed database must consider the location of the data and the partitioning of the data into database fragments. We will examine these issues in this chapter.

E

N

 TEN

10.1 THE EVOLUTION OF DISTRIBUTED DATABASE MANAGEMENT SYSTEMS

A **distributed database management system (DDBMS)** governs the storage and processing of logically related data over interconnected computer systems in which both data and processing functions are distributed among several sites. To understand how and why the DDBMS is different from the DBMS, it is useful to briefly examine the changes in the database environment that set the stage for the development of the DDBMS.

During the 1970s, corporations implemented centralized database management systems to meet their structured information needs. Structured information is usually presented as regularly issued formal reports in a standard format. Such information, generated by 3GL programming languages, is created by specialists in response to precisely channeled requests. Structured information needs are thus well served by centralized systems.

Basically, the use of a centralized database required that corporate data be stored in a single central site, usually a mainframe or midrange computer. Data access was provided through dumb terminals. The centralized approach, illustrated in Figure 10.1, worked well to fill the structured information needs of corporations, but it fell short when quickly moving events required faster response times and equally quick access to information. The slow progression from information request to approval, to specialist, to user simply did not serve decision makers well in a dynamic environment. What was needed was quick, unstructured access to databases, using ad hoc queries to generate on-the-spot information.

FIGURE 10.1 CENTRALIZED DATABASE MANAGEMENT SYSTEM

Database management systems based on the relational model could provide the environment in which unstructured information needs would be met by employing ad hoc queries. End users would be given the ability to access data when needed. Unfortunately, the early relational model implementations did not yet deliver acceptable throughput when compared to the well-established hierarchical or network database models.

The 1980s gave birth to a series of crucial social and technological changes that affected database development and design. Among these changes were:

- Business operations became more decentralized geographically.
- Competition increased at the global level.
- Customer demands and market needs favored a decentralized management style.

- Rapid technological change created low-cost microcomputers with mainframe-like power. As a result, a fast-growing number of corporations adopted local area networks (LANs) as the basis for their computerized solutions. The LAN's rise to prominence was also aided by the increasing cost of mainframe solutions.

- The large number of applications based on DBMSs and the need to protect investments in centralized DBMS software made the notion of data sharing attractive.

These factors created a dynamic business environment in which companies had to respond quickly to competitive and technological pressures. As large business units restructured to form leaner-and-meaner, quickly reacting, dispersed operations, two database requirements became obvious:

- Rapid ad hoc data access became crucial in the quick-response decision-making environment.

- The decentralization of management structures based on the decentralization of business units made decentralized multiple-access and multiple-location databases a necessity.

During the 1990s the factors we have just described became even more firmly entrenched. However, the way these factors were addressed was strongly influenced by:

- The growing acceptance of the Internet and, particularly, the World Wide Web (WWW), as the platform for data access and distribution. The WWW is, in effect, the *repository* for distributed data. (We'll examine the Internet's impact on data access and distribution in Chapter 13, "Databases in Electronic Commerce.")

- The increased focus on data analysis that led to data mining and data warehousing. Although a data warehouse is not usually a distributed database, it does rely on techniques—such as data replication and distributed queries—that facilitate its data extraction and integration. (We'll examine data warehouse design, implementation, and use in Chapter 12, "The Data Warehouse.")

At this point, the long-term impact of the WWW on *distributed* database design and management is unclear. Perhaps the WWW's success may even foster the use of distributed databases, as bandwidth becomes a more troublesome bottleneck. Perhaps the resolution of bandwidth problems will simply confirm the centralized database standard. In any case, distributed databases exist today, and many distributed database operating concepts and components are likely to find a place in future database developments.

The decentralized database is especially desirable because centralized database management is subject to problems such as:

- *Performance degradation* due to a growing number of remote locations over greater distances.

- *High costs* associated with maintaining and operating large central (mainframe) database systems.

- *Reliability problems* created by dependence on a central site.

The dynamic business environment and the centralized database's shortcomings spawned a demand for applications based on data access from different sources at multiple locations. Such a multiple-source/multiple-location database environment is managed by a distributed database management system (DDBMS).

10.2 DDBMS ADVANTAGES

Distributed database management systems deliver several advantages over traditional systems:

- *Data are located near the "greatest demand" site*. The data in a distributed database system are dispersed to match business requirements.

- *Faster data access*. End users often work with only a subset of the company's data. If such data subsets are locally stored and accessed, the database system will deliver faster data access than is possible with remotely located centralized data.

- *Faster data processing.* A distributed database system makes it possible to process data at several sites, thereby spreading out the system's workload.

- *Growth facilitation.* New sites can be added to the network without affecting the operations of other sites. Such flexibility enables the company to expand relatively easily and rapidly.

- *Improved communications.* Because local sites are smaller and located closer to customers, local sites foster better communications among departments and between customers and company staff. Quicker and better communication often helps to improve information systems. For example, a local accounts receivable operation uses sales department data directly, without having to depend on delayed reports from the central office.

- *Reduced operating costs.* It is much more cost-effective to add workstations to a network than to update a mainframe system. The cost of dedicated data communication lines and mainframe software is reduced proportionally. Development work is done more cheaply and more quickly on low-cost PCs than on mainframes. In fact, most corporations prohibit mainframe development work. Mainframes are not destined to fade from the database picture; their power cannot be discounted. Nevertheless, the existence of PC-based networks makes it possible to distribute workloads more prudently and to reserve the expensive mainframes for more specialized end-processing tasks.

- *User-friendly interface.* PCs and workstations are usually equipped with an easy-to-use graphical user interface (GUI). The GUI simplifies use and training for end users.

- *Less danger of a single-point failure.* In a centralized system, the mainframe's failure brings down all the system's operations. In contrast, a distributed system is able to shift operations when one of the computers fails. The system workload is picked up by other workstations, because one feature of the distributed system is that data exist at multiple sites.

- *Processor independence.* The end user is able to access any available copy of the data, and an end user's request is processed by any processor at the data location. In other words, requests do not depend on a specific processor; any available processor can handle the user's request.

10.3 DDBMS DISADVANTAGES

Current distributed database management systems are subject to some problems. These include:

- *Complexity of management and control.* Management of distributed data is a more complex task than management of centralized data. Applications must recognize data location, and they must be able to stitch together data from different sites. Database administrators must have the ability to coordinate database activities to prevent database degradation due to data anomalies. Transaction management, concurrency control, security, backup, recovery, query optimization, access path selection, and so on, must all be addressed and resolved. In short, keeping the various components of a distributed database synchronized is a daunting task.

- *Security.* The probability of security lapses increases when data are located at multiple sites. The responsibility of data management will be shared by different people at several sites, and LANs do not yet have the sophisticated security of centralized mainframe installations.

- *Lack of standards.* Although distributed databases depend on effective communication, there are no standard communication protocols *at the database level.* (Although TCP/IP is the de facto standard at the network level, there is no standard at the application level.) In fact, few official standards exist in any of the distributed database protocols, whether they deal with communication or data access control. For

example, different database vendors employ different—and often incompatible—techniques to manage the distribution of data and processing in a DDBMS environment. Consequently, distributed database users must wait (and lobby!) for the definitive emergence of standard protocols before distributed databases can deliver all their potential benefits.

- *Increased storage requirements.* The distributed database stores multiple copies of data required at different sites, thus requiring additional disk storage space. This disadvantage is a minor one, because disk storage space is relatively cheap, and it is becoming cheaper.

- *Greater difficulty in managing the data environment.* Disk access and storage in a widely dispersed data storage environment are more complex than they would be in a centralized database. Therefore, managing this data environment becomes more difficult from both software and human perspectives.

- *Increased training cost.* Training costs are generally higher in a distributed model than they would be in a centralized model, sometimes even to the extent of offsetting operational and hardware savings.

Distributed databases are used successfully, but we have a long way to go before they will yield the full flexibility and power of which they are theoretically capable. The inherently complex distributed data environment increases the urgency for standard protocols governing transaction management, concurrency control, security, backup, recovery, query optimization, access path selection, and so on. Such issues must be addressed and resolved before DDBMS technology is widely embraced.

The remainder of this chapter explores the distributed database's basic components and concepts. Because the distributed database is usually based on the relational database model, we will use relational terminology to explain the basic distributed concepts and components.

10.4 DISTRIBUTED PROCESSING AND DISTRIBUTED DATABASES

In **distributed processing**, a database's logical processing is shared among two or more physically independent sites that are connected through a network. For example, the data input/output (I/O), the data selection, and data validation might be performed on one computer, and a report based on that data might be created on another computer.

A basic distributed processing environment is illustrated in Figure 10.2. Figure 10.2 shows that a distributed processing system shares the database processing chores among three sites connected through a communications network. Although the database resides at only one site (Miami), each site can access the data and update the database. The database is located on Computer A, a network computer known as the *database server*.

A **distributed database**, on the other hand, stores a logically related database over two or more physically independent sites. The sites are connected via a computer network. In contrast, the distributed processing system uses only a single-site database but shares the processing chores among several sites. In a distributed database system, a database is composed of several parts known as **database fragments**. The database fragments are located at different sites, and can also be replicated among various sites. An example of a distributed database environment is shown in Figure 10.3.

FIGURE 10.2 DISTRIBUTED PROCESSING ENVIRONMENT

The database in Figure 10.3 is divided into three database fragments (E1, E2, and E3) located at different sites. The computers are connected through a network system. In a fully distributed database, the users Betty, Hernando, and Alan do not need to know the name or location of each database fragment in order to access the database. Also, the users may be located at sites other than Miami, New York, or Atlanta and still be able to access the database as a single logical unit.

As you examine and contrast Figures 10.2 and 10.3, you should keep the following points in mind:

- Distributed processing does not require a distributed database, but a distributed database requires distributed processing.

- Distributed processing may be based on a single database located on a single computer. In order to manage distributed data, copies or parts of the database processing functions must be distributed to all data storage sites.

- Both distributed processing and distributed databases require a network to connect all components.

FIGURE 10.3 DISTRIBUTED DATABASE ENVIRONMENT

10.5 CHARACTERISTICS OF DISTRIBUTED DATABASE MANAGEMENT SYSTEMS

A distributed database management system (DDBMS) governs the storage and processing of logically related data over interconnected computer systems in which both data and processing functions are distributed among several sites. A DBMS must have at least the following functions to be classified as distributed:

- *Application interface* to interact with the end user or application programs and with other DBMSs within the distributed database.

- *Validation* to analyze data requests.

- *Transformation* to determine which data request components are distributed and which ones are local.

- *Query optimization* to find the best access strategy. (Which database fragments must be accessed by the query and how must data updates, if any, be synchronized?)

- *Mapping* to determine the data location of local and remote fragments.

- *I/O interface* to read or write data from or to permanent local storage.

- *Formatting* to prepare the data for presentation to the end user or an application program.

- *Security* to provide data privacy at both local and remote databases.

- *Backup and recovery* to ensure the availability and recoverability of the database in case of a failure.

- *DB administration* features for the database administrator.

- *Concurrency control* to manage simultaneous data access and ensure data consistency across database fragments in the DDBMS.

- *Transaction management* to ensure that the data move from one consistent state to another. This activity includes the synchronization of local and remote transactions as well as transactions across multiple distributed segments.

A fully distributed database management system must perform all the functions of a centralized DBMS:

1. Receive an application's (or an end user's) request.

2. Validate, analyze, and decompose the request. The request may include mathematical and/or logical operations, such as: *Select all customers with balance greater than $1,000.* The request may require data from only a single table, or it may require access to several tables.

3. Map the request's logical-to-physical data components.

4. Decompose the request into several disk I/O operations.

5. Search for, locate, read, and validate the data.

6. Ensure database consistency, security, and integrity.

7. Validate the data for the conditions, if any, specified by the request.

8. Present the selected data in the required format.

In addition, a distributed DBMS must handle all necessary functions imposed by the distribution of data and processing. And it must perform these additional functions *transparently* to the end user. The DDBMS's transparent data access features are illustrated in Figure 10.4.

FIGURE 10.4 A FULLY DISTRIBUTED DATABASE MANAGEMENT SYSTEM

The single logical database in Figure 10.4 consists of two database fragments, A1 and A2, located at sites 1 and 2, respectively. Mary can query the database as if it were a local database, and so can Tom. Both users "see" only one logical database and *do not need to know the names of the fragments*. In fact, the end users need not even know that the database is divided into separate fragments, *nor do they need to know where the fragments are located*.

To better understand the different types of distributed database scenarios, we must first define the distributed database system's components.

10.6 DDBMS COMPONENTS

The DDBMS must include (at least) the following components:

- *Computer workstations* (sites or nodes) that form the network system. The distributed database system must be independent of the computer system hardware.

- *Network hardware and software* components that reside in each workstation. The network components allow all sites to interact and exchange data. Because the components—computers, operating systems, network hardware, and so on—are likely to be supplied by different vendors, it is desirable to ensure that distributed database functions can be run on multiple platforms.

- *Communications media* that carry the data from one workstation to another. The DDBMS must be communications-media-independent; that is, it must be able to support several types of communications media.

- The **transaction processor** (**TP**), which is the software component found in each computer that requests data. The transaction processor receives and processes the application's data requests (remote and local). The TP is also known as the **application processor** (**AP**) or the **transaction manager** (**TM**).

- The **data processor** (**DP**), which is the software component residing on each computer that stores and retrieves data located at the site. The DP is also known as the **data manager** (**DM**). A data processor may even be a centralized DBMS.

Figure 10.5 illustrates the placement of and interaction among the components. The communication among TPs and DPs shown in Figure 10.5 is made possible through a specific set of rules, or *protocols*, used by the DDBMS.

The protocols determine how the distributed database system will:

- Interface with the network to transport data and commands between data processors (DPs) and transaction processors (TPs).

- Synchronize all data received from DPs (TP side) and route retrieved data to the appropriate TPs (DP side).

- Ensure common database functions in a distributed system. Such functions include security, concurrency control, backup, and recovery.

DPs and TPs can be added to the system without affecting the operation of the other components. A TP and a DP can reside on the same computer, allowing the end user to access local as well as remote data transparently. In theory, a DP can be an independent centralized DBMS with proper interfaces to support remote access from other independent DBMSs in the network.

FIGURE 10.5 DISTRIBUTED DATABASE SYSTEM COMPONENTS

Note: Each TP can access data on any DP, and
each DP handles all requests for local data from any TP.

10.7 LEVELS OF DATA AND PROCESS DISTRIBUTION

We can classify current database systems on the basis of how process distribution and data distribution are supported. For example, a DBMS may store data in a single site (centralized DB) or in multiple sites (distributed DB) and may support data processing at a single site or at multiple sites. Table 10.1 uses a simple matrix to classify database systems according to data and process distribution. These types of processes are discussed in the sections that follow.

TABLE 10.1 DATABASE SYSTEMS: LEVELS OF DATA AND PROCESS DISTRIBUTION

	SINGLE-SITE DATA	MULTIPLE-SITE DATA
Single-site process	Host DBMS (Mainframes)	Not applicable (Requires multiple processes)
Multiple-site process	File server Client/server DBMS (LAN DBMS)	Fully distributed Client/server DDBMS

10.7.1 SINGLE-SITE PROCESSING, SINGLE-SITE DATA (SPSD)

In the **single-site processing**, **single-site data** (**SPSD**) scenario, all processing is done on a single CPU or host computer (mainframe, midrange, or PC), and all data are stored on the host computer's local disk. Processing cannot be done on the end user's side of the system. Such a scenario is typical of most mainframe and midrange computer DBMSs. The DBMS is located on the host computer, which is accessed by dumb terminals connected to it. (See Figure 10.6.) This scenario is also typical of the first generation of single-user microcomputer databases.

FIGURE 10.6 SINGLE-SITE PROCESSING, SINGLE-SITE DATA (CENTRALIZED)

Using Figure 10.6 as an example, the functions of the TP and the DP are embedded within the DBMS located on a single computer. The DBMS usually runs under a time-sharing, multitasking operating system, which allows several processes to run concurrently on a host CPU accessing a single DP. All data storage and data processing are handled by a single CPU.

10.7.2 MULTIPLE-SITE PROCESSING, SINGLE-SITE DATA (MPSD)

Under the **multiple-site processing**, **single-site data** (**MPSD**) scenario, multiple processes run on different computers sharing a single data repository. Typically, the MPSD scenario requires a network file server running conventional applications that are accessed through a LAN. Many multiuser accounting applications, running under a personal computer network, fit such a description. (See Figure 10.7.)

FIGURE 10.7 MULTIPLE-SITE PROCESSING, SINGLE-SITE DATA

As you examine Figure 10.7, note that:

- The TP on each workstation acts only as a redirector to route all network data requests to the file server.
- The end user sees the file server as just another hard disk. Because only the data storage input/output (I/O) is handled by the file server's computer, the MPSD offers only very limited distributed processing capabilities.
- The end user must make a direct reference to the file server in order to access remote data. All record- and file-locking activity is done at the end-user location.
- All data selection, search, and update functions take place at the workstation, thus requiring that entire files travel through the network for processing at the workstation. Such a requirement increases network traffic, slows response time, and increases communication costs.

The inefficiency of the last condition can be illustrated easily. For example, suppose that the file server computer stores a CUSTOMER table containing 10,000 data rows, 50 of which have balances greater than $1,000. If site A issues the SQL query:

```
SELECT *
    FROM CUSTOMER
        WHERE CUS_BALANCE > 1000;
```

All 10,000 CUSTOMER rows must travel through the network to be evaluated at site A.

A variation of the multiple-site processing, single-site data approach is known as **client/server architecture**. This architecture, discussed in detail in Appendix G, "Multi-Tier Client/Server Systems," located on the CD accompanying this book, is similar to that of the network file server, *except that all database processing is done at the server site, thus reducing the network traffic*. Although both the network file server and the client/server systems perform multiple-site processing, the latter's processing is distributed. Note that the network file server approach requires the database to be located at a single site. In contrast, the client/server architecture is capable of supporting data at multiple sites.

ADDITIONAL CD CONTENT

Appendix G, "Multi-Tier Client/Server Systems," is located on the student CD.

10.7.3 MULTIPLE-SITE PROCESSING, MULTIPLE-SITE DATA (MPMD)

The **multiple-site processing, multiple-site data (MPMD)** scenario describes a fully distributed database management system with support for multiple data processors and transaction processors at multiple sites. Depending on the level of support for different types of centralized DBMSs, DDBMSs are classified as either homogeneous or heterogeneous.

Homogeneous DDBMSs integrate only one type of centralized DBMS over a network. Thus, the same DBMS will be running on different mainframes, minicomputers, and microcomputers. In contrast, **heterogeneous DDBMSs** integrate different types of centralized DBMSs over a network (see Figure 10.8). A **fully heterogeneous DDBMS** will support different DBMSs that may even support different data models (relational, hierarchical, or network) running under different computer systems, such as mainframes and microcomputers.

FIGURE 10.8 HETEROGENEOUS DISTRIBUTED DATABASE SCENARIO

Platform	DBMS	Operating System	Network Communications Protocol
IBM 3090	DB2	MVS	APPC LU 6.2
DEC/VAX	VAX rdb	MVS	DECnet
IBM AS/400	SQL/400	OS/400	3270
RISC computer	Informix	UNIX	TCP/IP
Pentium CPU	Oracle	Windows XP	NetBIOS

No DDBMS currently provides full support for the scenario depicted in Figure 10.8, nor for the fully heterogeneous environment. Some DDBMS implementations support several platforms, operating systems, and networks, and allow remote data access to another DBMS. However, such DDBMSs still are subject to certain restrictions. For example:

- Remote access is provided on a read-only basis and does not support write privileges.
- Restrictions are placed on the number of remote tables that may be accessed in a single transaction.
- Restrictions are placed on the number of distinct databases that may be accessed.
- Restrictions are placed on the database model that may be accessed. Thus, access may be provided to relational databases but not to network or hierarchical databases.

The preceding list of restrictions is by no means exhaustive. The DDBMS technology continues to change rapidly, and new features are added frequently. Managing data at multiple sites leads to a number of issues that must be addressed and understood. Therefore, next we will examine several key features of distributed database management systems.

10.8 DISTRIBUTED DATABASE TRANSPARENCY FEATURES

A distributed database system requires functional characteristics that can be grouped and described as transparency features. DDBMS transparency features have the common property of allowing the end user to feel like the database's only user. In other words, the user believes that (s)he is working with a centralized DBMS; all the complexities of a distributed database are hidden, or transparent, to the user.

The DDBMS transparency features are:

- **Distribution transparency**, which allows a distributed database to be treated as a single logical database. If a DDBMS exhibits distribution transparency, the user does not need to know:
 - That the data are partitioned.
 - That the data can be replicated at several sites.
 - The data location.
- **Transaction transparency**, which allows a transaction to update data at several network sites. Transaction transparency ensures that the transaction will be either completed entirely or aborted, thus maintaining database integrity.
- **Failure transparency**, which ensures that the system will continue to operate in the event of a node failure. Functions that were lost because of the failure will be picked up by another network node.
- **Performance transparency**, which allows the system to perform as if it were a centralized DBMS. The system will not suffer any performance degradation due to its use on a network or due to the network's platform differences. Performance transparency also ensures that the system will find the most cost-effective path to access remote data.
- **Heterogeneity transparency**, which allows the integration of several different local DBMSs (relational, network, and hierarchical) under a common, or global, schema. The DDBMS is responsible for translating the data requests from the global schema to the local DBMS schema.

We will examine distribution, transaction, and performance transparency features in greater detail in the next few sections.

10.9 DISTRIBUTION TRANSPARENCY

Distribution transparency allows us to manage a physically dispersed database as though it were a centralized database. The level of transparency supported by the DDBMS varies from system to system. Three levels of distribution transparency are recognized:

- **Fragmentation transparency** is the highest level of transparency. The end user or programmer does not need to know that a database is partitioned. Therefore, neither fragment names nor fragment locations are specified prior to data access.

- **Location transparency** exists when the end user or programmer must specify the database fragment names but does not need to specify where these fragments are located.

- **Local mapping transparency** exists when the end user or programmer must specify both the fragment names and their locations.

We have summarized the transparency features in Table 10.2.

TABLE 10.2 A SUMMARY OF TRANSPARENCY FEATURES

IF THE SQL STATEMENT REQUIRES:			
FRAGMENT NAME?	**LOCATION NAME?**	**THEN THE DBMS SUPPORTS**	**LEVEL OF DISTRIBUTION TRANSPARENCY**
Yes	Yes	Local mapping	Low
Yes	No	Location transparency	Medium
No	No	Fragmentation transparency	High

As you examine Table 10.2, you might ask why there is no reference to a situation in which the fragment name is "No" and the location name is "Yes." The reason for not including this scenario is simple: You cannot have a location name that fails to reference an existing fragment. (If you don't need to specify a fragment name, its location is clearly irrelevant.)

To illustrate the use of various transparency levels, let us suppose we have an EMPLOYEE table containing the attributes EMP_NAME, EMP_DOB, EMP_ADDRESS, EMP_DEPARTMENT, and EMP_SALARY. The EMPLOYEE data are distributed over three different locations: New York, Atlanta, and Miami. The table is divided by location; that is, all New York employee data are stored in fragment E1, Atlanta employee data are stored in fragment E2, and Miami employee data are stored in fragment E3. (See Figure 10.9.)

FIGURE 10.9 FRAGMENT LOCATIONS

Now suppose that the end user wants to list all the employees with a date of birth prior to January 1, 1940. To focus on the transparency issues, let us also suppose that the EMPLOYEE table is fragmented and that each fragment is unique. The **unique fragment** condition indicates that each and every row is unique, regardless of the fragment in which it is located. Finally, let us assume that no portion of the database is replicated at any other site on the network.

Depending on the level of distribution transparency support, we may examine three query cases.

Case 1: The Database Supports Fragmentation Transparency

The query conforms to a nondistributed database query format; that is, it does not specify fragment names or locations. The query reads:

SELECT *
 FROM EMPLOYEE
 WHERE EMP_DOB < '01-JAN-1940';

Case 2: The Database Supports Location Transparency

Fragment names must be specified in the query, but fragment location is not specified. The query reads:

SELECT *
 FROM E1
 WHERE EMP_DOB < '01-JAN-1940';
UNION
SELECT *
 FROM E2
 WHERE EMP_DOB < '01-JAN-1940';
UNION
SELECT *
 FROM E3
 WHERE EMP_DOB < '01-JAN-1940';

Case 3: The Database Supports Local Mapping Transparency

Both the fragment name and location must be specified in the query. Using pseudo-SQL:

SELECT *
 FROM E1 NODE NY
 WHERE EMP_DOB < '01-JAN-1940';
UNION
SELECT *
 FROM E2 NODE ATL
 WHERE EMP_DOB < '01-JAN-1940';
UNION
SELECT *
 FROM E3 NODE MIA
 WHERE EMP_DOB < '01-JAN-1940';

> **NOTE**
>
> NODE indicates the location of the database fragment. NODE is used for illustration purposes and is not part of the standard SQL syntax.

As you examine the preceding query formats, you can see how distribution transparency affects the way end users and programmers interact with the database.

Distribution transparency is supported by a **distributed data dictionary (DDD)**, or a **distributed data catalog (DDC)**. The DDC contains the description of the entire database as seen by the database administrator. The database description, known as the **distributed global schema**, is the common database schema used by local TPs to translate user requests into subqueries (remote requests) that will be processed by different DPs. The DDC is itself distributed, and it is replicated at the network nodes. Therefore, the DDC must maintain consistency through updating at all sites.

Keep in mind that some of the current DDBMS implementations impose limitations on the level of transparency support. For instance, you might be able to distribute a database, but not a table, across multiple sites. Such a condition indicates that the DDBMS supports location transparency but not fragmentation transparency.

10.10 TRANSACTION TRANSPARENCY

Transaction transparency is a DDBMS property that ensures that database transactions will maintain the distributed database's integrity and consistency. Remember that a DDBMS database transaction can update data stored in many different computers connected in a network. Transaction transparency ensures that the transaction will be completed only if all database sites involved in the transaction complete their part of the transaction.

Distributed database systems require complex mechanisms to manage transactions and ensure the database's consistency and integrity. In order to understand how the transactions are managed, you should know the basic concepts governing remote requests, remote transactions, distributed transactions, and distributed requests.

10.10.1 DISTRIBUTED REQUESTS AND DISTRIBUTED TRANSACTIONS[1]

Whether or not a transaction is distributed, it is formed by one or more database requests. The basic difference between a nondistributed transaction and a distributed transaction is that the latter can update or request data from several different remote sites on a network. To illustrate the distributed transaction concepts more easily, we will begin by establishing the difference between remote and distributed transactions, using the BEGIN WORK and COMMIT WORK transaction format. We will assume the existence of location transparency to avoid having to specify the data location.

A **remote request**, illustrated in Figure 10.10, lets a single SQL statement access data to be processed by a single remote database processor. In other words, the SQL statement (or request) can reference data at only one remote site.

[1] The details of distributed requests and transactions were originally described in David McGoveran and Colin White, "Clarifying Client/Server," *DBMS* 3(12), November 1990, pp. 78–89.

FIGURE 10.10 A REMOTE REQUEST

Similarly, a **remote transaction**, composed of several requests, accesses data at a single remote site. A remote transaction is illustrated in Figure 10.11.

FIGURE 10.11 A REMOTE TRANSACTION

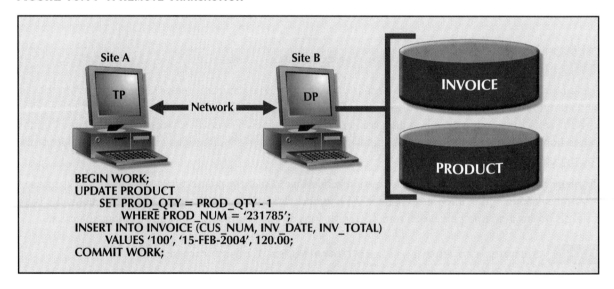

As you examine Figure 10.11, note the following remote transaction features:

- The transaction updates the PRODUCT and INVOICE tables (located at site B).
- The remote transaction is sent to and executed at the remote site B.
- The transaction can reference only one remote DP.
- Each SQL statement (or request) can reference only one (the same) remote DP at a time, and the entire transaction can reference and can be executed at only one remote DP.

A **distributed transaction** allows a transaction to reference several different (local or remote) DP sites. Although each single request can reference only one local or remote DP site, the transaction as a whole can reference multiple DP sites, because each request can reference a different site. The distributed transaction process is illustrated in Figure 10.12.

FIGURE 10.12 A DISTRIBUTED TRANSACTION

Note the following features in Figure 10.12:

- The transaction references two remote sites (B and C).

- The first two requests (UPDATE PRODUCT and INSERT INTO INVOICE) are processed by the DP at the remote site C, and the last request (UPDATE CUSTOMER) is processed by the DP at the remote site B.

- Each request can access only one remote site at a time.

The third characteristic may create problems. For example, suppose that the table PRODUCT is divided into two fragments, PRODl and PROD2, located at sites B and C, respectively. Given this scenario, the preceding distributed transaction cannot be executed, because the request:

SELECT *
 FROM PRODUCT
 WHERE PROD_NUM = '231785';

cannot access data from more than one remote site. Therefore, the DBMS must be able to support a distributed request.

A **distributed request** lets a single SQL statement reference data located at several different local or remote DP sites. Because each request (SQL statement) can access data from more than one local or remote DP site, a transaction can access several sites. The ability to execute a distributed request provides fully distributed database processing capabilities, because we are able to:

- Partition a database table into several fragments.

- Reference one or more of those fragments with only one request. In other words, we have fragmentation transparency.

The location and partition of the data should be transparent to the end user. Figure 10.13 illustrates a distributed request. As you examine Figure 10.13, note that the transaction uses a single SELECT statement to reference two tables, CUSTOMER and INVOICE. The two tables are located at two different sites, B and C.

FIGURE 10.13 **A DISTRIBUTED REQUEST**

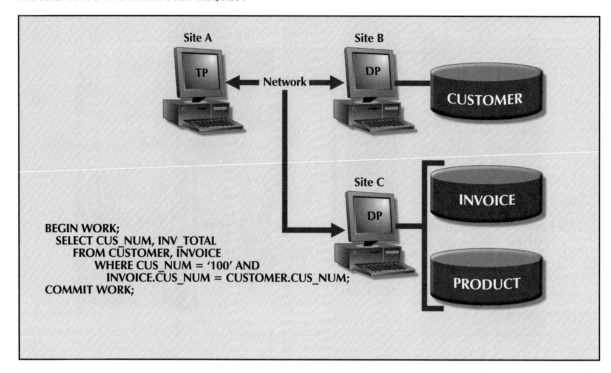

The distributed request feature also allows a single request to reference a physically partitioned table. For example, suppose that a CUSTOMER table is divided into two fragments, C1 and C2, located at sites B and C, respectively. Let us further suppose that the end user wants to obtain a list of all customers whose balances exceed $250. The request is illustrated in Figure 10.14. Full fragmentation transparency support is provided only by a DDBMS that supports distributed requests.

FIGURE 10.14 **ANOTHER DISTRIBUTED REQUEST**

Understanding the different types of database requests in distributed database systems helps us to address the transaction transparency issue more effectively. Transaction transparency ensures that distributed transactions are treated as centralized transactions, ensuring serializability of transactions. (Review Chapter 9, "Transaction Management and Concurrency Control," if necessary.) That is, the execution of concurrent transactions, whether or not they are distributed, will take the database from one consistent state to another. We will address distributed concurrency control next.

10.10.2 DISTRIBUTED CONCURRENCY CONTROL

Concurrency control becomes especially important in the distributed database environment because multisite, multiple-process operations are much more likely to create data inconsistencies and deadlocked transactions than are the single-site systems. For example, the TP component of a DDBMS must ensure that all parts of the transaction, at all sites, are completed before a final COMMIT is issued to record the transaction.

Suppose that each transaction operation was committed by each local DP, but one of the DPs could not commit the transaction's results. Such a scenario would yield the problems illustrated in Figure 10.15: The transaction(s) would yield an inconsistent database, with its inevitable integrity problems, because we cannot uncommit committed data! The solution for the problem illustrated in Figure 10.15 is a *two-phase commit protocol*, which we will explore next.

FIGURE 10.15 THE EFFECT OF A PREMATURE COMMIT

10.10.3 TWO-PHASE COMMIT PROTOCOL

Centralized databases require only one DP. All database operations take place at only one site, and the consequences of database operations are immediately known to the DBMS. In contrast, distributed databases make it possible for a transaction to access data at several sites. A final COMMIT must not be issued until all sites have

committed their parts of the transaction. The **two-phase commit protocol** guarantees that, if a portion of a transaction operation cannot be committed, all changes made at the other sites participating in the transaction will be undone to maintain a consistent database state.

Each DP maintains its own transaction log. The two-phase commit protocol requires that each individual DP's transaction log entry be written before the database fragment is actually updated. (See Chapter 9, "Transaction Management and Concurrency Control," Section 9.6.) Therefore, the two-phase commit protocol requires a DO-UNDO-REDO protocol and a write-ahead protocol.

The **DO-UNDO-REDO protocol** is used by the DP to roll back and/or roll forward transactions with the help of the system's transaction log entries. The DO-UNDO-REDO protocol defines three types of operations:

1. DO performs the operation and records the "before" and "after" values in the transaction log.

2. UNDO reverses an operation, using the log entries written by the DO portion of the sequence.

3. REDO redoes an operation, using the log entries written by the DO portion of the sequence.

To ensure that the DO, UNDO, and REDO operations can survive a system crash while they are being executed, a write-ahead protocol is used. The **write-ahead protocol** forces the log entry to be written to permanent storage before the actual operation takes place.

The two-phase commit protocol defines the operations between two types of nodes: the **coordinator** and one or more **subordinates**, or *cohorts*. The protocol is implemented in two phases:

Phase 1: Preparation

1. The coordinator sends a PREPARE TO COMMIT message to all subordinates.

2. The subordinates receive the message, write the transaction log using the write-ahead protocol, and send an acknowledgment (YES/PREPARED TO COMMIT or NO/NOT PREPARED) message to the coordinator.

3. The coordinator makes sure that all nodes are ready to commit, or it aborts the action.

If all nodes are PREPARED TO COMMIT, the transaction goes to Phase 2. If one or more nodes reply NO or NOT PREPARED, the coordinator broadcasts an ABORT message to all subordinates.

Phase 2: The Final COMMIT

1. The coordinator broadcasts a COMMIT message to all subordinates and waits for the replies.

2. Each subordinate receives the COMMIT message, then updates the database using the DO protocol.

3. The subordinates reply with a COMMITTED or NOT COMMITTED message to the coordinator.

If one or more subordinates did not commit, the coordinator sends an ABORT message, thereby forcing them to UNDO all changes.

The objective of the two-phase commit is to ensure that all nodes commit their part of the transaction; otherwise, the transaction is aborted. If one of the nodes fails to commit, then the information necessary to recover the database is in the transaction log, and the database can be recovered with the DO-UNDO-REDO protocol. (Remember that the log information was updated using the write-ahead protocol.)

10.11 PERFORMANCE TRANSPARENCY AND QUERY OPTIMIZATION

One of the most important functions of a database is its ability to make data available. Because all data reside at one single site in a centralized database, the DBMS must evaluate every data request and find the most efficient way to access the local data. In contrast, the DDBMS makes it possible to partition a database into several fragments, thereby rendering the query translation more complicated because the DDBMS must decide which fragment of the

database to access. In addition, the data may also be replicated at several different sites. The data replication makes the access problem even more complex, because now the database must decide which copy of the data to access. The DDBMS uses query optimization techniques to deal with such problems and to ensure acceptable database performance.

The objective of a query optimization routine is to minimize the total cost associated with the execution of a request. The costs associated with a request are a function of the:

- Access time (I/O) cost involved in accessing the physical data stored on disk.
- Communication cost associated with the transmission of data among nodes in distributed database systems.
- CPU time cost associated with the processing overhead of managing distributed transactions.

Although costs are often classified as either communication or processing costs, it is difficult to separate the two. Not all query optimization algorithms use the same parameters, and the algorithms do not all assign the same weight to each parameter. For example, some algorithms minimize total time, others minimize the communication time, and still others do not factor in the CPU time, considering it insignificant relative to other cost sources.

To evaluate query optimization, keep in mind that the TP must receive data from the DP, synchronize it, assemble the answer, and present it to the end user or an application. Although this process is standard, you should consider that a particular query may be executed at any one of several different sites. The response time associated with remote sites cannot be easily predetermined, because some nodes are able to finish their part of the query in less time than others.

One of the most important characteristics of query optimization in distributed database systems is that it must provide distribution transparency as well as *replica* transparency. (Distribution transparency was explained earlier in this chapter.) **Replica transparency** refers to the DDBMS's ability to hide the existence of multiple copies of data from the user.

Most of the algorithms proposed for query optimization are based on two principles:

1. The selection of the optimum execution order.
2. The selection of sites to be accessed to minimize communication costs.

Within these two principles, a query optimization algorithm can be evaluated on the basis of its *operation mode* or the *timing of its optimization*.

Operation modes can be classified as manual or automatic. **Automatic query optimization** means that the DDBMS finds the most cost-effective access path without user intervention. **Manual query optimization** requires that the optimization be selected and scheduled by the end user or programmer. Automatic query optimization is clearly more desirable from the end user's point of view, but the cost of such convenience is the increased overhead that it imposes on the DDBMS.

Query optimization algorithms can also be classified according to when the optimization is done. Within this timing classification, query optimization algorithms can be classified as *static* or *dynamic*.

- **Static query optimization** takes place at compilation time. In other words, the best optimization strategy is selected when the query is compiled by the DBMS. This approach is common when SQL statements are embedded in procedural programming languages, such as COBOL or Pascal. When the program is submitted to the DBMS for compilation, it creates the plan necessary to access the database. When the program is executed, the DBMS uses that plan to access the database.
- **Dynamic query optimization** takes place at execution time. Database access strategy is defined when the program is executed. Therefore, access strategy is dynamically determined by the DBMS at run time, using the most up-to-date information about the database. Although dynamic query optimization is efficient, its cost is measured by run-time processing overhead. The best strategy is determined every time the query is executed; this could happen several times in the same program.

Finally, query optimization techniques can be classified according to the type of information that is used to optimize the query. For example, queries may be based on *statistically based* or *rule-based* algorithms.

- A **statistically based query optimization algorithm** uses statistical information about the database. The statistics provide information about database characteristics such as size, number of records, average access time, number of requests serviced, and number of users with access rights. These statistics are then used by the DBMS to determine the best access strategy.

- The statistical information is managed by the DDBMS and is generated in one of two different modes: dynamic or manual. In the **dynamic statistical generation mode**, the DDBMS automatically evaluates and updates the statistics after each access. In the **manual statistical generation mode**, the statistics must be updated periodically through a user-selected utility, such as IBM's RUNSTAT command used by DB2 DBMSs.

- A **rule-based query optimization algorithm** is based on a set of user-defined rules to determine the best query access strategy. The rules are entered by the end user or database administrator, and they typically are very general in nature.

10.12 DISTRIBUTED DATABASE DESIGN

Whether the database is centralized or distributed, the design principles and concepts described in Chapter 3, "The Relational Database Model," Chapter 4, "Entity Relationship (ER) Modeling," and Chapter 5, "Normalization of Database Tables" are still applicable. However, the design of a distributed database introduces three new issues:

1. How to partition the database into fragments.
2. Which fragments to replicate.
3. Where to locate those fragments and replicas.

Data fragmentation and *data replication* deal with the first two issues, and *data allocation* deals with the third issue.

10.12.1 DATA FRAGMENTATION

Data fragmentation allows us to break a single object into two or more segments or fragments. The object might be a user's database, a system database, or a table. Each fragment can be stored at any site over a computer network. Information about data fragmentation is stored in the distributed data catalog (DDC), from which it is accessed by the TP to process user requests.

Data fragmentation strategies, as discussed here, are based at the table level and consist of dividing a table into logical fragments. We will explore three types of data fragmentation strategies: *horizontal*, *vertical*, and *mixed*. (Keep in mind that a fragmented table can always be re-created from its fragmented parts by a combination of unions and joins.)

- **Horizontal fragmentation** refers to the division of a relation into subsets (fragments) of tuples (rows). Each fragment is stored at a different node, and each fragment has unique rows. However, the unique rows all have the same attributes (columns). In short, each fragment represents the equivalent of a SELECT statement, with the WHERE clause on a single attribute.

- **Vertical fragmentation** refers to the division of a relation into attribute (column) subsets. Each subset (fragment) is stored at a different node, and each fragment has unique columns—with the exception of the key column, which is common to all fragments. This is the equivalent of the PROJECT statement in SQL.

- **Mixed fragmentation** refers to a combination of horizontal and vertical strategies. In other words, a table may be divided into several horizontal subsets (rows), each one having a subset of the attributes (columns).

To illustrate the fragmentation strategies, let's use the CUSTOMER table for the XYZ Company, depicted in Figure 10.16. The table contains the attributes CUS_NUM, CUS_NAME, CUS_ADDRESS, CUS_STATE, CUS_LIMIT, CUS_BAL, CUS_RATING, and CUS_DUE.

ADDITIONAL CD CONTENT

The database used to illustrate the material in this chapter is found on the CD bundled with this book in the \Databases\Student\Ch10 folder. For example, the source of the sample CUSTOMER table shown in Figure 10.16 is the Ch10_Text database located in the above folder.

FIGURE 10.16 A SAMPLE CUSTOMER TABLE

Table name: CUSTOMER

CUS_NUM	CUS_NAME	CUS_ADDRESS	CUS_STATE	CUS_LIMIT	CUS_BAL	CUS_RATING	CUS_DUE
10	Sinex, Inc.	12 Main St.	TN	$3,500.00	$2,700.00	3	$1,245.00
11	Martin Corp.	321 Sunset Blvd.	FL	$6,000.00	$1,200.00	1	$0.00
12	Mynux Corp.	910 Eagle St.	TN	$4,000.00	$3,500.00	3	$3,400.00
13	BTBC, Inc.	Rue du Monde	FL	$6,000.00	$5,890.00	3	$1,090.00
14	Victory, Inc.	123 Maple St.	FL	$1,200.00	$550.00	1	$0.00
15	NBCC Corp.	909 High Ave.	GA	$2,000.00	$350.00	2	$50.00

Horizontal Fragmentation

Suppose that XYZ Company's corporate management requires information about its customers in all three states, but company locations in each state (TN, FL, and GA) require data regarding local customers only. Based on such requirements, we decide to distribute the data by state. Therefore, we define the horizontal fragments to conform to the structure shown in Table 10.3.

TABLE 10.3 HORIZONTAL FRAGMENTATION OF THE CUSTOMER TABLE BY STATE

FRAGMENT NAME	LOCATION	CONDITION	NODE NAME	CUSTOMER NUMBERS	NUMBER OF ROWS
CUST_H1	Tennessee	CUS_STATE = 'TN'	NAS	10, 12	2
CUST_H2	Georgia	CUS_STATE = 'GA'	ATL	15	1
CUST_H3	Florida	CUS_STATE = 'FL'	TAM	11, 13, 14	3

Each horizontal fragment may have a different number of rows, but each fragment MUST have the same attributes. The resulting fragments yield the three tables depicted in Figure 10.17.

FIGURE 10.17 TABLE FRAGMENTS IN THREE LOCATIONS

Table name: CUST_H1 **Location: Tennessee** **Node: NAS**

	CUS_NUM	CUS_NAME	CUS_ADDRESS	CUS_STATE	CUS_LIMIT	CUS_BAL	CUS_RATING	CUS_DUE
▶	10	Sinex, Inc.	12 Main St.	TN	$3,500.00	$2,700.00	3	$1,245.00
	12	Mynux Corp.	910 Eagle St.	TN	$4,000.00	$3,500.00	3	$3,400.00

Table name: CUST_H2 **Location: Georgia** **Node: ALT**

	CUS_NUM	CUS_NAME	CUS_ADDRESS	CUS_STATE	CUS_LIMIT	CUS_BAL	CUS_RATING	CUS_DUE
▶	15	NBCC Corp.	909 High Ave.	GA	$2,000.00	$350.00	2	$50.00

Table name: CUST_H3 **Location: Florida** **Node: TAM**

	CUS_NUM	CUS_NAME	CUS_ADDRESS	CUS_STATE	CUS_LIMIT	CUS_BAL	CUS_RATING	CUS_DUE
▶	11	Martin Corp.	321 Sunset Blvd.	FL	$6,000.00	$1,200.00	1	$0.00
	13	BTBC, Inc.	Rue du Monde	FL	$6,000.00	$5,890.00	3	$1,090.00
	14	Victory, Inc.	123 Maple St.	FL	$1,200.00	$550.00	1	$0.00

Vertical Fragmentation

We may also divide the CUSTOMER relation into vertical fragments that are composed of a collection of attributes. For example, suppose that the company is divided into two departments: the service department and the collections department. Each department is located in a separate building, and each has an interest in only a few of the CUSTOMER table's attributes. In this case, the fragments are defined as shown in Table 10.4.

TABLE 10.4 VERTICAL FRAGMENTATION OF THE CUSTOMER TABLE

FRAGMENT NAME	LOCATION	NODE NAME	ATTRIBUTE NAMES
CUST_V1	Service Bldg.	SVC	CUS_NUM, CUS_NAME, CUS_ADDRESS, CUS_STATE
CUST_V2	Collection Bldg.	ARC	CUS_NUM, CUS_LIMIT, CUS_BAL, CUS_RATING, CUS_DUE

Each vertical fragment must have the same number of rows, but the inclusion of the different attributes depends on the key column. The vertical fragmentation results are displayed in Figure 10.18. Note that the key attribute (CUS_NUM) is common to both fragments CUST_V1 and CUST_V2.

FIGURE 10.18 VERTICALLY FRAGMENTED TABLE CONTENTS

Table name: CUST_V1 **Location: Service Bldg.** **Node: SVC**

	CUS_NUM	CUS_NAME	CUS_ADDRESS	CUS_STATE
▶	10	Sinex, Inc.	12 Main St.	TN
	11	Martin Corp.	321 Sunset Blvd.	FL
	12	Mynux Corp.	910 Eagle St.	TN
	13	BTBC, Inc.	Rue du Monde	FL
	14	Victory, Inc.	123 Maple St.	FL
	15	NBCC Corp.	909 High Ave.	GA

Table name: CUST_V2 **Location: Collection Bldg.** **Node: ARC**

	CUS_NUM	CUS_LIMIT	CUS_BAL	CUS_RATING	CUS_DUE
▶	10	$3,500.00	$2,700.00	3	$1,245.00
	11	$6,000.00	$1,200.00	1	$0.00
	12	$4,000.00	$3,500.00	3	$3,400.00
	13	$6,000.00	$5,890.00	3	$1,090.00
	14	$1,200.00	$550.00	1	$0.00
	15	$2,000.00	$350.00	2	$50.00

Mixed Fragmentation

The XYZ Company's structure requires that the CUSTOMER data be fragmented horizontally to accommodate the different company locations; within the locations, the data must be fragmented vertically to accommodate the different departments (service and collection). In short, the CUSTOMER table requires mixed fragmentation.

Mixed fragmentation requires a two-step procedure. First, we introduce horizontal fragmentation for each site, based on the location within a state (CUS_STATE). The horizontal fragmentation yields the subsets of customer tuples (horizontal fragments) that are located at each site. Because the departments are located in different buildings, we use vertical fragmentation within each horizontal fragment to divide the attributes, thus meeting each department's information needs at each subsite. Mixed fragmentation yields the results displayed in Table 10.5.

TABLE 10.5 MIXED FRAGMENTATION OF THE CUSTOMER TABLE

FRAGMENT NAME	LOCATION	HORIZONTAL CRITERIA	NODE NAME	RESULTING ROWS AT SITE	VERTICAL CRITERIA ATTRIBUTES AT EACH FRAGMENT
CUST_M1	TN-Service	CUS_STATE = 'TN'	NAS-S	10, 12	CUS_NUM, CUS_NAME CUS_ADDRESS, CUS_STATE
CUST_M2	TN-Collection	CUS_STATE = 'TN'	NAS-C	10, 12	CUS_NUM, CUS_LIMIT, CUS_BAL, CUS_RATING, CUS_DUE
CUST_M3	GA-Service	CUS_STATE = 'GA'	ATL-S	15	CUS_NUM, CUS_NAME CUS_ADDRESS, CUS_STATE
CUST_M4	GA-Collection	CUS_STATE = 'GA'	ATL-C	15	CUS_NUM, CUS_LIMIT, CUS_BAL, CUS_RATING, CUS_DUE
CUST_M5	FL-Service	CUS_STATE = 'FL'	TAM-S	11, 13, 14	CUS_NUM, CUS_NAME CUS_ADDRESS, CUS_STATE
CUST_M6	FL-Collection	CUS_STATE = 'FL'	TAM-C	11, 13, 14	CUS_NUM, CUS_LIMIT, CUS_BAL, CUS_RATING, CUS_DUE

Each fragment displayed in Table 10.5 contains customer data by state and, within each state, by department location, to fit each department's data requirements. The tables corresponding to the fragments listed in Table 10.5 are shown in Figure 10.19.

10.12.2 DATA REPLICATION

Data replication refers to the storage of data copies at multiple sites served by a computer network. Fragment copies can be stored at several sites to serve specific information requirements. Because the existence of fragment copies can enhance data availability and response time, data copies can help to reduce communication and total query costs.

Suppose database A is divided into two fragments, A1 and A2. Within a replicated distributed database, the scenario depicted in Figure 10.20 is possible: fragment A1 is stored at sites S1 and S2, while fragment A2 is stored at sites S2 and S3.

FIGURE 10.19 TABLE CONTENTS AFTER THE MIXED FRAGMENTATION PROCESS

Table name: CUST_M1 Location: TN-Service Node: NAS-S

	CUS_NUM	CUS_NAME	CUS_ADDRESS	CUS_STATE
▶	10	Sinex, Inc.	12 Main St.	TN
	12	Mynux Corp.	910 Eagle St.	TN

Table name: CUST_M2 Location: TN-Collection Node: NAS-C

	CUS_NUM	CUS_LIMIT	CUS_BAL	CUS_RATING	CUS_DUE
▶	10	$3,500.00	$2,700.00	3	$1,245.00
	12	$4,000.00	$3,500.00	3	$3,400.00

Table name: CUST_M3 Location: GA-Service Node: ATL-S

	CUS_NUM	CUS_NAME	CUS_ADDRESS	CUS_STATE
▶	15	NBCC Corp.	909 High Ave.	GA

Table name: CUST_M4 Location: GA-Collection Node: ATL-C

	CUS_NUM	CUS_LIMIT	CUS_BAL	CUS_RATING	CUS_DUE
▶	15	$2,000.00	$350.00	2	$50.00

Table name: CUST_M5 Location: FL-Service Node: TAM-S

	CUS_NUM	CUS_NAME	CUS_ADDRESS	CUS_STATE
▶	11	Martin Corp.	321 Sunset Blvd.	FL
	13	BTBC, Inc.	Rue du Monde	FL
	14	Victory, Inc.	123 Maple St.	FL

Table name: CUST_M6 Location: FL-Collection Node: TAM-C

	CUS_NUM	CUS_LIMIT	CUS_BAL	CUS_RATING	CUS_DUE
▶	11	$6,000.00	$1,200.00	1	$0.00
	13	$6,000.00	$5,890.00	3	$1,090.00
	14	$1,200.00	$550.00	1	$0.00

FIGURE 10.20 DATA REPLICATION

Replicated data are subject to the mutual consistency rule. The **mutual consistency rule** requires that all copies of data fragments be identical. Therefore, to maintain data consistency among the replicas, the DDBMS must ensure that a database update is performed at all sites where replicas exist.

Although replication has some benefits, it also imposes additional DDBMS processing overhead, because each data copy must be maintained by the system. To illustrate the replica overhead imposed on a DDBMS, consider the processes that the DDBMS must perform to use the database:

- If the database is fragmented, the DDBMS must decompose a query into *subqueries* to access the appropriate fragments.

- If the database is replicated, the DDBMS must decide which copy to access. A Read operation selects the *nearest copy* to satisfy the transaction. A Write operation requires that *all copies* be selected and updated to satisfy the mutual consistency rule.

- The TP sends a data request to each selected DP for execution.

- The DP receives and executes each request and sends the data back to the TP.

- The TP assembles the DP responses.

The problem becomes more complex if you consider additional factors such as network topology and communication throughputs.

Three replication scenarios exist: a database can be *fully replicated*, *partially replicated*, or *unreplicated*.

- A **fully replicated database** stores multiple copies of *each* database fragment at multiple sites. In this case, all database fragments are replicated. A fully replicated database can be impractical due to the amount of overhead it imposes on the system.

- A **partially replicated database** stores multiple copies of *some* database fragments at multiple sites. Most DDBMSs are able to handle the partially replicated database well.

- An **unreplicated database** stores each database fragment at a single site. Therefore, there are no duplicate database fragments.

Several factors influence the decision to use data replication:

- Database size.

- Usage frequency.

- Costs—performance, software overhead, and management—associated with synchronizing transactions and their components vs. fault-tolerance benefits that are associated with replicated data.

When the usage frequency of remotely located data is high and the database is large, data replication can reduce the cost of data requests. Data replication information is stored in the distributed data catalog (DDC), whose contents are used by the TP to decide which copy of a database fragment to access. The data replication makes it possible to restore lost data.

10.12.3 DATA ALLOCATION

Data allocation describes the process of deciding where to locate data. Data allocation strategies are as follows:

- With **centralized data allocation**, the entire database is stored at one site.

- With **partitioned data allocation**, the database is divided into several disjointed parts (fragments) and stored at several sites.

- With **replicated data allocation**, copies of one or more database fragments are stored at several sites.

Data distribution over a computer network is achieved through data partition, through data replication, or through a combination of both. Data allocation is closely related to the way a database is divided or fragmented. Most data allocation studies focus on one issue: *which* data to locate *where*.

Data allocation algorithms take into consideration a variety of factors, including:

- Performance and data availability goals.

- Size, number of rows, and the number of relations that an entity maintains with other entities.

- Types of transactions to be applied to the database, the attributes accessed by each of those transactions, and so on.

Some algorithms include external data, such as network topology or network throughput. No optimal or universally accepted algorithm exists yet, and very few algorithms have been implemented to date.

10.13 CLIENT/SERVER VS. DDBMS

Because the trend toward distributed databases is firmly established, many database vendors have used the *client/server* label to indicate distributed database capability. However, the label does not always accurately fit the required client/server characteristics.

Client/server architecture refers to the way in which computers interact to form a system. The client/server architecture features a *user* of resources, or a client, and a *provider* of resources, or a server. The client/server architecture can be used to implement a DBMS in which the client is the TP and the server is the DP.

Client/server interactions in a DDBMS are carefully scripted. The client (TP) interacts with the end user and sends a request to the server (DP). The server receives, schedules, and executes the request, *selecting only those records that are needed by the client*. The server then sends the data to the client *only* when the client requests the data.

Client/server applications offer several advantages:

- Client/server solutions tend to be less expensive than alternate minicomputer or mainframe solutions.

- Client/server solutions allow the end user to use the microcomputer's GUI, thereby improving functionality and simplicity.

- There are more people with PC skills than with mainframe skills in the job market.

- The PC is well established in the workplace.

- Numerous data analysis and query tools exist to facilitate interaction with many of the DBMSs that are available in the PC market.

- There is a considerable cost advantage to offloading applications development from the mainframe to powerful PCs.

Client/server applications are also subject to some disadvantages:

- The client/server architecture creates a more complex environment, in which different platforms (LANs, operating systems, and so on) are often difficult to manage.

- An increase in the number of users and processing sites often paves the way for security problems.

- The client/server environment makes it possible to spread data access to a much wider circle of users. Such an environment increases the demand for people with broad knowledge of computers and software applications. The burden of training increases the cost of maintaining the environment.

ADDITIONAL CD CONTENT

Refer to **Appendix G, "Multi-Tier Client/Server Systems,"** for complete coverage of client/server computing concepts, components, and managerial implications.

10.14 C. J. DATE'S TWELVE COMMANDMENTS FOR DISTRIBUTED DATABASES

No discussion of distributed databases is complete unless it includes C. J. Date's distributed database commandments.[2] Date's commandments describe a fully distributed database, and, although no current DDBMS conforms to all of them, the rules do constitute a useful distributed database target. The twelve rules are as follows:

1. *Local site independence.* Each local site can act as an independent, autonomous, centralized DBMS. Each site is responsible for security, concurrency control, backup, and recovery.

2. *Central site independence.* No site in the network relies on a central site or any other site. All sites have the same capabilities.

3. *Failure independence.* The system is not affected by node failures. The system is in continuous operation even in the case of a node failure or an expansion of the network.

4. *Location transparency.* The user does not need to know the location of the data in order to retrieve those data.

5. *Fragmentation transparency.* The user sees only one single logical database. Data fragmentation is transparent to the user. The user does not need to know the name of the database fragments in order to retrieve them.

6. *Replication transparency.* The user sees only one single logical database. The DDBMS transparently selects the database fragment to access. The DDBMS manages all fragments transparently to the user.

7. *Distributed query processing.* A distributed query may be executed at several different DP sites. Query optimization is performed transparently by the DDBMS.

8. *Distributed transaction processing.* A transaction may update data at several different sites. The transaction is transparently executed at several different DP sites.

9. *Hardware independence.* The system must run on any hardware platform.

10. *Operating system independence.* The system must run on any operating system software platform.

11. *Network independence.* The system must run on any network platform.

12. *Database independence.* The system must support any vendor's database product.

S U M M A R Y

A distributed database stores logically related data in two or more physically independent sites connected via a computer network. The database is divided into fragments, which can be horizontal (a set of rows) or vertical (a set of attributes). Each fragment can be allocated in a different network node.

Distributed processing is the division of logical database processing among two or more network nodes. Distributed databases require distributed processing. A distributed database management system (DDBMS) governs the processing and storage of logically related data through interconnected computer systems.

The main components of a DDBMS are the transaction processor (TP) and the data processor (DP). The transaction processor component is the software that resides on each computer node that requests data. The data processor component is the software that resides on each computer that stores and retrieves data.

Current database systems can be classified by the extent to which they support processing and data distribution. We use three major categories to classify distributed database systems: (1) single-site processing, single-site data (SPSD); (2) multiple-site processing, single-site data (MPSD); and (3) multiple-site processing, multiple-site data (MPMD).

[2] Date, C. J. "Twelve Rules for a Distributed Database," *Computerworld*, June 8, 1987, 2(23) pp. 77–81.

A homogeneous distributed database system integrates only one particular type of DBMS over a computer network. A heterogeneous distributed database system integrates several different types of DBMSs over a computer network.

DDBMS characteristics are best described as a set of transparencies: distribution, transaction, failure, heterogeneity, and performance transparency. All transparencies share the common objective of making the distributed database behave as though it were a centralized database system; that is, the end user sees the data as part of a single logical centralized database and is unaware of the system's complexities.

Distribution transparency allows us to manage a physically dispersed database as though it were a centralized database. There are three levels of distribution transparency: fragmentation, location, and local mapping transparency.

Transaction transparency ensures that the execution of a distributed transaction maintains the integrity and consistency of the distributed database.

Failure transparency ensures the continuous operation of the DDBMS if there is a failure in one or more nodes.

Performance transparency ensures that the system performs as a single centralized DBMS and that the overall performance of the system is not significantly affected by the distribution of the data among several remote locations.

Heterogeneity transparency ensures that the DDBMS integrates several different types of local DBMSs into a single common, or global, database schema.

A transaction is formed by one or more database requests. An undistributed transaction updates or requests data from a single site. A distributed transaction can update or request data from multiple sites.

Distributed concurrency control is required in a network of distributed databases. A two-phase COMMIT protocol is used to ensure that all parts of a transaction are completed.

A distributed DBMS evaluates every data request to find the optimum access path in a distributed database. The DDBMS must optimize the query to reduce access, communications, and CPU costs associated with the query.

The design of a distributed database must consider the fragmentation and replication of data. The designer must also decide how to allocate each fragment or replica to obtain better overall response time and to ensure data availability to the end user.

A database can be replicated over several different sites on a computer network. The replication of the database fragments has the objective of improving data availability, thus decreasing access time. A database can be partially, fully, or not replicated. Data allocation strategies are designed to determine the location of the database fragments or replicas.

Database vendors often label software as client/server database products. The client/server architecture label refers to the way in which two computers interact over a computer network to form a system. Client/server systems have become successful in the past few years, mainly because of the increasing power of microcomputers, the growth of local area networks, and the relatively low cost of these systems compared to similar mainframe solutions.

K E Y T E R M S

application processor (AP)	data allocation	data fragmentation
automatic query optimization	centralized	horizontal
client/server architecture	partitioned	mixed
coordinator	replicated	vertical
		data manager (DM)

data processor (DP)

data replication

database fragments

distributed database

distributed database management system (DDBMS)

distributed data catalog (DDC)

distributed data dictionary (DDD)

distributed global schema

distributed processing

distributed request

distributed transaction

distribution transparency

DO-UNDO-REDO protocol

dynamic query optimization

dynamic statistical generation mode

fragmentation transparency

failure transparency

fully heterogeneous DDBMS

fully replicated database

heterogeneity transparency

heterogeneous DDBMS

homogeneous DDBMS

local mapping transparency

location transparency

manual query optimization

manual statistical generation mode

multiple-site processing, multiple-site data (MPMD)

multiple-site processing, single-site data (MPSD)

mutual consistency rule

partially replicated database

performance transparency

remote request

remote transaction

replica transparency

rule-based query optimization algorithm

single-site processing, single-site data (SPSD)

static query optimization

statistically based query optimization algorithm

subordinates

transaction manager (TM)

transaction processor (TP)

transaction transparency

two-phase commit protocol

unique fragment

unreplicated database

write-ahead protocol

REVIEW QUESTIONS

1. Describe the evolution from centralized DBMSs to distributed DBMSs.
2. List and discuss some of the factors that influenced the evolution of the DDBMS.
3. What are the advantages of the DDBMS?
4. What are the disadvantages of the DDBMS?
5. Explain the difference between a distributed database and distributed processing.
6. What is a fully distributed database management system?
7. What are the components of a DDBMS?
8. Explain the transparency features of a DDBMS.
9. Define and explain the different types of distribution transparency.
10. Describe the different types of database requests and transactions.
11. Explain the need for the two-phase commit protocol. Then describe the two phases.
12. What is the objective of the query optimization functions?
13. To which transparency feature are the query optimization functions related?
14. What are the different types of query optimization algorithms?
15. Describe the three data fragmentation strategies. Give some examples.
16. What is data replication, and what are the three replication strategies?
17. Explain the difference between file server and client/server architecture.

PROBLEMS

The following problem is based on the DDBMS scenario in Figure P10.1.

FIGURE P10.1 **THE DDBMS SCENARIO FOR PROBLEM 1**

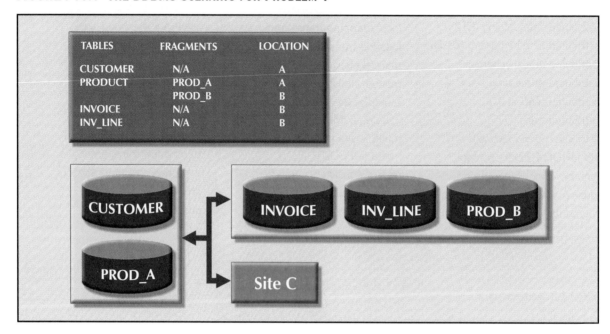

1. Specify the minimum type(s) of operation(s) the database must support (remote request, remote transaction, distributed transaction, or distributed request) in order to perform the following operations:

 At Site C

 a. SELECT *
 FROM CUSTOMER;

 b. SELECT *
 FROM INVOICE
 WHERE INV_TOT > 1000;

 c. SELECT *
 FROM PRODUCT
 WHERE PROD_ QOH < 10;

 d. BEGIN WORK;
 UPDATE CUSTOMER
 SET CUS_BAL = CUS_BAL + 100
 WHERE CUS_NUM = '10936';
 INSERT INTO INVOICE(INV_NUM, CUS_NUM, INV_DATE, INV_TOTAL)
 VALUES ('986391', '10936', '15-FEB-2002', 100);
 INSERT INTO LINE(INV_NUM, PROD_NUM, LINE_PRICE)
 VALUES ('986391', '1023', 100);
 UPDATE PRODUCT
 SET PROD_QOH = PROD_ QOH –1
 WHERE PROD_NUM = '1023';
 COMMIT WORK;

 e. BEGIN WORK;
 INSERT INTO CUSTOMER(CUS_NUM, CUS_NAME, CUS_ADDRESS, CUS_BAL)
 VALUES ('34210', 'Victor Ephanor', '123 Main St.', 0.00);

```
        INSERT INTO INVOICE(INV_NUM, CUS_NUM, INV_DATE, INV_TOTAL)
            VALUES ('986434', '34210', '10-AUG-1999', 2.00);
        COMMIT WORK;
```

At Site A

f. SELECT CUS_NUM,CUS_NAME,INV_TOTAL
 FROM CUSTOMER, INVOICE
 WHERE CUSTOMER.CUS_NUM = INVOICE.CUS_NUM;

g. SELECT *
 FROM INVOICE
 WHERE INV_TOTAL > 1000;

h. SELECT *
 FROM PRODUCT
 WHERE PROD_QOH < 10;

At Site B

i. SELECT *
 FROM CUSTOMER;

j. SELECT CUS_NAME, INV_TOTAL
 FROM CUSTOMER,INVOICE
 WHERE INV_TOTAL > 1000 AND CUSTOMER.CUS_NUM = INVOICE.CUS_NUM;

k. SELECT *
 FROM PRODUCT
 WHERE PROD_QOH < 10;

2. The following data structure and constraints exist for a magazine publishing company:

 a. The company publishes one regional magazine in each region: Florida (FL), South Carolina (SC), Georgia (GA), and Tennessee (TN).

 b. The company has 300,000 customers (subscribers) distributed throughout the four states listed in Part a.

 c. On the first of each month, an annual subscription INVOICE is printed and sent to each customer whose subscription is due for renewal. The INVOICE entity contains a REGION attribute to indicate the state (FL, SC, GA, TN) in which the customer resides:

 CUSTOMER (CUS_NUM, CUS_NAME, CUS_ADDRESS, CUS_CITY, CUS_ZIP, CUS_SUBSDATE)

 INVOICE (INV_NUM, INV_REGION, CUS_NUM, INV_DATE, INV_TOTAL)

 The company's management is aware of the problems associated with centralized management and has decided that it is time to decentralize the management of the subscriptions into the company's four regional subsidiaries. Each subscription site will handle its own customer and invoice data. The management at company headquarters, however, will have access to customer and invoice data to generate annual reports and to issue ad hoc queries, such as:

 • List all current customers by region.
 • List all new customers by region.
 • Report all invoices by customer and by region.

 Given these requirements, how must you partition the database?

3. Given the scenario and the requirements in Problem 2, answer the following questions:

 a. What recommendations will you make regarding the type and characteristics of the required database system?

 b. What type of data fragmentation is needed for each table?

 c. What criteria must be used to partition each database?

 d. Design the database fragments. Show an example with node names, location, fragment names, attribute names, and demonstration data.

 e. What type of distributed database operations must be supported at each remote site?

 f. What type of distributed database operations must be supported at the headquarters site?

11

OBJECT ORIENTED DATABASES

In this chapter, you will learn:

❖ What basic concepts govern OO systems

❖ How OO features are related to the more traditional relational and ER models

❖ What the basic features of an OO database management system (OODBMS) are

❖ What effect OO concepts are likely to have on data modeling and design

❖ About the advantages and disadvantages of OODBMSs

❖ How OO concepts have influenced the relational model

❖ How Oracle9i implements object extensions

The data management and application development environments have become far more complex than envisioned by the creators of the hierarchical, network, and relational DBMSs. Although current databases can handle complex data at both the data management and application development levels, such capability is a relatively recent phenomenon. In the late 1980s and early 1990s, database experts faced increasingly complex data requirements that were difficult to handle with existing database technology. The changing composition of the data being modeled—the database might include graphics, video, audio, maps, diagrams, fingerprints, and sound, as well as numbers and text—invited a reorganization of existing database systems. This reorganization effort led to a new wave of database technologies based on object oriented programming concepts, and to the addition of new features to relational databases that enabled them to better handle complex data.

Object oriented (OO) technology draws its strength from powerful programming and modeling techniques and advanced data-handling capabilities. Because OO technology has become an important contributor to the evolution of database systems, we explore the characteristics of OO systems and how these characteristics affect data modeling and design.

We also will investigate how relational database vendors have responded to the demand for databases capable of handling increasingly complex data types, by creating what are known as extended relational or object/relational databases. You will see how some of these features are implemented in Oracle.

E L E V E N

11.1 OBJECT ORIENTATION AND ITS BENEFITS

Object orientation is a modeling and development methodology based on object oriented (OO) concepts. More precisely, we define **object orientation** as a set of design and development principles based on conceptually autonomous computer structures known as objects. Each object represents a real-world entity with the ability to act upon itself and interact with other objects.

Considering this definition, it takes little imagination to see that using objects makes modularity almost inevitable. Object orientation concepts have been widely applied to many computer-related disciplines, especially those involving complex programming and design problems. Table 11.1 summarizes some object orientation contributions to computer-related disciplines.

TABLE 11.1 OBJECT ORIENTATION CONTRIBUTIONS

COMPUTER-RELATED AREA	OO CONTRIBUTIONS
Programming languages	Reduces the number of lines of code Decreases development time Enhances code reusability Makes code maintenance easier Enhances programmer productivity
Graphical user interfaces (GUIs)	Enhances ability to create easy-to-use interfaces Improves system end-user friendliness Makes it easier to define standards
Databases	Supports abstract data types Supports complex objects Supports multimedia data types
Design	Captures more of the data model's semantics Represents the real world more accurately Supports complex data manipulations in specialized applications that target graphics, imaging, mapping, financial modeling, telecommunications, geospatial applications, medical applications, and so on
Operating systems	Enhances system portability by creating layers of abstractions to handle hardware-specific issues Facilitates system extensibility through the use of inheritance and other object oriented constructs

11.2 THE EVOLUTION OF OBJECT ORIENTED CONCEPTS

Object oriented concepts stem from **object oriented programming (OOP)**, which was developed as an alternative to traditional programming methods. In an OOP environment, the programmer creates or uses objects, which are self-contained, reusable modules that contain data as well as the procedures used to operate on such data.

OO concepts first appeared in programming languages such as Ada, ALGOL, LISP, and SIMULA. These programming languages set the stage for the introduction of more refined OO concepts. Smalltalk, C++, and Java are popular **object oriented programming languages (OOPLs)**. Java is used to create Web applications that run on the Internet and are independent of operating systems.

OOPLs were developed to:

- Provide an easy-to-use software development environment.

- Provide a powerful software modeling tool for application development.

- Decrease development time by reducing the amount of code.

- Improve programmer productivity by making that code reusable.

OOP changes not only the way in which programs are written but also how those programs behave. In the object oriented view of the world, each object can manipulate the data that is part of that object. In addition, each object can send messages to change the data of other objects. Consequently, the OO environment has several important attributes:

- The data set is no longer passive.

- Data and procedures are bound together, creating an object.

- The object has an innate ability to act on itself.

An object can interact with other objects to create a system. Because such objects carry their own data and code, it becomes easier to produce reusable modular systems. It is precisely this characteristic that makes OO systems seem natural to those with little programming experience, but confusing to many whose traditional programming expertise has trained them to split data and procedures. It is not surprising that OO notions became much more viable with the advent of personal computers, because personal computers are typically operated by end users rather than by programmers and systems design specialists.

OO programming concepts have also had an effect on most computer-based activities, including those based on databases. Because a database is designed to capture data about a business system, it can be viewed as a set of interacting objects. Each object has certain characteristics (attributes) and has relationships with other objects (methods). Given such a structure, OO systems have an intuitive appeal for those doing database design and development. As database designers rather than programmers, we cannot afford to ignore the OO revolution.

11.3 OBJECT ORIENTED CONCEPTS

Although OO concepts have their roots in programming languages, and the programmers among you will recognize basic programming elements, *you do not need to know anything about programming to understand these concepts.*

11.3.1 OBJECTS: COMPONENTS AND CHARACTERISTICS

In OO systems everything we deal with is an object, whether it be a student, an invoice, an airplane, an employee, a service, a menu panel, a report, or so forth. Some objects are tangible and some are not. We can define an **object** as an abstract representation of a real-world entity that has a unique identity, embedded properties, and the ability to interact with other objects and act upon itself.

Note the difference between *object* and *entity*. An entity has data components and relationships but lacks manipulative ability. We'll notice other differences later.

A defining characteristic of an object is its *unique identity*. To emphasize this point, let us examine the real-world objects displayed in Figure 11.1. As you examine these simple objects, note that the student named J. D. Wilson has a unique (biological) identity and therefore constitutes a different object from M. R. Gonzalez or V. K. Spelling. Note also that, although they share common *general* characteristics such as name, Social Security number, address, and date of birth, each object exists independently in time and space.

FIGURE 11.1 REAL-WORLD STUDENT OBJECTS

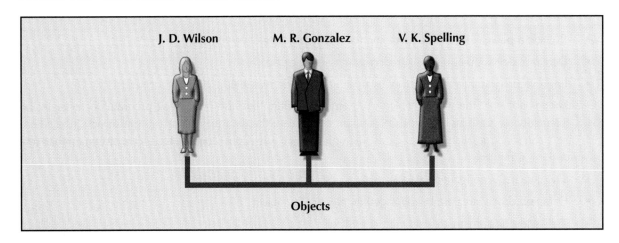

11.3.2 OBJECT IDENTITY

The object's identity is represented by an **object ID** (**OID**), which is unique to that object. The OID is assigned by the system at the moment of the object's creation *and cannot be changed under any circumstances.*

Do not confuse the relational model's primary key with an OID. In contrast to the OID, a primary key is based on *user-given* values of selected attributes and can be changed at any time. The OID is assigned by the system, does not depend on the object's attribute values, and cannot be changed. *The OID can be deleted only if the object is deleted, and that OID can never be reused.*

11.3.3 ATTRIBUTES (INSTANCE VARIABLES)

Objects are described by their attributes, known as **instance variables** in an OO environment. For example, the student John D. Smith may have the attributes (instance variables) shown in Table 11.2. Each attribute has a unique name and a data type associated with it. In Table 11.2 the attribute names are SOCIAL_SECURITY_NUMBER, FIRST_NAME, MIDDLE_INITIAL, LAST_NAME, and so on. Traditional data types, also known as **base data types** or **conventional data types**, are used in most programming languages and include real, integer, string, and so on.

TABLE 11.2 OBJECT ATTRIBUTES

ATTRIBUTE NAME	ATTRIBUTE VALUE
SOCIAL_SECURITY_NUMBER	414-48-0944
FIRST_NAME	John
MIDDLE_INITIAL	D
LAST_NAME	Smith
DATE_OF_BIRTH	11/23/1966
MAJOR *	Accounting
SEMESTER_GPA	2.89
OVERALL_GPA	3.01

TABLE 11.2 OBJECT ATTRIBUTES (CONTINUED)

ATTRIBUTE NAME	ATTRIBUTE VALUE
COURSES_TAKEN *	ENG201;MATH243;HIST201;ACCT211;ECON210;ECON212;ACCT212; CIS220;ENG202;MATH301;HIST202;CIS310;ACCT343;ACCT345;ENG242; MKTG301;FIN331;ACCT355
ADVISOR*	Dr. W. R. Learned
* Represents an attribute that references one or more other objects	

Attributes also have a domain. The **domain** logically groups and describes the set of all possible values that an attribute can have. For example, the possible values of SEMESTER_GPA (see Table 11.2) can be represented by the real number base data type. But that does not mean that any real number is a valid GPA. Keep in mind that base data types define base domains; that is, *real* represents all real numbers, *integer* represents all integers, *date* represents all possible dates, *string* represents any combination of characters, etc. However, base data type domains are the building blocks that are used to construct more restrictive *named* domains at a higher logical level. For example, to define the domain for the GPA attribute precisely, we must create a domain named "GPA." Every domain has a name and a description, including the base data type, size, format, and constraints for the domain's values. We can define the "GPA" domain, therefore, as "any positive number between 0.00 and 4.00, with only two decimal places." In this case, we have a domain name "GPA," a base data type "real," a constraint rule "any positive number between 0.00 and 4.00," and a format "with only two decimal places." The "GPA" domain will provide the values for the SEMESTER_GPA and OVERALL_GPA attributes. Domains can also be defined as lists of possible values separated by commas. For example the "GENDER" domain can be defined as "Male, Female" or as "M, F."

It is important to note that the relational database model also supports domains. In fact, C. J. Date, one of the relational database model's "parents," presents domains as the way in which relational systems are able to support abstract data types, thus providing the same functionality as object oriented databases.[1]

Just as in the ER model, an object's attribute can be *single-valued* or *multivalued*. Therefore, the object's attribute can draw a single value or multiple values from its domain. For example, the SOCIAL_SECURITY_NUMBER attribute takes only one value from its domain, because the student can have only one Social Security number. In contrast, an attribute such as LANGUAGE or HOBBY can have many values, because a student might speak many languages or have many hobbies.

Object attributes may reference one or more other objects. For example, the attribute MAJOR refers to a Department object, the attribute COURSES_TAKEN refers to a list (or collection) of Course objects, and the attribute ADVISOR refers to a Professor object. At the implementation level, the OID of the referenced object is used to link both objects, thus allowing the implementation of relationships between two or more objects. Using the example in Table 11.2, the MAJOR attribute contains the OID of a Department object (Accounting), and the ADVISOR attribute contains the OID of a Professor object (Dr. W. R. Learned). The COURSES_TAKEN attribute contains the OID of an object that contains a list of Course objects; such an object is known as a **collection object**.

[1] See C. J. Date and Hugh Darwen, *Foundation for Object/Relational Databases: The Third Manifesto*, Addison Wesley, 1998.

N O T E

Observe the difference between the relational and OO models at this point. In the relational model a table's attribute may contain only a value that may be used to join rows in different tables. The OO model does not need such joins to relate objects to one another.

11.3.4 OBJECT STATE

The **object state** is the set of values that the object's attributes have at a given time. Although the object's state can vary, its OID remains the same. If we want to change the object's state, we must change the values of the object's attributes. To change the object's attribute values, we must send a message to the object. This *message* will invoke a *method*.

11.3.5 MESSAGES AND METHODS

A **method** is the code that performs a specific operation on the object's data. Methods protect data from direct and unauthorized access by other objects. To help you understand messages and methods, imagine the object to be a nutshell. The nutshell's nucleus (the nut) represents the object's data structure, and the shell represents its methods. (See Figure 11.2.)

FIGURE 11.2 DEPICTION OF AN OBJECT

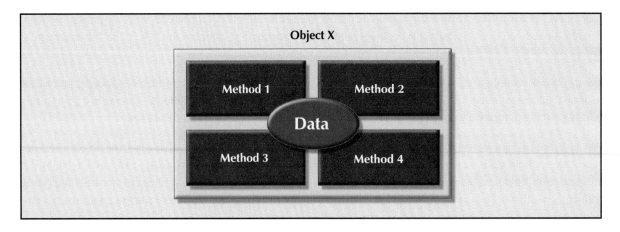

Every operation performed on an object must be implemented by a method. Methods are used to change the object's attribute values or to return the value of selected object attributes. Methods represent real-world actions, such as changing a student's major, adding a student to a course, or printing a student's name and address. In effect, *methods are the equivalent of procedures in traditional programming languages*. In OO terms, methods represent the object's behavior.

Every method is identified by a *name* and has a *body*. The body is composed of computer instructions written in some programming language to represent a real-world action. For example, using the object attributes described in Table 11.2, we can define a method *Avegpa* that will return the average GPA of a student by using the object's attributes SEMESTER_GPA and OVERALL_GPA. The method named *Avegpa* may thus be represented by the transformation shown in Figure 11.3.

FIGURE 11.3 METHOD COMPONENTS

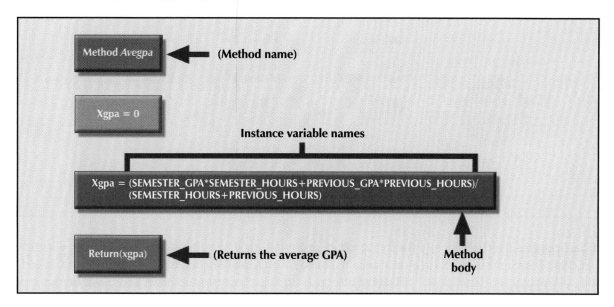

As you examine Figure 11.3, note that the Return(xgpa) would yield (3.2 * 15) + (3.0 * 60)/(15 + 60) = (48 + 180)/75 = 3.04 for a student with the following characteristics:

- Current semester GPA is 3.2.
- Current class load is 15 semester hours.
- Previous GPA was 3.0 earned for a total of 60 previous hours.

As you examine this example, note that a method can access the instance variables (attributes) of the object for which the method is defined.

To invoke a method you send a message to the object. A **message** is sent by specifying a receiver object, the name of the method, and any required parameters. The internal structure of the object cannot be accessed directly by the message sender, which is another object. Denial of access to the structure ensures the integrity of the object's state and hides the object's internal details. The ability to hide the object's internal details (attributes and methods) is known as **encapsulation**.

An object may also send messages to change or interrogate another object's state. (To **interrogate** means to ask for the interrogated object's instance variable value or values.) To perform such object-change and interrogation tasks, the method's body can contain references to other objects' methods (send messages to other objects), as depicted in Figure 11.4.

11.3.6 CLASSES

OO systems classify objects according to their similarities and differences. Objects that share common characteristics are grouped into classes. In other words, a **class** is a collection of similar objects with shared structure (attributes) and behavior (methods).

FIGURE 11.4 OBJECTS SEND MESSAGES TO EACH OTHER

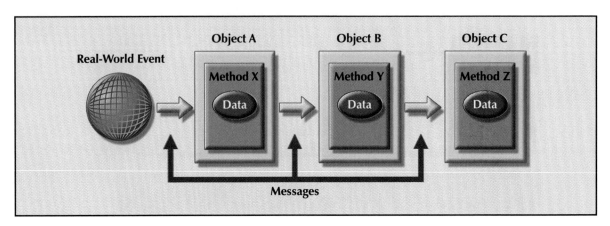

A class contains the description of the data structure and the method implementation details for the objects in that class. Therefore, all objects in a class share the same structure and respond to the same messages. In addition, a class acts as a "storage bin" for similar objects. Each object in a class is known as a **class instance** or **object instance** (see Figure 11.5).

FIGURE 11.5 CLASS ILLUSTRATION

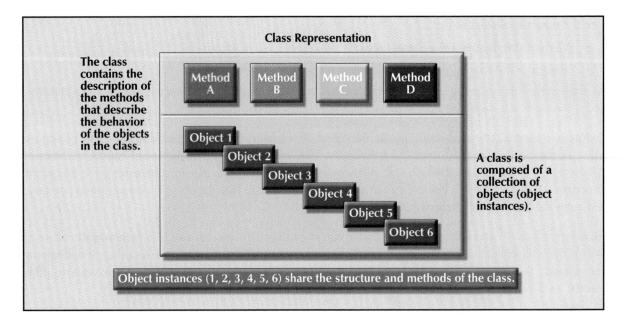

Using the example shown earlier in Table 11.2, we can define a class named Student to store student objects. All objects of the class Student, shown in Figure 11.6, share the same structure (attributes) and respond to the same messages (implemented by methods). Note that we defined the *Avegpa* method earlier; the *Enroll* and *Grade* methods shown in Figure 11.6 have been added. Each instance of a class is an object with a unique OID, and each object knows to which class it belongs.

FIGURE 11.6 **REPRESENTATION OF THE CLASS STUDENT**

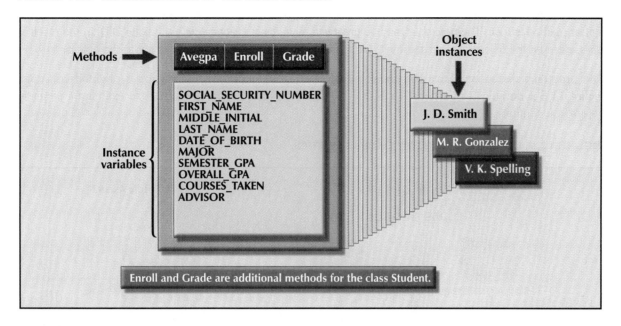

11.3.7 PROTOCOL

The class's collection of messages, each identified by a message name, constitutes the object or class *protocol*. The **protocol** represents an object's public aspect; that is, how it is known by other objects as well as end users. In contrast, the implementation of the object's structure and methods constitutes the object's *private aspect*. Both are illustrated in Figure 11.7. For example, Figure 11.6 shows three methods (Avegpa, Enroll, and Grade) that represent the public aspect of the Student object. Because these methods are public, other objects communicate with the student object using any of these methods. The internal representation of the methods (for example, see Figure 11.3) yields the private aspect of the object. The private aspect of an object is not available for use by other objects.

FIGURE 11.7 **PUBLIC AND PRIVATE ASPECTS OF AN OBJECT**

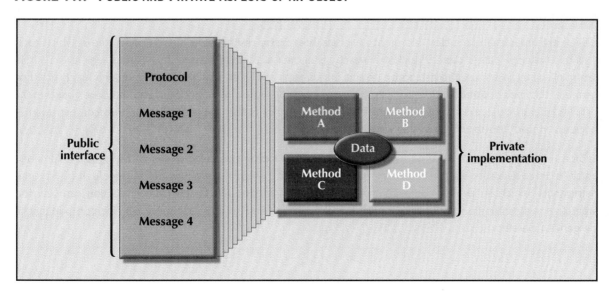

Usually, a message is sent to an object instance. However, it is also possible to send a message to the class rather than to the object. When the receiver object is a class, the message will invoke a *class method*. One example of a class method is *new*. The *new* class method creates a new object instance (with a unique OID) in the receiver class. Because the object does not exist yet, the message *new* is sent to the class and not to the object.

The preceding discussions have laid the foundation for your understanding of object oriented concepts. Figure 11.8 is designed to put together all the pieces of this part of the OO puzzle, so examine it carefully before you continue.

FIGURE 11.8 OO SUMMARY: OBJECT CHARACTERISTICS

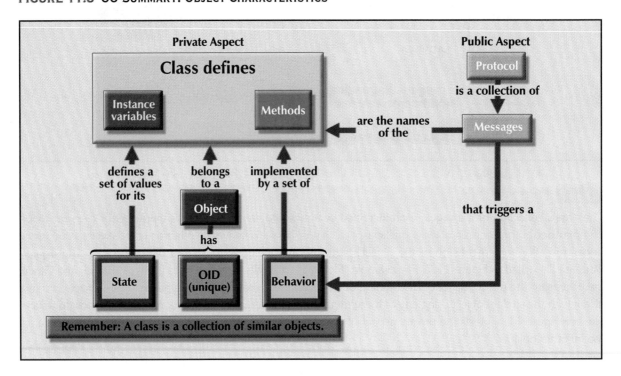

11.3.8 SUPERCLASSES, SUBCLASSES, AND INHERITANCE

Classes are organized into a class *hierarchy*. A **class hierarchy** resembles an upside-down tree in which each class has only one parent class. The class hierarchy is known as a **class lattice** if its classes can have multiple parent classes. Class is used to categorize objects into groups of objects that share common characteristics. For example, the class *automobile* includes large luxury sedans as well as compact cars, and the class *government* includes federal, state, and local governments. Figure 11.9 illustrates that the generalization *musical instruments* includes stringed instruments as well as wind instruments.

As you examine Figure 11.9, note that Piano, Violin, and Guitar are **subclasses** of Stringed instruments, which is, in turn, a subclass of Musical instruments. Musical instruments defines the **superclass** of Stringed instruments, which is, in turn, the superclass of the Piano, Violin, and Guitar classes. As you can see, the superclass is a more general classification of its subclasses, which, in turn, are more specific components of the general classification.

FIGURE 11.9 MUSICAL INSTRUMENTS CLASS HIERARCHY

The class hierarchy introduces a powerful OO concept known as *inheritance*. **Inheritance** is the ability of an object within the hierarchy to inherit the data structure and behavior (methods) of the classes above it. For example, the Piano class in Figure 11.9 inherits its data structure and behavior from the superclasses Stringed instruments and Musical instruments. Thus Piano inherits the strings and its sounding board characteristic from the Stringed instruments superclass and the musical scale from its Musical instruments superclass. *It is through inheritance that OO systems can deliver code reusability.*

In OO systems, all objects are derived from the superclass Object, or the Root class. Therefore, all classes share the characteristics and methods of the superclass Object. The inheritance of data and methods goes from top to bottom in the class hierarchy. There are two types of inheritance: single and multiple inheritance.

Single Inheritance

Single inheritance exists when a class has only one immediate (parent) superclass above it. Such a condition is illustrated by the Stringed instruments and Wind instruments classes in Figure 11.9. Most of the current OO systems support single inheritance. When the system sends a message to an object instance, the entire hierarchy is searched for the matching method, using the following sequence:

1. Scan the class to which the object belongs.
2. If the method is not found, scan the superclass.

The scanning process is repeated until either one of the following occurs:

1. The method is found.
2. The top of the class hierarchy is reached without finding the method. The system then generates a message to indicate that the method was not found.

To illustrate the scanning process, let's examine the Employee class hierarchy shown in Figure 11.10. If we send the *monthPay* message to a pilot's object instance, the object will execute the *monthPay* method defined in its Employee superclass. Note the code reusability benefits obtained through object oriented systems: the *monthPay* method's code is available to both the Pilot and Mechanic subclasses.

FIGURE 11.10 SINGLE INHERITANCE

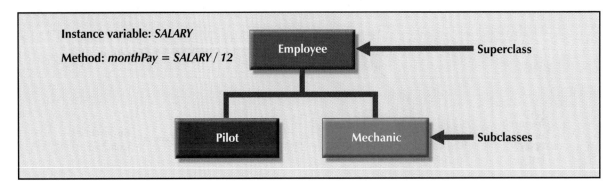

Multiple Inheritance

Multiple inheritance exists when a class can have more than one immediate (parent) superclass above it. This condition is illustrated in Figure 11.11. Figure 11.12 provides an example of multiple inheritance, illustrating that the Motorcycle subclass inherits characteristics from both the Motor Vehicle and Bicycle superclasses. From the Motor Vehicle superclass, the Motorcycle subclass inherits:

- Characteristics, such as fuel requirements, engine pistons, and horsepower.
- Behaviors, such as start motor, fill gas, and depress clutch.

From the Bicycle superclass, the Motorcycle subclass inherits:

- Characteristics, such as two wheels and handlebars.
- Behaviors, such as straddle the seat and move the handlebar to turn.

FIGURE 11.11 MULTIPLE INHERITANCE

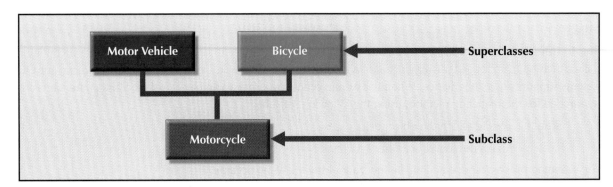

The assignment of instance variable or method names must be treated with some caution in a multiple inheritance class hierarchy. For example, if you use the same name for an instance variable or method in each of the superclasses, the OO system must be given some way to decide which method or attribute to use. To illustrate this point, let's suppose that both the Motor Vehicle and Bicycle superclasses shown in Figure 11.12 use a MAXSPEED instance variable.

FIGURE 11.12 MOTOR VEHICLE AND BICYCLE INSTANCE VARIABLES

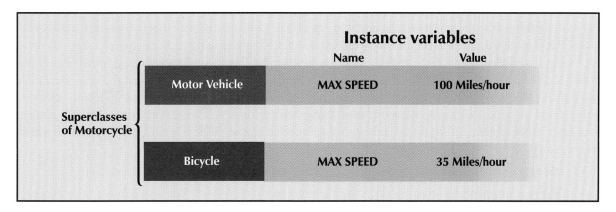

Which version of the MAXSPEED instance variable will be inherited by the Motorcycle's method in this case? A human being would use judgment to correctly assign the 100 miles/hour value to the motorcycle. The OO system, however, cannot make such value judgments and might:

- Produce an error message in a pop-up window explaining the problem.
- Ask the end user to supply the correct value or to define the appropriate action.
- Yield an inconsistent or unpredictable result.
- Use user-defined inheritance rules for the subclasses in the class lattice. These inheritance rules govern a subclass's inheritance of methods and instance variables.

11.3.9 METHOD OVERRIDING AND POLYMORPHISM

We may override a superclass's method definition by redefining the method at the subclass level. To illustrate the method override, we will use the Employee class hierarchy depicted in Figure 11.13.

FIGURE 11.13 EMPLOYEE CLASS HIERARCHY METHOD OVERRIDE

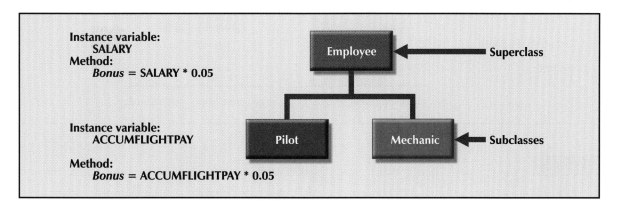

As you examine the summary presented in Figure 11.13, note that we have defined a *Bonus* method to compute a Christmas bonus for all employees. The specific bonus computation depends on the type of employee. In this case, *with the exception of pilots*, an employee receives a Christmas bonus equal to 5 percent of his or her salary. Pilots receive a Christmas bonus based on accumulated flight pay rather than on annual salary. By defining the *Bonus* method in the Pilot subclass, we are overriding the Employee Bonus method for all objects that belong to the Pilot subclass. However, the Pilot subclass bonus redefinition does not affect the bonus computation for the Mechanic subclass. In contrast to method overriding, polymorphism allows different objects to respond to the same message in different ways. Polymorphism is a very important feature of OO systems because it allows objects to behave according to their specific characteristics. In OO terms, **polymorphism** means that:

- We may use the same name for a method defined in different classes in the class hierarchy.
- The user may send the same message to different objects that belong to different classes and yet always generate the correct response.

To illustrate the effect of polymorphism, let's examine the expanded Employee class hierarchy shown in Figure 11.14. Using the class hierarchy in Figure 11.14, the system computes a pilot's or mechanic's monthly pay by sending the same message, *monthPay*, to the pilot or mechanic object. The object will return the correct monthly pay amount even though the *monthPay* includes *flyPay* for the pilot object and *overtimePay* for the mechanic object. The computation of the regular monthly salary payment for both subclasses (Pilot and Mechanic) is the same: the annual salary divided by 12 months.

FIGURE 11.14 EMPLOYEE CLASS HIERARCHY POLYMORPHISM

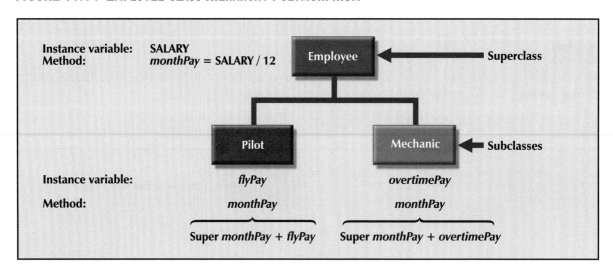

> **N O T E**
>
> In Figure 11.14 we used Smalltalk syntax: **Super** *monthPay* in the pilot's *monthPay* method indicates that the object inherits its superclass *monthPay* method. Other OOPLs, such as C++, use the *Employee.monthPay* syntax.

As you examine the polymorphism example in Figure 11.14, note that:

- The Pilot *monthPay* method definition overrides and expands the Employee *monthPay* method defined in the Employee superclass.
- The *monthPay* method that was defined in the Employee superclass is reused by the Pilot and Mechanic subclasses.

Polymorphism thus augments method override to enhance the code reusability so prized in modular programming and design.

11.3.10 ABSTRACT DATA TYPES

A data type describes a set of objects with similar characteristics. All conventional programming languages use a set of predefined base data types (real, integer, and string or character). Base data types are subject to a predefined set of operations. For example, the integer base data type allows operations such as addition, subtraction, multiplication, and division.

Conventional programming languages also include type constructors, the most common of which is the *record type constructor*. For example, a programmer can define a CUSTOMER record type by describing its data fields. The CUSTOMER record represents a new data type that will store CUSTOMER data, and the programmer can directly access such a data structure by referencing the record's field names. A record data type allows operations such as WRITE, READ, or DELETE. However, new operations cannot be defined for base data types.

Like conventional data types, an **abstract data type (ADT)** describes a set of similar objects. However, an abstract data type differs from a conventional data type in that:

- The ADT's operations are user-defined.
- The ADT does not allow direct access to its internal data representation or method implementation. In other words, the ADT encapsulates its definition, thereby hiding its characteristics.

Some OO systems, such as Smalltalk, implement base data types as ADTs.

To create an abstract data type you must define:

- Its name.
- The data representation or instance variables of the objects belonging to the abstract data type; each instance variable has a data type that may be a base data type or another ADT.
- The abstract data type operations and constraints, both of which are implemented through methods.

You might have noted that the abstract data type definition resembles a class definition. Some OO systems differentiate between class and type, using *type* to refer to the *class data structure and methods,* and using *class* to refer to the *collection of object instances.* A type is a more static concept, while a class is a more run-time concept. In other words, when you define a new class, you are actually defining a new type. The type definition is used as a pattern or template to create new objects belonging to a class at run time.

A simple example will help you understand the subtle distinction between OO type and class. Suppose you bought a cross-stitch pattern with which to create pillow covers. The pattern you bought includes the *description* of its structure as well as *instructions* about its use. That pattern will be the type definition. The collection of all the actual pillow covers, each with a unique serial number or OID, that you create with the help of that pattern, constitutes the class.

Abstract data types together with inheritance provide support for complex objects. A **complex object** is formed by combining other objects in a set of complex relations. An example of such a complex object might be found in a security system that uses different data types, such as:

- Conventional (tabular) employee data; for example name, phone, or date of birth.
- Bitmapped data to store the employee's picture.
- Voice data to store the employee's voice pattern.

The ability to deal in a relatively easy manner with such a complex data environment gives OO credibility in today's database marketplace.

11.3.11 OBJECT CLASSIFICATION

An object can be classified according to the characteristics (simple, composite, compound, hybrid, and associative) of its attributes. A **simple object** is an object that contains only single-valued attributes and has no attributes that refer to another object. For example, an object that describes the current semester can be defined as having the following single-valued attributes: SEM_ID, SEM_NAME, SEM_BEGIN_DATE, and SEM_END_DATE.

A **composite object** is an object that contains at least one multivalued attribute and has no attributes that refer to another object. An example of a composite object would be a MOVIE object in a movie rental system. For example, MOVIE might be defined as having the following attributes: MOVIE_ID, MOVIE_NAME, MOVIE_PRICE, MOVIE_TYPE, and MOVIE_ACTORS. In this case, MOVIE_ACTORS will be a multivalued attribute that tracks the many performers in the movie.

A **compound object** is an object that contains at least one attribute that references another object. An example is the STUDENT object in Table 11.2. In this example, the ADVISOR attribute refers to the PROFESSOR object.

A **hybrid object** is an object that contains a repeating group of attributes, at least one of which refers to another object. A typical example of a hybrid object would be the invoice example introduced in Chapter 3, "The Relational Database Model," Figure 3.30. In this case, an invoice contains many products, and each product has a quantity sold and a unit price. The object representation of the invoice contains a repeating group of attributes to represent the product, quantity sold, and unit price (PROD_CODE, LINE_UNITS, and LINE_PRICE) for each product sold. Therefore, the object representation of the invoice does not require a new INV_LINE object as in the ER model representation.

Finally, an **associative object** is an object used to represent a relationship between two or more objects. The associative object can contain its own attributes to represent specific characteristics of the relationship. A good example of an associative object is the Enroll example in Chapter 3, Figure 3.26. In this case, the ENROLL object relates to a STUDENT and a CLASS object, and includes an ENROLL_GRADE attribute that represents the grade earned by the student in the class.

In real-world data models, you find fewer simple and composite objects and more compound, hybrid, and associative objects. We will discuss these types of objects in greater detail in Section 11.4.3.

11.4 CHARACTERISTICS OF AN OBJECT ORIENTED DATA MODEL

The object oriented concepts described in previous sections represent the core characteristics of an **object oriented data model** (**OODM**), also known as an object data model or ODM. At the very least, an object oriented data model must:

- Support the representation of complex objects.
- Be **extensible**; that is, it must be capable of defining new data types as well as the operations to be performed on them.
- Support encapsulation; that is, the data representation and the method's implementation must be hidden from external entities.
- Exhibit inheritance; an object must be able to inherit the properties (data and methods) of other objects.
- Support the notion of object identity (OID) described earlier in this chapter.

For instructional purposes we will use OODM component descriptions and definitions, to the extent possible, to correspond to the entity relationship model components described in Chapter 3. Although most of the basic OODM components were defined earlier in this chapter, a quick summary may help you understand the subsequent material more easily:

- The OODM models real-world entities as *objects*.
- Each object is composed of *attributes* and a set of *methods*.
- Each attribute can *reference* another object or a set of objects.
- The attributes and the methods' implementation are hidden, *encapsulated*, from other objects.
- Each object is identified by a unique *object ID (OID)*, which is independent of the value of its attributes.
- Similar objects are described and grouped in a *class* that contains the description of the data (attributes or instance variables) and the methods' implementation.
- The class describes a *type* of object.
- Classes are organized in a *class hierarchy*.
- Each object of a class *inherits* all properties of its superclasses in the class hierarchy.

Armed with this summarized OO component description, note the comparison between the OO and ER model components presented in Table 11.3.

TABLE 11.3 COMPARING THE OO AND ER MODEL COMPONENTS

OO DATA MODEL	ER MODEL
Type	Entity definition
Object	Entity
Class	Entity set
Instance variable	Attribute
N/A	Primary key
OID	N/A
Method	N/A
Class hierarchy	ER diagram

11.4.1 OBJECT SCHEMAS: THE GRAPHICAL REPRESENTATION OF OBJECTS

A graphical representation of an object resembles a box, with the instance variable names inside the box. Generally speaking, the object representation is shared by all objects in the class. Therefore, you will discover that the terms *object* and *class* are often used interchangeably in the illustrations. With this caveat in mind, let us begin by examining the illustration based on the Person class, shown in Figure 11.15. In this case, the instance variables NAME, ADDRESS, DOB, and SEX use a string base data type, and the AGE instance variable uses an integer base data type.

FIGURE 11.15 SHARED REPRESENTATION FOR ALL OBJECTS OF THE CLASS PERSON

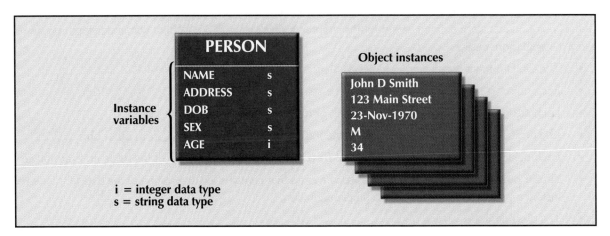

Next, let's examine the *state* of a Person object instance (see Figure 11.16). As you examine Figure 11.16, note that the AGE instance variable can also be viewed as a *derived* attribute. Derived attributes may be implemented through *methods*. For instance, we can create a method named *Age* for the Person class. This method will return the difference in years between the current date and the date of birth (DOB) for a given object instance. Aside from the fact that methods can generate derived attribute values, methods have the added advantages of encapsulation and inheritance.

FIGURE 11.16 STATE OF A PERSON OBJECT INSTANCE

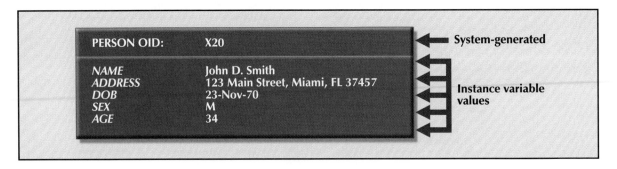

Keep in mind that the OO environment allows us to create abstract data types from base data types. For example, NAME, ADDRESS, and DOB are composite attributes that can be implemented through classes or ADTs. To illustrate this point, we have defined Name, Address, and DOB to be abstract data types in Figure 11.17.

FIGURE 11.17 DEFINING THREE ABSTRACT DATA TYPES

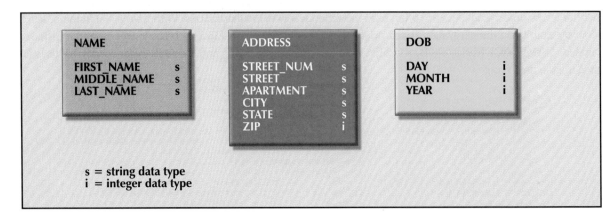

As you examine Figure 11.17, note that the Person class now contains attributes that point to objects of other classes or abstract data types. The new data types for each instance variable of the class Person are shown in Figure 11.18.

FIGURE 11.18 OBJECT REPRESENTATION FOR INSTANCES OF THE CLASS PERSON WITH ADTs

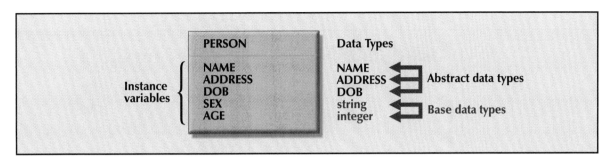

The **object space**, or **object schema**, is used to represent the composition of the state of an object at a given time.

The object's state for an instance of class Person is illustrated in Figure 11.19. As you examine Figure 11.19, note the use of OIDs to reference other objects. For example, the attributes NAME, ADDRESS, and DOB now contain an OID of an instance of their respective class or ADT instead of the base value. The use of OIDs for object references avoids the data consistency problem that would appear in a relational system if the primary key value were changed by the end user when changing the object's state. This is because the OID is independent of the object's state.

FIGURE 11.19 OBJECT STATE FOR AN INSTANCE OF THE CLASS PERSON, USING ADTs

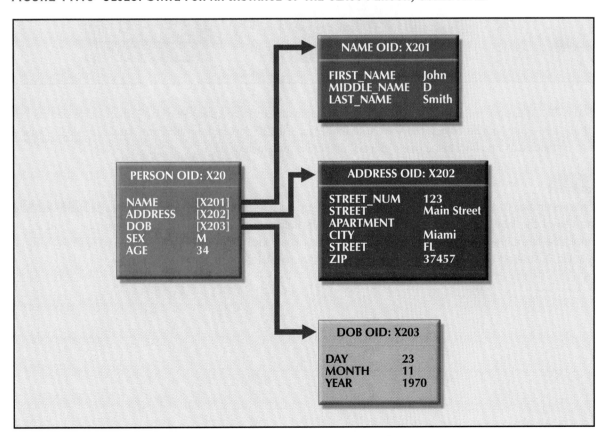

To illustrate this point further, we will use a rental property application by which we track many rental properties and the persons living in them. In this case, two people living at the same address are likely to reference the same Address object instance (see Figure 11.20). This condition is sometimes labeled as **referential object sharing**. A change in the Address object instance will be reflected in both Person instances.

FIGURE 11.20 REFERENTIAL OBJECT SHARING

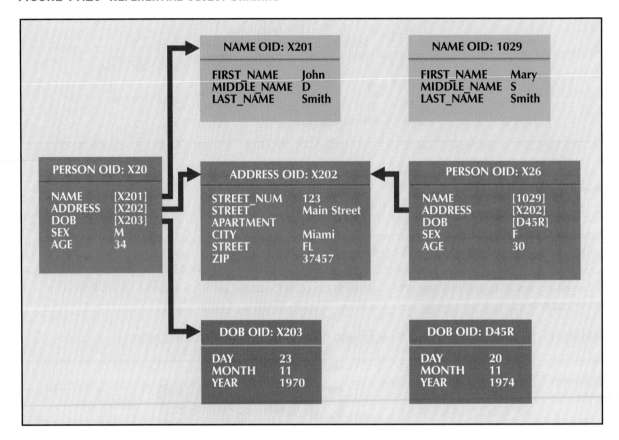

Figure 11.20 illustrates the state of two different object instances of the class Person; both object instances reference the same Address object instance. Note that Figure 11.20 depicts four different classes or ADTs: Person (two instances), Name (two instances), Address, and DOB (two instances).

11.4.2 CLASS-SUBCLASS RELATIONSHIPS

Remember that classes inherit the properties of their superclasses in the class hierarchy. This property leads to the use of the label "is a" to describe the relationship between the classes within the hierarchy. That is, an employee *is a* person, and a student *is a* person. This basic idea is sufficiently important to warrant a more detailed illustration based on the class hierarchy. (See Figure 11.21.)

FIGURE 11.21 **CLASS HIERARCHY**

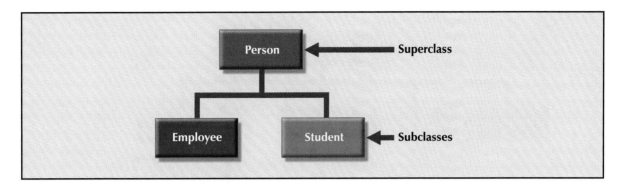

In the hierarchy shown in Figure 11.21, the Employee object is described by seven attributes, shown in Figure 11.22. Social Security number (SS#) is recorded as a string base data type, and SALARY is recorded as an integer base data type. The NAME, ADDRESS, DOB, SEX, and AGE are all inherited from the Person superclass.

FIGURE 11.22 **EMPLOYEE OBJECT REPRESENTATION**

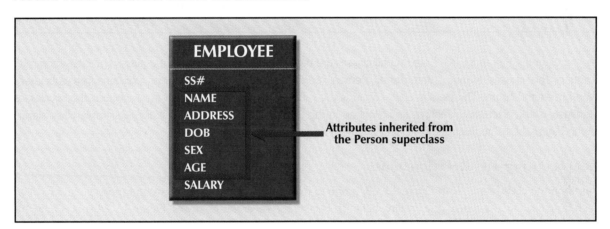

This example is based on the fact that the OODM supports the *class-subclass relationship*, for which it enforces the necessary integrity constraints. Note that the relationship between a superclass and its subclasses is 1:M; that is, if we assume single inheritance, each superclass can have many subclasses, and each subclass is related to only one superclass.

11.4.3 **INTEROBJECT RELATIONSHIPS: ATTRIBUTE-CLASS LINKS**

In addition to supporting the class-subclass relationship, the OODM supports the attribute-class relationship. An attribute-class relationship, or **interobject relationship**, is created when an object's attribute references another object of the same or different class.

The interobject relationship is different from the class-subclass relationship we explored earlier. To illustrate this difference, let's examine the class hierarchy for the EDLP (Every Day Low Prices) Retail Corporation, shown in Figure 11.23.

FIGURE 11.23 CLASS HIERARCHY FOR THE EDLP RETAIL CORPORATION

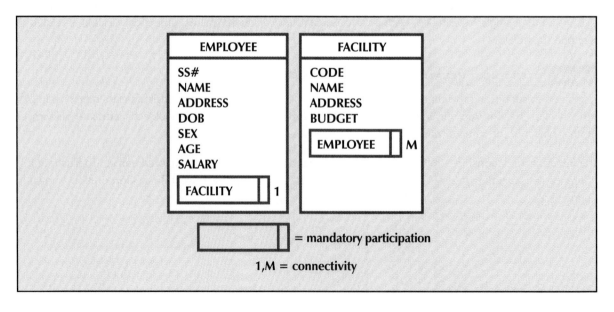

As you examine Figure 11.23, note that all classes are based on the Root Object superclass. The class hierarchy contains the classes Manufacturer, Item, Person, and Facility. The Facility class contains the subclasses Warehouse and Store. The Person class contains the subclass Employee, which in turn contains the subclasses Manager, Clerk, Secretary, Cashier, and Stocker. In the following discussion we will use the simple class hierarchy in Figure 11.23 to illustrate basic 1:M and M:N relationships.

Representing 1:M Relationships

Based on the class hierarchy in Figure 11.23, a one-to-many relationship exists between Employee and Facility: each Employee works in only one Facility, but each Facility has several Employees. Figure 11.24 shows how this relationship may be represented.

FIGURE 11.24 REPRESENTING A 1:M RELATIONSHIP

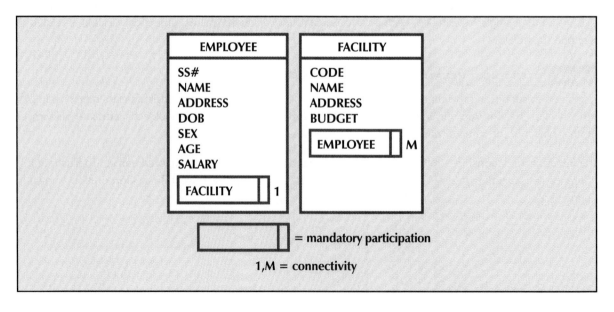

As you examine the relationship between Employee and Facility portrayed in Figure 11.24, note that the Facility object is included within the Employee object and vice versa; that is, the Employee object is also included within the Facility object. To examine the relationships in greater detail, we will use the following techniques:

- Related classes are enclosed in boxes to make relationships more noticeable.

- The double line on the box's right side indicates that the relationship is mandatory.

- Connectivity is indicated by labeling each box. In this case, we put a 1 next to Facility in the Employee object to indicate that each employee works in only one facility. The M beside Employee in the Facility object indicates that each facility has many employees.

Note that we use the ER notation to represent a mandatory entity and to indicate the connectivity of a relationship (1:M). The purpose of this notation is to maintain consistency with earlier diagrams.

Rather than just include the object box within the class, we prefer to use a name that is descriptive of the class *characteristic* that we are trying to model. Such a procedure is especially useful when two classes are involved in more than one relationship; in such cases we suggest that the attribute's name be written above the class box and that the class box be indented to indicate that the attribute will reference that class. For example, we can represent two relationships between Employee and Facility by using WORKS_AT and WORKERS, as indicated in Figure 11.25. Note that two relationships exist: the 1:M relationship is based on the business rule "each facility employs many employees, and each employee is employed by only one facility," and the 1:1 relationship is based on the business rule "each facility is managed by only one employee, and each manager manages only one facility."

FIGURE 11.25 REPRESENTING 1:1 AND 1:M RELATIONSHIPS

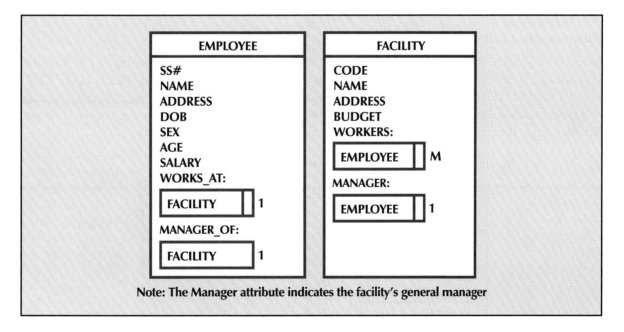

Note: The Manager attribute indicates the facility's general manager

As you examine Figure 11.25, note that the relationships are represented in both participating classes. This condition allows us to invert the relationship, if necessary. For example, the Facility object within the Employee object represents the "Manager_of" relationship. In this case, the Facility object is optional and has a maximum connectivity of 1. The Employee and Facility objects are examples of compound objects. Another type of 1:M relationship can be illustrated by examining the relationship between employees and their dependents. To establish this relationship we first create a Dependent subclass, using Person as its superclass. *Note that we cannot create a Dependent subclass by using Employee as its superclass, because the class hierarchy represents an "is a" relationship.* In other words, each Manager is an Employee, each Employee is a Person, each Dependent is a Person, and each Person is an Object in our object space—but each Dependent is not an Employee. Figure 11.26 shows the proper presentation of the relationship between Employee and Dependent.

FIGURE 11.26 **EMPLOYEE-DEPENDENT RELATIONSHIP**

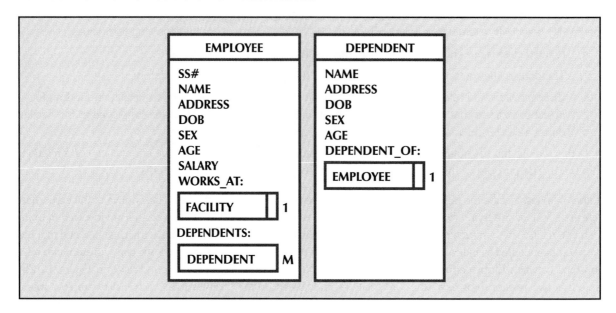

As you examine Figure 11.26, note that Dependent is optional to Employee and that Dependent has a 1:M relationship with Employee. However, Employee is mandatory to Dependent. *The weak entity concept disappears in the OODM because each object instance is identified by a unique OID.*

Representing M:N Relationships

Using the same EDLP Retail Corporation class hierarchy, a many-to-many (M:N) relationship can be illustrated by exploring the relationship between Manufacturer and Item, as represented in Figure 11.27. Figure 11.27 depicts a condition in which each Item may be produced by many Manufacturers, and each Manufacturer may produce many Items. Figure 11.27 thus represents a conceptual view of the M:N relationship between Item and Manufacturer. In this representation, Item and Manufacturer are both compound objects.

FIGURE 11.27 **REPRESENTING THE M:N RELATIONSHIP**

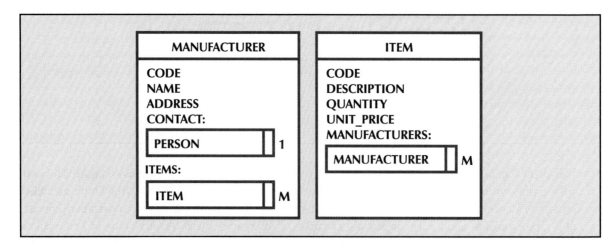

Also, note that the CONTACT attribute in the Manufacturer class in Figure 11.27 references only one instance of the Person class. A slight complication arises at this point: it is likely that each contact (person) has a phone

number, yet we did not include a phone number attribute in the Person class. In this case, the designer may add the attribute so that it will be available to all Person subclasses.

Representing M:N Relationships with an Intersection Class

Suppose that we add a condition to the just-explored ITEM class that allows us to track additional data for each item. For example, let's represent the relationship between Item and Facility so that each Facility may contain several Items and each Item may be located at several Facilities. In addition, we want to track the quantity and location (aisle and row) of an Item at each Facility. We have illustrated these conditions in Figure 11.28. The right square bracket "]" in Figure 11.28 indicates that the included attributes are treated as one logical unit. Therefore, each Item instance may contain several occurrences of Facility, each accompanied by related values for the AISLE, ROW, and QTY_ON_HAND attributes. The inverse case is true for each instance of Facility. The Item and Facility objects in this relationship are hybrid objects with a repeating group of attributes. Note that the semantic requirements for this relationship indicate that the Item or Facility objects are accessed first in order to know the aisle, row, and quantity on hand for each item.

FIGURE 11.28 REPRESENTING THE M:N RELATIONSHIP WITH ASSOCIATED ATTRIBUTES

To translate the preceding discussion to a more relational view of such an M:N scenario, we would have to define an **intersection** (bridge) **class** to connect both Facility and Item and to store the associated attributes. In this case, we might create a Stocked_Item associative object class to contain the Facility and Item object instances and the values for each of the AISLE, ROW, and QTY_ON_HAND attributes. Such a class is equivalent to the Interclass_Connection construct of the Semantic Data Model. Figure 11.29 shows how the Item, Facility, and Stocked_Item object instances might be represented.

Having examined the depiction of the basic OO relationships, we can represent the object space as shown in Figure 11.30.

FIGURE 11.29 REPRESENTING THE M:N RELATIONSHIP WITH INTERSECTION CLASS

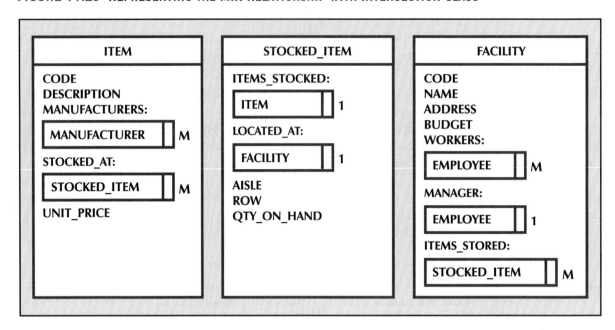

FIGURE 11.30 OBJECT SPACE REPRESENTATION

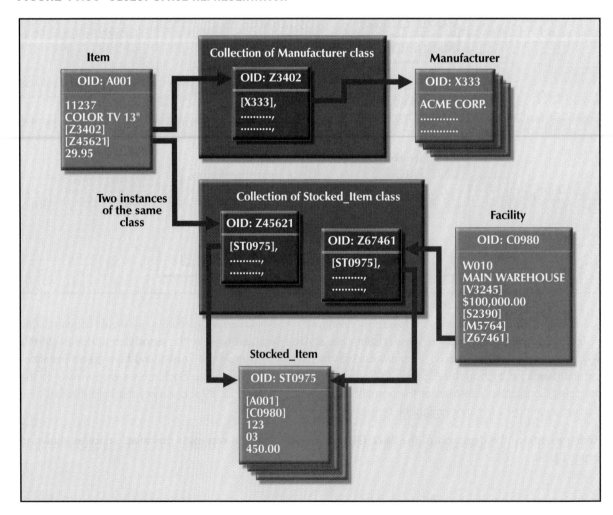

Figure 11.30 contains much critical design information, and you should particularly examine the following points:

- The Stocked_Item associative object instance contains references to an instance of each related (Item and Facility) class. The Stocked_Item intersection class is necessary only when we must keep track of the additional information we referred to earlier.

- The Item object instance in this object schema contains the collection of Stocked_Item object instances, each one of which contains a Facility object instance. The inverse of this relationship is also true: a Facility object instance contains the collection of Stocked_Item object instances, each one of which contains an Item object instance. *You should realize that these two relationships represent two different application views of the same object schema.* It is desirable for a data model to provide such flexibility.

- The interobject references use the OID of the referenced objects in order to *access* and *include* them in the object space.

- The values inside square brackets [] represent the OID of an object instance of some class. *The "collection of" classes represent a class of objects in which each object instance contains a collection of objects of some class.* For example, the Z3402 and Z45621 OIDs reference objects that constitute a collection of Manufacturers and a collection of Stocked_Items, respectively.

- In the relational model, the ITEM table would not contain any data regarding the MANUFACTURERs or the STOCKED_ITEMs in its structure. To provide (combined) information about ITEM, STOCKED_ITEM, and FACILITY, we would have to perform a relational join operation. The OODM does not need joins to combine data from different objects, because the Item object *contains* the references to the related objects; these references are automatically brought into the object space when the Item object is accessed.

11.4.4 LATE AND EARLY BINDING: USE AND IMPORTANCE

A very desirable OODM characteristic is its ability to let any object's attribute contain objects that define different data types (or classes) at different times. With this feature an object can contain a *numeric value* for a given instance variable, and the next object (of the same class) can contain a *character value* for the same instance variable. This characteristic is achieved through late binding. With **late binding** the data type of an attribute is not known until execution time or runtime. Therefore, two different object instances of the same class can contain values of different data types for the same attribute.

In contrast to the OODM's ability to use late binding, a conventional DBMS requires that a base data type be defined for each attribute at the time of its creation. For example, suppose that you want to define an INVENTORY to contain the following attributes: ITEM_TYPE, DESCRIPTION, VENDOR, WEIGHT, and PRICE. In a conventional DBMS you create a table named INVENTORY and assign a base data type to each attribute, as shown in Figure 11.31.

Recall from earlier chapters that, when working with conventional database systems, the designer must define the data type for each attribute *when the table structure is defined*. This approach to data type definition is called early binding. **Early binding** allows the database to check the data type for each of the attribute's values at compilation or definition time. For instance, the ITEM_TYPE attribute in Figure 11.31 is limited to numeric values. Similarly, the VENDOR attribute may contain only numeric values to match the primary key of some row in a VENDOR table with the same numeric value restriction.

FIGURE 11.31 INVENTORY TABLE WITH PREDETERMINED (BASE) DATA TYPES

Table: INVENTORY

Attributes	Conventional (Base) Data Type
ITEM_TYPE	Numeric
DESCRIPTION	Character
VENDOR	Numeric
WEIGHT	Numeric
PRICE	Numeric

Now let's take a look at Figure 11.32 to see how an OODM would handle this early-binding problem. As was true in the conventional database environment, the OODM allows the data types to be defined at creation time. However, quite *unlike* the conventional database, the OODM allows the data types to be user-defined ADTs. In this example of early binding, the abstract data types Inv_type, String_of_characters, Vendor, Weight, and Money are associated with the instance variables at definition time. Therefore, the designer may define the required operations for each data type. For example, the Weight data type can have methods to show the weight of the item in pounds, kilograms, and so on. Similarly, the Money data type may have methods to return the price as numbers or letters, denominated in U.S. dollars, Euros, or British pounds. (Remember that abstract data types are implemented through classes.)

FIGURE 11.32 INVENTORY CLASS WITH EARLY BINDING

Class: INVENTORY

Instance variables	Type
ITEM_TYPE	Inv_type
DESCRIPTION	String_of_characters
VENDOR	Vendor
WEIGHT	Weight
PRICE	Money

In a late-binding environment, the object's attribute data type is not known prior to its use. Therefore, an attribute can have any type of value assigned to it. Using the same basic data set described earlier, Figure 11.33 shows the attributes (instance variables) INV_TYPE, DESCRIPTION, VENDOR, WEIGHT, and PRICE without a prior data

type definition. Because no data types are predefined for the class instance variables, two different objects of the Inventory class may have different value types for the same attribute. For example, INV_TYPE can be assigned a character value in one object instance and a numeric value in the next one. Late binding also plays an important role in polymorphism, to allow the object to decide which method's implementation to use at run time.

FIGURE 11.33 **OODM INVENTORY CLASS WITH LATE BINDING**

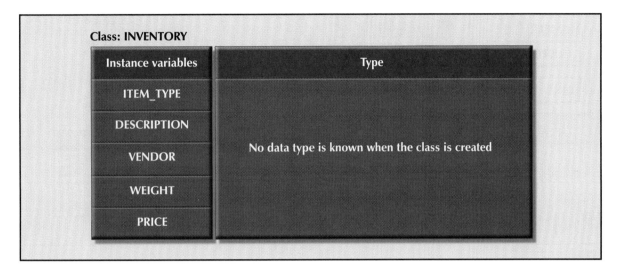

11.4.5 SUPPORT FOR VERSIONING

Versioning is an OODM feature that allows us to track the history of change in the state of an object. Versioning is a very powerful modeling feature, especially in computer-aided design (CAD) environments. For example, an engineer using CAD can load a machine component design in his/her workstation, make some changes, and see how those changes affect the component's operation. If the changes do not yield the expected results, the engineer can undo those changes and restore the component to its original state. Versioning is one of the reasons that the OODBMS is such a strong player in the CAD and computer-aided manufacturing (CAM) arenas.

11.5 OODM AND PREVIOUS DATA MODELS: SIMILARITIES AND DIFFERENCES

Although the OODM has much in common with relational or ER data models, the OODM introduces some fundamental differences. The following summary offers some detailed comparisons to help clarify the OODM characteristics we introduced in this chapter.

11.5.1 OBJECT, ENTITY, AND TUPLE

The OODM concept of *object* extends well beyond the concept of *entity* or *tuple* in other data models. Although an OODM object resembles the entity and the tuple in the ER and relational models, an OODM object has additional characteristics, such as behavior, inheritance, and encapsulation. Such OODM characteristics make OO modeling much more natural than ER and relational modeling. In fact, the ER and relational models often force the designer to create new artificial entities in order to represent real-world entities. For example, in the ER model an invoice is usually represented by two separate entities; the second (LINE) entity is usually weak because its existence depends on the first (INVOICE) entity and its primary key is partially derived from the INVOICE entity. (See Figure 11.34.)

FIGURE 11.34 AN INVOICE REPRESENTATION

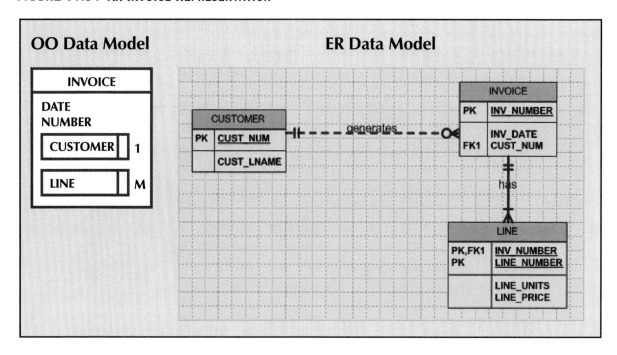

As you examine Figure 11.34, note that the ER approach requires the use of two different entities to model a single real-world INVOICE entity. Such an artificial construct is imposed by the relational model's inherent limitations. The ER model's artificial representation introduces additional overhead in the underlying system. In contrast, the OODM's INVOICE object is directly modeled as an object into the object space, or object schema.

11.5.2 CLASS, ENTITY SET, AND TABLE

The concept of *class* can be associated with the ER and relational models' concepts of *entity set* and *table*, respectively. However, class is a more powerful concept that allows not only the description of the data structure but also the description of the behavior of the class objects. A class also allows for both the concept and the implementation of abstract data types in the OODM. The ADT is a very powerful modeling tool because it allows the end user to create new data types and use them like any other base data type that accompanies a database. The ADT thus yields an increase in the *semantic* content of the objects being modeled.

11.5.3 ENCAPSULATION AND INHERITANCE

ADT brings us two other OO features that are not supported in previous models: encapsulation and inheritance. Classes are organized in class hierarchies. An object belonging to a class inherits all the properties of its superclasses. Encapsulation means that the data representation and the method's implementation are hidden from other objects and from the end user. In an OODM only the methods can access the instance variables. In contrast, the conventional system's data components or fields are directly accessible from the external environment.

Conventional models do not incorporate the methods found in the OODM. The closest thing to methods is the use of triggers and stored procedures in SQL databases. However, because triggers do not include the encapsulation and inheritance benefits that are typical of the object model's methods, triggers do not yield the same functionality as methods.

11.5.4 OBJECT ID

The object ID (OID) is not supported in either the ER or the relational model. Although database users may argue that Oracle Sequences and MS Access AutoNumber provide the same functionality as an OID, this argument is true *only* to the extent that they can be used to uniquely identify data elements. However, unlike the object model, in which the relationships are implicit, the relational model still uses value-based relationships such as

```
SELECT *
    INVOICE, INV_LINE
        WHERE INVOICE.INV_ID = INV_LINE.INV_ID;
```

The hierarchical and CODASYL models support some form of ID that can be considered similar to the OID, thus supporting the argument presented by some researchers who insist that the OO evolution is a step back on the road to the old pointer systems. Therefore, OO-based systems return us to the modeling and implementation complexities that were typical of the hierarchical and network models.

11.5.5 RELATIONSHIPS

The main property of any data model is found in its representation of relationships among the data components. The relationships in an OODM can be of two types: interclass references or class hierarchy inheritance. The ER and the relational models use a *value-based* relationship approach. Using a value-based approach means that a relationship among entities is established through a common value in one or several of the entity attributes. In contrast, the OODM uses the object ID, which is identity-based, to establish relationships among objects, and *such relationships are independent of the state of the object.* (While this property makes it easy to deal with the database objects at the end-user applications level, you may have concluded at this point that the price of this convenience is greater conceptual complexity.)

11.5.6 ACCESS METHODS

The ER and relational data models depend on the use of SQL to retrieve data from the database. SQL is a set-oriented query language that is based on a formally defined mathematical model. Given its set-oriented heritage, SQL uses associative access methods to retrieve related information from a database, based on the value of some of its attributes. For example, to retrieve a list of customer records based on the value of their year-to-date purchases, SQL would use:

```
SELECT *
    FROM CUSTOMER
        WHERE CUS_YTD_BUYS >= 5000;
```

If no CUS_YTD_BUYS value parameter is specified, SQL "understands" that condition to mean "any value," thus reducing the query statement to:

```
SELECT *
    FROM CUSTOMER;
```

As a consequence of having more semantics in its data model, the OODM produces a schema in which relations form part of the structure of the database. Accessing the structured object space resembles the record-at-a-time access of the old structured hierarchical and network models, especially if you use a 3GL or even the OOPL supported by the OODBMS. The OODM is suited to support both navigational (record-at-a-time) and set-oriented access. The navigational access is provided and implemented directly by the OODM through the OIDs. The OODM uses the OIDs to navigate through the object space structure developed by the designer.

Associative set-oriented access in the OODM must be provided through explicitly defined methods. Therefore, the designer must implement operations to manipulate the object instances in the object schema. The implementation of such operations will have an effect on performance and the database's ability to manage data. This is where the main problem of the object model appears: the lack of a universally accepted underlying mathematical model for data manipulation. Not having such a universal access standard hampers the OODM, because it forces each implementation to create its own version of an **object query language (OQL)**. The OQL is the database query language used by an OODBMS. Of course, different vendors will create different versions of OQL, and this in turn limits true interoperability. However, there are several groups, such as the Object Management Group (OMG, *www.omg.org*)[2], that are currently working on the development of standards for object oriented technology. For example, to facilitate modeling in an OO environment, the OMG has developed the Unified Modeling Language (UML) standard. UML represents an attempt to create a universal modeling notation to facilitate such activities as system development, data modeling, and network design. (See **Appendix B** on the student CD to learn more about UML.)

ADDITIONAL CD CONTENT

Appendix B provides an introduction to the Unified Modeling Language (UML). This supplement is located on the student CD, in the \Appendixes\Appendix B folder.

Although there is no standard way to manipulate sets with an OQL, the relational model's SQL2 standard did not provide ways to manipulate objects in an object oriented database, either. However, the mismatch between OQL and SQL has been reduced with the publication of the SQL3 or SQL-99 standard in 1999 by the American National Standards Institute (ANSI). SQL3 paves the way toward the integration of object oriented extensions within relational databases. For more information about this standard, see document number ANSI/ISO/IEC 9075, Parts 1 through 5, at *www.ansi.org*.

In previous sections we discussed the object oriented concepts that were derived from OOPLs. We used those concepts to establish the characteristics of the object oriented data model (OODM) and study its graphical representation. In this section, we compared the OODM to previous data models, and we explained that one of the major problems of the OODM is that it fails to conform to a universally accepted standard. Yet, in spite of the lack of standards, there is agreement about *minimal* OODBMS characteristics. We will explore those characteristics in the next section.

11.6 OBJECT ORIENTED DATABASE MANAGEMENT SYSTEMS

During the past few years, the data management and application environment has become far more complex than the one envisioned by the creators of the hierarchical, network, or relational DBMSs. Such complex application environments may best be served by an **object oriented database management system (OODBMS)**. The OODBMS is a database management system that integrates the benefits of typical database systems with the more powerful modeling and computational (programming) characteristics of the object oriented data model. (See Figure 11.35.)

[2] Check the OMG main Web site (*www.omg.org*) for the most recent object standards developments. Although the OMG develops standards, we see little evidence that the OMG standards are being widely adopted in the database modeling marketplace, which is the focus of our interest.

FIGURE 11.35 OBJECT ORIENTED DATABASE MANAGEMENT SYSTEMS

NOTE

A DBMS based on the object model may be labeled an object oriented database management system (OODBMS) or an object database management system (ODBMS). Given the frequent use of "OO" and "OODBMS" labels in the early stages of object oriented research, we will use the OODBMS label as a matter of personal preference.

OODBMS products are used to develop complex systems such as:

- Medical applications that handle digitized data such as x-rays, MRI scans, and ultrasounds, together with textual data used for medical research and patient medical history analysis.

- Financial applications in portfolio and risk management. Such applications yield a real-time view of data that is based on multiple computations and aggregations applied to data that are acquired from complex stock transactions across the world. These applications can handle "time series" data as a user-defined data type with its own internal representation and methods.

- Telecommunications applications such as network configuration management applications that automatically monitor, track, and reconfigure communications networks based on hundreds of parameters in real time. Companies such as Ericsson, Ameritech, and Bay Networks use OODBMSs to support their telecommunications management applications. Motorola's Iridium Global Communications System manages their complex network of satellites and ground stations using an OODBMS.

- The BaBar Physics experiment at the Stanford Linear Accelerator Center enters one terabyte of data per day into an OODBMS.

- Computer-aided design (CAD) and computer-aided manufacturing (CAM). Such applications make use of complex data relations as well as multiple data types.

- Computer-assisted software engineering (CASE) applications, which are designed to handle very large amounts of interrelated data.

- Multimedia applications, such as geographic information systems (GIS), that use video, sound, and high-quality graphics that require specialized data-management features such as intersect, inside, within, point, line, and polygon.

Many OODBMSs use a subset of the object oriented data model features. Therefore, those who create the OODBMS tend to select the OO features that best serve the particular OODBMS's purpose, such as support for early or late binding of the data types and methods, and support for single or multiple inheritance. Whatever the choices, the critical factor for a successful OODBMS implementation appears to be finding the best mix of OO and conventional DBMS features that will not sacrifice the benefits of either one.

11.6.1 FEATURES OF AN OBJECT ORIENTED DBMS

As shown in Figure 11.35, an OODBMS is the result of combining OO features, such as class inheritance, encapsulation, and polymorphism, with database features such as data integrity, security, persistence, transaction management, concurrency control, backup, recovery, data manipulation, and system tuning. The "Object oriented Database System Manifesto" (Atkinson et al., 1989)[3] was the first comprehensive attempt to define OODBMS features, and included 13 mandatory features as well as optional characteristics of the OODBMS. The 13 rules are divided into two sets: the first eight characterize an OO system, and the last five characterize a DBMS. The 13 rules are listed in Table 11.4. We will briefly discuss each of these rules.

TABLE 11.4 THE THIRTEEN OODBMS RULES

	RULES THAT MAKE IT AN OO SYSTEM
Rule 1	The system must support complex objects.
Rule 2	Object identity must be supported.
Rule 3	Objects must be encapsulated.
Rule 4	The system must support types or classes.
Rule 5	The system must support inheritance.
Rule 6	The system must avoid premature binding.
Rule 7	The system must be computationally complete.
Rule 8	The system must be extensible.
	RULES THAT MAKE IT A DBMS
Rule 9	The system must be able to remember data locations.
Rule 10	The system must be able to manage very large databases.
Rule 11	The system must accept concurrent users.
Rule 12	The system must be able to recover from hardware and software failures.
Rule 13	Data query must be simple.

- *Rule 1. The system must support complex objects.* It must be possible to construct complex objects from existing objects. Examples of such object constructors are sets, lists, and tuples that allow the user to define aggregations of objects as attributes.

[3] Malcolm Atkinson et al., "The Object Oriented Database System Manifesto." This white paper first presented at the First International Conference on Deductive and Object Oriented Databases, Kyoto, Japan, in 1989 may be downloaded from the Object Data Management Group Web site, at *www.odmg.org*. Select White papers, then Database, then OO Database System Manifesto. You will also find a series of follow-up white papers that explore the 1997–1999 OO rule extensions and standards. These extensions and standards are designed to augment the manifesto, and none claims to replace any part thereof.

- *Rule 2. Object identity must be supported.* The OID must be independent of the object's state. This feature allows the system to compare objects at two different levels: comparing the OID (identical objects) and comparing the object's state (equal or shallow equal objects).

> **NOTE**
>
> If the objects have different OIDs but their attribute values are equal, the objects are not identical, but they are considered to be "shallow equal." To use an analogy, identical twins are alike, yet different.

- *Rule 3. Objects must be encapsulated.* Objects have a public interface, but private implementation of data and methods. The encapsulation feature ensures that only the public aspect of the object is seen, while the implementation details are hidden.

- *Rule 4. The system must support types or classes.* This rule allows the designer to choose whether the system supports types or classes. Types are used mainly at compile time to check type errors in attribute value assignments. Classes are used to store and manipulate similar objects at execution time. In other words, class is a more dynamic concept, and type is a more static one.

- *Rule 5. The system must support inheritance.* An object must inherit the properties of its superclasses in the class hierarchy. This property ensures code reusability.

- *Rule 6. The system must avoid premature binding.* This feature allows us to use the same method's name in different classes. The OO system decides which implementation to access at run time, on the basis of the class to which the object belongs. This feature is also known as late binding or dynamic binding.

- *Rule 7. The system must be computationally complete.* The basic notions of programming languages are augmented by features common to the database data manipulation language (DML), thereby allowing us to express any type of operation in the language.

- *Rule 8. The system must be extensible.* The final OO feature concerns its ability to define new types. There is no management distinction between user-defined types and system-defined types.

- *Rule 9. The system must be able to remember data locations.* The conventional DBMS stores its data permanently on disk; that is, the DBMS displays data persistence. OO systems usually keep the entire object space in memory; once the system is shut down, the entire object space is lost. Much of the OODBMS research has focused on finding a way to permanently store objects and retrieve them from secondary storage (disk).

- *Rule 10. The system must be able to manage very large databases.* Typical OO systems limit the object space to the amount of primary memory available. For example, Smalltalk cannot handle objects larger than 64K. Therefore, a critical OODBMS feature is to optimize the management of secondary storage devices by using buffers, indexes, data clustering, and access path selection techniques.

- *Rule 11. The system must support concurrent users.* Conventional DBMSs are especially capable in this area. The OODBMS must support the same level of concurrency as conventional systems.

- *Rule 12. The system must be able to recover from hardware and software failures.* The OODBMS must offer the same level of protection from hardware and software failures that the traditional DBMS provides; that is, the OODBMS must provide support for automated backup and recovery tools.

- *Rule 13. Data query must be simple.* Efficient querying is one of the most important features of any DBMS. Relational DBMSs have provided a standard database query method through SQL, and the OODBMS must provide an object query language (OQL) with similar capability.

Optional OODBMS features include:

- *Support for multiple inheritance.* Multiple inheritance introduces greater complexity by requiring the system to manage potentially conflicting properties between classes and subclasses.

- *Support for distributed OODBMSs.* The trend toward systems application integration constitutes a powerful argument in favor of distributed databases. If the OODBMS is to be integrated seamlessly with other systems through networks, the database must support some degree of distribution.

- *Support for versioning.* Versioning is a new characteristic of the OODBMS that is especially useful in such applications as CAD or CAM. Versioning allows us to maintain a history that tracks all object transformations. Therefore, we can browse through all the different object states, in effect letting us walk back and forth in time.

11.6.2 ORACLE OBJECT EXAMPLES

Oracle databases (since version 8) support object oriented extensions. These extensions allow users to create object types using DDL commands. Such object types are the equivalent of classes (see Section 11.3.6) or abstract data types (ADT) in the object model (see Section 11.3.10). Oracle supports various object types:

- *Column type.* This object type provides data type extensibility by allowing the user to define his or her own data types. A column object type is the equivalent of an abstract data type (ADT). An ADT can be used when defining a column data type within a relational table. An ADT can also have methods associated with it; such methods are implemented using PL/SQL, C++, or Java.

- *Row type.* A row type object is used to define an object table object. An **object table** is the equivalent of a relational table composed of many rows, where each row is an object of the same type. Each row object has a unique system generated object ID (OID) or object identifier (see Section 11.3.2).

- *Collection objects.* Oracle also provides support for two types of collection objects (note the discussion that accompanies Figure 11.29, in Section 11.4.3).
 - *Variable length arrays (VARRAY).* Enables the user to create an object type as an array of objects of a given type.
 - *Nested tables.* Allows the creation of a relational table in which one of the attributes is a table. Specifically, a relational table contains an attribute with an object table type.

To illustrate creating and using various object types, we will use Oracle9*i* in the following discussion.

Column Type

A column type is basically a new data type (abstract data type) you can use when you define an attribute in a table. By creating a column type you are defining a new class with shared attributes and methods. To illustrate the use of column types, let's create two column types named: T_ADDRESS and T_JOB, respectively. The T_ADDRESS column type will contain the street, city, state, and zip code attributes. The T_JOB column type will be used to store data about a job (company name, start date, end date, and monthly salary). The T_JOB column type will have two methods: *monthsonjob*, which returns the number of months spent in a given job, and *totalearned*, which returns the result of the multiplication of the number of months employed and the monthly salary. Figure 11.36 shows use of the CREATE TYPE command to create the two column types.

FIGURE 11.36 CREATION OF THE **T_ADDRESS** AND **T_JOB** COLUMN DATA TYPES

```
Oracle SQL*Plus
File  Edit  Search  Options  Help
SQL> CREATE OR REPLACE TYPE T_ADDRESS AS OBJECT (
  2   STREET     VARCHAR2(20),
  3   CITY       VARCHAR2(15),
  4   STATE      CHAR(2),
  5   ZIP        CHAR(5)
  6  );
  7  /

Type created.

SQL> CREATE OR REPLACE TYPE T_JOB AS OBJECT (
  2   CNAME   VARCHAR(15),
  3   START_DATE  DATE,
  4   END_DATE    DATE,
  5   MONTHLYSALARY NUMBER(7,2),
  6   MEMBER FUNCTION MONTHSONJOB RETURN NUMBER,
  7   MEMBER FUNCTION TOTALEARNED RETURN NUMBER
  8  );
  9  /

Type created.

SQL>
```

In Figure 11.36, note that the T_JOB column data type creation command includes references to the methods that are to be created. By using the MEMBER FUNCTION clause, you define the name of the method to be created, any optional parameters that may be required (shown in parentheses), and the type of value (such as number or character) to be returned. To actually create the methods, you use the CREATE TYPE BODY command, as shown in Figure 11.37.

FIGURE 11.37 CREATION OF THE **T_JOB** METHODS

```
Oracle SQL*Plus
File  Edit  Search  Options  Help
SQL> CREATE OR REPLACE TYPE BODY T_JOB AS
  2   MEMBER FUNCTION MONTHSONJOB RETURN NUMBER IS
  3    BEGIN
  4     RETURN TRUNC(((SELF.END_DATE - SELF.START_DATE)/30),1);
  5    END;
  6   MEMBER FUNCTION TOTALEARNED RETURN NUMBER IS
  7    BEGIN
  8     RETURN(MONTHSONJOB*SELF.MONTHLYSALARY);
  9    END;
 10  END ;
 11  /

Type body created.

SQL>
```

As you examine Figure 11.37, note that the method definition is done using standard PL/SQL commands. You might also define methods using other languages, such as C++ or Java. Each method definition starts with the MEMBER FUNCTION keywords. The actual method code is contained within the BEGIN and END clauses.

Figure 11.38 shows the creation of a WORKER table that uses the T_ADDRESS and T_JOB column types that were defined earlier.

FIGURE 11.38 CREATION OF THE WORKER TABLE USING T_ADDRESS AND T_JOB COLUMN TYPES

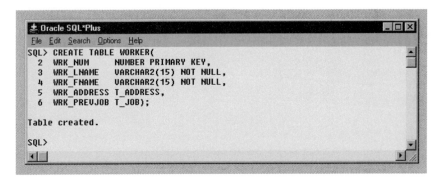

Once you have created the table, you use standard SQL commands to insert data. However, to enter data in a column type attribute, you must use the column type name as indicated in Figure 11.39, with the INSERT statements.

FIGURE 11.39 WORKING WITH COLUMN TYPES IN THE WORKER TABLE

```
Oracle SQL*Plus
File  Edit  Search  Options  Help
SQL> INSERT INTO WORKER
  2  VALUES (1,'VILLEGAS','ROBERTO',
  3        T_ADDRESS('123 Main St.','London','OH','76987'),
  4        T_JOB('GLOBALCOM','01-AUG-1999','03-DEC-2002',3000));

1 row created.

SQL>
SQL> INSERT INTO WORKER
  2  VALUES (2,'GORTTY','JANE',
  3        T_ADDRESS('453 Sulla St.','London','OH','76987'),
  4        T_JOB('SURESTART','15-MAY-2000','15-OCT-2002',2500));

1 row created.

SQL>
SQL> INSERT INTO WORKER
  2  VALUES (3,'SMITH','MARGO',
  3        T_ADDRESS('7854 Court Av.','London','OH','76984'),
  4        T_JOB('TELCO-R-US','01-JUL-2001','15-SEP-2002',4000));

1 row created.

SQL> SELECT   WRK_FNAME || ', ' || WRK_LNAME   AS NAME,
  2        W.WRK_ADDRESS.STREET           AS STREET,
  3        W.WRK_ADDRESS.CITY    AS CITY,
  4        W.WRK_ADDRESS.STATE   AS STATE,
  5        W.WRK_PREVJOB.MONTHSONJOB() AS MONTHS,
  6        W.WRK_PREVJOB.TOTALEARNED() AS TOTAL
  7  FROM     WORKER W;

NAME                          STREET               CITY             ST    MONTHS      TOTAL
----------------------------- -------------------- ---------------- --  ---------- ----------
ROBERTO, VILLEGAS             123 Main St.         London           OH      40.6     121800
JANE, GORTTY                  453 Sulla St.        London           OH      29.4      73500
MARGO, SMITH                  7854 Court Av.       London           OH      14.7      58800

SQL>
```

To retrieve data from a column type-based attribute using a SELECT statement, you must first declare an alias for the table. In this case we have used the alias W. Next, you can refer to a column type attribute or method using a dot-separated notation such as W.WRK_ADDRESS.STREET or W.WRK_PREVJOB.TOTALEARNED(), as shown in Figure 11.39.

Row Type

A row type enables you to create a table in which each row is an object instance. Such a table is called an object table to differentiate it from a relational table. To demonstrate the use of row types, let's create the OTBL_BAND object table in which each row is a musician object. To accomplish this task, let's first create a T_MUSICIAN column type. (Remember that a column type is an object.) The T_MUSICIAN column type will have an AGE method that uses the system date and the musician's date of birth to return the age of each musician. To create the object table, you use the CREATE TABLE OF command, as shown in Figure 11.40.

FIGURE 11.40 CREATION OF THE OTBL_BAND OBJECT TABLE

```
Oracle SQL*Plus
File  Edit  Search  Options  Help
SQL> CREATE TYPE T_MUSICIAN AS OBJECT {
  2   NAME              VARCHAR(15),
  3   DOB               DATE,
  4   INSTRUMENT        CHAR(8),
  5   MEMBER FUNCTION AGE RETURN NUMBER
  6   );
  7   /

Type created.

SQL> CREATE OR REPLACE TYPE BODY T_MUSICIAN AS
  2   MEMBER FUNCTION AGE RETURN NUMBER IS
  3     BEGIN
  4       RETURN TRUNC(((SYSDATE - SELF.DOB)/365),1);
  5     END;
  6   END ;
  7   /

Type body created.

SQL> CREATE TABLE OTBL_BAND OF T_MUSICIAN OBJECT ID IS SYSTEM GENERATED;

Table created.

SQL>
```

To insert data to the newly created object table, you use the INSERT command, as shown in Figure 11.41. Note that you do not have to identify the column type as was required with the WORKER table. The difference is that the WORKER table is a relational table containing attributes with abstract data types. In such cases you need to specify the abstract data (column type). In the case of the OTBL_BAND table, you are adding objects to an *object table*. However, you still must use an alias to invoke a method (see Figure 11.41).

VARRAY Collection Type

The variable length array (VARRAY) creates a new object type that represents a collection of objects of a similar type (objects or base data types). For example, an employee may have multiple dependents. In this case, you can store all the dependents in an array for each of the employees. Figure 11.42 shows the commands required to create the T_DEPENDLIST variable array object type and the EMP3 table containing the E_DEPENDENTS attribute, which uses the T_DEPENDLIST data type. Note that we have defined the variable array to hold a maximum of 10 dependent names.

FIGURE 11.41 WORKING WITH THE **OTBL_BAND** OBJECT TABLE

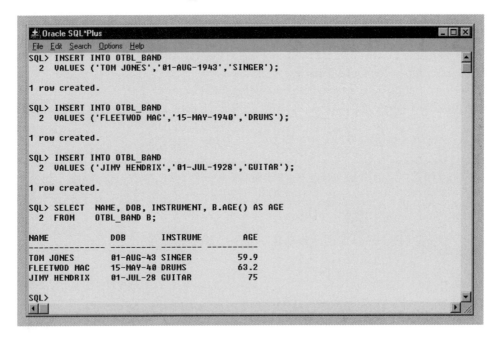

FIGURE 11.42 CREATING AND WORKING WITH THE **VARRAY** OBJECT TYPE

```
Oracle SQL*Plus
File  Edit  Search  Options  Help

SQL> CREATE TYPE T_DEPENDLIST AS VARRAY(10) OF VARCHAR(10);
  2  /

Type created.

SQL> CREATE TABLE EMP3 (
  2    E_LNAME         VARCHAR(15),
  3    E_FNAME         VARCHAR(15),
  4    E_DEPENDENTS    T_DEPENDLIST
  5  );

Table created.

SQL> INSERT INTO EMP3
  2  VALUES ('SMITH','ALAN',T_DEPENDLIST('STAN','MARY'));

1 row created.

SQL> INSERT INTO EMP3
  2  VALUES ('PRATT','ALYSSA',T_DEPENDLIST('CARLOS','PETER','DEB'));

1 row created.

SQL> SELECT  E_LNAME, E_FNAME, E_DEPENDENTS
  2  FROM     EMP3 E;

E_LNAME         E_FNAME
--------------- ---------------
E_DEPENDENTS
--------------------------------------------------------------------------------
SMITH           ALAN
T_DEPENDLIST('STAN', 'MARY')

PRATT           ALYSSA
T_DEPENDLIST('CARLOS', 'PETER', 'DEB')

SQL>
```

Nested Table Collection Type

If you have related data that are more extensive than you would expect to find in an array, you could use a nested table. A nested table is created when an attribute within a relational table definition (CREATE TABLE) is assigned a table data type. For example, Figure 11.43 shows the creation of the EMP4 table containing the E_DEPENDENTS attribute, which uses the T_DEPTAB data type. In turn, the T_DEPTAB data type is defined as a table type. Conceptually speaking, the attribute is, in effect, a table.

FIGURE 11.43 CREATING AND WORKING WITH A NESTED TABLE OBJECT TYPE

```
Oracle SQL*Plus
File  Edit  Search  Options  Help
SQL> CREATE TYPE T_DEPDNT AS OBJECT (
  2  NAME    VARCHAR(12)
  3  )
  4  /

Type created.

SQL> CREATE TYPE T_DEPTAB AS TABLE OF T_DEPDNT;
  2  /

Type created.

SQL> CREATE TABLE EMP4 (
  2  E_LNAME          VARCHAR(15),
  3  E_FNAME          VARCHAR(15),
  4  E_DEPENDENTS    T_DEPTAB)
  5  NESTED TABLE E_DEPENDENTS STORE AS DEPTABLE;

Table created.

SQL> INSERT INTO EMP4
  2  VALUES ('SMITH','ALAN',T_DEPTAB(T_DEPDNT('STAN'),T_DEPDNT('BILL')));

1 row created.

SQL> INSERT INTO EMP4
  2  VALUES ('PRATT','ALYSSA',T_DEPTAB(T_DEPDNT('CARLOS'),T_DEPDNT('PETER'),T_DEPDNT('DEB')));

1 row created.

SQL> SELECT * FROM EMP4;

E_LNAME          E_FNAME
---------------  ---------------
E_DEPENDENTS(NAME)
---------------------------------------------------------------------------
SMITH            ALAN
T_DEPTAB(T_DEPDNT('STAN'), T_DEPDNT('BILL'))

PRATT            ALYSSA
T_DEPTAB(T_DEPDNT('CARLOS'), T_DEPDNT('PETER'), T_DEPDNT('DEB'))

SQL> |
```

11.7 HOW OBJECT ORIENTATION AFFECTS DATABASE DESIGN

A conventional relational database design process involves the application of ER modeling and normalization techniques to develop and implement a design. During such a design process, the emphasis is placed on modeling real-world objects through simple tabular relations, usually presented in 3NF. Unfortunately, as we have already seen, the relational and ER models sometimes cannot adequately represent some objects. Consequently, the ER model makes use of such constructs as bridge (composite) entities that widen the semantic gap between the real-world objects and their corresponding representations.

You may have noticed that our database design process generally focuses on the identification of the data elements, rather than including the data operations as part of the process. In fact, the definition of data constraints and data transformations is usually considered late in the database design process. Such definitions are implemented by external application program code. In short, operations are not a part of the database model.

Why does the conventional model tolerate and even require the existence of the data/procedures dichotomy? After all, the idea of object oriented design had been contemplated even in the classical database environment. The reason is simple: until recently, database designers simply had no access to tools that bonded data and procedures.

The object oriented database design approach answers the problem of a split between data and procedures by providing both the data identification and the procedures or manipulations to be performed on that data. Object oriented database design forces us to think of data and procedures as a self-contained entity. Specifically, the OO design requires the database description to include the objects and their data representation, constraints, and operations. Such a design can produce a much more complete and meaningful description of the database than was possible with conventional database design.

OO design is iterative and incremental in nature. The database designer identifies each real-world object and defines its internal data representation, semantic constraints, and operations. Next, the designer groups similar objects in classes and implements the constraints and operations through methods. At this point the designer faces two major challenges:

1. Build the class hierarchy or the class lattice (if multiple inheritance is allowed), using base data types and existing classes. This task will define the superclass-subclass relationships.

2. Define the interclass relationships (attribute-class links), using both base data types and ADTs.

The importance of these tasks cannot be overestimated, because the better the use of the class hierarchy and the treatment of the interclass relationships, the more flexible and closer to the real world the final model will be.

Code reusability does not come easy. One of the hardest tasks in OODB design is creating the class hierarchy—using existing classes to construct new ones. Future DBAs will have to develop specialized skills to properly perform this task and to incorporate code that represents data behavior. Thus DBAs are likely to become surrogate database programmers who must define data-intrinsic behavior. The role of DBAs is likely to change when they take over some of the programming burden of defining and implementing operations that affect the data.

Both DBAs and designers face additional problems: in contrast to the relational or ER design processes, there are few computerized OODB design tools, and, if the design is to be implemented in any of the conventional DBMSs, it must be translated carefully, because conventional databases do not support abstract data types, non-normalized data, inheritance, encapsulation, or other OO features.

As is true in any of the object oriented technologies, the lack of standards also affects OO database design. There is neither a widely accepted standard methodology to guide the design process, nor a set of rules (like the normalization rules in the relational model) to evaluate the design. This situation is improving. The Object Management Group (OMG), mentioned earlier, produces vendor-independent standards and specifications for object-based systems and components. The OMG created the Unified Modeling Language (UML), a graphical language for the modeling, design, and visualization of object oriented systems. UML is used to model not only the database component of a system, but also its processes, modules, network components, and the interaction among all of them. OMG also created object standards that define the Object Management Architecture (OMA), which allows the interoperation of objects across diverse systems and platforms. The OMA standard includes the Common Object Request Broker Architecture (CORBA) and Common Object Services Specifications (COSS). This framework is used by OODBMS vendors and developers to implement systems that are highly interoperable with other OODBMSs, as well as RDBMSs and older DBMS systems.

Some vendors are already offering products, such as IBM's System Object Model (SOM) and HP's Object Request Broker (ORB), that comply with the OMG's CORBA and COSS specifications. Other object architectures have emerged as alternatives to the concerted standards efforts, especially Microsoft's object linking and embedding (OLE) and Component Object Model (COM). Although the OLE/COM specification is not a standards-based effort, the sheer established market volume is making this option the de facto object standard for the Microsoft Windows environment.

11.8 OODBMS: ADVANTAGES AND DISADVANTAGES

Compared to the RDBMS market share, OODBMSs have a long way to go before they can claim double-digit market percentage. In fact, at this writing the OODBMS occupies a strong niche market, as the Apple "Mac" does in the microcomputing arena. Like the Mac's impact on microcomputing, the OODBMS has been the vehicle for technological innovation, but has not been the beneficiary of market share growth based on its technological innovations. Yet, in spite of its lack of market share, the OODBMS is worth examining, especially because its OO features drive the changes in database technology that define today's object/relational DBMS.

Part of the OODBMS's lack of market acceptance is that the RDBMS has incorporated many OO features while retaining its conceptual simplicity, thus diminishing the OODBMS's allure. Nevertheless, as long as the RDBMS does not incorporate C. J. Date's recommended domain implementation, the OODBMS offers benefits that are worth examining.[4] Most of those benefits are expressed in terms of the complex object management capabilities we have explored in some detail. To obtain those benefits, the OODBMS depends on the use of an OOPL. That is why we examine some of the OODBMS's benefits with reference to programming issues.

Advantages

- OODBMSs allow the inclusion of more semantic information in the database, thus providing a more natural and realistic representation of real-world objects.

- OODBMSs provide an edge in the support for complex objects, which makes them especially desirable in specialized application areas. Conventional databases simply lack the ability to provide efficient applications in CAD, CAM, medical imaging, spatial imaging, and specialized multimedia environments.

- OODBMSs permit the extensibility of base data types, thereby increasing both the database functionality and its modeling capabilities.

- *If the platform allows efficient caching,* OODBMSs provide dramatic performance improvements compared to relational database systems when managing complex objects.

- Versioning is a useful feature for specialized applications such as CAD, CAM, medical imaging, spatial imaging, engineering, text management, and desktop publishing.

- The reusability of classes allows for faster development and easier maintenance of the database and its applications.

- Faster application development time is obtained through inheritance and reusability. This benefit is obtained only after mastering the use of OO development features such as:

 - The proper use of the class hierarchy; for example, how to use existing classes to create new classes.

 - OO design methodology.

- The OODBMS provides a possible solution to the problem of integrating existing and future DBMSs into a single environment. This solution is based on its strong data-abstraction capabilities and its promise of portability.

[4] See C. J. Date's "Back to the Relational Future," *http://www.dpbd.com/vault/9808date.html.*

Disadvantages

- OODBMSs face strong and effective opposition from the firmly established RDBMSs, especially when those RDBMSs—such as IBM's DB2 Universal Database and Oracle9i—incorporate many OO features that would otherwise have given the OODBMS the clear competitive edge in a complex data environment. Therefore, the OODBMS's design and implementation complexities become more difficult to justify.

- The OODBMS is based on the object model—which lacks the solid theoretical foundation of the relational model on which the RDBMS is built.

- In some sense, OODBMSs are considered a throwback to the traditional pointer systems used by hierarchical and network models. This criticism is not quite on the mark when it associates the pointer system with the navigational data manipulation style and fixed access paths that led to the relational system's dominance. Nevertheless, the *complexity* of the OODBMS pointer systems cannot be denied.

- OODBMSs do not provide a standard ad hoc query language, as relational systems do. At this point, the development of the object query language (OQL) is far from complete. Some OODBMS implementations are beginning to provide extensions to the relational SQL to make the integration of the OODBMS and RDBMS possible.

- The relational DBMS provides a comprehensive solution to business database design and management needs, supplying both a data model and a set of fairly straightforward normalization rules for designing and evaluating relational databases. OODBMSs do not yet provide a similar set of tools.

- The initial learning curve for the OODBMS is steep. If you consider the direct training costs and the time it takes to fully master the uses and advantages of object orientation, you will appreciate why OODBMSs seldom are rated as the first option when solutions are sought for noncomplex business-oriented problems.

- The OODBMS's low market presence, combined with its steep learning curve, means that there are few people who are qualified to use the presumed power of OO technology. Most of the technology is currently focused on engineering application areas of software development. Therefore, only companies with the right mix of resources (money, time, and qualified personnel) can afford to invest in OO technology.

- The lack of compatibility between different OODBMSs makes switching from one piece of software to another very difficult. With RDBMSs, different products are very similar, and switching from one to another is relatively easy.

A few years ago, we speculated that future systems would manage objects with embedded data and methods, rather than with records, tuples, or files. We also suggested that, although the portability details were not clear yet, they would have a major and lasting impact on how we would design and use databases. Given the benefit of hindsight, we now know that the OODBMS's reach has been limited by the object-relational DBMS's successful integration of many OO concepts. In any case, the OODBMS has had a major impact on how we view and manage databases, and the battle of the relational and object titans is far from over. Finally, because the object concepts are likely to remain the focus for future DBMS developments, they continue to be worth understanding.

11.9 HOW OO CONCEPTS HAVE INFLUENCED THE RELATIONAL MODEL

Most relational databases are designed to serve general business applications that require ad hoc queries and easy interaction. The data types encountered in such applications are well defined and are easily represented in common tabular formats with equally common short and well-defined transactions. However, RDBMSs are not as well suited as OODBMSs to the complex requirements of some applications, and the RDBMS is beginning to reach its limits in a business data environment that is changing with the advent of mixed-media data storage and retrieval.

The fast-changing data environment has forced relational model advocates to respond to the OO challenge by extending the relational model's conceptual reach. The result of their efforts is usually referred to as the extended relational model (ERM) or, more appropriately, the object/relational model (O/RM). Although this O/RM effort is still a work in progress, its basic features provide support for:

- Extensibility of new user-defined (abstract) data types
- Complex objects
- Inheritance
- Procedure calls (rules or triggers)
- System-generated identifiers (OID surrogates)

This is not an exhaustive list of all the extensions added to the relational model, nor do all extended relational models incorporate all the listed additions. However, this list contains the most crucial and desirable extended relational features.

NOTE

It's worth noting again that C. J. Date's "Third Manifesto" is based on Date's observation that the relational model already contains the desired capabilities through its support of domains. Therefore, the implementation of such domain support will yield the benefits now claimed for the OO "extensions" of the relational model. (See C. J. Date's April, 1999 "Back to the Relational Future: The Third Manifesto is Ready for Prime Time," *http://www.dbpd.com/vault/9808date.html*.) However, at this point, the relational domain implementations have not (yet?) been developed commercially, while the OO "extensions" to the relational database model are a commercial fact of life.

These enhancements to the relational model enhancements are based on the following concepts:

- Semantic and object oriented concepts are necessary to support the new generation of applications— especially if those applications will be deployed through the Internet.
- These concepts can and must be added to the relational model.
- The benefits of the relational model must be preserved to protect the investment in relational technology and to provide downward compatibility.

Most current extended relational DBMSs conform to the notions expressed in C. J. Date's "Third Manifesto" (see preceding Note). They also provide additional and very useful features, such as:

- Oracle Corporation and IBM have developed suites of products marketed as Universal Database Servers. Although the Universal Database Server is not a pure object oriented DBMS—it lacks the object storage component—this product supports complex data types such as multimedia data and spatial data, and it's Internet-ready. The Internet feature allows users to query the database using the World Wide Web (WWW). Oracle9*i* also includes support for object oriented extensions and storage. IBM's DB2 Universal Database Server has similar capabilities.
- IBM's DB2 Universal Database system is a proven database that is used by many Fortune 1000 corporations. IBM's system supports digitized data (video, audio) as well as user-defined data types and procedures. The Universal Database is also being positioned as a key player in the Internet area with its support for Web access and Java programming interfaces.

11.10 THE NEXT GENERATION OF DATABASE MANAGEMENT SYSTEMS

The adaptation of OO concepts in several computer-related areas has changed both systems design and system behavior. The next generation of DBMSs is very likely to incorporate features borrowed from object oriented database systems, artificial intelligence systems, expert systems, distributed databases, and the Internet.

OODBMSs represent only one step toward the next generation of database systems. The use of OO concepts will enable future DBMSs to handle more complex problems with both normalized and nonnormalized data. The extensibility of database systems is one of the many major object oriented contributions that enable databases to support new data types such as sets, lists, arrays, video, bitmap pictures, voice, and maps. The SQL3 standard provides such extensibility by supporting user-defined data types in addition to its predefined data types (numeric, integer, string, and so on). For example, in SQL3 a DBA can create a new abstract data type that represents a collection of objects, then use that data type in a table definition. This procedure enables a database column to contain a *collection* of values instead of a single value.

Recent market history indicates that the OODBMS will probably continue to occupy a niche within the database market. This niche will be characterized by applications that require very large amounts of data with several complex relations and with specialized data types. For example, the OODBMS seems likely to maintain its standing in CAD, CAM, computer-integrated manufacturing, specialized multimedia applications, medical applications, architectural applications, mapping applications, simulation modeling, and scientific applications.

However, current market conditions seem to dictate that the object/relational databases will become dominant in most complex business applications, based on the need to maintain compatibility with existing systems, the universal acceptance of the relational model as a standard, and the sheer weight of its considerable market share.

SUMMARY

Object oriented concepts are derived from OO programming languages, which were an alternative to traditional programming methods. These concepts allowed their users to closely model the real world, thereby improving programmer productivity.

An object is a conceptual representation of a real-world entity that has embedded data representation (attributes) and behavior (methods). Each object has a unique ID, which is independent of the object's attributes. An object has attributes (instance variables), and each attribute can reference another object. The state of an object is the set of values that the object has at a given time. Methods implement the behavior of objects. Methods are invoked through messages. The methods and the data representation are encapsulated to be hidden from external sources.

Similar objects are grouped in classes. The class is a collection of like objects with shared structure (data) and behavior (methods). Each object in a class is a class instance or object instance. The collection of messages that an object can receive is the object's protocol. Classes are organized into a class hierarchy. An object belongs to a class in the class hierarchy. The object inherits the instance variables (attributes) and methods of all the superclasses above it. Single inheritance allows an object to have only one parent superclass. Multiple inheritance allows an object to have two or more parent superclasses. Polymorphism is the characteristic by which different objects can respond to the same message in different ways.

Abstract data types are implemented through classes. An ADT allows the end user to describe and manipulate a set of similar objects with encapsulated data representation and behavior. A complex object is formed by several different objects in complex relations.

The OO data model allows the designer to create a more authentic representation of real-world objects. The OODM supports the notions of encapsulation, inheritance, OID, complex objects, and extensibility of data types. The object schema, or object space, describes the database. The OODM uses the OID to relate objects and to navigate through the object space of classes and subclasses. The OODM does not need JOINs to relate data, as the relational model does.

OODBMSs combine OO features with database features. The critical factor of an OODBMS is attaining the right balance between object oriented and DBMS features without losing the benefits of either.

OODBMSs yield several benefits over conventional DBMSs. The benefits include more semantic information in the database, support for complex objects, better performance in complex applications, reusability of classes, and extensibility of the data types supported by the database. The OODBMS's current shortcomings include the lack of a standard ad hoc query language, a steep learning curve, and a lack of people with the right skills.

Relational database advocates have extended the relational model to support several OO features. The extended relational model is based on the relational model and provides support for complex objects, extensibility of new user-defined data types, inheritance, methods, and system-generated IDs or OID surrogates.

KEY TERMS

abstract data type (ADT)	inheritance	object oriented programming (OOP)
associative object	instance variable	object oriented programming language (OOPL)
base data type	interobject relationship	object query language (OQL)
class	interrogate	object space (object schema)
class hierarchy	intersection class	object state
class instance	late binding	object table
class lattice	message	polymorphism
collection object	method	protocol
complex object	multiple inheritance	referential object sharing
composite object	object	simple object
compound object	object ID (OID)	single inheritance
conventional data type	object instance	subclass
domain	object orientation	superclass
early binding	object oriented data model (OODM)	versioning
encapsulation	object oriented database management system (OODBMS)	
extensible		
hybrid object		

REVIEW QUESTIONS

1. Discuss the evolution of object oriented concepts. Then explain how those concepts have affected computer-related activities.

2. How would you define object orientation? What are some of its benefits? How are OO programming languages related to object orientation?

3. Define and describe the following:
 a. Object
 b. Attributes
 c. Object state
 d. Object ID (OID)

4. Define and contrast the concepts of method and message. What OO concept provides the differentiation between a method and a message? Give examples.

5. Explain how encapsulation provides a contrast to traditional programming constructs such as record definition. What benefits are obtained through encapsulation? Give an example.

6. Using an example, illustrate the concepts of class and class instances.

7. What is a class protocol, and how is it related to the concepts of methods and classes? Draw a diagram to show the relationships among these OO concepts: object, class, instance variables, methods, object state, object ID, behavior, protocol, and messages.

8. Define the concepts of class hierarchy, superclasses, and subclasses. Then explain the concept of inheritance and the different types of inheritance. Use examples in your explanations.

9. Define and explain the concepts of method overriding and polymorphism. Use examples in your explanations.

10. Explain the concept of abstract data types and how they differ from traditional or base data types. What is the relationship between a type and a class in OO systems?

11. What are the five minimum attributes of an OO data model?

12. Describe the difference between early and late binding and how each affects the object oriented data model. Give examples.

13. What is an object space? Using a graphic representation of objects, depict the relationship(s) that exist between a student taking several courses and a course taken by several students. What type of object is needed to depict this relationship?

14. Compare and contrast the OODM with the ER and relational models. How is a weak entity represented in the OODM? Give examples.

15. Name and describe the 13 mandatory features of an OODBMS.

16. What are the advantages and disadvantages of an OODBMS?

17. Explain how OO concepts affect database design. How does the OO environment affect the DBA's role?

18. What are the essential differences between the relational database model and the object database model?

19. Using a simple invoicing system as your point of departure, explain how its representation in an entity relationship model (ERM) differs from its representation in an object data model (ODM). (*Hint:* Check Figure 11.34.)

20. What are the essential differences between an RDBMS and an OODBMS?

21. Discuss the object/relational model's characteristics.

P R O B L E M S

1. Convert the following relational database tables to the equivalent OO conceptual representation. Explain each of your conversions with diagram. (*Note:* The RRE Trucking Company database includes the three tables shown in Figure P11.1.)

A D D I T I O N A L C D C O N T E N T

The databases used in this problem section are located in the **Databases\Student\Ch11** folder on the student CD.

FIGURE P11.1 THE RRE TRUCKING COMPANY DATABASE

Table name: TRUCK **Database name: Ch11_RRE_Trucking**

VEHICLE_NUM	BASE_CODE	TYPE_CODE	VEHICLE_MILES	VEHICLE_BUY_DATE	VEHICLE_VIN
5001	101	1	162,123.5	08-Nov-01	AA-322-12212-W11
5002	102	1	276,984.3	23-Mar-99	AC-342-22134-Q23
5003	101	2	212,346.6	27-Dec-00	AC-445-78656-Z99
5004	101	1	99,894.3	21-Feb-01	WQ-112-23144-T34
5005	103	2	245,673.1	15-Apr-00	FR-998-32245-W12
5006	101	6	293,245.7	30-Aug-99	AD-456-00845-R45
5007	102	3	132,012.3	01-Dec-00	AA-341-96573-Z84
5008	102	3	144,213.6	21-Sep-00	DR-559-22189-D33
5009	103	2	80,932.9	16-Jan-02	DE-887-98456-E94
5010	104	1	34,213.4	12-Apr-03	FD-221-21100-F32
5011	101	1	42,326.8	09-Nov-02	DT-324-04056-H22
5012	104	6	152,339.4	23-May-01	GF-657-22134-K48
5013	101	3	298,145.8	11-Apr-99	HR-344-54560-J92
5014	104	2	8,122.2	09-Sep-03	RW-289-38956-H87
5015	103	1	154,667.9	26-Mar-01	HH-231-55498-K37
5016	105	1	1,200.6	18-Apr-04	FR-332-23459-G55
5017	105	1	3,345.5	23-Feb-04	DD-545-78896-X39

Table name: BASE

BASE_CODE	BASE_CITY	BASE_STATE	BASE_AREA_CODE	BASE_PHONE	BASE_MANAGER
101	Nashville	TN	615	123-4567	Andrea D. Gallagher
102	Lexington	KY	606	234-5678	George H. Delarosa
103	Kansas City	MO	573	345-6789	Maria J. Talindo
104	Athens	GA	901	456-7890	Peter F. McAfee
105	Gainesville	FL	904	678-6543	Lee A. Chau

Table name: TYPE

TYPE_CODE	TYPE_DESCRIPTION
1	Single box, double-axle
2	Single box, single-axle
3	Tandem trailer, single-axle
4	Tandem trailer, double-axle
5	Utility
6	Dump truck, single-axle
7	Dump truck, double-axle

2. Using the tables in Figure P11.1 as a source of information:
 a. Define the implied business rules for the relationships.
 b. Using your best judgment, choose the type of participation of the entities in the relationship (mandatory or optional). Explain your choices.
 c. Develop the conceptual object schema.

3. Using the data presented in Problem 1, develop an object space diagram representing the object's state for the instances of Truck listed below. Label each component clearly with proper OIDs and attribute names.
 a. The instance of the class Truck with TRUCK_NUM = 5001.
 b. The instances of the class Truck with TRUCK_NUM = 5003 and 5004.

4. Given the information in Problem 1, define a superclass Vehicle for the Truck class. Redraw the object space you developed in Problem 3, taking into consideration the new superclass that you just added to the class hierarchy.

5. Assume the following business rules:
 - A course contains many sections, but each section has only one course.
 - A section is taught by one professor, but each professor may teach one or more different sections of one or more courses.
 - A section may contain many students, and each student is enrolled in many sections, but each section belongs to a different course. (Students may take many courses, but they cannot take many sections of the same course!)
 - Each section is taught in one room, but each room may be used to teach several different sections of one or more courses.
 - A professor advises many students, but a student has only one advisor.

Based on these business rules:

a. Identify and describe the main classes of objects.

b. Modify your description in (a) to include the use of abstract data types such as Name, DOB, and Address.

c. Use object representation diagrams to show the relationships between:
 • Course and Section
 • Section and Professor
 • Professor and Student

d. Use object representation diagrams to show the relationships between:
 • Section and Student
 • Room and Section

What type of object is necessary to represent these relationships?

e. Using an OO generalization, define a superclass Person for Student and Professor. Describe this new superclass and its relationship to its subclasses.

6. Convert the following relational database tables to the equivalent OO conceptual representation. Explain each of your conversions with a diagram. (*Note:* The R&C Stores database includes the three tables shown in Figure P11.6.)

FIGURE P11.6 THE **R&C** STORES DATABASE

Table name: REGION **Database name: Ch11_RC_Stores**

REGION_CODE	REGION_LOCATION
1	East
2	West

Table name: STORE (selected attributes)

STORE_CODE	STORE_NAME	STORE_YTD_SALES	REGION_CODE	EMP_CODE
1	Access Junction	$1,403,455.76	2	108
2	Database Corner	$1,821,987.39	2	112
3	Tuple Charge	$1,386,783.22	1	107
4	Attribute Alley	$1,344,568.56	2	103
5	Primary Key Point	$3,330,098.45	1	115

Table name: EMPLOYEE (partial listing)

EMP_CODE	EMP_TITLE	EMP_LNAME	EMP_FNAME	EMP_INITIAL	STORE_CODE
101	Mr.	Williamson	John	W	1
102	Ms.	Ratula	Nancy		2
103	Ms.	Greenboro	Lottie	R	4
104	Mrs.	Rumpersfro	Jennie	S	5
105	Mr.	Smith	Robert	L	3
106	Mr.	Renselaer	Cary	A	1
107	Mr.	Ogallo	Roberto	S	3
108	Ms.	Johnsson	Elizabeth	I	1
109	Mr.	Eindsmar	Jack	W	2
110	Ms.	Jones	Rose	R	4
111	Mr.	Broderick	Tom		3
112	Mr.	Washington	Alan	Y	2
113	Mr.	Smith	Robert	N	3
114	Mrs.	Smith	Sherry	H	4
115	Mr.	Olenko	Howard	U	5
116	Mr.	Archialo	Barry	V	5
117	Ms.	Grimaldo	Jeanine	K	4
118	Mr.	Rosenberg	Andrew	D	4
119	Mr.	Rosten	Peter	F	4
120	Mr.	Zack	Robert	S	1
121	Ms.	Mcbee	Jennifer	A	1
122	Mr.	Ryan	Herman	G	3

7. Convert the following relational database tables to the equivalent OO conceptual representation. Explain each of your conversions with a diagram. (*Note:* The Avion Sales database includes the tables shown in Figure P11.7.)

FIGURE P11.7 THE AVION SALES DATABASE

Table name: CUSTOMER **Database name: Ch11_AvionSales**

	CUS_NUM	CUS_LNAME	CUS_FNAME	CUS_INITIAL	CUS_AREACODE	CUS_PHONE	CUS_YTD_BUY	CUS_CREDIT	CUS_BALANCE
	10010	Ramas	Alfred	A	615	844-2573	$2,011.56	$5,000.00	$0.00
	10011	Dunne	Leona	K	713	894-1238	$4,613.29	$5,000.00	$1,708.51
	10012	Smith	Kathy	W	615	894-2285	$3,217.94	$5,000.00	$0.00
	10013	Olowski	Paul	F	615	894-2180	$10,020.53	$5,000.00	$9,562.67
	10014	Orlando	Myron		615	222-1672	$14,234.85	$7,500.00	$4,358.55
	10015	O'Brian	Amy	B	713	442-3381	$3,005.77	$5,000.00	$0.00
	10016	Brown	James	G	615	297-1228	$5,432.76	$5,000.00	$1,217.84
	10017	Williams	George		615	290-2556	$1,629.95	$1,000.00	$1,629.95
	10018	Farriss	Anne	G	713	382-7185	$3,856.31	$2,500.00	$1,232.42
	10019	Smith	Olette	K	615	297-3809	$8,938.43	$7,500.00	$854.76

Table name: INVOICE

	INV_NUM	CUS_NUM	EMP_NUM	INV_DATE	INV_SUB	INV_TAX8%	INV_TOTAL	INV_PYMT	INV_BALANCE
	10001	10015	103	14-Jan-04	$2,325.00	$186.00	$2,511.00	$2,511.00	$0.00
	10002	10018	105	14-Jan-04	$1,599.00	$127.92	$1,726.92	$1,726.92	$0.00
	10003	10010	105	15-Jan-04	$7,284.00	$582.72	$7,866.72	$5,000.00	$2,866.72

Table name:

INV_NUM	INVLINE_NUM	PROD_CODE	INVLINE_UNITS	INVLINE_PRICE	INVLINE_TOTAL
10001	1	GPS-55AVD	1	$625.00	$625.00
10001	2	TMA-350D	2	$850.00	$1,700.00
10002	1	KX-155	1	$1,599.00	$1,599.00
10003	1	TNL1000(DC)	1	$1,695.00	$1,695.00
10003	2	KMA-24	1	$599.00	$599.00
10003	3	CP-136M	4	$980.00	$3,920.00
10003	4	TRT-250D	1	$1,070.00	$1,070.00

Table name: PRODUCT (selected attributes)

	PROD_CODE	PROD_COST	PROD_PRICE	PROD_QOH	PROD_MIN_QOH	PROD_LAST_ORDER
	ADF-841	$1,699.00	$2,595.00	11	5	11-Mar-2004
	AIRMAP	$619.00	$895.00	34	15	07-Sep-2003
	APOLLO2001	$1,529.00	$2,245.00	17	10	12-Dec-2003
	APOLLO360	$1,229.00	$1,995.00	32	20	19-Jan-2004
	APOLLO-920	$816.00	$1,225.00	26	12	16-Feb-2004
	AT-150	$649.00	$950.00	16	10	24-Jun-2003
	CP-136M	$605.00	$980.00	15	5	26-May-2003
	EC-10X	$1,545.00	$1,999.00	12	5	08-Jun-2003
	FLTPRO	$499.00	$725.00	12	5	20-May-2003
	GPS-150	$1,349.00	$1,995.00	29	15	20-Nov-2003
	GPS-155	$3,888.00	$5,995.00	14	8	14-Jan-2004
	GPS-55AVD	$439.00	$625.00	31	12	11-Dec-2003
	GPS-95XL	$699.00	$1,075.00	27	12	10-Feb-2004
	KLN-89	$2,289.00	$3,195.00	18	12	11-Dec-2003
	KLN-89B	$3,899.00	$5,895.00	14	8	11-Dec-2003
	KLN-90B	$6,311.00	$8,400.00			
	KMA-24	$399.00	$599.00			
	KMA-24H	$459.00	$699.00			
	KN-62A	$1,344.00	$1,899.00			
	KX-155	$1,119.00	$1,599.00			
	KX-165	$1,299.00	$1,695.00			
	KY-196A/197A	$659.00	$999.00			
	SKYNAV5000	$1,249.00	$1,799.00			
	TMA-350D	$499.00	$850.00			
	TNL1000(DC)	$1,250.00	$1,695.00			
	TNL-2000	$3,999.00	$6,650.00			
	TNL-2000A	$1,999.00	$2,695.00			
	TRT-250D	$699.00	$1,070.00			

Record: 1 of 28

Table name: EMPLOYEE

	EMP_NUM	EMP_TITLE	EMP_LNAME	EMP_FNAME	EMP_INITIAL	EMP_HIRE_DATE
	100	Mr.	Kolmycz	Andrew	H	15-Mar-85
	101	Dr.	Lewis	Barbara	K	25-Apr-90
	102	Ms.	Vandam	Marie	L	20-Dec-93
	103	Mr.	Jones	Robert		28-Aug-96
	104	Dr.	Lange	William	B	20-Oct-97
	105	Mr.	Williams	George	D	08-Nov-97
	106	Mrs.	Duzak	Judith	E	05-Jan-89
	107	Ms.	Diante	Melanie	K	02-Jul-97
	108	Ms.	Wiesenbach	Roberta	F	18-Nov-95
	109	Mr.	Travis	Peter		14-Apr-89
	110	Mr.	Genkazi	George	R	01-Dec-03

8. Using the ERD shown in Appendix E, "The University Lab: Conceptual Design Verification, Logical Design, and Implementation," Figure E.22 (the Check_Out component), create the equivalent OO representation. (Appendix E is on the student CD accompanying this book.)

9. Using the contracting company's ERD in Chapter 5, "Normalization of Database Tables," Figure 5.13, create the equivalent OO representation.

12

THE DATA WAREHOUSE

In this chapter, you will learn:

- ❖ How operational data and decision support data differ
- ❖ What a data warehouse is and how data for it are prepared
- ❖ What star schemas are and how they are constructed
- ❖ What steps are required to successfully implement a data warehouse
- ❖ What data mining is and what role it plays in decision support

The preceding 11 chapters are dedicated to the fact that data are crucial raw material in the information age. As a result, data storage and management have become the focus of database design and implementation. You learned that good database design produces effective operational databases through which we can track customers, sales, inventories, and other variables of interest. Such operational databases are critical to the success of almost any organization.

You also learned that, ultimately, the reason for generating, storing, and managing data is to create information that becomes the basis for rational decision making. To facilitate the decision-making process, decision support systems (DSSs) were developed. However, information requirements have become so complex that it is difficult for a DSS to extract all the necessary information from the data structures typically found in an operational database. Therefore, a new data storage facility, called a data warehouse, has been developed. The data warehouse extracts or obtains its data from operational databases as well as from external sources, providing a more comprehensive data pool. Also, the data in a data warehouse are stored in structures that simplify information generation and make it possible to generate a type and extent of information that otherwise might not be available. It's no surprise that data warehouses have become the main data source for modern DSSs. This chapter explores the main concepts and components, and the various architectures, of the data warehouse, and examines the tools that make using data warehouse information simpler and more effective.

In parallel with data warehouse development, advances in hardware, client/server technology, and DBMS software have provided the means to implement a new DSS component called an online analytical processing (OLAP) system. OLAP provides advanced data analysis tools (including multidimensional data analysis) to extract information from a data warehouse, a relational DBMS, or a multidimensional DBMS. In addition to OLAP, data mining has become a significant and growing extension to DSSs. Data mining employs automated tools to analyze the wealth of data now available through data warehouses and other sources and to identify possible relationships and anomalies. Data mining is a proactive methodology that is being used to uncover business problems and opportunities.

12.1 THE NEED FOR DATA ANALYSIS

Organizations tend to grow and prosper as they gain a better understanding of their environment. Typically, business managers must be able to track daily transactions to evaluate how the business is performing. By tapping into the operational database, management can develop strategies to meet organizational goals. In addition, data analysis can provide information about short-term tactical evaluations and strategies, such as: Are our sales promotions working? What market percentage are we controlling? Are we attracting new customers? Tactical and strategic decisions are also shaped by constant pressure from external and internal forces, including globalization, the cultural and legal environment, and, perhaps most important, technology.

Given the many and varied competitive pressures, managers are always looking for competitive advantages through product development and maintenance, service, market positioning, sales promotion, and so on. Managers understand that the business climate is very dynamic, and thus mandates their prompt reaction to change in order to remain competitive. In addition, the modern business climate requires that managers approach increasingly complex problems involving a rapidly growing number of internal and external variables. It's also no surprise that there is a growing interest in creating support systems dedicated to facilitating quick decision making in a complex environment.

Different managerial levels require different decision support needs. For example, transaction-processing systems, based on operational databases, are tailored to serve the information needs of people who deal with short-term inventory, accounts payable, or purchasing. Middle-level managers, general managers, vice presidents, and presidents focus on strategic and tactical decision making. Such managers require detailed information designed to help them make decisions in a complex data and analysis environment. To support such decision making, information systems (IS) departments have created decision support systems, or DSSs.

Table 12.1 gives some real-world examples of companies that have implemented data warehouse, data mart, OLAP, and/or data-mining tools; the table shows the original problem and how the use of these tools benefited the companies.

TABLE 12.1 SOLVING BUSINESS PROBLEMS AND ADDING VALUE WITH DATA WAREHOUSE-BASED SOLUTIONS

COMPANY	PROBLEM	BENEFIT
MOEN Manufacturer of bathroom and kitchen fixtures and supplies Source: Cognos Corp. www.cognos.com	• Information generation very limited and time-consuming. • Only five people knew how to extract data using a 3GL. • Response time unacceptable for Managers' decision-making purposes	• Provided quick answers to ad hoc questions for decision making. • Provided access to data for decision-making purposes. • Received in-depth view of product performance and customer margins.
Pacific Gas Transmission Co. Natural gas provider in Pacific Northwest Source: Oracle Corp. www.oracle.com	• Rapid changes in markets due to deregulation. • Diminishing profits in traditional services.	• Managers able to analyze data quickly. • Positioned company to quickly identify market trends. • Created new services and pricing structures.
SEGA Interactive entertainment systems and video games Source: Oracle Corp. www.oracle.com	• Needed way to rapidly analyze great amount of data. • Needed to track advertising, coupons, and rebates associated with effects of pricing changes. • Formerly used Excel spreadsheets, leading to human errors.	• Eliminated data-entry errors. • Identified successful marketing strategies to dominate interactive entertainment niches. • Used product analysis to identify better markets/product offerings.

TABLE 12.1 SOLVING BUSINESS PROBLEMS AND ADDING VALUE WITH DATA WAREHOUSE-BASED SOLUTIONS (CONTINUED)

COMPANY	PROBLEM	BENEFIT
Owens and Minor, Inc. Medical and surgical supply distributor Source: CFO Magazine *www.cfomagazine.com*	• Lost its largest customer, which represented 10% of its annual revenue ($360 million). • Stock plunged 23%. • Cumbersome process to access information from antiquated mainframe system.	• In just five months increased earnings per share. • Gained more business because the data warehouse was opened to its clients. • Managers gained quick access to data for decision-making purposes.
LA Cellular Cellular telephone company in Los Angeles area Source: *PC Week Online* *www.zdnet.com/pcweek/stories*	• Needed reduced response time to business questions. • Needed to identify which promotions were working, and which were not. • Needed to identify which customers to call—out of a database containing millions of customers—in order to offer new promotions.	• Achieved a 20% increase in subscribers, as a result of properly matching customers with promotions. • Could identify which promotions were effective. • Response times were cut from 14 minutes to 1 minute.

12.2 DECISION SUPPORT SYSTEMS

Decision support is a methodology (or a series of methodologies) designed to extract information from data and to use such information as a basis for decision making. A **decision support system (DSS)** is an arrangement of computerized tools used to assist managerial decision making within a business. A DSS usually requires extensive data "massaging" to produce information.

The DSS is used at all levels within an organization and is often tailored to focus on specific business areas or problems such as finance, insurance, healthcare, banking, sales, and manufacturing. The DSS is interactive and provides ad hoc query tools to retrieve data and to display data in different formats. For example, a DSS operator might want to:

- Compare the relative rates of productivity growth by company division over some specified period of time.

- Define the relationship between advertising types and sales levels. This relationship might then be used for forecasting.

- Define relative market shares by selected product lines.

How does the DSS help answer such questions? By combining historical operational data with business models that reflect the business activities.

N O T E

Although the DSS has an unquestionably important role in modern business operations, keep in mind that the *manager* must initiate the decision support process by asking the appropriate questions. The DSS exists to support the manager; it does *not* replace the management function. If the manager fails to ask the appropriate questions, problems will not be identified and solved, and opportunities will be missed. In spite of the very powerful DSS presence, the human component is still at the center of DSS technology.

A DSS is usually composed of four main components: a data store component, a data extraction and filtering component, an end-user query tool, and an end-user presentation tool.

- The **data store** component is basically a DSS database. The data store contains two main types of data: business data and business model data. The business data are extracted from the operational database and from external data sources; they represent a *snapshot* of the company situation. The business data are not merely a copy of the operational data. Instead, they summarize and arrange the operational data in structures that are optimized for data analysis and query speed. The external data sources provide data that cannot be found within the company but that are relevant to the business, such as stock prices, market indicators, marketing information (for example, demographics), and competitors' data. Business models are generated by special algorithms to model the business to identify and enhance the understanding of business situations and problems. For example, to define the relationships among advertising types, expenditures, and sales, for forecasting purposes the DSS might use some type of regression model, then use the results to perform a time-series analysis.

- The **data extraction** and **data filtering** component is used to extract and validate the data taken from the operational database and the external data sources. For example, to determine the relative market share by selected product line, the DSS requires data from competitors' products. Such data can be located in external databases provided by industry groups or by companies that market such data. As the name implies, this component extracts the data, filters the extracted data to select the relevant records, and packages the data in the right format to be added to the DSS data store component. You'll see later that the data in the DSS database have characteristics that are quite distinct from those in the operational database.

- The **end-user query tool** is used by the data analyst to create the queries that access the database. Depending on the DSS implementation, the query tool accesses either the operational database or, more commonly, the DSS database. This tool advises the user on which data to select and how to build a reliable business data model.

- The **end-user presentation tool** is used by the data analyst to organize and present the data. This tool helps the end user select the most appropriate presentation format, such as summary report, map, pie or bar graph, or mixed graphs. The query tool and the presentation tool are the front end to the DSS.

Each DSS component shown in Figure 12.1 has generated a fast-growing market for specialized DSS tools. And, thanks to the advancement of client/server technologies, those components are able to interact with other components to form a truly open DSS architecture.

Although the DSS is used at strategic and tactical managerial levels within organizations, *its effectiveness depends on the quality of data gathered at the operational level.* Yet operational data are seldom well suited to the decision support tasks. The differences between operational data and decision support data will be examined in the next section.

12.2.1 OPERATIONAL DATA VERSUS DECISION SUPPORT DATA

Operational data and DSS data serve different purposes. Therefore, it is not surprising to learn that their formats and structures differ.

FIGURE 12.1 MAIN COMPONENTS OF A DECISION SUPPORT SYSTEM (DSS)

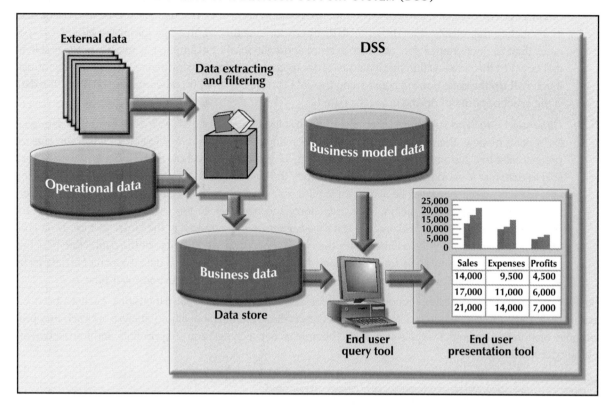

Most operational data are stored in a relational database in which the structures (tables) tend to be highly normalized. Operational data storage is optimized to support transactions that represent daily operations. For example, each time an item is sold, it must be accounted for. Customer data, inventory data, and so on are in a frequent update mode. To provide effective update performance, operational systems store data in many tables, each with a minimum number of fields. Thus, a simple sales transaction might be represented by five or more different tables (for example, invoice, invoice line, discount, store, and department). Although such an arrangement is excellent in an operational database, it is not query friendly. For example, to extract a simple invoice you would have to join several tables. Whereas operational data capture daily business transactions, DSS data give tactical and strategic business meaning to the operational data. From the data analyst's point of view, DSS data differ from operational data in three main areas: timespan, granularity, and dimensionality.

- *Timespan.* Operational data represent current (atomic) transactions. Such transactions might define a purchase order, a sales invoice, an inventory movement, and so on. Operational data cover a short time frame. In contrast, DSS data tend to cover a longer time frame. Managers are seldom interested in a specific sales invoice to customer X; rather, they tend to focus on sales generated during the last month, the last year, or the last five years. Rather than concern themselves with a single customer purchase, they might be interested in the buying patterns of such a customer or groups of customers. In short, DSS data tend to be historic in nature. That is, the DSS data represent company transactions up to a given point in time: yesterday, last week, last month, and the like. Data analysts are (or should be!) aware that the sales invoice generated two minutes ago is not likely to be found in the DSS database.

- *Granularity (level of aggregation).* DSS data must be presented at different levels of aggregation, from highly summarized to near-atomic. Managers at different levels in the organization require data with different levels of aggregation. It is also possible that a single problem requires data with different summarization levels. For example, if a manager must analyze sales by region, (s)he must be able to access data

showing the sales by region, by city within the region, by store within the city within the region, and so on. In this case, the manager requires summarized data to compare the regions, but (s)he also needs data in a structure that enables him or her to **drill down**, or decompose, the data into more atomic components (that is, finer-grained data at lower levels of aggregation). For example, it is necessary to be able to drill down to the stores within the region in order to compare store performance by region. In contrast, if you **roll up** the data, you are aggregating the data to a higher level. (In other words, rolling up the data is the exact opposite of drilling down the data.)

- *Dimensionality.* This is probably the most distinguishing characteristic of DSS data. From the data analyst's point of view, the data are always related in many different ways. For example, when we analyze product sales to a customer during a given timespan, we are likely to ask how many widgets of type X were sold to customer Y during the last six months. In fact, the question tends to expand quickly to include many different data slices. For instance, we might want to know how product X fared relative to product Z during the past six months, by region, state, city, store, and customer. In this case, both place and time are part of the picture. In fact, data analysts are always interested in developing the larger picture. Data analysts tend to include many data dimensions; they have a multidimensional view of the data. Note how this view is different from the single-transaction view that is typical of operational data. Operational data focus on representing atomic transactions, rather than on the effects of the transactions over time.

Figure 12.2 shows how decision support data can be examined from multiple dimensions (such as product, region, and year), using a variety of filters to produce each dimension. The ability to analyze, extract, and present information in meaningful ways is one of the differences between decision support data and transaction-at-a-time operational data.

A D D I T I O N A L C D C O N T E N T

The operational data in Figure 12.2 are found in the **Excel\Student\Ch12** folder on the student CD. The decision support data in Figure 12.2 show the output for the solution to Problem 2 at the end of this chapter.

From the DSS designer's point of view, we can summarize the differences between operational and DSS data as follows:

- Operational data represent transactions as they happen, in real time. DSS data are a snapshot of the operational data at a given point in time. Therefore, DSS data are historic, representing a time slice of the operational data.

- Operational and DSS data are different in terms of transaction type and transaction *volume*. Whereas operational data are characterized by update transactions, DSS data are mainly characterized by *query* (read-only) transactions. DSS data also require *periodic* updates to load new data that are summarized from the operational data. Finally, the transaction volume in operational data tends to be very high when compared with the low-to-medium levels found in DSS data.

- Operational data are commonly stored in many tables, and the stored data represent the information about a given transaction only. DSS data are generally stored in a few tables that store data derived from the operational data. The DSS data do not include the details of each operational transaction. Instead, DSS data represent transaction *summaries*, and, therefore, the DSS stores data that are integrated, aggregated, and summarized for decision support purposes.

FIGURE 12.2 **TRANSFORMING OPERATIONAL DATA INTO DECISION SUPPORT DATA**

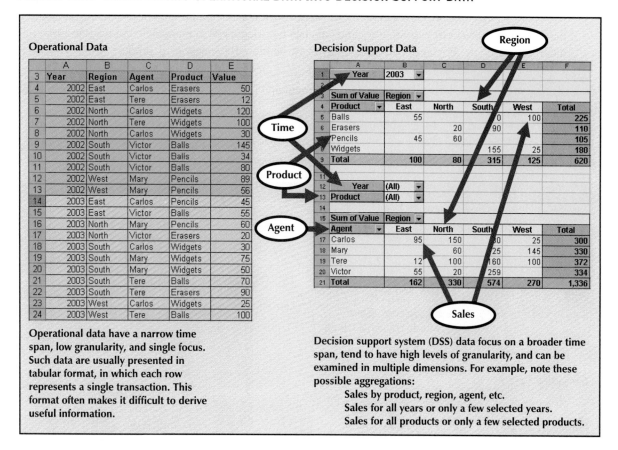

Operational Data

	A	B	C	D	E
3	Year	Region	Agent	Product	Value
4	2002	East	Carlos	Erasers	50
5	2002	East	Tere	Erasers	12
6	2002	North	Carlos	Widgets	120
7	2002	North	Tere	Widgets	100
8	2002	North	Carlos	Widgets	30
9	2002	South	Victor	Balls	145
10	2002	South	Victor	Balls	34
11	2002	South	Victor	Balls	80
12	2002	West	Mary	Pencils	89
13	2002	West	Mary	Pencils	56
14	2003	East	Carlos	Pencils	45
15	2003	East	Victor	Balls	55
16	2003	North	Mary	Pencils	60
17	2003	North	Victor	Erasers	20
18	2003	South	Carlos	Widgets	30
19	2003	South	Mary	Widgets	75
20	2003	South	Mary	Widgets	50
21	2003	South	Tere	Balls	70
22	2003	South	Tere	Erasers	90
23	2003	West	Carlos	Widgets	25
24	2003	West	Tere	Balls	100

Operational data have a narrow time span, low granularity, and single focus. Such data are usually presented in tabular format, in which each row represents a single transaction. This format often makes it difficult to derive useful information.

Decision Support Data

Region **Time** **Product** **Agent** **Sales**

	A	B	C	D	E	F
1	Year	2003 ▾				
2						
3	Sum of Value	Region ▾				
4	Product ▾	East	North	South	West	Total
5	Balls	55		0	100	225
6	Erasers		20	90		110
7	Pencils	45	60			105
8	Widgets			155	25	180
9	Total	100	80	315	125	620
11						
12	Year	(All) ▾				
13	Product	(All) ▾				
14						
15	Sum of Value	Region ▾				
16	Agent ▾	East	North	South	West	Total
17	Carlos	95	150	30	25	300
18	Mary		60	25	145	330
19	Tere	12	100	160	100	372
20	Victor	55	20	259		334
21	Total	162	330	574	270	1,336

Decision support system (DSS) data focus on a broader time span, tend to have high levels of granularity, and can be examined in multiple dimensions. For example, note these possible aggregations:

Sales by product, region, agent, etc.
Sales for all years or only a few selected years.
Sales for all products or only a few selected products.

- The degree to which DSS data are summarized is very high when contrasted with operational data. Therefore, you will see a great deal of derived data in DSS databases. For example, rather than storing all 10,000 sales transactions for a given store on a given day, the DSS database might simply store the total number of units sold and the total sales dollars generated during that day. DSS data might be collected to monitor such aggregates as total sales for each store, for each product, and so on. The purpose for the summaries is simple: they are to be used to establish and evaluate sales trends, product sales comparisons, and so on, that serve decision needs. (How well are sales performing? Should we discontinue this product? Has the advertising been effective as measured by sales changes?)

- The data models that govern operational data and DSS data are different. The operational database's frequent and rapid data updates make data anomalies a potentially devastating problem. Therefore, the data requirements in a typical relational transaction (operational) system generally require normalized structures that yield many tables, each of which contains the minimum number of attributes. In contrast, the DSS database is not subject to such transaction updates, and the focus is on querying capability. Therefore, DSS databases tend to be non-normalized and include few tables, each of which contains a large number of attributes.

- Query activity (frequency and complexity) in the operational database tends to be low, to allow more processing cycles for the more crucial update transactions. Therefore, queries against operational data typically are narrow in scope, low in complexity, and speed-critical. In contrast, DSS data exist for the sole purpose of serving query requirements. Queries against DSS data typically are broad in scope, high in complexity, and less speed-critical.

- Finally, DSS data are characterized by very large amounts of data. The large data volume is the result of two factors: First, data are stored in nonnormalized structures that are likely to display many data redundancies and duplications. Second, the same data can be categorized in many different ways to represent different snapshots. For example, sales data might be stored in relation to product, store, customer, region, manager, and so on.

Table 12.2 summarizes the differences between operational and DSS data from the database designer's point of view.

TABLE 12.2 CONTRASTING OPERATIONAL AND DSS DATA CHARACTERISTICS

CHARACTERISTIC	OPERATIONAL DATA	DSS DATA
Data currency	Current operations Real-time data	Historic data Snapshot of company data Time component (week/month/year)
Granularity	Atomic-detailed data	Summarized data
Summarization level	Low; some aggregate yields	High; many aggregation levels
Data model	Highly normalized Mostly relational DBMS	Nonnormalized Complex structures Some relational, but mostly multidimensional DBMS
Transaction type	Mostly updates	Mostly query
Transaction volumes	High update volumes	Periodic loads and summary calculations
Transaction speed	Updates are critical	Retrievals are critical
Query activity	Low to medium	High
Query scope	Narrow range	Broad range
Query complexity	Simple to medium	Very complex
Data volumes	Hundreds of megabytes and up to gigabytes	Hundreds of gigabytes to terabytes

The many differences found between operational data and decision support data are good indicators of the requirements of the DSS database. If operational data are not well matched to decision support requirements, we must create databases to hold data in a format that is better suited to decision support. So, what are the requirements of a DBMS that will be used to store decision support data? We will answer that question in the next section.

12.2.2 DSS DATABASE REQUIREMENTS

A DSS database is a specialized DBMS tailored to provide fast answers to complex queries. We'll address the four main DSS database requirements: the database schema, data extraction and loading, the end-user analytical interface, and database size.

Database Schema

The DSS database schema must support complex (nonnormalized) data representations. As we noted earlier, the DSS database must contain data that are aggregated and summarized, and, most important, that maintain relations with many other data elements. In addition to meeting these requirements, the queries must be able to

extract multidimensional time slices. If you are using an RDBMS, these conditions suggest using nonnormalized and even duplicated data. To see why this must be true, let's take a look at the 10-year sales history for a single store containing a single department. At this point, the data are fully normalized within the single table, as shown in Table 12.3.

TABLE 12.3 TEN-YEAR SALES HISTORY FOR A SINGLE DEPARTMENT, IN MILLIONS OF DOLLARS

YEAR	SALES
1994	8,227
1995	9,109
1996	10,104
1997	11,553
1998	10,018
1999	11,875
2000	12,699
2001	14,875
2002	16,301
2003	19,986

This structure works well if we have only one store with only one department. However, it is very unlikely that there is much need for a DSS in such a simple environment. One would suppose that a DSS becomes a factor when dealing with more than one store, each of which has more than one department. To support all the DSS requirements, the database must contain data for all the stores and all their departments—and the database must be able to support multidimensional queries that track sales by stores, by departments, and over time. For simplicity, let's say that there are only two stores (A and B), and two departments (1 and 2) within each store. We will also change the time dimension to include yearly data. Table 12.4 shows the sales figures under the specified conditions. We show only 1994, 1998, and 2003, using ellipses (...) to indicate that data values were omitted. If you examine Table 12.4, you can see that the number of rows and attributes already multiplies quickly and that the table exhibits multiple redundancies.

TABLE 12.4 YEARLY SALES SUMMARIES, TWO STORES AND TWO DEPARTMENTS PER STORE, IN MILLIONS OF DOLLARS

YEAR	STORE	DEPARTMENT	SALES
1994	A	1	1,985
1994	A	2	2,401
1994	B	1	1,879
1994	B	2	1,962

TABLE 12.4 YEARLY SALES SUMMARIES, TWO STORES AND TWO DEPARTMENTS PER STORE, IN MILLIONS OF DOLLARS (CONTINUED)

YEAR	STORE	DEPARTMENT	SALES
...	...	...	...
1998	A	1	3,912
1998	A	2	4,158
1998	B	1	3,426
1998	B	2	1,203
...	...	...	...
2003	A	1	7,683
2003	A	2	6,912
2003	B	1	3,768
2003	B	2	1,623

Now suppose that we have grown to 10 departments per store and 20 stores nationwide. And suppose we want to access *yearly* sales summaries. Now we are dealing with 200 rows and 12 attributes per row. (Actually, there are 13 attributes per row if we add each store's sales total for each year.)

The DSS database schema must also be optimized for query (read-only) retrievals. To optimize query speed, the DBMS must support features such as bitmap indexes and data partitioning to increase search speed. In addition, the DBMS query optimizer must be enhanced to support the nonnormalized and complex structures found in decision support databases.

Data Extraction and Filtering

The DSS database is created largely by extracting data from the operational database and by importing additional data from external sources. Thus, the DBMS must support advanced data extraction and filtering tools. The data extraction capabilities should allow batch and scheduled data extraction, to minimize the impact on the operational database. The data extraction capabilities should also support different data sources: flat files and hierarchical, network, and relational databases, as well as multiple vendors. Data filtering capabilities must include the ability to check for inconsistent data or data validation rules. Finally, the DBMS must support advanced data integration, aggregation, and classification, in order to filter and integrate the operational data into the DSS database.

Using data from multiple external sources also usually means having to solve data-formatting conflicts. For example, dates and Social Security numbers might occur in different formats, measurements can be based on different scales, the same data elements might have different names, and so on. In short, data must be filtered and purified to ensure that only the pertinent decision support data are stored in the database and that they are stored in a standard format.

End-User Analytical Interface

The DSS DBMS must support advanced data modeling and data presentation tools. Using these tools makes it easier for the data analysts to define the nature and extent of business problems. Once the problems have been defined, the DSS DBMS must generate the necessary queries to retrieve the appropriate data from the DSS database. If necessary, the query results may then be evaluated with data-analysis tools supported by the DSS

DBMS. Because queries yield crucial information for decision makers, the queries must be optimized for speedy processing. In short, the DSS DBMS must support:

- Data modeling and presentation tools.
- Data analysis tools.
- Query generation and optimization components.

The end-user analytical interface is one of the most critical DSS DBMS components. If properly implemented, an analytical interface permits the user to navigate through the data to simplify and accelerate the decision-making process.

Database Size

DSS databases tend to be very large; gigabyte and terabyte ranges are not unusual. (As mentioned earlier, the DSS database typically contains redundant and duplicated data in order to improve data retrieval and simplify information generation.) Therefore, the DBMS must be capable of supporting **very large databases** (**VLDB**s). To adequately support a VLDB, the DBMS might be required to use advanced hardware, such as multiple disk arrays and, even more important, to support multiple-processor technologies, such as a symmetric multiprocessor (SMP) or a massively parallel processors (MPP).

The complex information requirements and the ever-growing demand for sophisticated data analysis sparked the creation of a new type of data repository. This repository contains data in formats that facilitate data extraction, data analysis, and decision making. This DSS-friendly data repository is known as a data warehouse and has become the foundation for a new generation of decision support systems.

12.3 THE DATA WAREHOUSE

Bill Inmon, the acknowledged father of the **data warehouse**, defines it as an *integrated, subject-oriented, time-variant, nonvolatile* database (italics added for emphasis) that provides support for decision making.[1] To understand this definition, let's take a more detailed look at its components:

- *Integrated.* The data warehouse is a centralized, consolidated database that integrates data derived from the entire organization and from multiple sources with diverse formats. Data integration implies that all business entities, data elements, data characteristics, and business metrics *are described in the same way throughout the enterprise.* Although this requirement sounds logical, you would be amazed to discover how many different measurements for "sales performance" can exist within an organization; the same scenario holds true for any other business element. For instance, the status of an order might be indicated with text labels such as "open," "received," "cancel," or "closed" in one department and as "1," "2," "3," or "4" in another department. A student's status might be defined as "freshman," "sophomore," "junior," or "senior" in the Accounting Department and as "FR," "SO," "JR," or "SR" in the Computer Information Systems Department. To avoid the potential format tangle, the data in the data warehouse must conform to a common format acceptable throughout the organization. This integration can be time-consuming but, once accomplished, it enhances decision making and helps managers better understand the company's operations. This understanding can be translated into recognition of strategic business opportunities.

- *Subject-oriented.* Data warehouse data are arranged and optimized to provide answers to questions coming from diverse functional areas within a company. Data warehouse data are organized and summarized by topic, such as sales, marketing, finance, distribution, and transportation. For each of these topics the data warehouse contains specific subjects of interest—products, customers, departments, regions, promotions, and so on. This form of data organization is quite different from the more functional or process-oriented

[1] Inmon, Bill and Chuck Kelley, "The Twelve Rules of Data Warehouse for a Client/Server World," *Data Management Review*, 4(5), May 1994, pp. 6–16.

organization of typical transaction systems. For example, an invoicing system designer concentrates on designing normalized data structures (relational tables) to support the business process by storing invoice components in two tables: INVOICE and INVLINE. In contrast, the data warehouse has a *subject* orientation. Data warehouse designers focus specifically on the data, rather than on the processes that modify the data. (After all, data warehouse data are not subject to numerous real-time data updates!) Therefore, instead of storing an invoice, the data warehouse stores its product and customer components, because decision support activities require the retrieval of sales summaries by product or customer.

- *Time-variant.* In contrast to operational data, which focus on current transactions, warehouse data represent the flow of data through time. The data warehouse can even contain projected data generated through statistical and other models. It is also time-variant in the sense that, once data are periodically uploaded to the data warehouse, all time-dependent aggregations are recomputed. For example, when data for previous weekly sales are uploaded to the data warehouse, the weekly, monthly, yearly, and other time-dependent aggregates for products, customers, stores, and other variables are also updated. Because data in a data warehouse constitute a snapshot of the company history as measured by its variables, the time component is crucial. The data warehouse contains a time ID that is used to generate summaries and aggregations by week, month, quarter, year, and so on. Once the data enter the data warehouse, the time ID assigned to the data cannot be changed.

- *Nonvolatile.* Once data enter the data warehouse, they are never removed. Because the data in the warehouse represent the company's history, the operational data, representing the near-term history, are always added to it. Because data are never deleted and new data are always added, the data warehouse is always growing. That's why the DSS DBMS must be able to support multigigabyte and even multiterabyte databases and multiprocessor hardware.

Table 12.5 summarizes the differences between data warehouses and operational databases.

TABLE 12.5 A COMPARISON OF DATA WAREHOUSE AND OPERATIONAL DATABASE CHARACTERISTICS

CHARACTERISTIC	OPERATIONAL DATABASE DATA	DATA WAREHOUSE DATA
Integrated	Similar data can have different representations or meanings. For example, Social Security numbers may be stored as ###-##-#### or as #########, and a given condition may be labeled as T/F or 0/1 or Y/N. A sales value may be shown in thousands or in millions.	Provide a unified view of all data elements with a common definition and representation for all business units.
Subject-oriented	Data are stored with a functional, or process, orientation. For example, data may be stored for invoices, payments, credit amounts, and so on.	Data are stored with a subject orientation that facilitates multiple views of the data and facilitates decision making. For example, sales may be recorded by product, by division, by manager, or by region.
Time-variant	Data are recorded as current transactions. For example, the sales data may be the sale of a product on a given date, such as $342.78 on 12-MAY-2004.	Data are recorded with a historical perspective in mind. Therefore, a time dimension is added to facilitate data analysis and various time comparisons.
Nonvolatile	Data updates are frequent and common. For example, an inventory amount changes with each sale. Therefore, the data environment is fluid.	Data cannot be changed. Data are only added periodically from historical systems. Once the data are properly stored, no changes are allowed. Therefore, the data environment is relatively static.

In summary, the data warehouse is usually a read-only database optimized for data analysis and query processing. Typically, data are extracted from various sources and are then transformed and integrated—in other words, passed through a data filter—before being loaded into the data warehouse. Users access the data warehouse via front-end tools and/or end-user application software to extract the data in usable form. Figure 12.3 illustrates how a data warehouse is created from the data contained in an operational database.

FIGURE 12.3 CREATING A DATA WAREHOUSE

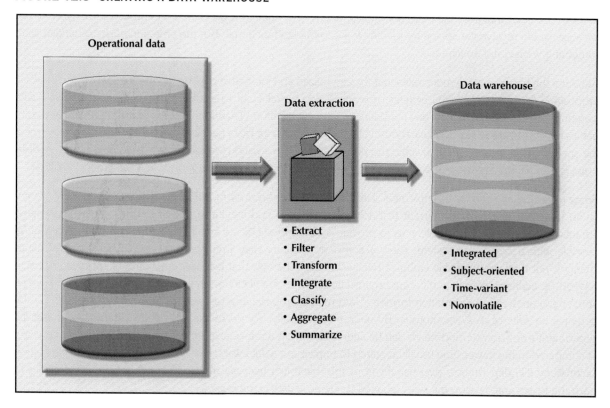

Creating a high-quality data warehouse is seldom simple, because its data are typically derived from a variety of different sources that use different data formats and structures. To illustrate the challenges that must be faced when integrating data from multiple sources, we will introduce you to a few of the data inconsistency problems found during the University Computer Lab design implementation presented in **Appendix D, "The University Lab: Conceptual Design," and in Appendix E, "The University Lab: Conceptual Design Verification, Logical Design, and Implementation."**

ADDITIONAL CD CONTENT

Appendix D and Appendix E are found on the student CD that accompanies this book.

In the case of the university lab, to deal with budgeting requirements and justify funding for special projects, we wanted to obtain data about the computer lab's usage rate. This project had two external constraints:

1. The data had to be generated by a newly installed, university-supplied magnetic card reader system. This system ran on a Sun Microsystems server, located elsewhere on campus. The system used a flat file and contained data about the access date, time, and the student ID only. The system did not yield any data from which we could generate information concerning student classification or department.

2. The student classification and department data were located in another flat-file-based system, stored in a different database on another computer. University-wide standardization requirements dictated the use of data segments contained in this database.

Following the dictates of Murphy's Law (that is, anything that can go wrong will go wrong), the card reader used data structures that did not match those found in the student database. We faced the typical scenario that leads to the types of problems encountered when creating a data warehouse: the information to be used during critical budget hearings came from different data sources and was subject to different data formats. Consequently, we had to create a separate database (relational, this time!) in which we could easily integrate data from both sources. This job turned out to be a major challenge. First, we had to deal with several layers of bureaucracy. Then we had to download the data from the diverse systems, import them into a desktop database, apply filters to clean the data, integrate the data, and finally, create the required reports. The job proved to be worthwhile, because we did receive the necessary funding to keep the lab functional. But the time and effort expended were something we had not anticipated.

Although the lab example demonstrates that the centralized and integrated data warehouse can be a very attractive proposition that can yield many benefits, managers may be reluctant to embrace this strategy. Creating a data warehouse requires time, money, and considerable managerial effort. Therefore, it is not surprising that many companies begin their foray into data warehousing by focusing on more manageable data sets that are targeted to meet the special needs of small groups within the organization. These small data stores are called data marts. A **data mart** is a small, single-subject data warehouse subset that provides decision support to a small group of people.

Some organizations choose to implement data marts not only because of the lower cost and shorter implementation time, but also because current technological advances and inevitable "people issues" make data marts attractive. Powerful computers allow us to provide a customized DSS to small groups in ways that might not be possible with a centralized system. Also, a company's culture may predispose its employees to resist major changes, but they might quickly embrace relatively minor changes that lead to demonstrably improved decision support. In addition, people at different organizational levels are likely to require data with different summarization, aggregation, and presentation formats. Data marts can serve as a test vehicle for companies exploring the potential benefits of data warehouses. By migrating gradually from data marts to data warehouses, a specific department's decision support needs can be addressed within a reasonable time frame (six months to one year), as compared to the longer time usually required to implement a data warehouse (one to three years). Information Technology (IT) departments also benefit from this approach because their personnel get the opportunity to learn the issues and develop the skills required to create a data warehouse.

The difference between a data mart and a data warehouse is only the size and scope of the problem being solved. Therefore, the problem definitions and data requirements are essentially the same for both. Fortunately, thanks to client/server technology, the hardware and software components may also be common to both the data mart and the data warehouse, thus making the growth path from one to the other less onerous. We will next examine how client/server technology has helped in the evolution of data warehouses.

12.3.1 DSS ARCHITECTURAL STYLES

Several DSS database architectural styles are available. These architectures provide advanced decision support features, and some are capable of providing access to multidimensional data analysis. The DSS tools that use multidimensional data analysis techniques are known as online analytical processing (OLAP) tools. (We discuss OLAP in detail in Section 12.4.) Table 12.6 summarizes the main architectural styles that are likely to be encountered in the DSS database environment.

You may be tempted to think that the data warehouse is just a big, summarized database. Our previous discussion indicates that a good data warehouse is much more than that. A complete data warehouse architecture includes support for a DSS data store, a data extraction and integration filter, and a specialized presentation interface. To be useful, the data warehouse must conform to uniform structures and formats to avoid data conflicts and to support decision making. In fact, before we can refer to a DSS database as a true data warehouse, it must conform to the rules described in the next section.

TABLE 12.6 **DSS ARCHITECTURAL STYLES**

SYSTEM TYPE	SOURCE DATA	DATA EXTRACTION/ INTEGRATION PROCESS	DSS DATA STORE	END-USER QUERY TOOL	END-USER PRESENTATION TOOL
Traditional mainframe-based OLTP	Operational data	None Reports, reads, and summarizes data directly from operational data	None Temporary files used for reporting purposes	Very basic Predefined reporting formats. Basic sorting, totaling, and averaging	Very basic Menu-driven, predefined reports, text and numbers only
Managerial information system (MIS) with third-generation language (3GL)	Operational data	Basic extraction and aggregation Reads, filters, and summarizes operational data into intermediate data store	Lightly aggregated data in RDBMS	Same as above, plus some ad hoc reporting using SQL	Same as above, plus some ad hoc columnar report definitions
First-generation departmental DSS	Operational data	Data extraction and integration process to populate a DSS data store; run periodically	1st DSS database generation Usually RDBMS	Query tool with some analytical capabilities and predefined reports	Advanced presentation tools with plotting and graphics capabilities
First-generation enterprise data warehouse using RDBMS	Operational data External data (census data)	Advanced data extraction and integration tools Features include access to diverse data sources, transformations, filters, aggregations, classifications, scheduling, and conflict resolution	Data warehouse integrated DSS database to support the entire organization Uses RDBMS technology optimized for query purposes Star schema model	Same as above, plus support for more advanced queries and analytical functions with extensions	Same as above, plus additional multidimensional presentation tools with drill-down capabilities
Second-generation data warehouse using MDBMS	Operational data External data (industry group data)	Same as first generation enterprise data warehouse using RDBMS	Data warehouse stores data using multidimensional database (MDBMS) technology based on data structures; referred to as "cubes" with multiple dimensions	Same as above, but uses different query interface to access MDBMS (proprietary)	Same as above, but uses "cubes" and multidimensional matrixes Limited in terms of cube size

12.3.2 TWELVE RULES THAT DEFINE A DATA WAREHOUSE

In 1994, William H. Inmon and Chuck Kelley created a list of 12 rules defining a data warehouse, which summarizes many of the points we've made about data warehouses.[2]

1. The data warehouse and operational environments are separated.

2. The data warehouse data are integrated.

3. The data warehouse contains historical data over a long time horizon.

4. The data warehouse data are snapshot data captured at a given point in time.

5. The data warehouse data are subject oriented.

6. The data warehouse data are mainly read-only with periodic batch updates from operational data. No online updates are allowed.

7. The data warehouse development life cycle differs from classical systems development. The data warehouse development is data-driven; the classical approach is process-driven.

8. The data warehouse contains data with several levels of detail: current detail data, old detail data, lightly summarized data, and highly summarized data.

9. The data warehouse environment is characterized by read-only transactions to very large data sets. The operational environment is characterized by numerous update transactions to a few data entities at a time.

10. The data warehouse environment has a system that traces data sources, transformations, and storage.

11. The data warehouse's metadata are a critical component of this environment. The metadata identify and define all data elements. The metadata provide the source, transformation, integration, storage, usage, relationships, and history of each data element.

12. The data warehouse contains a chargeback mechanism for resource usage that enforces optimal use of the data by end users.

Note how these 12 rules capture the complete data warehouse life cycle, from its introduction as an entity separate from the operational data store, to its components, functionality, and management processes. The current generation of OLAP provides a comprehensive infrastructure to design, develop, implement, and use decision support systems within an organization. We will next describe these systems in greater detail.

12.4 ONLINE ANALYTICAL PROCESSING

The need for more intensive decision support prompted the introduction of a new generation of tools. These new tools, called **online analytical processing (OLAP)**, create an advanced data analysis environment that supports decision making, business modeling, and operations research. OLAP systems share four main characteristics; they:

- Use multidimensional data analysis techniques.

- Provide advanced database support.

- Provide easy-to-use end-user interfaces.

- Support client/server architecture.

Let's examine each of these characteristics.

[2] Inmon, Bill and Chuck Kelley, "The Twelve Rules of Data Warehouse for a Client/Server World," *Data Management Review*, 4 (5), May 1994, pp. 6-16.

12.4.1 MULTIDIMENSIONAL DATA ANALYSIS TECHNIQUES

The most distinct characteristic of OLAP tools is their capacity for multidimensional analysis. (Recall that multidimensionality is also a requirement for decision support.) In multidimensional analysis, data are processed and viewed as part of a multidimensional structure. This type of data analysis is particularly attractive to business decision makers, because they tend to view business data as data that are related to other business data.

To better understand this view, let's examine how a business data analyst might investigate sales figures. In this case, (s)he probably is interested in the sales figures as they relate to other business variables such as customers and time. In other words, customers and time are viewed as different dimensions of sales. Figure 12.4 illustrates how the operational (one-dimensional) view differs from the multidimensional view of sales.

FIGURE 12.4 OPERATIONAL VS. MULTIDIMENSIONAL VIEW OF SALES

Database name: Ch12_Text

Table name: DW_INVOICE

	INV_NUM	INV_DATE	CUS_NAME	INV_TOTAL
+	2034	15-May-04	Dartonik	$1,400.00
+	2035	15-May-04	Summer Lake	$1,200.00
+	2036	16-May-04	Dartonik	$1,350.00
+	2037	16-May-04	Summer lake	$3,100.00
+	2038	16-May-04	Trydon	$400.00

Table name: DW_LINE

INV_NUM	LINE_NUM	PROD_DESCRIPTION	LINE_PRICE	LINE_QUANTITY	LINE_AMOUNT
2034	1	Optical Mouse	$45.00	20	$900.00
2034	2	Wireless RF remote and laser pointer	$50.00	10	$500.00
2035	1	Everlast Hard Drive, 60 GB	$200.00	6	$1,200.00
2036	1	Optical Mouse	$45.00	30	$1,350.00
2037	1	Optical Mouse	$45.00	10	$450.00
2037	2	Roadster 56KB Ext. Modem	$120.00	5	$600.00
2037	3	Everlast Hard Drive, 60 GB	$205.00	10	$2,050.00
2038	1	NoTech Speaker Set	$50.00	8	$400.00

Multidimensional View of Sales

Customer Dimension	Time Dimension		Totals
	15-May-04	16-May-04	
Dartonik	$1,400.00	$1,350.00	$2,750.00
Summer Lake	$1,800.00	$3,100.00	$4,900.00
Trydon		$400.00	$400.00
Totals	$3,200.00	$4,850.00	$8,050.00

Sales are located in the intersection of a customer row and time column

Aggregations are provided for both dimensions

As you examine Figure 12.4, note that the tabular (operational) view of sales data is not well suited to decision support, because the relationship INVOICE → LINE between INVOICE and LINE does not provide a business perspective of the sales data. On the other hand, the end user's view of sales data *from a business perspective* is more closely represented by the multidimensional view of sales than by the tabular view of separate tables. Note also that the multidimensional view allows end users to consolidate or aggregate data at different levels: total sales figures by customers and by date. Finally, the multidimensional view of data allows a business data analyst to easily switch business perspectives (dimensions) from sales by customer to sales by division, by region, and so on.

Multidimensional data analysis techniques are augmented by the following functions:

- Advanced data presentation functions: 3-D graphics, pivot tables, crosstabs, data rotation, three-dimensional cubes, and so on. Such data presentation facilities are compatible with desktop spreadsheets, statistical packages, and query and report-writer packages.

- Advanced data aggregation, consolidation, and classification functions that allow the business data analyst to create multiple data aggregation levels, slice and dice data (see Section 12.5.3), and drill down and roll up data across different dimensions and aggregation levels. For example, aggregating data across the time dimension (by week, month, quarter, and year) allows the business data analyst to drill down and roll up across time dimensions.

- Advanced computational functions: business-oriented variables (market share, period comparisons, sales margins, product margins, percentage changes, etc.), financial and accounting ratios (profitability, overhead, cost allocations, returns, etc.), statistical and forecasting functions, and so on. These functions are provided automatically, and the end user does not need to redefine their components each time they are accessed.

- Advanced data modeling functions: support for "what-if" scenarios, variable assessment, variable contributions to outcome, linear programming, and other modeling tools.

Because many analysis and presentation functions are common to existing desktop spreadsheet packages, most OLAP vendors have closely integrated their systems with desktop spreadsheets such as Microsoft Excel and IBM Lotus 1-2-3. Using the features available in graphical end-user interfaces such as Windows, the OLAP menu option simply becomes another option within the spreadsheet menu bar (see Figure 12.5). This seamless integration is a plus for OLAP systems and for spreadsheet vendors, because end users gain access to advanced data analysis features by using familiar programs and interfaces. Therefore, additional training and development costs are minimized.

12.4.2 ADVANCED DATABASE SUPPORT

To deliver efficient decision support, OLAP tools must have advanced data access features. Such features include:

- Access to many different kinds of DBMSs, flat files, and internal and external data sources.

- Access to aggregated data warehouse data as well as to the detail data found in operational databases.

- Advanced data navigation features such as drill-down and roll-up.

- Rapid and consistent query response times.

- The ability to map end-user requests, expressed in either business or model terms, to the appropriate data source and then to the proper data access language (usually SQL). The query code must be optimized to match the data source, regardless of whether the source is operational or data warehouse data.

- Support for very large databases. As we already explained, the data warehouse can easily and quickly grow to multiple gigabytes and even terabytes.

FIGURE 12.5 INTEGRATION OF OLAP WITH A SPREADSHEET PROGRAM

To provide a seamless interface, OLAP tools map the data elements from the data warehouse and from the operational database to their own data dictionaries. These metadata are then used to translate end-user data analysis requests into the proper (optimized) query codes, which are then directed to the appropriate data source(s).

12.4.3 EASY-TO-USE END-USER INTERFACE

Advanced OLAP features become much more useful if access to them is kept simple. OLAP tool vendors learned this lesson early and have equipped their sophisticated data extraction and analysis tools with easy-to-use graphical interfaces. Many of the interface features are "borrowed" from previous generations of data analysis tools that are already familiar to end users. This familiarity makes OLAP easily accepted and readily used.

12.4.4 CLIENT/SERVER ARCHITECTURE

Client/server architecture provides a framework within which new systems can be designed, developed, and implemented. (If necessary, review the coverage in **Appendix F, "Multi-tier Client/Server Systems,"** on the student CD, which provides an in-depth look at the characteristics of client/server system architecture and principles.)

The client/server environment enables us to divide an OLAP system into several components that define its architecture. These components can then be placed on the same computer or they can be distributed among several computers. Thus OLAP is designed to meet ease-of-use, as well as system flexibility, requirements.

12.4.5 OLAP ARCHITECTURE

OLAP operational characteristics can be divided into three main modules:

- Graphical user interface (GUI)
- Analytical processing logic
- Data-processing logic

These three OLAP modules, in the client/server environment, make possible the defining features of OLAP: multidimensional data analysis, advanced database support, and an easy-to-use interface. Figure 12.6 illustrates OLAP's client/server components and attributes.

FIGURE 12.6 OLAP CLIENT/SERVER ARCHITECTURE

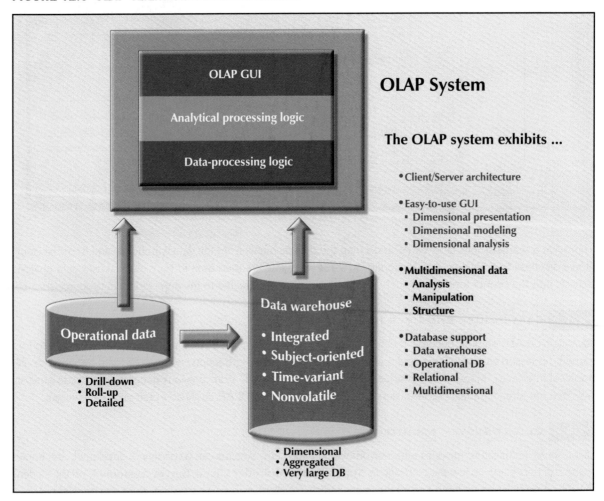

As Figure 12.6 illustrates, OLAP systems are designed to use both operational and data warehouse data. Although Figure 12.6 shows the OLAP system components located on a single computer, this single-user scenario is only one of many. In fact, one problem with the installation shown here is that each data analyst must have a powerful computer to store the OLAP system and perform all data processing locally. In addition, each analyst uses a separate copy of the data. Therefore, the data copies must be synchronized to ensure that analysts are working with the same data. In other words, each end user must have his/her own "private" copy (extract) of the data and programs, thus returning us to the *islands of information* problems discussed in **Appendix F**. This approach does not provide the benefits of a single business image shared among all users.

A more common and more practical architecture is one in which the OLAP GUI runs on client workstations, while the OLAP engine, or server, composed of the OLAP analytical processing logic and OLAP data-processing logic, runs on a shared computer. In this case, the OLAP server will be a front end to the data warehouse's decision support data. This front end or middle layer (because it sits between the data warehouse and the end-user GUI) accepts and processes the data-processing requests generated by the many end-user analytical tools. The end-user GUI might be a custom-made program or, more likely, a plug-in module that is integrated with Lotus 1-2-3, Microsoft Excel, or a third-party data analysis and query tool. Figure 12.7 illustrates such an arrangement.

FIGURE 12.7 OLAP SERVER ARRANGEMENT

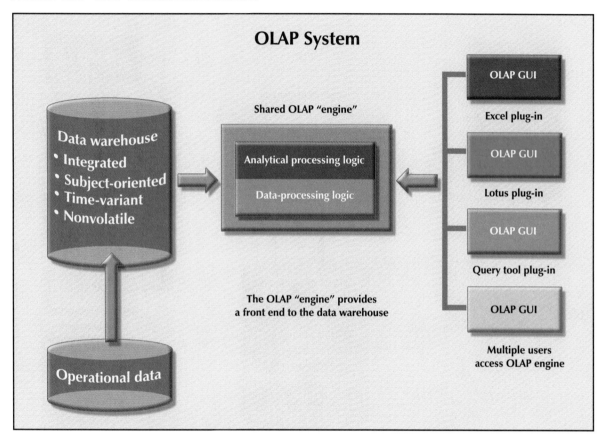

As you examine Figure 12.7, note that the data warehouse is created and maintained by a process or software tool that is independent of the OLAP system. This independent software performs the data extraction, filtering, and integration necessary to transform operational data into data warehouse data. This scenario reflects the fact that, in most cases, the data warehousing and data analysis activities are handled separately.

At this point you might ask why you need a data warehouse, if OLAP provides the necessary multidimensional data analysis of operational data. The answer lies in our definition of OLAP. We defined OLAP as an "advanced data analysis environment that supports decision making, business modeling, and an operation's research activities." The key word here is "environment," which includes client/server technology. Environment is defined as "surroundings or atmosphere." And an atmosphere surrounds a nucleus. *In this case, the nucleus is composed of all business activities within an organization as represented by the operational data.* Just as there are several layers within the atmosphere, there are several layers of data processing, each outer layer representing a more aggregated data analysis. The fact is that an OLAP system might access both data storage types (operational or data warehouse) or only one; it depends on the vendor's implementation of the product you selected. In any case, multidimensional data analysis requires some type of multidimensional data representation, which is normally provided by the OLAP engine.

In most implementations, the data warehouse and OLAP are interrelated and complementary environments. While the data warehouse represents integrated, subject-oriented, time-variant, and nonvolatile decision support data, the OLAP system provides the front end through which end users access and analyze such data. Yet an OLAP system can also directly access operational data, transforming it and storing it in a multidimensional structure. In other words, the OLAP system can provide a multidimensional data store component, as shown in Figure 12.8.

FIGURE 12.8 OLAP SERVER WITH MULTIDIMENSIONAL DATA STORE ARRANGEMENT

Figure 12.8 represents a scenario in which the OLAP engine extracts data from an operational database and then stores it in a multidimensional structure for further data analysis. The extraction process follows the same conventions used with data warehouses. Therefore, the OLAP provides a mini-data-warehouse component that looks remarkably like the data mart mentioned in previous sections. In this scenario, the OLAP engine has to perform all the data extraction, filtering, integration, classification, and aggregation functions that the data warehouse normally provides. In fact, when properly implemented, the data warehouse performs all data preparation functions instead of letting OLAP perform those chores, so there is no duplication of functions. Better yet, the data warehouse handles the data component much more efficiently than OLAP does, so you can appreciate the benefits of having a central data warehouse serve as the large enterprise decision support database.

To provide better performance, some OLAP systems merge the data warehouse and data mart approaches by storing small extracts of the data warehouse at end-user workstations. The objective is to increase the speed of data access and data visualization (the graphic representations of data trends and characteristics). The logic behind this approach is the assumption that most end users usually work with fairly small, stable data warehouse

data subsets. For example, the sales analyst is most likely to work with sales data, whereas a customer representative is likely to work with customer data, and so on. Figure 12.9 illustrates this scenario.

FIGURE 12.9 OLAP SERVER WITH LOCAL MINI DATA MARTS

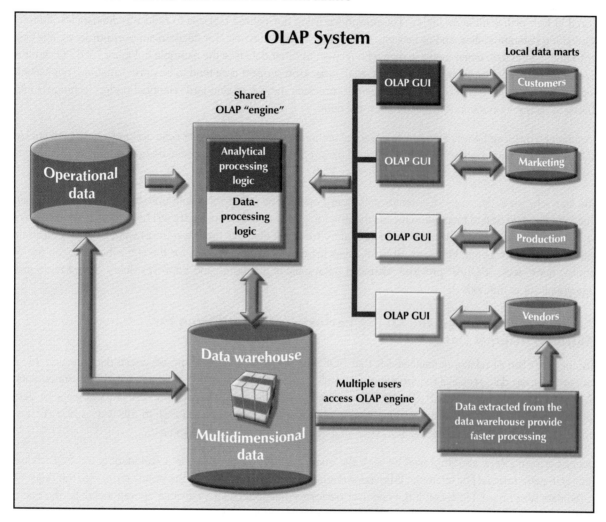

We have now summarized the main OLAP architectures you are likely to encounter. Whatever the arrangement of the OLAP components, one thing is certain: multidimensional data must be used. But how are such multidimensional data to be stored and managed best? OLAP proponents are sharply divided. Some favor the use of relational databases to store the multidimensional data; others argue for the superiority of specialized multidimensional databases to store multidimensional data. We will next examine the basic characteristics of each approach.

12.4.6 RELATIONAL OLAP

Relational online analytical processing (ROLAP) provides OLAP functionality by using relational databases and familiar relational query tools to store and analyze multidimensional data. This approach builds on existing relational technologies and represents a natural extension to all those companies that already use relational database management systems within their organizations. ROLAP adds the following extensions to traditional RDBMS technology:

- Multidimensional data schema support within the RDBMS.

- Data access language and query performance are optimized for multidimensional data.

- Support for very large databases (VLDBs).

Multidimensional Data Schema Support within the RDBMS

Relational technology uses normalized tables to store data. The reliance on normalization as the design methodology for relational databases is seen as a stumbling block to its use in OLAP systems. Normalization divides business entities into smaller pieces to produce the normalized tables. For example, sales data components might be stored in four or five different tables. The reason for using normalized tables is to reduce redundancies, thereby eliminating data anomalies, and to facilitate data updates. Unfortunately, for decision support purposes, it is easier to understand data when they are seen with respect to other data (see the example in Figure 12.4). Given this view of the data environment, we have stressed that decision support data tend to be nonnormalized, duplicated, and preaggregated. These characteristics seem to preclude the use of standard relational design techniques and RDBMSs as the foundation for multidimensional data.

Fortunately for those heavily invested in relational technology, ROLAP uses a special design technique to enable RDBMS technology to support multidimensional data representations. This special design technique is known as a star schema; we will cover it in detail in Section 12.5.

The star schema is designed to optimize data query operations rather than data update operations. Naturally, changing the data design foundation means that the tools used to access such data will have to change. End users who are familiar with the traditional relational query tools will discover that these tools will not work efficiently with the new star schema. However, ROLAP saves the day by adding support for the star schema to the use of familiar query tools. ROLAP provides advanced data analysis functions, and improves query optimization and data visualization methods.

Data Access Language and Query Performance Optimized for Multidimensional Data

Another criticism of relational databases is that SQL is not suited for performing advanced data analysis. Most decision support data requests require the use of multiple-pass SQL queries or multiple nested SQL statements. To answer this criticism, ROLAP extends SQL so that it can differentiate between access requirements for data warehouse data (based on the star schema) and operational data (normalized tables). In this way, a ROLAP system is able to properly generate the SQL code required to access the star schema data.

Query performance is also improved because the query optimizer is modified so it can identify the SQL code's intended query targets. For example, if the query target is the data warehouse, the optimizer passes the requests to the data warehouse. However, if the end user performs drill-down queries against operational data, the query optimizer identifies this operation and properly optimizes the SQL requests before passing them through to the operational DBMS.

Another source of improved query performance is the use of advanced indexing techniques such as bitmapped indexes within relational databases. As the name suggests, a bitmapped index is based on 0 and 1 bits to represent a given condition. For example, if the REGION attribute in Figure 12.2 has only four outcomes—North, South, East, and West—these outcomes may be represented as shown in Table 12.7. (Only the first 10 rows from Figure 12.2 are represented in Table 12.7. The "1" represents "bit on," and the "0" represents "bit off." For example, to represent a row with a REGION attribute = "East," only the "East" bit would be on. Note that each row must be represented in the index table.)

As you examine Table 12.7, note that the index takes a minimum amount of space. Therefore, bitmapped indexes are much more efficient at handling large amounts of data than are the indexes typically found in many relational databases. But do keep in mind that bitmapped indexes are primarily used in situations where the number of possible values for an attribute—in other words, the attribute domain—is fairly small. For example, REGION only has four outcomes in this example. Marital status—married, single, widowed, divorced—would be another good bitmapped index candidate, as would be gender—M or F.

TABLE 12.7 BITMAP REPRESENTATION OF REGION VALUES

NORTH	SOUTH	EAST	WEST
0	0	1	0
0	0	1	0
1	0	0	0
1	0	0	0
1	0	0	0
0	1	0	0
0	1	0	0
0	1	0	0
0	0	0	1
0	0	0	1

ROLAP tools are mainly client/server products in which the end-user interface, the analytical processing, and the data processing take place on different computers. Figure 12.10 shows the interaction of the client/server ROLAP components.

ADDITIONAL CD CONTENT

If you want to learn more about client/server technology, architecture, components, and uses, look in the \Appendixes\Appendix F folder.

Support for Very Large Databases

Recall that support for VLDBs is a mandatory requirement for DSS databases. Therefore, if the relational database is used in a DSS role, it also must be able to store very large amounts of data. Both the storage capability and the process of loading data into the database are crucial. Therefore, the RDBMS must have the proper tools to import, integrate, and populate the data warehouse with operational data. Decision support data are normally loaded in bulk (batch) mode from the operational data. However, batch operations require that both the source and the destination databases be reserved (locked). The speed of the data-loading operations is important, especially when you realize that most operational systems run 24 hours a day, 7 days a week, 52 weeks a year. Therefore, the window of opportunity for maintenance and batch loading is open only briefly, typically during slack periods.

With an open client/server architecture, ROLAP provides advanced decision support capabilities that are scalable to the entire enterprise. Clearly, ROLAP is a logical choice for companies that already use relational databases for their operational data. Given the size of the relational database market, it is hardly surprising that most current RDBMS vendors have extended their products to support data warehouses.

FIGURE 12.10 TYPICAL ROLAP CLIENT/SERVER ARCHITECTURE

ROLAP System

ROLAP server

ROLAP analytical processing logic

ROLAP data-processing logic

ROLAP GUI

ROLAP GUI

ROLAP GUI

ROLAP GUI

Data warehouse data

Operational data

ROLAP server interprets end-user requests and builds complex SQL queries required to access the data warehouse. If an end user requests a drill-down operation, the ROLAP server builds the required SQL code to access the operational database.

GUI front end runs on client computer and passes data-analysis requests to the ROLAP server. The GUI receives data replies from the ROLAP server and formats them according to the end user's presentation needs.

An RDBMS is used to manage both data warehouse and operational data. The RDBMS executes data requests and sends back the data sets.

12.4.7 MULTIDIMENSIONAL OLAP

Multidimensional online analytical processing (MOLAP) extends OLAP functionality to **multidimensional database management systems (MDBMSs)**. (An MDBMS uses special proprietary techniques to store data in matrix-like n-dimensional arrays.) MOLAP's premise is that multidimensional databases are best suited to manage, store, and analyze multidimensional data. Most of the proprietary techniques used in MDBMSs are derived from engineering fields such as computer-aided design/computer-aided manufacturing (CAD/CAM) and geographic information systems (GIS).

Conceptually, MDBMS end users visualize the stored data as a three-dimensional cube known as a **data cube**. The location of each data value in the data cube is a function of the x, y, and z axes in a three-dimensional space. The x, y, and z axes represent the dimensions of the data value. The data cubes can grow to n-number of dimensions, thus becoming hypercubes. Data cubes are created by extracting data from the operational databases or from the data warehouse. One important characteristic of data cubes is that they are static; that is, they are not subject to change and must be created before they can be used. Data cubes cannot be created by ad hoc queries. Instead, you query pre-created cubes with defined axes; for example, a cube for sales will have the product, location, and time dimensions, and you can query only those dimensions. Therefore, the data cube creation process is critical and requires in-depth front-end design work. This front-end design work may be well justified because MOLAP databases are known to be much faster than their ROLAP counterparts, especially when dealing with small to medium data sets. To speed data access, data cubes are normally held in memory, in what is called the

cube cache. (A data cube is only a window to a predefined subset of data in the database. A data *cube* and a data*base* are not the same thing.) Because MOLAP also benefits from a client/server infrastructure, the cube cache can be located at the MOLAP server, at the MOLAP client, or in both locations. Figure 12.11 shows the basic MOLAP architecture.

FIGURE 12.11 MOLAP CLIENT/SERVER ARCHITECTURE

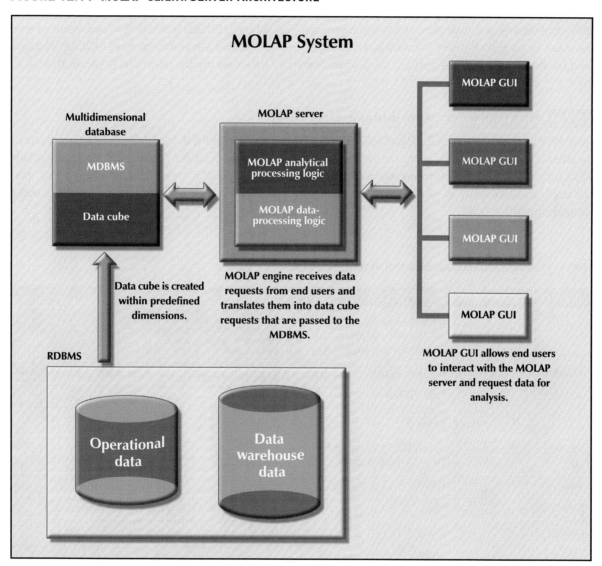

Because the data cube is predefined with a set number of dimensions, the addition of a new dimension requires that the entire data cube be re-created. This re-creation process is a time-consuming operation. Therefore, if data cubes are created too often, the MDBMS loses some of its speed advantage over the relational database. And, although MDBMSs have performance advantages over relational databases, the MDBMS is best suited to small and medium data sets. Scalability is somewhat limited, because the size of the data cube is restricted to avoid lengthy data access times caused by having less work space (memory) available for the operating system and the application programs. In addition, the MDBMS makes use of proprietary data storage techniques that, in turn, require proprietary data access methods using a multidimensional query language.

Multidimensional data analysis is also affected by how the database system handles sparsity. **Sparsity** is a measurement of the density of the data held in the data cube. Sparsity is computed by dividing the total number of

12.5 STAR SCHEMAS

The **star schema** is a data modeling technique used to map multidimensional decision support data into a relational database. In effect, the star schema creates the near equivalent of a multidimensional database schema from the existing relational database. The star schema was developed because existing relational modeling techniques, ER, and normalization did not yield a database structure that served advanced data analysis requirements well.

Star schemas yield an easily implemented model for multidimensional data analysis, while still preserving the relational structures on which the operational database is built. The basic star schema has four components: facts, dimensions, attributes, and attribute hierarchies. We will describe each one of these components in detail in the next sections.

12.5.1 FACTS

Facts are numeric measurements (values) that represent a specific business aspect or activity. For example, sales figures are numeric measurements that represent product and/or service sales. Facts commonly used in business data analysis are units, costs, prices, and revenues. Facts are normally stored in a fact table that is the center of the star schema. The **fact table** contains facts that are linked through their dimensions (covered in the next section).

Facts can also be computed or derived at run time. Such computed or derived facts are sometimes called **metrics** in order to differentiate them from stored facts. The fact table is updated periodically (daily, weekly, monthly, etc.) with data from operational databases.

12.5.2 DIMENSIONS

Dimensions are qualifying characteristics that provide additional perspectives to a given fact. Earlier in the chapter we stated that dimensions are of interest because *DSS data are almost always viewed in relation to other data.* For instance, sales might be compared by product from region to region, and from one time period to the next. The kind of problem typically addressed by a DSS might be: "Make a comparison of the sales of unit X by region, for the first quarters of 1990 through 1999." In this example, sales have product, location, and time dimensions. In effect, dimensions are the magnifying glass through which we study the facts. Such dimensions are normally stored in **dimension tables**. Figure 12.12 depicts a star schema for sales with product, location, and time dimensions.

12.5.3 ATTRIBUTES

Each dimension table contains attributes. Attributes are often used to search, filter, or classify facts. *Dimensions provide descriptive characteristics about the facts through their attributes.* Therefore, the data warehouse designer must define common business attributes that will be used by the data analyst to narrow a search, group information, or describe dimensions. Using a sales example, we can identify some possible attributes for each dimension, illustrated in Table 12.9.

FIGURE 12.12 **SIMPLE STAR SCHEMA**

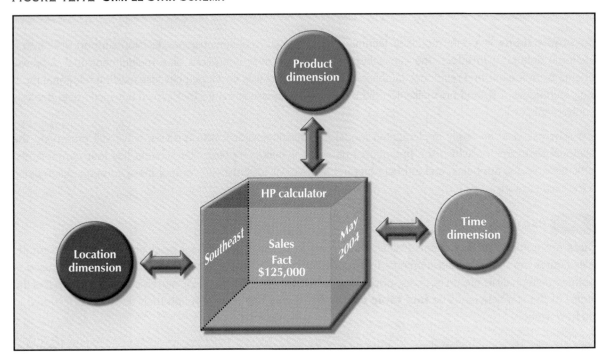

TABLE 12.9 **POSSIBLE ATTRIBUTES FOR SALES DIMENSIONS**

DIMENSION NAME	DESCRIPTION	POSSIBLE ATTRIBUTES
Location	Anything that provides a description of the location Example: Nashville, Store 101, South Region, TN, etc.	Region, state, city, store, etc.
Product	Anything that provides a description of the product sold For example, hair care product, shampoo, Natural Essence brand, 5.5 oz. bottle, blue liquid, etc.	Product type, product ID, brand, package, presentation, color, size, etc.
Time	Anything that provides a time frame for the sales fact. For example, the year 1999, the month of July, the date 07/29/1999, the time 4:46 p.m., etc.	Year, quarter, month, week, day, time of day, etc.

These product, location, and time dimensions add a business perspective to the sales facts. The data analyst can now group the sales figures for a given product, in a given region, and at a given time. The star schema, through its facts and dimensions, can provide the data when needed and in the required format. And, it can do this without imposing the burden of the additional and unnecessary data (such as order number, purchase order number, status, etc.) that commonly exist in operational databases.

Conceptually, the sales example's multidimensional data model is best represented by a three-dimensional cube. Of course, we are not implying that there is a limit on the number of dimensions that can be associated to a fact table. There is no mathematical limit to the number of dimensions we can use. However, the use of a three-dimensional model makes it easier for us to visualize the problem. In this three-dimensional example, using the multidimensional data analysis jargon, the cube illustrated in Figure 12.13 represents a view of sales dimensioned by product, location, and time.

FIGURE 12.13 THREE-DIMENSIONAL VIEW OF SALES

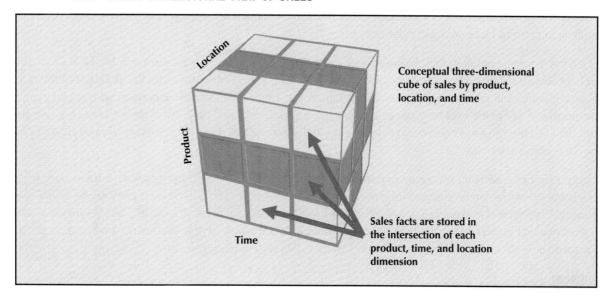

As you examine Figure 12.13, note that each sales value stored in the cube is associated with the location, product, and time dimensions. However, keep in mind that this cube is only a *conceptual* representation of multidimensional data, and it does not show how the data are physically stored in a data warehouse. A ROLAP engine stores data in an RDBMS and uses its own data analysis logic and the end-user GUI to perform multidimensional analysis. A MOLAP system stores data in an MDBMS, using proprietary matrix and array technology to simulate this multidimensional cube.

Whatever the underlying database technology, one of the main features of multidimensional analysis is its ability to focus on specific "slices" of the cube. For example, the product manager might be interested in examining the sales of a product, whereas the store manager is interested in examining the sales made by a particular store. Using multidimensional jargon, the ability to focus on slices of the cube to perform a more detailed analysis is known as **slice and dice**. Figure 12.14 illustrates the slice and dice concept. As you look at Figure 12.14, note that each cut across the cube yields a slice. Intersecting slices produce small cubes that constitute the "dice" part of the "slice and dice" operation.

FIGURE 12.14 SLICE AND DICE VIEW OF SALES

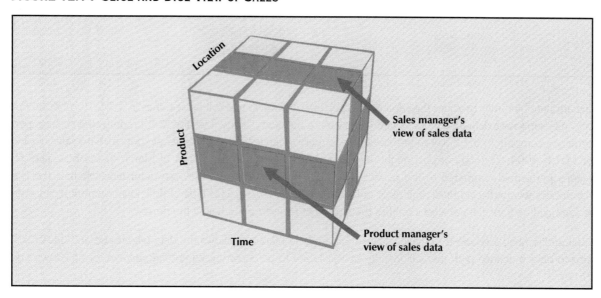

To slice and dice, we must be able to identify each slice of the cube. This is done by using the values of each attribute in a given dimension. For example, to use the location dimension we might need to define a STORE_ID attribute in order to focus on a particular store.

Given the requirement for attribute values in a slice and dice environment, let's reexamine Table 12.9. Note that each attribute adds an additional perspective to the sales facts, thus setting the stage for finding new ways to search, classify, and possibly aggregate information. For example, the location dimension adds a geographic perspective of where the sales took place: in which region, state, city, and store. All of the attributes are selected with the objective of providing decision support data to the end user so that (s)he can study sales by each one of the dimension's attributes.

Remember that time is an especially important dimension. The time dimension provides a framework from which sales patterns can be analyzed and, possibly, predicted. Also, the time dimension plays an important role when the data analyst is interested in looking at sales aggregates by week, month, quarter, and so on. Given the importance and universality of the time dimension from a data analysis perspective, many vendors have added automatic time dimension management features to their data warehousing products.

12.5.4 ATTRIBUTE HIERARCHIES

Attributes within dimensions can be ordered in a well-defined attribute hierarchy. The **attribute hierarchy** provides a top-down data organization that is used for two main purposes: aggregation and drill-down/roll-up data analysis. For example, Figure 12.15 shows how the location dimension attributes can be organized in a hierarchy by region, state, city, and store.

FIGURE 12.15 LOCATION ATTRIBUTE HIERARCHY

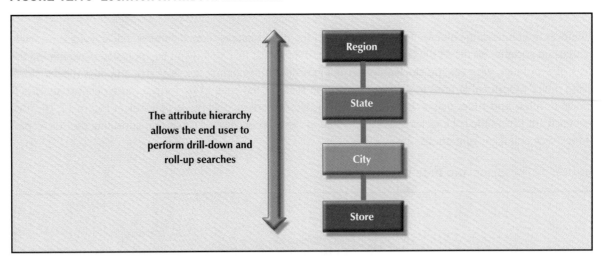

The attribute hierarchy provides the capability to perform drill-down and roll-up searches in a data warehouse. For example, suppose a data analyst looks at the answers to the query "How does the 2003 month-to-date sales performance compare to the 2004 month-to-date sales performance?" The data analyst spots a sharp sales decline for March 2004. The data analyst might decide to drill down inside the month of March to see how sales by regions performed compared to the previous year. By doing this the analyst can determine whether the low March sales were reflected in all regions or only in a particular region. This type of drill-down operation can even be extended until the data analyst identifies the store that is performing below the norm.

The just-described scenario is possible because the attribute hierarchy allows the data warehouse and OLAP systems to have a defined path that will identify how data are to be decomposed and aggregated for drill-down and

roll-up operations. It is not necessary for all attributes to be part of an attribute hierarchy; some attributes exist merely to provide narrative descriptions of the dimensions. But keep in mind that the attributes from different dimensions can be grouped to form a hierarchy. For example, after you drill down from city to store, you may want to drill down using the product dimension so the manager can identify slow products in the store. The product dimension can be based on the product group (dairy, meat, etc.) or on the product brand (Brand A, Brand B, etc.).

Figure 12.16 illustrates a scenario in which the data analyst studies sales facts using the product, time, and location dimensions. In this example, the product dimension is set to "All products," meaning that the data analyst will see all products on the y axis. The time dimension (x axis) is set to "Quarter," meaning that the data are aggregated by quarters (for example, total sales of products A, B, and C in Q1, Q2, Q3, and Q4). Finally, the location dimension is initially set to "Region," thus ensuring that each cell contains the total regional sales for a given product in a given quarter.

FIGURE 12.16 ATTRIBUTE HIERARCHIES IN MULTIDIMENSIONAL ANALYSIS

The simple data analysis scenario illustrated in Figure 12.16 provides the data analyst with three different information paths. On the product dimension (the y axis), the data analyst can request to see all products, products grouped by type, or just one product. On the time dimension (the x axis), the data analyst can request time-variant data at different levels of aggregation: year, quarter, month, or week. Each sales value initially shows the total sales, by region, of each product. When using a GUI, clicking on the region cell will enable the data analyst to drill down to see sales by states within the region. Clicking again on one of the state values will yield the sales for each city in the state, and so forth.

As the preceding examples illustrate, attribute hierarchies determine how the data in the data warehouse are extracted and presented. The attribute hierarchy information is stored in the DBMS's data dictionary and is used by the OLAP tool to properly access the data warehouse. Once such access is ensured, query tools must be closely integrated with the data warehouse's metadata, and they must support powerful analytical capabilities.

12.5.5 STAR SCHEMA REPRESENTATION

Facts and dimensions are normally represented by physical tables in the data warehouse database. The fact table is related to each dimension table in a many-to-one (M:1) relationship. In other words, many fact rows are related to each dimension row. Using our sales example, we can conclude that each product appears many times in the sales fact table.

Fact and dimension tables are related by foreign keys and are subject to the familiar primary key/foreign key constraints. The primary key on the one side, the dimension table, is stored as part of the primary key on the many side, the fact table. *Because the fact table is related to many dimension tables, the primary key of the fact table is a composite primary key.* Figure 12.17 illustrates the relationships among the sales fact table and the product, location, and time dimension tables. To show you how easily the star schema can be expanded, we have added a customer dimension to the mix. Adding the customer dimension merely required including the CUST_ID in the SALES fact table and adding the CUSTOMER table to the database.

FIGURE 12.17 STAR SCHEMA FOR SALES

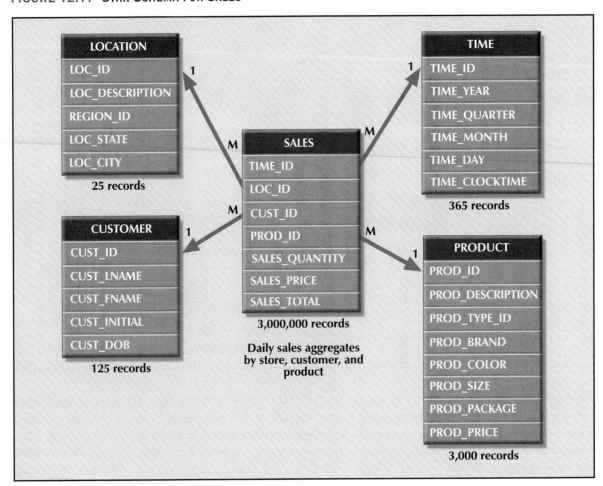

The composite primary key for the sales fact table is composed of TIME_ID, LOCATION_ID, CUST_ID, and PRODUCT_ID. Each record in the sales fact table is uniquely identified by the combination of values for each of the fact table's foreign keys. *By default, the fact table's primary key is always formed by combining the foreign keys pointing to each dimension table to which it is related.* In this case, each sales record represents each product sold to a specific customer, at a specific time, and in a specific location. In this schema, the time dimension table represents daily periods, so the sales fact table represents daily sales aggregates by product and by customer. Because fact tables contain the actual values used in the decision support process, these values are repeated many times in the fact tables. Therefore, the fact tables are always the largest tables in the star schema. Because the dimension tables contain only nonrepetitive information (all unique salespersons, all unique products, etc.), the dimension tables are always smaller than the fact tables.

In a typical star schema, each dimension record is related to thousands of fact records. For example, "widget" appears only once in the product dimension, but it has thousands of corresponding records in the sales fact table. This characteristic of the star schema facilitates data retrieval functions, because most of the time the data analyst will look at the facts through the dimension's attributes. Therefore, a DSS-optimized data warehouse DBMS first searches the smaller dimension tables before accessing the larger fact tables.

Data warehouses usually have many fact tables. Each fact table is designed to answer specific decision support questions. For example, suppose that we develop a new interest in orders, while maintaining our original interest in sales. In this scenario, we should maintain an orders fact table and a sales fact table in the same data warehouse. If orders are considered to be an organization's key interest, the orders fact table should be the center of a star schema that might have vendor, product, and time dimensions. In this case, an interest in vendors yields a new vendor dimension represented by a new VENDOR table in our database. The product dimension is represented by the same product table that we used in our initial sales star schema. However, given the interest in orders as well as sales, the time dimension now requires special attention. If the orders department uses the same time periods as the sales department, time can be represented by the same time table. If different time periods are used, we must create another table, perhaps named ORDER_TIME, to represent the time periods used by the orders department. In Figure 12.18, the Orders star schema shares the product, vendor, and time dimensions.

Multiple fact tables also can be created for performance and semantic reasons. The following section explains several performance-enhancing techniques that can be used within the star schema.

12.5.6 STAR SCHEMA PERFORMANCE-IMPROVING TECHNIQUES

The creation of a database that provides fast and accurate answers to data analysis queries is the data warehouse design's prime objective. Therefore, performance-enhancement actions might target query speed through the facilitation of SQL code as well as through better semantic representation of business dimensions. Four techniques are often used to optimize data warehouse design:

- Normalization of dimensional tables
- Multiple fact tables representing different aggregation levels
- Denormalization of fact tables
- Table partitioning and replication

FIGURE 12.18 ORDERS STAR SCHEMA

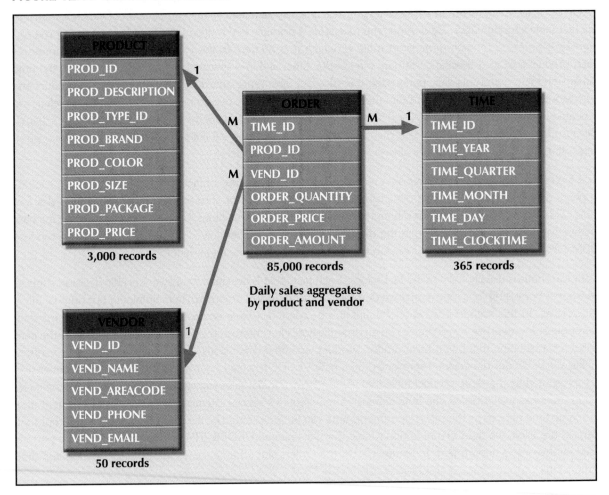

Normalization of Dimensional Tables

Normalization is done to achieve semantic simplicity and facilitate end-user navigation through the dimensions. For example, if the location dimension table contains transitive dependencies among region, state, and city, we can revise these relationships to the 3NF (third normal form), as shown in Figure 12.19. (If necessary, review normalization techniques in Chapter 5, "Normalization of Database Tables.")

By normalizing the dimension tables, we simplify the data-filtering operations related to the dimensions. In this example, the region, state, city, and location contain very few records compared to the sales fact table. Only the location table is directly related to the sales fact table.

> **N O T E**
>
> Although we gain structural simplicity by using the dimension tables shown in Figure 12.19, there is a price to pay for this simplicity. For example, if you want to aggregate the data by region, you must use a four-table join, thus increasing the complexity of the SQL statements. The star schema in Figure 12.17 uses a LOCATION dimension table that greatly facilitates the data retrieval by eliminating multiple join operations. Here is yet another example of the trade-offs that designers must consider.

FIGURE 12.19 NORMALIZED DIMENSION TABLES

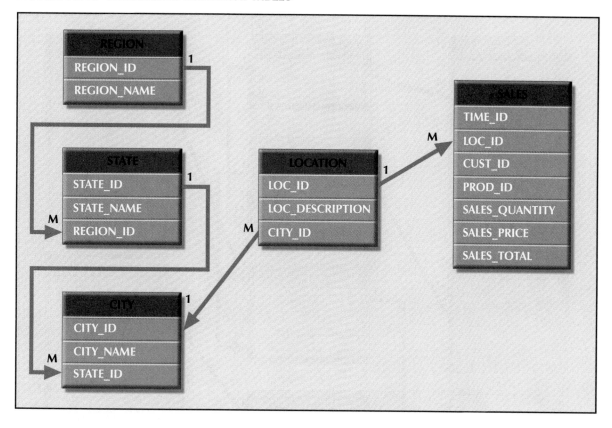

Multiple Fact Tables Representing Different Aggregation Levels

We can also speed up query operations by creating and maintaining multiple fact tables related to each level of aggregation (region, state, and city) in the location dimension. These aggregate tables are precomputed at the data-loading phase, rather than at run time. The purpose of this technique is to save processor cycles at run time, thereby speeding up data analysis. An end-user query tool optimized for decision analysis then properly accesses the summarized fact tables, instead of computing the values by accessing a lower level of detail fact table. This technique is illustrated in Figure 12.20. Using the initial sales example, we have added the region, state, and city aggregate fact tables.

The data warehouse designer must identify which levels of aggregation to precompute and store in the database. These multiple aggregate fact tables are updated during each load cycle in batch mode. And, because the objective is to minimize access and processing time, the data warehouse designer must select which aggregation fact tables to create, according to the expected frequency of use and the processing time required to calculate a given aggregation level at run time.

Denormalizing the Fact Tables

Denormalizing fact tables improves data access performance and saves data storage space. The latter objective, however, is becoming less of an issue. Data storage costs decrease almost daily, and DBMS limitations that restrict database and table size limits, record size limits, and the maximum number of records in a single table, have far more negative effects than raw storage space costs.

FIGURE 12.20 **MULTIPLE FACT TABLES**

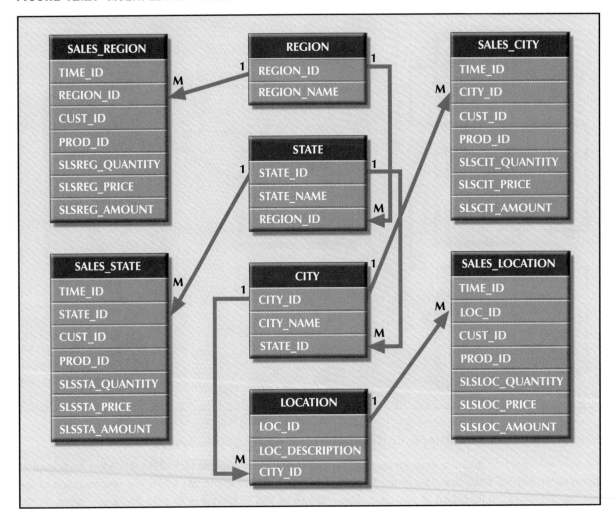

Denormalization improves performance by using one single record to store data that normally take many records. For example, to compute the total sales for all products in all regions we might have to access the region sales aggregates and summarize all the records in this table. If we have 300,000 product sales, we could be summarizing at least 300,000 rows. Although this might not be a very taxing operation for a DBMS, a comparison of, let's say, 10 years' worth of previous sales begins to bog down the system. In such cases, it is useful to have special aggregate tables that are denormalized. For example, a YEAR_TOTALS table might contain the following fields: YEAR_ID, MONTH_1, MONTH_2...MONTH_12, and each year's total. Such tables easily can be used to serve as a basis for year-to-year comparisons at the top month level, the quarter level, or the year level. Here again, design criteria, such as frequency of use and performance requirements, are evaluated against the possible overload placed on the DBMS to manage these denormalized relations.

Table Partitioning and Replication

Because we already covered table partitioning and replication in detail in Chapter 10, "Distributed Database Management Systems," we will examine these techniques only as they specifically relate to the data warehouse. Table partitioning and replication are particularly important when a DSS is implemented in widely dispersed geographic areas. **Partitioning** splits a table into subsets of rows or columns and places the subsets close to the client computer to improve data access time. **Replication** makes a copy of a table and places it in a different location, also to improve access time.

No matter which performance-enhancement scheme is used, time is the most common dimension used in business data analysis. Therefore, it is very common to have one fact table for each level of aggregation defined within the time dimension. For example, in our sales example we might have five aggregate sales fact tables: daily, weekly, monthly, quarterly, and yearly. These fact tables must have an implicit or explicit periodicity defined. **Periodicity**, usually expressed as current year only, previous years, or all years, provides information about the timespan of the data stored in the table.

At the end of each year, daily sales for the current year are moved to another table that contains previous years' daily sales only. This table actually contains all sales records from the beginning of operations, with the exception of the current year. The data in the current year and previous years tables thus represent the complete sales history of the company. We can have the previous years sales table replicated at several locations to avoid remote access to the historic sales data, which can cause slow response time. The possible size of this table is enough to intimidate all but the bravest of query optimizers. Here is one case in which denormalization would be of value!

12.6 IMPLEMENTING A DATA WAREHOUSE

Organization-wide information system development is subject to numerous constraints. Some of these constraints are based on available funding. Others are a function of management's view of the role played by an IS department and of the extent and depth of the information requirements. Add the constraints imposed by corporate culture, and you understand why no single formula can describe perfect data warehouse development. Therefore, rather than proposing a single data warehouse design and implementation methodology, we identify a few factors that appear to be common to data warehousing.

12.6.1 THE DATA WAREHOUSE AS AN ACTIVE DECISION SUPPORT FRAMEWORK

Perhaps the first thing to remember is that a data warehouse is not a static database. Instead, it is a dynamic framework for decision support that is, almost by definition, always a work in progress. Because it is the foundation of a modern DSS, the design and implementation of the data warehouse means that you are involved in the design and implementation of a complete database-system-development infrastructure for company-wide decision support. Although it is easy to focus on the data warehouse database as the DSS central data repository, you must remember that the decision support infrastructure includes hardware, software, people, and procedures, as well as data. The argument that the data warehouse is the only *critical* DSS success component is as misleading as the argument that a human being needs only a heart or a brain to function. The data warehouse is a critical component of a modern DSS, but it is certainly not the only critical component. Therefore, its design and implementation must be examined in the light of the entire infrastructure.

12.6.2 A COMPANY-WIDE EFFORT THAT REQUIRES USER INVOLVEMENT

Designing a data warehouse means being given an opportunity to help develop an integrated data model that captures the data that are considered to be essential to the organization, from both end-user and business perspectives. Data warehouse data cross departmental lines and geographical boundaries. Because the data warehouse represents an attempt to model all the organization's data, you are likely to discover that organizational components (divisions, departments, support groups, etc.) often have conflicting goals, and it certainly will be easy to find data inconsistencies and damaging redundancies. Information is power, and the control of its sources and uses is likely to trigger turf battles, end-user resistance, and power struggles at all levels. Building the perfect

data warehouse is not just a matter of knowing how to create a star schema; it requires managerial skills to deal with conflict resolution, mediation, and arbitration. In short, the designer must:

- Involve end users in the process.

- Secure end users' commitment from the beginning.

- Create continuous end-user feedback.

- Manage end-user expectations.

- Establish procedures for conflict resolution.

12.6.3 SATISFY THE TRILOGY: DATA, ANALYSIS, AND USERS

Great managerial skills are not, of course, solely sufficient. The technical aspects of the data warehouse must be addressed as well. The old adage of input-process-output repeats itself here. The data warehouse designer must satisfy:

- Data integration and loading criteria.

- Data analysis capabilities with acceptable query performance.

- End-user data analysis needs.

The foremost technical concern in implementing a data warehouse is to provide end-user decision support with advanced data analysis capabilities—at the right moment, in the right format, using the right data, and at the right cost.

12.6.4 APPLY DATABASE DESIGN PROCEDURES

We covered the database life cycle and the database design process in Chapter 8, "Database Design," so perhaps it is wise to begin with a review of the traditional database design procedures. These design procedures must then be adapted to fit the data warehouse requirements. Remember that the data warehouse derives its data from operational databases, and you will understand why a solid foundation in operational database design is important. (It's difficult to produce good data warehouse data when the operational database data are corrupted.) Figure 12.21 depicts a simplified process for implementing the data warehouse.

As noted, developing a data warehouse is a company-wide effort that requires many resources: human, financial, and technical. Providing company-wide decision support requires a sound architecture based on a mix of people skills, technology, and managerial procedures that is often difficult to find and implement. For example:

- The sheer and often mind-boggling quantity of decision support data is likely to require the latest hardware and software—that is, advanced computers with multiple processors, advanced database systems, and very large-capacity storage units. In the not-too-distant past, these requirements usually prompted the use of a mainframe-based system. Today's client/server technology offers many other choices to implement a data warehouse.

- It is also imperative to have very detailed procedures to orchestrate the flow of data from the operational databases to the data warehouse. Data flow control includes data extraction, validation, and integration.

- To implement and support the data warehouse architecture, you also need people with advanced database design, software integration, and management skills.

FIGURE 12.21 DATA WAREHOUSE IMPLEMENTATION ROAD MAP

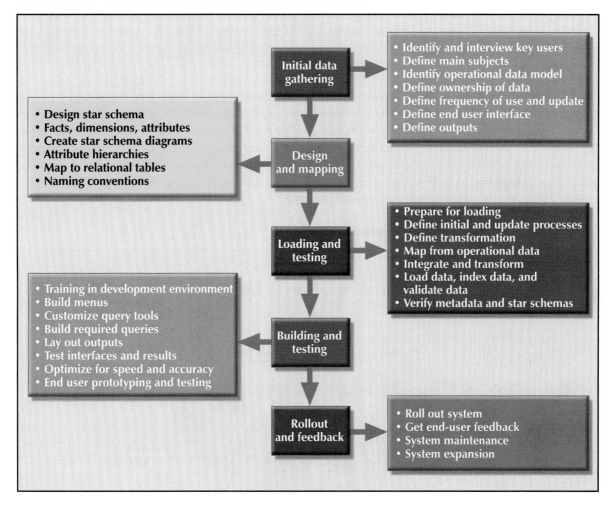

12.7 DATA MINING

The purpose of data analysis is to discover previously unknown data characteristics, relationships, dependencies, or trends. Such discoveries then become part of the information framework on which decisions are built. *A typical data analysis tool relies on the end users to define the problem, select the data, and initiate the appropriate data analyses to generate the information that helps model and solve problems those end users uncover.* In other words, the end user reacts to an external stimulus—the discovery of the problem itself. If the end user fails to detect a problem, no action is taken. Given this limitation, some current DSSs now support various types of automated alerts. These alerts are software agents that constantly monitor certain parameters, such as sales indicators and inventory levels, and then perform specified actions (send e-mail or alert messages, run programs, etc.) when such parameters reach predefined levels.

In contrast to the traditional (reactive) DSS tools, data mining is *proactive*. Instead of having the end user define the problem, select the data, and select the tools to analyze such data, *data-mining tools automatically search the data for anomalies and possible relationships, thereby identifying problems that have not yet been identified by the end user*. In other words, **data-mining** tools analyze the data, uncover problems or opportunities hidden in the data relationships, form computer models based on their findings, and then use the models to predict business behavior—requiring minimal end-user intervention. The end user is, therefore, able to use the

system's findings to gain knowledge that may yield competitive advantages. Data mining describes a new breed of specialized decision support tools that automate data analysis. In short, data-mining tools *initiate* analyses to create knowledge. Such knowledge can be used to address any number of business problems. For example, banks and credit card companies use knowledge-based analysis to detect fraud, thereby decreasing fraudulent transactions.

Given its premise, we might describe data mining as a methodology designed to perform knowledge-discovery expeditions over the database data with only minimal end-user intervention during the actual knowledge-discovery phase. To put this description in perspective, let's use the pyramid in Figure 12.22 to represent three stages: data, information, and knowledge. Data form the pyramid base and represent what most organizations collect in their operational databases. The second level contains information that represents the purified and processed data. Information forms the basis for decision making and business understanding. Knowledge is found at the pyramid's apex and represents highly specialized information.

FIGURE 12.22 EXTRACTION OF KNOWLEDGE FROM DATA

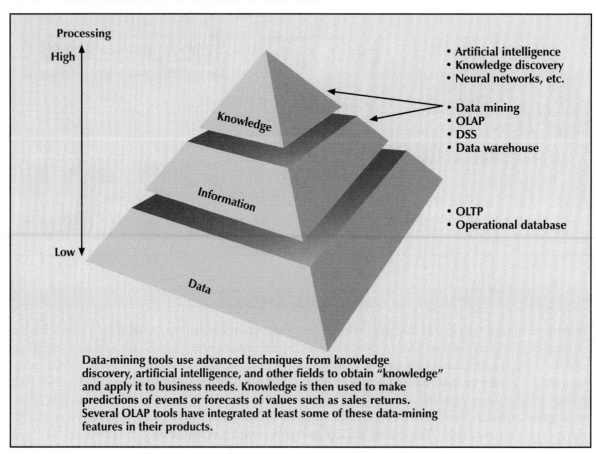

It is difficult to provide a precise list of data-mining tool characteristics. For one thing, the current generation of data-mining tools contains many design and application variations to fit data-mining requirements. Additionally, the many variations exist because there are no established standards that govern the creation of data-mining tools. In short, each data-mining tool seems to be governed by a different approach and focus, thus generating families of data-mining tools that focus on market niches such as marketing, retailing, finance, healthcare, investments, insurance, and banking. Within a given niche, data-mining tools can use certain algorithms, and those algorithms can be implemented in different ways and/or applied over different data.

In spite of the lack of precise standards, we can conclude that data mining is subject to four phases:

1. Data preparation
2. Data analysis and classification
3. Knowledge acquisition
4. Prognosis

In the *data preparation phase*, the main data sets to be used by the data-mining operation are identified and cleansed of any data impurities. Because the data in the data warehouse are already integrated and filtered, the data warehouse usually is the target set for data-mining operations.

The *data analysis and classification phase* studies the data to identify common data characteristics or patterns. During this phase the data-mining tool applies specific algorithms to find:

- Data groupings, classifications, clusters, or sequences.
- Data dependencies, links, or relationships.
- Data patterns, trends, and deviations.

The *knowledge-acquisition phase* uses the results of the data analysis and classification phase. During this phase, the data-mining tool (with possible intervention by the end user) selects the appropriate modeling or knowledge-acquisition algorithms. The most typical algorithms used in data mining are based on neural networks, decision trees, rules induction, genetic algorithms, classification and regression trees, memory-based reasoning, or nearest neighbor and data visualization. A data-mining tool may use many of these algorithms in any combination to generate a computer model that reflects the behavior of the target data set.

Although many data-mining tools stop at the knowledge-acquisition phase, others continue to the *prognosis phase*. In this phase, the data-mining findings are used to predict future behavior and forecast business outcomes. Examples of data-mining findings can be:

- Sixty-five percent of customers who did not use a particular credit card in the last six months are 88 percent likely to cancel that account.
- Eighty-two percent of customers who bought a 27-inch or larger TV are 90 percent likely to buy an entertainment center within the next four weeks.
- If age < 30, and income <= 25,000, and credit rating < 3, and credit amount > 25,000, then the minimum loan term is 10 years.

The complete set of findings can be represented in a decision tree, a neural net, a forecasting model, or a visual presentation interface that is then used to project future events or results. For example, the prognosis phase might project the likely outcome of a new product rollout or a new marketing promotion. Figure 12.23 illustrates the different phases of the data-mining techniques.

Because data-mining technology is still in its infancy, some of the data-mining findings might fall outside the boundaries of what business managers expect. For example, a data-mining tool might find a close relationship between a customer's favorite brand of soda and the brand of the tires on the customer's car. Clearly, this relationship might not be held in high regard among sales managers. (In regression analysis, such relationships are commonly described by the label "idiot correlation.") Fortunately, data mining usually yields more meaningful results. In fact, data mining has proved to be very helpful in finding practical relationships among data that help define customer buying patterns, improve product development and acceptance, reduce healthcare fraud, analyze stock markets, and so on.

FIGURE 12.23 DATA-MINING PHASES

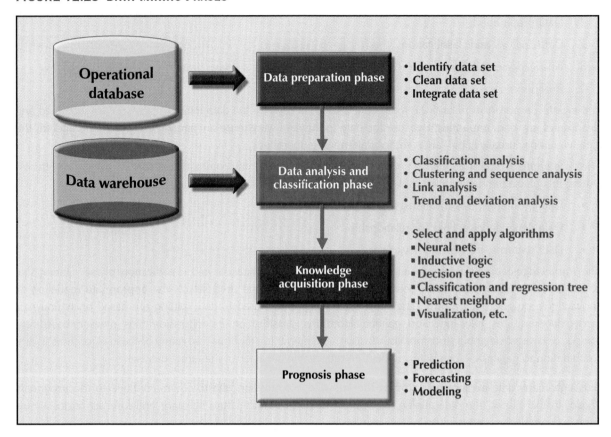

Ideally, we can expect the development of databases that not only store data and various statistics about data usage, but also have the ability to learn about and extract knowledge from the stored data. Such database management systems, also known as inductive or intelligent databases, are the focus of intense research in many laboratories. Although such databases have yet to lay claim to substantial commercial market penetration, both "add-on" and DBMS-integrated data-mining tools have proliferated in the data warehousing database market. Table 12.10 contains a sample of data warehouse and data-mining software vendors.

TABLE 12.10 A SAMPLE OF CURRENT DATA WAREHOUSING AND DATA-MINING VENDORS

VENDOR	PRODUCT	WEB ADDRESS
SAS Institute	SAS Data Warehouse	www.sas.com
IBM	DB2 Business Intelligence	www.software.ibm.com
Business Objects	Business Objects	www.businessobjects.com
Oracle	Oracle9i	www.oracle.com
Cognos	Business Intelligence Series	www.cognos.com
MicroStrategy	MicroStrategy 7i	www.strategy.com
Hyperion	Essbase XTD	www.hysoft.com

SUMMARY

To survive and prosper in an increasingly dynamic business environment, tactical and strategic business decision making requires managers to have quick and accurate access to information. Data analysis is used to derive and interpret information from data. Therefore, data analysis plays an important supporting—some would argue crucial—role in successful decision making. Given the importance of data analysis, many businesses have organized their data analysis practitioners into formal "decision support" groups.

Decision support refers to a methodology (or a series of methodologies) designed to extract information from data and to use such information as a basis for decision making. A decision support system (DSS) is an arrangement of computerized tools used to assist managerial decision making within a business. The DSS is usually composed of four main high-level components: a data store component, a data extraction and filtering component, an end-user query tool, and an end-user presentation tool. The DSS is used at strategic and tactical managerial levels within organizations, but its usefulness depends directly on the quality of data gathered at the operational level. Operational data are not best suited for decision support. From the end-user point of view, DSS data differ from operational data in three main areas: timespan, granularity, and dimensionality.

The requirements for a DSS DBMS are divided into four main categories: database schema, data extraction and loading, end-user analytical interface, and database size requirements. The DSS database schema must support complex (nonnormalized) data representations. The DSS database contains data that are aggregated and summarized. Although the DSS database may contain externally derived data, it is most often created by extracting data from the operational database. The DBMS must support advanced data extraction and filtering capabilities. The DSS DBMS provides hooks to support advanced data-modeling and data presentation tools.

The data warehouse is an integrated, subject-oriented, time-variant, nonvolatile database that provides support for decision making. The data warehouse is usually a read-only database optimized for data analysis and query processing. Business data analysts access the data warehouse through front-end tools and end-user application software to extract the data in usable forms. A data mart is a small, single-subject data warehouse subset that provides decision support to a small group of people.

Online analytical processing (OLAP) refers to an advanced data analysis environment that supports decision making, business modeling, and operations research. OLAP systems have four main characteristics: use of multidimensional data analysis techniques, advanced database support, easy-to-use end-user interfaces, and client/server architecture. OLAP systems have three main modules: OLAP graphical user interface (GUI), OLAP analytical processing logic, and OLAP data-processing logic. Relational online analytical processing (ROLAP) provides OLAP functionality using relational databases and familiar relational query tools to store and analyze multidimensional data. This approach is based on relational technologies and represents their natural extension for organizations that already use relational database management systems. Multidimensional online analytical processing (MOLAP) provides OLAP functionality using multidimensional databases (MDBMS) to store and analyze multidimensional data. Multidimensional database systems use special proprietary techniques to store data in matrix-like arrays of n-dimensions.

The star schema is a data-modeling technique used to map multidimensional decision support data into a relational database with the purpose of performing advanced data analysis. Star schemas provide an easily understood model for multidimensional data analysis. The basic star schema has four components: facts, dimensions, attributes, and attribute hierarchies. Facts are numeric measurements or values representing a specific business aspect or activity. Dimensions are general qualifying categories that provide additional perspectives to a given fact. Each dimension table is composed of attributes that are often used to search, filter, or classify facts. Conceptually, the multidimensional data model is best represented by a three-dimensional cube. Attributes can

be ordered in well-defined attribute hierarchies. The attribute hierarchy provides a top-down organization that is used for two main purposes: to permit aggregation and to provide drill-down/roll-up data analysis. Facts and dimensions are normally represented with physical tables in the data warehouse database. The fact table is related to each dimension table in a many-to-one relationship. The primary key of the fact table is formed by combining the foreign keys of each dimension table to which it is related.

Four techniques are generally used to optimize data warehouse design: normalization of dimensional tables, multiple fact tables representing different aggregation levels, denormalization of fact tables, and table partitioning and replication.

The implementation of any company-wide information system is subject to conflicting organizational and behavioral factors. Because the data warehouse is built to model all data in an entire organization—whose departments might have conflicting goals—it is likely to find data inconsistencies and redundancy. Therefore, extensive end-user input is required to develop the data warehouse, to minimize conflicts. To resolve conflicts, managers must develop skills for conflict resolution, mediation, and arbitration.

Data mining employs a new breed of specialized decision support tools that automate the analysis of operational data with the intention of finding previously unknown data characteristics, relationships, dependencies, and/or trends. The data-mining process has four phases: data preparation, data analysis and classification, knowledge acquisition, and prognosis.

The data warehouse is the prime storage location for decision support data, and it promotes integration with two new decision support components: online analytical processing systems and data-mining systems. OLAP tools access the data warehouse to provide advanced multidimensional data analysis. Data-mining tools use the data warehouse to extract the target data set. Then they apply data-mining algorithms to obtain highly specialized information or knowledge about the data and, therefore, about the organization's business or operations. By working together, these components constitute a source of tactical and strategic business decision support.

KEY TERMS

attribute hierarchy

cube cache

data cube

data extraction

data filtering

data mart

data mining

data store

data warehouse

decision support

decision support system (DSS)

dimensions

dimension table

drill down

end-user presentation tool

end-user query tool

facts

fact table

hypercube

main memory database (MMDB)

metrics

multidimensional online analytical processing (MOLAP)

multidimensional database management system (MDBMS)

online analytical processing (OLAP)

partitioning

periodicity

relational online analytical processing (ROLAP)

replication

roll up

slice and dice

sparsity

star schema

very large database (VLDB)

actual values in the cube by the total number of cells in the cube. Because the data cube's dimensions are predefined, not all cells are populated. In other words, some cells are empty. Returning to our sales example, there might be many products that are not sold during a given time period in a given location. In fact, you will often find that fewer than 50 percent of the data cube's cells are populated. In any case, multidimensional databases must handle sparsity effectively in order to reduce processing overhead and resource requirements.

Relational proponents also argue that using proprietary solutions makes it difficult to integrate the MDBMS with other data sources and tools used within the enterprise. Yet, although it takes a substantial investment of time and effort to integrate the new technology and the existing information systems architecture, MOLAP may be a good solution for those shops in which small- to medium-sized databases are the norm and application software speed is critical.

12.4.8 RELATIONAL VS. MULTIDIMENSIONAL OLAP

Table 12.8 summarizes some OLAP and MOLAP pros and cons. Keep in mind, too, that the selection of one or the other often depends on the evaluator's vantage point. For example, a proper OLAP evaluation must include price, supported hardware platforms, compatibility with the existing DBMS, programming requirements, performance, and availability of administrative tools. Nevertheless, the summary in Table 12.8 provides a useful comparison starting point.

TABLE 12.8 RELATIONAL VS. MULTIDIMENSIONAL OLAP

CHARACTERISTIC	ROLAP	MOLAP
Schema	Uses star schema Additional dimensions can be added dynamically	Uses data cubes Additional dimensions require re-creation of the data cube
Database size	Medium to large	Small to medium
Architecture	Client/server Standards-based Open	Client/server Proprietary
Access	Supports ad hoc requests Unlimited dimensions	Limited to predefined dimensions
Resources	High	Very high
Flexibility	High	Low
Scalability	High	Low
Speed	Good with small data sets; average for medium to large data sets	Faster for small to medium data sets; average for large data sets

ROLAP and MOLAP vendors are working toward the integration of their respective solutions within a unified decision support framework. Many OLAP products are able to handle tabular and multidimensional data with the same ease. For example, if you are using Excel OLAP functionality, as shown in Figure 12.5, note that you can access relational OLAP data in an SQL server, as well as cube (multidimensional data) in the local computer. In the meantime, relational databases use the star schema design successfully to handle multidimensional data, and their market share makes it unlikely that their popularity will fade anytime soon.

REVIEW QUESTIONS

1. What are decision support systems, and what role do they play in the business environment?
2. Explain how the main components of a DSS interact to form a system.
3. What are the most relevant differences between operational and decision support data?
4. What is a data warehouse, and what are its main characteristics?
5. Give three examples of problems likely to be encountered when operational data are integrated into the data warehouse.

Use the following scenario to answer Questions 6 through 12:

While working as a database analyst for a national sales organization, you are asked to be part of its data warehouse project team.

6. Prepare a high-level summary of the main requirements for evaluating DBMS products for data warehousing.
7. Your data warehousing project group is arguing about prototyping a data warehouse before its implementation. The project group members are especially concerned about the need to acquire some data warehousing skills before implementing the enterprise-wide data warehouse. What would you recommend? Explain your recommendations.
8. Suppose you are selling the data warehouse idea to your users. How would you define multidimensional data analysis to them and explain its advantages?
9. The data warehousing project group has invited you to provide an OLAP overview before making a commitment. The group's members are particularly concerned about the OLAP client/server architecture requirements and how OLAP will fit the existing environment. Your job is to explain to them the main OLAP client/server components and architectures.
10. One of your vendors recommends using an MDBMS. How would you explain this recommendation to your project leader?
11. The project group is ready to make a final decision between ROLAP and MOLAP. What should be the basis for this decision? Why?
12. The data warehouse project is in the design phase. Explain to your fellow designers how you would use a star schema in the design.
13. Trace the evolution of DSS from its origins to today's advanced analytical tools. What major technologies influenced this evolution?
14. What is OLAP, and what are its main characteristics?
15. Explain ROLAP, and give the reasons you would recommend its use in the relational database environment.
16. Explain the use of facts, dimensions, and attributes in the star schema.
17. Explain multidimensional cubes and describe how the slice and dice technique fits into this model.
18. In the star schema context, what are attribute hierarchies and aggregation levels, and what is their purpose?
19. Discuss the most common performance improvement techniques used in star schemas.
20. Explain some of the most important issues in data warehouse implementation.
21. What is data mining, and how does it differ from traditional DSS tools?
22. How does data mining work? Discuss the different phases in the data-mining process.

PROBLEMS

ADDITIONAL CD CONTENT

The databases used in the problem set are found in the \Databases\Student\Ch12 folder on your CD. These databases are stored in Microsoft Access 2000 format. The databases, named Ch12_P1.mdb, Ch12_P3.mdb, and Ch12_P4.mdb, contain the data for Problems 1, 3, and 4, respectively. The data for Problem 2 are stored in Microsoft Excel format in the \Excel\Student\Ch12 folder on your CD. The spreadsheet filename is Ch12_P2.xls.

1. The University Computer Lab's director keeps track of lab usage, measured by the number of students using the lab. This particular function is very important for budgeting purposes. The computer lab director assigns you the task of developing a data warehouse in which to keep track of the lab usage statistics. The main requirements for this database are to:

 • Show the total number of users by different time periods.

 • Show usage numbers by time period, by major, and by student classification.

 • Compare usage for different majors and different semesters.

 Use the **Ch12_P1.mdb** database, which includes the following tables:

USELOG	contains the student lab access data.
STUDENT	is a dimension table containing student data.

 Given the (bulleted) requirements and using the DW-P1.MDB data, complete Problems 1a through 1g:
 a. Define the main facts to be analyzed. (*Hint*: These facts become the source for the design of the fact table.)
 b. Define and describe the appropriate dimensions. (*Hint*: These dimensions become the source for the design of the dimension tables.)
 c. Draw the lab usage star schema, using the fact and dimension structures you defined in Problems 1a and 1b.
 d. Define the attributes for each of the dimensions in Problem 1b.
 e. Recommend the appropriate attribute hierarchies.
 f. Implement your data warehouse design, using the star schema you created in Problem 1c and the attributes you defined in Problem 1d.
 g. Create the reports that will meet the requirements listed in this problem's introduction.

2. Ms. Victoria Ephanor manages a small product distribution company. Because the business is growing fast, Ms. Ephanor recognizes that it is time to manage the vast information pool to help guide the accelerating growth. Ms. Ephanor, who is familiar with spreadsheet software, currently employs a small sales force of four people. She has asked you to develop a data warehouse application prototype that enables her to study sales figures by year, region, salesperson, and product. (This prototype is to be used as the basis for a future data warehouse database.)

Using the data supplied in the **Ch12-P2.xls** file, complete the following seven problems:

a. Identify the appropriate fact table components.
b. Identify the appropriate dimension tables.
c. Draw a star schema diagram for this data warehouse.
d. Identify the attributes for the dimension tables that will be required to solve this problem.
e. Using a Microsoft Excel spreadsheet—or any other spreadsheet capable of producing pivot tables—generate a pivot table to show the sales by product and by region. The end user must be able to specify the display of sales for any given year. (The sample output is shown in the first pivot table in Figure P12.2E.)

FIGURE P12.2E USING A PIVOT TABLE

f. Using Problem 2e as your base, add a second pivot table (see Figure P12.2E) to show the sales by salesperson and by region. The end user must be able to specify sales for a given year or for all years, and for a given product or for all products.

g. Create a 3-D bar graph to show sales by salesperson, by product, and by region. (See the sample output in Figure P12.2G.)

FIGURE P12.2G 3-D BAR GRAPH SHOWING THE RELATIONSHIPS AMONG AGENT, PRODUCT, AND REGION

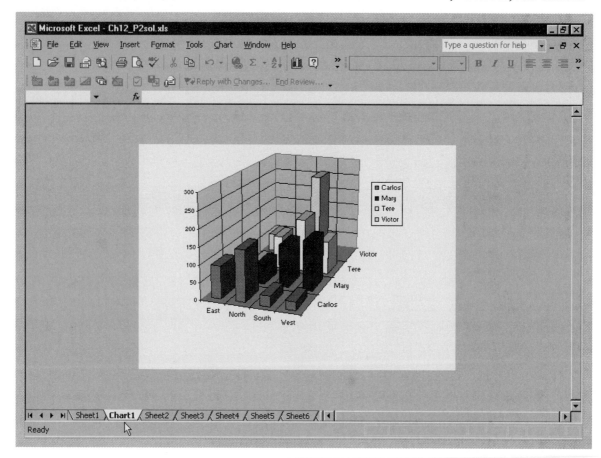

3. Mr. David Suker, the inventory manager for a marketing research company, is interested in studying the use of supplies within the different company departments. Mr. Suker has heard that his friend, Ms. Ephanor, has developed a small spreadsheet-based data warehouse model (see Problem 2) that she uses to analyze sales data. Mr. Suker is interested in developing a small data warehouse model like Ms. Ephanor's, in order to analyze orders by department and by product. He will use Microsoft Access as the data warehouse DBMS and Microsoft Excel as the analysis tool.

 a. Develop the order star schema.

 b. Identify the appropriate dimensions attributes.

 c. Identify the attribute hierarchies required to support the model.

 d. Develop a crosstab report (in Microsoft Access), using a 3-D bar graph to show orders by product and by department. (The sample output is shown in Figure P12.3.)

FIGURE P12.3 CROSSTAB REPORT: ORDERS BY PRODUCT AND DEPARTMENT

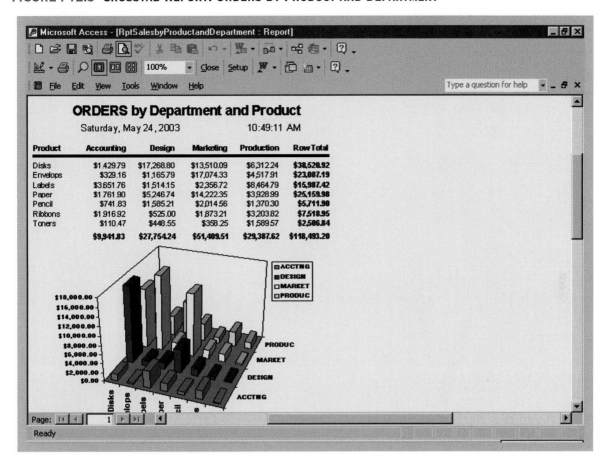

4. ROBCOR, Inc., whose sample data are contained in the database named **DW-P4.mdb**, provides "on demand" aviation charters, using a mix of different aircraft and aircraft types. Because ROBCOR, Inc. has grown rapidly, it has hired you to be its first database manager. (The company's database, developed by an outside consulting team, already has a charter database in place to help manage all of its operations.) Your first and critical assignment is to develop a decision support system to analyze the charter data. (Please review Problems 28–30 in Chapter 3, "The Relational Database Model," in which the operations have been described.) The charter operations manager wants to be able to analyze charter data such as cost, hours flown, fuel used, and revenue. She would also like to be able to drill down by pilot, type of airplane, and time periods.

Given these requirements, complete the following:

a. Create a star schema for the charter data.

b. Define the dimensions and attributes for the charter operation's star schema.

c. Define the necessary attribute hierarchies.

d. Implement the data warehouse design, using the design components you developed in Problems 4a through 4c.

e. Generate the reports that will illustrate that your data warehouse is able to meet the specified information requirements.

PART

V

DATABASES & THE INTERNET

13 DATABASES IN ELECTRONIC COMMERCE

14 WEB DATABASE DEVELOPMENT

13

DATABASES IN ELECTRONIC COMMERCE

In this chapter, you will learn:

- ❖ What electronic commerce (e-commerce) is

- ❖ About the effect of e-commerce and its benefits and disadvantages

- ❖ About the different electronic commerce styles

- ❖ About the Internet architectural components required to conduct electronic commerce

- ❖ About database design and implementation issues for electronic commerce applications

- ❖ What Extensible Markup Language (XML) is and why it is important

PREVIEW

Unless you have been living on a *very* remote island, it is highly probable that the Internet and the World Wide Web (WWW) have touched your life. The Internet has changed how organizations of all types and origins operate. Buying goods and services through the Internet is becoming commonplace. In fact, all over the world, companies are using the Internet in ways that could not have been imagined just a few years ago.

Electronic commerce (e-commerce) enables organizations—whether they are public or private, for profit or not-for-profit—to market and sell products and services to a global market of millions of users. Intranets (networks that use Internet technology but operate within an organization) have likewise streamlined internal business operations.

E-commerce companies not only sell products or services to consumers or end-users, but also to other companies. The Internet has brought about new technologies that facilitate the exchange of business documents and data among business partners. Companies are using the Internet to create new types of systems that integrate their data to increase efficiency and reduce costs.

Online databases are critical components of many e-commerce applications. This chapter introduces the world of e-commerce, and illustrates some special concerns that must be addressed in designing e-commerce databases.

Orbitz: The Birth of an E-Commerce Business

In June 2001, Orbitz.com was launched, and has since become a major player in the online travel service industry. Orbitz enables users to access thousands of flight itineraries and research airfares from most major U.S. airlines, as well as plan vacations, rent cars, book hotel rooms, and access other travel-related services (see Figure 13.1).

Orbitz is owned by five of the largest U.S. airlines: American, Continental, Delta, Northwest, and United. This joint ownership by competitors created much press coverage, which fueled a rising level of expectation in the Internet user community, and spurred an antitrust investigation by the U.S. Justice Department.

The e-commerce Web site quickly established a major presence. During its first 48 hours online, Orbitz sold over $4 million in tickets. During its first week online, Orbitz averaged over 10,000 tickets sold per day. But its initial success was accompanied by problems. Orbitz underestimated the need for customer service agents. Because most visitors were new to the Internet, the customer service department was overwhelmed by customer calls. Also, technical cabling problems slowed down response time on a part of its Web site. Orbitz has so far been able to meet its challenges and use them to improve its services and market share. In August 2003, the Justice Department closed its antitrust investigation, concluding that the site hasn't reduced competition or hurt buyers. To make it on the Internet, companies must put together good business planning with superb technological implementation.

(*Sources*: Cox, Beth, "Orbitz Finds a Madding Crowd," *ecommerce-guide.com*, June 7, 2001; "Antitrust Probe Clears Orbitz," *Salt Lake Tribune*, August 1, 2003.)

FIGURE 13.1 ORBITZ E-COMMERCE WEB SITE

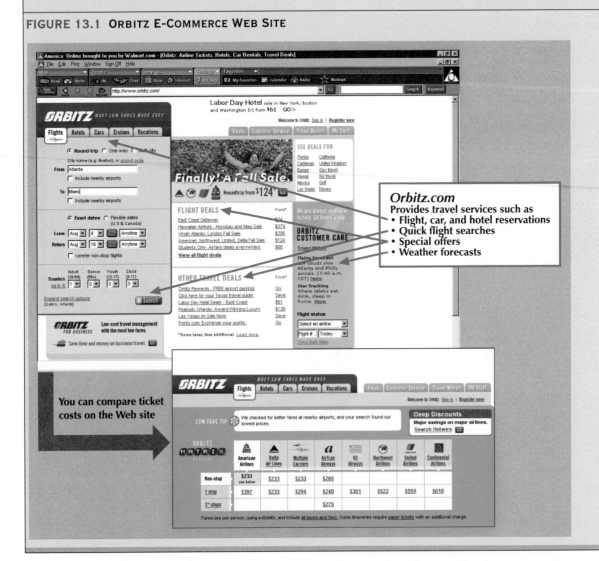

13.1 WHAT IS ELECTRONIC COMMERCE?

There are many definitions of e-commerce—short for electronic commerce. The Webopedia online technology dictionary (*webopedia.com*) defines e-commerce in general as "conducting business online." However, even a cursory examination of professional publications makes it painfully obvious that the definition of e-commerce changes according to whom you ask—and the definitions appear to evolve as fast as does the underlying technology. In this book, we define **electronic commerce (e-commerce)** as the use of electronic networked computer-based technology to:

- Bring new products, services, or ideas to market.

- Support and enhance business operations (including sales of products/services over the Web).

Although businesses have always used many different types of electronic technology, e-commerce is mainly identified with the use of the Internet as a medium to transact business (buy, sell, and trade products and services) and to add value to an organization. Because the Internet, and in particular the Web, plays such a crucial role in enabling the development and execution of e-commerce, some experts argue that e-commerce should be called Internet commerce (I-commerce), instead.

Although it is easy to view e-commerce as just an online extension of common customer activities, most e-commerce transactions actually take place among businesses. Companies use the Internet and **intranets**, internal networks which use Internet technology but are used within an organization, to streamline their production and distribution processes, and to enhance their internal and external operations. In fact, a main selling point of e-commerce is that it provides a competitive advantage to the organizations that use it. Because e-commerce operations have become so embedded in the business environment, many people refer to e-commerce as electronic business (e-business).

The external evidence of a company's e-commerce activities is the company's Web site. This Web site can be as simple as a few static pages that provide product line and contact information, or as complex as a complete online database-driven product catalog with dynamic ordering and credit card payment processing.

E-commerce is now recognized as a prime revenue source. Putting your products online makes them immediately available for millions of potential buyers. Companies are competing for a share of the online market by attracting (and keeping!) customers focused on their Web sites. But the Internet is a challenging place; many companies are discovering that competing in the online market is much more difficult than creating and using Web pages. In short, e-commerce is much more than just another marketing channel; e-commerce is the kernel of a new business model dedicated to rapidly bringing products, ideas, or services to large markets in relatively inexpensive ways through the use of Internet technology.

E-commerce is not an end in itself; it is a road that is traveled to compete and survive in the 21^{st} century business environment. On this road, Internet technology has played—and will continue to play—a crucial role. To know where you are going, it is useful to know where you have been. Therefore, the next section provides a brief glimpse of the major milestones on the e-commerce road.

13.2 THE ROAD TO ELECTRONIC COMMERCE

Businesses have been using technology for many decades to enhance their operations. For example, phones and fax machines are long-standing business uses of technology. Should such technologies be considered part of the e-commerce model? The short answer is no. (We're assuming that the phone is not connected to a computer modem!) The just-mentioned examples are not strictly e-commerce because they depend heavily on human intervention to conduct business transactions. Phones and fax machines simply connect the sender with the receiver by transmitting sound or images. A human at the receiving end of the transmission still has to *manually* process the information in order to complete the business transaction. The key to e-commerce is using computer

networks, especially the Internet, to automate and streamline business transactions. The development of the Internet is closely allied with the development of e-commerce.

The Internet was born as an extension of the DARPA network, started in the early 1960s by the U.S. government as a military project to ensure computer communications in case of nuclear attack. Almost immediately after its creation, it was extended to include higher education institutions to facilitate critical research. The Internet's reach did not extend into the business arena until many years later.

The following list summarizes a few of the major technological milestones in business. As you examine the list, note that all of the technological milestones share the common feature of drastically reducing human intervention:

- In the early 1960s, banks created a private telephone network to do electronic funds transfers (EFT). This service allowed two banks to exchange funds electronically in a fast, efficient, and secure manner. This service was restricted to the participating banks and these banks were required to cover the costs associated with using and maintaining the system.

- In the early 1970s, banks created services such as the Automated Teller Machine (ATM) to provide after-hours services to their customers. In the beginning, ATMs were installed by a few banks nationwide and customers were allowed only a limited number of transactions. As the ATM's popularity grew, companies were created to provide ATM service to most banks.

- In the late 1970s and early 1980s, **Electronic Data Interchange (EDI)** emerged. EDI is a communications protocol that enabled companies to exchange business documents over private phone networks. The use of EDI facilitated the coordination of business operations between business partners. The use of EDI became especially well established in the automotive industry. For example, Nissan used EDI to send requests for parts (car seats, tires, or other components) to subcontractors as soon as a car entered the assembly line. EDI's problem was that its maintenance and implementation costs were high, and the EDI formats varied from company to company. Given these drawbacks, EDI became a secondary (niche) player as Internet communications capabilities grew. However, EDI recently has been adapted to Internet technology, which has drastically reduced the investment in communication infrastructure required by previous generations of EDI.

- During the early 1980s and through the 1990s, the personal computer (PC) facilitated the rapid expansion of the Internet and ultimately provided the spark that led to the explosive use of the World Wide Web, usually referred to as the Web. The Web, perhaps the best-known of many Internet services, made the transfer of information among multiple organizations as simple as a mouse click. The Web also became the basis for the exploration and exploitation of new Internet-based technologies that led to the enhancement of business processes within and between corporations.

- In the late 1990s and early 2000s, networking technologies blossomed and expanded the reach, speed, and in some cases, the security of Internet-based communications and transactions. Many companies began to take advantage of Virtual Private Networks (VPNs), which are private, relatively secure networks that "tunnel" through the Internet using Internet technology. Wide area networks (WANs) that incorporate satellite and microwave links, and wireless technologies such as cell phones and wireless networks have also increased the scope and possibilities for businesses large and small. New developments, such as Extensible Markup Language (XML), are being used to facilitate the dynamic exchange of data among geographically dispersed applications. XML has introduced a new dimension in the provision of Internet-based services.

N O T E

If you want to examine a more detailed history of the Internet, visit the following Web sites:
- *http://www.pbs.org/opb/nerds2.0.1/timeline/*
- *http://www.zakon.org/robert/internet/timeline/*

As an example of how e-commerce has developed, it is worthwhile to take a look at one well-known e-commerce success story. In 1992, Jeff Bezos launched *Amazon.com* as a Web site to sell books directly to consumers over the Internet. Five years later, sales exceeded $131 million. Soon, Amazon began expanding into many other retailing areas, selling everything from baby goods to music and electronics to clothing and home goods (see Figure 13.2). It pioneered the "personalized" Web page based on tracking customers' buying habits and preferences, and established alliances and cross-links with other major retailers such as Target. By 2003, it was an international presence with net sales of nearly $5 billion and climbing. In just a few years, *Amazon.com* had become one of the most successful retail Web sites and a model for many other Internet-based businesses.

FIGURE 13.2 AMAZON.COM HOME PAGE

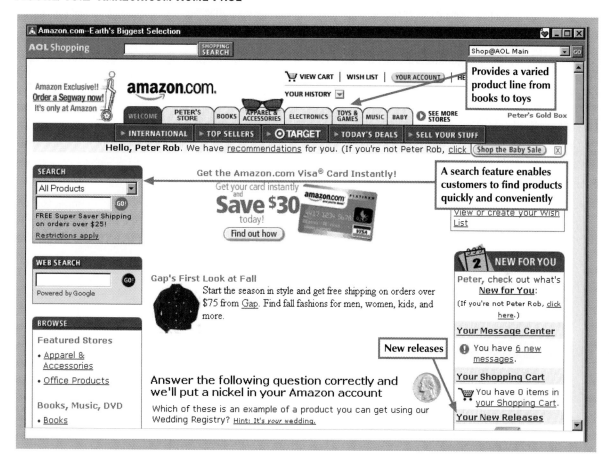

Why did businesses jump on the Internet bandwagon with such abandon? Previous technologies took much longer to find a place at the business table. In the next section we'll examine why the Web's acceptance is so broad and deep.

13.3 THE IMPACT OF E-COMMERCE

In the 1990s, economists coined the term "the new economy" to refer to the business marketplace based on computer technologies and delivered through the Internet. And although the "tech bubble burst" of 2000 scrubbed some of the luster off many e-commerce businesses, many of those businesses quickly adapted to the new market realities and continued to expand. Today, a majority of businesses, even "traditional" or so-called "brick-and-mortar" businesses, have a presence on the Web. (Can you think of a major business that does not have a substantial Web presence?)

Quite aside from the marketing aspects that are enhanced by e-commerce, there are many other reasons for implementing e-commerce within organizations. These reasons—including rapid response to competitive pressures and customer service requirements, facilitation of transaction management, and inventory management—have an added urgency in today's world of global markets, mergers, and acquisitions.

The Internet economy works within a global market of interconnected consumers and sellers. Businesses no longer compete just with businesses down the street or across town or even in the same country. Small organizations also have discovered that there is, at least potentially, a global market for their products, services, and ideas. This global market has attracted millions of businesses and organizations to the Internet and e-commerce. The Internet has reached critical mass with hundreds of millions of consumers worldwide, and it has become the new frontier for organizations in their quest for profits and enhanced public services.

For IS departments, the new frontier is the use of Internet technologies to provide services to customers, partners, employees, and the general public. The Internet is driving the development of a new generation of information systems. For many IS applications, Internet technologies facilitate sharing heterogeneous information in an environment that provides multiple benefits at a minimum cost.

The next sections highlight the advantages and disadvantages of e-commerce.

13.3.1 E-COMMERCE BENEFITS

E-commerce has benefits both for buyers and for sellers, including:

- *Easy comparison shopping.* Consumers can quickly compare prices for just about anything under the sun at sites like *www.mysimon.com* or *www.dealtime.com*. Such sites have had a major impact on industries like insurance (*www.quotesmith.com*), automobiles (*www.autobytel.com*), and air travel (*www.qixo.com*).

- *Reduced costs and increased competition.* Online comparison-shopping means intense competition among the suppliers of goods and services, which means lower costs for consumers. For businesses, cost reductions are reflected in a lower cost per transaction. Even though initial Internet infrastructure and implementation costs are high, the cost per transaction tends to be much lower because of the marginal cost of each additional transaction and the growing volume of customers using the Internet.

- *Convenience.* Online shoppers can purchase products conveniently from their home, without the need to drive a car to a store, thus saving substantial time and transportation costs.

- *24 × 7 × 365 operation.* Online stores, unlike their brick-and-mortar counterparts, remain open year-round, including weekends and holidays. (Most transactions are automated to the extent that they do not require full-time staffing.) The benefit to businesses is clear, too—they can gain new customers year-round and at any time.

- *Global access.* Through the Internet, businesses have access to millions of users worldwide. Companies can develop national and international exposure. Such exposure creates brand awareness and, businesses hope, customer loyalty.

- *Lower entry barriers.* An organization looking to establish an online presence has to expend less for physical location start-up, hiring and training a large sales and managerial staff, and pays fewer expensive rental, utilities, and advertising fees. Because there is no need to build and furnish attractive office and/or store interiors, the total time required to launch an online store is reduced. E-commerce has revolutionized the concepts of business time and place.

- *Increased market (customer) knowledge.* The Internet environment makes it relatively easy to compile and track customer information. Businesses can create extensive customer profiles that include purchasing preferences, demographic and geographic information, and online behavior. Such information can be used to design Web sites that attract more users, target specific markets, and display features customized to individuals' interests. Businesses benefit by increasing customer loyalty and generating repeat sales.

13.3.2 E-COMMERCE DISADVANTAGES

Although the list of e-commerce advantages is long, the e-commerce environment is far from perfect. In fact, some e-commerce disadvantages cause consumers and businesses both to suffer considerable anxiety:

- *Hidden costs.* Online purchases are often accompanied by high shipping and restocking fees, a lack of warranty coverage, and unacceptable delivery times. In fact, excessive shipping fees is one of the most frequently cited reasons shoppers choose not to buy online.

- *Vulnerability to technical failure.* When an e-commerce Web site cannot service customers because of a network failure, a software virus, or internal hardware failure, the organization loses sales, credibility, and even customers.

- *Cost of staying in business.* Although getting *into* business is relatively easier in an e-commerce environment, *staying* in business may be more difficult. Increased competition means businesses operate with very thin profit margins. To be profitable, e-businesses must maintain high sales volumes, which means developing and maintaining large and loyal customer bases. Also, to survive and remain competitive, businesses must invest heavily in often-costly technology.

- *Lack of security.* One of the main roadblocks to the wide acceptance of e-commerce by businesses and consumers alike is the perceived lack of adequate security for online transactions. For example, many consumers are wary of providing credit card information over the Internet. Several credit card companies are currently working on a set of standards to make credit card transactions on the Web more secure.

- *Invasion of privacy.* The incredible capacity for online data collection is a mixed blessing to customers. Information about sales, customer purchasing habits, demographic data, credit information, and so on is stored in databases connected to Web servers, thus potentially exposing the information to cyber criminals. Another concern is that customer information is often sold to marketing companies, who then engage in e-mail campaigns to attract new customers. It doesn't take long for the customer's e-mail inbox to be filled with unwanted, unsolicited e-mail, or spam.

- *Low service levels.* Another common complaint about doing business online is that online companies tend to provide minimal customer service. Although technology has automated business transactions to a large extent, there remains a real need for the human touch. Therefore, customer service has become a major differentiating factor.

- *Legal issues.* Legal problems encountered in the e-commerce environment include: software and copyright infringements; credit card fraud and stolen identities, and online fraud (failure to deliver products and/or services to the customers who paid for them).

13.4 E-COMMERCE STYLES

E-commerce transactions can be grouped according to who are the sellers and the buyers. Using this distinction, we can classify the principal e-commerce styles as:

- **Business to business (B2B)**: electronic commerce between businesses.

- **Business to consumer (B2C)**: electronic commerce between business and consumers.

- **Intra-business**: internal electronic commerce activities, most of which involve interactions between employers and their employees.

Although some may argue that we should include **government to business (G2B)** and **government to consumer (G2C)** as additional e-commerce styles, we consider them special cases of B2B and B2C. Consumer to consumer (C2C) transactions, particularly C2C transactions facilitated by an intermediary business (C2B2C), are also growing, facilitated by Web sites like eBay and Amazon.com that provide avenues for consumers to buy and sell from each other. Figure 13.3 identifies the primary e-commerce players and their interactions.

FIGURE 13.3 **E-Commerce Styles**

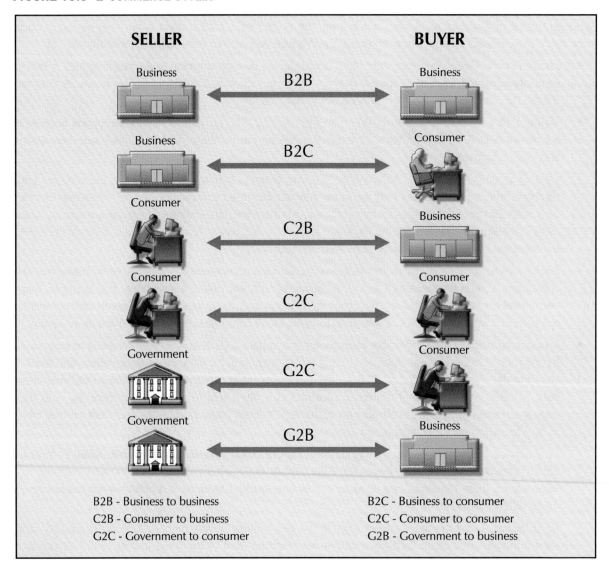

13.4.1 **Business to Business (B2B)**

Although B2B transactions include intangibles such as transmitting contracts and moving accounts, most transactions involve sellers and buyers of products and/or services. Generally, the seller is any company that sells a product and/or service using electronic exchanges such as the Internet or EDI. Examples of B2B transactions are:

- A Nissan Corporation manufacturing plant issues an electronic order for a number of tires of a given model and size. That order is received by one of the approved tire providers and the order is fulfilled. When the tires are received, Nissan issues a payment electronically through a bank funds transfer.

- Using the Web, the University of Tennessee makes hotel reservations for a group of researchers who will conduct a seminar.

- The Red Cross uses the Web to compare hardware and software prices and subsequently orders hardware and software through the Web.

B2B is the biggest and fastest-growing component of the e-commerce market. Some estimate that B2B transactions will grow to over $1 trillion by 2004. (Check the *www.emarketer.com* site for current information and estimates.)

Regardless of which estimates are used, it is clear that the Internet is a dominant player in the B2B economy. In this economy, the main focus is using technology to automate the value chain of a business. The **value chain** refers to all activities required to design, plan, manufacture, market, sell, and support a product or service. By examining the value chain components with reference to Internet technologies, businesses can automate and enhance their operations. For example, many companies now use Enterprise Resource Planning (ERP) to manage and enhance all aspects of the company's value chain, from raw material procurement to promoting customer satisfaction. Figure 13.4 illustrates the use of the value chain to automate B2B transactions.

FIGURE 13.4 E-COMMERCE AUTOMATION OF SUPPLY CHAIN

B2B transactions are subject to different types of implementation. The two most important versions are:

- *B2B integration.* In this scenario, companies establish partnerships to reduce costs and time and to increase business and competitiveness. For example, a company that manufactures computers will partner with its suppliers for hard disks, memory, and other components. Such a partnership allows the company to automate its purchasing system, integrating it with its suppliers' ordering systems, which then will link their respective inventory systems. In this case, when a component in company "C" gets below the minimum quantity-on-hand requirement, it will automatically generate an order to supplier "S". Both systems would be integrated and would exchange business data, generally using XML through the Internet. (We will examine XML later in the chapter.) In addition, the company may integrate its distribution system

with that of its distributors. Finally, the distributors may integrate their activities with those of their retailers. And the retailers may, in turn, integrate their operations with those of their customers. Given such integration, sellers align their operations with those of buyers, thereby achieving levels of efficiency that make it difficult to switch to other buyers and/or sellers.

- *B2B marketplace.* In this scenario, the objective is to provide a way in which businesses can easily search, compare, and purchase products and services from other businesses. The Web-based system will basically work as an online broker that will service both buyers and sellers. Within this system, the focus shifts to attracting new members, either sellers or buyers. The "broker" offers sellers a way to market their product or service to other businesses, while buyers are attracted because they can compare products from different buyers and get access to special deals only offered to the members. In this scenario, the broker obtains revenue through membership and transaction fees. Figure 13.5 shows an example of B2B Web marketplaces for the automotive industry.

FIGURE 13.5 COVISINT.COM: AUTOMOTIVE B2B MARKETPLACE

One important aspect involved in implementing B2B solutions is the integration of the *databases* to support information (data) exchanges with other database systems. Later in the chapter, you will learn about Internet technologies to facilitate such data exchange.

13.4.2 BUSINESS TO CONSUMERS (B2C)

A business to consumer (B2C) operation is one that uses the Internet to sell products and/or services directly to consumers or end users. In B2C e-commerce, the Internet—and in particular the Web—is the marketing, sales, and post-sales support channel. B2C is oriented to attracting customers to the Web sites and offering products and services in new and innovative ways. Table 13.1 lists a sample of B2C e-commerce Web sites.

TABLE 13.1 SAMPLE B2C WEB SITES

INDUSTRY	B2C WEB SITES	INDUSTRY	B2C WEB SITES
Travel	Travelocity.com Expedia.com CheapTickets.com	Computer	Dell.com IBM.com Pricewatch.com
Retailing	Landsend.com Spiegel.com JCPenney.com	Health Services	HealthNet.com WebMD.com
Financial	Fidelity.com Etrade.com	Auctions	Ebay.com
Banking	Netbank.com 1st-online-banking.com	Reverse Auctions	Priceline.com LendingTree.com
Music	CDNow.com	Insurance	Quotesmith.com
Government	Internal Revenue Service (www.irs.com)	Education	University of Phoenix Cardean University

Two variations of the B2C marketplace exist:

- *Consumer to business to consumer (C2B2C).* A consumer offers items for sale to other consumers through a third-party Web site. The Web's many auction sites like *www.ebay.com* are good examples.

- *Business to business to consumer (B2B2C).* A business offers products or services to consumers through a third-party Web site. A typical example of B2B2C is a reverse auction such as LendingTree.com (*www.lendingtree.com*), where consumers request bids for loans, and financial institutions compete for consumers' business.

13.4.3 INTRANETS AND EXTRANETS

An intranet is a company-owned and -operated Internet that can be accessed only by the computers within the company's computer network. Companies use intranets to support the internal business activities of their many departments such as procurement, manufacturing, marketing, sales, distribution, and customer support. The focus in this case is on the automation and enhancement of the company's internal business processes and on increasing communication among the employees. For example, a company can use an intranet for its human resources Web site to enable employees to sign up for company benefits such as dental coverage and automatic payroll deposit.

If an intranet extends beyond a single corporate entity, it is known as an **extranet**. An extranet extends the intranet to the corporation's value chain. Extranets facilitate B2B integration. For example, a large retailer may use an extranet to extend the internal product and ordering information to its suppliers and distributors, thereby helping to automate its supplier and distributor functions. Such an extranet enhances the intra-organizational workflow.

Intranets and extranets use relatively simple Internet tools, such as Web browsers and HTML-formatted documents, to facilitate creating and distributing information, as well as interdepartmental business transactions. (HTML stands for Hypertext Markup Language. HTML is the standard document-formatting language for Web pages. In Chapter 14, "Web Database Development," you will learn how to create database-driven dynamic Web pages.) The same Internet tools and architecture can be used both internally and externally with relatively minor changes. For example, the automation of inventory control using a Web-based front-end tied to a central database repository can serve internal business processes—such as price updates and stock queries—as well as external e-commerce transactions such as online sales. Internet technology also allows companies to control which computers (inside and outside the organization) have access to the network.

We'll examine the basic Internet architectural components that are required to support e-commerce in the next section.

13.5 E-COMMERCE ARCHITECTURE

Companies embracing e-commerce must deal with managerial and technological issues. Managerial issues range from establishing partnerships with suppliers, distributors, and vendors to the design and development of well-orchestrated business plans. (Although managerial issues are critical to the success of an e-commerce initiative, they are beyond the scope of this book.) Technological issues include the hardware and software components that provide the backbone for reliable and secure e-commerce transactions. Regardless of the type and extent of an organization's information structure before it embraces e-commerce, the proper design, development, and deployment of a well-planned architecture is necessary to support e-commerce business transactions, both internal and external. How those issues are confronted depends on whether the information systems architecture is already established and on the e-commerce style.

In this section, we describe the basic architectural components that must exist to support e-commerce transactions.

To better understand e-commerce architecture, we can divide it into a series of layers. Each layer provides services to the layer below it. Those layers are:

- Basic Internet services.
- Business-enabling services.
- E-commerce business services.

Figure 13.6 offers a bird's-eye view of e-commerce architecture.

FIGURE 13.6 **E-Commerce Architecture**

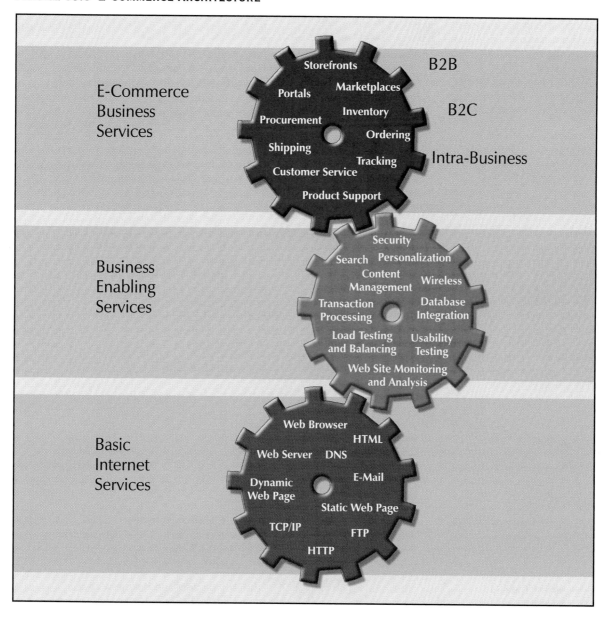

13.5.1 Basic Internet Services

The Internet provides the basic services that facilitate the transmission of data and information between computers. The terms Internet and World Wide Web are often used interchangeably, but they are not synonyms. The World Wide Web functions as one of the many services of the Internet. Table 13.2 describes the basic building blocks and services provided by both the Internet and the World Wide Web.

TABLE 13.2 INTERNET BUILDING BLOCKS AND BASIC SERVICES

BUILDING BLOCK/SERVICE	DESCRIPTION
Internet	A worldwide network of networks. The Internet acts as a "super network" that connects thousands of smaller networks across the world. You can think of the Internet as the "highway" on which data travel, as in the phrase "the information superhighway." To connect thousands of heterogeneous networks, the Internet uses a standard network protocol known as *TCP/IP* and devices known as *routers*.
TCP/IP	Transmission Control Protocol/Internet Protocol. The basic network protocol that determines the rules used to create and route "packets" of data between computers in the same or different networks. Each computer connected to the Internet has a unique TCP/IP address. The TCP/IP address is divided into two parts used to identify the network and the computer (or host).
Router	Special hardware/software equipment that connects multiple and diverse networks. The router is in charge of delivering packets of data from a local network to a remote network. Routers are the traffic cops of the Internet, monitoring all traffic and moving data from one network to another.
World Wide Web (WWW or the Web)	Worldwide network collection of specially formatted and interconnected documents known as *Web pages*. The Web is just *one* of *many* services provided by the Internet.
Web page	A document containing text and special commands (or tags) written in *Hypertext Markup Language (HTML)*. A Web page can contain text, graphics, video, audio, and other elements.
Hypertext Markup Language (HTML)	Standard document-formatting language for Web pages. HTML allows documents to be presented in a Web browser in a standard manner.
Hyperlink	Web pages are linked to each other—that is, each Web page calls other Web pages—creating the effect of a "web." Because a link can connect to different types of documents such as text, graphics, animated graphics, video, and audio, it is known as a "hyperlink." A hyperlink is generally expressed as a *URL* in an HTML-formatted Web page.
Uniform Resource Locator (URL) or Web address	A URL identifies the address of a resource on the Internet. The URL is an abbreviation (ideally easily remembered) that uniquely identifies an Internet resource. Examples of URLs include *www.dell.com*, *www.ford.com*, *www.amazon.com*, *www.faa.gov*, and *www.mtsu.edu*.
Hypertext Transfer Protocol (HTTP)	The standard protocol used by the *Web browser* and *Web server* to communicate—to send requests and replies between servers and browsers. HTTP uses TCP/IP to transmit the data between computers on the Internet.
Domain Name System (Service)	DNS translates the "English-like" domain names (such as whitehouse.org or ebay.com) to the appropriate TCP/IP addresses. The DNS service lies at the heart of the Internet because most hyperlinks use URLs to refer to other Web pages.
Web browser	The end-user application used to *browse* or *navigate* (move from page to page) through the Internet. The browser is a graphical application that runs on the client computer, and its main function is to display Web pages. A client uses the Web browser (for example, Netscape Navigator, Microsoft Internet Explorer, and Opera) to request Web pages from a Web *server*.
Web server	A specialized application whose only function is to "listen" for client requests, process them, and send the requested Web page back to the client browser. The Web server and the Web client communicate using a special protocol known as *Hypertext Transfer Protocol* or *HTTP*.

TABLE 13.2 INTERNET BUILDING BLOCKS AND BASIC SERVICES (CONTINUED)

BUILDING BLOCK/SERVICE	DESCRIPTION
Web site	Term used to refer to the Web server and the collection of Web pages stored on the local hard disk of the server computer or an accessible shared directory.
Static Web page	A Web page whose contents remain the same (when viewed in a browser) unless the page is manually edited. An example of a static Web page is a standard pricelist posted by a manufacturer for inspection by the manufacturer's customers.
Dynamic Web page	A Web page whose contents are automatically created and tailored to an end user's needs each time the end user requests the page. For example, an end user can access a Web page that displays the latest stock prices for the companies selected by the end user.
File Transfer Protocol (FTP)	Protocol used to provide file transfer capabilities among computers in the Internet. An FTP client requests a file to an FTP server. The FTP server listens for clients' requests, processes them, and sends the requested files back to the client.
Electronic mail (e-mail)	Messages transmitted electronically among computers on the Internet. A mail server stores e-mail messages in end-user mailboxes. Mail clients retrieve e-mail from the mail server. When a client sends an e-mail, it is temporarily stored on the mail server, which in turn delivers the e-mail to the correct destination.
News and discussion group services	Specialized services that allow the creation of "virtual communities" in which users exchange messages regarding specific topics, for example, aviation, sports, and computers. This service allows end users to post information on shared bulletin boards for public access.

Figure 13.7 shows the relationships among the components defined in Table 13.2.

As you examine Figure 13.7, note that the client enters a Web address or URL in the Web browser. In turn, the Web browser issues an HTTP page request to the Web server. The request is handled through TCP/IP for transmission over the network. TCP/IP transforms the request into packets and sends them to the router, which in turn routes them to the Internet. Once on the Internet, the TCP/IP packets travel across several routers until they reach the destination server. The server receives the Web page request, fetches the page, and sends it to the client's Web browser in the same way. The client receives the HTML-formatted Web page and displays it on the screen.

The client and the server can be located in the same building or across the globe. As you can see, this makes the Web a great medium to deliver information across geographic boundaries. The Web browser integrates all the services provided by the Web site. The end user does not care or know about all the different applications running on the server side.

Web pages may be static or dynamic. Static Web pages display information that does not change much over time, or is not time–critical. Dynamic Web pages are used mainly for pages with contents that change over time and cannot be anticipated, for example, an online ordering system. Static Web pages are adequate to display information such as product catalogs or contact information; dynamic Web pages are better suited to e-commerce applications such as online ordering with product-customization options. For example, you can visit the Dell home computer e-commerce site (*dell4me.com*) and view how you can dynamically configure your computer. Dynamic Web pages are at the heart of most B2B and B2C Web transactions.

FIGURE 13.7 BASIC INTERNET SERVICES

13.5.2 BUSINESS-ENABLING SERVICES

The Internet services described in the previous section are sufficient to operate a basic Web site. However, they do not provide the support required to conduct even rudimentary business transactions. Business-enabling services are implemented by hardware and software components working together to provide the additional functionality not provided by basic Internet services. Table 13.3 describes the services that enhance Web sites by providing the ability to perform searches, authenticate and secure business data, manage Web site contents, and more. The list in Table 13.3 is not comprehensive; technological advances continue to enable new services, which in turn are used to enable yet additional services.

TABLE 13.3 BUSINESS-ENABLING SERVICES

SERVICE	DESCRIPTION
Search services	Search services provide Web sites with the ability to perform searches on their contents. These services can be used in intranets to search for payroll information, benefits, vacation times, contract information, etc. A B2C Web site could use this feature to search for product return information, customer support data, etc. Search services are a "must have" for all e-commerce Web sites.
Security	Services oriented to ensure the security and privacy of the data by providing encryption, digital certificates, SSL, S-HTTP, firewalls, and proxy servers. (These are covered in greater detail in Section 13.6.)

TABLE 13.3 BUSINESS-ENABLING SERVICES (CONTINUED)

SERVICE	DESCRIPTION
Site monitoring and data analysis	Ensures the Web site is performing at an optimal level. Monitors the main indicators of system and network performance. This service also includes the identification of network bottlenecks, often noticed by the end user when Web pages load very slowly. Analysis features study network traffic to determine which pages are attracting many visitors and which ones are not. This feedback answers questions about the effectiveness of the pages.
Load testing, balancing, and Web caching	Load testing is performed before an e-commerce site goes online. Its main objective is to test the Web site to ensure it will support the load imposed by having thousands of users accessing it. When the expected transaction load is too much for a single server to handle, multiple servers are required. Load balancing ensures that the processing load is distributed evenly among multiple servers. These services provide performance-enhancing techniques to operate the system at optimum speed.
Usability testing	In an e-commerce environment, not having a Web site is bad enough, but having a badly designed Web site may even be worse. Usability testing ensures that the Web site features and services are presented in a user-friendly manner. Is the search function hard to find? Are the colors chosen for the Web site difficult to read on portables' screens? Are the options presented logically?
Personalization	Personalization features allow customizing Web pages for individual users. Personalization serves to make the site user-friendly to attract more users and keep them coming back. To see personalization in action, go to *yahoo.com* and click "My Yahoo." You can then specify your interests and your zip code and get a Web page tailored for you with your local weather, news, etc.
Web development	E-commerce applications require business logic to be integrated into the Website. Web development tools provide the means to add business logic to Web pages. HTML is not a programming language, it is a document-formatting specification created to properly present documents in a Web browser. Business logic can be added to Web sites by using one of many Web-based programming environments, such as Java, JavaScript, JScript, or VBScript. Such programming environments allow the creation of dynamic Web pages that form the basis of e-commerce Web sites.
Database integration	Business transaction data are normally stored in databases. The integration of enterprise databases with the Web is a *requirement* for e-commerce success. Many DBMSs, such as Oracle and Microsoft SQL Server, already come with Web development environments that integrate the database with the Internet. Third-party vendors provide solutions that allow corporate databases (including legacy data) to be integrated into a company's Web site. (We will examine this topic in greater detail in Chapter 14, " Web Database Development.")
Transaction processing	As you might imagine, transaction processing services are enormously important in e-commerce. In Chapter 9, "Transaction Management and Concurrency Control," we examined the importance of transaction management—in an e-commerce environment the problem is vastly magnified, because transactions may originate from customers all over the world.
Content management	Content management automates the creation and management of a Web site's contents and provides a flexible and consistent way for many different individuals and departments to create Web pages. Content management is critical for companies that are in the business of providing information.
Messaging	The Internet is a highway through which applications communicate. An e-commerce application sends messages from a client to a server, and vice versa. Messaging ensures the proper routing and delivery of applications-oriented messages among multiple services.
Wireless device support	Wireless and mobile devices promise to become big players in e-commerce in the near future. Services to support wireless communication will become increasingly important in e-commerce.

13.5.3 E-COMMERCE BUSINESS SERVICES

E-commerce business services form the top layer of the e-commerce architecture (look at Figure 13.6 again). This layer uses the services of the two layers below it to map business logic and to automate business processes. Automating business processes enhances the business unit, department, and intra-business operations. At this layer, business-enabling services (layer 2) interact with basic Internet services (layer 1) to provide support for the front-end e-commerce services. Note that it is the business functions that drive the operations of the services below it—not the other way around. This layer takes longer to implement, because care must be taken to develop a thorough business understanding, establish business partnerships, and understand customer behavior.

When the Web front end is designed and implemented, a decision must be made about what business services to provide and how to provide them. For B2C e-commerce sites, the front end could be as simple as an online product catalog or as sophisticated as a database-based "storefront" with three-dimensional (3D) views of the available products, plus a shopping cart application that enables users to order items while they browse the products.

Common services provided by e-commerce Web sites include the automation of the supply chain for B2B purposes. Such automation covers online procurement, just-in-time inventory, online ordering, order tracking, product delivery, product support, customer satisfaction management, and so on.

The next two sections introduce two essential e-commerce features that enable complete online transactions: security and payment processing.

13.6 SECURITY

For e-commerce to be successful, it must ensure the security and privacy of all business transactions and the data associated with those transactions. In an e-commerce context, **security** encompasses all the activities related to protecting the data and other e-commerce components against accidental or intentional (usually illegal) access or use by unauthorized users. **Privacy** deals with the rights of individuals and organizations to determine the "who, what, when, where, and how" of data use.

Even before the emergence of e-commerce, information privacy and security was a major concern in business organizations. In some cases these concerns merited the creation of a Data Security Officer position to oversee data security and privacy standards and procedures. Integrating databases with the Web and using the Internet as a transmission medium for business transactions has made the need for such security even greater.

The World Wide Web was originally created with the objective of sharing data easily, rather than securely. For this purpose, the Web is a good fit for nonprofit organizations and business/consumer advocacy groups that are more concerned with information distribution than with secure transactions. However, private, for-profit organizations require security and privacy to engage in e-commerce business transactions. For example, who would use a credit card to pay online if there were no measures to protect the credit card information against theft? How can messages traveling on the Internet be protected against modifications? How can we verify the identity of the sender? How can we protect the Web storefront against attacks from cybervandals? There is no foolproof way to protect against all possible threats, but there are several technologies that address security problems. Although Internet security is a vast topic, the focus here will be on the principal mechanisms used to protect electronic business transactions, Web-store front-ends, and associated data.

E-commerce data must be secured from a transaction's beginning to its end. Let's examine the following online purchasing scenario, illustrated in Figure 13.8:

1. A customer orders products online, entering order and credit card information on a merchant's Web page.

2. The information travels from the customer's computer over the Internet to the merchant's Web server.

3. The merchant uses a third-party company to process payment authorization.

4. The payment-processing company contacts the customer's credit-card issuing company to authorize the transaction.

5. The customer's credit card issuer authorizes the transaction.

6. The merchant receives authorization, stores the order and payment data in a database, and sends order confirmation to the customer.

7. The seller uses a third-party shipping company to deliver the products.

8. The customer receives order and shipping confirmation.

FIGURE 13.8 A SAMPLE E-COMMERCE TRANSACTION

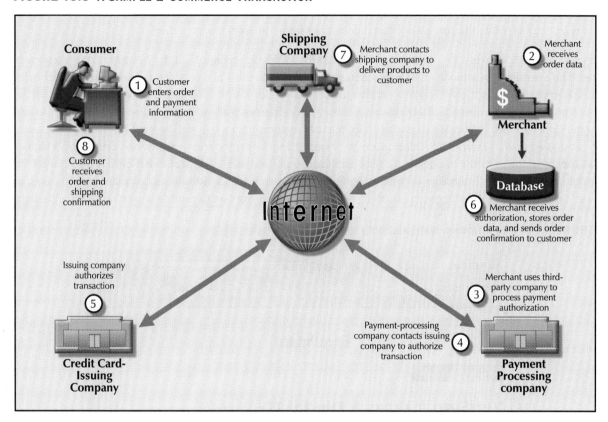

Given the just-described scenario, security (procedures and technology) must be maintained to:

- *Authenticate the identity of the transaction's participants* by ensuring that both the buyer and the seller are who they say they are. In other words, there must be a secure way to properly identify transaction participants and the authenticity of the messages.

- *Protect the transaction data from unauthorized modifications* while it travels on the Internet. The Internet is formed by millions of interconnected networks. E-commerce data has to pass through several different networks to travel from the client to the server, thereby increasing the risks of data being stolen, modified, or forged.

- *Protect the resources (data and computers).* This includes protecting the end user and the business data stored on the Web server and in the databases from unauthorized access. It also includes securing the Web server against attacks from hackers wanting to break into the system to modify or steal data, or to impair normal operations by limiting resource availability.

13.6.1 AUTHENTICATION

Authentication refers to the process of properly and uniquely identifying entities. Such an entity could be a user in a computer system or a database, a computer in a network, or participants in an e-commerce transaction. You probably are familiar with the authentication used in local area network systems in which you are given a unique

user ID and password. For example, on most systems that use Microsoft Windows, you must enter your user ID and password to log on. In this way, Windows authenticates who you are. Based on your user ID (identity), Windows will assign you access rights (read, write, etc.) to resources such as printers or files.

In the case of e-commerce, authentication concerns properly identifying a user on the Internet. Suppose that you have placed an online order for a Sony receiver. After a week, your order still has not arrived. But when you call the merchant, you are informed that the item was shipped several days ago to an address that is not yours. Apparently, somebody changed the shipping address while it traveled on the Internet prior to its delivery to the merchant's Web site. From this admittedly simplified scenario, you can see why verifying the identity of participants, as well as hiding the contents of communications from intruders are important.

Internet authentication is a bit more complicated than the Windows scenario. On the Internet, digital certificates issued by a certification authority (CA) are used to authenticate both users and vendors. A **certification authority (CA)** is a private entity or company that certifies the user or vendor is who (s)he claims to be. Certification authority companies work in concert with credit card verification companies or other financial institutions in order to verify the identity of the certificate's requesters. When a user registers with a CA such as VeriSign, (s)he will be asked to provide appropriate verification information (see Figure 13.9). Using the end-user-provided data, the CA verifies the identity of the requester and issues a digital certificate to the end user. A **digital certificate** is a unique identifier given to an entity. The holder of the certificate may be an end user, a Web site, a computer, a Web page, or even a program. Digital certificates are used in combination with encryption to provide security and authentication.

FIGURE 13.9 REGISTERING WITH A CERTIFICATION AUTHORITY

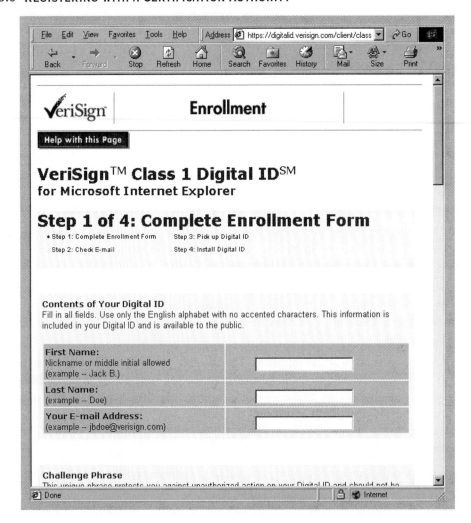

13.6.2 ENCRYPTION

Digital certification does not totally guard against improper use of data. So how can we further ensure that the original sender sent the document and that it has not been forged? This is where encryption comes into the picture. **Encryption** is a process of inputting data in "plain text" to yield an output "encoded" version of the data, making the data unintelligible to unauthorized users. To secure e-commerce transactions in the Web, the client Web browser must encrypt the data before it sends it over the Internet to the merchant's Web server. The merchant's Web server receives the encrypted transaction data and decrypts it. All sensitive communications between client and server must be encrypted. Encryption works to promote:

- Data privacy to ensure that the data cannot be understood if intercepted.

- Data authenticity to ensure that the data were not forged.

Most encryption algorithms use mathematical formulas and an encryption key to encode the data. The **encryption key** is a very large number used to encrypt and decrypt the data. The length of the key—the number of digits in it—determines how secure the data will be: the longer the key, the more secure the data. Most encryption algorithms use key lengths that range from 40 bits to 128 bits or more. You can download Web browsers such as Netscape and Internet Explorer with support for either 40-bit or 128-bit encryption.

Encryption algorithms are of two types: symmetric (private-key) or asymmetric (public-key). **Symmetric** or **private-key encryption** uses a *single* numeric key to encode and decode data. Both sender and receiver must know the encryption key. The most widely used standard for private-key encryption is **Data Encryption Standard (DES)**, used by the U.S. government. Because each client/server combination would require a different key, this type of encryption is basically used by government entities and is rarely used in the Internet for e-commerce transactions.

Asymmetric or **public-key encryption** uses two numeric keys:

- The public key is available to anyone wishing to communicate securely with the key's owner.

- The private key is available only to the owner.

Both keys are able to encrypt and decrypt each other's messages.

An example of public-key encryption is **Pretty Good Privacy (PGP)** by Pretty Good Privacy Inc. PGP is a fairly popular and inexpensive method for encrypting e-mail messages on the Internet. The most commonly used public-key encryption technology is **RSA** encryption by RSA Data Security Inc. RSA has become the de facto standard for Internet encryption. Figure 13.10 shows security settings for the most common Web browsers.

Digital certificates use public-key encryption techniques to create digital signatures. A **digital signature** is an encrypted attachment added to the electronic message to verify the sender's identity. The digital certificate received by the user includes a copy of its public key. The digital certificate's owner makes its public-key available to anyone wanting to send encrypted documents to the certificate's owner.

13.6.3 TRANSACTION SECURITY

As you know, Web pages are plain-text documents that are created through the use of HTML. In fact, the vast majority of Web pages that travel within the Web are transferred in plain text. Such plain text transmissions are acceptable for many applications, but they are clearly unacceptable when sensitive e-commerce data such as credit card and bank account information are transmitted. In this section, we will examine two standard protocols that are used to securely transmit HTML documents over the Internet: Secure Sockets Layer and Secure Hypertext Transfer Protocol.

FIGURE 13.10 WEB BROWSER SECURITY SETTINGS

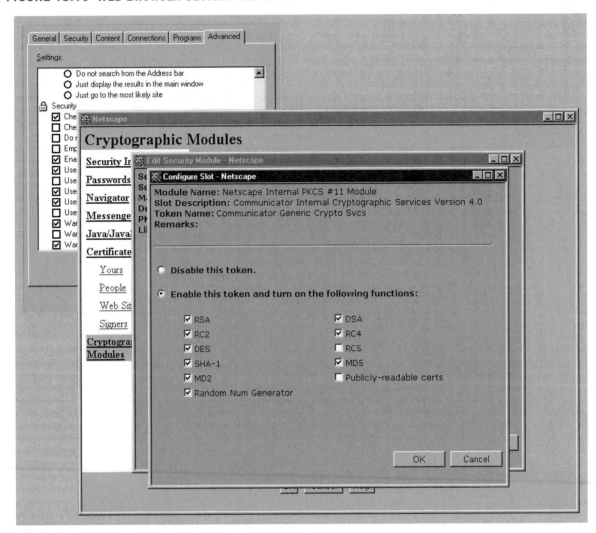

Secure Sockets Layer (SSL) is a protocol used to implement secure communication channels between client and server computers on the Internet. SSL was originally created by a group of companies led by Netscape Communications, IBM, and Microsoft. The SSL protocol requires that the following conditions be met:

- The client accesses a merchant's Web server that supports SSL.
- The server sends its digital certificate, including the server's public-key, to the client.
- The client uses the certificate to verify the server with the certification authority (CA).
- The client generates a private-key for the session, encrypts the key using the server's public-key, and sends it over the Internet to the server.
- The server receives the data, decrypts the data with its private-key, and extracts the SSL session private-key.

When the client and the server each have the SSL session private-key, they communicate by sending encrypted data back and forth.

The advantage of SSL is that it can be used with many different Internet services (FTP, Telnet, and HTTP), which means that it is widely supported on the Internet. To request a secure connection, just use an https prefix to the server's Web address. For example, to access a secure connection to Microsoft, enter *https://www.microsoft.com* in your browser. As you examine Figure 13.11, note the closed padlock on the lower right that indicates a secure session.

FIGURE 13.11 WEB BROWSER SECURE SESSION

Secure Hypertext Transfer Protocol (S-HTTP) is used to securely transfer Web documents over the Internet. S-HTTP supports use of private and public keys for authentication and encryption. S-HTTP has not been widely used, because it only supports encrypted HTTP data and does not support other Internet protocols as SSL does.

In the next section, we will look at the technologies that are used to protect the integrity of the resources that are connected to the Internet.

13.6.4 RESOURCE SECURITY

Resource security refers to the protection of the resource(s) connected to the Internet from external and internal threats. Specifically, resource security means protecting the computers connected to the Internet from viruses, intrusion by hackers, and denial of service attacks.

The most common threat is probably the virus attack. A **virus** is a malicious program that affects the normal operation of a computer system. Client and server computers both must be protected against this type of threat by the use of a virus protection program. Virus protection programs must be constantly updated with the latest signature files to identify newer viruses. For more information about viruses, visit the Web sites of vendors such as Symantec (*www.symantec.com*) and McAfee (*www.mcafee.com*).

A **hacker** is a person who maliciously and illegally accesses a Web site with the intention of stealing data, changing Web pages, or impairing Web site operations. Illegal access to Web sites with the intention of defacing or changing Web page contents is very common. Because e-commerce Web servers connect directly to the Internet, they are prime hacker targets. One of the most common hacker activities is to render Web sites unusable with a denial of service attack. A **denial of service attack** overloads Web servers and routers with millions of requests for service, rendering the services unavailable to legitimate users. A distributed denial of service attack originates from many different computers at the same time. At this point, there is no methodology that is totally effective in protecting Web sites against distributed denial of service attacks. However, one of the most effective and commonly used defenses against viruses and hacker attacks is a *firewall*.

A **firewall** is used to protect the network from unauthorized access from the outside world (public Internet). Specifically, a firewall is a hardware and/or software component that is used to limit and control Internet traffic going into the company's network infrastructure and data that are allowed to be moved outside the company's network.

Firewalls can be classified as:

- *Packet filter firewall.* Works at the TCP/IP packet level, examining every packet that moves into and out of the company's private network.

- *Gateway server firewall.* Works at the application level, examining every application request (HTTP, FTP, Telnet, etc.) that moves into or out of the company's network.

- *Proxy server firewall.* A proxy server is an intermediary between client computers inside a private network and the Internet. The proxy server accepts requests from clients and forwards them to the Internet. When replies come back from the Internet, the proxy server hands the data back to the client that made the request. The client computer application must be configured to work with the proxy server. (Figure 13.12 shows an example of the Internet Explorer proxy settings.)

- *State inspection firewall.*[1] Firewall technology that compares only parts of incoming packets to parts of the related outgoing packets.

FIGURE 13.12 WEB BROWSER PROXY SETTINGS

[1] For additional information about this type of firewall, please visit CheckPoint Software at *www.checkpoint.com*.

Firewalls, proxy servers, and virus protection programs are just a few of the technologies used to secure and protect e-commerce sites. No single technology is able to protect an e-commerce Web server against viruses, hacker intrusions, or denial of service attacks. Implementing resource security includes a mixture of the technologies examined in this section, plus others—such as intrusion detection systems, vulnerability assessment tools, and secure operating systems—that are not within the scope of this book.

13.7 WEB PAYMENT PROCESSING

A key function of e-commerce Web sites is their ability to process online payments for products and/or services. In a traditional business transaction such as buying a CD at a music store, you pay for your purchase using cash, check, money order, or credit card. On the Web, the most common method of payment is credit card, but there are some less-common alternatives. In this section we briefly introduce three technologies used to aid in processing electronic payments: digital cash, online credit card processing, and electronic wallets.

13.7.1 DIGITAL CASH

One of the risks of paying with credit cards is theft. If your credit card information is stolen, you are very likely to have extra charges in your next credit card bill. If somebody steals a $5 bill from your wallet, you only lose five dollars. The use of cash thus has certain advantages. **Digital cash** is the digital equivalent of hard currency (coins or bills). Digital cash uses digital certificates to verify the identity of the transaction's participants and requires the existence of a bank or financial institution from which the user will buy digital cash. When the user buys digital cash, (s)he receives digital bills or coins backed up by a financial institution and represented by binary IDs. The user will later transfer the digital cash to a merchant to pay for products and/or services. The merchant submits the digital cash to the bank to get the account credited. Digital cash remains an evolving technology, with a number of competing companies and few standards.

One of the main advantages of digital cash is its lower cost per transaction when compared to bank checks and credit cards. In fact, given relatively high transaction costs, some online merchants have imposed a minimum purchase for credit card transactions. Nevertheless, at this point digital cash usage still pales in comparison to credit card transactions. Reasons for the reluctance to embrace digital cash include the following:

- Difficulty in getting merchants, banks, and customers to change the way they do business.
- Regulatory obstacles.
- Customers regard credit cards as the most convenient way to pay for online purchases.

Given the current lack of general acceptance, the future of digital cash providers is uncertain.

A more successful approach to Web payment processing has been taken by companies like PayPal, which sets up secure transfers from and to users' credit card or bank accounts. Registered PayPal users can shop at thousands of PayPal-enabled sites, plus they can safely send or request money from anyone with an e-mail address. PayPal has been popular because it also offers business services, and is widely used as a payment system for online auctions. eBay acquired PayPal in 2002. By 2003 it had more than 31 million members in 38 countries.

13.7.2 CREDIT CARD PROCESSING

Most online purchases are paid with a credit card. Although Web merchants accept credit cards as a form of payment, not all merchants use identical methods to process credit card payments. In previous sections you learned about different security techniques that are used to protect credit card information. We will now briefly explore some of the ways in which merchants can process credit card payments.

The simplest payment processing method is manual processing. In this scenario, a seller collects credit card information from the customer through the Web site, using a secure connection. (In some cases, the buyer must place the actual order by phone.) The seller then contacts the credit card issuer—by phone or through a credit card verification device—to get authorization information. If problems arise—such as insufficient credit, a wrong card number, or a stolen card—the merchant e-mails the customer to inform him/her about the problem. Although this multi-step manual process continues to survive, thanks to the existence of many small Web companies, its inefficiency has caused most sellers to modify or abandon it.

A more efficient variation of the manual method is for the merchant to use third-party credit card processing software such as IC Verify (*www.icverify.com*) or to employ the services of a third-party processing company such as InternetSecure Inc. (*www.internetsecure.com*). In this scenario, the client customer computer, the merchant's Web server, the payment processing company server, the customer, and the merchant's bank work together to process the payment. In its simplest form, the procedure requires the following steps:

1. The customer sends credit card information through the Web to the seller's Web server.
2. The seller's Web server verifies the order and sends the credit card information to the credit card processing company.
3. The credit card processing company verifies the credit card with the customer bank and obtains authorization information.
4. The merchant's Web server receives confirmation of the transaction.
5. The merchant's Web server e-mails the purchase order confirmation to the client.
6. When the merchandise is delivered, the seller and the credit card company settle the payment.

One important feature of all credit card payment processing systems is that the charge is only issued after the seller ships the products and/or services to the client.

Two of the major credit card companies, VISA and MasterCard, have created a system known as *Secure Electronic Transactions*. The **Secure Electronic Transactions (SET)** standard provides for the security and privacy of the credit card information through the use of digital certificates, digital signatures, and public-key encryption. Security is automatically enforced for all communications between customer, merchant, and financial institutions. One of the main advantages of SET is that it provides a way for the buyer to transfer his/her credit card information to the credit card issuer *without the seller being able to see the credit card* information—by requiring merchants to encrypt all credit card information. SET has been widely accepted among the most prominent Internet companies and vendors such as Netscape, Microsoft, IBM, and GTE. SET also defines the minimum network and security infrastructure requirements for companies that want to perform electronic payment verification on the Web. This standard includes the use of firewalls, encryption, digital signatures, encrypted database data, antivirus software, operating system patches, end-user authentication, auditing, and other security policies.

13.7.3 ELECTRONIC WALLETS

In the real world, it is very likely you keep your coins, bills, and credit cards in your wallet. An **electronic wallet** is the equivalent of a physical wallet—it can contain your credit card information, digital cash, and other personal information such as shipping addresses, phone numbers, and e-mail addresses. The electronic wallet is a small program that is used in conjunction with your Web browser and the merchant's Web site to automatically enter payment information when online purchases are made. Figure 13.13 shows the configuration of the Microsoft Wallet.

FIGURE 13.13 SETTING UP A MICROSOFT ELECTRONIC WALLET

Clearly, payment processing is a critical e-commerce Web site component. Merchants who want to succeed in e-commerce must be able to accept credit cards for online purchasing. (Customers are much more likely to buy with merchants that accept credit cards than with any other payment method.) While security and privacy of e-commerce transactions are still a concern, new standards are emerging that create a framework for facilitating credit card transactions in more secure ways.

Now that you've had an overview of the world of electronic commerce, let's turn to database design for e-commerce applications.

13.8 DATABASE DESIGN FOR E-COMMERCE APPLICATIONS

Experience tells us that we do not have to reinvent a heat source to make hot water each time we want a cup of coffee. Similarly, to design e-commerce databases you do not need to invent "new" design techniques. To the contrary, the design techniques you have learned in this book (ER modeling, normalization, SDLC, DBLC, transaction management, and so on) provide the basic tools and the knowledge required to build successful e-commerce databases. However, e-commerce databases have a few requirements that you have not encountered before. That is why we illustrate a basic database design for an e-commerce application in this section. In Chapter 14, you will actually set up a Web-enabled database and learn more about Web database development.

Let's start by defining the scope of our database. Our simple e-commerce Web site must include at least the core features that facilitate the sale of products and/or services. Therefore, the database must support the Web site's ability to show the available products and/or services, and to conduct basic sales transactions. In addition, e-commerce Web sites should offer features that focus on customer service, product returns, and Web customer profiling, which make the customer Web experience a pleasant one. To accomplish this, an e-commerce database design must include a few additional support tables. However, we will focus mainly on the database entities that directly support the sale of products in an e-commerce database.

To start the design process, let's establish some basic business rules and their effect(s) on the design:

- The objective of the e-commerce design is to sell products to customers. Therefore, our database's first two tables will be PRODUCT and CUSTOMER.

- Each customer may place one or more orders. Each order is placed by one customer. Therefore, there is a 1:M relationship between CUSTOMER and ORDER.

- Each order contains one or more order lines. Each order line is contained within an order. Therefore, there is a 1:M relationship between ORDER and ORDLINE.

- Each order line references one product. Each product may appear on many order lines. (The company can sell more than one HP ink-jet printer!) Therefore, there is a 1:M relationship between PRODUCT and ORDLINE.

- Customers who browse the product catalog would like to see products grouped by category or type. (For example, it would be useful to see product lists broken down as computers, printers, application software, operating systems, etc.) Therefore, each PRODUCT belongs to one PRODTYPE, and each PRODTYPE has one or many PRODUCTs associated with it.

- Customers who browse the Web catalog must be able to select products and store them in an electronic shopping cart. The shopping cart temporarily holds the products until the customer checks out. Therefore, our next entity is SHOPCART. Each SHOPCART belongs to one CUSTOMER and references one or more PRODUCTs.

- When the customer checks out, (s)he enters credit card and shipping information. This information is added to the ORDER. (Note that the business rule identifies required attributes.)

- When the credit card authorization is received, an order is placed for the products found in the shopping cart. The SHOPCART information is used to create an ORDER, which contains one or more ORDLINEs. *After the order is placed and the customer leaves the Web site, the shopping cart data are deleted.*

- Because the merchant offers many shipping options, we create a SHIPOPTION table to store the details of each shipping option, that is, Land, 2-day, Next day, UPS, FedEx, etc.

- Because the merchant offers many payment options, we create a PMTOPTION table to store the details of each payment option, that is, MasterCard, Visa, American Express, etc.

- Because each state may have a different tax rate, we create two tables, that is, STATE and TAXRATE, to keep track of the states (and countries) and the tax rate for each one.

The entities based on these business rules are shown in Table 13.4.

TABLE 13.4 MAIN TABLES FOR E-COMMERCE DATABASE

TABLE NAME	TABLE DESCRIPTIONS
CUSTOMER	Contains details for each registered customer. This table will contain general customer data, shipping preference, credit card data, and billing data (for customer accounts). Some customers may prefer not to register and they will have to enter the customer details each time they place an order.
PRODUCT	Contains product details such as stock IDs, prices, quantity on hand, etc.
PRODTYPE	Identifies the main product type classifications.
ORDER	Contains general orders details, for example, date, number, customer, etc.
ORDLINE	Contains the products ordered for each order.

SUPPORT TABLES	
SHOPCART	This table contains the purchase quantity for each product selected by the customer. This is a "working" table—*whose contents are deleted when the customer exits the Web site or closes the browser.*
PMTTYPE	The different payment options offered by the merchant.
SHIPTYPE	The different shipping options offered by the merchant.
TAXRATE	The tax rate for each state and/or country.
STATE	The list of each state and/or country for which the tax is charged.
PROMOTION	Special promotions such as vouchers or sales discounts.
PRICEWATCH	Customers who want to be notified if a product's price reaches a certain level.
PRODPRICE	Optional table used to manage multiple price levels

NOTE

To save space, and to recognize that you learned to translate the business rules into an ERD in Chapter 4, "Entity Relationship (ER) Modeling," and Chapter 5, "Normalization of Database Tables," we will not develop the ERD for this design.

After defining the tables required to support the e-commerce activities, let's identify the basic attributes for each table. Please note that the following attribute summary is only a sample of the most important—and most commonly used—attributes, and it is not meant to be comprehensive. (Your specific environment will determine what attributes are relevant and/or what attributes might be added.)

NOTE

In the course of our discussion, we will mention some programming and design practices that may be used to build e-commerce Web sites. For example, because the data stored in the tables exist to support Web transactions, it would be advisable to define some fields—such as product names or descriptions—for Web page display purposes. Also, fields such as credit card numbers and passwords should be stored in an encrypted format. Even the fields themselves may be encrypted, thus making it unlikely that even a field's contents can be improperly accessed! (Databases such as Oracle9*i* and IBM's DB2 support encryption.)

The CUSTOMER Table

The CUSTOMER table contains the details for each registered customer. Keep in mind that some customers would prefer not to register, because they feel uncomfortable providing such registration information. Therefore, we will have to decide whether to require registration for all purchases:

- With mandatory registration, the Web site will need a form to register new customers. Also, a login form is required for returning customers before they start browsing the product catalog. Given the registration data, the customer's shipping and credit card data can be automatically placed in the order when the customer checks out. (One of the benefits of registration is that customers may be rewarded with discounted prices.)

- With optional registration, there is no need to generate a registration or login form for each sale or visit. The customer will have to enter all shipping and credit card information with each order.

Obviously, the simplest option is not to require registration. Therefore, we have decided to make registration optional. The main CUSTOMER table attributes are shown in Table 13.5.

TABLE 13.5 **CUSTOMER Table**

ATTRIBUTE NAME	DESCRIPTION	PK/FK
CUST_ID	Customer ID—automatically generated	PK
CUST_DATEIN	Date the customer was added to the table	
CUST_LNAME	Last Name	
CUST_FNAME	First Name	
CUST_ADDR1	Address line 1	
CUST_ADDR2	Address line 2	
CUST_CITY	City	
CUST_STATE	State or Region if international customer	FK
CUST_ZIP	Zip code	
CUST_CNTRY	Country	
CUST_PHONE	Phone	
CUST_EMAIL	E-mail address	
CUST_LOGINID	Login ID for registered customers	
CUST_PASSWD	Password for login—encrypted field	
CUST_CCNAME	Name as it appears on credit card	
CUST_CCNUM	Credit card number—encrypted field	
CUST_CCEXDATE	Credit card expiration date in mm/yy format	
CUST_ACRNUM	Account Receivable number—to interface with the internal account receivable system or a reference PO number for clients set up in net 30 terms	
CUST_BLLADDR1	Billing address line 1	
CUST_BLLADDR2	Billing address line 2	
CUST_BLLCITY	Billing address city	
CUST_BLLSTATE	Billing address state	FK

TABLE 13.5 **CUSTOMER** TABLE (CONTINUED)

ATTRIBUTE NAME	DESCRIPTION	PK/FK
CUST_BLLZIP	Billing address zip	
CUST_BLLCNTRY	Billing address country	
SHIP_ID	Favorite shipping type	FK
CUST_SHPADDR1	Shipping address line 1	
CUST_SHPADDR2	Shipping address line 2	
CUST_SHPCITY	Shipping address city	
CUST_SHPSTATE	Shipping address state	FK
CUST_SHPZIP	Shipping address zip	
CUST_SHPCNTRY	Shipping address country	
CUST_TAXID	Tax ID for tax-exempt customers	
CUST_MBRTYPE	Membership type—used for special promotions and to determine product pricing according to membership level, for example: regular price, member price, or gold member price	

The PRODUCT Table

The PRODUCT table is the central entity in our database. The PRODUCT table contains the relevant product information about all of the products offered on the Web site. This table (see Table 13.6) is related to the PRODTYPE, ORDLINE, and PROMOTION tables.

TABLE 13.6 **PRODUCT** TABLE

ATTRIBUTE NAME	DESCRIPTION	PK/FK
PROD_ID	Product ID—automatically generated	PK
PROD_NAME	Product short name—shown in promotions, invoices, etc. For example: Verbatim CD-R	
PROD_DESCR	Product description—a long description of the product, used in Web pages for product information	
PROD_OPTIONS	Product options, for example: colors, sizes, style. (There are many ways to handle sizes or colors for apparel and shoe industries; several of them require separate product entries or the creation of other tables in 1:M relationships.)	
PROD_IMAGE_1	URL of the product's image file. Could occur many times (front view, back view, side view, top view).	
PROD_SKU	Stock number used by the vendor or supplier	
PROD_PARTNUM	Part number from manufacturer, for example, VBTM 34563	
VEND_ID	Vendor—The vendor ID for the product, for example, Global Suppliers	FK
PTYPE_ID	Product type (category), for example, Storage	FK
PROD_UNIT_SIZE	Unit size of the product: box, case, each	

TABLE 13.6 **PRODUCT TABLE** (CONTINUED)

ATTRIBUTE NAME	DESCRIPTION	PK/FK
PROD_UNIT_QTY	Unit quantity in unit size: 12, 6, 1	
PROD_QOH	Quantity on hand in the warehouse per each product	
PROD_QORDER	Quantity on order—items that are ordered, but not yet shipped. To determine if an item is in stock, subtract the quantity on order from quantity on hand.	
PROD_REORD_LEVEL	Reorder level—when the quantity on hand is equal to this amount, the product is reordered	
PROD_REORD_QTY	How much to reorder from vendor	
PROD_REORD_DATE	Estimated date the order will arrive from the vendor	
PROD_PRICE	Regular price per unit quantity (each): $1.05 per CD-R	
PROD_MSRP	Manufacturer's suggested retail price—to show savings	
PROD_PRICE_D1	Price discount 1—for members or order quantity level: 3%	
PROD_PRICE_D2	Price discount 2—for gold members or order quantity level: 6%	
PROD_TAX	Yes or No—is the product taxable?	
PROD_ALTER_1	Alternate product if not in stock. Could occur many times. This is a foreign key to the same product table. Again, this could be implemented by creating another separate table in 1:M relationship.	**FK**
PROD_PROMO	Yes/No. Product participates in promotions? Default is Yes	
PROD_WEIGHT	Weight of product, used for shipping purposes	
PROD_DIMEN	Product dimensions, used for shipping purposes	
PROD_NOTES	Notes about the product shipping, handling instructions, etc.	
PROD_ACTIVE	Yes or No. If not active, the product is not available to customers. Useful for product recalls or when we want to stop sales of a given product	

Generally, there is one row for each product, except for those products that come in various sizes, colors, or styles, such as shoes, shirts, etc. In those cases, we have two options:

- The customer enters the size, color, and style as additional attributes in the order.

- Create unique product entries for each product size, color, and style combination.

- Create a new product option (PRODOPT) table in a 1:M relationship with the PRODUCT table. This table will have one record for each combination of color, size, and style for a given product.

The PRODTYPE Table

This table describes the different product categories. The categories could be limited to just one level or could be multiple levels. In this example, we have decided to use two levels. This decision will permit the use of categories, such as "printer"; within that category we could reference subcategories, such as "laser" or "inkjet." (See Table 13.7.)

TABLE 13.7 **PRODTYPE** **TABLE**

ATTRIBUTE NAME	DESCRIPTION	PK/FK
PTYPE_ID	Product type ID—automatically generated	**PK**
PTYPE_NAME	Product type name—"Inkjet Printer"	
PTYPE_PARENT	Product type parent—such as "Printer"	**FK**

The ORDER Table

This table contains all of the customer orders. After the credit card company approves the transaction, the order is added to the ORDER table. If the credit card is rejected (invalid number, expired, stolen, etc.), the order is not added. There will be one ORDER row for each new customer order, regardless of the number of products ordered. If a registered customer places an order, the credit card and shipping information could be automatically entered in the ORDER table. The ORDER table is in a 1:M relationship with ORDLINE table. (See Table 13.8.)

TABLE 13.8 **ORDER** **TABLE**

ATTRIBUTE NAME	DESCRIPTION	PK/FK
ORD_ID	Order ID—automatically generated	**PK**
ORD_DATE	Date the order was added	
CUST_ID	Customer ID (optional)—some customers will not register. If this were a registered customer, the CUS_ID would be automatically added by Web system.	**FK**
PMT_ID	Payment type ID—selected by the customer	**FK**
ORD_CCNAME	Name as it appears on credit card—copied from CUSTOMER data, manually entered by an unregistered customer, or by electronic wallet software.	
ORD_CCNUM	Credit card number—encrypted field—copied from CUSTOMER data, manually entered by an unregistered customer or by electronic wallet software.	
ORD_CCEXDATE	Credit card expiration date in mm/yy format—copied from CUSTOMER data, manually entered by an unregistered customer or by electronic wallet software.	
SHIP_ID	Selected shipping type—automatically or manually entered. This is used if only one company or shipment is used to fulfill the order.	**FK**
ORD_SHIPADDR1	Shipping address line 1—automatically or manually entered	
ORD_SHIPADDR2	Shipping address line 2—automatically or manually entered	
ORD_SHIPCITY	Shipping address city—automatically or manually entered	
ORD_SHIPSTATE	Shipping address state—automatically or manually entered	**FK**
ORD_SHIPZIP	Shipping address zip—automatically or manually entered	
ORD_SHIPCNTRY	Shipping address country—automatically or manually entered	
ORD_SHIPDATE	Date the order shipped—if complete shipment. If partial shipment, see ORDLINE for shipment dates for each product line.	

TABLE 13.8 **ORDER TABLE (CONTINUED)**

ATTRIBUTE NAME	DESCRIPTION	PK/FK
ORD_SHIPCOST	Total shipment cost—estimated shipment cost for order. This is the result of applying a given shipment cost formula according to the shipment method.	
ORD_PRODCOST	Total product cost—the sum of all product prices * quantity ordered	
ORD_TAXCOST	Total cost of sales tax—computed by adding the taxes for each individual product ORDLINE table.	
PROM_ID	Promotion ID applied to order (optional)	
ORD_TOTCOST	Total cost of order: PRODCOST + SHIPCOST + TAXCOST – PRO_AMT (from promotion table)	
ORD_TRXNUM	Transaction confirmation number from credit card company	
ORD_STATUS	Status of the order: Open, Shipped, or Paid	

The ORDLINE Table

This table contains one or more products related to each order. Each ORDLINE row is related to one PRODUCT row and to one ORDER row. The ORDLINE contains the quantity and price for each product ordered. The ORDLINE table gets the PROD_ID and ORL_QTY from the SHOPCART table. (See Table 13.9.)

TABLE 13.9 **ORDLINE TABLE**

ATTRIBUTE NAME	DESCRIPTION	PK/FK
ORL_ID	Order line ID—automatically generated	**PK**
ORD_ID	Order ID from ORDER table	**FK**
PROD_ID	Product ID	**FK**
ORL_QTY	Quantity ordered	
ORL_PRICE	Product price—after all promotions and discounts	
ORL_TAX	Percentage tax rate applied to this product. Some products or customers may be tax exempt. If the product/customer is taxable, the tax rate is obtained according to the STATE in the shipping address.	
SHIP_ID	Shipping company and type used to ship this product—for cases in which partial shipment is required	**FK**
ORL_SHIPDATE	Date this product shipped	

The SHOPCART Table

The SHOPCART table is a special table that is used by the Web site to temporarily store the products during the customer's shopping activities. To understand how the SHOPCART table works, you need to understand the online ordering process:

- When a customer first visits our Web site, (s)he may or may not log in. If (s)he logs in, the Web server keeps his/her customer ID (CUST_ID) in memory.

- The customer browses the product catalog. If the customer is a registered customer, (s)he sees the member prices, otherwise (s)he sees regular prices (PROD_PRICE minus the respective discount percentage, PROD_PRICE_D1 or PROD_PRICE_D2).

- A shopping cart is automatically assigned to the user the first time (s)he orders a product by clicking the "Add to Shopping Cart" or the "Order Now" button. A unique shopping cart ID is negotiated between the client's browser and the merchant's Web server, using a secure session. This ID only lasts until the customer successfully checks out, cancels the order, exits the Web site, or closes the browser.

- The shopping cart stores the PROD_ID and the quantity for each product selected by the customer.

- When the customer clicks the Check Out button, an Order confirmation screen is shown. This screen shows all the product details for the product(s) placed in the shopping cart.

- When the customer accepts the order, (s)he is shown another screen in which to enter the shipping and payment information. The customer then confirms the order.

- After the customer has confirmed the order, the Web server requests a transaction confirmation from the credit card company. The completion of this process may take from 10 seconds to one minute.

- Once the confirmation has been received, the ORDER data and the ORDLINE data are saved.

- The SHOPCART data are deleted.

The SHOPCART structure is shown in Table 13.10.

TABLE 13.10 **SHOPCART** TABLE

ATTRIBUTE NAME	DESCRIPTION	PK/FK
CART_ID	Shopping cart unique ID—automatically generated	PK
CART_PROD_ID	Product ID—this is a copy of the PROD_ID value. Because the SHOPCART table could potentially become a high-traffic table with many add and delete operations, we do not want to relate it to the PRODUCT table for performance reasons. Remember, these values will be automatically copied to the ORDLINE table when the customer transaction is processed. However, a relationship could be established if desired.	
CART_QTY	Quantity ordered	

The PMTTYPE Table

This table contains one row for each payment method accepted by the merchant. (See Table 13.11.)

TABLE 13.11 **PMTTYPE** TABLE

ATTRIBUTE NAME	DESCRIPTION	PK/FK
PMT_ID	Payment type ID—automatically generated	PK
PMT_NAME	Name: VISA, MasterCard, American Express, Net30	
PMT_MCHNT_ID	Merchant ID—used by payment processing systems. This ID is given to the merchant when it registers with a credit card company	
PMT_NOTES	Additional notes	

The SHIPTYPE Table

This table contains one row for each shipping method supported by the merchant. (See Table 13.12.)

TABLE 13.12 **SHIPTYPE TABLE**

ATTRIBUTE NAME	DESCRIPTION	PK/FK
SHIP_ID	Shipping type ID—automatically generated	**PK**
SHIP_NAME	Name: UPS Next Day, UPS Three Days, FedEx Overnight, etc.	
SHIP_COST	Shipping cost per weight unit—this depends on the formula used by the shipping company. Most are based on the shipping zip code and the size and weight of the products being shipped. Therefore, it's very likely that you are going to need additional attributes in this table as you develop the design.	
SHIP_NOTES	Additional shipping notes	

The TAXRATE Table

This table contains the different sales tax rates used in each state. Note that the sales tax is not applied to every order, but only in those states in which the merchant is required by law to collect taxes on products sold in that state. How is the sales tax requirement determined? Federal and state regulations determine this requirement. Normally, the sales tax determination could be based on the shipping address. However, the customer billing address or the credit card billing address may also be used for this purpose. This table is related to the STATE table. Also, tax-exempt institutions are not charged tax. (See Table 13.13.)

TABLE 13.13 **TAXRATE TABLE**

ATTRIBUTE NAME	DESCRIPTION	PK/FK
STATE_ID	State ID from the STATE table—required	**PK, FK**
TAX_RATE	Percentage sales tax rate applied—required	
TAX_NOTES	Additional notes—like reason for the tax charge, etc.	

The STATE Table

This table contains one entry for each state or country. This table is related to the TAXRATE table. The table may contain modified entries for countries that use postal regions or other identifiers. This table can also contain a COUNTRY field for businesses that have multinational offices. (See Table 13.14.)

TABLE 13.14 **STATE TABLE**

ATTRIBUTE NAME	DESCRIPTION	PK/FK
STATE_ID	State ID—automatically generated	**PK**
STATE_NAME	Name of the state—required	

The PROMOTION Table

This table is used to represent special sales promotions. It contains one entry for each sale or promotion offered by the merchant. All promotions have a starting and ending date. Some promotions may apply to one product line or to a specific product. Some promotions may offer a discount percentage and other promotions—such as vouchers—will have a specific face amount. You could combine the attributes shown in Table 13.15 in many ways to represent different promotions.

TABLE 13.15 **PROMOTION** TABLE

ATTRIBUTE NAME	DESCRIPTION	PK/FK
PROMO_ID	Promotion ID—automatically generated	**PK**
PROMO_NAME	Name of the promotion: Summer sale, Christmas sale, Voucher	
PROMO_DATE	Date the promotion was introduced	
PROMO_BEGDATE	Date promotion begins	
PROD_ENDDATE	Date promotion ends	
PTYPE_ID	Product type ID—optional. The type of product(s) affected by the promotion	**FK**
PROD_ID	The product(s) affected by the promotion—optional	**FK**
PROMO_MINQTY	Minimum purchase quantity required for promotion—optional	
PROMO_MAXQTY	Maximum purchase quantity to which the promotion applies	
PROMO_MINPUR	Minimum total purchase cost required for promotion—optional	
PROMO_PCTDISC	Percent discount of promotion—optional	
PROMO_DOLLAR	Dollar amount of promotion—optional	
PROMO_CEILING	Maximum value of promotion (if percentage)—optional	

The PRICEWATCH Table

Many e-commerce Web sites offer a "pricewatch" service. This service sends an e-mail to a customer when the price of a product is below or equal to a price that was preselected by the customer. The PRICEWATCH table implements this feature. (See Table 13.16.) A variation of this table can also be used for reverse bids.

TABLE 13.16 **PRICEWATCH** TABLE

ATTRIBUTE NAME	DESCRIPTION	PK/FK
PW_ID	Price watch ID—automatically generated	**PK**
PW_DATE	Date timestamp when the row was inserted in the table	
CUST_ID	Customer ID—optional. We want to offer this service to all prospective visitors as well as to registered customers.	**FK**
PW_CUST_NAME	Name of customer—required Manually entered or automatically copied from CUSTOMER data	
PW_CUST_EMAIL	Customer e-mail—required E-mail address to send email notification.	
PW_ENDDATE	Date the price watch stops—optional Customer enters this date if (s)he wants	
PROD_ID	The product for which the price watch is done—required	**FK**
PW_LOWPRICE	The price at which the customer wants to be informed. If the product price is equal to or less than this value, the system will send an e-mail to the customer.	

The PRODPRICE Table

The PRODPRICE table is used to manage multilevel pricing. Some e-commerce sites offer several different prices, depending on the order quantity. For example, if you purchase one to five (inclusive) pairs of shoes, the price per pair might be $39.95. However, if you purchase six or more pairs, the price per pair might drop to $35.95. This table is in a 1:M relationship with the PRODUCT table. (See Table 13.17.) If multilevel pricing is used, the PROD_PRICE in the PRODUCT table is not used.

TABLE 13.17 **PRODPRICE TABLE**

ATTRIBUTE NAME	DESCRIPTION	PK/FK
PROD_ID	Product ID from the PRODUCT table	**PK, FK**
PROD_QTYFROM	Product purchase quantity from in range—required For example: 1 or 6 or 11	**PK**
PROD_QTYTO	Product purchase quantity to in range—required For example: 5 or 10 or 9999	**PK**
PROD_PRICE	Price for the quantity range—required	

13.9 EXTENSIBLE MARKUP LANGUAGE (XML)

You learned earlier in the chapter that most e-commerce transactions take place between businesses. Because B2B e-commerce integrates business processes among companies, it requires the transfer of business information among different business entities. But the way in which businesses represent, identify, and use data tends to differ substantially from company to company. (Is a *product code* the same thing as an *item-ID*?) Until recently, the expectation was that a purchase order traveling through the Web would be in the form of an HTML document, containing tags and data. The HTML Web page displayed on the Web browser would include formatting tags as well as the order details. If an application wants to get the order data from the Web page, there is no easy way to extract the order details (such as the order number, the date, the customer number, the item, the quantity, the price, or payment details) from the HTML document. (In effect, the HTML approach requires going through the document one line at a time.) The HTML document can only describe how to display the order in a Web browser; it does not permit the manipulation of the order's data elements, that is, date, shipping information, payment details, product information, and so on. To solve this problem, a new standard markup language, known as Extensible Markup Language or XML was developed.

Extensible Markup Language (XML) is a meta-language used to represent and manipulate data elements. XML is designed to facilitate the exchange of structured documents such as orders or invoices over the Internet. The **World Wide Web Consortium (W3C)**[2] published the first XML 1.0 standard definition in 1998. This standard sets the stage for giving XML the real-world appeal of being a true vendor-independent platform. Therefore, it is not surprising that XML is rapidly becoming the data exchange standard for e-commerce applications.

The XML meta-language allows the definition of new *tags* (such as <ProdPrice>) to describe the data elements used in an XML document. Given this feature, XML is said to be an *extensible* language. XML is derived from the Standard Generalized Markup Language. **Standard Generalized Markup Language (SGML)** is an international standard for the publication and distribution of highly complex technical documents—such as those used by the aviation industry and the military services—that are too complex and unwieldy for the Web. Just like

[2] You can visit the W3C Web page, located at *www.w3.org*, to get additional information about the efforts that were made to develop the XML standard.

HTML—which was also derived from SGML—an XML document is a text file, but it has a few—but very important—additional characteristics:

- XML allows the definition of new tags to describe data elements, such as <ProductId>.

- XML is case sensitive: <ProductID> is not the same as <Productid>.

 - XML tags must be well formed, that is, each opening tag has a corresponding closing tag. For example, the product identification would require the format <ProductId>2345-AA</ProductId>.

 - XML tags must be properly nested. For example, a properly nested XML tag might look like this: <Product><ProductId>2345-AA</ProductId></Product>.

- You can use the <— and —> symbols to enter comments in the XML document.

- The "XML" or "xml" prefixes are reserved for XML tags only.

XML is *not* a new version or replacement for HTML. XML is concerned with the description and representation of the data, rather than how the data are displayed. (The data display remains the job of HTML.) XML provides the semantics that facilitate the sharing, exchange, and manipulation of structured documents over organizational boundaries. In short, XML and HTML perform complementary, rather than overlapping, functions. To illustrate the use of XML, let's assume a B2B example in which Company A uses XML to exchange product data with Company B over the Internet. Figure 13.14 shows the contents of the ProductList.xml document.

FIGURE 13.14 CONTENTS OF THE PRODUCTLIST.XML DOCUMENT

Using the XML example shown in Figure 13.14, we can illustrate several important XML features:

- The first line represents the XML document declaration, and it is mandatory.

- Every XML document has a *root element*. In our example, the second line declares the ProductList root element.

- The root element contains *child elements* or subelements. In our example, line three declares Product as a child element of ProductList.

- Each element can contain *subelements*. For example, each Product element is composed of several child elements, represented by P_CODE, P_DESCRIPT, P_INDATE, P_ONHAND, P_MIN, and P_PRICE.

- The XML document reflects a hierarchical tree structure where elements are related in a parent-child relationship; each parent element can have many child elements. For example, the root element is ProductList. Product is the child element of ProductList. Product has six child subelements: P_CODE, P_DESCRIPT, P_INDATE, P_ONHAND, P_MIN, and P_PRICE.

Once Company B receives the ProductList.xml document, it can process it—assuming it can understand the tags created by Company A. The meaning of the XML tags in the example shown in Figure 13.14 is fairly self-evident, but there is no easy way to validate the data, or to check whether the data are complete. For example, we could encounter a P_INDATE value of "25/14/2004"—but is that value correct? And what happens if Company B expects a Vendor element as well? How can companies share data descriptions about their business data elements? In the next section, we will show how document type definitions and XML schemas are used to address these concerns.

13.9.1 DOCUMENT TYPE DEFINITIONS (DTDS) AND XML SCHEMAS

B2B solutions require a high degree of business integration between companies. Companies that use B2B transactions must have a way to understand and validate each other's tags. One way to accomplish that task is through the use of Document Type Definitions. A **Document Type Definition (DTD)** is a file with a .dtd extension that describes XML elements—in effect, a DTD file provides the composition of the database's logical model, and defines the syntax rules or valid tags for each type of XML document. (The DTD component is very similar to having a public data dictionary for business data.) Companies that intend to engage in e-commerce business transactions must develop and share DTDs. Figure 13.15 shows the ProductList.dtd document for the ProductList.xml document shown earlier in Figure 13.14.

FIGURE 13.15 CONTENTS OF THE PRODUCTLIST.DTD DOCUMENT

As you examine Figure 13.15, note that the ProductList.dtd file provides definitions of the elements in the ProductList.xml document. In particular, note that:

- The first line declares the ProductList root element.
- The ProductList root element has one child, the Product element.
- The plus "+" symbol indicates that Product occurs one or more times within ProductList.
- An asterisk "*" would mean that the child element occurs zero or more times.
- A question mark "?" would mean that the child element is optional.
- The second line describes the Product element.
- The question mark "?" after the P_INDATE and P_MIN indicates that they are optional elements.
- The third through eighth lines show that the Product element has six child subelements.
- The #PCDATA keyword represents the actual text data.

To use a DTD file to define elements within an XML document, the DTD must be referenced from within that XML document. Figure 13.16 shows the ProductListv2.xml document that includes the reference to the Product List.dtd in the second line.

FIGURE 13.16 CONTENTS OF THE PRODUCTLISTV2.XML DOCUMENT

```
productlistv2.xml - Notepad
File  Edit  Search  Help
<?xml version ="1.0"?>
<!DOCTYPE ProductList SYSTEM "ProductList.dtd">
<ProductList>
        <Product>
                <P_CODE>23109-HB</P_CODE>
                <P_DESCRIPT>Claw hammer</P_DESCRIPT>
                <P_ONHAND>23</P_ONHAND>
                <P_PRICE>5.95</P_PRICE>
        </Product>
        <Product>
                <P_CODE>23114-AA</P_CODE>
                <P_DESCRIPT>Sledge hammer, 12 lb.</P_DESCRIPT>
                <P_ONHAND>8</P_ONHAND>
                <P_MIN>5</P_MIN>
                <P_PRICE>14.40</P_PRICE>
        </Product>
</ProductList>
```

As you examine Figure 13.16, note that the P_INDATE and P_MIN do not appear in all Product definitions because they were declared to be optional elements. The DTD can be referenced by many XML documents of the same type. For example, if Company A routinely exchanges product data with Company B, it needs to create the DTD only once. All subsequent XML documents will refer to the DTD and Company B will be able to verify the data being received.

To further demonstrate the use of XML and DTD for e-commerce business data exchanges, let's assume the case of two companies exchanging order data. Figure 13.17 shows the DTD and XML documents for such a scenario.

Although the use of DTDs is a great improvement for data sharing over the Web, a DTD provides only descriptive information to understand how the elements—root, parent, child, mandatory or optional—relate to each other. A DTD provides limited additional semantic value, such as data type support or data validation rules. Such information is very important for database administrators who are in charge of large e-commerce databases. To solve this problem, the W3C published an XML Schema standard in May 2001, to provide a better way to describe XML data.

The **XML schema** is an advanced data definition language that is used to describe the structure (elements, data types, relationship types, ranges, and default values) of XML data documents. One of the main advantages of an XML schema is that it more closely maps to database terminology and features. For example, an XML schema can define common database types, such as date, integer or decimal, minimum and maximum values, list of valid values, and required elements. Using the XML schema, a company would be able to validate the data for values that may be out of range, incorrect dates, valid values, and so on. For example, a university application must be able to specify that a GPA value must be between zero and 4.0 and it must be able to detect an invalid birth date, such as "14/13/1987." (There is no 14th month.) Many vendors are rapidly adopting this new standard and are supplying tools to translate DTD documents into XML Schema Definition (XSD) documents. It is widely expected that XML schemas will replace DTD as the method to describe XML data.

FIGURE 13.17 DTD AND XML DOCUMENTS FOR ORDER DATA

Unlike a DTD document, which uses a unique syntax, an **XML Schema Definition (XSD)** file uses a syntax that resembles an XML document. Figure 13.18 shows the XML Schema Definition document for our OrderData XML document.

The code shown in Figure 13.18 is a simplified version of the XML schema document. As you can see, the XML schema syntax is similar to the XML document syntax. In addition, the XML schema introduces additional semantic information for the OrderData XML document, such as string, date, and decimal data types, required elements, and minimum and maximum cardinalities for the data elements.

FIGURE 13.18 THE XML SCHEMA DOCUMENT FOR THE ORDER DATA

OrderData.xsd

```
<xsd:schema xmlns:xsd="http://www.company.com/xmlschema">

<xsd:element name="OrderData" type="Order"/>

<xsdcomplexType name="Order">
        <xsd:element name="ORD_ID"        type="xsd:string"/>
        <xsd:element name="ORD_DATE"      type="xsd:date"/>
        <xsd:element name="CUS_NAME"      type="xsd:string"/>
        <xsd:element name="ORD_SHIPTO"    type="xsd:string"/>
        <xsd:element name="ORD_PRODS"     type="productlist"/>
        <xsd:element name="ORD_TOT"       type="xsd:decimal"/>
<xsd:complexType name="Order">

<xsd:complexType name="productlist">
        <xsd:element name="product"       type="aproduct" ninOccurs="1" maxOccurs="unbounded"/>
</xsd:complexType>

<xsd:complexType name="aproduct">
        <xsd:element name="P_CODE"        type="xsd:string"          use="required"/>
        <xsd:element name="P_DESCRIPT"    type="xsd:string"          use="required"/>
        <xsd:element name="P_QTY"         type="xsd:positiveInteger" use="required"/>
        <xsd:element name="P_PRICE"       type="xsd:decimal"         use="required"/>
</xsd:complexType>

</xsd:schema>
```

13.9.2 XML PRESENTATION

One of the main benefits of XML is that it separates data structure from its presentation and processing. By separating data and presentation, you are able to present the same data in different ways—which is similar to having views in SQL. But what are the mechanisms that are used to present data?

The Extensible Style Language specification provides the mechanism to display XML data. **Extensible Style Language (XSL)** is a specification used to define the rules by which XML data are formatted and displayed. The XSL specification is divided into two parts: Extensible Style Language Transformations (XSLT) and XSL style sheets.

- **Extensible Style Language Transformations (XSLT)** describe the general mechanism that is used to extract and process data from one XML document and enable its transformation within another document. Using XSLT, we can extract data from an XML document and convert it into a text file, an HTML Web page, or a Web page that is formatted for a mobile device. What the user sees in these cases is actually a view or HTML representation of the actual XML data. XSLT can also be used to extract certain elements from an XML document, such the product codes and product prices, to create a product catalog. We can even use XSLT to transform an XML document into another XML document.

- **XSL style sheets** define the presentation rules applied to XML elements—something like presentation templates. The XSL style sheet describes the formatting options to apply to XML elements when they are displayed on the browser, cellular phone display, PDA screen, and so on.

Figure 13.19 illustrates the framework used by the different components to translate XML documents into viewable Web pages, an XML document, or some other document.

Browser support for XML, XLST, and XSL is limited—recent versions of Microsoft Internet Explorer (MSIE) support only some of the XML standards, and earlier versions of Netscape Navigator (4.7 or earlier) do not support XML.

FIGURE 13.19 FRAMEWORK FOR XML TRANSFORMATIONS

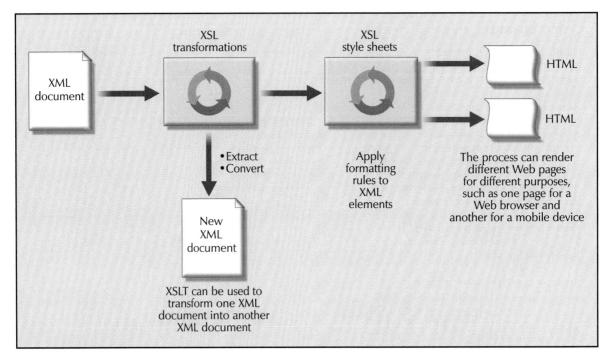

To display the XML document with MS Internet Explorer (version 5.0 or later), just enter the URL of the XML document in the browser's address bar. Figure 13.20 is based on the ProductList.xml document created earlier. As you examine Figure 13.20, note that MSIE shows the XML data in a color-coded, collapsible, tree-like structure. (Actually, this is the MSIE default style sheet that is used to render XML documents.)

FIGURE 13.20 DISPLAYING XML DOCUMENTS IN MSIE 5.0

MSIE 5.0 also provides *data binding* of XML data to HTML documents. Figure 13.21 shows the HTML code that is used to bind an XML document to an HTML table. The example uses the <xml> tag to include the XML data in the HTML document in order to later bind it to the HTML table. This example only works in MSIE 5.0 or later.

FIGURE 13.21 **XML DATA BINDING IN MSIE 5.0**

13.9.3 XML APPLICATIONS

Now that you have some idea what XML is, the next question is: how can you use it? What kinds of applications lend themselves particularly well to XML? In this section, we will list some of the uses of XML. Keep in mind that the future use of XML is limited only by the imagination and creativity of developers, designers, and programmers.

- *B2B exchanges.* As we noted earlier, XML enables the exchange of B2B data, providing the standard to be used by all organizations that need to exchange data with partners, competitors, government, or customers. In particular, XML is positioned to replace EDI as the standard for the automation of the supply chain, because it is less expensive and more flexible.

- *Legacy systems integration.* XML provides the "glue" to integrate legacy system data with modern e-commerce Web systems. Web and XML technologies could be used to inject some new life into "old but trusted" legacy applications. Another example would be the use of XML to import transaction data from multiple operational databases to a data warehouse database.

- *Web page development.* XML provides several features that make it a good fit for certain Web development scenarios. For example, Web portals with large amounts of personalized data can use XML to pull data from multiple external sources (such as news, weather, and stocks). and apply different presentation rules to format pages on desktop computers as well as mobile devices.

- *Database support.* Databases are at the heart of e-commerce applications. A DBMS that supports XML exchanges will be able to integrate with external systems (Web, mobile data, legacy systems, etc.) and thus enable the creation of new types of systems. IBM, Oracle, and Microsoft already offer database products that support XML. Such databases can import or export data in XML format, or generate XML documents from SQL queries, etc.—while still storing the data using its native data model format. Alternately, a DBMS can also support a XML data type to store XML data in its native format. The implications of these capabilities are far-reaching—you would even be able to store a hierarchical-like tree structure inside a relational structure. Of course, such activities would also require that the query language be extended to support queries on XML data.

- *Database meta-dictionaries.* XML can also be used to create meta-dictionaries or vocabularies for databases. Such meta-dictionaries can be used by applications that need to access other external data sources. (Until now, each time an application wanted to exchange data with another application, a new interface had to be specifically built for that purpose.) DBMS vendors can publish meta-dictionaries to facilitate data exchanges and the creation of data views from multiple applications—hierarchical, relational, object-oriented, object-relational, or extended-relational. The meta-dictionaries will all use a common language, regardless of the DBMS type. We expect that the development of industry-specific meta-dictionaries will enable the development of complex B2B interactions, such as those likely to be found in the aviation, automotive, and pharmaceutical industries. We also expect that application-specific initiatives will take place. For example, such initiatives would create XML meta-dictionaries for data warehousing, system management, and complex statistical applications. Even the United Nations and a not-for-profit standards-promoting organization named Oasis are working on a new specification called *ebXML* that will create a standard XML vocabulary for e-business (the most current information about this development can be found on at *www.ebxml.org*).

- *XML databases.*[3] Given the huge number of expected XML-based data exchanges, businesses are already looking for ways to better manage and utilize such data. Currently, there are many different products in the market that address this problem. Their approaches range from simple middleware XML software to object databases with XML interfaces to full XML database engines and servers. The current generation of relational databases is tuned for the storage of normalized rows, that is, manipulating one row of data at a time. We know that business data do not always conform to such a requirement. XML databases provide for the storage of data in complex relationships. For example, an XML database would be well suited to store the contents of a book. (The book's structure would dictate its database structure: a book typically consists of chapters, sections, paragraphs, figures, charts, footnotes, end notes, etc.) Examples of XML databases are dbXML from The dbXML Group (*www.dbxml.org*), Ipedo XML Database from Ipedo (*www.ipedo.com*), and Tamino from Software AG (*www.softwareag.com*).

- *XML services.* Many companies are already working to develop a new breed of services around XML and Web technologies. Those services promise to break down the interoperability barriers among systems and companies alike. XML provides the infrastructure that facilitates heterogeneous systems to work together across the desk, the street, and the world. Services would use XML and other Internet technologies to publish their interfaces. Other services, wanting to interact with existing services, will locate them and learn their vocabulary (service request and replies) to establish a "conversation."

[3] For a comprehensive analysis of XML database products, see "XML Database Products" by Ronald Bourret at *www.rpbourret.com*.

SUMMARY

E-commerce is the use of electronic computer-based technology—in particular, the Internet—to bring new products, services, or ideas to market, and to support enhanced business operations. The Internet has dramatically affected the way in which companies conduct business. The Internet has also had a major impact on not-for-profit and government institutions.

E-commerce enables companies to market and sell products and/or services to a global market of millions of users. Companies engaged in e-commerce use the Internet to integrate their systems with those of other companies in order to increase efficiency and reduce costs.

The adoption of e-commerce technologies enables companies to provide rapid responses to competitive pressures by enhancing internal operations and facilitating business transactions. Some of the customer advantages of using e-commerce include; comparison shopping, reduced costs, increased competition, $24 \times 7 \times 365$ operations, and overall convenience. Business advantages include: access to a global market of million of users, lower entry barriers, and increased market knowledge. However, e-commerce also has some drawbacks for customers and sellers—hidden costs, lack of network reliability, higher costs of staying in business, lack of security, lack of privacy, low service levels, and legal issues such as copyright infringement and fraud.

E-commerce can be classified as business to business (B2B), business to consumer (B2C), and intra-business. B2B e-commerce refers to all transactions that are conducted among businesses, such as manufacturers and suppliers, manufacturers and wholesalers, wholesalers and retailers, and retailers and shipping companies. B2C e-commerce refers to all transactions that take place between retailers and customers. Intra-business e-commerce refers to business processes conducted within an organization.

E-commerce architecture provides a framework that describes the technical components of e-commerce Web sites and their interactions. The architecture is divided into three separate, yet interrelated, layers of services and applications. The bottom layer, Basic Internet Services, provides the "plumbing" required for the transmission of data and information between computers. The middle layer, Business Enabling Services, provides additional services to better support business transactions. These services add the security, reliability, integrity, and performance features required by businesses to operate in the Internet. The top layer of the architecture, E-Commerce Business Services, uses the bottom layers to map business logic in order to represent and automate business processes.

The security and privacy of business transactions and data are crucial for the success of e-commerce. E-commerce data must be secure from a transaction's beginning to its end; the storage of the transaction details and results must also be secure.

Credit card processing and digital cash are the most common ways to process payments on the Internet. Electronic wallets allow the customer to store his/her personal, credit card, and digital cash information in a secure place that can later be used to pay for online purchases.

E-commerce databases contain normalized structures that are similar to typical operational databases. In this chapter we developed the foundation for an e-commerce database structure.

Extensible Markup Language (XML) facilitates the exchange of B2B data over the Internet. XML provides the semantics that facilitate the exchange, sharing, and manipulation of structured documents across organizational boundaries. With XML, organizations of all kinds can exchange data in XML-formatted documents. XML enables the systems to produce the description and the representation of data, thus setting the stage for data manipulation in ways that were not possible before. XML documents can be validated through the use of Document Type Definition (DTD) documents and XML Schema Definition (XSD) documents. The use of DTD, XML schemas, and XML documents permits a much greater level of integration between diverse systems than was possible before. XML impacts B2B data exchanges, legacy systems integration, Web page development, database systems interactions, and documents storage and management.

KEY TERMS

asymmetric encryption
authentication
business to business (B2B)
business to consumer (B2C)
certification authority (CA)
content management
Data Encryption Standard (DES)
database integration
denial of service attack
digital cash
digital certificate
digital signature
Document Type Definition (DTD)
dynamic Web page
electronic commerce (e-commerce)
Electronic Data Interchange (EDI)
electronic mail (e-mail)
electronic wallet
encryption
encryption key
Extensible Markup Language (XML)
extranet
File Transfer Protocol (FTP)
Firewall
government to business (G2B)
government to consumer (G2C)
hacker

hyperlink
Hypertext Markup
 Language (HTML)
Hypertext Transfer Protocol (HTTP)
Internet
intra-business
intranet
load balancing
load testing
messaging
news and discussion group services
personalization
Pretty Good Privacy (PGP)
privacy
private-key encryption
public-key encryption
resource security
RSA
router
static Web page
search services
Secure Electronic Transaction (SET)
Secure Hypertext Transport
 Protocol (S-HTTP)
Secure Sockets Layer (SSL)
security
site monitoring and data analysis

Standard Generalized Markup
 Language (SGML)
static Web page
symmetric encryption
TCP/IP
transaction processing
Uniform Resource Locator (URL)
value chain
virus
Web browser
Web caching
Web development
Web page
Web server
Web site
wireless device support
World Wide Web (WWW or
 the Web)
World Wide Web
 Consortium (W3C)
XML schema
XML Schema Definition (XSD)
XSL (Extensible Style Language)
XSL style sheets
XSLT (Extensible Style Language
 Transformation)

REVIEW QUESTIONS

1. What does e-commerce mean and how did it evolve?
2. Identify and briefly explain five benefits and five disadvantages of e-commerce.
3. Define and contrast B2B and B2C e-commerce styles.
4. Describe and give an example of each of the two principal B2B forms.
5. Describe e-commerce architecture, then briefly describe each one of its components.
6. What types of services are provided by the bottom layer of the e-commerce architecture?
7. Name and explain the operation of the main building blocks of the Internet and the Internet's basic services.
8. What does business-enabling do? What services layer does it provide? Give six examples of such services.
9. What is the definition of security? Explain why security is so important for e-commerce transactions.
10. Give an example of an e-commerce transaction scenario. With what three things should security be concerned in this e-commerce transaction?

11. You are hired as a resource security officer for an e-commerce company. Briefly discuss what technical issues must be addressed in your security plan.

12. What is XML and why it is important?

13. What are Document Type Definition (DTD) documents and what do they do?

14. What are XML Schema Definition (XSD) documents and what do they do?

PROBLEMS

1. Use the Internet in your university computer lab or home to research the scenarios described in Problems 1–9, then work through the following problems:
 a. What Web sites did you visit?
 b. Classify each site (B2B, B2C, etc.).
 c. What information did you collect? Was this information useful? Why?
 d. What decision(s) did you make based on the information you collected?

2. Research—and document—the purchase of a new car. Explain why you plan you buy this car, based on the research you have done.

3. Research—and document—the purchase of a new house.

4. You are in the market for a new job. Search the Web for your ideal next job. Document your job search and your job selection.

5. You need to do your taxes; download IRS form 1040 and look for online tax processing help. Document what you have done.

6. Research the purchase of a 20-year level term life insurance policy and report your findings.

7. Research—and document—the purchase of your new computer.

8. Vacation time is here! Research—and document—your next summer vacation's destination(s) and activities.

9. You have some money to invest. Research—and document—mutual funds information for investment purposes. Report your investment decision(s), based on the research you have done.

10. Create a sample XML document and DTD for the exchange of:
 a. customer data
 b. product and pricing data
 c. order data
 d. promotion data
 e. pricewatch data
 f. student transcript data
 (*Hint*: Use Section 13.9 as your map.)

11. Modify the SaleCo database in Chapter 6 to include support for Internet transactions. (*Hint*: Use the table examples in Section 13.8 as your guide.)

14

WEB DATABASE DEVELOPMENT

In this chapter, you will learn:

- ❖ How Internet databases are typically used

- ❖ About the architecture of Web-to-database middleware

- ❖ How Web-to-database middleware (ColdFusion) is used to integrate databases with the Internet

- ❖ What special considerations govern Web database development

P R E V I E W

In Chapter 13, "Databases in Electronic Commerce" we described the building blocks of the Internet. We also discussed the mechanisms required to provide a secure e-commerce infrastructure for delivery of products and services to the mass market of Internet users. Additionally, we explored some critical issues that affect database design and development within an e-commerce framework.

A database is a central data repository for critical business transactions. Such transactions may be generated through traditional business channels or through the Internet. In this chapter, we will explore how a database can be connected to the Web.

Given the growing relationship between the Web and databases, database professionals must know how to create, use, and manage Web interfaces to those databases. This chapter examines the basics of Web database development. In keeping with our hands-on approach, you will have a chance to try out ColdFusion, an important Web-application server tool for creating Web database front ends.

F O U R T E E N

14.1 INTERNET TECHNOLOGIES AND DATABASES

Millions of people all over the world now use their computers and Web browser software to access the Internet and connect to databases over the Web. Web database connectivity opens the door to new innovative services that:

- Allow rapid response to competitive pressures by bringing new services and products to market quickly.
- Increase customer satisfaction through the creation of Web-based support services.
- Yield fast and effective information dissemination through universal access from across the street or across the globe.

Given such advantages, many IS departments face the need to create universal data access architectures, based on Internet standards, to streamline operations and to facilitate decision making. Table 14.1 shows a sample of Internet technology characteristics and the benefits that they provide.

TABLE 14.1 CHARACTERISTICS AND BENEFITS OF INTERNET TECHNOLOGIES

INTERNET CHARACTERISTIC	BENEFIT
Hardware and software independence	Savings in equipment/software acquisition Runs on most existing equipment Platform independence and portability No need for multiple platform development
Common and simple user interface	Reduced training time and cost Reduced end-user support cost No need for multiple platform development
Location independence	Global access through Internet infrastructure Reduced requirements (and costs!) for dedicated connections
Rapid development at manageable costs	Multiple development tools available Plug and play development tools (open standards) Development is more interactive Reduced development times Tools are relatively inexpensive Free client access tools (Web browsers) Low entry costs—free Web servers are often available Reduced costs of maintaining private networks Distributed processing and scalability using multiple servers

In the current business and global information environment, it's easy to see why many database professionals consider the DBMS "connection" to the Internet to be a critical element in IS development. As you will learn in the following sections, database application development and, in particular, the creation and management of user interfaces and database connectivity, are profoundly affected by the Web. However, having a Web-based database interface does not negate the database design and implementation issues that were addressed in the previous chapters. In the final analysis, whether you make a purchase online or by standing in line, the system-level transaction details are essentially the same, and they require the same basic database structures and relationships. If there is any immediate lesson to be learned it is this: *The effects of bad database design, implementation, and management are multiplied in an environment in which transactions may be measured in millions per day, rather than in hundreds per day.*

14.2 TYPICAL USES OF INTERNET DATABASES

The Internet is rapidly changing the way information is generated, accessed, and distributed. At the core of this change are the Web's ability to access data in databases (local and remote), the simplicity of the interface, and cross-platform (heterogeneous) functionality. The Web has helped create a new information dissemination standard.

Although e-commerce and its complement of transaction management and security issues are growing exponentially, the most typical uses of databases on the Internet involve *sharing information you are willing to make public*. Because such information is public, it is available to your competitors as well as to your customers. Therefore, the shared information is likely to be based on non-critical data. For example, the U.S government's Securities and Exchange Commission's EDGAR database holds financial information about thousands of businesses that is available to anyone through the Internet (*http://www.sec.gov/edgar/searchedgar/webusers.htm*). Also, a business may use the Internet to provide marketing and customer support by making available searchable technical databases, product catalogs, contact information, and so on. Internet technologies are now also used to develop systems that provide corporate information to users inside the company (intranets), and systems that make relevant information available to users of *another* company with which the company does business (extranets). Internet technology is also used to develop applications that range from "front-line" business applications such as customer services to "back-line" business applications such as manufacturing. Table 14.2 provides a summary of common Internet-based applications.

TABLE 14.2 A SAMPLE OF APPLICATIONS THAT USE WEB-DATABASE TECHNOLOGY

TYPE OF APPLICATION	INTERNALLY AVAILABLE	EXTERNALLY AVAILABLE
General Corporate Information		
Phone Directories	X	X
Suggestion Boxes	X	X
Manufacturing Information		
Product Design	X	
Inventory Costing	X	
Project Tracking	X	
Sales Information		
Product Catalogs	X	X
Order Entry	X	X
Customer Support	X	X

The following sections will be devoted to examining Web-to-database middleware and the interfaces that enable end users to interact with databases over the Web.

14.3 WEB-TO-DATABASE MIDDLEWARE: SERVER-SIDE EXTENSIONS

In Chapter 13 you learned about the basic interaction between a client Web browser and a Web server. In general, the Web server is the main hub through which all Internet services are accessed. For example, when an end user uses a Web browser to dynamically query a database, the client browser requests a Web page. When the Web server receives the page request, it looks for the page on the hard disk and, when it finds this page (for example, a stock quote, product catalog information, or an airfare listing), the server sends it back to the client.

You also learned how dynamic Web pages are at the heart of e-commerce Web sites. In this database-query scenario, the Web server generates the Web page contents before it sends the page to the client Web browser. The only problem with the preceding query scenario is that the Web server must first include the database query result on the page *before* it sends that page back to the client. Unfortunately, neither the Web browser nor the Web server knows how to connect to and read data from the database (remember, the Web server and the Web browser can understand only HTML-formatted pages). Therefore, to support this type of (database-query) request, the Web server's capability must be extended so it can understand and process database requests. This job is done through a server-side extension.

A **server-side extension** is a program that interacts directly with the Web server to handle specific types of requests. In the preceding database query example, the server-side extension program retrieves the data from databases and passes the retrieved data to the Web server, which in turn sends it to the client's browser for display purposes. The server-side extension makes it possible to retrieve and present the query results, but what's more important is that *it provides its services to the Web server in a way that is totally transparent to the client browser*. In short, the server-side extension adds significant functionality to the Web server and, therefore, to the Internet.

A database server-side extension program is also known as **Web-to-database middleware**. (To explore middleware in greater detail, review **Appendix F, "Multi-Tier Client/Server Systems"**). Figure 14.1 shows the interaction between the browser, the Web server, and the Web-to-database middleware (ColdFusion in this case).

ADDITIONAL CD CONTENT

Appendix F is located on the student CD in the \Appendixes\Appendix F folder.

As you examine Figure 14.1, trace the Web-to-database middleware actions:

1. The client browser sends a page request to the Web server.

2. The Web server receives and validates the request. In this case, the server will pass the request to the Web-to-database middleware for processing. Generally, the requested page contains some type of scripting language to enable the database interaction.

3. The Web-to-database middleware reads, validates, and executes the script. In this case, it connects to the database, queries the database, dynamically generates an HTML-formatted page that includes the data retrieved from the database, and sends it to the Web server.

4. The Web server returns the just-created HTML page—which now includes the query result—to the client browser for display.

FIGURE 14.1 WEB-TO-DATABASE MIDDLEWARE (COLDFUSION)

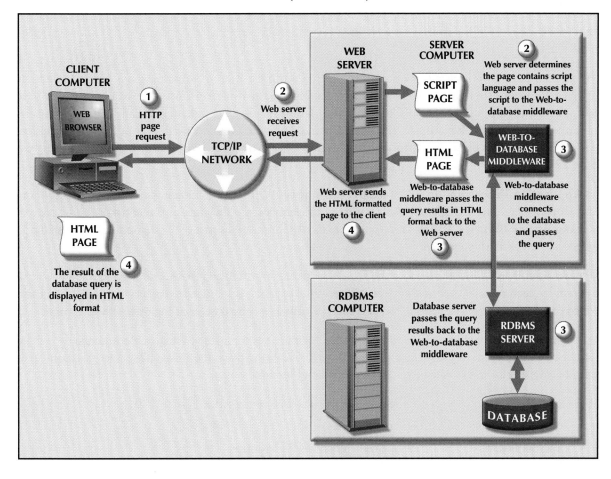

The interaction between the Web server and the Web-to-database middleware is crucial to the development of a successful Internet database implementation. Therefore, the middleware must be well integrated with the other Internet services and the components that are involved in its use. For example, when installing Web-to-database middleware, the middleware must verify the type of Web server being used and install itself to match that Web server's requirements. In addition, how well the Web server and the Web-to-database service interact will depend on the Web server interfaces that are supported by the Web server.

14.3.1 WEB SERVER INTERFACES

Extending Web server functionality implies that the Web server and the Web-to-database middleware will properly communicate with each other. (Database professionals often use the word "interoperate" to indicate that each party to the communication can respond to the communications of the other. Our use of "communicate" will assume interoperation.) If a Web server is to successfully communicate with an external program, both programs must use a standard way to exchange messages and respond to requests. A Web server interface defines how a Web server communicates with external programs. Currently there are two well-defined Web server interfaces:

- Common Gateway Interface (CGI)
- Application Programming Interface (API)

The **Common Gateway Interface (CGI)** uses script files that perform specific functions based on the client's parameters that are passed to the Web server. The script file is a small program containing commands written in a

programming language, usually PERL, C++, or Visual Basic. The script file's contents can be used to connect to the database and to retrieve data from it, using the parameters passed by the Web server. Next, the script converts the retrieved data to HTML format and passes the data to the Web server, which sends the HTML-formatted page to the client.

The main disadvantage of using CGI scripts is that the script file is an external program that is individually executed for each user request. Such a scenario decreases system performance. For example, if you have 200 concurrent requests, the script is loaded 200 *different* times, which takes significant CPU and memory resources away from the Web server. The language and method used to create the script also can affect system performance. For example, performance is degraded by using an interpreted language or by writing the script inefficiently.

An **Application Programming Interface (API)** is a newer Web server interface standard that is much more efficient and faster than a CGI script. APIs are more efficient because they are implemented as shared code or as **dynamic-link libraries (DLLs)**. This means that the API is treated as part of the Web server program that is dynamically invoked when needed.

APIs are faster than CGI scripts because the code is resident in memory and there is no need to run an external program for each request. Instead, the same API serves all requests. Another advantage is that an API can use a shared connection to the database instead of creating a new one every time, as is the case with CGI scripts.

Although APIs are more efficient in handling requests, their use has some disadvantages. Because the APIs share the same memory space as the Web server, an API error can bring down the server. The other disadvantage is that APIs are specific to the Web server and the operating system.

At the time of this writing, there are three well-established Web server APIs:

- Netscape API (NSAPI) for Netscape servers
- Internet Server API (ISAPI) for Microsoft Windows Web servers
- WebSite API (WSAPI) for O'Reilly Web servers

The various types of Web interfaces are illustrated in Figure 14.2.

Regardless of the type of Web server interface used, the Web-to-database middleware program must be able to connect with the database. This connection can be accomplished in one of two ways:

- Use the native SQL access middleware provided by the vendor. For example, you can use SQL*Net if you are using Oracle.
- Use the services of **Open DataBase Connectivity (ODBC)** or the Object Linking and Embedding Database (OLE DB) interface if running in a Windows platform.

Most database vendors support ODBC as the standard way to connect to the database. Therefore, we will next examine how ODBC is used as database access middleware.

FIGURE 14.2 WEB SERVER INTERFACES

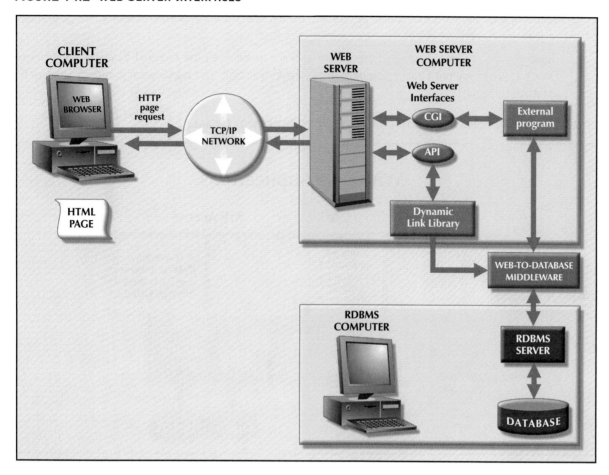

14.3.2 OPEN DATABASE CONNECTIVITY (ODBC)

To see how the server-extension program and the DBMS interact, let's examine what is required in the most commonly encountered operating system environment, Microsoft Windows. Within the Windows environment, most productivity applications provide "hooks" that are used to access databases such as Access, DB2, Oracle, or SQL Server. Such database access is achieved through ODBC database access middleware. ODBC is Microsoft's implementation of a superset of the SQL Access Group **Call Level Interface** (**CLI**) standard for database access. ODBC allows any Windows application to access data sources that range from relational database systems to some flat file systems, using standard SQL. Figure 14.3 illustrates how Windows applications use ODBC and other Windows services.

Figure 14.3 conveys the following information:

- All Windows applications use shared services such as memory management, spooling, and user interface management.

- One of those shared services is ODBC. For example, an application such as MS Access can use ODBC to directly access data from external sources. Such data can be located in the same computer or in another computer located on the network.

- Shared Windows services are implemented as shared code that is dynamically linked into the Windows operating environment (DLLs, stored as files with the .dll extension). By running as a DLL, the code speeds up load and run times.

- Databases usually offer "native" database middleware to establish the required database connections. For example, if you use Oracle, you must install the Oracle SQL*Net middleware in order to access the database.

FIGURE 14.3 WINDOWS APPLICATIONS USE ODBC TO ACCESS DATABASES

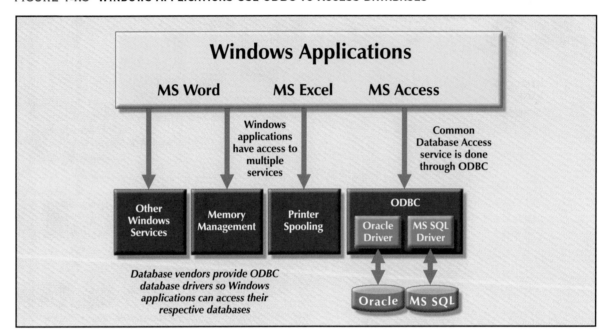

Remember that the Web browser cannot directly access databases. The Web browser's main function is to *request Web pages from the Web server.* This characteristic is important if the browser is to remain independent of the operating system. For example, ODBC is only available in Windows and in some UNIX implementations. The proper use of ODBC in a Web environment is illustrated in Figure 14.4.

As you examine Figure 14.4, note in particular the following interactions:

1. The Web browser requests a page from a Web server.
2. The Web server passes the request to the Web-to-database middleware using CGI or API.
3. The Web-to-database middleware uses ODBC to connect to the database.
4. The middleware receives the query result and creates the HTML-formatted page.
5. The Web-to-database middleware sends the page back to the Web server, using the Web server interface (CGI or API) transmission standard.
6. The Web server sends the page to the browser.
7. The HTML output is displayed on the client computer's Web browser.

FIGURE 14.4 WEB-TO-DATABASE MIDDLEWARE USES ODBC TO ACCESS DATABASES

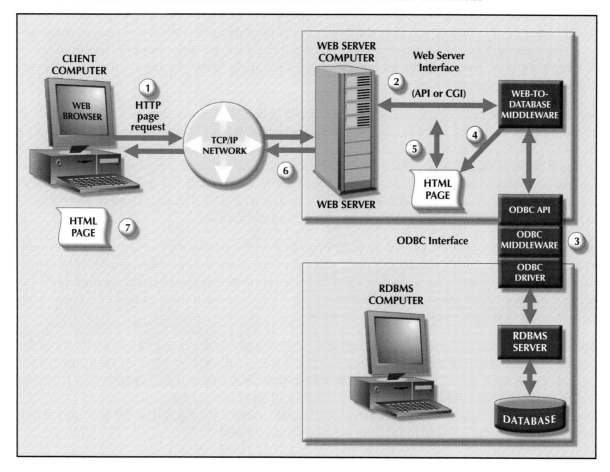

14.4 THE WEB BROWSER

As you learned in Chapter 13, a Web browser is software, such as Microsoft Internet Explorer or Netscape Navigator, that lets you navigate (browse) the Web. The Web browser is located in the client computer and is the end-user interface to the World Wide Web. (Remember that the browser's main function is to request pages and to present the pages to the end user.) Each time the end user clicks a hyperlink, the browser generates an HTTP GET page request that is sent to the designated Web server, using the TCP/IP Internet protocol.

The Web browser's job is to *interpret* the HTML code that it receives from the Web server and to present the different page components in a standard way. Unfortunately, the browser's interpretation and presentation capabilities are not sufficient to develop Web-based applications. To improve upon these capabilities, you must use plug-ins and other client-side extensions.

14.4.1 CLIENT-SIDE EXTENSIONS

Client-side extensions add functionality to the Web browser. Although client-side extensions are available in various forms, the most commonly encountered extensions are:

- Plug-ins
- Java and JavaScript
- ActiveX and VBScript

A **plug-in** is an external application that is automatically invoked by the browser when needed. Because it is an *external* application, the plug-in is operating-system specific. The plug-in is associated with a data object—generally using the file extension—to allow the Web server to properly handle data that are not originally supported. For example, if one of the page components is a .pdf document, the Web server will receive the data, it will recognize it as a "portable document format" object, and launch Adobe Acrobat Reader to present and manipulate the document on the client computer.

Java is an object-oriented programming language developed by Sun Microsystems that runs on top of the Web browser software. Java applications are compiled and stored in the Web server. (In many respects, Java resembles C++.) Calls to Java routines are embedded inside the HTML page. When the browser finds this call, it downloads the Java code from the Web server and runs it in the client computer. Java's main advantage is that it enables application developers to develop their applications once and run them in many environments. (For developing Web applications, interoperability is a very important issue. Unfortunately, Microsoft and Netscape client browsers are not 100 percent interoperable, thus limiting portability.)

JavaScript, developed by Netscape, is a scripting language (one that enables the running of a series of commands or macros) that allows Web authors to design interactive sites. Because JavaScript is much simpler to generate than Java, it is easier to learn. JavaScript code is embedded in the Web pages. It is downloaded with the Web page and is activated when a specific event takes place—like a mouse click on an object or a page being loaded from the server into memory.

ActiveX is Microsoft's alternative to Java. ActiveX is a specification for writing programs that will run inside the Microsoft client browser (Internet Explorer). Because ActiveX is oriented mainly to Windows applications, it has low portability. ActiveX extends the Web browser by adding "controls" to Web pages. (Examples of such controls are drop-down windows, a slider, a calendar, and a calculator.) These controls let you manipulate data inside the browser, and are downloaded from the Web server when needed. ActiveX controls can be created in several programming languages, but C++ and Visual Basic are most commonly used.

VBScript is another Microsoft product that is used to extend browser functionality. VBScript is derived from Microsoft Visual Basic. Like JavaScript, VBScript code is embedded inside an HTML page and is activated by triggering events such as clicking a link.

From the developer's point of view, using routines that permit data validation on the client side is an absolute necessity. For example, when data are entered on a Web form and no data validation is done on the client side, the entire data set must be sent to the Web server. This scenario requires the server to perform all data validation, thus wasting valuable CPU processing cycles. Therefore, client-side data input validation is one of the most basic requirements for Web applications. Most of the data validation routines are done in Java, JavaScript, ActiveX, or VBScript.

14.5 USING A WEB-TO-DATABASE PRODUCTION TOOL: COLDFUSION

To understand how Web-to-database interfaces work, you need to know how they are created and to see them in action. In this section you will have a chance to try Macromedia ColdFusion, one of a new breed of products known as Web application servers. A **Web application server** is a middleware application that expands the functionality of Web servers by linking them to a wide range of services, such as databases, directory systems, and search engines. The Web application server also provides a consistent run-time environment for Web applications.

ColdFusion application middleware can be used to:

- Connect to and query a database from a Web page.
- Present database data in a Web page, using various formats.
- Create dynamic Web search pages.

- Create Web pages to insert, update, and delete database data.

- Define required and optional relationships.

- Define required and optional form fields.

- Enforce referential integrity in form fields.

- Use simple and nested queries and form select fields to represent business rules.

> **N O T E**
>
> Although ColdFusion has a wide range of features, the purpose of this section is to show you how to create and use a simple, yet useful, Web-to-database interface. You can learn additional ColdFusion features by tapping into its detailed and well-organized online documentation (*www.macromedia.com*).

ColdFusion has several important characteristics:

- It is a powerful and stable software product that can be used to produce and support even the most complex Web-to-database access solutions.

- In spite of its power, it is a developer- and user-friendly product. ColdFusion has a strong and growing corporate presence. Using ColdFusion, you can get some hands-on experience with the Web-to-database environment, while improving the marketability of your knowledge.

- Macromedia offers a 30-day free evaluation version of the latest ColdFusion software, which may be downloaded from ***www.macromedia.com***. Because ColdFusion includes a complete set of online documentation with full working demo applications that illustrate all the functionality of the product, you will incur no documentation charges.

Some of the many other Web application servers, as of this writing, include Oracle Application Server by Oracle Corp., WebLogic by BEA Systems, NetDynamics by Sun Microsystems, NetObjects' Fusion, Microsoft's Visual Studio.NET, and WebObjects by Apple.

Web application servers provide features such as:

- An integrated development environment with session management and support for persistent application variables.

- Security and authentication of users through user IDs and passwords.

- Computational languages to represent and store business logic in the application server.

- Automatic generation of HTML pages integrated with Java, JavaScript, VBScript, ASP, and so on.

- Performance and fault-tolerant features.

- Database access with transaction management capabilities.

- Access to multiple services, such as file transfers (FTP), database connectivity, electronic mail, and directory services.

All of these products offer the ability to connect Web servers to multiple data sources and other services. These products vary in terms of the range of available features, robustness, scalability, ease of use, compatibility with other Web and database tools, and the extent of the development environment.

14.5.1 HOW COLDFUSION WORKS

We have described ColdFusion as a full-fledged Web application server that provides hooks to databases, e-mail systems, directory systems, search engines, and so on. To do its job, ColdFusion provides a server-side markup language (HTML extensions or **tags**) known as the **ColdFusion Markup Language** (**CFML**), which is used to create ColdFusion application pages known as *scripts*. A **script** is a series of instructions executed in *interpreter*

mode. The script is a plain text file that is not compiled like COBOL, C++, or Java. ColdFusion scripts contain the code that is required to connect, query, and update a database from a Web front end.

ColdFusion scripts contain a combination of HTML, ColdFusion tags, and, when necessary, Java, JavaScript, or VBScript. ColdFusion tags start with <CF and may include an opening and closing component such as <CFQUERY> (begin a query) and </CFQUERY> (end a query). ColdFusion scripts are saved in files with a .cfm extension. When a client browser requests a .cfm page from the Web server, the ColdFusion application server executes the .cfm script instructions, and sends the resulting output—in HTML format—to the Web server. The Web server then sends the document to the client computer for display. Figure 14.5 shows the application server components and actions.

FIGURE 14.5 HOW COLDFUSION WORKS

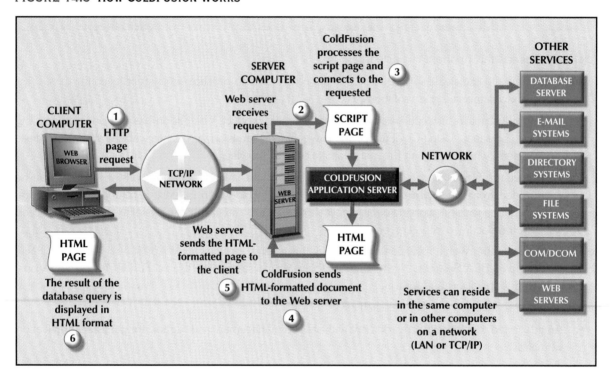

14.5.2 THE ROBCOR SAMPLE DATABASE

To illustrate how ColdFusion can be used to provide the Web-to-database interface, we will use a small Microsoft Access database named RobCor. In the following sections, we will guide you through the creation of several ColdFusion scripts designed to select, insert, update, and delete data from the RobCor database.

N O T E

To focus our efforts on the use of CFML to access databases, these exercises assume that you are familiar with basic HTML tags and the HTML editing process. The examples shown in this chapter can be created using any standard text editor such as Notepad.exe.

The RobCor database, whose relational schema is shown in Figure 14.6, was designed to track the purchase orders placed by users in a multidepartment company.

ADDITIONAL CD CONTENT

The database used in the ColdFusion scripts is located on the student CD in the **\ColdFusion\robcor\Databases** folder. The database name is **orderdb.mdb**. Note that this database has been saved in MS Access 97 format. Please do not change the database format to a newer version. Unless you have the latest MS Access ODBC/OLE drivers, changing the format could render the database inaccessible to the ColdFusion scripts.

FIGURE 14.6 **THE CH14_ROBCOR DATABASE'S RELATIONAL SCHEMA**

As you examine Figure 14.6, note that the database contains the following tables: USER, DEPARTMENT, VENDOR, INVTYPE, ORDERS, and ORDER_LINE. The relationships between these tables are derived from the following business rules:

- A department employs many users.
- A department may be managed by one of those users.
- Each user belongs to one department, and each department can have many users.
- Each department may have a department manager. (That is, a department manager is optional.)
- Each order is placed to only one vendor, and each vendor can receive many orders.
- Each order contains one or more order lines.
- Each order line refers to one inventory type.

USER_1 is a virtual component created by MS Access when *multiple relationships* between USER and DEPARTMENT are set. MS Access created the USER_1 virtual table to represent the "USER *manages* DEPARTMENT" relationship. This is a one-to-one *optional* relationship, thus allowing the USR_ID field in the DEPARTMENT table

to be null. (We created this relationship to later illustrate how you can manage optional relationships within a Web interface.)

14.5.3 CREATING A SIMPLE QUERY WITH CFQUERY AND CFOUTPUT

Let's begin by creating a simple script to produce a query that will list all the vendors in the VENDOR table. This script will perform two tasks:

1. Query the database, using standard SQL to retrieve a data set that contains all records found in the VENDOR table.

2. Format all of the records generated in Step 1 in HTML to let them be included in the page that is returned to the client browser.

The script shown in Script 14.1 contains the required code.

A D D I T I O N A L C D C O N T E N T

The script filenames are shown with each script. For example, the script file for Script 14.1 is named **rc-1.cfm**. The script files are on your student CD in the **\ColdFusion\robcor** folder.

SCRIPT 14.1 (RC-1.CFM): A SIMPLE QUERY USING CFQUERY AND CFOUTPUT

```
1   <HTML>
2   <HEAD>
3   <TITLE>Rob & Coronel -  ColdFusion Examples</TITLE>
4   <CFQUERY NAME="venlist" DATASOURCE="RobCor">
5        SELECT * FROM VENDOR ORDER BY VEN_CODE
6   </CFQUERY>
7   </HEAD>
8   <BODY BGCOLOR="LIGHTBLUE">
9   <H1>
10  <CENTER><B>Simple Query using CFQUERY and CFOUTPUT</B></CENTER>
11  <CENTER><B>(Vertical Output)</B></CENTER>
12  </H1>
13  <BR>
14  <HR>
15  <CFOUTPUT>
16       Your query returned #venlist.RecordCount# records
17  </CFOUTPUT>
18  <CFOUTPUT QUERY="venlist">
19  <PRE><B>
20   VENDOR CODE:        #VEN_CODE#
21   VENDOR NAME:        #VEN_NAME#
22   CONTACT PERSON:     #VEN_CONTACT_NAME#
23   ADDRESS:            #VEN_ADDRESS#
24   CITY:               #VEN_CITY#
25   STATE:              #VEN_STATE#
26   ZIP:                #VEN_ZIP#
27   PHONE:              #VEN_PH#
28   FAX:                #VEN_FAX#
29   E-MAIL:             #VEN_EMAIL#
30   CUSTOMER ID:        #VEN_CUS_ID#
31   SUPPORT ID:         #VEN_SUPPORT_ID#
32   SUPPORT PHONE:      #VEN_SUPPORT_PH#
33   VENDOR WEB PAGE:    #VEN_WEB_PAGE#
34  <HR></B></PRE>
35  </CFOUTPUT>
36  <FORM ACTION="rc-0.cfm" METHOD="post">
37       <INPUT TYPE="submit" VALUE="Main Menu ">
38  </FORM>
39  </BODY>
40  </HTML>
```

N O T E

If you installed the ColdFusion demo, you can run this script—and all subsequent scripts—by pointing the browser to the Web server address. If your computer is the Web server, you can use *http://127.0.0.1/RobCor/ rc-1.cfm* as your Web address. If your Web server is a different computer selected by your instructor, use the address supplied by your instructor.

As you examine Script 14.1, note that its ColdFusion tags are CFQUERY (to query a database) and CFOUTPUT (to display the data returned by the query). (Note that the CFML and HTML tags are shown in different colors.) Let's take a closer look at these two CFML tags:

- *<CFQUERY>* tag (lines 4–6). This tag sets the stage for the database connection and the execution of the enclosed SQL statement. You should include all query statements *before* or *within* the document's HTML header (<HEAD>) section. Using this procedure, the page will display the output on the client side *after all queries have been executed*. If you do not use this procedure, the browser will be perceived as "slow" because the page will start to display output and then pause to wait for additional data to arrive from the Web server. The CFQUERY tag requires the following parameters:

 - *NAME* = *"queryname"*. This is a mandatory parameter, whose name uniquely identifies the record set returned by the database query—in this case, *venlist*. You can have multiple queries, each with a unique name, in a single script.

 - *DATASOURCE* = *"datasourcename"*. This, too, is a mandatory parameter that uses the database name as defined in ODBC. Keep in mind that the database name is case-sensitive. Therefore, if the database name is *"RobCor"*, do not use "ROBCOR" or "robcor". You use the ColdFusion Administrator interface to define all data sources. Data sources can use ODBC, a native driver such as Oracle SQL*Net, or Microsoft OLE (object linking and embedding).

 - *SQL statement* (line 5) is another mandatory parameter. In this case, the parameter is defined by the query:

 SELECT * FROM VENDOR ORDER BY VEN_CODE

 but you could use any ODBC SQL-compliant statement.

- *<CFOUTPUT>* tag (lines 15–17 and 18–35). This tag is used to display the results from a CFQUERY or to call other ColdFusion variables or functions. Its parameters are:

 - QUERY = *"queryname"*. This is an optional parameter (see line 18). If you use a query name for a query that returns 10 rows, this tag will execute all commands between the opening and closing CFOUTPUT tags 10 times—one per row. In short, this tag works like a loop that is executed as many times as the number of rows in the named query set.

 - You can include any valid HTML tags or text within the opening and closing CFOUTPUT tags. ColdFusion uses pound signs (#) to reference query fields in the resulting query set or to call other ColdFusion functions or variables. When ColdFusion encounters a name enclosed within pound signs, it will evaluate this named variable to verify that it is a field name of the named query, an internal variable, or a function. Following this evaluation, ColdFusion will replace the named variable with the value that corresponds to the query, the internal variable, or the function. In this case, line 16 is a call to a ColdFusion internal variable. When the query is executed, this variable's value is the number of rows in

the query output. The name of the query must precede the *RecordCount* keyword, and the two components must be separated by a period. For example, we used *#venlist.RecordCount#* to name the variable in line 16.

- Lines 19–34: These lines are repeated as a loop, one for each record returned in the named query. In this example, the loop is defined by *QUERY* ="venlist". Note that—in lines 20–33—the field names (enclosed in pound signs) will be replaced by the actual values of the fields that are returned by the query.

The output produced by rc-1.cfm is shown in Figure 14.7.

FIGURE 14.7 **THE RC-1.CFM SCRIPT OUTPUT (VERTICAL LISTING)**

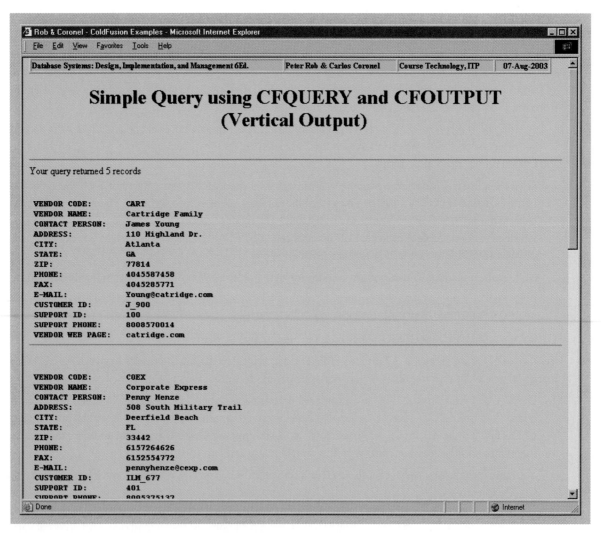

A variation of the just-described approach is found in Script 14.2, in which the output is preformatted (using the HTML <PRE> tag) one-row-per-record, using the CFOUTPUT tag.

SCRIPT 14.2 (RC-2.CFM): CFQUERY WITH TABULAR CFOUTPUT

```
1   <HTML>
2   <CFQUERY NAME="VENDATA" DATASOURCE="RobCor">
3         SELECT * FROM VENDOR ORDER BY VEN_CODE
4   </CFQUERY>
5   <HEAD>
6   <TITLE>Rob & Coronel -  ColdFusion Examples</TITLE>
7   </HEAD>
8   <BODY BGCOLOR="LIGHTBLUE">
9   <H1>
10  <B><CENTER>Simple Query using CFQUERY and CFOUTPUT</CENTER></B>
11  <B><CENTER>(Horizontal Output)</CENTER></B>
12  </H1>
13  <BR>
14  <HR>
15  <PRE>
16  <B>
17  VENDOR CODE  VENDOR NAME  CONTACT PERSON     ADDRESS        CITY       STATE       ZIP        PHONE
18  </B>
19  <HR>
20  <CFOUTPUT QUERY="VENDATA">
21  #VEN_CODE#   #VEN_NAME#   #VEN_CONTACT_NAME# #VEN_ADDRESS#  #VEN_CITY# #VEN_STATE# #VEN_ZIP#  #VEN_PH#<BR>
22  </CFOUTPUT>
23  </PRE>
24  <FORM ACTION="rc-0.cfm" METHOD="post">
25      <INPUT TYPE="submit" VALUE="Main Menu ">
26  </FORM>
27  </BODY>
28  </HTML>
```

The output of Script 14.2 is shown in Figure 14.8.

FIGURE 14.8 THE RC-2.CFM SCRIPT OUTPUT (HORIZONTAL LISTING)

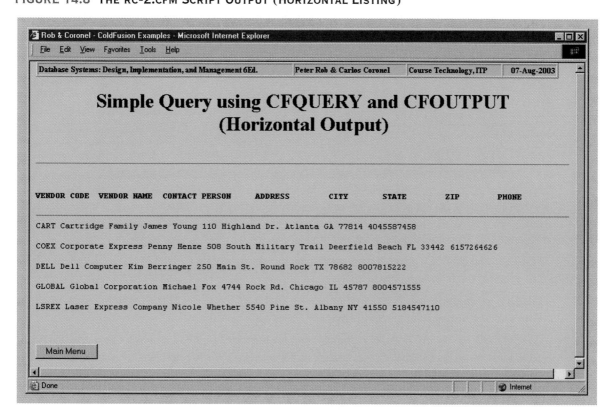

14.5.4 CREATING A SIMPLE QUERY WITH CFQUERY AND CFTABLE

As you can see in Figure 14.8, the output produced by CFOUTPUT is not aligned. To give the output a more polished look, the vendor list can be displayed in tabular format, based on the CFTABLE tag. This tag will automatically create a tabular output in which each row in the data set is placed in a row in the table. The source code for this example is stored in Script 14.3.

SCRIPT 14.3 (RC-3.CFM): CFQUERY WITH CFTABLE

```
<HTML>
<HEAD>
<TITLE>Rob & Coronel - ColdFusion Examples</TITLE>
<CFQUERY NAME="VENDATA" DATASOURCE="RobCor">
        SELECT * FROM VENDOR ORDER BY VEN_CODE
</CFQUERY>
</HEAD>
<BODY BGCOLOR="LIGHTBLUE">
<H1>
<B><CENTER>Simple Query Using CFQUERY and CFTABLE</CENTER></B>
</H1>
<BR>
<TABLE BGCOLOR="Silver" BORDERCOLOR="Fuchsia" FRAME="BORDER">
<TR><HR></TR>
<TR>
<CFTABLE QUERY="VENDATA" STARTROW="1" COLSPACING="1" COLHEADERS>
<CFCOL HEADER="<B>CODE</B>" WIDTH="8" TEXT="#VEN_CODE#">
<CFCOL HEADER="<B>VENDOR_NAME</B>" WIDTH="25" TEXT="#VEN_NAME#">
<CFCOL HEADER="<B>CONTACT_PERSON</B>" WIDTH="14" TEXT="#VEN_CONTACT_NAME#">
<CFCOL HEADER="<B>ADDRESS</B>" WIDTH="20" TEXT="#VEN_ADDRESS#">
<CFCOL HEADER="<B>CITY</B>" WIDTH="10" TEXT="#VEN_CITY#">
<CFCOL HEADER="<B>STATE</B>" WIDTH="2" TEXT="#VEN_STATE#">
<CFCOL HEADER="<B>ZIP</B>" WIDTH="5" TEXT="#VEN_ZIP#">
<CFCOL HEADER="<B>PHONE</B>" WIDTH="10" TEXT="#VEN_PH#">
<CFCOL HEADER="<B>FAX</B>" WIDTH="10" TEXT="#VEN_FAX#">
<CFCOL HEADER="<B>E-MAIL</B>" WIDTH="10" TEXT="#VEN_EMAIL#">
<CFCOL HEADER="<B>CUSTOMER_ID</B>" WIDTH="8" TEXT="#VEN_CUS_ID#">
<CFCOL HEADER="<B>SUPPORT_PHONE</B>" WIDTH="6" TEXT="#VEN_SUPPORT_ID#">
<CFCOL HEADER="<B>WEB PAGE</B>" WIDTH="14" TEXT="#VEN_WEB_PAGE#">
</CFTABLE>
</TR>
<TR>
<FORM ACTION="rc-0.cfm" METHOD="post">
    <INPUT TYPE="submit" VALUE="Main Menu ">
</FORM>
</TR>
</TABLE>
</BODY>
</HTML>
```

The output of Script 14.3 is shown in Figure 14.9.

The rc-3.cfm script's CFTABLE ColdFusion tag contents are as follows:

- <CFTABLE> tag (line 16). This tag, used to display the results from a CFQUERY (lines 4–6) in a tabular format, requires the following parameters:

 - QUERY = "queryname". A required parameter that uses the name of the query that generated the data set to be displayed in tabular format.

 - STARTROW = "1". An optional parameter that is used to tell ColdFusion which query row will be the table's first row. This parameter is particularly useful when your query returns many rows and you do not want to display them all in one long page. Instead, you can display them in multiple pages, each page displaying the number of rows defined by the parameter. For example, if you want to list 10 rows per page, the starting row will be 1 for the first page, 11 for the second page, 21 for the third, and so on.

- *COLSPACING = "1"*. An optional parameter that is used to define the number of spaces to be placed between columns.

- *COLHEADERS*. An optional parameter that will make the first row in the table a row header. This row header contains the name of each column, using the values defined in the CFCOL tags.

• The CFCOL tag (lines 17–29) is used to define each table column, using the following parameters:

- *HEADER = "header text"*. This is the header text that will appear in the table's header row for each of the displayed columns. The header text can include HTML tags.

- *WIDTH = "x"*. This parameter defines the column width.

- *TEXT = "#queryfieldname#"*. This is the actual value to be placed in the selected column. For example, line 17 will cause ColdFusion to replace *#VEN_CODE#* with the actual values retrieved by the query for this field.

FIGURE 14.9 THE RC-3.CFM SCRIPT OUTPUT (FORMATTED HORIZONTAL LISTING)

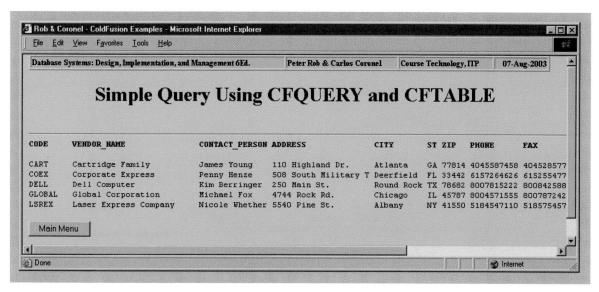

14.5.5 CREATING A DYNAMIC SEARCH PAGE

At this point, you have seen how the CFQUERY, CFOUTPUT, and CFTABLE tags are used to send the SQL statement to the database to retrieve a data set. Given the script files used thus far, the query will always retrieve the same records from the VENDOR table. (Because the SQL statement is "hard-coded" inside the page, it cannot be changed—unless the SQL code is manually edited each time it is to be used to generate a different query output.) Such a static output display clearly limits the query's usefulness.

To create a practical dynamic query environment, you can create a dynamic search query in which the query search condition can be changed at the user's option—without requiring script page editing. To demonstrate the process, we will use two fields, vendor code and vendor state, to search for user-specified vendor records. (Naturally, you can create a search form that uses as many fields as you think are necessary.)

To perform a dynamic query over the Web, you must complete these two steps:

1. Create a script that will generate a form which will be used to enter the criteria to be used in the search. In other words, the form allows the user to enter the parameter values that are to be used in the query statement.

2. Create a script that will execute the query and display the results based on the parameters that are passed to it by the script created in Step 1.

The script required to complete Step 1 is listed in Script 14.4a.

SCRIPT 14.4A (RC-4A.CFM): DYNAMIC SEARCH QUERY: CRITERIA ENTRY FORM

```
1   <HTML>
2   <CFQUERY NAME="STATELIST" DATASOURCE="RobCor">
3       Select VEN_STATE from VENDOR Order by VEN_STATE
4   </CFQUERY>
5   <HEAD>
6   <TITLE>Rob & Coronel -  ColdFusion Examples</TITLE>
7   </HEAD>
8   <BODY BGCOLOR="LIGHTBLUE">
9   <H1>
10  <CENTER><B>Dynamic Search Query: Criteria Entry Form</B></CENTER>
11  </H1>
12  <FORM ACTION="rc-4b.cfm" METHOD=POST>
13  <TABLE ALIGN="CENTER" BGCOLOR="Silver">
14      <TR>
15      <TD ALIGN="right">VEN_CODE</TD>
16      <TD>
17      <INPUT TYPE ="text" NAME="VEN_CODE" SIZE="10" MAXLENGTH="10"></TD>
18      </TR>
19      <TR>
20      <TD ALIGN="right">VEN_STATE</TD>
21      <TD><SELECT NAME="VEN_STATE" SIZE=1>
22          <OPTION SELECTED VALUE="ANY">ANY
23           <CFOUTPUT QUERY="STATELIST">
24          <OPTION VALUE="#STATELIST.VEN_STATE#">#VEN_STATE#
25          </CFOUTPUT>
26          </SELECT>
27      </TD>
28      </TR>
29      <TR>
30      <TD ALIGN="right" VALIGN="middle">
31          <INPUT TYPE="submit" VALUE="Search">
32          </FORM>
33      </TD>
34      <TD ALIGN="right" VALIGN="middle">
35          <FORM ACTION="rc-0.cfm" METHOD="post">
36              <INPUT TYPE="submit" VALUE="Main Menu ">
37          </FORM>
38      </TD>
39      </TR>
40  </TABLE>
41  </BODY>
42  </HTML>
```

The rc-4a.cfm script output is shown in Figure 14.10.

FIGURE 14.10 THE RC-4A.CFM SCRIPT OUTPUT (STATE SEARCH CRITERIA ENTRY FORM)

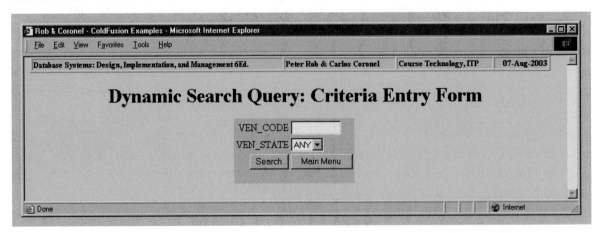

As you examine Figure 14.10, note that the rc-4a.cfm script generates a form in which the user enters the search parameters. Let's follow the rc-4a.cfm script to see how it works:

- In lines 2–4 we retrieve all existing values from the *VEN_STATE* field in the VENDOR table. This query effectively tells us what states are represented in the vendor table. If there is no vendor for a given state, it will not be shown in the returned data set. This query is named "*STATELIST*", and it will later be used in our form.

- In lines 12–32 we define the form. Note that, when the user clicks the "submit" button, the rc-4b.cfm script—shown later—will be called to receive the form's variables, *VEN_CODE* and *VEN_STATE*.

- Line 17 presents the first input text box to let the user enter a value for the *VEN_CODE* form variable. This value will be used in the SQL statement to search for all records with matching VEN_CODE values.

- Lines 21–27 create a drop-down SELECT box to let the user pick the state to be used in the vendor table query. This selection will later be passed to the rc-4b.cfm script.

- In lines 23–25 the CFOUTPUT tag is used to build the selection options, using the states that occur in the VENDOR table. The default ("SELECTED") option shown in line 22 gives the user the ability to search without selecting any particular state. If line 22 is not included, there will be no way to limit the search to only a vendor code, nor will the user be able to display all vendors from all states. To limit the search, the VEN_STATE form field must have the option to contain a null "value." In short, line 22's default condition provides crucial flexibility!

When the user clicks the Search button, the rc-4b.cfm script is executed, and the two form variables (VEN_CODE and VEN_STATE) are passed to the rc-4b.cfm script. The rc-4b.cfm script is shown next.

SCRIPT 14.4B (RC-4B.CFM): THE VENDOR SEARCH RESULTS

```
1   <HTML>
2   <CFQUERY NAME="SearchVendor" DATASOURCE="RobCor">
3       SELECT VEN_CODE, VEN_NAME, VEN_CONTACT_NAME, VEN_ADDRESS, VEN_CITY, VEN_STATE, VEN_PH
4       FROM VENDOR
5       WHERE 0=0
6       <CFIF #FORM.VEN_CODE# IS NOT "">
7           AND VENDOR.VEN_CODE LIKE '%#FORM.VEN_CODE#%'
8       </CFIF>
9       <CFIF #FORM.VEN_STATE# IS NOT "ANY">
10          AND VENDOR.VEN_STATE LIKE '%#FORM.VEN_STATE#%'
11      </CFIF>
12      ORDER BY VEN_CODE
13  </CFQUERY>
14  <HEAD>
15  <TITLE>Rob & Coronel -  ColdFusion Examples</TITLE>
16  </HEAD>
17  <BODY BGCOLOR="LIGHTBLUE">
18  <H1>
19  <CENTER><B>Vendor Search Results</B></CENTER>
20  </H1>
21  <BR>
22  <CFTABLE QUERY="SearchVendor" STARTROW="1" COLSPACING="1" COLHEADERS>
23  <CFCOL HEADER="<B>VENDOR CODE</B>" WIDTH="14" TEXT="#VEN_CODE#">
24  <CFCOL HEADER="<B>VENDOR NAME</B>" WIDTH="20" TEXT="#VEN_NAME#">
25  <CFCOL HEADER="<B>CONTACT PERSON</B>" WIDTH="20" TEXT="#VEN_CONTACT_NAME#">
26  <CFCOL HEADER="<B>ADDRESS</B>" WIDTH="20" TEXT="#VEN_ADDRESS#">
27  <CFCOL HEADER="<B>CITY</B>" WIDTH="20" TEXT="#VEN_CITY#">
28  <CFCOL HEADER="<B>STATE</B>" WIDTH="2" TEXT="#VEN_STATE#">
29  <CFCOL HEADER="<B>PHONE</B>" WIDTH="20" TEXT="#VEN_PH#">
30  </CFTABLE>
31  <CFIF #SEARCHVENDOR.RECORDCOUNT# IS 0>
32      <H2>No records were found matching your criteria </H2>
33  <CFELSE>
34      <CFOUTPUT>Your seach returned #SearchVendor.RecordCount# record(s).</CFOUTPUT>
35  </CFIF>
36  <FORM ACTION="rc-0.cfm" METHOD="post">
37      <INPUT TYPE="submit" VALUE="Main Menu ">
38  </FORM>
39  </BODY>
40  </HTML>
```

The rc-4b.cfm script output is shown in Figure 14.11.

FIGURE 14.11 THE RC-4B.CFM SCRIPT OUTPUT (VENDOR SEARCH RESULTS: ALL STATES)

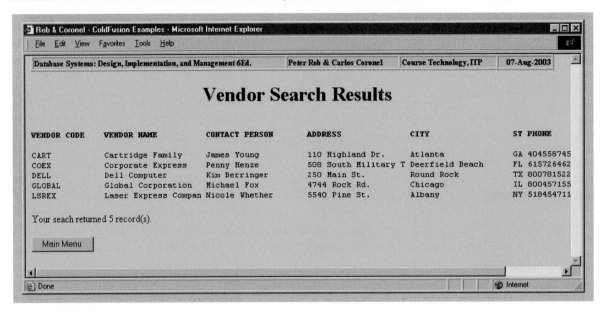

As you examine Script rc-4b, note how it references the form variables received from the calling script (rc-4a.cfm) and how it uses the CFIF tag to dynamically build SQL statements. Let's take a closer look at this script:

- Lines 2–13 are used to create the SQL statement and to query the database.

- Line 5 creates a dummy WHERE clause that will be used to anchor the query's conditional criteria. The "0=0" conditional criterion is the default condition that will list all records. This line is required to form the basis for additional conditions, using SQL's logical operators. Naturally, if no additional conditions are specified, the default value ensures that the query will list all records.

- Lines 6–8 use a CFIF tag to evaluate the VEN_CODE form field passed from the calling page. (In this case, the rc-4a.cfm script created the calling page.) If the field specified in line 6 is not null (""), the condition specified in line 7 is added to the query.

- Lines 9–11 use the CFIF tag to evaluate the VEN_STATE form field passed from the calling page (rc-4a.cfm). If this field is not "ANY", line 10 adds the "state" condition to the query. (Remember that "ANY" is the default selection for the rc-4a.cfm script's VEN_STATE field.) This default selection ensures that it is not necessary to search for any specific state. Therefore, the user can select a state to limit the query output to that state; or by not selecting a state, the user can generate the query output for all states.

- Lines 22–30 indicate that the CFTABLE and CFCOL tags are used to display the query result set in a tabular format.

- Finally, a CFIF statement is used to evaluate the number of records in the resulting record set and display the appropriate message. For example, line 31 ensures that, if the record set returns zero (0) rows, the "*No records were found matching your search criteria*" message is displayed in line 32. If the record count is not zero, line 34 generates a message that indicates the number of records returned in the query result set.

Figure 14.12 shows the output for a dynamic search based on Vendors for the condition VEN_STATE = "GA".

FIGURE 14.12 THE VENDOR LIST FOR THE CONDITION VEN_STATE = "GA"

The dynamic search process clearly makes the Web a viable end-user information-generation tool. However, to conduct *transactions*, you must be able to modify the database's table contents. (For example, if you make a withdrawal from inventory, you must be able to update the inventory table. If you make a purchase, you must be able to generate an invoice record, and so on.) Therefore, we will shift our attention to the basic procedures that may be used to insert, update, and delete data through Web interfaces. Before you can develop Web-based transaction applications, you need to know why and how the shortcomings of HTML and browsers affect data manipulation activities. These shortcomings are a function of the Web's basic structure, which is often described as a so-called *stateless system*.

14.5.6 THE WEB AS A STATELESS SYSTEM

The Web is said to be a stateless system. Simply put, the label **stateless system** indicates that, at any given time, a Web server does not know the status of any of the clients communicating with it. That is, there is no open communications line between the server and each client accessing it—that, of course, is impractical in a *world-wide* Web! Instead, client and server computers interact in very short "conversations" that follow the request-reply model. For example, the browser is only concerned with the *current* page, so there is no way for the second page to know what was done in the first page. The only time the client and server computers communicate

is when the client requests a page—when the user clicks a link—and the server sends the requested page to the client. Once the client receives the page and its components, the client/server communication is ended. Therefore, although you may be browsing a page and *think* that the communication is open, you are actually just browsing the HTML document stored in the local cache (temporary directory) of the client browser. The server does not have any idea what the end user is doing with the document, what data is entered in a form, what option is selected, and so on. In the Web, if we want to act on a client's selection, we need to jump to a new page (go back to the Web server), therefore losing track of whatever was done before!

Not knowing what was done before, or what a client selected before it got to this page, makes adding business logic to the Web cumbersome. For example, suppose that you need to write a program that performs the following steps: display a data entry screen, capture data, validate data, and save data. This entire sequence can be completed in a single COBOL program because COBOL uses a working storage section that holds in memory all variables used in the program. Now imagine the same COBOL program—but *each* section (PERFORM statement) is now a separate program! That is precisely how the Web works. In short, the Web's stateless nature means that extensive processing required by a program's execution cannot be done directly in a single Web page; the client browser's processing ability is limited by the lack of processing ability and the lack of a working storage area to hold variables used by all pages in a Web site.

Keep in mind that a Web browser's function is to display a page on the client computer. The browser—through its use of HTML—does not have computational abilities beyond formatting output text and accepting form field inputs. Even when the browser accepts form field data, there is no way to perform immediate data entry validation. Therefore, to perform such crucial processing in the client, the Web defers to other Web programming languages such as Java, JavaScript, and VBScript. The browser most resembles a dumb terminal that displays only data and can perform only rudimentary processing such as accepting form data inputs. Most of the processing takes place at the server end—in this case, the Web application server.

To circumvent the above-mentioned shortcomings of the Web environment, most Web application servers have session management capabilities that allow a Web server to maintain status information for each client session in the server's memory. Each Web server vendor has its own way to maintain this information. In the case of ColdFusion, session variables are declared using the <CFSET session.variablename=value> command syntax. For example, to declare a user type session variable, you would say: <CFSET session.usertype = "ADMIN">. Then, the variable would be available to all pages in the same client session. The client session starts the first time the client's browser connects to the Web server, and ends when the client accesses a page outside the Web site, closes his/her browser, or stays idle for a given time-out period.

Knowing and understanding the significance of the Web's limitation—its stateless nature—means that alternate processing strategies can be developed and used. We'll next examine how the Web's interface can, in spite of its limitations, be used effectively to accept and manage the crucial transactions that are required by most business processes.

14.5.7 INSERTING DATA

In this section you will create a data entry form to insert data in the DEPARTMENT table. In the following example, the DEPARTMENT table contains three fields: department ID and department description, which are required, and an optional manager user ID that references the USER table. Of course, *if the user enters a user ID, that ID must match a user ID in the USER table.* Given this basic scenario, let's see how ColdFusion can be used to establish basic server-side data validation for the required fields and how ColdFusion can implement and manage data entry for an optional relationship.

We need at least two pages to accomplish the just-described tasks. The first page, generated by a script named rc-5a.cfm, creates a form to get the data. The second page, generated by a script named rc-5b.cfm, inserts the data in the table. Data validation will take place at the server side. Let's first look at Script 14.5a. The script is followed by its output in Figure 14.13.

SCRIPT 14.5A (RC-5A.CFM): INSERT QUERY: DATA ENTRY SCRIPT

```
1   <HTML>
2   <HEAD>
3   <TITLE>Rob & Coronel - ColdFusion Examples</TITLE>
4   <CFQUERY NAME="USRLIST" DATASOURCE="RobCor">
5   SELECT USR_ID, USR_LNAME, USR_FNAME, USR_MNAME
6         FROM USER
7         WHERE USR_ID NOT IN (SELECT USR_ID FROM DEPARTMENT WHERE USR_ID > 0)
8         ORDER BY USR_LNAME, USR_FNAME, USR_MNAME
9   </CFQUERY>
10  </HEAD>
11  <BODY BGCOLOR="LIGHTBLUE">
12  <H1>
13  <CENTER><B>Insert Query: Data Entry Screen</B></CENTER></H1>
14  <FORM ACTION="rc-5b.cfm" METHOD="post">
15  <CENTER><H1>Department Data</H1></CENTER><!--- the following code defines required fields --->
16      <INPUT TYPE="hidden" NAME="DEPT_ID_required" VALUE="You must enter a DEPT_ID">
17      <INPUT TYPE="hidden" NAME="DEPT_DESC_required" VALUE="You must enter a description">
18      <TABLE ALIGN="CENTER" BGCOLOR="Silver">
19          <TR>
20              <TD>
21              <PRE>
22  Department ID:
23              <INPUT TYPE="text" NAME="DEPT_ID" SIZE="10" MAXLENGTH="10"><BR>Description:
24              <INPUT TYPE="text" NAME="DEPT_DESC" SIZE="35" MAXLENGTH="35"><BR>Manager    :
25              <SELECT NAME="USR_ID" SIZE="1"><!--- select user from list --->
26                  <OPTION VALUE="">----------
27                  <CFOUTPUT QUERY="USRLIST">
28                      <OPTION VALUE="#UsrList.USR_ID#">[#USR_ID#] #USR_LNAME#, #USR_FNAME#, #USR_MNAME#
29                  </CFOUTPUT>
30              </SELECT></PRE>
31          </TD>
32          <TD>
33              <INPUT TYPE="submit" VALUE="Add Record">
34  </FORM>
35  <FORM ACTION="rc-0.cfm" METHOD="post">
36      <INPUT TYPE="submit" VALUE=" Main Menu ">
37  </FORM>
38          </TD>
39      </TR>
40  </TABLE>
41  </BODY>
42  </HTML>
```

FIGURE 14.13 THE RC-5A.CFM SCRIPT OUTPUT (DEPARTMENT DATA ENTRY SCREEN)

Script 14-5a displays a data entry form to allow the end user to enter the new department data. To show you how the form works, we'll add a new Transportation department (TRANS) and assign a manager to run this new department. The rc-5a.cfm script is designed to perform a data validation check in the USR_ID field, using a query and a select box. To see how the script accomplishes these tasks, let's look at some key lines:

- Lines 4–9 execute a nested query to find all user IDs for employees who are not already department managers. By performing this portion of the data validation procedure, we ensure that there are no duplicate user IDs in the DEPARTMENT table. Therefore, it will not be possible to have a manager manage more than one department. Also, because a department might not (yet) have a manager assigned to it, the USR_ID might not have a value in it. To perform the requisite checks, we start with a nested query, using the *USR_ID > 0* condition. This condition will be true only for those records in which a manager (USR_ID) exists.

- Lines 14–34 define the form that will be used to enter the data. Note that the rc-5b.cfm script will be called when the user clicks the Add Record button.

- Lines 16 and 17 are special form tags ColdFusion uses to perform data validation at the server side. In this case, we will validate the entries for the two *required* fields, the department identification code, DEPT_ID, and the department description, DEPT_DESC. This task is performed by using an INPUT form tag with the following parameters:

 - *TYPE* = "*hidden*". This parameter ensures that the field will not be displayed on the screen.

 - *NAME* = "*fieldname_required*". This parameter specifies the field to be checked, followed by the word "_required". Other parameter options are:

 _integer to check for integer values only.

 _date to check for valid dates only in the format mm/dd/yy.

 _range to check for a value within a range. The range is given in the value = parameter, for example: value = "MIN=10 MAX=20".

 - *VALUE* = "*error message*". This parameter contains the error message to be displayed when the constraint is violated (in our case, when the field is empty). If the parameter type = _range, this field contains the maximum and minimum values used in the validation.

- Lines 24–30 are used to create a drop-down select box to show all the *available* users who can be selected to be department manager. Note in particular the following lines:

 - Line 26 defines a null option, represented on the screen by a dotted line to indicate that the department does not have a manager assigned. This line is used to implement the optionality of the USR_ID field. If we do not include this line, there will be no way to assign a null to the USR_ID field.

 - Lines 27–29 are used to generate a list of all users who are eligible to manage this new department. Remember that the USRLIST query returns only the USR_ID of users who are not already in the DEPARTMENT table. In other words, the USRLIST query lists only those users *who are still available to become department managers*. (Remember that the business rule specifies that a manager can manage only one department.)

When the user clicks the Add Record button, the form is sent to the Web server for processing. There, ColdFusion will evaluate the required fields, sending an error message if one of the required fields is empty. (See Figure 14.14.)

FIGURE 14.14 THE INSERT QUERY FORM: SERVER-SIDE VALIDATION ERROR MESSAGE

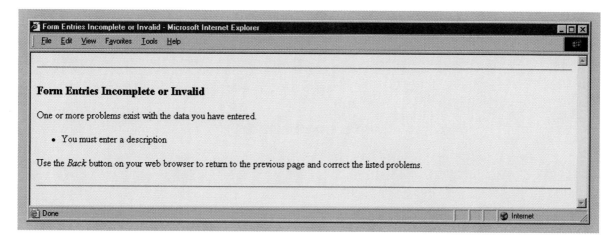

If the server-side data validation yields the conclusion that the data entry is valid, the second script, rc-5b.cfm, is executed. This script receives the form fields from the rc-5a.cfm script and uses the CFINSERT tag to add the record to the database. Once the record has been added, the rc-5b.cfm script presents a confirmation screen to the end user.

SCRIPT 14.5B (RC-5B.CFM): INSERT QUERY: CONFIRMATION SCRIPT

```
1   <HTML>
2   <HEAD>
3   <TITLE>Rob & Coronel - ColdFusion Examples</TITLE>
4   <!--- inserting record in table --->
5   <CFINSERT DATASOURCE="RobCor" TABLENAME="DEPARTMENT">
6   </HEAD>
7   <BODY BGCOLOR="LIGHTBLUE">
8   <H1>
9   <CENTER><B>Insert Query: Result Confirmation</B></CENTER></H1>
10  You have added the following record:
11  <CFOUTPUT>
12  <PRE>
13      DEPT_ID  : #DEPT_ID#
14      DEPT_DESC: #DEPT_DESC#
15      USR_ID   : #USR_ID#
16  </PRE>
17  </CFOUTPUT>
18  <FORM ACTION="rc-0.cfm" METHOD="post">
19      <INPUT TYPE="submit" VALUE="Main Menu ">
20  </FORM>
21  </BODY>
22  </HTML>
```

The execution of Script rc-5b is shown in Figure 14.15.

FIGURE 14.15 THE RC-5B.CFM SCRIPT OUTPUT (DEPARTMENT DATA ENTRY CONFIRMATION)

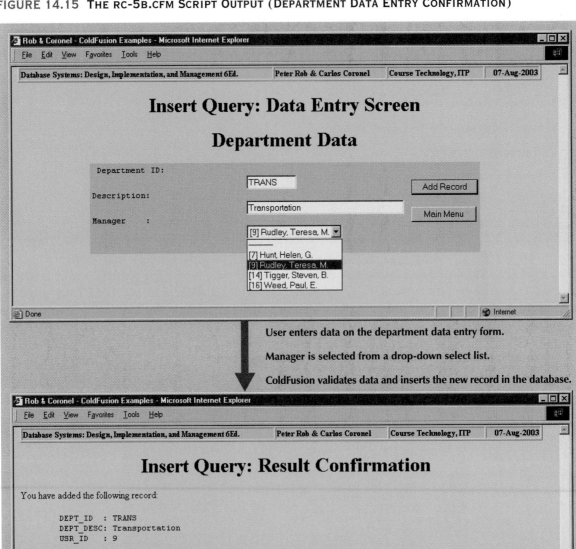

To see how the rc-5b.cfm script uses the CFINSERT tag to add the record to the database, check line 5. Note that this tag uses two parameters:

- *DATASOURCE* = *"datasource_name"*. The name of the ODBC database connection.
- *TABLENAME* = *"table_name"*. The name of the table to be updated.

How does the CFINSERT tag know what fields to insert in the table? And where does it get the values to insert into the table columns? The answer to both questions is found by examining the rc-5b.cfm script CFINSERT tag. This tag uses the field names that were defined on the form generated by the rc-5a.cfm script. Recall that the rc-5a.cfm script produced a form containing the DEPT_ID, DEPT_DESC, and USER_ID fields. The CFINSERT tag compares these form field names with the names of the table columns in order to do the insert. To avoid an error condition, the form that was created in the calling page must have form field names that match the table column names.

The MS Access database enforces entity integrity for the RobCor database's DEPARTMENT table. Therefore, entering an *existing* department ID on the input form automatically triggers an ODBC database

integrity violation error, as shown in Figure 14.16. (Remember that entity integrity is violated when a primary key value is duplicated; your error message may look slightly different depending on your version of ColdFusion.)

FIGURE 14.16 THE INSERT QUERY: ODBC INTEGRITY VIOLATION ERROR

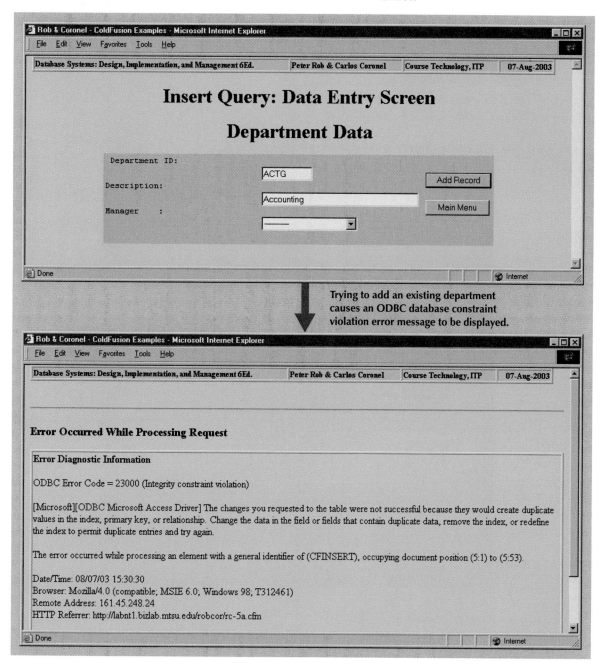

14.5.8 DATA UPDATES

Data updates require multiple pages. For example, if you want to produce a simple update in the DEPARTMENT table's records, the update process requires *three* different pages:

- The first page, produced by Script 14.6a, allows the end user to select the record to be updated. When the user clicks this page's Edit button, the second page, produced by Script 14.6b, is called and the first page's search field value is passed to this second page. (To keep the process as simple as possible, we will

use the primary key, DEPT_ID, to ensure a unique match. If you use secondary search fields, you may find more than one record, and you would need to use an *additional* page to select one of the multiple records to generate a unique match.)

- The second page (rc-6b.cfm) reads the selected record, then displays a data entry form to enable the end user to modify the data. When the end user clicks the Update button, this page calls the third script and passes the second page's form fields to the third page.

- The third and last page, generated by Script 14.6c, updates the data in the database and presents a confirmation message to the end user.

Although this process seems simple enough, there are several issues that must be addressed, as we will see next. Let's begin by taking a close look at Script 14.6a.

SCRIPT 14.6A (RC-6A.CFM): UPDATE QUERY: RECORD SELECTION SCRIPT

```
1    <HTML>
2    <HEAD>
3    <TITLE>Rob & Coronel - ColdFusion Examples</TITLE>
4    <CFQUERY NAME="Deptlist" DATASOURCE="RobCor">
5        SELECT * FROM DEPARTMENT ORDER BY DEPT_ID
6    </CFQUERY>
7    </HEAD>
8    <BODY BGCOLOR="LIGHTBLUE">
9    <H1>
10   <CENTER><B>Update Query: Record Selection Screen</B></CENTER>
11   </H1>
12   <TABLE ALIGN="CENTER" BGCOLOR="Silver">
13   <TR VALIGN="TOP">
14   <TD>
15   <FORM ACTION="rc-6b.cfm" METHOD="post">
16       <SELECT NAME="DEPT_ID" SIZE=1>
17       <CFOUTPUT QUERY="Deptlist">
18           <OPTION VALUE="#DEPT_ID#">[#DEPT_ID#] - #DEPT_DESC#
19       </CFOUTPUT>
20       </SELECT>
21       <INPUT TYPE=HIDDEN NAME="DEPT_ID_required" VALUE="DEPT_ID is required">
22   </TD>
23   <TD>
24   <INPUT TYPE="submit" VALUE=" Edit ">
25   </FORM>
26   </TD>
27   <TD>
28   <FORM ACTION="rc-0.cfm" METHOD="post">
29       <INPUT TYPE="submit" VALUE="Main Menu ">
30   </FORM>
31   </TD>
32   </TR>
33   </TABLE>
34   </BODY>
35   </HTML>
```

The rc-6a.cfm script output is shown in Figure 14.17.

FIGURE 14.17 THE RC-6A.CFM SCRIPT OUTPUT (RECORD SELECTION SCREEN)

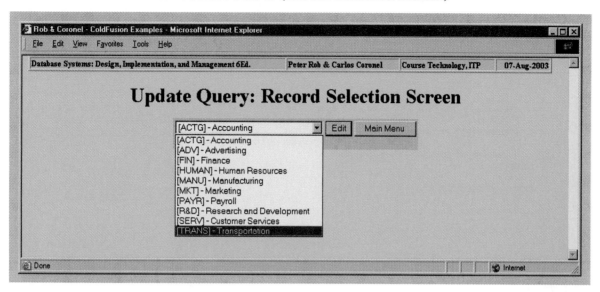

The rc-6a.cfm script produces the form that lets the end user select the record to be updated. This record selection process requires the completion of the following steps:

- Lines 4–6 execute a query ("Deptlist") to retrieve all DEPARTMENT table records. This record set will later be used to create a drop-down selection box.

- Lines 17–19 use a CFOUTPUT tag to produce a list of available options. In this example, ColdFusion uses an OPTION tag for each department in the "Deptlist" query.

- Lines 15–25 produce the record selection form. This form uses an HTML form tag to pick the department to be updated.

- Line 21 defines the "DEPT_ID" form field to be a required field. This definition ensures that the end user will not generate "variable not defined" errors when the next page is called.

When the end user clicks the form's Edit button, Script 14-6b is called. The rc-6b.cfm script will receive the DEPT_ID form field as a parameter, using it to find the matching department table record. It will then prepare and display the data edit form.

SCRIPT 14.6B (RC-6B.CFM): UPDATE QUERY: EDIT RECORD SCRIPT

```
1   <HTML>
2   <HEAD>
3   <TITLE>Rob & Coronel -  ColdFusion Examples</TITLE>
4   <CFQUERY NAME="DeptData" DATASOURCE="RobCor">
5     SELECT * FROM DEPARTMENT WHERE (DEPARTMENT.DEPT_ID='#form.DEPT_ID#')
6   </CFQUERY>
7   <CFQUERY NAME="USRLIST" DATASOURCE="RobCor">
8     SELECT USR_ID, USR_LNAME, USR_FNAME, USR_MNAME
9       FROM USER
10      WHERE USR_ID NOT IN (SELECT USR_ID FROM DEPARTMENT
11                             WHERE USR_ID > 0 AND DEPT_ID <> '#form.DEPT_ID#')
12      ORDER BY USR_LNAME, USR_FNAME, USR_MNAME
13  </CFQUERY>
14  </HEAD>
15  <BODY BGCOLOR="LIGHTBLUE">
16  <H1>
17  <CENTER><B>Update Query: Edit Record Screen</B></CENTER>
18  </H1>
19  <FORM ACTION="rc-6c.cfm" METHOD="post">
20  <TABLE ALIGN="CENTER" BGCOLOR="Silver" BORDERCOLOR="Blue">
21  <TR>
22  <TD>
23  <CFOUTPUT QUERY="DeptData">
24  <PRE>
25  <INPUT TYPE="hidden" NAME="DEPT_ID" VALUE="#DEPTDATA.DEPT_ID#">
26  Department ID: <B>#DEPT_ID#</B><BR>
27  Description  : <INPUT TYPE="text" NAME="DEPT_DESC" VALUE="#DEPT_DESC#"SIZE=35 MAXLENGTH=35><BR>
28  Manager      : <SELECT NAME="USR_ID" SIZE=1><!--- select user from list --->
29                  <OPTION <CFIF #DEPTDATA.USR_ID# EQ "">SELECTED</CFIF> VALUE="">----------
30  </CFOUTPUT>
31  <CFOUTPUT QUERY="USRLIST">
32                  <OPTION <CFIF #DEPTDATA.USR_ID# EQ #USRLIST.USR_ID#>SELECTED</CFIF> VALUE="#UsrList.USR_ID#">
33                  [#USR_ID#] #USR_LNAME#, #USR_FNAME#, #USR_MNAME#
34  </CFOUTPUT>
35                  </SELECT>
36  </PRE>
37  </TD>
38  <TD VALIGN="TOP">
39  <INPUT TYPE="submit" VALUE=" Update ">
40  </FORM>
41  <FORM ACTION="rc-0.cfm" METHOD="post">
42     <INPUT TYPE="submit" VALUE="Main Menu">
43  </FORM>
44  </TD>
45  </TR>
46  </TABLE>
47  </BODY>
48  </HTML>
```

The rc-6b script output is shown in Figure 14.18.

FIGURE 14.18 THE RC-6B.CFM SCRIPT OUTPUT (EDIT RECORD SCREEN)

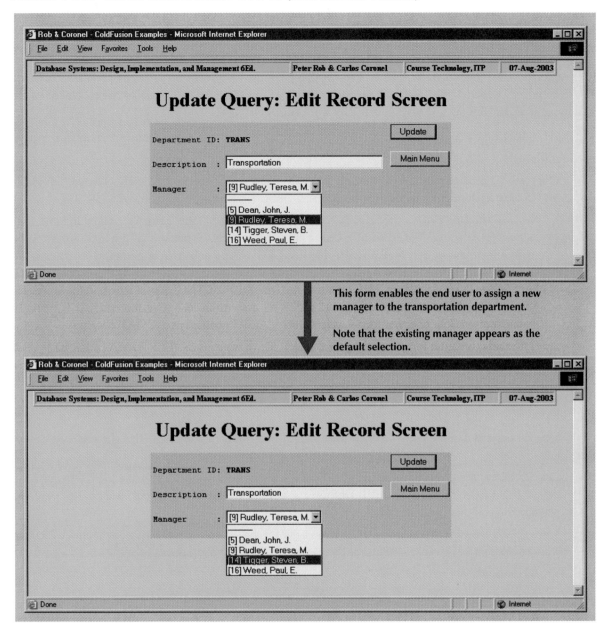

This form enables the end user to assign a new manager to the transportation department.

Note that the existing manager appears as the default selection.

As you compare the rc-6b.cfm script and the sequence shown in Figure 14.19, note that the script generates the following actions:

- Lines 4–6 read the Department data, using the *"#form.dept_id#"* parameter received from the rc-6a.cfm script.

- Lines 7–13 specify and execute a key query. When a new department is to be inserted, we use a "USRLIST" query to list all users who are not already managers of a department. This query applies only to *new* record inserts; it cannot be used to edit an *existing* record.

To see why it is necessary to modify the original USRLIST query, let's suppose that USR_ID = 13 is the current ACTG department's manager. If we try to *edit* the ACTG department record using the *original* USRLIST query, the list of "available users" will yield all users who are *not* already in the DEPARTMENT table. Therefore, because

the ACTG department's current manager *is* listed in the DEPARTMENT table's USR_ID column, the current manager will not appear on the list of users who are available as ACTG department managers. In short, if we try to edit (update) a record using the original USRLIST query, we will be forced to select a manager other than the current one. This is clearly not what's intended!

To avoid the just-described dilemma, the nested query criteria of the USRLIST query must be modified to exclude the "to be edited" department's DEPT_ID from the subquery. Therefore, line 11 specifies the nested query criteria to be WHERE USR_ID > 0 AND DEPT_ID <> '#form.DEPT_ID#'.

Given this modification, the ACTG department's current manager (USR_ID = 13) will appear on the list of available user IDs.

- Lines 19–40 produce the user edit form. This form allows the end user to modify only two DEPARTMENT table fields: department description, DEPT_DESC, and department manager, USR_ID.

 - *Because the DEPT_ID is the DEPARTMENT table's primary key, we cannot allow the end user to modify its value.* The reason for this restriction is simple. Suppose that the end user edits the ACTG department, that is, the DEPT_ID = 'ACTG'. If the end user is permitted to change the DEPT_ID form 'ACTG' to 'ACTNG', the department data update will pass the DEPT_ID = 'ACTNG' form field to the update script. Unfortunately, the DEPT_ID=ACTNG' does not exist in the DEPARTMENT table. Therefore, the database will return an error to indicate that the end user is trying to update a record that does not exist—which is true: 'ACTG' exists, but 'ACTNG' does not.

- Line 25 creates an input form variable named DEPT_ID, to which the "#DEPTDATA.DEPT_ID#" value is assigned. In other words, the script assigns the current record value to the edit mode. Note that this variable is hidden, so it will not be shown on the screen. This value assignment ensures that the DEPT_ID is passed to the rc-6c.cfm script. (Remember that all form variables that are defined with an INPUT or SELECT form tag are passed to the called program.)

- Line 26 ensures that the current department ID is shown on the screen. Note that the end user cannot edit this value.

- Line 27 allows the end user to modify the department's description. Note that the INPUT tag sets the default value for this field to "#DEPT_DESC#", thus ensuring that the previous field contents are displayed. The end user can then modify the displayed values.

- Lines 28–35 allow the end user to choose a manager for the selected department. These lines will create a select box that lists all valid options for the manager field. The valid options are:

 - All users who are not managers.

 - If the department has a manager, the existing manager.

 - A null manager option to indicate that no manager has (yet) been assigned to the selected department.

 Given these options, the end user can set the manager to null, leave the current manager unchanged, or choose another manager. If the department being edited already has a manager, that manager must appear as the default (SELECTED) option.

- Line 29 allows the manager ID to be set to null. (Note that VALUE = "".) This line also contains a CFIF tag to evaluate the current value of the department's USR_ID field. If the USR_ID is "" (that is, the selected department does not have a manager), this null will be the SELECTED option. Otherwise, this option will appear on the list of options, but will not be preselected. This null option must be available to ensure that a department's manager can be removed.

- Lines 31–34 use the CFOUTPUT tag to create OPTION entries for each of the user IDs in the "USRLIST" query. (Remember that the CFOUTPUT tag will loop through each record in the named query.) For each added option, a CFIF tag compares the existing USR_ID in the department table

(#DEPTDATA.USR_ID#) with the USR_ID being added (#USRLIST.USR_ID#). If the two are equal, the "SELECTED" keyword is added to the OPTION tag.

When the user clicks the form's Update button, Script 14.6b calls Script 14.6c, passing its variable values.

SCRIPT 14.6C (RC-6C.CFM): UPDATE QUERY: RESULT CONFIRMATION SCRIPT

```
1   <HTML>
2   <HEAD>
3   <TITLE>Rob & Coronel -  ColdFusion Examples</TITLE>
4   <CFUPDATE DATASOURCE="RobCor" TABLENAME="Department">
5   </HEAD>
6   <BODY BGCOLOR="LIGHTBLUE">
7   <H1>
8   <CENTER><B>Update Query: Result Confirmation</B></CENTER>
9   </H1>
10  <CFOUTPUT>
11  You have successfully Updated the following data
12  <PRE>
13  DEPARTMENT ID:  <B>#DEPT_ID#</B>
14  DESCRIPTION  :  <B>#DEPT_DESC#</B>
15  MANAGER      :  <B>#USR_ID#</B>
16  </PRE>
17  </CFOUTPUT>
18  <FORM ACTION="rc-0.cfm" METHOD="post">
19      <INPUT TYPE="submit" VALUE="Main Menu ">
20  </FORM>
21  </BODY>
22  </HTML>
```

The rc-6c.cfm script's output is shown in Figure 14.19.

FIGURE 14.19 THE RC-6C.CFM SCRIPT OUTPUT (UPDATE QUERY RESULT CONFIRMATION)

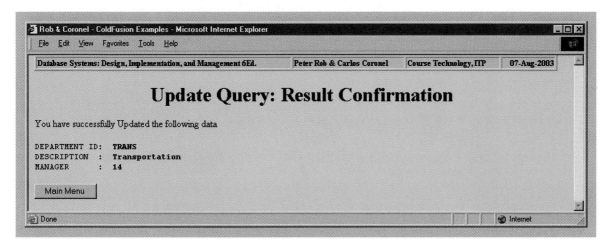

The rc-6c.cfm script uses the CFUPDATE tag to update the DEPARTMENT table. The parameters for this tag are:

- *DATASOURCE = "datasource_name"*. The name of the ODBC database connection.
- *TABLENAME = "table_name"*. The name of the table to be updated.

The CFUPDATE tag uses the form fields passed from the calling page (DEPT_ID, DEPT_DESC, and USER_ID) to update the named table. As was true with the other .cfm pages, the form created in the calling page must name its form fields to match the table column names. Failure to adhere to this naming requirement will generate a "variable not found" error.

The most basic Web-based data management process requires at least three actions: create a new record, modify an existing record, and delete a record. We have discussed the first two, so it's time to examine the "delete" action.

14.5.9 DELETING DATA

The "delete" query we'll examine in this section enables the end user to delete a department record. As was true for the update query, the delete query's implementation requires three pages:

- The first page (Script 14.7a) allows the end user to select the record to be deleted. When the user clicks the form's Delete button, Script 14.7b is invoked, and the DEPT_ID form field value is passed to it.

- The second page (Script 14.7b) reads the selected record and displays its data on the screen. This query also performs a referential integrity check to ensure that the end user cannot delete a department that still contains users. When the user clicks the Delete button, this page calls the third page, passing the DEPT_ID form's field value to that page.

- The last page (Script 14.7c) deletes the department row from the database table, using the DEPT_ID form field value passed from its calling program, rc-7b.cfm.

Let's look at the first of these three scripts.

SCRIPT 14.7A (RC-7A.CFM): DELETE QUERY: RECORD SELECTION SCRIPT

```
<HTML>
<HEAD>
<TITLE>Rob & Coronel - ColdFusion Examples</TITLE>
<CFQUERY NAME="Deptlist" DATASOURCE="RobCor">
    SELECT * FROM DEPARTMENT ORDER BY DEPT_ID
</CFQUERY>
</HEAD>
<BODY BGCOLOR="LIGHTBLUE">
<H1>
<CENTER><B>Delete Query: Record Selection Screen</B></CENTER>
</H1>
<TABLE ALIGN="CENTER" BGCOLOR="Silver">
<TR VALIGN="TOP">
<TD>
<FORM ACTION="rc-7b.cfm" METHOD="post">
    <SELECT NAME="DEPT_ID" SIZE=1>
    <CFOUTPUT QUERY="Deptlist">
        <OPTION VALUE="#DEPT_ID#">[#DEPT_ID#] - #DEPT_DESC#
    </CFOUTPUT>
    </SELECT>
    <INPUT TYPE=HIDDEN NAME="DEPT_ID_required" VALUE="DEPT_ID is required">
</TD>
<TD>
    <INPUT TYPE="submit" VALUE="Delete">
</FORM>
</TD>
<TD>
<FORM ACTION="rc-0.cfm" METHOD="post">
    <INPUT TYPE="submit" VALUE="Main Menu">
</FORM>
</TD>
</TR>
</TABLE>
</BODY>
</HTML>
```

The rc-7a.cfm script output is shown in Figure 14.20.

FIGURE 14.20 THE RC-7A.CFM SCRIPT OUTPUT (RECORD SELECTION SCREEN)

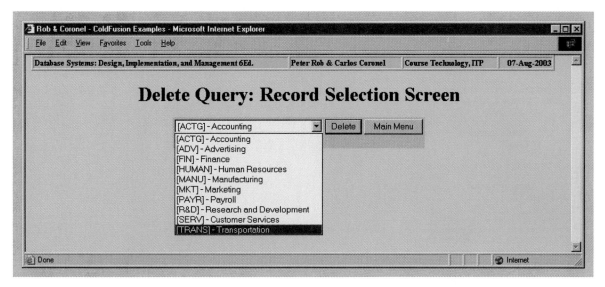

The rc-7a.cfm script allows the end user to select a record to be deleted. To trace its operations, let's examine the following lines:

- Lines 4–6 perform a query ("Deptlist") to retrieve all records from the DEPARTMENT table. This query result set will be used to display a record selection form.

- Lines 15–25 define the record selection form. When the user clicks the form's Delete button, the rc-7a.cfm script calls Script 14.7b. As before, the rc-7a script form passes its DEPT_ID form field to the rc-7b.cfm script.

- Lines 16–20 create the SELECT form control.

- Lines 17–19 use a CFOUTPUT tag to dynamically create the OPTION list.

- Line 21 uses the INPUT tag to define the DEPT_ID field as a required field. ColdFusion uses this command to perform sever-side validation on this field. If this field is left blank, ColdFusion will return the error message text entered in the "VALUE" parameter. Therefore, line 21 ensures that the end user selects a record before trying to delete it. (You can't delete a record without first specifying which one it is!) If the end user fails to select a record first, clicking the Delete button produces an ODBC database error message to indicate the attempted deletion of a nonexistent record—or, even worse, it could delete all the table rows!

The second script (rc-7b.cfm) performs two important functions:

- It reads the record to be deleted and presents its data on the screen to let the end user confirm that this is the record (s)he wants to delete.

- It performs the referential integrity validation. Remember that the DEPARTMENT and USER tables maintain a 1:M relationship expressed by "each department may have one or more users." Therefore, the end user cannot be allowed to delete a department if it still contains users.

Given its importance, let's next examine the details of Script 14.7b.

SCRIPT 14.7B (RC-7B.CFM): DELETE QUERY: SHOW RECORD SCRIPT

```
1   <HTML>
2   <HEAD>
3   <TITLE>Rob & Coronel -  ColdFusion Examples</TITLE>
4   </HEAD>
5   <CFQUERY NAME="DeptData" DATASOURCE="RobCor">
6       SELECT * FROM DEPARTMENT WHERE (DEPARTMENT.DEPT_ID='#form.DEPT_ID#')
7   </CFQUERY>
8   <CFIF #DEPTDATA.USR_ID# IS NOT "">
9       <CFQUERY NAME="Usrdata" DATASOURCE="RobCor">
10          SELECT USR_ID, USR_LNAME, USR_FNAME, USR_MNAME FROM USER
11          WHERE (USER.USR_ID = #deptdata.usr_id#)
12      </CFQUERY>
13  </CFIF>
14  <CFQUERY NAME="UsrTot" DATASOURCE="RobCor">
15      SELECT COUNT(*) AS T1 FROM USER
16      WHERE (USER.DEPT_ID = '#form.DEPT_ID#')
17  </CFQUERY>
18  </HEAD>
19  <BODY BGCOLOR="LIGHTBLUE">
20  <H1><CENTER><B>Delete Query: Show Record Screen</B></CENTER></H1>
21  <FORM ACTION="rc-7c.cfm" METHOD="post">
22  <CFOUTPUT QUERY="DeptData">
23  <INPUT TYPE="hidden" NAME="DEPT_ID" VALUE="#deptdata.DEPT_ID#">
24  <INPUT TYPE="hidden" NAME="DEPT_DESC" VALUE="#deptdata.DEPT_DESC#">
25  <INPUT TYPE="hidden" NAME="USR_ID" VALUE=#DEPTDATA.USR_ID#>
26  <TABLE ALIGN="CENTER" BGCOLOR="Silver" BORDERCOLOR="Blue">
27  <TR>
28  <TD>
29  <PRE>
30  Department ID: #DEPT_ID#<BR>
31  Description  : #DEPT_DESC#<BR>
32  Manager     : <CFIF #DEPTDATA.USR_ID# IS NOT "">#Usrdata.USR_LNAME# #Usrdata.USR_FNAME# #Usrdata.USR_MNAME#</CFIF>
33  </CFOUTPUT>
34  </PRE>
35  </TD>
36  <TD VALIGN="TOP">
37  <CFIF #USRTOT.T1# EQ 0>
38      <INPUT TYPE="submit" VALUE=" Delete ">
39  <CFELSE>
40      <SMALL><B>We cannot delete this record <BR>because there are dependent <BR>users assigned to this department</B></SMALL>
41  </CFIF>
42  </FORM>
43  <FORM ACTION="rc-0.cfm" METHOD="post"><INPUT TYPE="submit" VALUE="Main Menu"></FORM>
44  </TD>
45  </TR>
46  </TABLE>
47  </BODY>
48  </HTML>
```

The rc-7b.cfm script output is shown in Figure 14.21.

FIGURE 14.21 THE RC-7B.CFM SCRIPT OUTPUT (SHOW RECORD SCREEN)

Let's examine the rc-7b.cfm script to understand how it works:

- Lines 5–7 use the CFQUERY tag to read the selected department data. The query uses the "*#form.DEPT_ID#*" form field passed from the rc-7a.cfm script.

- Lines 8–13 retrieve the department manager data from the USER table. Because this is an optional field, we first check to see if the user ID is not null. If this non-null condition is met, we read the user data from the USER table using the "*#deptdata.usr_id#*" value. If the user ID is null, there is no need to read the user data.

- Lines 14–17 perform referential integrity validation checks. The process starts by executing a query to see if there are users still assigned to the department to be deleted. Note that the SQL query in lines 15 and 16 counts the number of users assigned to the department. (If this count yields a value greater than zero, the department contains at least one user.) Note that we stored the count in variable T1.

- Lines 21–42 define a form to display the department data and to confirm the record deletion. When the user clicks the Delete button, the rc-7c.cfm script is called and the three variables (DEPT_ID, DEPT_DESC, and USR_ID) are passed to it.

- Line 23 defines the form's DEPT_ID field as "hidden", and the to-be-deleted DEPARTMENT table's DEPT_ID value is assigned to this hidden field. (Although this hidden input field does not show on the screen, it is passed to the next script.)

- Lines 24 and 25 perform the same function as line 23, defining the remaining form fields (DEPT_DESC and USER_ID) as "hidden" and assigning the corresponding department field values to these hidden form fields. These hidden form field values, too, are passed to the next script.

- Lines 30–32 display the department data for the record to be deleted. This action enables the end user to see the record to confirm that this is, in fact, the department record (s)he wants to delete. Note that the fields specified in lines 30 and 31 use the "Deptdata" query source as specified in line 22's CFOUTPUT tag. In contrast, note that line 32's field name prefix indicates that it uses fields from the "Usrdata" query.

- Line 32 uses a CFIF tag to check whether there are user data. If the "*#deptdata.usr_id#*" is not null, the user data are displayed.

- Lines 37–41 check to see if there are any users assigned to the department. If user records do not exist (#usrtot.t1# EQ 0), the Delete button is shown. (Check line 15 again and note that its count is now the basis for the condition check.) If the count is anything other than zero, the form displays a message to indicate that there are still users assigned to this department, and the Delete button is not shown.

 If the record can be deleted and the user clicks the Delete button, Script 14.7c is called and the (hidden) form fields are passed to it.

SCRIPT 14.7c (RC-7C.CFM): DELETE QUERY: RESULT CONFIRMATION SCRIPT

```
<HTML>
<HEAD>
<TITLE>Rob & Coronel - ColdFusion Examples</TITLE>
<CFQUERY NAME="DeleteDept" DATASOURCE="RobCor">
    DELETE FROM DEPARTMENT WHERE (DEPT_ID = '#FORM.DEPT_ID#')
</CFQUERY>
</HEAD>
<BODY BGCOLOR="LIGHTBLUE">
<H1>
<CENTER><B>Delete Query: Result Confirmation</B></CENTER>
</H1>
<CFOUTPUT>
You have successfully Deleted the following data
<PRE>
DEPARTMENT ID:   <B>#DEPT_ID#</B>
DESCRIPTION:     <B>#DEPT_DESC#</B>
MANAGER    :     <B>#USR_ID#</B>
</PRE>
</CFOUTPUT>
<FORM ACTION="rc-0.cfm" METHOD="post">
    <INPUT TYPE="submit" VALUE="Main Menu ">
</FORM>
</BODY>
</HTML>
```

The rc-7c.cfm script output is shown in Figure 14.22.

FIGURE 14.22 THE RC-7C.CFM SCRIPT OUTPUT (DELETE QUERY CONFIRMATION SCREEN)

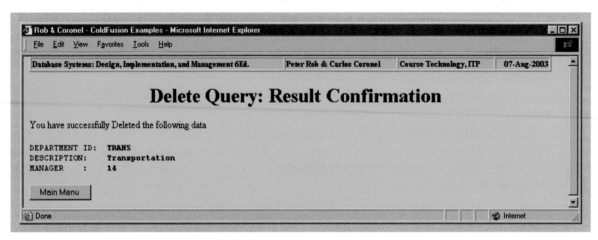

The rc-7c.cfm script deletes the department record from the database and displays a confirmation screen. To see how this script accomplishes these tasks, check the following lines:

- Lines 4–6 perform a query used to delete the department record from the database. This query executes the SQL "Delete" statement, using the form's DEPT_ID field received from the rc-7b.cfm script.

- Lines 10–17 are used to confirm that the record has been deleted and to display the deleted data.

Figure 14.23 shows how the delete sequences work. As you examine the screens, note that the attempt failed to delete a department that still contains users. Also, note that the Delete button is not shown in the second screen. There is, after all, no reason to display a Delete button if the selected department record cannot be deleted.

FIGURE 14.23 **THE DELETE RECORD VALIDATION**

NOTE

The ColdFusion techniques presented in this chapter represent just the tip of the proverbial iceberg in the development of database-enabled Web applications. At the time of this writing, ColdFusion provides hundreds of additional tags and functions to help you properly develop professional Web applications. Although the preceding examples are far from exhaustive, they do provide a compelling illustration of the Web interface's power and flexibility.

In today's increasingly Web-driven business environment, there is little doubt that you will work with databases that are Web-enabled. Given the clear competitive advantages provided by database Web access, it is tempting to focus on the Web side of the Web-database equation. Yet it is important to realize that a Web interface to a badly designed database is a recipe for epic-scale database disasters. On the other hand, good database design and implementation, coupled with sound Web development techniques, yield countless business tactical and strategic advantages, virtually boundless professional opportunities, and personal satisfaction.

14.6 INTERNET DATABASE SYSTEMS: SPECIAL CONSIDERATIONS

Internet database systems involve more than just the development of database-enabled Web applications. In addition, there are issues to be addressed when Web interfaces are used as the gateway to corporate and institutional databases. For example, there are data security, transaction management, client-side data validation, and many other operational and management challenges to meet. Although we have discussed many of these issues in detail in previous chapters, they are particularly relevant to Web database development, so we revisit some of them here.

Whether you are talking about databases in a conventional client/server environment or in the latest Internet arena, database systems development requires sound database design and implementation. The database system must exist within a secure environment that is well suited to maintain well-monitored and protected data access, robust transaction management with a focus on data integrity maintenance, and solid data recovery. Finally, from the end user's and business manager's points of view, the database is not particularly useful unless its front end is characterized by user-friendly, information-capable, and transaction-supportive end-user applications.

Production database and data warehouse designs are not affected—*at the conceptual level*—by the change from the conventional client/server to the Internet's client/server environment. Therefore, we need not revisit the basic design processes and procedures. However, Internet database application development is quite different from that found in the traditional client/server environment. And, given the vastness of the Internet, development issues such as security, backup, and transaction volume are even more critical than they were in the traditional environment.

Clearly, concurrent database access by multiple heterogeneous clients affects how transactions are defined and managed. The existence of support for multiple data sources and types, the advent of increasing platform independence and portability, process distribution and scalability, and open standards have a major effect on how applications are developed, installed, and managed.

Arguably, database application development is most affected by the Internet. The characteristics of the Internet—and particularly its Web service—fundamentally change the way that applications work. The stateless nature of the Web has a major impact on how database queries are presented and executed. Just think of the Web's request-reply model and how different it is from the conventional programmer's view.

If database systems are to be developed and managed intelligently, today's database administrator must understand the Internet-based business environment to successfully cope with the issues that drive the development, use, and management of Web-to-database interfaces. Since it all begins and ends with data, we will begin by looking at the incredibly diverse data types that are supported in the Internet environment.

14.6.1 WHAT DATA TYPES ARE SUPPORTED?

Web development requires the concurrent management of many and quite different data types. Typically, conventional databases support such data types as Julian dates, various types of numbers (integer, floating point, and currency) and text (fixed and variable-length). Most leading RDBMS vendors support extended data types such as binary or OLE (object linking and embedding) objects in the conventional database environment.

Because interactive Web sites tend to integrate data from multiple sources—word-processor documents, spreadsheets, presentations, pictures, movies, sounds, and even holograms—some Web and database designers use extended data types to store page components (in binary format) for later incorporation into the Web pages.

Although the page arrangement may provide better data organization from the database point of view, several issues must be addressed:

- How does one store and extract data objects such as documents, pictures, and movies through a Web browser? Remember that the Web client expects every page component to be a file stored in the Web server's directory. Therefore, the DBMS or the Web-to-database middleware must provide special functions or subroutines that allow objects to be dynamically extracted from a database field to the Web server's directory, and vice versa.

- How much overhead will be created by the storage of binary objects in the database? How robust must the DBMS be to handle binary object transactions? What are the limitations for extended or OLE data types? How many extended or OLE data type fields can tables have?

- Does the client browser support the data type of the object being accessed? Are the necessary plug-ins available? Is there a way to automatically translate documents from their native format to HTML? For example, a PowerPoint presentation can be viewed within an Internet Explorer browser using the PowerPoint Viewer plug-in, but Netscape Navigator cannot do that.

- Finally, storing pictures or multimedia presentations in the database can very quickly increase the size of the database. Does the DBMS support very large databases? What about transaction speed? The concurrent insertion and extraction of binary objects in database fields could take quite a toll in database transaction performance. How many users are going to access the database? How frequently?

Web-to-database interface design must juggle all these issues and find the right balance to ensure that the database does not become the Web-based system's bottleneck.

14.6.2 DATA SECURITY

In earlier sections we noted that security is one of the key issues when databases are accessible through the Internet. Most DBMS vendors provide interfaces to manage database security. When you create a database Web interface, security can be implemented in the Web server, the database, and in the networking infrastructure. In many ways, building multiple firewalls is the essence of Internet database security.

At the Web server level, most Web clients and servers can perform secure transactions by using encryption routines at the TCP/IP protocol level. Clients and servers can exchange security certificates to ensure that the clients and servers are who they say they are. Therefore you must ensure that the clients and servers are properly registered and that they have compatible encryption protocols. Also, the Web administrator can use TCP/IP addresses and firewalls to restrict access to the site. The firewalls will ensure that only authorized data travel outside the company.

All RDBMS vendors provide security mechanisms at the database end, providing some form of login authentication for users who are trying to access the database. At the SQL level, administrators can use the GRANT and REVOKE commands to assign access restrictions to tables and/or to specific SQL commands.

Web-to-database middleware vendors usually have several security mechanisms available to interface with databases. For example, when using ODBC data sources, the administrator can restrict end-user access to certain SQL statements such as SELECT, UPDATE, INSERT, or DELETE, or to some combination of these commands. And, although the Web pages operate in the request-reply model, the use of Web interfaces does not preclude the creation of algorithms to guarantee data entity and referential integrity requirements. Data security measures must also include logs to relate data manipulation activities to specific end users. Such logs ensure that each database update is directly associated with an authorized user.

Security must also be extended to support electronic commerce, or e-commerce. Such support is key to the Web site's ability to execute secure business transactions over the Internet. If you want to be able to take credit card orders over the Internet, such order processing, rooted in a production database environment, must have strict security mechanisms in place to safeguard the transactions. In addition, the order transaction must be able to *securely* interact with multiple sites—such as distributors and banks—making sure that the transaction information cannot be modified and that the information cannot be stolen.

14.6.3 TRANSACTION MANAGEMENT

Although the preceding comments focus on transaction-management issues that must be addressed to successfully conduct e-commerce, the concept of database *transactions* is foreign to the Web. Remember that the Web's request-reply model means that the Web client and the Web server interact by using very short messages. These messages are limited to the request for and delivery of pages and their components. (Page components may include pictures, multimedia files, and so on.) The dilemma created by the Web's request-reply model is that:

- The Web cannot maintain an open line between the client and the database server.
- The mechanics of a recovery from incomplete or corrupted database transactions *require* that the client must maintain an open communications line with the database server.

Clearly, the creation of mission-critical Web applications mandates support for database transaction management capabilities. Given the just-described dilemma, *the designers must ensure proper transaction-management support at the database server level.*

Many Web-to-middleware products provide transaction-management support. For example, ColdFusion provides this support through the use of its CFTRANSACTION tag. If the transaction load is very high, this function can be assigned to an independent computer. By using this approach, the Web application and database servers are free to perform other tasks, and the overall transaction load is distributed among multiple processors.

14.6.4 DENORMALIZATION OF DATABASE TABLES

When the Web is used to interact with databases, the application design must take into account that the Web forms cannot use the multiple data entry lines that are typical of parent-child (1:M) relationships. Yet these 1:M relationships are crucial in e-commerce. For example, think of order and order line, or invoice and invoice line. Most end users are familiar with the conventional GUI entry forms that support multitable (parent-child) data entry through a multiple-component structure composed of a main form and a subform. Using such main-form/subform forms, the end user can enter multiple purchases associated with a single invoice. All the data entry is done on a single screen.

Unfortunately, the Web environment does not support this very common type of data-entry screen. As illustrated in the ColdFusion script examples, the Web can easily handle single-table data entry. However, when multitable data entries or updates are needed—such as order with order lines, invoice with invoice lines, and reservation with reservation lines—the Web falls short. Although implementing the parent/child data entry is not impossible in a Web environment, its final outcome is less than optimum, usually counter-intuitive, less user-friendly, and prone to errors.

To see how the Web developer might deal with the parent/child data entry, let's briefly examine how you might deal with the ORDER and ORDER_LINE relationship used to store customer orders. Using an applications middleware server such as ColdFusion to create a Web front end to update orders, one or more of the following techniques might be used:

- Design HTML frames to separate the screen into order header and detail lines. An additional frame would be used to provide status information or menu navigation.
- Use recursive calls to pages to refresh and display the latest items added to an order.

- Create temporary tables or server-side arrays to hold the child table data while in the data entry mode. This technique is usually based on the bottom-up approach in which the end user first selects the products to order. When the ordering sequence is completed, the order-specific data, such as customer ID, shipping information, and credit card details, are entered. Using this technique, the order detail data are stored in the temporary tables or arrays.

- Use stored procedures or triggers to move the data from the temporary table or array to the master tables.

Although the Web itself does not support the parent/child data entry directly, it is possible to resort to the Web programming languages, such as Java, JavaScript, or VBScript, to create the required Web interfaces. The price of this approach is a steeper application development learning curve and a need to hone programming skills. And, while this augmentation works, it also means that complete programs are stored outside the HTML code that is used in your Web site.

S U M M A R Y

In just a few years, the Internet has become the de facto standard for global communications. The Internet is a worldwide network of computers running the TCP/IP suite of network protocols. The World Wide Web, known as the Web, is one of the many services that run on the Internet; it provides a GUI to access HTML-formatted documents. At this point, the Internet is host to millions of company and organizational Web sites that provide many different types of services. MIS departments have taken note of the Internet as the new paradigm for systems development because it provides multiple benefits, such as universal data access, a common GUI interface, and lower end-user training costs.

The Web server provides standard interfaces such as the Common Gateway Interface (CGI) and the application programming interface (API). All Web servers support CGI, but CGI creates more processing overhead because it uses external programs that are loaded multiple times. Web APIs are closely integrated to the Web server program and can, therefore, achieve better performance. However, the Web APIs can cause Web servers to crash if not implemented properly. Another problem with Web APIs is that they are vendor- and operating-system-specific. Therefore, there are many Web APIs, such as ISAPI for Microsoft Web servers, NSAPI for Netscape Web servers, and WSAPI for O'Reilly's Web Site.

Database access through the Web is achieved through database-to-middleware software. This middleware software resides between the Web server and the resource to be accessed. Its main function is to provide Web access to other external services such as databases, e-mail systems, directory services, and FTP services.

ColdFusion is a Web application server that provides database access (among other services) to the Web. ColdFusion can access databases through ODBC or through native database drivers. ODBC provides a common layer of abstraction to multiple data sources. ColdFusion uses script files containing ColdFusion Markup Language (CFML) to access databases. The most common database-oriented CFML tags are: CFQUERY, CFOUTPOUT, CFTABLE, CFINSERT, and CFUPDATE. ColdFusion also provides support for conditional processing tags such as CFIF.

Developing applications for the Internet/intranets/extranets means overcoming the limitations of the Web interface. Database administrators in today's Internet-based environment must understand the implications of Web-to-database development in terms of extended data type support, database security, database transaction management, and database design.

When evaluating the Internet's contribution to the database environment, it is appropriate to point out that we have completed a circle. That is, we have replaced the old "big iron," room-sized mainframes with Web servers that run on inexpensive boxes with hundreds of times the processing power of the old "big iron" boxes. On the

client side, we have replaced the old dreaded "green" text-based terminal with inexpensive GUI-based computers that have processing power hundreds of times greater than the old mainframes. Yet, in spite of the tremendous strides we have made in the microcomputer arena, we have, in some fundamental respects, returned to the old mainframe model in which the processing is done and managed on the server side.

KEY TERMS

ActiveX

application programming
 interface (API)

Call Level Interface (CLI)

client-side extension

ColdFusion Markup
 Language (CFML)

Common Gateway Interface (CGI)

dynamic-link library (DLL)

Java

JavaScript

Open DataBase
 Connectivity (ODBC)

plug-in

script

server-side extension

stateless system

tag (ColdFusion or HTML)

VBScript

Web application server

Web-to-database middleware

REVIEW QUESTIONS

1. What does the statement "the Web is a stateless system" mean? What implications does a stateless system have for database applications developers?

2. What are Web server interfaces used for? Give some examples.

3. What is ODBC, and what is its function?

4. What is a Web application server, and how does it work from a database perspective?

5. What are scripts, and what is their function? (Think in terms of database applications development!)

6. Describe the basic services provided by the ColdFusion Web application server.

7. Search the Internet for Web application servers other than ColdFusion and prepare a short presentation to your class about one of them.

8. Briefly describe the function of the following ColdFusion tags: CFQUERY, CFOUTPUT, CFTABLE, CFIF, CFINSERT, and CFUPDATE.

9. Discuss what issues must be addressed by Web-to-database interfaces in terms of data types, security, database transaction management, and denormalization of database tables.

10. Describe the Web page development problems related to database parent/child relationships.

PROBLEMS

In the following exercises students are required to create ColdFusion scripts. The teacher can assign individual or multiple scripts to each student. When you create these scripts, include one main script to show the records and the main options (for a total of five scripts for each table). Consider and document foreign key and business rules when creating your scripts.

1. Create ColdFusion scripts to search, add, edit, and delete records for the USER table in the RobCor database.

2. Create ColdFusion scripts to search, add, edit, and delete records for the INVTYPE table in the RobCor database.

3. Create ColdFusion scripts to search, add, edit, and delete records for the VENDOR table in the RobCor database.

4. Modify the insert scripts (rc-5a.cfm and rc-5b.cfm) for the DEPARTMENT table so that the users who can be manager of a department are only those who belong to that department.

5. Create an Order data-entry screen using the ORDERS and ORDER_LINE tables in the RobCor database. To do this you could use frames and other advanced ColdFusion tags. Please consult the online manual and review the demo applications.

PART

VI

DATABASE ADMINISTRATION

15 DATABASE ADMINISTRATION

15

DATABASE ADMINISTRATION

In this chapter, you will learn:

- ❖ That data are a valuable business asset that requires careful management
- ❖ Why and how the database plays a critical role in an organization
- ❖ That the introduction of a DBMS has important technological, managerial, and cultural organizational consequences
- ❖ What the database administrator's managerial and technical roles are
- ❖ About several database administration tools
- ❖ About various database administration strategies
- ❖ How various database administration technical tasks are performed with Oracle

PREVIEW

This chapter shows you the basis for a successful database administration strategy. Such a strategy requires that data be considered important and valuable resources, to be treated and managed as corporate assets.

We will explore how a database fits within an organization, what data views and requirements exist at different management levels, and how the DBMS supports each of these views and requirements. Database administration must be fully understood and accepted within an organization before a sound data administration strategy can be implemented. We will examine important data management issues by looking at the managerial and technical roles of the database administrator (DBA).

The technical aspects of database administration will be augmented by a discussion of database administration tools and the corporate-wide data architectural framework. The managerial aspects of database administration will be explained by showing you how the database administration function fits within classical organizational structures. Because Oracle is the current leader in mid- to high-level corporate database markets, we will use the Oracle DBMS software to illustrate some of the technical aspects of the DBA's role in managing the database environment.

15.1 DATA AS A CORPORATE ASSET

In Chapter 1, "Database Systems," you learned that data are the raw material from which information is produced. Therefore, it is not surprising to learn that in today's information-driven environment, data are accepted as a valuable asset that requires careful management.

To assess data's monetary value, just take a look at what's stored in a company database: data about customers, suppliers, inventory, operations, and so on. How many opportunities will be lost if the data in the database are lost? What will be the actual cost of data loss? For example, an accounting firm whose entire database is lost when a computer virus damages the files on its hard disks would incur significant direct and indirect costs. The accounting firm's problems would be magnified if the data loss occurred during the tax season! Data loss puts any company in a difficult position. The company might be unable to handle daily operations effectively, it might be faced with the loss of customers who require quick and efficient service, and it might lose the opportunity to gain new customers.

Data are now well understood as a valuable *resource* to be translated into *information*. If the information is accurate and timely, its use is likely to trigger actions that enhance the company's competitive position and generate wealth. In effect, an organization is subject to a *data-information-decision cycle*: The data *user* applies intelligence to *data* to produce *information* that is the basis of *knowledge* used in *decision making* by the user. This cycle is illustrated in Figure 15.1.

FIGURE 15.1 THE DATA-INFORMATION-DECISION-MAKING CYCLE

As you examine the flow in Figure 15.1, keep in mind that the decisions made by high-level managers trigger actions within the organization's lower levels. Such actions produce additional data to be used for monitoring company performance. In turn, the additional data must be recycled within the data/information/decision framework. Thus, data form the basis for decision making, strategic planning, control, and operations monitoring.

A critical success factor of an organization is efficient asset management. To manage data as a corporate asset, managers must understand the value of information—that is, processed data. In fact, there are companies (for example, those that provide credit reports) whose only product is information and whose success is solely a function of information management.

15.2 THE NEED FOR AND ROLE OF DATABASES IN AN ORGANIZATION

Data are used by different people in different departments for different reasons. Therefore, data management must address the concept of shared data. Chapter 1 showed how the need for data sharing made the electronic DBMS almost inevitable. Used properly, the DBMS facilitates:

- *Interpretation* and *presentation* of data in useful formats, by transforming raw data into information.

- *Distribution* of data and information to the right people at the right time.

- Data *preservation* and data usage *monitoring* for adequate periods of time.

- *Control* over data duplication and use, both internally and externally.

Whatever the type of organization, the database's predominant role is *to support managerial decision making at all levels in the organization.*

An organization's managerial structure might be divided into three levels: top, middle, and operational. Top-level management makes strategic decisions, middle management makes tactical decisions, and operational management makes daily, operational decisions.

The DBMS must provide tools that give each level of management a different view of the data and support the required level of decision making. We can get a better fix on the database support roles by summarizing the activities typical of each management level.

- *Top management.* At this level, the database must be able to:

 1. Provide the information necessary for strategic decision making, strategic planning, policy formulation, and goals definition.

 2. Provide access to external and internal data to identify growth opportunities and to chart the direction of such growth. (Direction refers to the nature of the operations: Will we become a service organization, a manufacturing organization, or some combination of the two?)

 3. Provide a framework for defining and enforcing organizational policies. (Remember that such polices are translated into business rules at lower levels in the organization.)

 4. Improve the likelihood of a positive return on investment for the company, by searching for new ways to reduce costs and/or by boosting productivity.

 5. Provide feedback to monitor whether the company is achieving its goals.

- *Middle management.* At this level, the database must be able to:

 1. Deliver the data necessary for tactical decisions and planning.

 2. Monitor and control the allocation and use of company resources.

 a. Specifically, how efficiently are resources allocated and used? Remember that data, too, are a company resource, so the database must provide the necessary information to monitor and control the use of data. For example, the database must enable the information systems department to ensure that all the database components are properly integrated.

 b. What potential or current operational problems exist? How well does the database environment allow us to define and study alternate solutions? Such solutions often require scenario playing and detailed forecasting information.

 c. Evaluate the performance of the different departments.

 3. Provide a framework for enforcing and ensuring the security and privacy of the data in the database. **Security** means protecting the data against accidental or intentional use by unauthorized users. **Privacy** deals with the rights of individuals and the organization to determine the "who, what, when, where, and how" of data usage.

- *Operational management.* At this level, the database must be able to:

 1. Represent and support the company operations as closely as possible. The data model must be flexible enough to incorporate all required present and expected data.

 2. Produce query results within specified performance levels. Keep in mind that the performance requirements increase for lower levels of management and operations. Thus, the database must support fast responses to a greater number of transactions at the operational management level.

 3. Enhance the company's short-term operational ability by providing:

 a. Timely information for customer support.

 b. Support for application development and computer operations.

A general database objective is to provide a seamless flow of information throughout the company.

The company's database is also known as the corporate or enterprise database. The **enterprise database** might be defined as the company's data representation that provides support for all present and expected future operations. Most of today's successful organizations depend on the enterprise database to provide support for all their operations—from design to implementation to sales and services, and from strategic planning to daily decision making.

15.3 INTRODUCTION OF A DATABASE: SPECIAL CONSIDERATIONS

Having a computerized database management system does not guarantee that the data will be properly used to provide the best solutions required by managers. A DBMS is just a tool for managing data and must be used effectively in order to produce the desired results. To use an analogy: In the hands of a carpenter a hammer can help produce furniture; in the hands of a child it might just do damage. The solution to company problems is not the mere existence of a computer system or its database but, rather, its effective management and use.

The introduction of a DBMS represents a big change and challenge; throughout the organization it is likely to have a profound impact, which might be positive or negative, depending on how it is administered. For example, one key consideration is adapting the DBMS to the organization, rather than forcing the organization to adapt to the DBMS. The main issue should be the organization's needs rather than the DBMS's technical capabilities. However, the introduction of a DBMS cannot be accomplished without affecting the organization. Just the flood of new DBMS-generated information has a profound effect on the way the organization functions and, therefore, on its corporate culture.

The introduction of a DBMS into an organization has been described as a process that includes three important aspects:[1]

- *Technological:* DBMS software and hardware.

- *Managerial:* Administrative functions.

- *Cultural:* Corporate resistance to change.

The *technological* aspect includes selecting, installing, configuring, and monitoring the DBMS to make sure that it handles data storage, access, and security efficiently. The person(s) in charge of addressing the technological aspect of the DBMS installation must have the technical skills necessary to provide or secure adequate support for all the different users of the DBMS: programmers, managers, and end users. Therefore, database administration *staffing* is a key technological consideration in the DBMS introduction. The selected personnel must exhibit the right mix of technical and managerial skills to provide a smooth transition to the new shared-data environment.

1 John P. Murray, "The Managerial and Cultural Issues of a DBMS," *370/390 Database Management* 1(8), September 1991, pp. 32–33.

The *managerial* aspect of the DBMS introduction should not be taken lightly. A high-quality DBMS does not guarantee a high-quality information system, just as having the best race car does not guarantee winning a race, without the best drivers and mechanics.

The introduction of a DBMS into an organization requires careful planning to create an appropriate organizational structure to accommodate the person(s) responsible for administering the DBMS. The organizational structure must also be subject to well-developed monitoring and controlling functions. The administrative personnel must have excellent interpersonal and communications skills, combined with broad organizational and business understanding. Top management must be committed to the new system and must define and support the data administration functions, goals, and roles within the organization.

The *cultural* impact of the introduction of a database system must be assessed carefully. The DBMS's existence is likely to have an effect on people, functions, and interactions. For example, additional personnel might be added, new roles might be allocated to existing personnel, and employee performance might be evaluated by new standards.

A cultural impact is likely because the database approach creates a more controlled and structured information flow. Department managers, who are used to handling their own data, must surrender their subjective ownership to the data administration function and must share such data with the rest of the company. Application programmers must learn and follow new design and development standards. Managers might be faced with what they consider to be an information overload and might require some time to adjust to the new environment.

When the new database comes online, people might be reluctant to use the information provided by the system and might question its value or accuracy. (Many will be surprised and possibly chagrined to discover that the information does not fit preconceived notions and strongly held beliefs!) The database administration department must be prepared to open its doors to end users, listen to their concerns, act on them when possible, and educate end users about the system's uses and benefits.

15.4 THE EVOLUTION OF THE DATABASE ADMINISTRATION FUNCTION

Data administration has its roots in the old, decentralized world of the file system. The cost of data and managerial duplication in such file systems gave rise to a centralized data administration function known as the **electronic data processing (EDP)** or **data processing (DP) department**. The DP department's task was to pool all computer resources to support all departments *at the operational level* with a minimal duplication of data and effort. The DP administration function was given the authority to manage all existing company file systems as well as resolve data and managerial conflicts created by the duplication and/or misuse of data.

The advent of the DBMS and its shared view of data produced a new level of data management sophistication and led the DP department to evolve into an **information systems (IS) department**. Its responsibilities were broadened to include:

- A *service* function to provide end users with active data management support.

- A *production* function to provide end users with specific solutions for their information needs through integrated application or management information systems.

The functional orientation of the IS department was reflected in its internal organizational structure. IS departments typically were structured as shown in Figure 15.2. As the demand for application development grew, the IS application development segment was subdivided by the type of supported system: accounting, inventory, marketing, and so on. However, this development meant that the database administration responsibilities were divided. The application development segment was in charge of gathering database requirements and logical database design, whereas the database operations segment took charge of implementing, monitoring, and controlling the DBMS operations.

FIGURE 15.2 THE IS DEPARTMENT INTERNAL ORGANIZATION

As the number of database applications grew, data management became an increasingly complex job, thus leading to the development of the database administration function. The person responsible for the control of the centralized and shared database was the **database administrator (DBA)**.

The size and role of the DBA function varies from company to company, as does its placement within a company's organizational structure. On the organization chart, the DBA function might be defined as either a staff or line position. Placing the DBA function in a staff position often creates a consulting environment, in which the DBA is able to devise the data administration strategy but does not have the authority to enforce it or to resolve possible conflicts.[2]

The DBA function in a line position has both the responsibility and the authority to plan, define, implement, and enforce the policies, standards, and procedures used in the data administration activity. We have illustrated the two possible DBA function placements in Figure 15.3.

The decision concerning DBA placement depends on the company's management style and on factors such as the size and complexity of company operations and the geographical distribution of company facilities. These factors also help determine the DBA function's internal structure.

There does not appear to be a DBA internal organization standard. In part, standards might be lacking because the DBA function itself is probably the most dynamic of any organization's functions. In fact, the fast-paced changes in DBMS technology dictate changing organizational styles. For example:

- The development of distributed databases can force an organization to decentralize the data administration function further. The distributed database requires the system DBA to define and delegate the responsibilities of each local DBA, thus imposing new and more complex *coordinating* activities on the system DBA.

- The growing use of Internet-ready and object oriented databases and the growing number of data warehousing applications are very likely to add to the DBA's data modeling and design activities, thus expanding and diversifying the DBA's job.

[2] For a historical perspective on the development of the DBA function and a broader coverage of its organizational placement alternatives, we recommend Jay-Louise Weldon's classic *Data Base Administration* (New York, Plenum Press, 1981). Although you may think that the book's publication date renders it obsolete, a surprising number of its topics are returning to the current operational database scene. (For example, a "thin" client in the client/server architecture reflects many of the "outdated" mainframe concepts, and the object model's conceptual structure incorporates many ideas that are traceable to the hierarchical database concepts.) Perhaps it is worthwhile to remember an old dictum: The more things change, the more they stay the same. Whatever you do today is likely to be built on some historical precedent, so you ignore history at your own peril!

FIGURE 15.3 THE PLACEMENT OF THE DBA FUNCTION

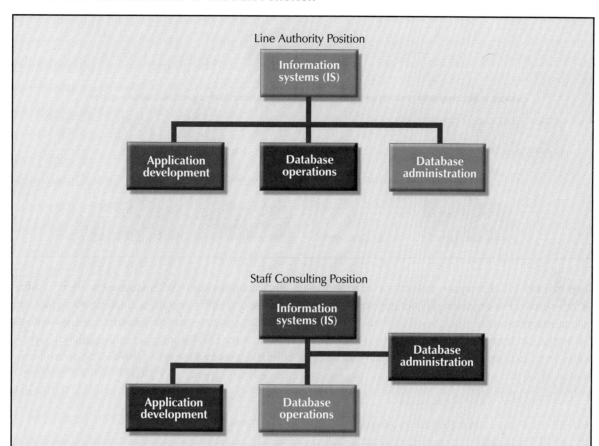

- The increasing sophistication and power of microcomputer-based DBMS packages provide an easy platform for the development of user-friendly, cost-effective, and efficient solutions to specific departmental information needs. But such an environment also invites data duplication, not to mention the problems created by people who lack the technical qualifications to produce good database designs. In short, the new microcomputer environment requires the DBA to develop a new set of technical and managerial skills.

Although no current standard exists, it is common practice to define the DBA function by dividing the DBA operations according to the Database Life Cycle (DBLC) phases. If this approach is used, the DBA function requires personnel to cover the following activities:

- Database planning, including the definition of standards, procedures, and enforcement.
- Database requirements gathering and conceptual design.
- Database logical design and transaction design.
- Database physical design and implementation.
- Database testing and debugging.
- Database operations and maintenance, including installation, conversion, and migration.
- Database training and support.

Figure 15.4 represents an appropriate DBA functional organization according to this model.

FIGURE 15.4 A DBA FUNCTIONAL ORGANIZATION

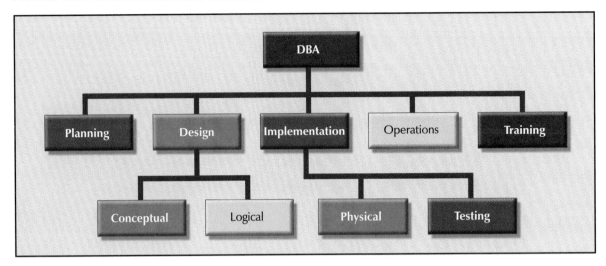

Keep in mind that a company might have several different and incompatible DBMSs installed to support different operations. For example, it is not uncommon to find corporations with a hierarchical DBMS to support the daily transactions at the operational level and a relational database to support middle and top management's ad hoc information needs. There can also be a variety of microcomputer DBMSs installed in the different departments. In such an environment, the company might have one DBA assigned for each DBMS. The general coordinator of all DBAs is sometimes known as the **systems administrator (SYSADM)**, whose position is illustrated in Figure 15.5.

FIGURE 15.5 MULTIPLE DATABASE ADMINISTRATORS IN AN ORGANIZATION

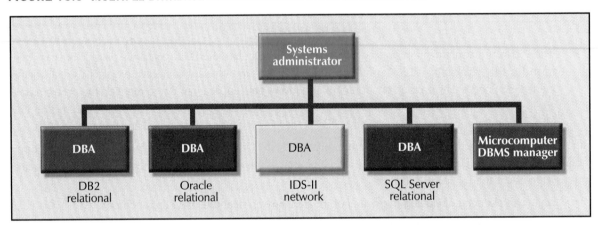

There is a growing trend toward specialization in the data management function. For example, the organization charts used by some of the larger corporations make a distinction between a DBA and the **data administrator (DA)**. The DA, also known as the **information resource manager (IRM),** usually reports directly to top management and is given a higher degree of responsibility and authority than the DBA, although their roles tend to overlap to some extent.

The DA is responsible for controlling the overall corporate data resources, both computerized and manual. The DA's job description thus covers a larger area of operations than that of the DBA, because the DA is in charge of controlling not only the computerized data but also the data outside the scope of the DBMS. The placement of the DBA within the expanded organizational structure may vary from company to company. Depending on this structure's components, the DBA might report to the DA, the IRM, the IS manager, or directly to the company's CEO.

A substantial portion of this book is devoted to relational database design and implementation. We have also examined DBMS features and characteristics. In short, we have thus far focused on the very important technical aspects of the database. However, there is another important side of the database coin waiting to be examined: even the most carefully crafted database system cannot operate without the human component. So, we must now find out how people perform the data administration activities that make a good database design truly useful.

Effective data administration requires both technical and managerial skills. In the previous section you learned that if an organization employs both DBAs and DAs, their functions tend to overlap somewhat. Nevertheless, we can make some useful generalizations. For example, the DA's job typically has a strong managerial orientation with company-wide scope. In contrast, the DBA's job tends to be more technically oriented and has a narrower, DBMS-specific scope. However, the DBA, too, must have a considerable store of people skills. After all, both the DA and the DBA perform "people" functions common to all departments in an organization. For example, both DA and DBA direct and control personnel staffing and training within their respective departments. Naturally, however, we intend to focus on the data aspects of their managerial roles.

Table 15.1 contrasts the general characteristics of both positions by summarizing the typical DA and DBA activities. All activities flowing from the characteristics shown in Table 15.1 are invested in the DBA if the organization does not employ both a DA and a DBA.

TABLE 15.1 CONTRASTING DA AND DBA ACTIVITIES AND CHARACTERISTICS

DATA ADMINISTRATOR (DA)	DATABASE ADMINISTRATOR (DBA)
Strategic planning	Control and supervision
Sets long-term goals	Executes plans to reach goals
Sets policies and standards	Enforces policies and procedures Enforces programming standards
Broad scope	Narrow scope
Long term	Short term (focus on daily operations)
Managerial orientation	Technical orientation
DBMS-independent	DBMS-specific

As you examine Table 15.1, note that the DA is responsible for providing a global and comprehensive administrative strategy for all the organization's data. In other words, the DA's data administration plans must always consider the entire data spectrum. The DA is thus responsible for the consolidation and consistency of both manual and computerized data.

The DA also must set data administration goals. These goals are defined by such issues as:

- Data "sharability" and time availability.
- Data consistency and integrity.
- Data security and privacy.
- Extent and character of data use.

Naturally, this list can be expanded to fit the organization's specific data needs. Regardless of how data management is conducted—and despite the fact that much authority is invested in the DA or DBA to define and control the way company data are used—the DA and DBA do not own the data. Instead, DA and DBA functions are defined to emphasize data as a shareable company asset.

The preceding discussion should not lead you to believe that there are universally accepted DA and DBA administrative standards. As a matter of fact, the style, duties, organizational placement, and internal structure of both functions vary from company to company. For example, many companies distribute DA duties between the DBA and the manager of information systems. For simplicity and to avoid confusion, we will use the label DBA as a general title that encompasses all appropriate data administration functions. Having made this point, we will move ahead to illustrate the DBA's role as an arbitrator between data and users.

The arbitration of interactions between the two most important assets of any organization, people and data, places the DBA in the dynamic environment portrayed in Figure 15.6.

FIGURE 15.6 A SUMMARY OF DBA ACTIVITIES

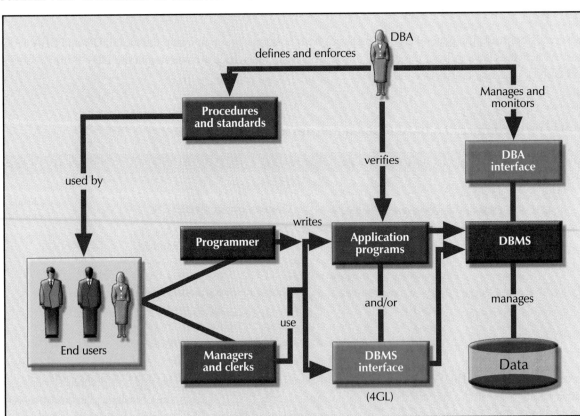

As you examine Figure 15.6, note that the DBA is the focal point for data/user interaction. The DBA defines and enforces the procedures and standards to be used by programmers and end users during their work with the DBMS. The DBA also verifies that programmer and end-user access meets the required quality and security standards.

Database users might be classified by the:

- Type of decision-making support (operational, tactical, or strategic) required.
- Degree of computer knowledge (novice, proficient, or expert).
- Frequency of access (casual, periodic, or frequent).

These classifications are not exclusive and usually overlap. For example, an operational user can be an expert with casual database access. Nevertheless, we might describe a typical top-level manager as a strategic novice user, with periodic database access. On the other hand, a database application programmer is an operational expert and frequent database user. Each organization thus employs people whose levels of database expertise span an entire spectrum. The DBA must be able to interact with all these people, understand their quite different needs, answer questions at all levels of the expertise scale, and communicate effectively.

The DBA activities portrayed in Figure 15.6 suggest the need for a diverse mix of skills. In large companies, such skills are likely to be distributed among several persons who work within the DBA function. In small companies, these skills may be the domain of just one individual. DBA skills are broad and varied, yet they are closely related to the nature of the job. We will classify these skills into two different categories: managerial and technical. Our classification is summarized in Table 15.2.

TABLE 15.2 DESIRED DBA SKILLS

MANAGERIAL	TECHNICAL
Broad business understanding	Broad data-processing background
Coordination skills	Systems development life cycle knowledge
Analytical skills	Structured methodologies: Data flow diagrams Structure charts Programming languages
Conflict resolution skills	Database life cycle knowledge
Communications skills (oral and written)	Database modeling and design skills Conceptual Logical Physical
Negotiation skills	Operational skills: database implementation, data dictionary management, security, etc.
Experience: 2–5 years in a large DP department	

As you examine Table 15.2, keep in mind that the DBA must perform two distinct roles. The DBA's managerial role is focused on personnel management and on interactions with the end-user community. The DBA's technical role covers activities involving the use of the DBMS—database design, development, and implementation—as well as the production, development, and use of application programs. We will examine the DBA's managerial and technical roles in greater detail in the following sections.

15.5.1 THE DBA'S MANAGERIAL ROLE

As a manager, the DBA must concentrate on the control and planning dimensions of the database administration function. Therefore, the DBA is responsible for:

- Coordinating, monitoring, and allocating database administration resources: people and data.
- Defining goals and formulating strategic plans for the database administration function.

More specifically, we can summarize the DBA's responsibilities as shown in Table 15.3.

TABLE 15.3 DBA ACTIVITIES AND SERVICES

DBA ACTIVITY			DBA SERVICE
Planning			End-user support
Organizing			Policies, procedures, and standards
Testing	← of →		Data security, privacy, and integrity
Monitoring			Data backup and recovery
Delivering			Data distribution and use

Table 15.3 illustrates that the DBA is generally responsible for planning, organizing, testing, monitoring, and delivering quite a few services. These services might be performed by the DBA or, more likely, by the DBA's personnel. To simplify the discussion and to make its points more broadly applicable, we will not make a distinction between the services rendered by the DBA and those provided by other IS personnel under the DBA's control. Instead, we will use the label DBA to encompass all service providers within the database administration function. So, within the confines of our DBA label, let's examine the DBA-provided services in greater detail.

End-User Support

The DBA interacts with the end user by providing data and information support services to the organization's departments. Because end users usually have very dissimilar computer backgrounds, end-user support services include:

- *User-requirements gathering.* The DBA must work within the end-user community to help gather data required to identify and describe the end users' problems. The DBA's communications skills are at a premium at this stage because the DBA works closely with persons who tend to have very different computer backgrounds and communication styles. User-requirements gathering requires the DBA to:

 - Develop a precise understanding of the users' views and needs. If the DBA does not develop a clear picture of the problem, the DBMS cannot deliver an appropriate and efficient solution. Remember, bad input produces bad output.

 - Identify present and future information needs. The end user can see only his or her portion of the information picture. In effect, the end user sees only the proverbial tip of the iceberg. In contrast to the end user, the DBA is in a position to see the organization's entire data schema.

 A good measure of overall data administration success is provided by how well all end-user information needs are met. If the DBA is proactive in anticipating problems and future end-user information needs, (s)he will create a positive climate for increasing end-user trust in the effectiveness of the DBA function.

- *Conflict and problem resolution.* The solution of end-user problems in one department might trigger conflicts with other departments. End users are typically concerned with their own specific data needs rather than with those of others, and they are not likely to consider how their data affect other departments within the organization. When data/information conflicts arise, the DBA function has the authority and responsibility to resolve them.

- *Finding solutions to information needs.* The ability and authority to resolve data conflicts enable the DBA to develop solutions that will properly fit within the existing data management framework. The DBA's primary objective is to provide solutions to the end users' information needs. Given the growing importance of the Internet, such solutions are likely to require the development and management of Web browsers to interface with the databases. In fact, the explosive growth of e-commerce requires the use of *dynamic* interfaces to facilitate interactive product queries and product sales. (See Chapter 13, "Databases in Electronic Commerce.")

- *Ensuring quality and integrity of applications and data.* Once a proper solution has been found, it must be properly implemented and used. Therefore, the DBA must work with both application programmers and end users to teach them the database standards and procedures required for data access and manipulation. The DBA must also make sure that the database transactions do not adversely affect the database's data quality. Certifying the quality of the application programs that access the database is crucial for the DBA function. Special attention must be given to the DBMS Internet interfaces, because such interfaces do not provide the transaction management features that are typically found in the DBMS-managed database environment. For example, if an "internal" application-based database transaction is required to fire a trigger, that trigger must also be fired when that transaction is generated via the Internet interface.

- *Building end-user confidence.* Finding adequate solutions to end users' problems increases end-user trust and confidence in the DBA function.

- *Managing the training and support of DBMS users.* One of the most time-consuming DBA activities is teaching end users how to use the database properly. The DBA must ensure that all users accessing the database have a basic understanding of the functions and the use of the DBMS software. The DBA coordinates and monitors all activities concerning end-user education.

Policies, Procedures, and Standards

A prime component of a successful data administration strategy is the continuous enforcement of the policies, procedures, and standards for correct data creation, usage, distribution, and deletion within the database. The DBA must define, document, and communicate the policies, procedures, and standards before they can be enforced. Basically:

- **Policies** are general statements of direction or action that communicate and support DBA goals.

- **Standards** are more detailed and specific than policies and describe the minimum requirements of a given DBA activity. In effect, standards are rules that are used to evaluate the quality of the activity. For example, standards define how application programs must be structured, the naming conventions to be used by programmers, and so on.

- **Procedures** are written instructions that describe a series of steps to be followed during the performance of a given activity. Procedures must be developed within existing working conditions, and they must support and enhance that environment.

To illustrate the distinctions among policies, standards, and procedures, let's look at the following examples:

Policies

All users must have passwords.

Passwords must be changed every six months.

Standards

A password must have a minimum of five characters.

A password may have a maximum of 12 characters.

Social Security numbers, names, and birth dates cannot be used as passwords.

Procedures

To create a password:

1. The end user sends a written request for an account creation to the DBA.

2. The DBA approves the request and forwards it to the computer operator.

3. The computer operator creates the account, assigns a temporary password, and sends the account information to the end user. A copy of the account information is sent to the DBA.

4. The user changes the temporary password to a permanent one.

Standards and procedures defined by the DBA are used by all end users who want to benefit from the database. Standards and procedures must complement each other and must constitute an extension of data administration policies. Procedures must facilitate the work of end users and the DBA. The DBA must define, communicate, and enforce procedures that cover areas such as:

- *End-user database requirements gathering.* What documentation is required? What forms must be used?

- *Database design and modeling.* What database design methodology is to be used (normalization, object-oriented methodology, etc.)? What tools are to be used (CASE tools, data dictionaries, ER diagrams, and so on)?

- *Documentation and naming conventions.* Must be used in the definition of all data elements, sets, and programs that access the database.

- *Design, coding, and testing of database application programs.* The DBA must define the standards for application program coding, documentation, and testing. The DBA standards and procedures are given to the application programmers, and the DBA must enforce such standards.

- *Database software selection.* The selection of the DBMS package and any other software related to the database must be properly managed. For example, the DBA might require that software be properly interfaced with existing software, that it have the features needed by the organization, and that it provide a positive return on investment. In today's Internet environment, the DBA must also work with Web administrators to find proper Web-to-database connectivity solutions.

- *Database security and integrity.* The DBA must define the policies governing security and integrity. Database security is especially crucial. Security standards must be clearly defined and strictly enforced. Security procedures must be designed to handle a multitude of security scenarios to ensure that security problems are minimized. Although no system can ever be completely secure, security procedures must be designed to meet critical standards. The growing use of Internet interfaces to databases opens the door to new security threats that are far more complex and more difficult to manage than those encountered with more traditional internally generated and controlled interfaces. Therefore, the DBA must work closely with Internet security specialists to ensure that the databases are properly protected from external attacks launched—either inadvertently or deliberately—by unauthorized users.

- *Database backup and recovery.* Database backup and recovery procedures must include the information necessary to guarantee proper execution and management of the backups.

- *Database maintenance and operation.* The DBMS's daily operations must be clearly documented. Operators must keep job logs, and they must write operator instructions and notes. Such notes are helpful in pinpointing the causes and solutions of problems. Operational procedures must also include precise information concerning backup and recovery procedures.

- *End-user training.* A full-featured training program must be established within the organization, and procedures governing such training must be clearly specified. The objective is to clearly indicate who does what, when, and how. Each end user must be aware of the type and extent of the available training methodology.

Procedures and standards must be revised at least annually to keep them up to date. Periodic revision of the standards and procedures ensures that the organization can adapt quickly to changes in the work environment caused by internal or external forces. Naturally, the introduction of new DBMS software, the discovery of security or integrity violations, the reorganization of the company, and similar changes require revision of the procedures and standards.

Data Security, Privacy, and Integrity

The security, privacy, and integrity of the data in the database are of great concern to DBAs who manage current DBMS installations. Technology has pointed the way to greater productivity through information management. Technology has also made it possible to distribute data across multiple sites, thus making it more difficult to maintain data control, security, and integrity. The multiple-site data configuration has made it imperative that the DBA use the security and integrity mechanisms provided by the DBMS to enforce the database administration policies defined in the previous section. In addition, DBAs must team up with Internet security experts to build firewalls, proxy services, and other security mechanisms to safeguard data from possible external attacks.

Protecting the security and privacy of the data in the database is a function of authorization management. **Authorization management** defines procedures to protect and guarantee database security and integrity. These procedures include, but are not limited to, user access management, view definition, DBMS utility access control, and DBMS usage monitoring.

- *User access management.* This function is designed to limit access to the database and likely includes at least the following procedures:

 - *Define each user to the database.* This is achieved at two levels: at the operating system level and at the DBMS level. At the operating system level, the DBA can request the creation of a logon user ID that allows the end user to log on to the computer system. At the DBMS level, the DBA can either create a different user ID or employ the same user ID to authorize the end user to access the DBMS.

 - *Assign passwords to each user.* This, too, can be done at both operating system and DBMS levels. The database passwords can be assigned with predetermined expiration dates. The use of expiration dates enables the DBA to screen end users periodically, and remind users to periodically change their passwords, thus making unauthorized access less probable.

 - *Define user groups.* Classifying users into user groups, according to common access needs, facilitates the DBA's job of controlling and managing the access privileges of individual users.

 - *Assign access privileges.* The DBA assigns access privileges or access rights to specific users to access specified databases. An access privilege describes the type of authorized access. For example, access rights may be limited to read-only, or the authorized access may include READ, WRITE, and DELETE privileges. Access privileges in relational databases are assigned through SQL GRANT and REVOKE commands. (See Chapter 8, "Database Design" and Appendix E, "The University Lab: Conceptual Design Verification, Logical Design, and Implementation," for examples of such commands.)

○ ADDITIONAL CD CONTENT

Appendix E is located on the student CD in the \Appendixes\Appendix E folder.

 - *Physical access control.* Physical security can prevent unauthorized users from directly accessing the DBMS installation and facilities. Some common physical security practices found in large database installations include: secured entrances, password-protected workstations, electronic personnel badges, closed-circuit video, and voice recognition technology.

- *View definition.* The DBA must define data views to protect and control the scope of the data that are accessible to an authorized user. The DBMS must provide the tools that allow the definition of views that are composed of one or more tables and the assignment of access rights to a user or a group of users. The SQL command CREATE VIEW is used in relational databases to define views.

- *DBMS utilities access control.* Database access can be controlled by placing limits on the use of the DBMS's query and reporting tools. The DBA must make sure that such tools are used properly and only by authorized personnel.

- *DBMS usage monitoring.* The DBA must also audit the use of the data in the database. Several DBMS packages contain features that allow the creation of an **audit log**, which automatically records a brief description of the database operations performed by all users. Such audit trails enable the DBA to pinpoint access violations. The audit trails can be tailored to record all database accesses or just failed database accesses.

> **N O T E**
>
> James Martin provides an excellent enumeration and description of the desirable attributes of a database security strategy that remains relevant today (James Martin, *Managing the Database Environment,* Englewood Cliffs, NJ: Prentice-Hall, 1977). Martin's security strategy is based on "the seven essentials of database security" and may be summarized as one in which:
>
Data are	Users are
> | Protected | Identifiable |
> | Reconstructable | Authorized |
> | Auditable | Monitored |
> | Tamperproof | |

Security breaches can yield a database whose integrity is either preserved or corrupted:

- *Preserved:* Action is required to avoid the repetition of similar security problems, but data recovery may not be necessary. As a matter of fact, most security violations are produced by unauthorized and unnoticed access for information purposes, and such snooping does not disrupt the database.

- *Corrupted:* Action is required to avoid the repetition of similar security problems, and the database must be recovered to a consistent state. Corrupting security breaches include database access by computer viruses and by hackers whose actions are designed to destroy or alter data.

The integrity of a database might be lost due to external factors beyond the DBA's control. For example, the database might be damaged or destroyed by an explosion, a fire, or an earthquake. Whatever the reason, the specter of database corruption or destruction makes backup and recovery procedures crucial to any DBA.

Data Backup and Recovery

We started this chapter by pointing out that data have value and must be managed like any other company asset. When data are not readily available, companies face potentially ruinous losses. Therefore, data backup and recovery procedures are critical in all database installations, and the DBA must ensure that the data in the database can be fully recovered in case of either physical data loss or loss of database integrity.

Data loss can be partial or total. A partial loss can be caused by the physical loss of part of the database or if part of the database has lost integrity. A total loss might mean that the database continues to exist but that its integrity is entirely lost—or that the entire database is physically lost. In any case, backup and recovery procedures are the cheapest database insurance you can buy.

The management of database security, integrity, backup, and recovery is so critical that many DBA departments have created a position staffed by the **database security officer (DSO)**. The DSO's sole job is to ensure database security and integrity. In large database shops, the DSO's activities are often classified as *disaster management.*

Disaster management includes all the DBA activities designed to secure data availability following a physical disaster or a database integrity failure. Disaster management includes all planning, organizing, and testing of database contingency plans and recovery procedures. The backup and recovery measures must include at least:

- *Periodic data and applications backups.* Some DBMSs include tools to ensure backup and recovery of the data in the database. The DBA should use these tools to render the backup and recovery tasks automatic. Products such as IBM's DB2 allow the creation of different backup types: full, incremental, and concurrent. (A **full backup**, also known as a **database dump**, produces a complete copy of the entire database. **Incremental backup** produces a backup of all data since the last backup date; **concurrent backups** take place while the user is working on the database.)

- *Proper backup identification.* Backups must be clearly identified through detailed descriptions and date information, thus enabling the DBA to ensure that the correct backups are used to recover the database. The most common backup medium is tape; the storage and labeling of such tapes must be done diligently by the computer operators, and the DBA must keep track of tape currency and location. (Other emerging backup/storage solutions include optical and disk-based backup devices. However, organizations that are large enough to hire a DBA do not typically use CDs and DVDs for enterprise backup.)

- *Convenient and safe backup storage.* There must be multiple backups of the same data, and each backup copy must be stored in a different location. The storage locations must include sites inside and outside the organization. (Keeping different backups in the same place defeats the purpose of having the multiple backups in the first place.) The storage locations must be properly prepared and may include fire-safe and quakeproof vaults, as well as humidity and temperature controls. The DBA must establish a policy to respond to two questions: (1) Where are the backups to be stored? and (2) How long are backups to be stored?

- *Physical protection of both hardware and software.* Protection might include the use of closed installations with restricted access, as well as the preparation of the computer sites to provide air conditioning, backup power, and fire protection. Physical protection also includes the provision of a backup computer and DBMS that may be used in case of emergency.

- *Personal access control to the software of a database installation.* Multilevel passwords and privileges, and hardware and software challenge/response tokens can be used to properly identify authorized users of resources.

- *Insurance coverage for the data in the database.* The DBA or security officer must secure an insurance policy to provide financial protection in the event of a database failure. Such insurance might be expensive, but it is less expensive than the disaster created by massive data loss.

Two additional points are worth making:

1. Data recovery and contingency plans must be thoroughly tested and evaluated, and they must be practiced frequently. So-called fire drills are not to be disparaged, and they require top-level management support and enforcement.

2. A backup and recovery program is not likely to cover all components of an information system. Therefore, it is appropriate to establish priorities concerning the nature and extent of the data recovery process.

Data Distribution and Use

Data are useful only when they reach the right users in a timely fashion. The DBA is responsible for ensuring that the data are distributed to the right persons, at the right time, and in the right format. The DBA's data distribution and use tasks can become very time-consuming, especially if the data delivery capacity is based on a typical applications programming environment, where users depend on programmers to deliver the programs to access the data in the database. Although the Internet and its intranet and extranet extensions have opened databases to corporate users, their use has also created a new set of challenges for the DBA.

Current data distribution philosophy makes it easier for *authorized* end users to access the database. One way to accomplish this task is to facilitate the use of a new generation of more sophisticated query tools and the new Internet Web front ends. These enable the DBA to educate end users to produce the required information without being dependent on applications programmers. Naturally, the DBA must ensure that appropriate standards and procedures are adhered to.

This distribution philosophy is common today, and it is likely that it will become more common as database technology marches on. Such an environment is more flexible for the end user. Clearly, enabling end users to become relatively self-sufficient in the acquisition and use of data can lead to the more efficient use of data in the decision process. Yet this "data democracy" can also produce some troublesome side effects. By letting end users micromanage their data subsets, we could inadvertently sever the connection between those users and the data administration function. The DBA's job under these circumstances might become sufficiently complicated to compromise the data administration function's efficiency. Data duplication might flourish again without checks at the organizational level to ensure the uniqueness of data elements. Thus, end users who do not completely understand the nature and sources of data might make improper use of the data elements.

15.5.2 THE DBA'S TECHNICAL ROLE

The DBA's technical role requires a broad understanding of DBMS functions, configuration, programming languages, data modeling and design methodologies, and other DBMS-related issues. For example, the DBA's technical activities include the selection, installation, operation, maintenance, and upgrading of the DBMS and utility software, as well as the design, development, implementation, and maintenance of the application programs that interact with the database.

Many of the DBA's technical activities are a logical extension of the DBA's managerial activities. For example, the DBA deals with database security and integrity, backup and recovery, training and support. Thus, the DBA's dual role might be conceptualized as a capsule whose technical core is covered by a clear managerial shell.

The technical aspects of the DBA's job are rooted in the following areas of operation:

- DBMS and utilities evaluation, selection, and installation.
- Design and implementation of databases and applications.
- Testing and evaluation of databases and applications.
- Operation of the DBMS, utilities, and applications.
- Training and supporting users.
- Maintenance of the DBMS, utilities, and applications.

The following sections explore the details of these operational areas.

DBMS and Utilities Evaluation, Selection, and Installation

One of the DBA's first and most important technical responsibilities is the selection of the database management system, utility software, and supporting hardware to be used in the organization. Therefore, the DBA must develop and execute an evaluation and acquisition plan for selecting the DBMS, utilities, and hardware. The evaluation and acquisition plan must be based primarily on the organization's needs, rather than on specific software and hardware features. The DBA must recognize that the search is for solutions to problems rather than for a computer or DBMS software. Put quite simply, a DBMS is a management tool and not a technological toy.

The evaluation and acquisition plan's first and most important step is to determine company needs. To establish a clear picture of these needs, the DBA must make sure that the entire end-user community, including top- and mid-level managers, is involved in this process. Once the needs are identified, the objectives of the data administration function can be clearly established, and the DBMS features and selection criteria may be defined.

To match DBMS capability to the organization's needs, the DBA would be wise to develop a checklist of desired DBMS features. Such a DBMS checklist should address at least these issues:

- *DBMS model.* Are the company's needs better served by a relational, object oriented, or object/relational DBMS? If this is a data warehouse application, should a relational or multidimensional DBMS be used?

- *DBMS storage capacity.* What maximum disk and database size is required? How many disk packages must be supported? How many tape units should be acquired?

- *Application development support.* Which 3GLs and 4GLs are supported? What application development tools (database schema design, data dictionary, performance monitoring, screen and menu painters) are available? Are end-user query tools provided? Does the DBMS provide Web front-end access?

- *Security and integrity.* Does the DBMS support referential and entity integrity rules, access rights, and so on? Does the DBMS support the use of audit trails to spot errors and security violations? Can the audit trail size be modified?

- *Backup and recovery.* Does the DBMS provide some automated backup and recovery tools? Does the DBMS support tape or optical disc-based backups? Does the DBMS automatically back up the transaction logs?

- *Concurrency control.* Does the DBMS support multiple users? What levels of isolation (table, page, row) does the DBMS offer? How much manual coding is needed in the application programs?

- *Performance.* How many transactions per second does the DBMS support? Are additional transaction processors needed?

- *Database administration tools.* Does the DBMS offer some type of DBA management interface? What type of information does the DBA interface provide? Does the DBMS provide alerts to the DBA when errors or security violations occur?

- *Interoperability and data distribution.* Can the DBMS work with other DBMS types in the same environment? What coexistence or interoperability level is achieved? Does the DBMS support READ and WRITE operations to and from other DBMS packages? Does the DBMS support a client/server architecture?

- *Portability and standards.* Can the DBMS run on different operating systems and platforms? Can the DBMS run on mainframes, minicomputers, and personal computers? Can the DBMS applications run without modification on all platforms? What national and industry standards does the DBMS follow?

- *Hardware.* What hardware does the DBMS require?

- *Data dictionary.* Does the DBMS have a data dictionary? If so, what information is kept in it? Does the DBMS interface with any data dictionary tool? Does the DBMS support any CASE tools?

- *Vendor training and support.* Does the vendor offer in-house training? What type and level of support does the vendor provide? Is the DBMS documentation easy to read and helpful? What is the vendor's upgrade policy?

- *Available third-party tools.* What additional tools are offered by third-party vendors (query tools, data dictionary, access management and control, storage allocation management tools, etc.)?

- *Cost.* What are the costs involved in the acquisition of the software and hardware? How many additional personnel are required, and what level of expertise is required of them? What are the recurring costs? What is the expected payback period?

Pros and cons of several alternate solutions must be evaluated during the selection process. Available alternatives are often restricted because software must be compatible with the organization's existing computer system. Remember that a DBMS is just part of the solution; it requires support from collateral hardware, application software, and utility programs. For example, the DBMS's use is likely to be constrained by the available CPU, a front-end processor, auxiliary storage devices, data communication devices, the operating system, a transaction processor system, and so on. The costs associated with such hardware and software components must be included in the estimations.

The selection process must also consider the site's preparation costs. For example, the DBA must include both one-time and recurring expenditures involved in the preparation and maintenance of the computer room installations.

The DBA must supervise the installation of all software and hardware designated to support the data administration strategy; must have a thorough understanding of the components being installed; and must be familiar with the installation, configuration, and startup procedures of such components. The installation procedures include details such as the location of backup and transaction log files, network configuration information, physical storage details, and so on.

Keep in mind that installation and configuration details are DBMS-dependent. Therefore, such details cannot be addressed in this book. You should consult the installation and configuration sections of your system's DBMS administration guide.

Design and Implementation of Databases and Applications

The DBA function also provides data modeling and design services to the end-user community. Such services are often coordinated with an application development group within the data-processing department. Therefore, one of the primary activities within the database administration function is to determine and enforce standards and procedures to be used. Once the appropriate standards and procedures framework is in place, the DBA must ensure that the database modeling and design activities are performed within this framework. The DBA then provides the necessary assistance and support during the design of the database at the conceptual, logical, and physical levels. (Remember that the conceptual design is both DBMS- and hardware-independent, the logical design is DBMS-dependent and hardware-independent, and the physical design is both DBMS- and hardware-dependent.)

The DBA function usually requires that several people be dedicated to database modeling and design activities. Such people might be grouped according to the organizational areas covered by the application. For example, database modeling and design personnel may be assigned to production systems, financial and managerial systems, executive and decision support systems, and so on. The DBA schedules the design jobs to coordinate the data design and modeling activities. Such coordination may require reassignment of available resources based on externally determined priorities.

The DBA also works with applications programmers to ensure the quality and integrity of database design and transactions. Such support services include reviewing the database application design to ensure that transactions are:

- *Correct:* The transactions mirror real-world events.
- *Efficient:* The transactions do not overload the DBMS.
- *Compliant with integrity and standards.*

These activities require personnel with broad database design and programming skills.

The implementation of the applications requires the implementation of the physical database. Therefore, the DBA must provide assistance and oversight during the physical design, including storage space determination and creation, data loading, conversion, and database migration services. The DBA's implementation tasks also include the generation, compilation, and storage of the application's access plan. An **access plan** is a stored procedure that, during compilation, predetermines how the application will access the database at run time. (In order to create and validate the access plan, the user must have the required rights to access the database.)

Before an application comes online, the DBA must develop, test, and implement the operational procedures required by the new system. Such operational procedures include the training, security, and backup and recovery plans, as well as assigning responsibility for database control and maintenance. Finally, the DBA must authorize application users to access the database from which the applications draw the required data.

The addition of a new database may require the fine-tuning and/or reconfiguration of the DBMS. Remember that the DBMS assists all applications by managing the shared corporate data repository. Therefore, if data structures are added or modified, the DBMS may require the assignment of additional resources to service the new and original users with equal efficiency.

Testing and Evaluation of Databases and Applications

The DBA must also provide testing and evaluation services for all the database and end-user applications. These services are the logical extension of the design, development, and implementation services described in the preceding section. Clearly, testing procedures and standards must already be in place before any application program can be approved for use in the company.

Although testing and evaluation services are closely related to the database design and implementation services, they usually are maintained independently. The reason for the separation is that applications programmers and designers often are too close to the problem being studied to detect errors and omissions.

Testing usually starts by loading the testbed database. Such a database contains test data for the applications, and its purpose is to check the data definition and integrity rules of the database and application programs.

The testing and evaluation of a database application cover all aspects of the system—from the simple collection and creation of data to its use and retirement. The evaluation process covers:

- Technical aspects of both the applications and the database; backup and recovery, security and integrity, use of SQL, and application performance must be evaluated.
- Evaluation of the written documentation to ensure that the documentation and procedures are accurate and easy to follow.
- Observance of standards for naming, documenting, and coding.
- Data duplication conflicts with existing data.
- The enforcement of all data validation rules.

Following the thorough testing of all applications, the database, and the procedures, the system is declared operational and can be made available to end users.

Operation of the DBMS, Utilities, and Applications

DBMS operations can be divided into four main areas:

- System support.
- Performance monitoring and tuning.
- Backup and recovery.
- Security auditing and monitoring.

System support activities cover all tasks directly related to the day-to-day operations of the DBMS and its applications. These activities range from filling out job logs to tape changing to checking and verifying the status of computer hardware, disk packages, emergency power sources, and so on. System-related activities include periodic and occasional tasks, such as running special programs and resource configurations for new and/or upgraded versions of database applications.

Performance monitoring and tuning require much of the DBA's attention and time. Such activities are designed to ensure that the DBMS, utilities, and applications maintain satisfactory performance levels. To carry out the performance-monitoring and tuning tasks the DBA must:

- Establish DBMS performance goals.
- Monitor the DBMS to evaluate whether the performance objectives are being met.

- Isolate the problem and find alternate solutions (if performance objectives are not met).

- Implement the selected performance solutions.

DBMSs often include performance-monitoring tools that allow the DBA to query database usage information. If the DBMS does not include performance-monitoring tools, they are available from many different sources: DBMS utilities are provided by third-party vendors, or they might be included in operating system utilities or transaction processor facilities. Most of the performance-monitoring tools allow the DBA to focus on selected system bottlenecks. The most common bottleneck in DBMS performance tuning is related to the use of indexes, query-optimization algorithms, and management of storage resources.

Because improper index selection can have a deleterious effect on system performance, most DBMS installations adhere to a carefully defined index creation and usage plan. Such a plan is especially important in a relational database environment.

The DBA is likely to spend much time trying to educate programmers and end users on the proper use of SQL statements to produce satisfactory performance. Typically, DBMS programmers' manuals and administration manuals contain useful performance guidelines and examples to demonstrate the proper use of SQL statements, both in the command-line mode and within application programs. Because relational systems do not give the user an index choice within a query, the DBMS makes the index selection for the user. Therefore, the DBA should create indexes that can be used to improve system performance. (See Appendix E for examples of database performance tuning.)

Query-optimization routines are usually integrated into the DBMS package, thereby allowing few tuning options. Query-optimization routines are oriented to improve concurrent access to the database. Several database packages let the DBA specify parameters to determine the desired level of concurrency. Concurrency is also affected by the types of locks used by the DBMS and requested by the applications. Because the concurrency issue is important to the efficient operation of the system, the DBA must be familiar with the factors that influence concurrency. (See Chapter 9, "Transaction Management and Concurrency Control," for more information on this subject.)

Available storage resources, in terms of both primary and secondary memory, also must be considered during DBMS performance tuning. The allocation of storage resources is determined when the DBMS is configured. Storage configuration parameters can be used to determine:

- The number of databases that may be opened concurrently.

- The number of application programs or users supported concurrently.

- The amount of primary memory (buffer pool size) assigned to each database and each database process.

- The size and location of the log files. (Remember that these files are used to recover the database. The log files can be located in a separate volume to reduce the disk's head movement and increase performance.)

Performance-monitoring issues are DBMS-specific. Therefore, the DBA must become familiar with the DBMS manuals to learn the technical details involved in this task.

Because data loss is likely to be devastating to the organization, backup and recovery activities are of primary concern during the DBMS operation. The DBA must establish a schedule for backing up database and log files at appropriate intervals. Backup frequency is dependent on the application type and on the relative importance of the data. All critical system components—the database, the database applications, and the transaction logs—must be backed up periodically.

Most DBMS packages include utilities that schedule automated database backups, be they full or incremental. Although incremental backups are naturally faster than full backups, an incremental backup requires the existence of an initial full backup to be useful for recovery purposes.

Database recovery after a media or systems failure requires the application of the transaction log to the correct database copy. The DBA must plan, implement, test, and enforce a "bulletproof" backup and recovery procedure.

Security auditing and monitoring assumes the appropriate assignment of access rights and the proper use of such access privileges by programmers and end users. The technical aspects of security auditing and monitoring are found in the creation of users, the assignment of access rights, the use of SQL commands to grant and revoke access rights to users and database objects, and the use of audit trails to determine the existence of security violations or attempted violations. The DBA must periodically generate an audit trail report to determine if there have been actual or attempted security violations—and, if so, from what locations and, if possible, by whom.

Training and Supporting Users

Training people to use the DBMS and its tools is included in the DBA's technical activities. In addition, the DBA provides or secures technical training in the use of the DBMS and its utilities for the applications programmers. Applications programmer training covers the use of the DBMS tools as well as the procedures and standards required for database programming.

Unscheduled, on-demand technical support for end users and programmers is also included in the DBA's activities. A technical troubleshooting procedure can be developed to facilitate such support. The technical procedure might include the development of a technical database used to find solutions to common technical problems.

Part of the support is provided by interaction with DBMS vendors. Establishing good relationships with software suppliers is one way to ensure that the company always has a good external support source. Vendors are the source for up-to-date information concerning new products and personnel retraining. Good vendor-company relations also are likely to give organizations an edge in determining the future direction of database development.

Maintenance of the DBMS, Utilities, and Applications

The maintenance activities of the DBA are an extension of the operational activities. Maintenance activities are dedicated to the preservation of the DBMS environment.

Periodic DBMS maintenance includes management of the physical or secondary storage devices. One of the most common maintenance activities is reorganizing the physical location of data in the database. (This is usually done as part of the DBMS fine-tuning activities.) The reorganization of a database might be designed to allocate contiguous disk-page locations to the DBMS, in order to increase performance. The reorganization process also might free the space allocated to deleted data, thus providing more disk space for new data.

Maintenance activities also include upgrading the DBMS and utility software. The upgrade might require the installation of a new version of the DBMS software or an Internet front-end tool. Or it might create an additional DBMS gateway to allow access to a host DBMS running on a different host computer. DBMS gateway services are very common in distributed DBMS applications running in a client/server environment. Also, new-generation databases include features such as spatial data support, data warehousing and star query support, and support for Java programming interfaces for Internet access.

Quite often, companies are faced with the need to exchange data in dissimilar formats or between databases. The maintenance efforts of the DBA include migration and conversion services for data in incompatible formats or for different DBMS software. Such conditions are common when the system is upgraded from one version to another, or when the existing DBMS is replaced by an entirely new DBMS. Database conversion services also include downloading data from the host (mainframe-based) DBMS to an end user's personal computer to allow that user to perform a variety of activities—spreadsheet analysis, charting, statistical modeling, and so on. Migration and conversion services can be done at the logical (DBMS- or software-specific) level or at the physical (storage media or operating-system-specific) level.

15.6 DATABASE ADMINISTRATION TOOLS

We cannot overstate the importance of the data dictionary as a prime DBA tool. Although we have already described the data dictionary's function and format in Chapter 3, "The Relational Database Model," and in Appendix E, its administrative use warrants additional discussion. This section examines the data dictionary as a data administration tool, as well as the DBA's use of computer-aided software engineering (CASE) tools to support database analysis and design activities.

15.6.1 THE DATA DICTIONARY

In Chapter 1 we defined a data dictionary as a DBMS component that stores the definition of data characteristics and relationships. You may recall that such "data about data" were labeled *metadata*. The DBMS data dictionary provides the DBMS with its self-describing characteristic. In effect, the data dictionary resembles an x-ray of the company's entire data set, and it is a crucial element in the data administration function.

Two main types of data dictionaries exist: *integrated* and *standalone*. An integrated data dictionary is included with the DBMS. For example, all relational DBMSs include a built-in data dictionary or system catalog that is frequently accessed and updated by the RDBMS. Other DBMSs, especially older types, do not have a built-in data dictionary; instead, the DBA may use third-party, *standalone data dictionary* systems.

Data dictionaries can also be classified as *active* or *passive*. An **active data dictionary** is automatically updated by the DBMS with every database access, thereby keeping its access information up to date. A **passive data dictionary** is not updated automatically and usually requires running a batch process. Data dictionary access information is normally used by the DBMS for query optimization purposes.

The data dictionary's main function is to store the description of all objects that interact with the database. Integrated data dictionaries tend to limit their metadata to the data managed by the DBMS. Standalone data dictionary systems are usually more flexible and allow the DBA to describe and manage all the organization's data, whether or not they are computerized. Whatever the data dictionary's format, its existence provides database designers and end users with a much-improved ability to communicate. In addition, the data dictionary is the tool that helps the DBA resolve data conflicts.

Although there is no standard format for the information stored in the data dictionary, several features are common. For example, the data dictionary typically stores descriptions of all:

- Data elements that are defined in all tables of all databases. Specifically, the data dictionary stores the names, data types, display format, internal storage format, and validation rules. The data dictionary tells where an element is used, by whom it is used, and so on.

- Tables defined in all databases. For example, the data dictionary is likely to store the name of the table creator, the date of creation, access authorizations, and the number of columns.

- Indexes defined for each database table. For each index, the DBMS stores at least the index name, the attributes used, the location, specific index characteristics, and the creation date.

- Defined databases: who created each database, the date of creation, where the database is located, who the DBA is, and so on.

- End users and administrators of the database.

- Programs that access the database, including screen formats, report formats, application programs, SQL queries, and so on.

- Access authorizations for all users of all databases.

- Relationships among data elements, including which elements are involved, whether the relationships are mandatory or optional, the connectivity and cardinality requirements, and so on.

If the data dictionary can be organized to include data external to the DBMS itself, it becomes an especially flexible tool for more general corporate resource management. The management of such an extensive data dictionary thus makes it possible to manage the use and allocation of all of the organization's information, regardless of whether it has its roots in the database data. This is why some managers consider the data dictionary to be a key element of information resource management. And this is also why the data dictionary might be described as the **information resource dictionary**.

The metadata stored in the data dictionary are often the basis for monitoring database use and the assignment of access rights to the database users. The information stored in the data dictionary is usually based on a relational table format, thus enabling the DBA to query the database with SQL commands. For example, SQL commands can be used to extract information about the users of a specific table or about the access rights of a particular user. We will illustrate such queries with the help of the following IBM DB2 system catalog tables:

SYSTABLES Stores one row for each table or view.
SYSCOLUMNS Stores one row for each column of each table or view.
SYSTABAUTH Stores one row for each authorization given to a user for a table or view in a database.

Examples of Data Dictionary Usage

Example 1

List the names and creation dates of all tables created by the user JONESVI in the current database.

SELECT NAME, CTIME
 FROM SYSTABLES
 WHERE CREATOR = 'JONESVI';

Example 2

List the names of the columns for all tables created by JONESVI in the current database.

SELECT NAME
 FROM SYSCOLUMNS
 WHERE TBCREATOR = 'JONESVI';

Example 3

List the names of all tables for which the user JONESVI has DELETE authorization.

SELECT TTNAME
 FROM SYSTABAUTH
 WHERE GRANTEE = 'JONESVI' AND DELETEAUTH = 'Y';

Example 4

List the names of all users who have some type of authority over the INVENTORY table.

SELECT DISTINCT GRANTEE
 FROM SYSTABAUTH
 WHERE TTNAME = 'INVENTORY';

Example 5

List the user and table names for all users who can alter the database structure for any table in the database.

```
SELECT GRANTEE, TTNAME
    FROM SYSTABAUTH
        WHERE ALTERAUTH = 'Y'
            ORDER BY GRANTEE, TTNAME;
```

As you can see by looking at the preceding examples, the data dictionary can be a tool for monitoring the security of data in the database by checking the assignment of data access privileges. Although the preceding examples targeted database tables and users, information about the application programs that access the database can also be drawn from the data dictionary.

The DBA can use the data dictionary to support data analysis and design activities. For example, the DBA can create: a report that lists all data elements to be used in a particular application; a list of all users who access a particular program; a report that checks for data redundancies, duplications, the use of homonyms and synonyms; and a number of other reports that describe data users, data access, and data structure. The data dictionary can also be used to ensure that applications programmers have met all the naming standards for the data elements in the database and that the data validation rules are correct. Thus, the data dictionary can be used to support a wide range of data administration activities.

Earlier in this section, we mentioned that data dictionaries are used to facilitate the design and implementation of information systems. Integrated data dictionaries are also essential to the use of computer-aided software engineering tools.

15.6.2 CASE TOOLS

CASE is the acronym for **computer-aided software engineering**. A CASE tool provides an automated framework for the Systems Development Life Cycle (SDLC). CASE is based on the use of structured methodologies and powerful graphical interfaces. Because they automate many tedious system design and implementation activities, CASE tools play an increasingly important role in information systems development.

CASE tools are usually classified according to the extent of support they provide for the SDLC. For example, **front-end CASE tools** provide support for the planning, analysis, and design phases; **back-end CASE tools** provide support for the coding and implementation phases. The benefits associated with CASE tools include:

- A reduction in development time and costs.
- The automation of the SDLC.
- Standardization of systems development methodologies.
- Easier maintenance of application systems developed with CASE tools.

One of the CASE tools' most important components is an extensive data dictionary, which keeps track of all objects created by the systems designer. For example, the CASE data dictionary stores descriptions of data flow diagrams, structure charts, all external and internal entities, data stores, data items, report formats, screen formats, and so on. A CASE data dictionary also describes the relationships among the components of the system.

Several CASE tools provide interfaces that interact with the DBMS. These interfaces allow the CASE tool to store its data dictionary information by using the DBMS. Such CASE/DBMS interaction demonstrates the interdependence that exists between systems development and database development, and it helps create a fully integrated development environment.

In a CASE development environment, the database and application designers use the CASE tool to store the description of the database schema, data elements, application processes, screens, reports, and other data relevant to the development process. The CASE tool integrates all systems development information in a common repository, which can be checked by the DBA for consistency and accuracy.

As an additional benefit, a CASE environment tends to improve the extent and quality of the communications among the DBA, the application designers, and the end users. The DBA can interact with the CASE tool to check the definition of the data schema for the application, the observance of naming conventions, the duplication of data elements, the validation rules for the data elements, and a host of other developmental and managerial variables. When the CASE tool indicates conflicts, rule(s) violations, and inconsistencies, it facilitates making corrections. Better yet, a correction is transported by the CASE tool to cascade its effects throughout the applications environment, thus greatly simplifying the job of the DBA and the application designer.

A typical CASE tool provides five components:

- Graphics designed to produce structured diagrams such as data flow diagrams, ER diagrams, class diagrams, and object diagrams.

- Screen painters and report generators to produce the information system's input/output formats (for example, the end-user interface).

- An integrated repository for storing and cross-referencing the system design data. This repository includes a comprehensive data dictionary.

- An analysis segment to provide a fully automated check on system consistency, syntax, and completeness.

- A program documentation generator.

Figure 15.7 illustrates how Microsoft Visio Professional can be used to produce an ER diagram.

FIGURE 15.7 AN EXAMPLE OF A CASE TOOL: VISIO PROFESSIONAL

One CASE tool, AllFusion ERwin Data Modeler, by Computer Associates, produces fully documented ER diagrams that can be displayed at different abstraction levels. In addition, ERwin can produce detailed relational designs: The user simply specifies the attributes and primary keys for each entity and describes the relations. ERwin then assigns foreign keys based on the specified relationships among the entities. Changes in primary keys are always updated automatically throughout the system.

Major relational DBMS vendors, such as Oracle, now provide fully integrated CASE tools for their own DBMS software as well as for RDBMSs supplied by other vendors. For example, Oracle's CASE tools can be used with IBM's DB2, SQL/DS, and Microsoft's SQL Server to produce fully documented database designs. Some vendors even take nonrelational DBMSs, develop their schemas, and produce the equivalent relational designs automatically.

There is no doubt that CASE has enhanced the database designer's and the applications programmer's efficiency. But, no matter how sophisticated the CASE tool, its users must be well versed in conceptual design ideas. In the hands of database novices, CASE tools simply produce impressive-looking bad designs.

15.7 DEVELOPING A DATA ADMINISTRATION STRATEGY

For a company to succeed, its activities must be committed to its main objectives or mission. Therefore, regardless of a company's size, a critical step for any organization is to ensure that its information system supports its strategic plans for each of the company's business areas.

The database administration strategy must not conflict with the information systems plans. After all, the information systems plans are derived from a detailed analysis of the company's goals, its condition or situation, and its business needs. Several methodologies are available to ensure the compatibility of data administration and information systems plans and to guide the strategic plan development. The most commonly used methodology is known as information engineering.

Information engineering (IE) allows us to translate the strategic goals of the company into the data and applications that will help the company achieve those goals. IE focuses on the description of the corporate data instead of the processes. The IE rationale is simple: Business data types tend to remain fairly stable and do not change much during their existence. In contrast, processes change often and thus require the frequent modification of existing systems. By placing the emphasis on data, IE helps decrease the impact on systems when processes change.

The output of the IE process is an **information systems architecture (ISA)** that serves as the basis for planning, development, and control of future information systems. Figure 15.8 shows the forces that affect ISA development.

Implementing IE methodologies in an organization is a costly process that involves much planning, commitment of resources, management liability, well-defined objectives, identification of critical factors, and control. An ISA provides a framework that includes the use of computerized, automated, and integrated tools such as a DBMS and CASE tools.

FIGURE 15.8 FORCES AFFECTING THE DEVELOPMENT OF THE ISA

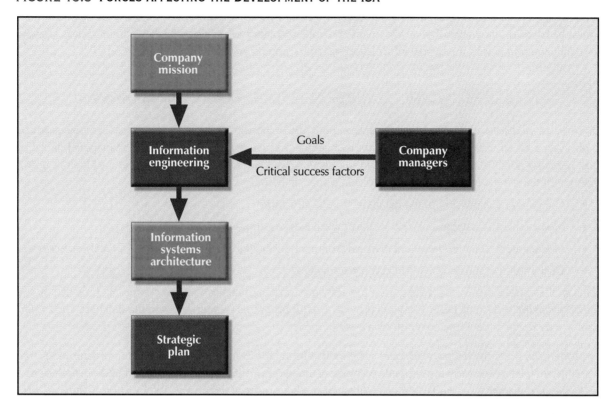

The success of the overall information systems strategy and, therefore, of the data administration strategy, depends on several critical success factors. Understanding the critical success factors helps the DBA develop a successful corporate data administration strategy. Critical success factors include managerial, technological, and corporate culture issues. Specifically, we identify the following:

- *Management commitment.* Top-level management commitment is necessary to enforce the use of standards, procedures, planning, and controls. The example must be set at the top.

- *Thorough company situation analysis.* The current situation of the corporate data administration must be analyzed to understand the company's position and to have a clear vision of what must be done. For example, how are database analysis, design, documentation, implementation, standards, codification, and other issues handled? Identify needs and problems, and then prioritize them.

- *End-user involvement.* End-user involvement is another aspect critical to the success of the data administration strategy. What is the degree of organizational change involved? Successful organizational change requires that people be able to adapt to such change. Users should be given an open communication channel to upper-level management to ensure the success of the implementation. Good communication channels are key in the overall process.

- *Defined standards.* Analysts and programmers must be familiar with appropriate methodologies, procedures, and standards. If they lack such familiarity, they may need to be trained in the use of such procedures and standards.

- *Training.* The vendor must train the DBA personnel in the use of the DBMS and other tools. End users must be trained to use the tools, standards, and procedures to obtain and demonstrate the maximum benefit, thereby increasing end-user confidence. Train key personnel first, so they can train others later.

- *A small pilot project.* A small project is recommended to ensure that the DBMS will work in the company, that the output is what was expected, and that the personnel have been trained properly.

This list of factors is not and cannot be comprehensive. Nevertheless, it does provide the initial framework for the development of a successful strategy. However, no matter how comprehensive the list of success factors is, it must be based on the notion that development and implementation of a successful data administration strategy are tightly integrated with the overall information systems planning activity of the organization.

15.8 THE DBA AT WORK: USING ORACLE FOR DATABASE ADMINISTRATION

Thus far, we've discussed the DBA's work environment and responsibilities in general terms. The objective was to give you a broad view of the wide variety of management activities, both technical and managerial, faced by most database administration professionals. In this section, we want to give you a more detailed look at how a DBA might handle the following technical tasks in a specific DBMS:

- Creation and expansion of the database storage structures.

- Management of database objects such as tables, indexes, triggers, procedures, and so on.

- Management of the end-user database environment, including the type and extent of database access.

- Customization of database initialization parameters.

Many of these tasks require the DBA to use the software tools and utilities that are commonly provided by the database vendor. In fact, all DBMS vendors provide a set of programs to interface with the database and to perform a wide range of database administrative tasks.

To properly illustrate the selected DBA tasks, we looked for a DBMS that is typically found in organizations that are sufficiently large and have a sufficiently complex database environment to require—and afford—the use of a DBA. We also wanted to use a DBMS with sufficient market presence to serve as the basis for a "typical" production database environment. Finally, we wanted to use a DBMS that was affordable even to small colleges and universities. Because Oracle met our stated requirements, we selected Oracle9*i* for Windows as the DBMS for the following examples.

> **N O T E**
>
> Although Microsoft Access is a superb DBMS, it is typically used in smaller organizations or in organizations and departments with relatively simple data environments. Access yields a superior database prototyping environment and, given its easy-to-use GUI tools, rapid front-end application development is a snap. And Access is one of the components in the MS Office suite, thus making end-user applications integration relatively simple and seamless. Finally, Access does provide some important database administration tools. However, an Access-based database environment does not typically require the services of a DBA. Therefore, MS Access did not fit this section's mission well.

Keep in mind that most of the tasks described in this section are encountered by DBAs regardless of their DBMS or their operating system. However, the *execution* of those tasks tends to be DBMS- and operating-system specific. Therefore, if you use IBM DB2 Universal Database or Microsoft SQL Server, you must adapt the procedures shown here to your DBMS. And, because we run our database server in the Windows operating system, some of the administration tasks will be specific to that OS. Therefore, if you use some other OS, you must adapt the procedures shown in this section to your OS.

Finally, we want to make clear that this section is not intended to serve as a database administration manual. Instead, it is a brief introduction to the way some typical DBA tasks would be performed in Oracle. Before we describe how to use Oracle to accomplish specific database administration tasks, you should become familiar with the tools Oracle uses for database administration, and how to log on, as discussed in the next two sections.

15.8.1 ORACLE DATABASE ADMINISTRATION TOOLS

All database vendors supply a set of database administration tools. In Oracle you perform most DBA tasks via the Oracle Enterprise Manager application. (See Figure 15.9.)

FIGURE 15.9 THE ORACLE ENTERPRISE MANAGER INTERFACE

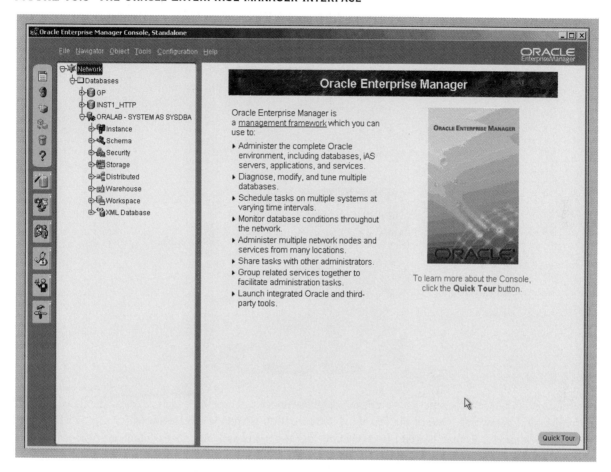

As you examine Figure 15.9, note that it shows all the available databases. In this section, we will use the ORALAB database. Note also the different database administration options under the ORALAB database in the Enterprise Manager window. For example:

- *Instance* allows the DBA to start and stop a database and to configure the database initialization parameters.

- *Schema* allows the DBA to manage database objects such as tables, indexes, triggers, procedures, and functions.

- *Security* allows the DBA to create users and to assign rights to users.

- *Storage* allows the DBA to manage the database storage and physical level components, such as tablespaces, datafiles, and rollback segments. (We'll define these Oracle terms in Section 15.8.4.)

The remaining options (Distributed, Warehouse, Workspace, and XML Database) enable the DBA to perform specialized tasks such as setting up a distributed database or creating a data warehouse. In the following sections we will examine the tasks most commonly encountered by a DBA.

15.8.2 THE DEFAULT LOGIN

To perform any administrative task you must connect to the database using a username with administrative (DBA) privileges. By default, Oracle automatically creates a SYSTEM user ID that has administrative privileges with every new database you create. You can define the preferred credentials for each database by selecting **Configuration** from the menu bar and then **Edit Local Preferred Credentials**. Figure 15.10 shows the Edit Local Preferred Credentials window that defines the user ID (SYSTEM) used to log on to the ORALAB database.

FIGURE 15.10 ORACLE EDIT LOCAL PREFERRED CREDENTIALS WINDOW

Keep in mind that usernames and passwords are database-specific. Therefore, each database can have different usernames and passwords. One of the first things you must do is change the password for the SYSTEM user. Immediately after you do that, you can start creating your users and assigning them database privileges.

15.8.3 ENSURING AN AUTOMATIC RDBMS START

One of the basic DBA tasks is to ensure that your database access is automatically started when you start the computer. Startup procedures will be different for each operating system. Because we use Oracle9i for Windows in this section's examples, we would identify the required services to ensure automatic database startup. (A **service** is the Windows system name for a special program that runs automatically as part of the operating system. This program ensures the availability of required services to the system and to end users on the local computer or over the network.) Figure 15.11 shows the required Oracle services that are started automatically when Windows starts up.

FIGURE 15.11 ORACLE RDBMS SERVICES

As you examine Figure 15.11, note the following Oracle services:

- *OracleOraHome92Agent* provides name resolution service for the DBMS. Each Oracle database must have a unique name that identifies it within the network. Such identification is crucial when you have multiple databases running in a distributed environment. (The DBA must create the names used to identify the databases in the network.) This service allows each user to refer to each database by name, without having to specify where the database is located.

- *OracleOraHome92TNSListener* is the process that "listens" and processes the end-user connection requests over the network. For example, when a SQL connection request such as "connect userid/password@ORALAB" is sent over the network, the listener service will take the request, validate it, and establish the connection.

- *OracleServiceORALAB* refers to the Oracle processes running in memory that are associated with the ORALAB **database instance**. You can think of a database instance as a separate location in memory that is reserved to run your database. Because you can have several databases (and, therefore, several instances) running in memory at the same time, you need to identify each database instance uniquely, using a different suffix. For example, note that OracleServiceGP starts the GP database instance, while OracleServiceORALAB starts the ORALAB database instance.

15.8.4 USING THE STORAGE MANAGER TO CREATE TABLESPACES AND DATAFILES

Each DBMS manages data storage differently. In this example, we will use the Oracle RDBMS to illustrate how the database manages data storage at the logical and the physical levels. In Oracle:

- A database is *logically* composed of one or more tablespaces. A **tablespace** is a logical storage space. Tablespaces are primarily used to logically group related data.

- The tablespace data are *physically* stored in one or more datafiles. A **datafile** physically stores the database's data. Each datafile is associated with one and only one tablespace. But each datafile can reside in a different directory on the hard disk or even on one or more different hard disks.

Given the preceding description of tablespaces and datafiles, we can conclude that a database has a one-to-many relationship with tablespaces and that a tablespace has a one-to-many relationship with datafiles. This set of 1:M hierarchical relationships isolates the end user from any physical details. However, *the DBA must be aware of these details in order to properly manage the database.*

To perform database storage management tasks—such as creating and managing tablespaces and datafiles—the DBA uses the Storage option in the Enterprise Manager window. (See Figure 15.12.)

FIGURE 15.12 THE ORACLE STORAGE MANAGER

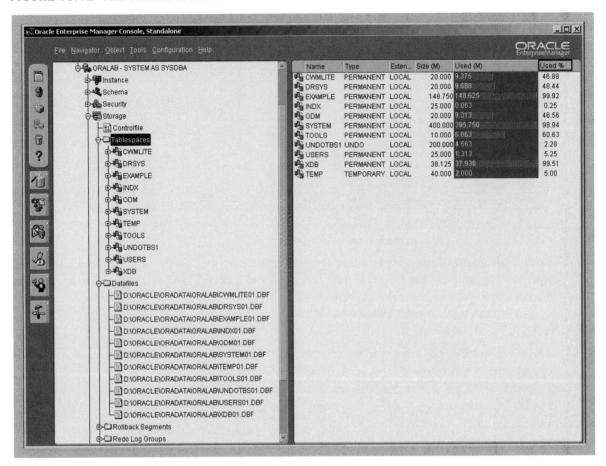

When the DBA creates a database, Oracle automatically creates the tablespaces and datafiles shown in Figure 15.12. We will describe just a few of them:

- The *SYSTEM* tablespace is used to store the data dictionary data. This tablespace is associated with the *SYSTEM01.DBF* datafile.

- The *USERS* tablespace is used to store the table data created by the end users. This tablespace is associated with the *USERS01.DBF* datafile.

- The *TEMP* tablespace is used to store the temporary tables and indexes created during the execution of SQL statements. For example, temporary tables are created if your SQL statement contains an ORDER BY, GROUP BY, or HAVING clause. This tablespace is associated with the *TMP01.DBF* datafile.

- The *UNDOTBS01* tablespace is used to store database transaction recovery information. If for any reason a transaction must be rolled back—to preserve database integrity—the UNDOTBS01 tablespace is used to store the undo information. This tablespace is associated with the *UNDOTBS01.DBF* datafile.

Using the Storage Manager, the DBA can:

- Create additional tablespaces to organize the data in the database. Therefore, if you have a database with several hundred users, you can create several user tablespaces to segment the data storage for different types of users. For example, you might create a teacher tablespace and a student tablespace.

- Create additional tablespaces to organize the different subsystems that may exist within the database. For example, you may create different tablespaces for human resources data, payroll data, accounting data, manufacturing data, and so on. Figure 15.13 shows the dialog box that was used to create a new tablespace called ROBCOR to hold the tables used in this book. This tablespace will be stored in the datafile named D:\ORACLE\ORADATA\ORALAB\ROBCOR.DBF, and its initial size is 5 megabytes. As you examine Figure 15.13, note that the tablespace will be immediately put online to be available to users for data storage purposes. Note also that at the bottom of the dialog box you can see the SQL code that was generated by Oracle to create the tablespace. (Actually, all DBA tasks can also be accomplished through the direct use of SQL commands. In fact, some diehard DBAs prefer writing their own SQL code, rather than using the "easy-way-out" GUI.)

FIGURE 15.13 CREATING A NEW TABLESPACE

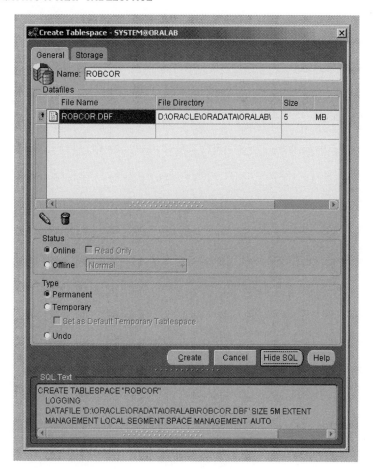

- Expand the tablespace storage capacity by creating additional datafiles. If you used the scroll bar shown at the bottom of the Datafiles window in Figure 15.13, you'd see that the Storage Manager shows the amount of space allocated to each tablespace along the window's right margin. Remember that the datafiles can be stored in the same directory or on different hard disks to increase access performance. For example, you could increase storage and access performance to the USERS tablespace by creating a new datafile in a different drive (for example, E:\ORACLE\ORADATA\ORALAB\USERS02.DBF).

15.8.5 MANAGING THE DATABASE OBJECTS: TABLES, VIEWS, TRIGGERS, AND PROCEDURES

Another important aspect of managing a database is monitoring the database objects that were created in the database. The Oracle Schema Manager gives the DBA a graphical user interface to create, edit, view, and delete database objects in the database. A **database object** is basically any object created by end users, for example, tables, views, indexes, stored procedures, and triggers. Figure 15.14 shows some of the different types of objects listed in the Oracle Schema Manager.

FIGURE 15.14 THE ORACLE SCHEMA MANAGER

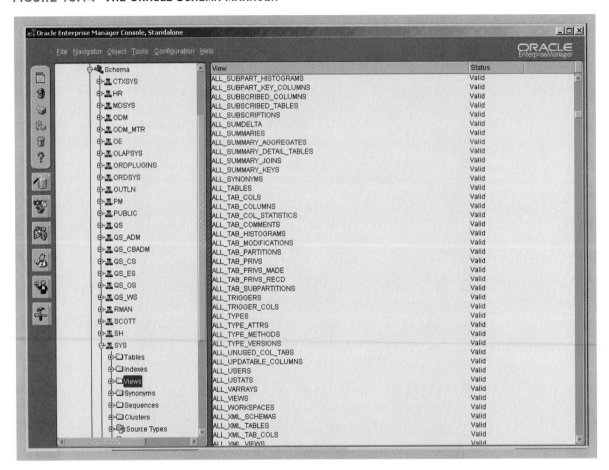

An Oracle **schema** is a logical section of the database that belongs to a given user, and such a schema is identified by the username. For example, if the user named SYSTEM creates a VENDOR table, the table will belong to the SYSTEM schema. Oracle prefixes the table name with your username. Therefore, the SYSTEM's VENDOR table name will be named SYSTEM.VENDOR by Oracle. Similarly, if the user PEROB creates a VENDOR table, that table will be created in the PEROB schema and will be named PEROB.VENDOR.

Within the schema, the user can create his/her own tables and other objects. You can have as many different schemas in the database as you have different users. Because each user sees only the object(s) created by him or her, each user may gain the impression that (s)he is the database's only user.

Normally, the user is only authorized to access the objects that belong to his/her schema. The user could, of course, give other users access to his/her data by changing access rights. In fact, all users with DBA authorization have access to all objects in all schemas in the database.

As you can see in Figure 15.14, the Schema Manager presents an organized view of all the objects in your database schema. With this program the DBA can create, edit, view, and delete tables, indexes, views, functions, triggers, procedures, and other specialized objects.

Figure 15.14 also shows some of the database tables that form the Oracle data dictionary. For example, the DBA can query the ALL_VIEWS table to list all defined views, or the ALL_USERS table to list all defined database users. (For a complete reference to all the tables that form the data dictionary, we recommend that you consult your Database Administrator's Guide.)

15.8.6 MANAGING USERS AND ESTABLISHING SECURITY

One of the most common database administration activities is creating and managing database users. (Actually, the creation of user IDs is just the first component of any well-planned database security function. As we indicated earlier in this chapter, database security is one of the most important database administration tasks.)

The Oracle Security Manager is the database administration utility that enables the DBA to create users, roles, and profiles:

- A **user** is a uniquely identifiable object that allows a given person to log on to the database. The DBA assigns privileges to access the objects in the database. Within the privilege assignment, the DBA may specify a set of limits that define how many of the database's resources the user can use.

- A **role** is a named collection of database access privileges that authorize a user to connect to the database and use the database system resources. Examples of roles are:
 - *CONNECT* allows a user to connect to the database and create and modify tables, views, and other data-related objects.
 - *RESOURCE* allows a user to create triggers, procedures, and other data management objects.
 - *DBA* gives the user database administration privileges.

- A **profile** is a named collection of settings that control how much of the database resource a given user can use. (If you consider the possibility that a runaway query could cause the database to lock up or to stop responding to the user's commands, you'll understand why it is important to limit access to the database resource!) By specifying profiles, the DBA can limit the amount of storage space a user can use, how long a user can be connected, how much idle time may be used before the user is disconnected, and so on. In an ideal world, all users would have unlimited access to all resources at all times, but in the real world we know that such access is neither possible nor desirable.

Figure 15.15 shows the Oracle Security Manager main screen. Note that the left window shows a graphical representation of the main types of objects—users, roles, and profiles—and the right window will show the actual objects within each type after you click the item in the left window.

FIGURE 15.15 **THE ORACLE SECURITY MANAGER**

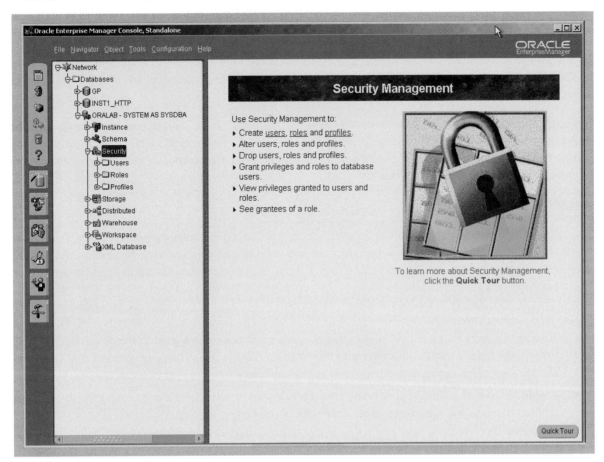

To create a new user, the DBA uses the Create User dialog box, shown in Figure 15.16.

The Create User dialog box contains many tabs; the most important are:

- The *General* tab allows the DBA to assign the name, password, and profile to the new user. Also, in this screen the DBA will define the default tablespace used to store table data and the temporary tablespace for temporary data.

- The *Role* tab allows the DBA to assign the roles for a user.

- The *Object* tab is used by the DBA to assign specific access rights to other database objects.

- The *Quota* tab allows the DBA to specify the maximum amount of storage that the user can have in each assigned tablespace.

FIGURE 15.16 THE CREATE USER DIALOG BOX

15.8.7 CUSTOMIZING THE DATABASE INITIALIZATION PARAMETERS

Fine-tuning a database is another important DBA task. This task usually requires the modification of some database configuration parameters, some of which can be changed in real time, using SQL commands, while others require the database to be shut down and restarted. Also, some parameters may affect only the database instance, while others affect the entire RDBMS and all instances running. So, it is very important that the DBA become familiar with database configuration parameters, especially those that affect performance.

Each database has an associated database initialization file that stores its run-time configuration parameters. The initialization file is read at instance startup and is used to set the working environment for the database. Oracle's Instance Manager allows the DBA to start up, shut down, and view/edit the database configuration parameters (stored in the initialization file) of a database instance. The Oracle Instance Manager program provides a GUI to modify this text file, shown in Figure 15.17.

FIGURE 15.17 THE ORACLE INSTANCE MANAGER

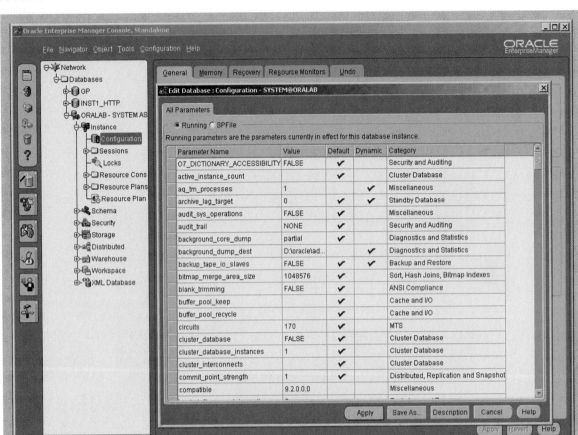

One of the important functions provided by the initialization parameters shown on the right side of Figure 15.17 is to reserve the resources that must be used by the database at run time. One of those resources is the primary memory to be reserved for database transaction caching. Such caching is used to fine-tune database performance. For example, the "shared_pool_size" parameter sets the amount of memory reserved for database transaction caching. This parameter should be set to a value that is sufficiently large to support all concurrent transactions.

Once you modify the initialization parameters, you may be required to restart the database. As you have seen in this brief section, the DBA is responsible for a wide range of tasks. The quality and completeness of the administration tools available to the DBA go a long way toward making his/her job easier. Even so, it is critical that the DBA become familiar with the tools and technical details of the RDBMS to properly and efficiently perform his/her job.

15.8.8 CREATING A NEW DATABASE

Although the general database creation format tends to be generic, its execution tends to be DBMS-specific. The leading RDBMS vendors offer the DBA the option to create databases manually, using SQL commands, or using a GUI-based process. Which option is selected depends on the DBA's sense of control and style.

Using the Oracle Database Configuration Assistant, it is much simpler to create a database. The DBA uses a wizard interface to answer a series of questions to establish the parameters for the database to be created. Figures 15.18 through 15.26 show how you create a database with the help of the Oracle Database Configuration Assistant.

FIGURE 15.18 CREATING A NEW DATABASE WITH THE DATABASE CONFIGURATION ASSISTANT

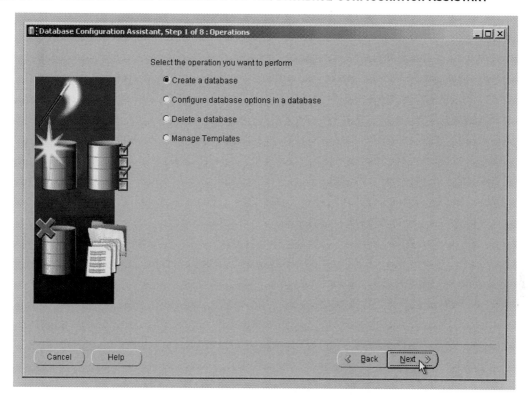

FIGURE 15.19 SELECTING THE NEW DATABASE TEMPLATE

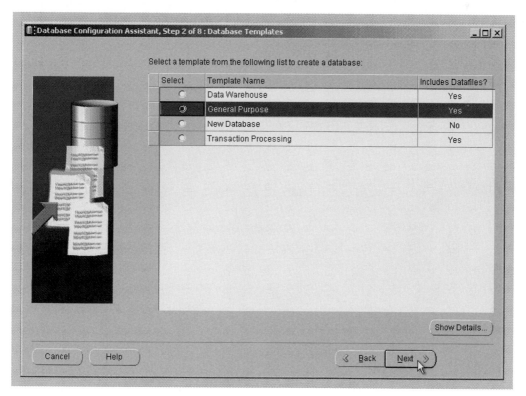

FIGURE 15.20 NAMING THE DATABASE

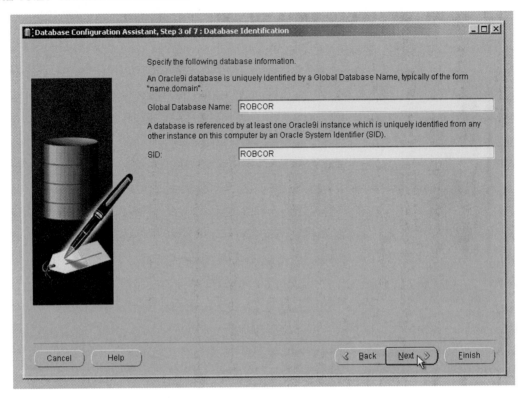

FIGURE 15.21 SELECTING THE DATABASE OPERATION MODE

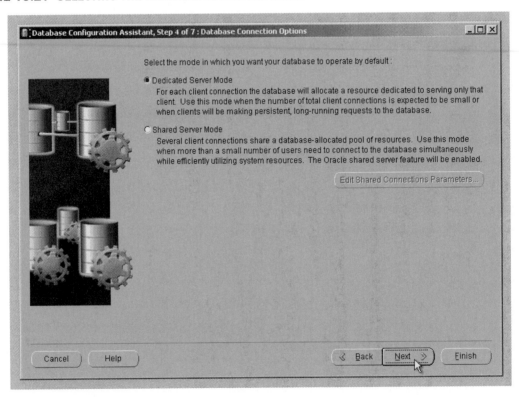

FIGURE 15.22 SPECIFYING INITIALIZATION PARAMETERS

FIGURE 15.23 SPECIFYING STORAGE PARAMETERS

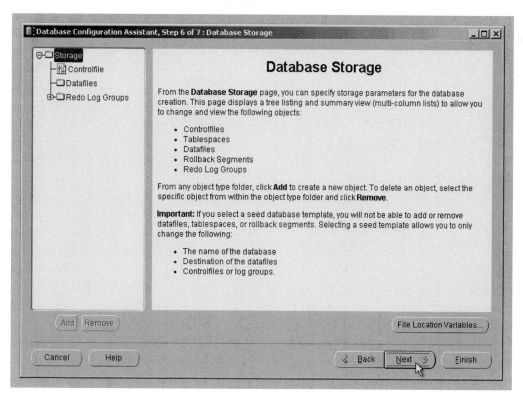

FIGURE 15.24 **DATABASE CREATION OPTIONS**

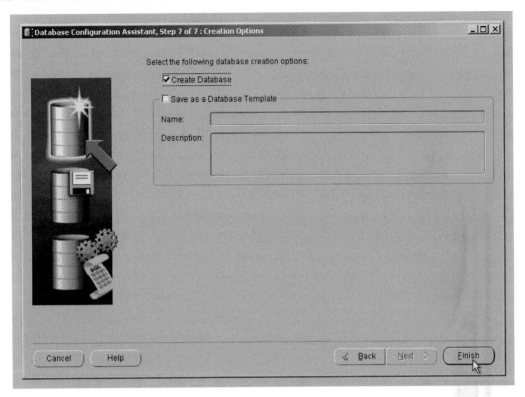

FIGURE 15.25 **SUMMARY OF DATABASE CREATION OPTIONS**

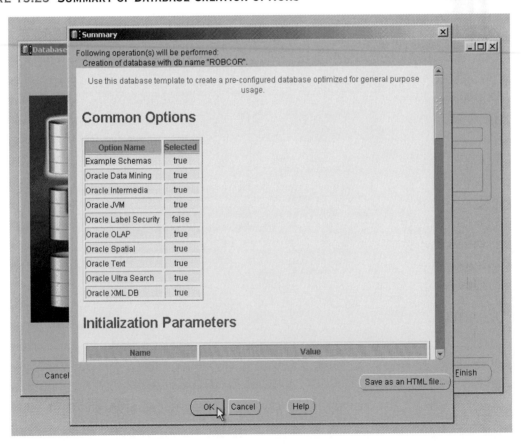

Finally, after confirming all the database options selected in Figures 15.18 through 15.24, the database creation process starts. This process creates the database structure, including the necessary data dictionary tables, the administrator user accounts, and other supporting processes required by the DBMS to manage the database. Figure 15.26 shows the rate at which the database creation process proceeds.

FIGURE 15.26 DATABASE CREATION PROGRESS

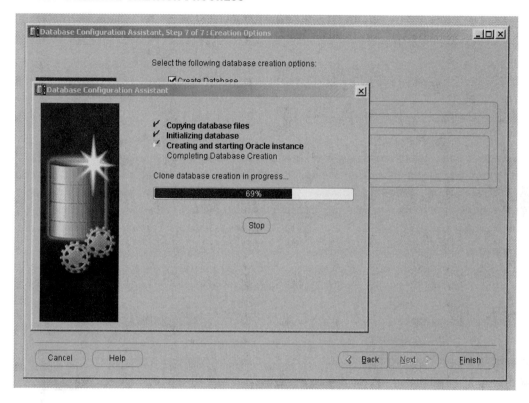

One of the disadvantages of using this graphical interface to create databases is that there are no records in the form of SQL scripts to document the creation of the database. Even given the GUI's help, the database creation process requires a solid understanding of the database's underlying structures and components.

SUMMARY

Data management is a critical activity for any organization. Data must be treated as a corporate asset that has a monetary value. The value of a data set is measured by the utility of the information derived from it. Good data management is likely to produce good information, which is the basis for better decisions.

The DBMS is the most commonly used electronic tool for corporate data management. The role of the DBMS is to support managerial decision making at all levels of the organization. Therefore, the DBMS must provide support for strategic, tactical, and operational decision making. The company data that are managed by the DBMS are stored in the corporate or enterprise database.

The introduction of a DBMS into an organization is a very delicate job. In addition to managing the technical details of DBMS introduction, the impact of the DBMS on the organization's managerial and cultural framework must be carefully examined.

The data administration function was developed in response to changes in data management that were fostered by technological advances in computer systems. The data administration function's development is based on the evolution from departmental data processing to the more centralized electronic data processing (EDP) department, to the more formal "data as a corporate asset" information systems (IS) department. Typical file systems were characterized by applications that tended to behave as distinct "islands of information." As applications began to share a common data repository, the need for centralized data management to control such data became clear, thus paving the way for the database administration function.

The database administrator (DBA) is responsible for the management of the corporate database. The internal organization of the database administration function varies from company to company. Although no standard exists, it is common practice to define the DBA function by dividing the DBA operations according to the database life cycle phases. Some companies have created a position with a broader data management mandate to manage both computerized and other data within the organization. This broader data management activity is handled by the data administrator (DA).

The DA and the DBA functions tend to overlap. Generally speaking, the DA is more managerially oriented than the more technically oriented DBA. Compared to the DBA function, the DA function is DBMS-independent, with a broader and longer-term focus. However, when the organizational chart does not include a DA position, the DBA executes all the functions of the DA. Because the DBA has both technical and managerial responsibilities, the DBA must have a diverse mix of skills.

The DBA's managerial role makes the DBA responsible for planning, organizing, testing, monitoring, and delivering services associated with the DBA function. The managerial services of the DBA function include at least:

- Supporting the end-user community.
- Defining and enforcing policies, procedures, and standards for the database function.
- Ensuring data security, privacy, and integrity.
- Providing data backup and recovery services.
- Monitoring the distribution and use of the data in the database.

The technical role of the DBA function is closely related to the DBA's managerial responsibilities. The technical role requires the DBA to be involved in at least these activities:

- DBMS and utilities evaluation, selection, and installation.
- Design and implementation of databases and applications.
- Testing and evaluation of databases and applications.
- Operation of the DBMS, utilities, and applications.
- Training and supporting users.
- Maintenance of the DBMS, utilities, and applications.

The development of the data administration strategy is closely related to the company's mission and objectives. Therefore, the development of an organization's strategic plan corresponds to that of data administration, requiring a detailed analysis of company goals, situation, and business needs. To guide the development of this overall plan, an integrating methodology is required. The most commonly used integrating methodology is known as information engineering (IE).

Information engineering allows us to translate strategic plans into a set of plan-conforming data and applications. Implementing the overall company strategy is a crucial process, which is influenced by managerial, technological, and cultural issues.

To help translate strategic plans into operational plans, the DBA has access to an arsenal of database administration tools. These tools include the data dictionary and computer-aided software engineering (CASE) tools.

This chapter also gave a very brief illustration of how some database administration functions are performed, using Oracle9i Server for Windows. This section covered several tasks performed by a typical Oracle DBA using the Oracle Database Administrator Tools. Since each DBMS vendor has a different set of tools to perform database administration tasks, you may need to consult the documentation for your DBMS for information on how to perform specific database administration tasks.

KEY TERMS

access plan

active data dictionary

audit log

authorization management

back-end CASE tool

CASE (computer-aided software engineering)

concurrent backup

data administrator (DA)

database administrator (DBA)

database dump

database instance (Oracle)

database object (Oracle)

database security officer (DSO)

datafile (Oracle)

data processing (DP) department

disaster management

electronic data processing (EDP) department

enterprise database

front-end CASE tool

full backup

incremental backup

information engineering (IE)

information resource dictionary

information resource manager (IRM)

information systems architecture (ISA)

information systems (IS) department

passive data dictionary

policies

privacy

procedures

profile (Oracle)

role (Oracle)

schema (Oracle)

security

service (Windows)

standards

systems administrator (SYSADM)

tablespace (Oracle)

user (Oracle)

REVIEW QUESTIONS

1. Explain the difference between data and information. Give some examples of raw data and information.

2. Explain the interactions among end user, data, information, and decision making. Draw a diagram and explain the interactions.

3. Suppose that you are a DBA staff member. What data dimensions would you describe to top-level managers to obtain their support for the data administration function?

4. How and why did database management systems become the organizational data management standard? Discuss some advantages of the database approach over the file-system approach.

5. Using a single sentence, explain the role of databases in organizations. Then explain your answer.

6. Define *security* and *privacy*. How are these two concepts related to each other?

7. Describe and contrast the information needs at the strategic, tactical, and operational levels in an organization. Use examples to explain your answer.

8. What special considerations must you take into account when contemplating the introduction of a DBMS into an organization?

9. Who is the database administrator? Describe the DBA's responsibilities.

10. How can the DBA function be placed within the organizational chart? What effect(s) will such placement have on the DBA function?

11. Why and how are new technological advances in computers and databases changing the DBA's role?

12. Explain the DBA department's internal organization, based on the DBLC approach.

13. Explain and contrast the differences and similarities between the DBA and DA.

14. Explain how the DBA plays an arbitration role between an organization's two main assets. Draw a diagram to facilitate your explanation.

15. Describe and characterize the skills desired for a DBA.

16. What are the DBA's managerial roles? Describe the managerial activities and services provided by the DBA.

17. What DBA activities are used to support the end-user community?

18. Explain the DBA's managerial role in the definition and enforcement of policies, procedures, and standards.

19. Protecting data security, privacy, and integrity are important database functions in authorization management. What activities are required in the DBA's managerial role of enforcing these functions?

20. Discuss the importance and characteristics of database backup and recovery procedures. Then describe the actions that must be detailed in backup and recovery plans.

21. Assume that your company assigned you the responsibility of selecting the corporate DBMS. Develop a checklist for the technical and other aspects involved in the selection process.

22. Describe the activities that are typically associated with the design and implementation services of the DBA technical function. What technical skills are desirable in the DBA's personnel?

23. Why are testing and evaluation of the database and applications not done by the same people who are responsible for their design and implementation? What minimum standards must be met during the testing and evaluation process?

24. Identify some bottlenecks in DBMS performance. Then propose some solutions used in DBMS performance tuning.

25. What are the typical activities involved in the maintenance of the DBMS, utilities, and applications? Would you consider application performance tuning part of the maintenance activities? Explain your answer.

26. Define the concept of a data dictionary. Discuss the different types of data dictionaries. If you were to manage an organization's entire data set, what characteristics would you look for in the data dictionary?

27. Using SQL statements, give some examples of how you would use the data dictionary to monitor the security of the database.

> **NOTE**
>
> If you use IBM DB2, the names of the main tables are SYSTABLES, SYSCOLUMNS, and SYSTABAUTH.

28. What characteristics do a CASE tool and a DBMS have in common? How can these characteristics be used to enhance the data administration function?

29. Briefly explain the concepts of information engineering (IE) and information systems architecture (ISA). How do these concepts affect the data administration strategy?

30. Identify and explain some of the critical success factors in the development and implementation of a successful data administration strategy.

31. What is the tool used by Oracle to create users?
32. In Oracle, what is a tablespace?
33. In Oracle, what is a database role?
34. In Oracle, what is a datafile, and how does it differ from a file systems file?
35. In Oracle, what is a database profile?
36. In Oracle, what is a database schema?
37. In Oracle, what role is required to create triggers and procedures?

2-tier client/server system—A system within which a client requests services directly from the server. See also *client/server*.

3-tier client/server system—A system design where the client's requests are handled by intermediate servers, which coordinate the execution of the client requests with subordinate servers. See also *client/server*.

A

abstract data types (ADT)—Data types that describe a set of similar objects with shared and encapsulated data representation and methods. An abstract data type is generally used to describe complex objects. See also *class*.

access plan—A stored procedure, generated at application compilation time, that is created and managed by the DBMS. The access plan predetermines the way an application's query will access the database at run-time.

access point—In the case of wireless networks, this allows you to connect wireless devices to a wired or wireless network.

active data dictionary—A data dictionary that is automatically updated by the database management system every time the database is accessed, thereby keeping its information current. See also *data dictionary*.

ActiveX—Microsoft's alternative to Java. A specification for writing programs that will run inside the Microsoft client browser (Internet Explorer). Oriented mainly to Windows applications, it is not portable. It adds "controls" such as drop-down windows and calendars to Web pages.

ad hoc queries—"Spur-of-the-moment" questions.

alias—An alternate name given to a column or table in any SQL statement.

ALTER—The SQL command used to make changes in the table structure. Followed by a keyword (ADD or MODIFY), it will add a column or change column characteristics.

American National Standards Institute (ANSI)—The group that accepted the DBTG recommendations and augmented database standards in 1975 through its SPARC committee.

AND—The SQL logical operator used to link multiple conditional expressions in a WHERE or HAVING clause. It requires that all conditional expressions evaluate to true.

anonymous PL/SQL block—A PL/SQL block that has not been given a specific name.

AppleTalk—The proprietary network communications protocol used by Apple Computers.

application processor—See *transaction processor*.

application programming interface (API)—Software through which programmers interact with middleware. Allows the use of generic SQL code, thereby allowing client processes to be database server independent.

application program-to-program communications (APPC)—A communications protocol used in IBM mainframe systems network architecture (SNA). Allows for communications between personal computers and IBM mainframe applications.

AREA—A named section of permanent storage space that is reserved to store the database.

asymmetric encryption—A form of encryption that uses two numeric keys—the public key and the private key. Both keys are able to encrypt and decrypt each other's messages. See also *public-key encryption*.

atomic transaction property—All portions of the transaction must be treated as a single logical unit of work, in which all operations must be applied and completed to produce a consistent database.

atomicity—See *transaction*.

attribute—A characteristic of an entity or object. An attribute has a name and a data type. See also *instance variable and entity*.

attribute domain—See *domain*.

attribute hierarchy—Provides a top-down data organization that is used for two main purposes: aggregation and drill-down/roll-up data analysis.

audit log—Database management system feature that automatically records a brief description of the database operations performed by all users.

authentication—The process through which the DBMS verifies that only registered users are able to access the database.

authorization management—All activities that are dedicated to enforcing the security, privacy, and integrity of the data in the database.

automatic query optimization—A method by which the DBMS takes care of finding the most efficient access path for the execution of a query.

AVG—An SQL aggregate function that outputs the mean average for the specified column.

B

back-end application—The process that provides service to clients. See also *server*.

back-end CASE tools—A computer-aided software tool that has been classified as "back-end" because it provides support for the coding and implementation phases of the SDLC. In comparison, front-end case tools provide support for the planning, analysis, and design phases.

base data types—A term used to describe the data types frequently used in traditional programming languages. Base data types include *real*, *integer*, and *string*.

base table—The table on which a view is based.

batch update routine—A routine that pools transactions into a single "batch" to update a master table in a single operation.

BETWEEN—A SQL special operator used to define range limits.

bidirectional physically paired logical relationships—In the hierarchical model, a relationship that links a logical child with its logical parent in two directions.

bidirectional virtually paired logical relationships—In the hierarchical model, a relationship created when a logical child segment is linked to its logical parent in two directions. The virtually paired relationship is different from the physically paired relationship in that no duplicates are created.

binary lock—A lock that has only two states: *locked* (1) or *unlocked* (0). If a data item is locked by a transaction, no other transaction can use that data item. See also *lock*.

binary relationship—An ER term used to describe an association (relationship) between two entities. Example: PROFESSOR teaches COURSE.

Boolean algebra—A branch of mathematics that deals with the use of the logical operators OR, AND, and NOT.

bottom-up—A design philosophy that begins by identifying individual design components and then aggregates those components into larger units. In database design, it is a process that begins by defining attributes and then groups those attributes into entities. Compare *top-down*.

boundaries—The external limits to which any proposed system is subjected. These include budget, personnel, and existing hardware and software.

Boyce-Codd normal form (BCNF)—A special form of third normal form (3NF) in which *every* determinant is a candidate key. A table that is in BCNF must be in 3NF.

bridge—A device that connects similar networks. Allows computers in one network to communicate with computers in another network.

bridge entity—See *composite entity*.

buffer—A reserved block of primary memory that temporarily stores data read from secondary memory. Used to take advantage of the computer's fast primary memory compared to the slower secondary memory, thereby minimizing the number of input/output (I/O) operations between the primary and secondary memories.

buffer cache—A shared, reserved memory area that stores the most recently accessed data blocks in RAM. Also called *data cache*.

bus topology—Network topology requiring that all computers be connected to a main network cable. Usually implemented using coaxial cable; it bears the disadvantage that a single lost computer can result in network segment breakdown.

business rules—Narrative descriptions of policies, procedures, or principles within an organization. Examples: A pilot cannot be on duty for more than ten hours during a twenty-four hour period. A professor may teach up to four classes during any one semester.

business to business (B2B)—Electronic commerce between businesses.

business to consumer (B2C)—Electronic commerce between a business and consumers.

C

call level interface (CLI)—A standard developed by the SQL Access Group for database access.

campus-wide network (CWN)—A typical college or university network in which buildings containing LANs are (usually) connected through a network backbone.

candidate key—See *key*.

cardinality—Assigns a specific value to connectivity. Expresses the specific number of allowed entity occurrences associated with a single occurrence of the related entity.

cascading order sequence—Refers to a sequencing at different levels. For example, a list in which all last names are alphabetically ordered and, within the last names, all first names are ordered, represents a cascading sequence.

CASE—See *computer-aided software engineering*.

centralized DBMS—A database management system that supports a database located at a single site.

centralized design—A process in which a single conceptual design is modeled to match an organization's database requirements. Typically used when the data component consists of a relatively small number of objects and procedures. Compare to *decentralized design*.

certification authority (CA)—A private entity or company that certifies the user or vendor is who (s)he claims to be.

checkpoint—In transaction management, an operation in which the database management system writes all of its updated buffers to disk.

Chen model—Entity relationship diagram described by Peter Chen, characterized by its use of diamonds to depict relationships and rectangles to depict entities.

class—A collection of like objects with shared structure (attributes) and behavior (methods). A class encapsulates the object's data representation and method's implementation. Classes are organized in a class hierarchy. See also *abstract data types*.

class hierarchy—The organization of classes in a hierarchical tree where each "parent" class is a *superclass* and each "child" class is a *subclass*.

class instance—Each individual object stored in a class. Each class instance must share the same structure and respond to the same messages if they are located in the same class. Also known as *object instance*.

class lattice—The class hierarchy is known as a class lattice if its classes can have multiple parent classes.

client—Any process that requests specific services from server processes. See *client/server architecture*.

client/server architecture—The client/server infrastructure composed of hardware and software to form a system composed of clients, servers, and middleware.

client-side extensions—These extensions add functionality to the Web browser. Although available in various forms, the most commonly encountered extensions are: plug-ins, Java, JavaScript, ActiveX, and VBScript.

cluster tables—Tables that store rows of different tables together in consecutive disk locations to speed up data access.

coaxial cable—Copper cables enclosed in two layers of insulation or shielding. Often referred to as "coax." Very similar to cable used for home cable TV.

cohesivity—The strength of the relationships between a module's components. Module cohesivity must be high.

ColdFusion Markup Language (CFML)—A server-side markup language (HTML extensions or tags) that is used to create ColdFusion application pages known as *scripts*.

collection object—An object that contains one or more objects.

COMMIT—The SQL command that permanently saves data changes to the database.

common gateway interface (CGI)—A Web server interface standard that uses script files to perform specific functions based on the client's parameters.

communications middleware—Any computer process(es) through which clients and servers communicate.

complex object—An object formed by several different objects in complex relationships. See also *abstract data types*.

composite attribute—An attribute that can be further subdivided to yield additional attributes. Example: A phone number (615-898-2368) may be divided into an area code (615), an exchange number (898), and a four-digit code (2368). Compare to *simple attribute*.

composite entity—An entity designed to transform an M:N relationship into two 1:M relationships. The composite entity's primary key comprises at least the primary keys of the entities that it connects. Also known as a *bridge entity*.

composite index—An index which is based on two or more attributes.

composite key—A multiple-attribute key.

composite object—An object that contains at least one multivalued attribute and has no attributes that refer to another object.

compound object—An object that contains at least one attribute that references another object.

computer-aided software engineering (CASE)—Tools used to automate part or all of the Systems Development Life Cycle.

concentrator—A device that takes multiple wires and combines them into a single method of transfer to allow multiple users to access the line simultaneously. It resembles a network wiring closet.

conceptual design—A process that uses data modeling techniques to create a model of a database structure that represents the real-world objects in the most realistic way possible. Both software- and hardware-independent.

conceptual model—An abstract view of the data as seen by high-level managers and database designers. Describes the main data objects, avoiding details.

concurrency control—A DBMS feature that is used to coordinate the simultaneous execution of transactions in a multiprocessing database system while preserving data integrity.

concurrent backups—A backup that takes place while one or more users are working on the database.

Conference on Data Systems Languages (CODASYL)—A group originally formed to help standardize COBOL; its DBTG subgroup helped to develop database standards in the early 1970s.

connectivity—Describes the classification of the relationship between entities. Classifications include 1:1, 1:M and M:N.

consistent database state—Database state in which all data-integrity constraints are satisfied.

content management—Automation of the creation and management of a Web site's contents.

conventional data types—See *base data types*.

coordinator—The transaction processor (TP) node that coordinates the execution of a two-phase COMMIT in a DDBMS. See also *data processor, transaction processor, two-phase commit protocol*.

correlated subquery—A subquery that executes once for each row in the outer query.

cost-based optimizer—A query optimizer technique that uses an algorithm based on statistics about the objects being accessed, that is, number of rows, indexes available, indexes sparsity, etc.

COUNT—An SQL aggregate function that outputs the number of rows containing not null values for the given column, sometimes used in conjunction with the DISTINCT clause.

CREATE INDEX—An SQL command that creates indexes on the basis of any selected attribute or attributes.

CREATE TABLE—An SQL command used to implement a table's structures using the characteristics and attributes given.

CREATE VIEW—An SQL command that creates a logical, "virtual" table which can be treated as a "real" table.

cross join—A join that performs a relational product (also known as the Cartesian product) of two tables.

Crow's Foot model—A notation of the entity relationship diagram using a three-pronged symbol to represent the "many" sides of the relationship.

cube cache—The cache in memory where data cubes are held. Using the cube cache assists in speeding up data access.

currency—This term indicates the position of the record pointer within the database and *refers to the most recently accessed record*.

cursor—A special construct used in procedural SQL to hold the data rows returned by a SQL query. You can think of a cursor as a reserved area of memory in which the output of the query is stored, like an array holding columns and rows. Cursors are held in a reserved memory area in the DBMS server, not in the client computer.

D

data—Raw facts; that is, facts that have not yet been processed to reveal their meaning to the end user.

data administrator (DA)—The person responsible for managing the entire data resource, whether computerized or not. The DA has broader authority and responsibility than the database administrator (DBA). See also *information resource manager*.

data anomaly—A data abnormality that exists when inconsistent changes to the database have been made. Example: An employee moves, but the address change is only corrected in one file and not across all files in the database.

data cache—A shared, reserved memory area that stores the most recently accessed data blocks in RAM. Also called *buffer cache*.

data cube—Data extracted from the operational databases or the data warehouse that helps MDBMS end users visualize the stored data as a multi-dimensional cube. The locations of each data value in the data cube is based on the *x*, *y* and *z* axes of the cube. They are static (must be created before they are used), so they can not be created by an *ad hoc* query.

data definition language (DDL)—The language that allows a database administrator to define the database structure, schema, and subschema.

data dependence—A data condition in which the data representation and manipulation are dependent on the physical data storage characteristics.

data dictionary—A DBMS component that stores metadata – data about data. Thus, the data dictionary contains the data definition as well as its characteristics and relationships. A data dictionary may also include data that are external to the DBMS. See also *active data dictionary* and *passive data dictionary*.

Data Encryption Standard (DES)—The most widely used standard for private-key encryption. DES is used by the U.S. government.

data extraction—A component used to extract and validate the data taken from the operational database and the external data sources. Also known as *data filtering*.

data filtering—See *data extraction*.

data inconsistency—A condition in which different versions of the same data yield different (inconsistent) results.

data independence—A condition that exists when data access is unaffected by changes in the physical data storage characteristics.

data integrity—A condition in which given data always yield the same result. Data integrity is mandatory in any database.

data management—A process that focuses on data collection, storage and retrieval. Common data management functions include addition, deletion, modification and listing.

data manager (DM)—See *data processing manager*.

data manipulation language (DML)—The language (set of commands) that allows the end user to manipulate the data in the database (Select, Insert, Update, Delete).

data mining—A method of employing automated tools to analyze the wealth of data now available through data warehouses and other sources and to identify possible relationships and anomalies. Data mining is a proactive methodology that is being used to uncover business problems and opportunities.

data model—A representation, usually graphic, of a complex "real-world" data structure. Data models are used in the database design phase of the database life cycle. See also *database models*.

data processing (DP) department—A centralized department created to coordinate and maintain the company's operational data and applications. See also *electronic data processing (EDP) department*.

data processing (DP) manager—A DP specialist who evolved into the department supervisor. Roles include: managing the technical and human resources, supervising the senior programmers, program troubleshooting.

data processing (DP) specialist—A now obsolete position formed in the conversion from manual filing systems to computer filing systems; once filled by an employee who created and programmed the necessary file structures, wrote the software that managed the data in those structures, and designed the application programs that produced reports from the file data.

data processor (DP)—The software component residing on each computer that stores and retrieves data through a DDBMS. The DP is responsible for managing the local data in the computer and coordinating access to such data. See also *transaction processor*.

data redundancy—A condition that exists when the data environment contains redundant—unnecessarily duplicated—data.

data replication—The storage of duplicated database fragments at multiple sites on a DDBMS. The duplication of such fragments is transparent to the end user.

data sparsity—A column distribution of values or the number of different values a column could have.

data store—The component of the decision support system that acts as a database for storage of business data and business model data. The data in the data store has already been extracted and filtered from the external and operational data, and will be stored for access by the end user query tool for the business data model.

data warehouse database—A database that focuses primarily on the storage of data used to generate information required to make tactical or strategic decisions.

database—A computer structure that houses a collection of related data. A database contains two types of data: end user data (raw facts) and metadata. The metadata consist of data about data, that is, the data characteristics and relationships.

database administrator control system (DBACS)—In the network model, the database definition processor that reads the database definition and validates the schema. (The DBACS works like a compiler.)

database administrator (DBA)—Person responsible for the planning, organization, control and monitoring of the centralized and shared corporate database. The DBA is the general manager of the database-administration department.

database description (DBD) statement database—In the hierarchical model, the series of commands that define the hierarchical tree structure and how the segments are stored in the database.

database design—The process that yields the description of the database structure. The database design process determines the database components. Database design is the second phase of the database life cycle.

database development—A term used to describe the process of database design and implementation.

database dump—A complete copy of the entire database saved and periodically updated in a separate memory location. Ensures a full recovery of all data in the event of a physical disaster or a database integrity failure. Also known as a *full backup*.

database fragments—Subsets of a distributed database. Although the fragments may be stored at different sites within the computer network, the set of all fragments is treated as a single database.

database instance—The separate location of memory that is reserved for running a specific database. (Usually part of Oracle.)

database integration—A business enabling service where a company's database is used in their Web page development.

database life cycle (DBLC)—The history of a database within the information system. Divided into six phases: initial study, design, implementation and loading, testing and evaluation, operation and maintenance and evolution.

database management system (DBMS)—Software that serves as an intermediary between the user and the database. The DBMS translates user requests into the computer code that is required to fulfill those requests. A DBMS manages the data stored within the database.

database object—Any object in a database created by end users, such as tables, views, indexes, stored procedures, and triggers. Must be monitored by the database manager.

database performance tuning—A set of activities and procedures designed to reduce the response time of the database system; that is, to ensure that an end-user query is processed by the DBMS in the minimum amount of time.

database recovery—Restoration of a database to its previous, consistent state. Recovery techniques are based on the atomic transaction property.

database request—The equivalent of a single SQL statement in an application program or transaction.

database security officer (DSO)—Responsible for the security, integrity, backup, and recovery of the database.

database system—An organization of components that define and regulate the collection, storage, management, and use of data in a database environment.

database task group (DBTG)—A CODASYL committee that helped develop database standards in the early 1970's. See also *CODASYL*.

database translator—A middleware component that translates generic SQL calls into specific database server syntax to create database server independence.

database-level lock—Prevents the database from accessing two tables simultaneously. The first transaction must be completed before the second is begun. Successful for batch processes, but unsuitable for online multi-user DBMSs.

datafile—A physical storage space that stores the database's data. It can reside in a different directory on the hard disk or on one or more different hard disks.

DBGEN—In the hierarchical model, the component that generates the physical database with all its necessary structures.

deadlock—A condition that exists when two or more transactions wait indefinitely for each other to release each other's lock on a previously locked data item. Also called a *deadly embrace*. See also *lock*.

deadly embrace—See *deadlock*.

decentralized design—A process in which conceptual design is used to model subsets of an organization's database requirements. After verification of the views, processes and constraints, the subsets are then aggregated into a complete design. Such modular designs are typical of complex systems, in which the data component consists of a relatively large number of objects and procedures. See also *centralized design*.

decision support—A methodology (or a series of methodologies) designed to extract information from data and to use such information as a basis for decision making.

decision support system (DSS)—An interactive arrangement of computerized tools tailored retrieve and display data regarding business problems and queries. The information is ultimately used to assist managers with problems such as finance, insurance, healthcare, banking, sales, and manufacturing. Consists of extracting and filtering external and operational data to the data store for use by the end user query tool and end user presentation tool.

deferred update—When the transaction operations do not immediately update the physical database. Also called *deferred write*.

deferred write—See *deferred update*.

degree—See *relationship*.

DELETE—An SQL command that allows you to eliminate specific data from a table.

denial of service attack—One of the most common hacker activities. This attack overloads Web servers and routers with millions of requests for service, rendering the services unavailable to legitimate users.

denormalization—A process that reduces a table's normal form by combining table's normal form from a higher to a lower normal form resulting in fewer tables. Sometimes done to increase processing speed. The price paid for such an increase in speed includes data redundancy and the resulting data anomalies. Use with caution. See also *normalization*.

dependency diagram—A representation of all data dependencies within a table.

dependent table—A condition in which a table's row are dependent in a related table's row. Typically, the table in the "many" side in a mandatory relationship is dependent on the table in the "one" side.

derived attribute—An attribute that does not physically exist within the entity and is derived via an algorithm. Example: Age = current date – birth date.

description of operations—A document that provides a precise, detailed, up-to-date, and thoroughly reviewed description of the activities that define an organization's operating environment.

desktop database—A single-user database that runs on a personal computer.

determinant—Any attribute in a specific row whose value directly effects and determines other values in that row. See also *Boyce-Codd normal form*.

determination—The role of the key. In the context of a database table, the statement "A determines B" indicates that knowing the value of attribute A means that you can look up (determine) the value of attribute B.

differential backup—A level of database backup in which only the last modifications to the database (when compared with a previous backup copy) are copied.

digital cash—The digital equivalent of hard currency (coins or bills of a given denomination).

digital certificate—A unique identifier given to an entity. The certificate holder may be an end user, a Web site, a computer, a Web page, or even a program. Digital certificates are used in combination with encryption to provide security and authentication.

digital signature—An encrypted attachment added to an electronic message to verify the sender's identity.

dimension tables—In a data warehouse, the tables that are used to search, filter or classify facts within a start schema. The fact table is in a one-to-many relationship with dimension tables.

dimensions—Qualifying characteristics that provide additional perspectives to a given fact.

disaster management—The set of database administrator activities dedicated to securing data availability following a physical disaster or a database integrity failure.

disjoint—A unique and non-overlapping subtype entity set.

diskblock—A referenced section of a disk that has a fixed size. Also called *diskpage* and *page*.

diskpage—See *diskblock*.

DISTINCT—An SQL clause designed to produce a list of only those values that are different from one another.

distributed data catalog (DDC)—A data dictionary that contains the description (fragment names, locations) of a distributed database.

distributed data dictionary (DDD)—See *distributed data catalog*.

distributed database—A logically related database that is stored over two or more physically independent sites.

distributed database management system (DDBMS)—A database management system that supports a database distributed across several different sites; governs the storage and processing of logically related data over interconnected computer systems in which both data and processing functions are distributed among several sites.

distributed database management system transparency—The feature that allows the end user to feel like the database's only user; that is, a distributed database can be treated as a single logical database. Distributed database transparency features include: distribution transparency, transaction transparency, failure transparency, performance transparency, and heterogeneity transparency.

distributed global schema—The database schema description of a distributed database as seen by the database administrator.

distributed processing—The activity of sharing (dividing) the logical processing of a database over two or more sites connected by a network.

distributed request—A database request that allows a single SQL statement to access data in several remote DPs in a distributed database.

distributed transaction—A database transaction that accesses data in several remote DPs in a distributed database.

distribution transparency—A DDBMS feature that allows a distributed database to be treated as though it were a single logical database.

document type definition (DTD)—A file with a .dtd filename extension that describes XML elements; in effect, a DTD file provides the description of the document's composition and defines the syntax rules or valid tags for each type of XML document.

domain—Used to organize and describe an attribute's set of possible values.

DO-UNDO-REDO protocol—Used by the data processor to roll back and/or roll forward transactions with the help of the system's transaction log entries.

drill down—To decompose data into more atomic components (that is, data at lower levels of aggregation). Used primarily in a decision support system to focus on specific geographic areas, business types etc. See also *roll up*.

durability—The transaction property indicating the permanence of the database's consistent state. Transactions that have been completed will not be lost in the event of a system failure if the database has proper durability.

dynamic SQL—A term used to describe an environment in which the SQL statement is not known in advance, but instead is generated at run-time. In a dynamic SQL environment, a program can generate the SQL statements at run-time that are required to respond to ad hoc queries.

dynamic statistical generation mode—The distributed database management system will automatically evaluate and update the statistical information after each access.

dynamic Web pages—Pages with contents that change over time and cannot be anticipated; for example, an online ordering system. The content of dynamic Web pages is not predetermined, but is generated at run-time.

dynamic-link libraries (DLLs)—Shared code libraries that are treated as part of the operating system or server process so they can be dynamically invoked at run-time.

E

early binding—A property by which the data type of an object's attribute must be known at definition time, bonding the data type to the object's attribute. Characteristic of an object oriented data model. See also *late binding*.

electronic commerce—See *e-commerce*.

e-commerce—The use of electronic computer-based technology to bring products, services, or ideas to market and to support or enhance buisness operations.

Electronic Data Interchange (EDI)—A communications protocol that enabled companies to exchange business documents over private phone networks.

electronic data processing (EDP) department—See *data processing (DP) department*.

electronic mail—The graphics and text messages sent to other, specific computer end users connected to the same network. Widely used for its low cost and ability to instantly send and receive information to one or more people. Also called e-mail.

electronic wallet—The equivalent of a physical wallet but for use in the virtual marketplace; it contains the end user's credit card(s), digital cash, and other personal information such as shipping addresses, phone numbers, and e-mail addresses.

embedded SQL—A term used to refer to SQL statements that are contained within an application programming language such as COBOL, C++, ASP, Java, or ColdFusion.

encapsulation—A feature by which the object can hide the internal data representation and method's implementation from external objects. Characteristic of an object oriented data model.

encryption—A process of inputting data in "plain text" to yield an output "encoded" version of the data, making the data unintelligible to unauthorized users.

encryption key—Used by encryption algorithms to encode data. The encryption key is a very large number used to encrypt and decrypt data.

end-user presentation tool—Used by the data analyst to organize and present the selected data compiled by the end user query tool.

end-user query tool—Used by the data analyst to create the queries that access the specific, desired information from the data store.

enterprise database—The overall company data representation, which provides support for present and expected future needs.

entity—Something about which you want to store data; typically a person, place, thing, concept or event. See also *attribute*.

entity instance—A term used in ER modeling to refer to a specific table row. Also known as an *entity occurrence*.

entity integrity—Property of a relational table that guarantees that each entity have a unique value in a primary key and that there are no null values in the primary key.

entity occurrence—See *entity instance*.

entity relationship (ER) diagram—A diagram that depicts an entity relationship model's entities, attributes and relations. It also displays connectivity and cardinality.

entity relationship (ER) model—A data model developed by P. Chen in 1975. It describes relationships (1:1, 1:M, M:N) among entities at the conceptual level with the help of ER diagrams.

entity set—A grouping of related entities.

equiJOIN—A join operator that links tables based on an equality condition that compares specified columns of the tables.

ERDM—See *extended relational data model*.

Ethernet—The dominant LAN standard used to interconnect computer systems. Ethernet is based on a bus or star topology that can use coaxial, twisted-pair, or fiber-optic cabling. Most Ethernet LANs transfer data at a speed of 100 Mbps (one hundred million bits per second).

exclusive lock—A lock that is reserved by a transaction. An exclusive lock is issued when a transaction requests permission to write (update) a data item and no locks are previously held on that data item by any other transaction. An exclusive lock does not allow any other transactions to access the database.

existence-dependent—A property of an entity whose existence depends on one or more other entities. In an existence-dependent environment, the existence-independent table must be created and loaded first, because the existence dependent key can not reference a table that does not yet exist. (Represented by a double–square box in the ER diagram.)

existence-independent—An entity that can exist apart from one or more related entities. It must be created first when referencing an existence-dependent table to it.

EXISTS—An SQL special operator that returns TRUE if a subquery returns any rows, or FALSE if it does not.

explicit cursor—In procedural SQL, a cursor created to hold the output of a SQL statement that may return two or more rows (but could return zero or only one row!).

extended relational data model (ERDM)—A model that includes the object oriented model's best features in an inherently simpler relational database structural environment.

extensible markup language (XML)—Derived from SGML, this is a meta-language used to represent and manipulate data elements. Unlike other markup languages, XML permits the manipulation of the document's data elements.

external model—The application programmer's view of the data environment. Given its business-unit focus, it works with a data subset of the global database schema.

extranet—An exclusive, company network that is accessible to certain individuals in the corporation's value chain (outside the organization), but not to the general public. Such an extranet enhances the intra-organizational workflow by allowing clients to place their own orders, process their own payments or complete other tasks usually performed my a member inside the organization.

F

fact table—In a data warehouse, refers to the star schema center table containing facts that are linked and classified through their common dimensions. A fact table is in a one-to-many relationship with each associated dimension table.

failure transparency—A DDBMS feature that allows continuous operation of a DDBMS, even in the event of a failure in one of the nodes of the network.

fat client—In a client/server system, a client that carries a relatively larger proportion of the processing load than compared to the server. Fat clients are always paired with thin servers and conversely, thin clients are always paired with fat servers.

fat server—In a client/server system, a server that carries a relatively larger proportion of the processing load than compared to the client. Fat servers are always paired with thin clients and conversely, thin servers are always paired with fat clients.

fiber-optic cable—Data transmission medium for computer networks. It uses light pulses to transmit the data from node to node and allows for the highest speed of information transfer available. Remains the most expensive method of information transfer due to the intricate laser equipment used for transferring the signal (data) to and from the cable line.

field—A character or group of characters (alphabetic or numeric) that defines a characteristic of a person, place or thing. For example, a person's social security, address, phone, bank balance, and so on all constitute fields.

field-level lock—Allows concurrent transactions to access the same row, so long as they require the use of different fields (attributes) within that row. Yields the most flexible multi-user data access but requires a high level of computer overhead.

file transfer protocol (FTP)—Used to provide file transfer capabilities among computers on the Internet/intranet using a public or known name for classifying and grouping the messages.

file—A named collection of related records.

firewall—Used to protect a network from unauthorized access from the outside world (public Internet). Specifically, a firewall is a hardware and/or software component that is used to limit and control Internet traffic going into a company's network infrastructure and data that are allowed to be moved outside a company's network.

first normal form (1NF)—The first stage in the normalization process. It describes a relation depicted in tabular format, with no repeating groups and in which a primary key has been identified. All non-key attributes in the relation are dependent on the primary key.

flags—Special codes implemented by designers to prevent nulls by bringing attention to the absence of a value in a table.

foreign key—See *key*.

fourth-generation language (4GL)—A nonprocedural language, such as SQL, that only requires the user to define what must be done; the details of *how* the user's commands are executed are handled by the 4GL.

fourth normal form (4NF)—A table is in 4NF if it is in 3NF and contains no multiple independent sets of multivalued dependencies.

fragmentation transparency—A DDBMS feature that allows a system to treat a distributed database as a single database even though the database is divided into two or more fragments.

frame—Created in the data-link layer to add control to the information by specifying the network and physical media being used. The information is added at the beginning (header) and at the end (trailer) of a network packet to enclose (frame) the packet data. It is secured by the data-link layer to ensure that safe frames are not lost in transition to the physical layer.

front-end application—Any process that the end-user interacts with to request services from a server process. See also *client*.

front-end CASE tools—A computer aided software tool that has been classified as "front-end" because it provides support for the planning, analysis and design phases of the SDLC. In comparison, back-end case tools provide support for the coding and implementation phases.

full backup—See *database dump*.

full functional dependence—A condition in which an attribute is functionally dependent on a composite key but not on any subset of that composite key.

fully heterogeneous distributed database system (fully heterogeneous DDBMS)—Integrates different types of database management systems (hierarchical, network and relational) over a network. It supports different database management systems that may even support different data models running under different computer systems, such as mainframes, minicomputers, and microcomputers. See also *heterogeneous DDBMS* and *homogeneous DDBMS*.

fully replicated database—Stores multiple copies of each database fragment at multiple sites.

functional dependence—Within a relation R, an attribute B is functionally dependent on an attribute A if and only if a given value of the attribute A determines exactly one value of the attribute B. The relationship "B is dependent on A" is equivalent to "A determines B" and is written as A➔B.

G

gateway—A type of middleware software that is used to translate client requests into the appropriate protocols needed to access specific services. Such software eliminates the requirement to use the same protocols in both clients and servers.

generalization hierarchy—Used in the entity relationship (ER) model to depict the relationship between a supertype entity set and one or more subtype entity sets. This hierarchy depicts the supertype entity as a parent of each of the subtype entities. See also *subtype* and *supertype*.

generalized update access method (GUAM)—Developed by North American Rockwell in the late 1960s to control file system redundancies; a precursor to database development.

Get Hold (GH)—In an IMS hierarchical DBMS, this statement is used to hold a segment for delete or replace operations. There are three different Get Hold statements: Get Hold Next (GHN), Get Hold Next within Parent (GHNP), and Get Hold Unique (GHU).

Get Unique (GU)—In an IMS hierarchical DBMS, a statement that is used to retrieve a database segment into the application program input area or record area.

government to business (G2B)—Special case of the Business to Business and Business to Commerce e-commerce styles. See also *government to consumer (G2C)*.

government to consumer (G2C)—Special case of the Business to Business and Business to Commerce e-commerce styles. See also *government to business (G2B)*.

GROUP BY—An SQL clause used to create frequency distributions when combined with any of the aggregate functions in a SELECT statement.

GSAM (generalized sequential access method)—In an IMS hierarchical DBMS, one of several complex storing schemes. (GSAM is a special case of the sequential access method.)

H

hacker—A person who maliciously and illegally accesses a Web site with the intention of stealing data, changing Web pages, or impairing Web site operations.

HAVING—A particularly useful extension of the GROUP BY feature. The HAVING clause is applied to the output of a GROUP BY operation to restrict the selected rows.

HDAM (hierarchical direct access method)—In an IMS hierarchical DBMS, one of several complex storing schemes. If direct-access pointers are required to keep track of the hierarchy of segments, HDAM is preferred.

heterogeneity transparency—A DDBMS feature that allows a system to integrate several different centralized DBMSs into one logical DDBMS.

heterogeneous distributed database systems (heterogeneous DDBMS)—Integrates different types of centralized database management systems over a network. See also *fully heterogeneous DDBMS* and *homogeneous DDBMS*.

HIDAM (hierarchical indexed direct access method)—In an IMS hierarchical DBMS, one of several complex storing schemes. If direct-access pointers are required to keep track of the hierarchy of segments, HIDAM is preferred.

hierarchical database model—No longer a major player in the current database market; important to know, however, because the basic concepts and characteristics form the basis for subsequent database development. This model is based on an "upside-down" tree structure in which each record is called a segment. The top record is the root segment. Each segment has a 1:M relationship to the segment directly below it. See also *child, parent, root* and *segment*.

hierarchical path—Ordered sequence of segments that must be accessed by the hierarchical DBMS in order to retrieve a given segment.

hierarchical structure—An ordered data arrangement that (logically) resembles an "upside-down" tree.

hierarchic sequence—See *preorder traversal*.

HISAM (hierarchical indexed sequential access method)—In an IMS hierarchical DBMS, a specific access method best-suited to particular kinds of applications. HISAM is particularly well suited for storing and retrieving data in hierarchic sequence, putting parent and children records in contiguous disk locations.

homogeneous distributed database systems (homogeneous DDBMS)—Integrates only one particular type of centralized database management system over a network. See also *heterogeneous DDBMS* and *fully heterogeneous DDBMS*.

homonyms—Indicates the use of the same name to label different attributes; should generally be avoided. Some relational software automatically checks for homonyms and either alerts the user to their existence or automatically makes the appropriate adjustments. See also *synonym*.

horizontal fragmentation—The distributed database design process that breaks up a table into subsets consisting of unique rows. See also *database fragments and vertical fragmentation*.

host language—A term used to describe any language that contains embedded SQL statements.

HSAM (hierarchical sequential access method)—In an IMS hierarchical DBMS, a specific access method best-suited to particular kinds of applications. HSAM is particularly well suited for storing and retrieving data in hierarchic sequence, putting parent and children records in contiguous disk locations.

hub—A warehouse of data packets housed in a central location on a local area network. Contains multiple *ports* that copy the data in the data packets to make it accessible to selected or all segments of the network.

hybrid object—The type of object classification that contains and represents a repeating group of attributes. One or more of the attributes reference another object that usually summarizes the contents of the hybrid object.

hypercubes—The data value in a data cube that has grown to *n*-number of (more than three) dimensions (represented by the *x*, *y* and *z* axes). Assists the MDBMS end users to visualize the data in a multi-dimensional form. See *data cube*.

hyperlink—Link between Web pages in hypertext or other electronic document types.

hypertext markup language (HTML)—Standard document-formatting language for Web pages.

hypertext transfer protocol (HTTP)—Standard protocol used by Web browser and Web server to communicate.

I

IDEF1X—An entity relationship (ER) modeling style derived from integrated computer-aided manufacturing (ICAM) studies that were conducted in the late 1970s.

identifying relationship—A relationship that exists when the related entities are existence-dependent. Also called a *strong relationship* or strong identifying relationship because the dependent entity's primary key contains the primary key of the parent entity.

IEEE 802.11—IEEE standards for wireless networks.

IEEE 802.3—IEEE standards for the Ethernet network.

IEEE 802.5—IEEE standards for token ring networks.

imaging server—A process that runs on a computer and provides image management services to client computers (scanning, searching, cataloging, comparisons, and so on).

immediate update—When the database is immediately updated by transaction operations during the transaction's execution, even before the transaction reaches its commit point.

implicit cursor—A cursor that is automatically created in procedural SQL when the SQL statement returns only one value.

IN—An SQL special comparison operator used to check whether an attribute value matches a value contained in a (sub)set of listed values.

inconsistent retrievals—Concurrency control problem that arises when a transaction calculating summary (aggregate) functions over a set of data—while other transactions are updating the data—yields erroneous results.

incremental backup—A process that makes a backup of only data that has changed in the database since the last incremental backup. Ensures a full recovery of all data in the event of a physical disaster or a database integrity failure.

index—An ordered array composed of index key values and row ID values (pointers). Indexes are generally used to speed up data retrieval.

index key—See *index*.

information—Facts (data) that are arranged in meaningful patterns. Information consists of *transformed* data and facilitates decision making.

information engineering (IE)—A methodology that translates the company's strategic goals into data and applications that will help to achieve the company's goals.

information management system (IMS)—A hierarchical database model based on GUAM and developed jointly by North American Rockwell and IBM. IMS became the hierarchical database standard in the 1970s.

information resource dictionary—See *data dictionary*.

information resource manager (IRM)—See *data administrator*.

information systems (IS)—A system that provides for data collection, storage and retrieval; facilitates the transformation of data into information and the management of both data and information. An information system comprises hardware, software (DMBS and applications), the database(s), people and procedures.

information systems (IS) department—An evolution of the data-processing department when responsibilities were broadened to include service and production functions.

inheritance—In the object oriented data model, the ability of an object to inherit the data structure nad methods of the classes above it in the class hierarchy. See also *class hierarchy* and *single inheritance*.

inner join—A join operation in which only rows that meet a given criteria are selected. The join criteria can be an equality condition (natural join or equijoin) or an inequality condition (theta join).

inner query—A query that is embedded (or nested) inside another query. Also known as a *nested query* or a *subquery*.

input/output (IO) request—A low-level (read or write) data access operation to/from computer devices (such as memory, hard disks, video, and printers).

INSERT—An SQL command that allows the insertion of data rows into a table, one row at a time or multiple rows at a time using a subquery.

instance variables—See *attribute*.

Institute of Electrical and Electronic Engineers (IEEE)—An organization that developed standards to provide uniformity among networks. These standards specify the technical details that define network topology and data transmission across shared media. In addition, the IEEE standards yield the rules that govern network cabling, cable distances between computers, devices used in the networks, and so on.

intelligent terminals—Term used to describe a device that provides enhanced I/O functions to a mainframe system such as a PC connected to a mainframe computer.

Interactive Database Processor (IDP)—In an IDS/II network DBMS, a processor that allows users to manipulate databases. The IDP front end is intended for users who have some programming knowledge and is not well-suited for most end users.

internal model—In database modeling, refers to a level of data abstraction that adapts the conceptual model to a specific DBMS model for implementation.

International Organization for Standardization (ISO)—An organization formed to develop standards for diverse network systems.

Internet—A global network of computers connected together through a standard network protocol known as Transmission Control Protocol/Internet Protocol (TCP/IP) You can think of the Internet as the "highway" on which the data travel. The terms *Internet* and *World Wide Web* are often used interchangeably, but they are *not* synonyms.

Internet Packet Exchange/Sequenced Packet Exchange (IPX/SPX)—A communications protocol, or set of rules, that determines how messages between computers are sent, interpreted and processed; often used in Novell networks.

interobject relationship—An attribute-class relationship created when an object's attribute references another object of the same or a different class.

interprocess communication (IPC) protocols—A capability, which is supported by various operating systems, that allows two processes to communicate with each other. The processes can be on the same computer or different computers that are connected through a network. IPC enables applications to share data without interfering with each other.

interrogate—To ask for the interrogated object's instance variable value or values. An object may send messages to interrogate another object's state.

intersection class—In the object oriented data model, a bridge class that connects classes to store the associated attributes. See *bridge*.

intranet—A company-owned and -operated computer network that is restricted to the company's internal use. Such systems can only be accessed by the computers inside the company's computer network. The purpose of such a system is to enhance company operations through improved data access management and communication.

IS NULL—A special SQL comparison operator used in conjunction with the WHERE clause to check whether an attribute is null or not. See also *BETWEEN, LIKE, IN, and EXISTS*.

islands of information—Independent data pools created and managed by different organizational components. Typical of old-style file systems.

isolation—A property of a database transaction that guarantees that a data item used by one transaction is not available to other transactions until the first transaction ends.

iterative process—A process based on repetition of processes and procedures.

J

Java—An object oriented programming language, developed by Sun Microsystems, that runs on top of the Web browser software. Java applications are compiled and stored in the Web server. Java's main advantage is its ability to let application developers develop their applications once and run them in many environments.

JavaScript—A scripting language (one that enables the running of a series of commands or macros) that was developed by Netscape and allows Web authors to design interactive Web sites. JavaScript code is embedded in Web pages. This JavaScript is downloaded with the page and is activated when a specific event takes place—like a mouse click on an object.

join columns—Term used to refer to the columns used to join two tables. The join columns generally share similar values.

K

key—An entity identifier based on the concept of functional dependence; may be classified as follows: *Superkey:* An attribute (or combination of attributes) that uniquely identifies each entity in the table. *Candidate key:* A minimal superkey, that is, one that does not contain a subset of attributes that is itself a superkey. *Primary key:* A candidate key selected as a unique entity identifier. *Secondary key:* A

key that is used strictly for data retrieval purposes. For example, a customer is not likely to know his or her customer number (primary key) but the combination of last name, first name, initial and telephone number is likely to make a match to the appropriate table row. *Foreign key:* An attribute (or combination of attributes) in one table whose values must match the primary key in another table or whose values must be null.

key attributes—The attribute(s) that form a primary key. See also *prime attribute*.

L

late binding—A characteristic in which the data type of an attribute is not known until execution time or run-time.

left outer join—In a pair of tables to be joined, a left outer join yields all the rows in the left table, including those that have no matching values in the other table. For example, a left outer join of Customer with Agent will yield all the Customer rows, including the ones that do not have a matching Agent row. See also *outer join*.

LIKE—A special operator allowed by ANSI-standard Structured Query Language (SQL) in conjunction with the WHERE clause to check for similar character strings. See also *BETWEEN, IS NULL, IN,* and *EXISTS*.

linking table—A table that implements a composite entity. See also *composite entity*.

load balancing—Services, provided by an intranet, that offer performance-enhancing techniques to maintain the intranet operating at optimum speed. In some cases, these systems also provide fault tolerance.

load testing—Services to ensure that a Web site will support the load imposed by having thousands of users access it.

local area network (LAN)—A network of computers that spans a small area, such as a single building. A LAN allows end user personal computers to share files via a central personal computer that acts as a network file server. With a LAN, multiple users can also share devices, such as printers.

local mapping transparency—A property of a DDBMS in which access to the data requires the end user to know both the name and the location of the fragments in order to access the database.

LOCATION MODE clause—In an IDS/II network DBMS, this clause determines where a record will be (physically) stored in the database and how the record will be retrieved.

location transparency—The property of a DDBMS in which access to the data requires that only the name of the database fragments need to be known. (Fragment locations need not be known.) See also *local mapping transparency*.

lock—A device that is employed to guarantee unique use of a data item to a particular transaction operation, thereby preventing other transactions from using that data item. A transaction requires a lock prior to data access, and such a lock is released *(unlocked)* after the operation's execution, to enable other transactions to lock the data item for their use.

lock manager—A DBMS component that is responsible for assigning and releasing locks.

logical data format—The way in which a human being views data.

logical design—A stage in the design phase that matches the conceptual design to the requirements of the selected DBMS an is, therefore, software-dependent. It is used to translate the conceptual design into the internal model for a selected database management system, such as DB2, SQL Server, Oracle, IMS, Informix, Access, and Ingress.

lost updates—A concurrency control problem in which data updates are lost during the concurrent execution of transactions.

M

main memory database (MMDB)—A new type of database management system that serves all query requests from memory, running 1,000 percent faster than traditional disk- and processor-bound relational databases.

mandatory—A term used to describe a relationship in which one entity occurrence must have a corresponding occurrence in another entity. Example: EMPLOYEE works in DIVISION. (A person cannot be an employee if he or she is not assigned to a company's division.)

manual query optimization—An operation mode that requires the end user or programmer to define the access path for the execution of a query.

manual statistical generation mode—One mode of generating statistical data access information used for query optimization. In this mode, the DBA must periodically run a routine to update the data access statistics, for example, running the RUNSTAT command in an IBM DB2 database.

many-to-many (M:N or M:M) relationships—One of three types of relationships (associations among two or more entities) in which one occurrence of an entity is associated with many occurrences of a related entity, and one occurrence of the related entity is associated with many occurrences of the first entity.

MAX—An SQL aggregate function that yields the maximum attribute value encountered in a given column.

member (network database)—One record type in network database terminology; a member record is equivalent to the hierarchical model's child.

message—In the OO data model, the name of a method sent to an object in order to perform an action. A message triggers the object's behavior. See *method.*

messaging—A service to ensure the proper routing and delivery of application-oriented data among multiple services.

metadata—Data about data; that is, data concerning data characteristics and relationships. See *data dictionary.*

method—In the OO data model, a named set of instructions to perform an action. Methods represent real-world actions. Methods are invoked through messages.

metrics—In a data warehouse, numeric facts that measure something of interest to the end user.

metropolitan area network (MAN)—Network type used to connect computers across a city or metropolitan area.

middleware—The computer software that allows clients and servers to communicate within the client/server architecture. Also known as the communication layer, composed of an Applications Programming Interface (API), a database translator, and a network translator. Middleware insulates client processes from the network protocols and the details of the server process protocols.

MIN—An SQL aggregate function that yields the minimum attribute value encountered in a given column.

minimal data rule—Defined as "All that is needed is there, and all that is there is needed." In other words, all data elements required by the database transactions must be defined in the model, and all data elements defined in the model must be used by at least one database transaction.

mixed fragmentation—Regarding data fragmentation, this refers to a combination of horizontal and vertical strategies: Meaning, a table may be divided into several rows, each one having a subset of the attributes (columns).

model—A description or analogy used to visualize something that cannot be directly observed; simplified abstractions of real-world events or conditions.

module—(1) A design segment that can be implemented as an autonomous unit, sometimes linked to produce a system. (2) An information system component that handles a specific function, such as inventory, orders, payroll, and so on.

module coupling—A description of the extent to which modules are independent of one another.

multidimensional database management system (MDBMS)—Database management system that uses proprietary techniques to store data in matrixlike arrays of *n*-dimensions, known as cubes.

multidimensional online analytical processing (MOLAP)—Extends online analytical processing functionality to multidimensional database management systems.

multiple access unit (MAU)—A wiring concentrator through which a token ring's computers are connected physically.

multiple inheritance—Exists when a class can have more than one immediate (parent) superclass above it.

multiple-site processing, multiple-site data (MPMD)—A scenario describing a fully distributed database management system with support for multiple data processors and transaction processors at multiple sites.

multiple-site processing, single-site data (MPSD)—A scenario in which multiple processes run on different computers sharing a single data repository.

multiuser DBMS—A database management system that supports multiple concurrent users.

multivalued attribute—Attributes that can have many values.

mutual consistency rule—A data replication rule that requires that all copies of data fragments be identical.

mutual exclusive rule—A condition in which only one transaction at a time can own an exclusive lock on the same object.

N

natural join—A relational operation that links tables by selecting only the rows with common values in their common attribute(s).

nested query—A query contained within a query. Usually consists of an outer query and one or more inner queries. The inner query generally executes first and the results are used as input for the outer query.

network backbone—The main network cabling system for one or more local area networks.

Network Basic Input/Output System (NetBIOS)—A network protocol originally developed by IBM and SYTEK Corporation in 1984.

network interface card (NIC)—Electronic circuit board that allows computers to communicate within a network.

network model—A data model created to represent complex data relationships more effectively than the hierarchical model could, to improve database performance, and to impose a database standard.

network operating system (NOS)—A computer operating system oriented toward providing server side services to clients (such as file and printer sharing and security management).

network protocols—A set of rules (at the physical level) that determines how messages between computers are sent, processed, and interpreted.

network segment—A single section of cable that connects several computers.

network translator—A middleware component that manages the network communications protocols.

news and discussion group services—Specialized services that allow the creation of "virtual communities" in which users exchange messages regarding specific topics.

non-identifying relationship—A relationship that occurs when the primary key of the dependent (many side) entity does not contain the primary key of the related parent entity. Also known as a *weak relationship*.

nonkey attribute—See *nonprime attribute*.

non-overlapping subtypes—Subtypes that contain a subset of the supertype entity set, but each entity instance (row) of the supertype can appear in only one of the non-overlapping subtypes. Also known as disjoint subtypes.

nonprime attribute—A nonkey attribute – that is, an attribute that is not even part of a key.

nonserialized items—Inventory items that do not require individual component tracking by an assigned serial number or code, such as laser printer paper or disks. Also called nondurable supplies.

normalization—A process that assigns attributes to entities in such a way that data redundancies are reduced or eliminated.

NOT—An SQL logical operator that negates a given predicate.

NULL—The absence of an attribute value. Note: a null is not a blank.

O

object—An abstract representation of a real-world entity that has a unique identity, embedded properties, and the ability to interact with other objects and with itself.

object ID (OID)—A system-generated object identifier that is independent of the object state and any physical address in memory.

object instance—Each particular object belonging to a class.

object orientation—A set of modeling and development principles focused on an autonomous entity with embedded intelligence to interact with other objects and itself.

object query language (OQL)—The database query language used by an object oriented database management system.

object space—The equivalent of the database schema, as seen by the designer in an object oriented database.

object state—The set of values that the object's attributes have at a given time.

object oriented data model (OODM)—A data model based on object oriented concepts. Provides support for new user-defined data types, inheritance, polymorphism, encapsulation, and so on.

object oriented database management system (OODBMS)—Software used to better manage data found within an object oriented database model.

object oriented database model (OODBM)—A data model whose basic modeling structure is an object.

object oriented programming (OOP)—An alternative to conventional programming methods based on object oriented concepts. It reduces programming time and lines of code, and increases programmers' productivity.

object oriented programming language (OOPL)—A programming language based on object oriented concepts.

object table—The equivalent of a relational table composed of many rows, where each row is an object of the same type. Each row object has a unique system generated object ID (OID) or object identifier.

one-to-many (1:M) relationship—One of three types of relationships (associations among two or more entities) which are used by data models. In a 1:M relationship, one entity instance is associated with many instances of the related entity.

one-to-one (1:1) relationship—One of three types of relationships (associations among two or more entities) which are used by data models. In a 1:1 relationship, one entity instance is associated with only one instance of the related entity.

online analytical processing (OLAP)—Decision support system (DSS) tools that use multidimensional data analysis techniques. OLAP creates an advanced data analysis environment that supports decision making, business modeling, and operations research activities.

Open DataBase Connectivity (ODBC)—Database middleware developed by Microsoft Corp. to provide a database access API to Windows applications.

optimistic approach—In concurrency control, this approach is based on the assumption that the majority of the database operations do not conflict and a transaction is executed without restriction until it is committed.

optimization algorithm—An algorithm used by the DBMS to determine the most efficient execution plan for a query.

optionality—Any condition in an entity-relationship model in which one or more optional relationships exist.

OR—The SQL logical operator used to link multiple conditional expressions in a WHERE or HAVING clause. It requires that only one of the conditional expressions be true.

ORDER BY—An SQL clause useful for ordering the output of a SELECT query (for example, in ascending or descending order.)

outer join—A relational-algebra JOIN operation that produces a table in which all unmatched pairs are retained; unmatched values in the related table are left null.

overlapping subtypes—Subtypes that contain a subset of the supertype entity set, but each entity instance (row) of the supertype can appear in more than one subtype.

owner (network database)—One record type in network database terminology; an owner record is equivalent to the hierarchical model's parent.

P

page-level lock—In this type of lock, the database management system will lock an entire diskpage, or section of a disk.

parent table—In a 1:M relationship between two tables, the one side is normally referred to as the parent table. In other words, the parent table is the table that has the primary key that is referenced by a foreign key in a related table.

partial dependency—In normalization, a condition in which an attribute is dependent on only a portion (subset) of the primary key.

partially replicated database—A distributed database in which copies of only some database fragments are stored at multiple sites. See also *fully replicated database*.

participants—An ER term used to label the entities that participate in a relationship. Example: PROFESSOR teaches CLASS. The *teaches* relationship is based on the participants PROFESSOR and CLASS.

partitioned data allocation strategy—In a DDBMS, the data allocation strategy in which the database is divided into fragments that are stored at several sites.

partitioning—The process of splitting a table into subsets of rows or columns.

passive data dictionary—A DBMS component that stores the definition of data characteristics and relationships which is not updated automatically and which usually requires a batch process to be run.

performance transparency—A DDBMS feature that allows a system to perform as though it were a centralized DBMS (no degradation of response times).

performance tuning—Activities that make the database perform more efficiently in terms of storage and access speed.

periodicity—Usually expressed as current year only, previous years, or all years; provides information about the time span of the data stored in the table.

persistent stored module (PSM)—A block of code (containing standard SQL statements and procedural extensions) that is stored and executed at the DBMS server.

personalization—Customization of a Web page for individual users.

physical data format—The way in which the computer "sees" the data.

physical design—A stage of database design that maps the data-storage and access characteristics of the database. Since these characteristics are a function of the types of devices supported by the hardware, the data-access methods supported by the system, and the selected DBMS, physical design is both hardware- and software- dependent. See also *physical model*.

physical model—A model in which the physical characteristics (location, path, and so on) are described for the data. Both hardware- and software-dependent. See also *physical design*.

plug-in—A client-side, external application that is automatically invoked by the browser when needed to manage specific types of data.

pointer—A reference device that "points" to the location of data within the computer data-storage medium.

policies—General statements of direction that are used to manage company operations through communicating and supporting the organization's objectives.

polymorphism—An object oriented data model characteristic by which different objects can respond to the same message in different ways.

preorder traversal—A "left-list" path within a hierarchical database. The left-list refers to the order (always starting from the left of the hierarchical tree) in which segments are referenced. Also known as the *hierarchical sequence*.

Pretty Good Privacy (PGP)—An example of public-key encryption by Pretty Good Privacy Inc. PGP is a fairly popular and inexpensive method for encrypting e-mail messages on the Internet.

primary key—See *key*.

prime attribute—A key attribute, that is, an attribute that is part of a key or the whole key.

privacy—Control of data usage dealing with the rights of individuals and of the organization to determine the "who, what, when, where, and how."

private-key encryption—Encryption that uses a *single* numeric key to encode and decode data. Both sender and receiver must know the encryption key. See also *symmetric encryption*.

procedural SQL (PL/SQL)—A type of SQL that allows the use of procedural code and SQL statements that are stored in the database as a single callable object that can be invoked by name.

procedure cache—A shared, reserved memory area that stores the most recently executed SQL statements or PL/SQL procedures (including triggers and functions). Also called the *SQL cache.*

procedures—A series of steps to be followed during the performance of a given activity or process.

processing option (PROCOPT)—A type of access granted to a program.

production database—The main database designed to keep track of the day-to-day operations of a company. See also *transaction database.*

profile—In Oracle, a named collection of settings that controls how much of the database resource a given user can use.

program commitment block (PCB)—In a hierarchical database, after the physical database has been defined through the DBD, a way through which application programs are given a subset of the physical database.

program specification block (PSB)—In a hierarchical database, this represents a logical view of a selected portion of the database and also defines the database(s), segments, and types of operations that can be performed by the application. Using PSBs yields better data security as well as improved program efficiency by allowing access to only the portion of the database that is required to perform a given function.

protocol—A specific set of rules to accomplish a specific function. In the object oriented data model, protocol refers to a collection of messages to which an object responds.

public-key encryption—A form of encryption that uses two numeric keys—the public key and the private key. Both keys are able to encrypt and decrypt each other's messages. See also *asymmetric encryption.*

Q

query—A question or task asked by an end user of a database in the form of SQL code.

query language—A nonprocedural language that lets the user specify *what* is to be done without specifying *how* it is to be done. An example of a query language is SQL.

query optimizer—A DBMS process that analyzes SQL queries and finds the most efficient way to access the data. The query optimizer generates the access or execution plan for the query.

R

record—A collection of related (logically connected) fields.

record at a time—This term indicates that the database commands affect a single record at a time.

RECORD NAME—In the IDS/II network DBMS, this clause initiates the record's definition by assigning it a unique name. An IDS/II network database must contain at least one record type.

recursive query—A nested query that joins a table to itself. For example, a recursive query joins the EMPLOYEE table to itself.

recursive relationship—An entity that displays relationships within itself.

redundant data—Duplicated data that are stored in more than one location.

redundant transaction logs—Most database management systems keep several copies of the transaction log to ensure that a physical failure of a disk will not impair the DBMS's ability to recover data.

referential integrity—A condition by which a dependent table's foreign key must have either a null entry or a matching entry in the related table. Even though an attribute may not have a *corresponding* attribute, it will be impossible to have an invalid entry.

referential object sharing—When an object instance is referenced by other objects. That is, two or more different objects point to the same object instance, a change in the referenced object instance values is automatically reflected in all other referring objects.

Rein85 model—Based on the same modeling conventions as the Crow's Foot model, its symbols are quite different, and it operates at a higher level of abstraction. Developed by D. Ranier in 1985.

relation—In the relation model, a term used to describe a set of associated attributes. Normally, relations will become relational tables.

relational algebra—A set of mathematical principles that form the basis of the manipulation of relational table contents; comprises eight main functions: SELECT, PROJECT, JOIN, INTERSECT, UNION, DIFFERENCE, PRODUCT, and DIVIDE.

relational database management system (RDBMS)—A collection of programs that manages the complexity of a relational database. The RDBMS software translates the user's logical requests (queries) into commands that physically locate and retrieve the requested data. A good RDBMS also creates and maintains a data dictionary (system catalog) to help provide data security, data integrity, concurrent access, easy access, and system administration to the data in the database through a query language (SQL) and application programs.

relational model—Developed by E. F. Codd (of IBM) in 1970, it represents a major breakthrough for users and designers because of its conceptual simplicity. The relational model produced an "automatic transmission" database to replace the "standard transmission" databases that preceded it.

relational online analytical processing (ROLAP)—Provides online analytical processing functionality by using relational databases and familiar relational query tools to store and analyze multidimensional data.

relational schema—The description of the organization of a database as seen by the database administrator that shows connecting fields and the relationship type.

relationship—An association between entities defined by a diamond-shaped symbol in an ER diagram. Its degree defines the number of related entities: unary, binary, ternary, or higher.

remote request—A DDBMS feature that allows a single SQL statement to access data in a single remote DP. See also *remote transaction*.

remote transaction—A DDBMS feature that allows a transaction (formed by several requests) to access data in a single remote DP. See also *remote request*.

repeater—A device used in Ethernet networks to add network segments to the network and to extend its signal reach.

repeating groups—The condition in which an entity characteristic can contain multiple values. In a relation, that characteristics can be represented by several attributes all of the same type, for example: an employee can have multiple dependents, a car can have many colors, a painter can paint many paintings.

replica transparency—Refers to the DDBMS's ability to hide the existence of multiple copies of data from the user.

replicated data allocation—Data allocation strategy by which copies of one or more database fragments are stored at several sites.

replication—The process of creating and managing duplicate versions of a database. Used to place copies in different locations and improve access time.

resource security—The protection of the resource(s) from external and internal threats. Specifically, resource security means protecting the resource from viruses, unauthorized access by hackers, or denial of service attacks.

right outer join—In a pair of tables to be joined, a right outer join yields all the rows in the right table, including the ones with no matching values in the other table. For example, a right outer join of CUSTOMER with AGENT will yield all the agent rows, including the ones that do not have a matching CUSTOMER row. See also *outer join*.

ring topology—Network topology in which computers are connected to one another via a cabling setup that, as the name implies, resembles a ring; it is more flexible than a bus topology, because addition or loss of a computer does not have a negative impact on other network activities.

role—In Oracle, a named collection of database access privileges that authorize a user to connect to the database and use the database system resources.

roll up—When you aggregating the data to a higher level. (Rolling up the data is the exact opposite of drilling down the data.) See *drill down*.

ROLLBACK—An SQL command that restores the database table contents to their original condition.

root—The starting component of a hierarchical database tree.

router—(1) An intelligent device used to connect dissimilar networks. (2) Hardware/software equipment that connects multiple and diverse networks.

row-level lock—A comparatively less restrictive database lock where the DBMS allows concurrent transactions to access different rows of the same table, even if the rows are located on the same page.

row-level trigger—A trigger that is executed once for each row affected by the triggering SQL statement. A row-level trigger requires the use of the FOR EACH ROW keywords in the trigger declaration.

RSA—The most commonly used public-key encryption technology, developed by RSA Data Security Inc. RSA is the de facto standard for Internet encryption.

rule of precedence—A basic algebraic rule that says conditions within the parentheses are executed first. For example, in the equation $2 + (3 \times 5)$ the multiplication portion would be calculated first, making the correct answer 17.

rule-based optimizer—A query optimization technique that uses a set of preset rules and points to determine the best approach to execute a query.

rule-based query optimization algorithm—A manual query optimization technique based on a set of user-defined rules to determine the best query access strategy.

S

scalability—A benefit of a server process that means it can be upgraded to run on more powerful platforms and provide services to more clients.

scheduler—The DBMS component that is responsible for establishing the order in which concurrent transaction operations are executed. The scheduler *interleaves* the executive of database operations in a specific order (sequence) to ensure *serializability*.

schema—A group of database objects (tables, indexes, views, queries, etc.) that are related to each other. Usually a schema belongs to a single user or application.

scope—That part of a system that defines the extent of the design, according to operational requirements.

script—A small program containing commands written in a programming language.

search services—A business enabling web service that allows Web sites to perform searches on their contents.

second normal form (2NF)—The second stage in the normalization process in which a relation is in 1NF and there are no partial dependencies (dependencies in only part of the primary key).

secondary key—A key that is used strictly for data retrieval purposes. For example, a customer is not likely to know his or her customer number (the primary key) but the combination of last name, first name, initial, and telephone number is likely to make a match to the appropriate table row.

Secure Electronic Transaction (SET)—Initiative to provide a standard for secure credit card transactions over the Internet.

Secure Hypertext Transfer Protocol (S-HTTP)—Protocol used to securely transfer Web documents over the Internet. S-HTTP supports use of private and public keys for authentication and encryption. S-HTTP has not been widely used, because it only supports encrypted HTTP data and does not support other Internet protocols as does SSL.

Secure Sockets Layer (SSL)—A protocol used to implement secure communication channels between client and server computers on the Internet.

security—(1) Protecting data against accidental or intentional use by unauthorized users. (2) Services designed to ensure safety and privacy of data through encryption, digital certificates, SSL, S-HTTP, firewalls, and proxy servers.

segment—The equivalent of a file record type in the hierarchical database model.

SELECT—An SQL command that yields the values of all or a subset of rows in a table. The SELECT statement is used to retrieve data from tables.

SENSEG (SENsitive SEGment)—In the IMS hierarchical database, this keyword declares the segments that will be available, starting with the root segment.

semantic data model (SDM)—The first of a series of data models that more closely represented the real world, modeling both data and their relationships in a single structure known as an object. The SDM, published in 1981, was developed by M. Hammer and D. McLeod.

sequence field—An attribute that contains values that are unique and sequential (ascending or descending). Some DBMS allows the explicit definition of sequence or auto-number attributes that are generally used to uniquely identify each row.

serializability—A property of a transaction that ensures that the result of the concurrent execution of transactions yields the same results as if executing the transactions in serial order.

serialized items—Inventory items that require careful tracking with serial numbers or codes, such as computers and printers. Also called durable inventory items.

server—Any process that provides requested services to clients. See *client/server architecture*.

server-side extension—A program that interacts directly with the server process to handle specific types of requests. They add significant functionality to Web servers and to intranets.

service—In Windows, the name for a special program that runs as part of the operating system. It ensures the availability of required services to the system and users.

set—In the network model, a description of a 1:M relationship between an owner record type and a member record type.

shared lock—A lock that is issued when a transaction requests permission to read data from the database and no exclusive locks are held on that data by another transaction. A shared lock allows other read-only transactions to access the database. See also *exclusive lock*.

SHISAM (simple hierarchical indexed sequential access method)—In the IMS hierarchical database, a specific access method best-suited to particular kinds of applications. SHISAM is particularly well-suited for storing and retrieving data in hierarchic sequence, putting parent and children records in contiguous disk locations.

SHSAM (simple hierarchical sequential access method)—In the IMS hierarchical database, a specific access method best-suited to particular kinds of applications. SHSAM is particularly well-suited for storing and retrieving data in hierarchic sequence, putting parent and children records in contiguous disk locations.

simple attribute—An attribute that cannot be subdivided into meaningful components. Compare to *composite attribute*.

simple object—An object that contains only single-valued attributes and has no attributes that refer to another object.

single inheritance—In the object oriented data model, the property of an object that allows it to have only one parent superclass from which it inherits its data structure and methods. See also *inheritance, multiple inheritance*.

single-site processing, single-site data (SPSD)—A scenario in which all processing is done on a single CPU or host computer (mainframe, minicomputer, or PC) and all data are stored on the host computer's local disk.

single-user DBMS—A database management system classification that depicts a DBMS that supports only one user at a time.

single-valued attribute—An attribute that can have only one value.

site monitoring and data analysis—Services to ensure that a Web site performs at an optimal level.

slice and dice—Using multidimensional jargon, the ability to cut slices off the data cube to perform a more detailed analysis.

sneakernet—One of the original ways to share data. When users needed to share data, they would simply copy the data to a disk and walk to the coworker's office, disk in hand.

software independence—A property of any model or application that does not depend on the software used to implement it. See also *software-dependent*.

sparsity—In multidimensional data analysis, a measurement of the density of the data held in the data cube.

SQL cache—A shared, reserved memory area that stores the most recently executed SQL statements or PL/SQL procedures (including triggers and functions). Also called the *procedure cache*.

SQL*Net—A proprietary solution limited to Oracle databases; used by Oracle to send SQL requests over a network.

SQL tuning—Activities oriented toward generating an SQL query that returns the correct answer in the least amount of time, using the minimum amount of resources at the server end.

Standard Generalized Markup Language (SGML)—An international standard for the publication and distribution of highly complex technical documents—such as those used by the aviation industry and the military services—that are too complex and unwieldy for the Web.

standards—A detailed and specific set of instructions that describe the minimum requirements for a given activity. Standards are used to evaluate the quality of the output.

Standards Planning and Requirements Committee (SPARC)—The ANSI committee that augmented database standards in 1975.

star schema—A data modeling technique used to map multidimensional decision support data into a relational database with the purpose of performing advanced data analysis.

star topology—Network topology with all computers connected to one another in a star configuration through a central computer or network hub. Like a ring topology, allows for computers to be added to or released from the network without having an impact on other computers.

stateless system—Describes the fact that, at any given time, a Web server does not know the status of any of the clients communicating with it. The Web does not reserve memory to maintain an open communications "state" between the client and the server.

statement-level trigger—A trigger that is assumed if you omit the FOR EACH ROW keywords. This type of trigger is executed once, before or after the triggering statement completes, and is the default case.

static query optimization—A query optimization method in which the access path to a database is predetermined at compilation time.

static SQL—A style of embedded SQL in which the SQL statements will not change while the application is running.

static Web pages—Web pages used to display information that does not change much over time or is not time-critical.

statistically based query optimization algorithm—A query optimization technique that uses statistical information about the database. These statistics are then used by the DBMS to determine the best access strategy.

stored function—A named group of procedural and SQL statements that returns a value, indicated by a RETURN statement in its program code.

stored procedure—(1) A named collection of procedural and SQL statements. (2) Business logic stored on the server in the form of SQL code or some other DBMS-specific procedural language.

strong (identifying) relationship—When two entities are existence-dependent; from a database design perspective, this exists whenever the primary key of the related entity contains a primary key component of the parent entity.

structural dependence—A data characteristic that exists when a change in the database schema affects data access, thus requiring changes in all access programs.

structural independence—A data characteristic that exists when changes in the database schema do not affect data access.

structured query language (SQL)—A powerful and flexible relational database software language composed of commands that enable users to create database and table structures, perform various types of data manipulation and data administration, and query the database to extract useful information.

subclass/superclass—See *class hierarchy*.

subordinate—In a Distributed Database Management System (DDBMS), a DP node that participates in a distributed transaction using the two-phase COMMIT protocol.

subquery—A query that is embedded (or nested) inside another query. Also known as a *nested query* or an *inner query*.

subschema—A subset of a schema. The subschema defines the portion of the database as "seen" by the application programs that use it. See also *schema*.

subtype (entity set)—An entity (set) that contains the unique characteristics (attributes) of an entity whose general characteristics are found in another, more broadly defined entity known as a supertype. In a generalization hierarchy, it is any entity that is found below a "parent" entity. Example: The subtype PILOT of the supertype EMPLOYEE.

SUM—An SQL aggregate function that yields the sum of all values for a selected attribute in a given column or expression.

superclass—This class is a more general classification of its subclasses, which, in turn, are more specific components of the general classification. Musical instruments defines the superclass of Stringed instruments, which is, in turn, the superclass of the Piano, Violin, and Guitar classes.

superkey—See *key*.

supertype (entity set)—An entity (set) that contains the general (commonly shared) characteristics of an entity (see *subtype*). If the entity set can include characteristics that are not common to all entities within the set, the supertype becomes the parent to one or more subtypes in a generalization hierarchy.

surrogate key—A system-assigned primary key, generally numeric and auto-incremented.

switch—An intelligent device that connects computers. Unlike a hub, a switch allows multiple simultaneous transmissions between two ports (computers). Therefore, switches have greater throughput and speed than regular hubs.

symmetric encryption—Encryption that uses a *single* numeric key to encode and decode data. Both sender and receiver must know the encryption key. See also *private-key encryption*.

synonym—The use of different names to identify the same object, such as an entity, an attribute, or a relationship; should (generally) be avoided. See also *homonym*.

system catalog—A detailed system data dictionary that describes all objects in the database.

systems administrator (SYSADM)—The person responsible for coordinating the activities of the data processing function. The systems administrator also manages the use of database software in a data processing (DP) department; solicits and evaluates database designs, coordinates the development of applications based on the data resource, and assigns the right to manage the database(s) to selected individuals. The systems administrator also coordinates the activities of all database administrators (DBAs) in a multi-database operation.

systems analysis—The process that establishes the need for and the extent of an information system.

systems development—The process of creating an information system.

Systems Development Life Cycle (SDLC)—The cycle that traces the history (life cycle) of an information system. The SDLC provides the "big picture" within which the database design and application development can be mapped out and evaluated.

Systems Network Architecture (SNA)—A network environment used by IBM mainframe computers.

T

table—A (conceptual) matrix comprising intersecting rows (entities) and columns (attributes) that represents an entity set in the relational model. Also called a *relation*.

table-level lock—A locking scheme that allows only one transaction to access a table at a time. A table level lock locks an entire table preventing access to any row by transaction T2 while transaction T1 is using the table.

tablespace—In an Oracle DBMS, a logical storage space used to group related data.

tag—In markup languages such as HTML and XML, a command inserted in a document to specify how the document should be formatted. Tags are used in server-side markup languages and interpreted by a Web browser for presenting data.

ternary relationship—An ER term used to describe an association (relationship) between three entities. Example: A CONTRIBUTOR contributes money to a FUND from which a RECIPIENT receives money. (Note that there is no direct relationship between CONTRIBUTOR and RECIPIENT.

theta join—A join operator that links tables using an inequality comparison operator (<, >, <=, >=) in the join condition.

thicknet—A more expensive and durable form of coaxial cable; allows for greater distances between computers than thinnet.

thin client—A client that carries a relatively smaller proportion of the processing load than compared to the server, thin clients are always paired with fat servers.

thin server—A server that carries a lesser processing load than the client processes, thin servers are always paired with fat clients.

thinnet—Cheaper form of coaxial cable that is easier to install; does not allow for the distance between networked computers that thicknet does.

third normal form (3NF)—A table is in 3NF if it is in 2NF and no nonkey attribute is functionally dependent on another nonkey attribute, that is, it cannot include transitive dependencies.

third-generation language (3GL)—A language that requires the programmer to specify both what must be done and how it is to be done. Examples include, COBOL, BASIC and FORTRAN.

time stamping—In transaction management, a technique used when scheduling concurrent transactions that assigns a global unique time stamp to each transaction.

token ring network standard (IEEE 802.5)—The standard for ring topology networks that can use shielded twisted-pair, unshielded twisted-pair, or fiber-optic cabling.

token—In a ring topology network, the marker that passes from computer to computer, similar to a baton in a relay race. Only the computer with the token can transmit at a given time.

top-down—A design philosophy that begins by defining the main (macro) structures of the system and then moves to define the smaller units within those structures. In database design, it is a process that first identifies entities and then defines the attributes within the entities. Compare *bottom-up*.

transaction—A sequence of database operations (one or more database requests) that access the database. A transaction is a logical *unit of work*, that is, it must be *entirely* completed or aborted – no intermediate ending states are accepted. All transactions must have the following properties: (1) *Atomicity* requires that, unless all operations (parts) of a transaction are completed, the transaction is aborted. A transaction is treated as a single, indivisible logical unit of work. (2) *Durability* assures that, once a transaction is completed, the database reaches a consistent state; that state cannot be lost, even in the event of a system failure. (3) *Serializability* ensures that the selected order of transaction operations creates a final database state the would have been produced if the transactions had been executed in a serial fashion. (4) *Isolation* assures that the data used during the execution of a transaction cannot be used by a second transaction until the first one is completed.

transaction log—A feature that is used by the DBMS to keep track of all transaction operations that update the database. The information stored in this log is used by the DBMS for recovery purposes.

transaction log backup—Backs up only the transaction log operations that are not reflected in a previous backup copy of the database.

transaction manager (TM)—Same as *transaction processor (TP)*.

transaction processing—In electronic-commerce, this is the series of actions, changes and/or functions among thousands of connected customers.

transaction processor (TP)—The software component on each computer that requests data through a DDBMS. The TP is responsible for the execution and coordination of all databases issued by a local application that access data on any data processor (DP). See also *data processor* and *transaction manager (TM)*.

transaction transparency—A DDBMS property that ensures that database transactions will maintain the distributed database's integrity and consistency. They ensure that the transaction will be completed only if all database sites involved in the transaction complete their part of the transaction.

transactional database—A database designed to keep track of the day to day transactions of an organization. See also *production database*.

transitive dependency—A condition in which an attribute is dependent on another attribute that is not part of the primary key.

Transmission Control Protocol/Internet Protocol (TCP/IP)—The official communications protocol of the Internet, a worldwide network of heterogeneous computer systems.

trigger—A procedural SQL code that is automatically invoked by the relational database management system upon the occurrence of a data manipulation event.

tuple—A table row.

twisted-pair—Network cable formed by pairs of wires that are twisted inside a protective insulating cover; choice cabling for most network installations because it is easy to install and carries a low price tag. It resembles typical telephone cable.

two-phase commit protocol—In a DDBMS, an algorithm used to ensure atomicity of transactions and database consistency, as well as integrity in distributed transactions.

two-phase locking—A set of rules that governs the way transactions acquire and relinquish locks. Two-phase locking guarantees serializability, but it does not prevent deadlocks. The two-phase locking protocol is divided into two phases: (1) the *growing phase*, in which the transaction acquires all the locks that is needs without unlocking any *existing* data locks. Once all locks have been acquired, the transaction is in its *locked* point. (2) A *shrinking phase*, in which the transaction releases all locks and cannot obtain a new lock.

U

unary relationship—An ER term used to describe an association *within* an entity. Example: A COURSE is a prerequisite to another COURSE.

uncommitted data—When trying to achieve concurrency control, uncommitted data causes data integrity and consistency problems. It occurs when two transactions are executed concurrently and the first transaction is rolled back after the second transaction has already accessed the uncommitted data, thus violating the isolation property of transactions.

unidirectional logical relationships—In a hierarchical database, relationships that are established by linking a logical child with a logical parent in a one-way arrangement.

Uniform Resource Locator (URL)—Identifier for a resource on the Internet. Also called Web address.

union-compatible—Two or more tables are union compatible if they share the same column names and the columns have compatible data types or domains.

unique fragment—In a DDBMS, a condition indicating that each and every row is unique, regardless of which fragment it is located in.

unique index—An index in which the index key can only have one pointer value (row) associated with it.

uniqueness—A property of time stamping that ensures that no equal time stamp values can exist.

universal database servers—Another name for an object/relational product used in the object/relational database model.

unreplicated database—A distributed database in which each database fragment is stored at a single site.

updatable view—A view that can be used to update attributes in base table(s) that are used in the view.

UPDATE—An SQL command that allows you to change attribute values in one or more rows of a table.

user—In Oracle, a uniquely identifiable object that allows a given person to log on to the database.

UWA (user work area)—A specific area of memory that contains several fields used to access and inform regarding the status of the database. The UWA also contains space for each record type defined in the subschema.

V

value chain—All activities required to design, plan, manufacture, market, sell, and support a product or service.

VBScript—A client-side extension in the form of a Microsoft language product used to extend the browser's functionality; derived from Visual Basic.

verification—The process whereby a conceptual model is validated against all required database transactions (INSERT, UPDATE, DELETE, and SELECT).

versioning—A property of an OODBMS that allows the database to keep track of the different transformations performed on an object.

vertical fragmentation—In distributed database design, the process that breaks a table up into subsets (fragments) consisting of a subset of columns from the original table. Fragments must share a common primary key. See also *database fragments, horizontal fragmentation.*

view—A virtual table based on a SELECT query.

virus—A malicious program that affects the normal operation of a computer system.

W

wait/die—A concurrency control scheme in which an older transaction waits and the younger transaction is rolled back and rescheduled.

weak entity—An entity that displays existence dependence and cannot exist by itself. Example: A DEPENDENT requires the existence of an EMPLOYEE. The weak entity usually inherits at least part of its key from other entities.

Web application server—A middleware application that expands the functionality of Web servers by linking them to a wide range of services, such as databases, directory systems, search engines, and so on.

Web browser—The end-user application used to navigate the Internet; runs on a client computer and requests services from a Web server.

Web caching—Performance-enhancing technique in which a temporary storage area is created to provide Web pages at an optimal speed.

Web development—The process of adding "business logic" to Web pages, thereby making them business-enabled.

Web page—A text document on the World Wide Web containing text and special commands written in Hypertext Markup Language.

Web server—A specialized application whose only function is to "listen" for client requests, process them, and send the requested Web resource back to the client browser.

Web site—Refers to the Web server and the collection of Web pages stored on the local hard disk of the server computer.

Web-to-database middleware—A database server-side extension program that retrieves data from databases and passes it on to the Web server, which sends it to the client's browser for display purposes.

wide area network (WAN)—Network type used to connect computer users across distant geographical areas; generally makes use of telephone or specialized communications companies.

wildcard character—A symbol that can be used as a general substitute for other characters or commands.

wireless adapter—In the case of wireless networks, this adapter, sometimes called a wireless NIC, allows a computer to communicate using a wireless network.

wireless device support—Services that facilitate data integration with mobile communication devices, such a smart phones; allow users to conduct business, make hotel reservations, purchase groceries, and pay bills from virtually anywhere.

wireless LAN (WLAN)—A local area network that is connected by wireless technology rather than wires.

workgroup database—A multiuser database that supports a relatively small number of users (usually fewer than 50) or for a specific department in an organization.

World Wide Web (WWW)—Worldwide network collection of specially formatted and interconnected documents known as Web pages. Also called the Web.

World Wide Web Consortium (W3C)—An international group involved with the Internet and Wed. Their purpose is to develop open standards for the web including the first XML 1.0 standard definition in 1998.

wound/wait—A concurrency control scheme in which the older transaction rolls back the younger transaction and reschedules it.

write-ahead protocol—See *write-ahead-log protocol*.

write-ahead-log protocol—Ensures that transaction logs are always written to permanent storage before any database data are actually updated. Also called *write-ahead protocol*.

write-through—Occurs when the database is immediately updated by transaction operations during the transaction's execution, even before the transaction reaches its commit point.

X

XSL (Extensible Style Language)—A specification used to define the rules by which XML data are formatted and displayed. The XSL specification is divided into two parts: Extensible Style Language Transformations (XSLT) and XSL style sheets.

XSL style sheets—Similar to presentation templates, these define the presentation rules applied to XML elements. The XSL style sheet describes the formatting options to apply to XML elements when they are displayed on the browser, cellular phone display, PDA screen, and so on.

XSLT (Extensible Style Language Transformations)—Term that describe the general mechanism that is used to extract and process data from one XML document and enable its transformation within another document.

INDEX